Engineering FLUID MECHANICS

For Mechanical, Civil, Electrical, Chemical, Aeronautical and Other Engineering and Design Students

In SI Units

Prof. K.L. KUMAR

Ph.D. (London), D.I.C., M.E. (Roorkee), B.E. (Delhi)

F.I.E. (India), M.I.S.T.E., M.S.F.M.F.P., M.I.S.T.A.M., M.A.S.E.E. (USA), I.M.A.I.A.E.T.

Professor, Faculty of Engineering and Technology,

University of Botswana, Gaborone

Formerly UNESCO Expert in Technical Education, Middle East

Professor & Head, Educational Technology and Coordinator, QIP

Professor & Head, Department of Applied Mechanics,

INDIAN INSTITUTE OF TECHNOLOGY, NEW DELHI

EURASIA PUBLISHING HOUSE (P) LTD.

RAM NAGAR, NEW DELHI-110 055

Sole Distributors :

S. CHAND & COMPANY PVT. LTD.

(An ISO 9001 : 2008 Company)

S. CHAND PUBLISHING
empowering minds

Head Office: 7361, RAM NAGAR, NEW DELHI - 110 055
Phone: 23672080-81-82, 9899107446, 9911310888 Fax: 91-11-23677446
www.schandpublishing.com; e-mail: helpdesk@schandpublishing.com

Branches

Ahmedabad	:	Ph: 27541965, 27542369, ahmedabad@schandpublishing.com
Bengaluru	:	Ph: 22268048, 22354008, bangalore@schandpublishing.com
Bhopal	:	Ph: 4274723, 4209587, bhopal@schandpublishing.com
Chandigarh	:	Ph: 2725443, 2725446, chandigarh@schandpublishing.com
Chennai	:	Ph. 28410027, 28410058, chennai@schandpublishing.com
Coimbatore	:	Ph: 2323620, 4217136, coimbatore@schandpublishing.com (Marketing Office)
Cuttack	:	Ph: 2332580; 2332581, cuttack@schandpublishing.com
Dehradun	:	Ph: 2711101, 2710861, dehradun@schandpublishing.com
Guwahati	:	Ph: 2738811, 2735640, guwahati@schandpublishing.com
Hyderabad	:	Ph: 27550194, 27550195, hyderabad@schandpublishing.com
Jaipur	:	Ph: 2219175, 2219176, jaipur@schandpublishing.com
Jalandhar	:	Ph: 2401630, 5000630, jalandhar@schandpublishing.com
Kochi	:	Ph: 2378740, 2378207-08, cochin@schandpublishing.com
Kolkata	:	Ph: 22367459, 22373914, kolkata@schandpublishing.com
Lucknow	:	Ph: 4026791, 4065646 lucknow@schandpublishing.com
Mumbai	:	Ph: 22690881, 22610885, mumbai@schandpublishing.com
Nagpur	:	Ph: 6451311, 2720523, 2777666, nagpur@schandpublishing.com
Patna	:	Ph: 2300489, 2302100, patna@schandpublishing.com
Pune	:	Ph: 64017298, pune@schandpublishing.com
Raipur	:	Ph: 2443142, raipur@schandpublishing.com (Marketing Office)
Ranchi	:	Ph: 2361178, ranchi@schandpublishing.com
Siliguri	:	Ph: 2520750, siliguri@schandpublishing.com (Marketing Office)
Visakhapatnam	:	Ph: 2782609 visakhapatnam@schandpublishing.com (Marketing Office)

First Edition 1976
Subsequent Editions and Reprints 1980, 84, 88, 90, 92, 95, 97, 98, 99, 2000, 2002, 2003, 2004, 2006, 2007, 2008, 2009, 2012, 2013, 2014
Reprint 2016

ISBN : 978-93-856-7648-2 **Code :** 1010 666

This book is for Sale in India & other SAARC Countries only.

PRINTED IN INDIA
By Vikas Publishing House Pvt. Ltd., Plot 20/4, Site-IV, Industrial Area Sahibabad, Ghaziabad-201010
and Published by Eurasia Publishing House (P) Ltd., 7361, Ram Nagar, New Delhi -110 055.

Preface to the Eighth Revised Edition

***I** wrote the first edition of the book in love for the subject of fluid mechanics; I have revised and brought the edition as a mark of affection for the students and younger staff members !*

With the passage of time since its first edition in 1976, my fascination for the subject of fluid mechanics has soared to new heights, perhaps with my own involvement with some new designs in the world full of new applications including bio-fluid mechanics, heart lung machines for heart transplants and prediction models of natural calamities such as hurricanes, tornadoes, wind storms and tsunamis!

Ever since the genius of Leonardo da Vinci drew the above concept drawing and our knowledge of fluid mechanics has realized his vision by way of aircraft, gliders, helicopters and other flying machines and a variety of fluid machines including pumps, compressors and turbines have become a reality in everyday life, the subject of fluid mechanics has become a core subject in all engineering disciplines.

I thank the Management of S. Chand & Company Ltd., for encouraging in every way to produce the book. Thanks are also due to Mr. Rupesh Gupta, Subject Editor for his editorial contribution.

I am pleased to dedicate the new edition to my past students who helped me learn the subject and to extend it to the present and future students and teachers in fluid mechanics all over the world!

(*Kris Kumar*)

Formerly of IIT Delhi; Currently, Visiting Professor, IIT Mumbai and New Delhi

On Sabbatical Leave from the University of Botswana

kumarkl@gmail.com and kumarkl@mopipi.ub.bw

Preface to the First Edition - 1976

Anything new ?

Anything special ?

Does it cover our syllabus ?

Solves numerical examples ?

Answers to the unsolved problems ?

The author has tried his best to place a positive and capital 'YES' to each of the above questions posed by a student of engineering in Mechanical, Civil, Electrical, Chemical or Aeronautical branch belonging to (*a*) a progressive engineering college in India or abroad, (*b*) an Indian Institute of Technology or (*c*) A.M.I.E. Section B (New Scheme), A.M.Ae.S.I. Section A (New Scheme) or any other engineering college in Asia, Africa or Europe. How has the author achieved this ?

The author has the experience of teaching Fluid Mechanics and allied subjects at the University of Roorkee, Imperial College of Science and Technology, London University and at the Indian Institute of Technology, Delhi. Through experience, contacts or otherwise, the syllabi of a number of universities were collected. This included the universities in India and abroad. These were examined in the light of the recommendations made by the Curriculum Development Cells set up by the Government of India. The data collection step was followed by data analysis. The largest common factor was determined, mean syllabus formulated and stretched to allow 10 to 20% overlap for any institution. What has thus emerged is indeed a series of fundamental topics on flow of paramount importance to all engineering students.

Realising that engineers and designers are more concerned with the applications than with theory, the author has woven the subject-matter with an adequate number of practical applications in engineering and a number of solved numerical examples. Unsolved problems of objective, brainstorming and numerical nature have been appended and answers to most of them have been given.

The authors acknowledges the general help received from his colleagues and friends notable among who are Professor Shankar Lal, Professor R.C. Malhotra, Professor Subhash Chander, Professor D.S. Kumar, Professor V. Seshadri, Miss Neelam Arora and Miss Sureshta Arora. The author wishes to express his gratitude to the National Book Trust of India for the generous subsidy in the publication of the text. The publishers M/s Eurasia Publishing House and in particular Shri R.C. Kumar and Shri Ravi Gupta have taken keen interest in the publication.

Acknowledgements would be incomplete without a word of appreciation for my family who have put up with my midnight lamps and general restlessness. My parents, my wife and little daughters provided an ideal bookwriting atmosphere at home throughout this venture. The students whom I have taught at home and abroad and for whose sake I have learnt the subject over the years are the contributors in absentia. My effort would be rewarded if the students and teachers of engineering accept this text and send their comments for further improvements.

Prof. K.L. Kumar, I.I.T. Delhi

(*kumarkl@gmail.com*)

Acknowledgement to Engineering Institutions

The author is grateful to the authorities in the following Institutions and Universities, who have made available their syllabi and whose faculty members have interacted in the preparation of the revised manuscript:

Indian Institutes of Technology :

(Delhi, Mumbai, Kanpur, Kharagpur, Roorkee, Guwahati and Chennai)

National Institutes of Technology:

(Agartala, Allahabad, Bhopal, Calicut, Durgapur, Goa, Hamirpur, Jalandhar, Jaipur, Jammu, Jamshedpur, Kurukshetra, Nagpur, Patna, Raipur, Rourkela, Silchar, Srinagar, Surat, Surathkal, Trichy, Warangal)

Indian Institute of Science, Bangalore; Jawaharlal Nehru Technological University, Hyderabad; Uttar Pradesh Technical University; Punjab Technical University, Jalandhar; Uttarakhand Technical University, Dehradun; Visveswariah Technological University, Bangalore; Vivekanand Technical University, Bhillai;

Birla Institute of Technology and Science : Pilani and Ranchi; College of Engineering, Anna University, Chennai; Dayalbagh Engineering Institute, Agra; Delhi College of Engineering, Delhi; Engineering Colleges of Bihar; Engineering Colleges of Pune; Engineering Colleges, Karnataka; Engineering Colleges, Nagpur; Faculty of Engineering and Technology, Jadavpur, Kolkata; Government Engineering College, Ujjain; Guindy Engineering College, Chennai; Hercourt Butler Technological Institute, Kanpur; Institute of Technology, BHU, Varanasi; Jamia Milia Islamia University, Jamia Nagar, New Delhi; Jodhpur University, Jodhpur; Karnataka University, Dharwar; Kerala University, Trivandrum; M.S. University of Baroda, Vadodra; Madhav Inst. of Sc. and Technology, Gwalior; Netaji Subhas Institute of Technology, Delhi; North Eastern Regional Institute of Science and Technology, Itanagar; Osmania University, Hyderabad; PSG College of Technology, Coimbatore; Punjab Engineering College, Chandigarh; SV Univ. College of Engineering, Tirupati; Thapar Institute of Engg. and Technology, Patiala; Tripura Engineering Colleges, Tripura; Ujjain Engineering Colleges; Vellore Institute of Technology, Vellore; Victoria Jubilee Technical Institute, Mumbai; Z.H. College of Engg. and Technology, AMU, Aligarh.

Institutions Abroad :

Arya Mehar University of Technology, Tehran, Iran; Asian Institute of Technology, Bangkok, Thailand; Foundation of Technical Institute, Iraq; Imperial College of Science and Technology, London., U.K.; London University, London, England; Massachusetts Institute of Technology, Boston, U.S.A.; Mehran University of Engineering and Technology, Nawabshah, Pakistan; Thangal Kunju Musaliar College of Engineering , Quilon; University of Asmara, Eritea; Azarabadegan, Tabriz, Iran; Baghdad, Iraq; Basrah, Iraq; Birmingham, England; Botswana, Gaborone, Botswana; Cairo, Egypt; Dar Es Salaam, Tanzania; Garyounis, Benghazi, Libya; Hong Kong; Ife-Ife, Nigeria; Kuwait; Mousl, Iraq; Riyadh, Saudi Arabia; Singapore; Colombo/Katubedda, Sri Lanka and University of Technology, Baghdad, Iraq.

Contents

Learning Objectives in Engineering Fluid Mechanics

On successful completion of all learning activities provided in each chapter, a student will be able to

1 — identify the different states of matter and define a fluid;
— differentiate between a system and a control volume in a fluid flow;
— define the stress and the rate of strain at a point in a fluid flow;
— apply the Newton's law of viscosity to simple engineering problems;
— explain the mechanism of surface tension in liquids and
— classify fluids vis-a-vis regimes of flow.

2 — state and prove the Pascal's law for a static fluid;
— derive the hydrostatic law in an incompressible and in a compressible fluid at rest;
— determine the hydrostatic force on a given plane or curved surface;
— determine the pressure at a desired point in a fluid by employing a single-tube, a U-tube (upright, inclined or inverted) or a differential manometer;
— compute the force of buoyancy on a partially or fully submerged body;
— identify the states of equilibrium for a floating body and
— determine the surface profile of a liquid in a vessel subjected to a constant acceleration or a constant rotation,

3 — explain the concepts of steady and uniform flow;
— determine the acceleration in a fluid flow;
— differentiate between streamlines, pathlines and streaklines;
— derive the continuity equation for an incompressible or a compressible flow;
— describe a real flow in terms of a stream function;
— explain the kinematics of a fluid element and establish the condition for irrotationality and
— plot simple flownets in terms of streamlines and velocity potential lines.

4 — establish the Euler's equation of motion for a two-dimensional steady flow of an incompressible fluid and extend it for three-dimensional viscous flow;
— obtain the Bernoulli equation by integrating the Euler's equation under appropriate conditions;
— establish the steady flow energy equation and obtain the Bernoulli equation from it;
— state the integral momentum equation and the angular momentum equation for simple engineering flows and
— determine the force exerted by a fluid flow on the passage confining it.

5 — determine the velocity at a point in a fluid by employing pitot probes;
— estimate the discharge through a pipe by using a venturimeter and
— estimate the discharge through an open channel by using a notch.

6 — describe the flownets and pressure distribution for simple ideal flows, i.e., uniform flow, source, sink, free vortex and their combinations and
— simulate the flow over a long rotating circular cylinder by superimposing a doublet and a free vortex on a uniform flow

7 — describe the conditions under which the flow in a round pipe is laminar or turbulent;
— establish the Hagen-Foiseuille equation for laminar flow in a pipe and
— determine the viscosity of a liquid by employing viscometers.

8 — explain the development of a boundary layer in an external and in an internal flow;
— describe the rationale for the different estimates of boundary layer thickness and
— demonstrate the phenomena of turbulence, separation and recirculation of flow and suggest some means to control the separation of flow.

9 — explain the origin of lift and drag forces for flow around an immersed body;
— classify different types of drag;
— describe the variation of lift and drag coefficients with .variations in flow;
— explain the D'Alembert paradox and the Magnus effect for flow over a cylinder and
— estimate the terminal velocity for free fall of a body.

10 — derive an expression for the frictional head loss in pipe flow;
— estimate minor losses in pipe transitions; -
— draw the energy gradient lines for flow through pipes and
— calculate the flow through pipes connected in series and in parallels.

11 — develop an expression for the discharge for uniform flow in an open channel;
— draw and interpret the specific energy diagram for a free-surface flow and
— describe the formation of a hydraulic jump and estimate the variation of Froude number and other parameters across a jump.

12 — differentiate between various processes of expansion and compression;
— determine the sonic velocity in a fluid;
— describe the effect of Mach number in flow;
— analyse a normal shock;
— describe the flow through a converging-diverging nozzle and
— estimate a supersonic velocity by employing a pilot-static probe.

13 — state and apply the Buckingham Pie theorem to arrange a set of given variables into dimensionless groups;
— employ the Rayleigh's indicial method to determine the functional form of a phenomenon in terms of dimensionless groups;
— state the advantages of the dimensional method, dimensionless groups and dimensional homogeneity in engineering and
— describe the principle of similitude and apply it for model testing.

14 — classify various types of fluid machines;
— calculate the efficiency of centrifugal pumps, axial flow machines and hydraulic turbines;
— describe the characteristic performance of a reaction turbine and centrifugal pump and
— calculate the head developed by centrifugal pumps operating in series and in parallels

and finally, apply the principles of fluid mechanics to analyse and to design simple fluid mechanical systems, subsystems and devices!

CHAPTER

Zero

The World of Fluid Mechanics

What, Why and How?

- Students want to know the importance and relevance of a course before they take interest! They wish to know why they should study Fluid Mechanics vis-a-vis natural phenomena and engineering applications.
- Lecturers need to inspire the students to learn a subject by showing Its real life applications all around us.
- This Chapter is, therefore, a window to the world of Fluid Mechanics; It unveils the fascinating nature of the subject!

Welcome to the world of fluid mechanics! Engineers and designers are fascinated with the subjcct of fluid mechanics for at least two good reasons. Firstly, it enables us to understand the interesting phenomena in nature and secondly it empowers us to design and make a variety of fluid flow equipment for the benefit of humankind. Let us look at a sample of the spectrum of natural phenomena and related applications.

Phenomena: How do birds fly in the air? How do fish swim in water? How do they steer themselves forward; up and down; left and right at will?

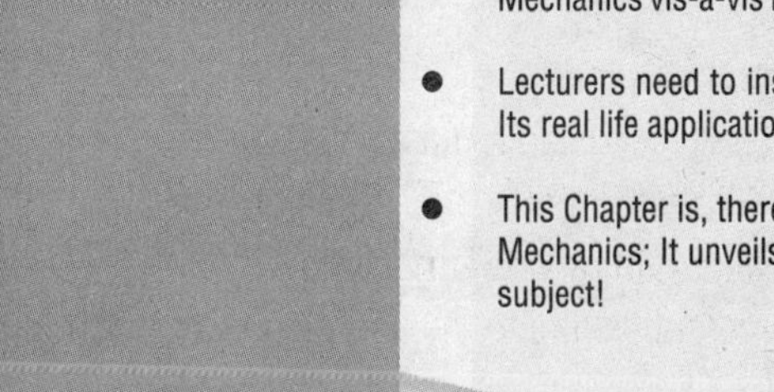

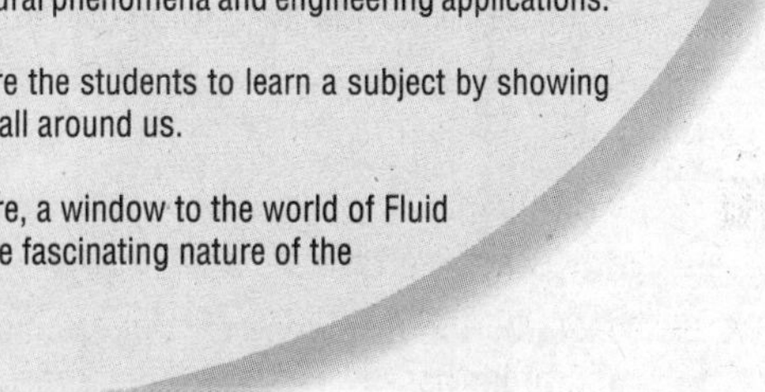

Understanding: Streamlining, floatation, lift and drag forces and propulsion by wings, etc.

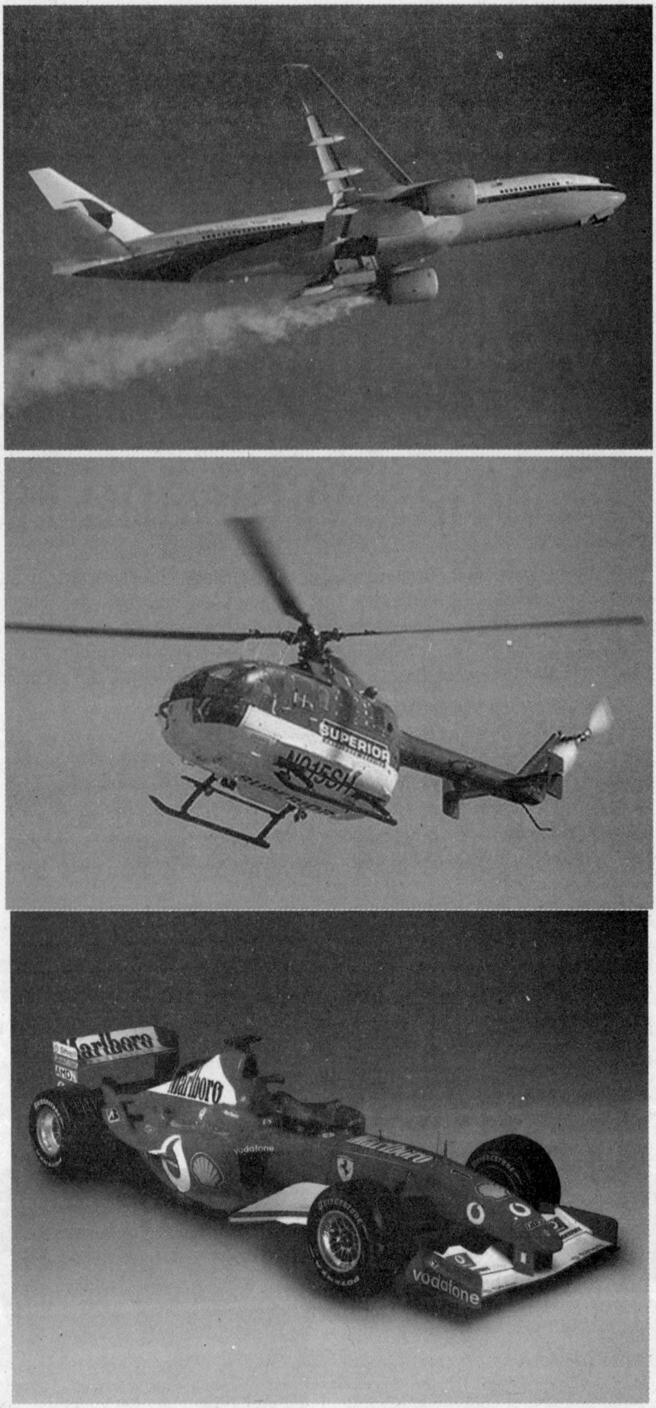

Applications: Aircraft, helicopters, gliders, jets, fighters, rockets and racing cars.

Applications: Boats, sail boats, ships, submarines, structures, cooling towers, etc.

Phenomena: Water falls, winds in coastal regions and fields, tides, waves, etc in the ocean.

Understanding: Natural sources of energy, environmental conservation and energy conversion.

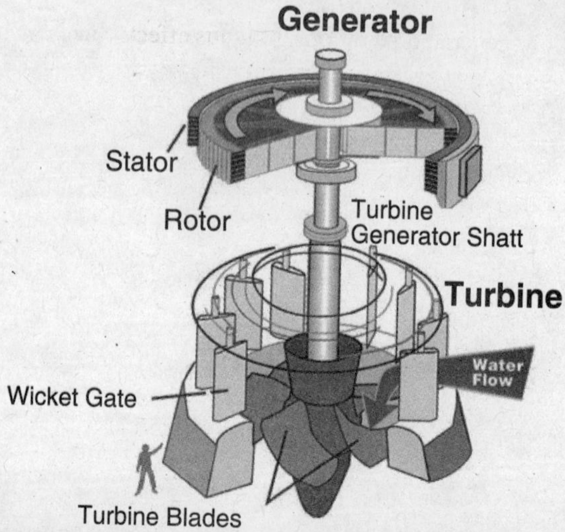

Applications: Design of hydraulic machines; turbines and pumps, wind mills and tidal power.

Phenomena: Laminar and turbulent flow, wakes and generation of *lift* by spinning a body in a free stream.

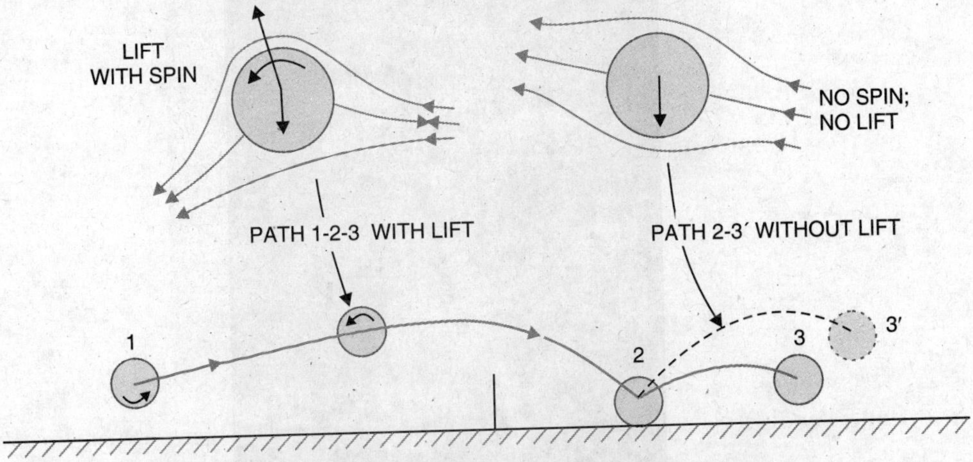

Understanding: Flow-generated vibrations, Magnus effect: change of path of spinning bodies.

Applications; Spinning the ball in cricket, table tennis and Frisbee disc. Conformal transformation to generate high lift-to-drag ratio Joukowski aerofoil for aircraft and axial flow compressors.

Phenomena: Human subsystems for breathing, blood flow; heart as a pump for blood and lungs for oxygenation.

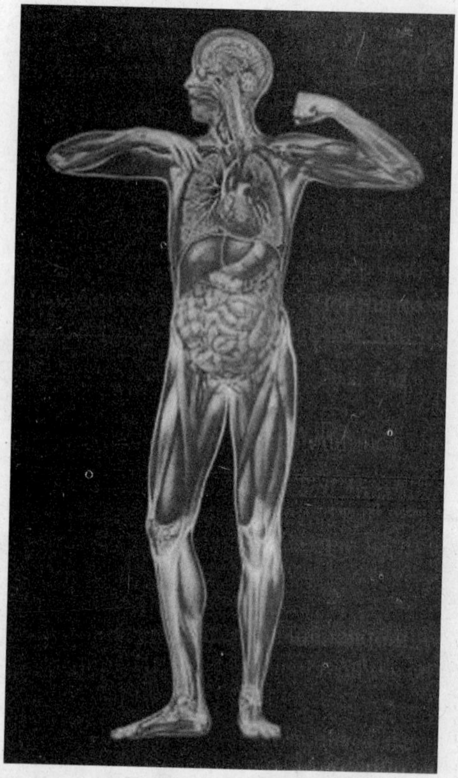

Understanding: Functioning of lungs, double-pump action of heart and several survival needs.

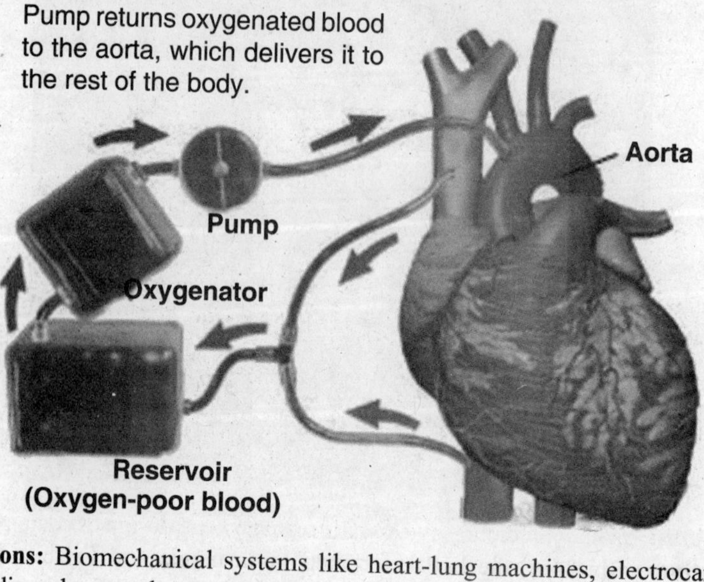

Applications: Biomechanical systems like heart-lung machines, electrocardiograms and oxygenator. Cardio-pulmonary bypass systems to enable the doctors to operate and install appliances.

Phenomena: Laminar and turbulent flow, hurricanes, wakes behind immersed bodies, etc.

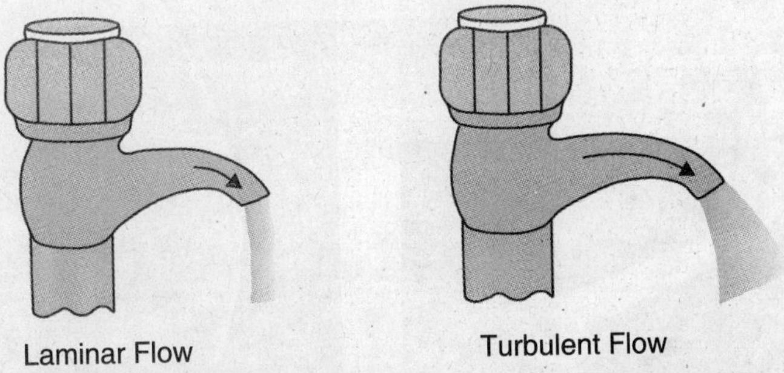

Understanding: Atmosphere, pressure gradients, critical Reynolds number, boundary layers and alternate shedding of Von Karman vortices

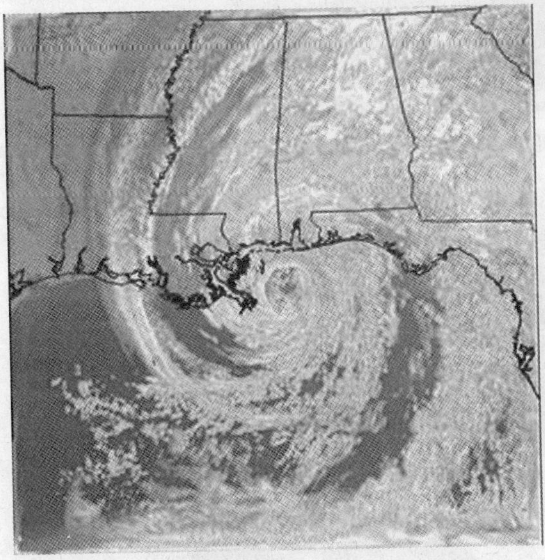

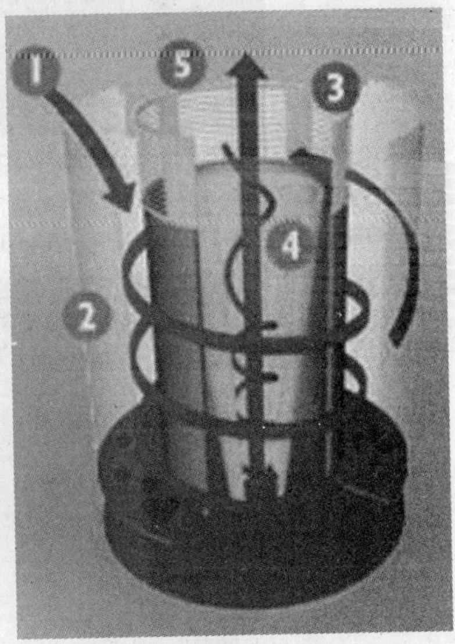

Applications: Weather prediction and disaster management, design of equipment without flow generated vibrations and Dyson's dual cyclone vacuum cleaner, etc.

Phenomenon: Abrupt changes in pressure and velocity beyond a certain speed of flow

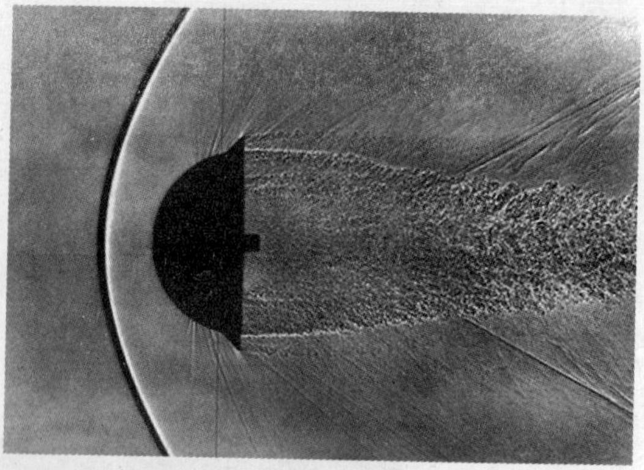

Understanding: Occurrence of hydraulic jump in water and other liquids and occurrence of normal and oblique shock waves beyond the speed of sound in a medium.

Applications: Supersonic flow passages and aircraft, e.g., Concord and supersonic fighter jets.

Phenomena: Fun with water and surfing, fountains, floatation, propulsion and sea beach sports.

Understanding: Playing with water and sand for relaxation, fun and learning.

Applications: Water sports and entertainment with water. Water bodies and fun.

All said and done, the study of fluid mechanics is so vast that we would feel ourselves like 'children playing at the sea shore, finding a pebble here and a pebble there' in the words of Newton!

Internet References and Acknowledgement

Some pictures have been taken from the Internet in order to promote web-surfing by students. Sincere acknowledgements are due to the following sites, among others:

Figure/ Photograph	Internet Site Acknowledged
Leonardo da Vinci Sketch	www.richarddawkins.net
Eagle Flying	http://swfloridawildlife.com/birds/Eagles
Bird Flying	thebirdpictures.blogspot.com
Fish Swimming	www.wetpixel.com
Aeroplane	www.webmail.groundreport.com
Helicopter	http://www.superiorambulance.com
Ferrari Formula 1 Car	cars.88000.org/30Ferrari_Formula_1
Sail Boat	www.sdmaritime.org
Nippon Maru Sailing	http://3.bp.blogspot.com
Submarine	www.dailymail.co.uk
Water Fall	http://i256.photobucket.com/albums
Francis Turbine	www.commons.wikimedia.org
Windmill	http://www.freefoto.com/images
Magnus Effect	www.juliantrubin.com
Harbhajan Singh	crickettamasha.googlepages.com
Human Systems	blogs.cgdev.org
Heart Lung Machine	happylife-smart49.blogspot.com
Hydraulic Jump	http://www.engineering.uiowa.edu/fluidslab
Shock Wave	www.psrd.hawaii.edu
Concode Flight	www.aerospaceweb.org
Dyson Dual Cyclone	www.espares.co.uk
Streamlined Car	us1.webpublications.com.au
Von Karman Vortices	panoramix.ift.uni.wroc.pl
Weather Forecasting	http://images.pennnet.com
Water Sport	www.sxmbeach.com
Water Entertainment	www.kamennydom.sk
Children Playing at a Beach	http://www.travelindiasmart.com/beaches

The introduction to Fluid Mechanics would be substantially supplemented and made interesting by viewing a video entitled FLOW VISUALISATION by Prof. K.L. Kumar, IIT Delhi. It is recommended to show the video to the students once at the beginning of the course and again, later during the course!

The video presents some interesting fluid flows through dye injection, hydrogen bubble, smoke injection, wool tuft and optical techniques. The phenomena include transition from laminar to turbulent flow, separation and shedding of vortices, creation of lift on rotating cylinders, spheres and aerofoils as also shockwaves in supersonic flow.

Available form :

Superintendent/Head of Educationul Technology Services Centre,
Indian Institute of Technology, Delhi,
New Delhi-110016; http:/www.iitd.ac.in/centre/etsc
Fax: +91–11- 2656 6917; Phone: +91– 11-2659 6415, 1339, 6539.

CHAPTER

1

Introductory Concepts

LEARNING OBJECTIVES

On successful completion of all learning activities planned and provided in the chapter, a student will be able to

- identify the different states of matter and define a fluid;
- differentiate between a system and a control volume in a fluid flow ;
- define the stress and the rate of strain at a point in a fluid flow;
- apply the Newton's law of viscosity to simple engineering problems ;
- explain the mechanism of surface tension in liquids and
- classify fluids vis-a-vis regimes of flow.

1.1 Introduction : Fluids and Non-fluids

There are two states of matter : fluid and non-fluid. The fluid state refers to liquid, vapour or gaseous phases and the non-fluid state means only the solid phase of matter. Existence of matter in these states is governed by the intermolecular attractive forces within a substance. Very strong intermolecular attractive forces exist in solids which give them the property of rigidity. Such forces are weaker in liquids and extremely weak in vapours and gases, so much so that liquids may change shape easily and acquire the shape of the container and that vapours and gases fill up the entire space of the container allotted to them !

It is more appropriate to classify substances as fluids and solids on the basis of their behaviour under the application of external forces. In particular, it is the effect of shear forces which distinguishes fluids from solids. *A fluid is a substance which deforms continuously when subjected to shear forces., however small the shear force may be.* The tendency of continuous deformation of a substance is called 'fluidity' and the act of continuous deformation is called 'flow'. A fluid would, therefore, flow when subjected to shear forces.

If a solid is subjected to shear forces, it has a tendency to resist deformation and on being deformed within the elastic limit, it has a tendency to regain its initial state of equilibrium when the applied shear forces are removed. A fluid, on the other hand, deforms continuously or flows when shear forces are applied to it and it has no tendency whatsoever to regain its initial state ; instead, it comes to rest as soon as the shear forces are removed. The state of rest of a fluid thus corresponds to the absence of shear forces.

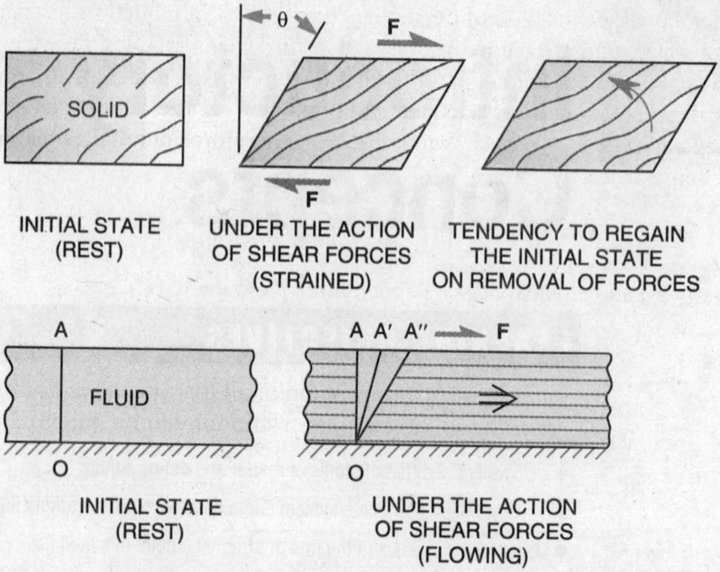

Fig. 1.1: Behaviour of a Solid and a Fluid under the Action and Removal of Shear Forces.

Behaviour of a typical solid and a typical fluid on the application and removal of shear forces are shown schematically in Fig. 1.1. It is interesting to note that a solid suffers strain when subjected to shear forces whereas a fluid suffers rate of strain, *i.e.*, it *flows* under similar circumstances.

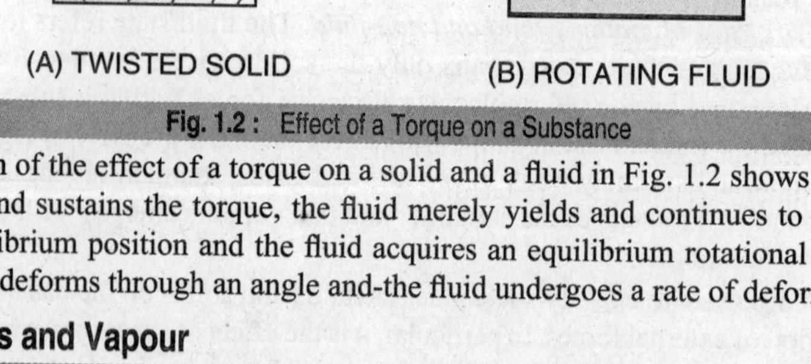

Fig. 1.2 : Effect of a Torque on a Substance

Comparison of the effect of a torque on a solid and a fluid in Fig. 1.2 shows that whereas the solid is twisted and sustains the torque, the fluid merely yields and continues to rotate. The solid acquires an equilibrium position and the fluid acquires an equilibrium rotational velocity. This is because the solid deforms through an angle and-the fluid undergoes a rate of deformation.

1.2. Liquid, Gas and Vapour

Fluids may exist as liquid, gas or vapour. Liquids are characterised by the presence of 'free surface' as a clear interface in the vicinity of an immiscible fluid. Gases and vapours, on the other hand, expand and adjust their shape and size so as to completely fill the space of the container. The same substance may exist in one or more phases. When a substance exists as a liquid, vapour or gas, it is said to have a single-phase existence. The term vapour is used for a gas when it is at a temperature below the critical temperature of the substance. Two-phase and multiphase existence refers to co-existence of two or more phases of the same substance.

1.3. Continuum

The mathematical idealisation of continuous distribution of matter, *i.e..*, when the properties of the matter are considered as continuous function of the space variables, then the matter is called a continuum. Hence, solid continuum and fluid continuum. In a continuum, any property function P (P may be scalar, vector or tensor) defined at a point (x, y, z) is a continuous and differentiable function of space variables x, y, z as shown in Fig. 1.3. Therefore, an infinitesimal increment in the function can be represented by

$$d(P) = \left(\frac{\partial P}{\partial x}\right) dx + \left(\frac{\partial P}{\partial y}\right) dy + \left(\frac{\partial P}{\partial z}\right) dz$$

The property P can be density (ρ), temperature (T), velocity ($\boldsymbol{U}$), etc.

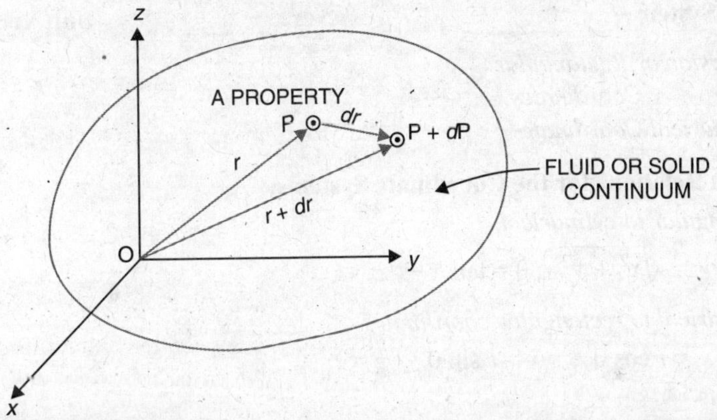

Fig 1.3: Property P(x, y, z) as a Continuous Function of the Space Coordinates (x, y, z).

1.4. Co-ordinate Systems

The position, displacement, velocity and acceleration of a fluid are referred to a system of co-ordinates and an origin. Following convention, a right-handed triad is used. In the Cartesian or Rectangular co-ordinates, the triad comprises of the x, y and z axes whereas in the cylindrical co-ordinates, the triad is made up of the r, θ and z co-ordinates as shown in Fig. 1.4. (*a*), (*b*) and (*c*). The transformation equations, unit vectors and relationship between the unit vectors for the two co-ordinate systems are tabulated.

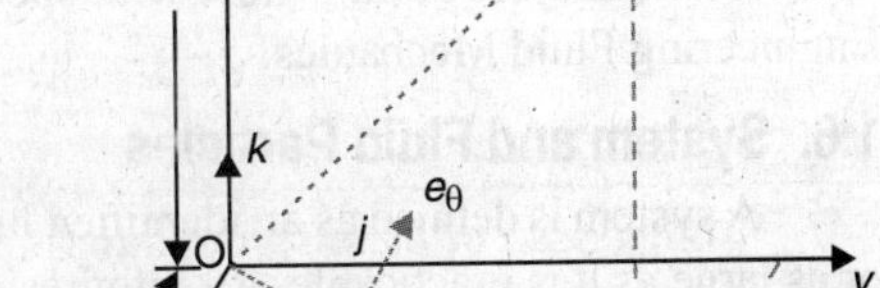

Fig. 1.4 (a): Coordinate Systems

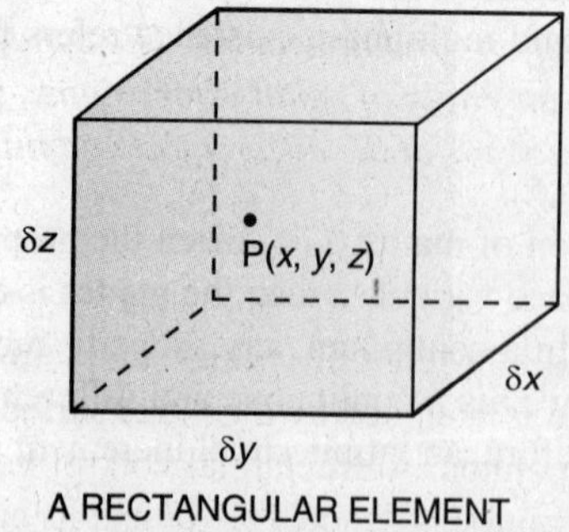

Fig. 1.4 (b)

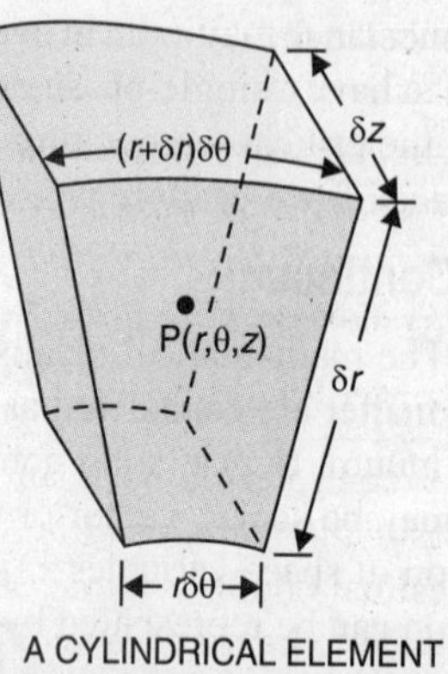

Fig. 1.4 (c)

Cartesian and Cylindrical Co-ordinate Systems

System	Axes	Unit Vectors
Cartesian or Rectangular Coordinates :	$x \; y \; z$	$i \; j \; k$
Cylindrical Coordinates :	$r \; \theta \; z$	$e_r \; e_\theta \; e_z$

Transformation Relations for the Coordinate Systems.

(A) *Rectangular to cylindrical.*

$$r = \sqrt{x^2 + y^2} \; ; \theta = \tan^{-1} \frac{y}{x}; \; z = z$$

(B) *Cylindrical to rectangular coordinates*

$$x = r \cos \theta \; ; \quad y = r \sin \theta \; ; \quad z = z$$

Unit vectors

$$e_r = \cos \theta \, i + \sin \theta \, j \; ; \; e_\theta = -\sin \theta \, i + \cos \theta \, j \; ; \; e_z = k$$

1.5. Fluid Mechanics

The subject of Fluid Mechanics deals with the study of the motion and general behaviour of a fluid governed by the law of conservation of mass, laws of mechanics and of thermodynamics. The laws of mechanics referred to are the Newton's laws of motion. The fluid under study may be flowing in a pipe or in a channel, in a pump or in a compressor, around an aircraft or a missile, in an ocean or in the atmosphere thus making the subject of *Fluid Mechanics as the most vital of all engineering studies.* The subject of fluid-flow with special emphasis to applications in engineering is termed as Engineering Fluid Mechanics.

1.6. System and Fluid Particles

A system is defined as an identified lump of fluid. The lump of fluid may either be infinitesimal or as large as it is practicable. An infinitesimal lump of fluid is called a 'fluid element' or a particle. A fluid particle is assumed to have very small but non-zero dimensions : its volume is taken as δV, mass δm and density ρ such that

$$\rho = \frac{\delta m}{\delta V} \qquad \text{... (1.1)}$$

A fluid particle may be chosen to be at any point. The stresses acting on the fluid particle are considered to be acting at that point in the fluid. The initial shape of a particle may be chosen arbitrarily. It is considered as a triangular prism or as a parallelopiped for the Cartesian coordinates, segment of a cylinder for the cylindrical system and segment of a sphere for the spherical system of co-ordinates for the sake of simplicity. The shape and size of a particle will, in general, change in the flow.

An element or a particle should not be confused with atoms and molecules. The concept of an element or a particle in a continuous fluid flow refers to an entity of small dimensions., small enough for a differential study in accordance with the requirements of the differential and integral calculus disregarding the atomic and molecular nature of matter.

1.7. Control Volume

A control volume is an identified volume fixed in space. A control volume may either be infinitesimally small or as large as desired. A control volume is bounded by a control surface. An infinitesimal control volume is called a differential control volume. A differential control volume is assumed to have very small but non-zero dimensions. Its volume is denoted as δV and its control surface as δS. The shape of the differential control volume may be chosen arbitrarily ; a parallelopiped for the rectangular co-ordinates and a segment of a hollow cylinder for the cylindrical co-ordinates.

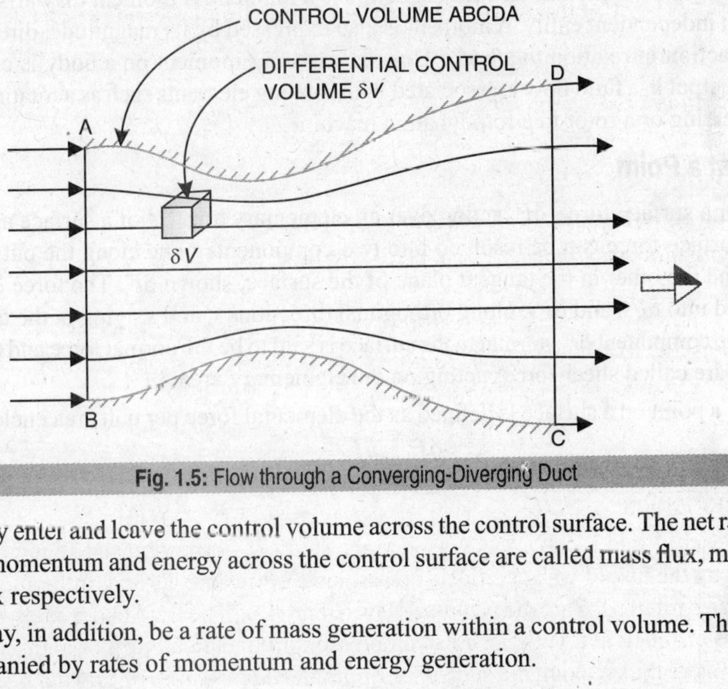

Fig. 1.5: Flow through a Converging-Diverging Duct

Fluid may enter and leave the control volume across the control surface. The net rates of change of fluid mass, momentum and energy across the control surface are called mass flux, momentum flux and energy flux respectively.

There may, in addition, be a rate of mass generation within a control volume. This may or may not be accompanied by rates of momentum and energy generation.

1.8. Force and Moment

Force is a manifestation of an external action on a body. A force changes or tends to change the state, orientation, shape or size of a body. A force acting on a body is specified by the magnitude, direction and sense as well as the line of action.

If a force is specified as a continuous function of space, it is said to constitute a force-field. In general,

$$F = F(x, y, z)$$

In particular, if a force is a function of one space co-ordinate it is said to constitute a linear force field. Examples of this are the spring force acting on a body in contact with it during compression and expansion of the spring and the gravitational force field directed towards the centre of the earth.

Forces are also classified as

(a) *Point forces*

(b) *Line forces*

(c) *Surface forces and*

(d) *Body forces*

A point force is assumed to act as a single concentrated force on a small but non-zero elemental area. A line force is assumed to act over a line element ; an example of this is the surface tension force. A force which acts distributed over the surface area is called a surface force and a force acting distributed over the volume of a body is termed as a body force. If the body in question is a fluid mass then the forces due to pressure and shearing action are surface forces whereas the force due to gravitational attraction is a body force. It is customary, to use the term 'force per unit length' for a line force, 'force per unit area' for a surface force and 'force per unit volume or per unit mass' for a body force.

The other manifestation of an external action is a moment. A moment may arise out of a set of forces or be an independent entity. A moment is also expressed by its magnitude, direction and sense as well as its action in relation to a body. Uniaxial twisting moment on a body is called as torque. Torque input/output in a fluid flow is associated with rotating elements such as a rotating disc, rotating journal in a bearing or a rotor in a rotodynamic machine.

1.9. Stress at a Point

Consider a surface force δF acting over an elementary area δA of a surface as shown in Fig. 1.6 (*a*). The surface force can be resolved into two components : one along the outward normal n, shown δF_n, and the other in the tangent plane of the surface, shown δF_t. The force δF_t may further be decomposed into δFs_1 and δFs_2 along orthogonal directions s_1 and s_2 lying in the tangent plane of the surface. The component δF_n normal to the surface is said to be the normal force and the components δFs_1 and δFs_2 are called shear forces acting on the elementary area δA.

Stress at a point on a surface is defined as the elemental force per unit area enclosing the point.

$$\text{Stress} = \underset{\delta A \to 0}{Lt.} \frac{\delta F}{\delta A} = \frac{dF}{dA} \quad \text{... (1.2)}$$

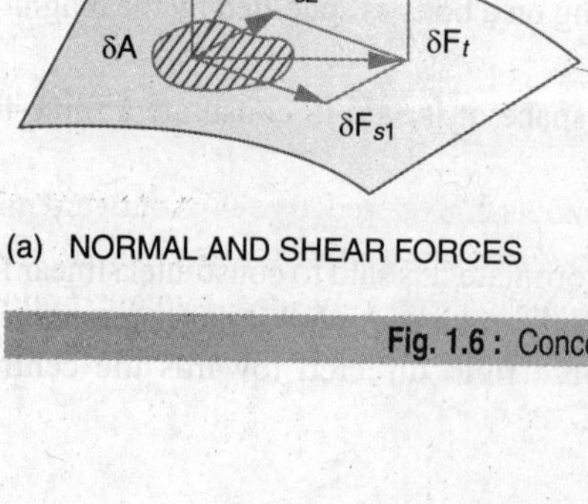

(a) NORMAL AND SHEAR FORCES

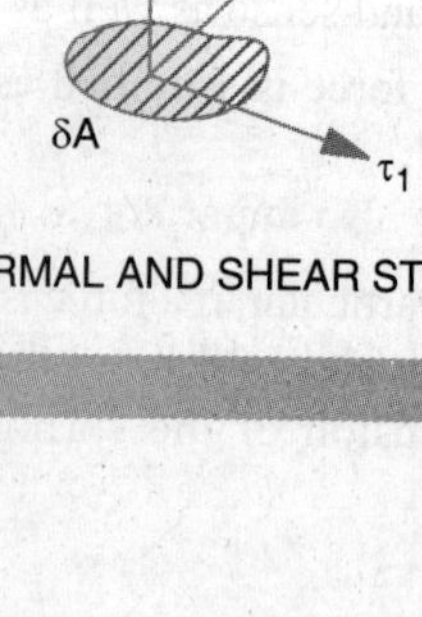

(b) NORMAL AND SHEAR STRESSES

Fig. 1.6 : Concept of Normal and Shear Forces

Normal stress σ at a point refers to the stress corresponding to the normal component, of the elemental force,

$$\sigma = \underset{\delta A \to 0}{Lt.} \frac{\delta F_n}{\delta A} = \frac{dF_n}{dA}$$

Shear stresses at a point refer to the stresses due to the shear forces,

$$\tau_1 = \frac{dFs_1}{dA} \text{ and } \tau_2 = \frac{dFs_2}{dA}$$

It follows that, at a point on a surface, there are orthogonal stress components : σ directed normal to the surface at that point and τ_1, τ_2 lying in the tangent plane of the surface at that point. This is shown in Fig. 1.6 (*b*).

Let us now consider the stress at a point *P* in space. The point is enclosed by an elementary parallelopiped with six surfaces. Each surface would be subjected to three components of shear stress, one normal to it and two lying in the plane of the surface. It is necessary to designate a stress component by two subscripts.

The first subscript stands for the direction of the normal to the plane on which it acts and the second subscript in the direction of the component of force producing it. For example, τ_{xy} refers to the shear stress on a face normal to *x* axis and the direction of the shear stress is along *y* axis. σ_{zz} refers to the normal stress on a face normal to *z* axis and due to a force along *z* axis: that is why it is normal stress. On a parallelopiped fluid element, there must be eighteen stresses, three per face as in Fig. 1.7. Of these, the nine stresses on the positive faces must equal the corresponding nine stresses on the three negative faces. Finally only six of these are independent stresses. *i.e.,* the three normal stresses σ_{xx}, σ_{yy}, σ_{zz} and only three of the six shear stresses $\tau_{xy}, \tau_{yz}, \tau_{zx}$ as a result of the complimentarity principle, whereby

$$\tau_{xy} = \tau_{yx},\ \tau_{yz} = \tau_{zy} \text{ and } \tau_{xz} = \tau_{zx},$$

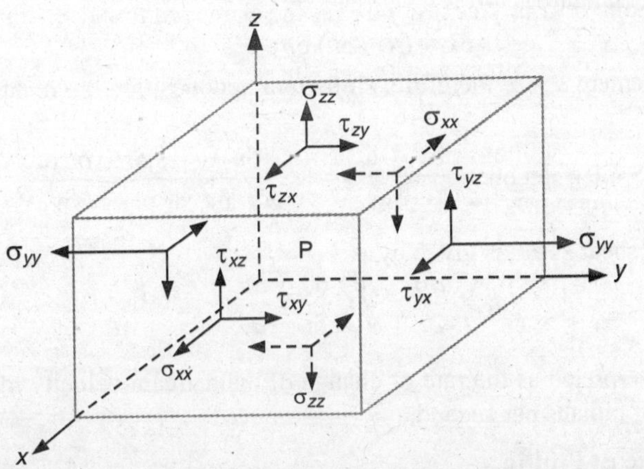

Fig. 1.7: Stress on the Faces of an Element Enclosing a Point P

It is also important to understand that an area element is considered + ve if its outward normal is in the + ve co-ordinate direction and – ve if otherwise. A stress is + ve if it acts along a + ve co-ordinate direction on a + ve area element or along a – ve co-ordinate on a – ve area element. Following this convention, all the 18 stresses have been shown + ve and most of them are labelled.

1.10. Rate of Strain

The rate of strain at a point in a fluid refers to the rate of shearing of the adjacent layers at that point. This may be interpreted in terms of the rate of change of angle θ of a line *a-a* joining the identified fluid elements in layers just above and just below the point *P.* Consider the changes that take place over an interval of time δt.

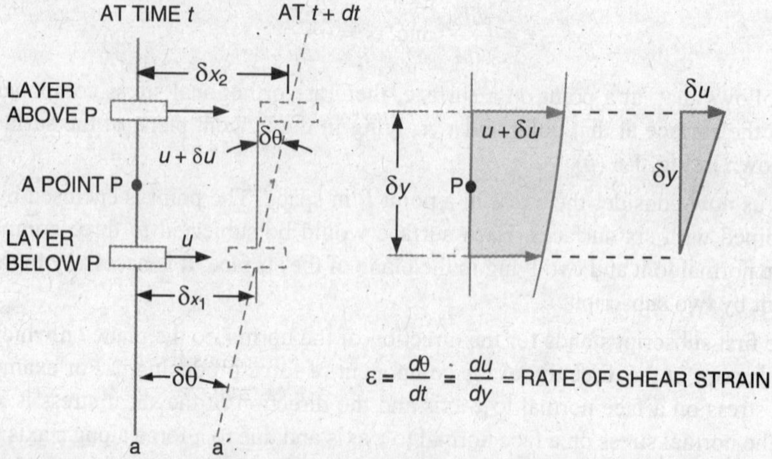

Fig. 1.8: Concept of Rate of Strain at a Point P

The element in the layer below *P,* moving with a velocity *u*, translates a distance δx_1 to the dotted position such that

$$\delta x_1 = u\,.\,\delta t$$

and the element in the layer above *P,* moving with a velocity $(u + \delta u)$ translates a distance δx_2 to the dotted position such that

$$\delta x_2 = (u + \delta u).\delta t$$

The line element $a - a$, therefore, shifts to a new position $a' - a'$ such that the inclination $\delta\theta$ is given by

$$\delta\theta \cong \tan\delta\theta = \frac{\delta x_2 - \delta x_1}{\delta y} = \frac{(u+\delta u).\delta t - u.\delta t}{\delta y} = \frac{\delta u.\delta t}{\delta y}$$

The rate of shear strain is given by

$$\varepsilon = \frac{d\theta}{dt} = \underset{\delta t \to 0}{Lt} \frac{\delta\theta}{\delta t} = \frac{du}{dy} \qquad ...(1.4)$$

which may be interpreted as the rate of change of mainstream velocity in the transverse direction. This is in terms of radians per second.

1.11. Properties of Fluids

1.11.1. Density ρ

Density, also called as mass density, is defined as the mass of a substance per unit volume. For a fluid continuum, density must be a continuous function of space. If a fluid element enclosing a point *P* has a volume d*V* and mass d*m,* then the density is given by

$$\rho = \underset{\delta V \to 0}{Lt} \frac{\delta m}{\delta V} = \frac{dm}{dV}$$

Obviously, density is a local fluid property which may vary from point to point as well as at a point with the passage of time.

The weight of a substance per unit volume is called the specific weight.

$$\text{Specific weight} = \underset{\delta V \to 0}{Lt} \frac{g\,\delta m}{\delta V} = \frac{g\,dm}{dV} = \rho g \qquad ...(1.5)$$

The specific gravity S of a liquid is the ratio of its density to that of water at 4°C and the specific gravity of a gas is the ratio of its density to that of air at STP ; water and air being accepted as the reference liquid and gaseous fluids respectively.

It is customary in the study of thermo-fluid dynamics to use the terms 'specific volume'. The specific volume is the volume per unit mass of the fluid. It follows, therefore, that the specific volume v is the inverse of the mass density ρ;

$$v = \frac{1}{\rho}$$

1.11.2. Pressure *p*

The pressure at a point in a fluid continuum is defined as the normal compressive force per unit area at that point. In other words, the compressive normal stress is referred to as the pressure ;

$$p = -\sigma \qquad ...(1.6)$$

It may be noted that fluids can withstand high compressive stresses or pressure but they are weak in tension. It is for this reason that the pressure in a fluid is always referred above the absolute value since negative absolute pressure would imply tension.

Atmospheric pressure may also serve as a suitable reference for pressure measurement. Pressure above the atmospheric value is called 'gauge pressure' and that below it is called 'vacuum' or sub-atmospheric pressure. Relationship between the pressure stated for different references is shown in Fig. 1.9.

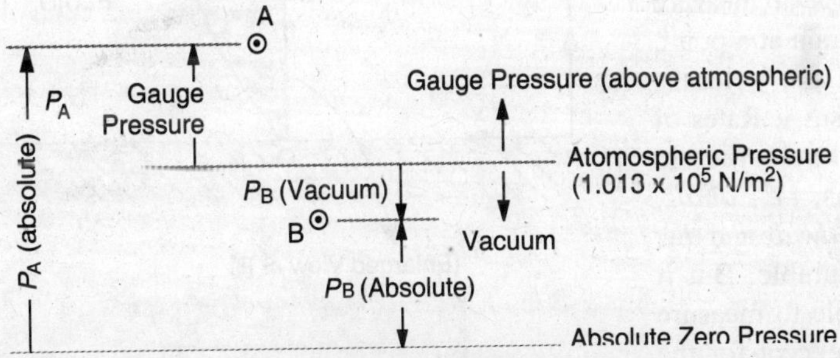

Fig. 1.9: Pressure References

Normal stress at a point may be different in different directions. This would mean that the pressure may be different in different directions. It is indeed so but the term 'thermodynamic pressure', or simply pressure is generally used to imply the 'mean pressure', i.e.,

$$p = -\frac{\sigma_{xx} + \sigma_{yy} + \sigma_{zz}}{3} \qquad ...(1.7)$$

In a fluid at rest or in motion in the absence of shear stresses, the normal stresses at a point are equal. Under such circumstances,

$$P = -\sigma_{xx} = -\sigma_{yy} = -\sigma_{zz}$$

1.11.3. Other Thermodynamic Variables

Temperature T, internal energy i, entropy s etc. for a gaseous fluid can be defined and estimated by involving them in the equations governing the flow. This can be done by invoking the thermodynamic relations which relate the thermodynamic variables.

For example, if the temperature T is to be included, the equation of state should be invoked. For a perfect gas,

$$p = \rho RT$$

where R is the characteristic gas constant.

$$R = \frac{\text{universal gas contant}}{\text{molecular weight of the gas}}$$

If, in addition to the temperature, internal energy is to be included, the definition of internal energy i should be invoked. For a perfect gas,

$$\delta i = C_v \delta T$$

where C_v is the specific heat at constant volume ; this is taken as constant during the flow. Since the ratio of specific heats

$$C_p / C_v = \gamma$$

is constant for a fluid, it follows that C_p should also be constant for a perfect gas.

$$\text{Also,}\quad R = C_p - C_v$$

1.11.4. Newton's Law of Viscosity

We have noticed that stress and 'rate of strain' at a point in a fluid are fundamental attributes for the flow of a fluid. Well, it is the application of a shear stress which brings about the rate of strain or flow of a fluid. It is easy to measure the rate of strain at a point, du/dy, from the velocity profile u versus y. Rates of strain for all velocities in all directions, *i.e.*, du/dz, dv/dz, dv/dx, dw/dx and dw/dy are measurable. But it is not possible to measure the stresses except for the normal stresses σ_{xx}, σ_{yy} and σ_{zz}. Newton succeeded in his attempt to relate the shear stress with the rate of strain at a point in a fluid. In fact, the Newton's law has a far reaching implication; it places all fluids on a common footing irrespective of their liquid, gaseous and vaporous states. This is by virtue of the property

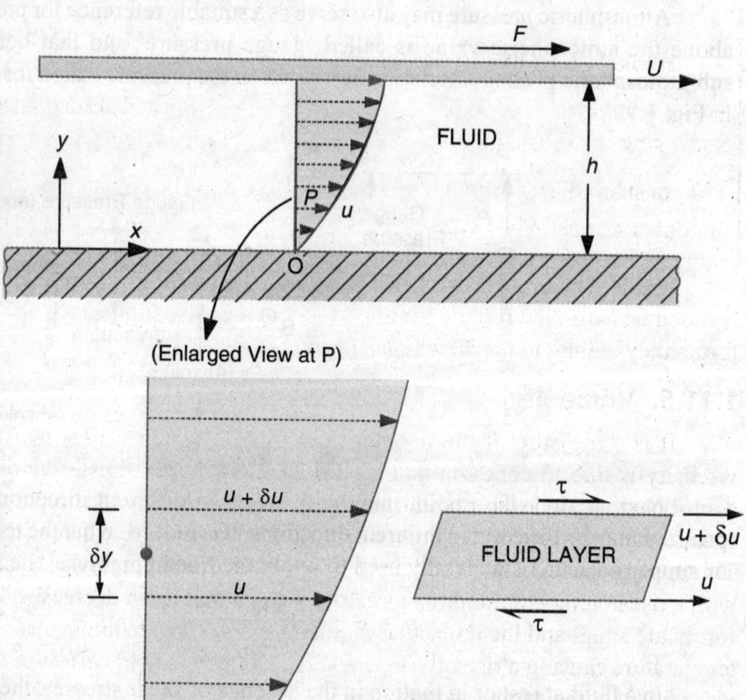

Fig. 1.10: Relationship between Stress and Rate of Strain

of viscosity common to all of them. Physical origin of viscosity in liquids and gases may be different but the fact that most of them obey Newton's law widens the scope of the subject of fluid mechanics.

According to the Newton's law of viscosity, the shear stress on a fluid element layer is directly proportional to the rate of strain, the constant of proportionality being called the coefficient of viscosity.

Consider, for example, a fluid contained between two parallel plates, distance h apart, as shown in Fig. 1.10. If the lower plate is held stationary and the upper plate is moved at a small but constant velocity u by the application of a force F, the fluid will be subjected to shearing forces and the velocity will vary across the gap. A typical velocity profile is shown. Consider the velocity profile, as shown by an enlarged view in the vicinity of a point P. The rate of strain at the point is given by

$$\varepsilon = \frac{du}{dy}$$

and the shear stresses across the layer is indicated as τ. According to the Newton's law of viscosity,

$$\tau \propto \varepsilon \text{ or } \tau \propto \frac{du}{dy}$$

which may be written as

$$\tau = \mu \frac{du}{dy} \qquad \text{... (1.8)}$$

where the constant of proportionality μ is called the coefficient of viscosity.

One may notice that the Newton's law of viscosity for fluids is analogous to the Hooke's law of elasticity for solids.

TABLE 1.1

Hooke's Law of Elasticity for Solids	Newton's Law of Viscosity for Fluids
Stress $\propto$ strain Shear stress $\propto$ shear strain	Shear stress $\propto$ rate of strain or velocity gradient.
The constant of proportionality is the modulus of elasticity.	The constant of proportionality is the coefficient of viscosity.

Fluids which obey this law are termed **Newtonian** and those which disobey it are broadly branded as **Non-Newtonian Fluids.** Most engineering fluids such as water, air and molten metals fortunately belong to the Newtonian class.

1.11.5. Viscosity

It is a measure of the internal fluid friction which causes resistance to flow. The origin of viscosity is due to cohesion and molecular momentum transfer across layers. These two factors contribute to provide the resistance to shearing reckoned by the property of viscosity. It is interesting to note that the viscosity of liquids drops with temperature whereas the reverse happens for gases. This is so because, in liquids, the cohesive forces predominate due to closely packed molecules and with a rise in temperature these forces decrease resulting in decrease of viscosity. In gases, cohesive forces are small and the molecular momentum transfer predominates which increases with a rise of temperature causing a rise of viscosity. Fig. 1.11 presents the effect of change in temperature on the viscosity of some engineering fluids.

The following expressions describe the variation of viscosity for liquids and gases adequately:

For a liquid, $\mu = \dfrac{\mu_0}{(1+\alpha t+\beta t^2)}$ ... (1.9) where μ_0, α and β are constants for the liquid.

For example, for water,

$$\mu = \frac{1.79\times10^{-3}}{(1+33.68\times10^{-2}t+2.21\times10^{-4}t^2)}$$

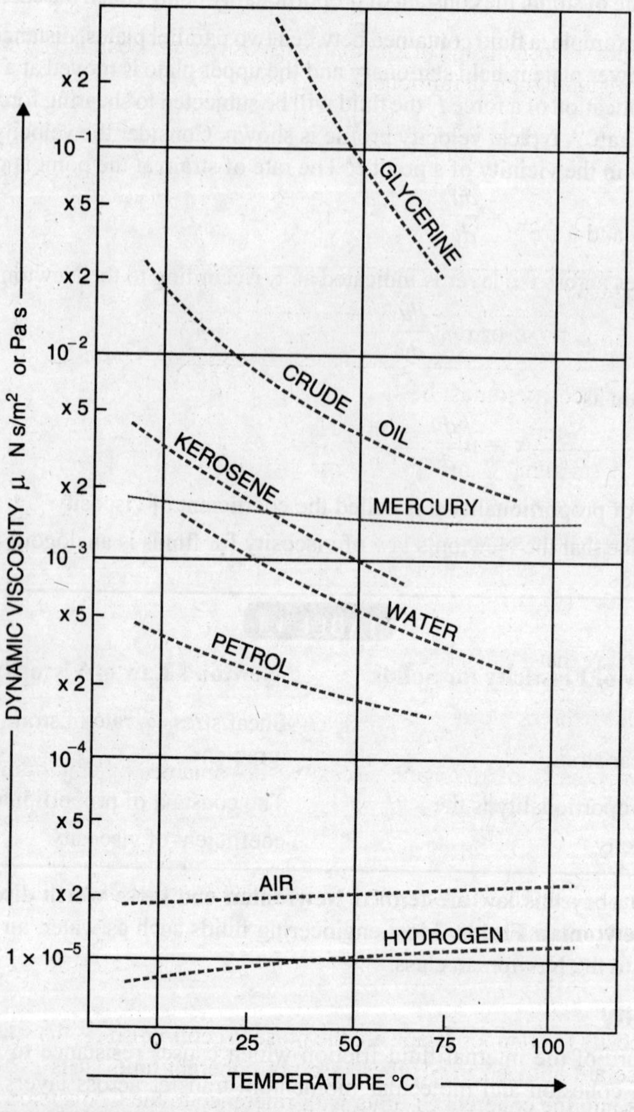

Fig. 1.11: Viscosity vs Temperature for Common Fluids

For a gas, on the other hand,

$$\mu = \mu_0 + \alpha\, t - \beta\, t^2 \qquad \text{... (1.10)}$$

For example, for air,

$$\mu = (17.040 + 56.02t - 0.1189\, t^2) \times 10^{-3}$$

The temperature t in the above expressions is taken in degrees celcius and the viscosity μ appears in SI units, *i.e.*, N s/m² or kg/m s or Pa s.

The viscosity μ as defined by Eq. (1.8) is frequently referred to as absolute or dynamic viscosity and the ratio of the dynamic viscosity to the fluid density is commonly termed as the kinematic viscosity v,

$$v = \frac{\mu}{\rho} \qquad \text{... (1.11)}$$

Units of dynamic viscosity μ and of kinematic viscosity v can be found by considering the units of the quantities involved in their definitions.

Since $$\mu = \frac{\tau}{\frac{du}{dy}}$$

and the units of τ, u and y are

Quantity	τ	u	y
Units	N/m²	m/s	m

the units of dynamic viscosity μ must be

$$\frac{\text{N/m}^2}{\text{(m/s)/m}} = \frac{\text{N}}{\text{m}^2}.\frac{\text{s}}{\text{m}}.\frac{\text{m}}{1} = \frac{\text{Ns}}{\text{m}^2}$$

Now that 1 N = 1 kg m/s²

the units of μ are expressible as

$$\frac{\text{Ns}}{\text{m}^2} = \frac{\text{kg m}}{\text{s}^2}\frac{\text{s}}{\text{m}^2} = \frac{\text{kg}}{\text{m s}}$$

Also, recognising that 1 Pa = 1 N/m²

the units of μ are also expressible as $\frac{\text{Ns}}{\text{m}^2}$ = Pa s

In the same way, units of kinematic viscosity v are obtained by employing the expression ;

$$v = \frac{\mu}{\rho}$$

Since the units of μ are N s/m² or kg/m s or Pa s, and the units of density ρ are kg/m³, the units of kinematic viscosity v must be

$$\frac{\text{N s/m}^2}{\text{kg/m}^2} = \frac{\text{kg/ms}}{\text{kg/m}^3} = \frac{\text{m}^2}{\text{s}}$$

Units of viscosity in common usage are the poise P, centi-poise cP, the stoke St and centistoke cSt. Although these are allowed in SI, these are not coherent units. It is important, therefore, to convert these units into the coherent SI units with reference to the Conversion Tables before using them in the calculations.

Multiply the Viscosity in		By	To Obtain the SI Units
Dynamic Viscosity	poise P	×0.1	Ns/m² ≡ kg/ms ≡ Pas
	centipoise cP	$\times 10^{-3}$	
Kinematic Viscosity	stoke St	$\times 10^{-4}$	m²/s
	centistoke cSt	$\times 10^{-6}$	

The effect of viscosity in a fluid flow is to give rise to shear stresses which oppose the motion of neighbouring elements in the flow. The velocity gradients or the rates of strain depend upon the boundary conditions. The velocity of a fluid in contact with a boundary must be the same as that of the boundary. For example, the velocity of a fluid flowing through a stationary pipe should be zero at the wall. This is known as the condition of 'no slip'. It may be noted that the velocity of a fluid must drop from the maximum value to zero continuously and monotonically.

Properties of some common fluids are listed in Table 1.2.

Table 1.2 **Properties of Common Fluids at S.T.P. (288.15 K and 1.01325×10^5 N/m²)**

Properties	Water	Air	Mercury	Lub. Oil	Glycerine
Mass Density ρ (kg/m³)	1000	1.22	13550	930	1260
Dynamic Viscosity μ (Ns/m²)	0.001	0.000 018	0.0016	0.1	0.0015
Kinematic Viscosity ν (m²/s)	1×10^{-6}	15×10^{-6}	$0.1\ 18 \times 10^{-6}$	1.08×10^{-6}	1.1×10^{-6}
Bulk modulus of Elasticity K (N/m²)	2.09×10^9	–	26×10^9	1.4×10^9	0.03×10^9
Surface Tension with Air σ (N/m)	0.073	–	0.505	0.035	0.06

1.11.6. Water as a Reference Fluid

Water, in its pure form, has been used as a reference fluid for the specifications of density, dynamic viscosity and kinematic viscosity of other fluids. The standard conditions of pure distilled water are that its freezing point should be 0°C, the boiling point 100°C and critical temperature 374°C and the critical pressure 221.2×10^5 N/m². The pure distilled water referred to may contain about one part in 6500 of heavy water (which has a dynamic viscosity approximately 1.25 times that of water). The standard reference of temperature and pressure conditions for water are 20°C and 101.325 kN/m² respectively. At these conditions the dynamic viscosity of water is

$$\mu = 1.0020 \text{ cP } (= 1.0020 \times 10^{-3} \text{ Ns/m}^2)$$

and the kinematic viscosity is

$$\nu = 1.0038 \text{ cSt } (= 1.0038 \times 10^{-6} \text{ m}^2/\text{s})$$

the density being

$$\rho = 998.23 \text{ kg/m}^3$$

The viscosity of water decreases with the rise of temperature.

Example 1.1. *The absolute viscosity μ of water., air and a lubricating oil at 20°C are 1 cP, 0.018 cP and 100 cP respectively. Express the values in SI units and determine the kinematic viscosity ν of these fluids. The densities are 1.0, 0.0012 and 0.930 gm/cm³ respectively.*

Solution.

Let us proceed in a tabular form as follows :

Property	Water	Air	Lube Oil
Absolute Viscosity μ			
centipoise cP	1.0	0.018	100.0
poise = g/cm s = P	0.01	0.000 18	1.0
kg/m s = N s/m²			
in SI units	0.001	18×10^{-6}	0.1
Mass Density ρ			
g/cm³	1.0	0.001	20.930
kg/m³ in SI units	1000.0	1.2	930.0
Kinematic Viscosity ν			
= μ/ρ			
≡ cm²/s ≡ stokes = St	0.01	0.15	1.08
(kg/ms)/(kg/m³)	1×10^{-6}	15×10^{-6}	108×10^{-6}
≡ m²/s in SI units			

Example 1.2. *A flat plate 0.1 m² area is pulled at 30 cm/s relative to another plate located at a distance of 0.01 cm from it, the fluid separating them being water with μ as 0.001 Ns/m². Find the force and power required to maintain the velocity.*

Solution.

For the configuration shown,

$$\delta u = 0.3 - 0 = 0.3 \text{ m/s},$$

$$\delta y = 0.0001 \text{ m}$$

$$\frac{du}{dy} \approx \frac{0.3}{0.0001} = 3000 \text{ rad/s}$$

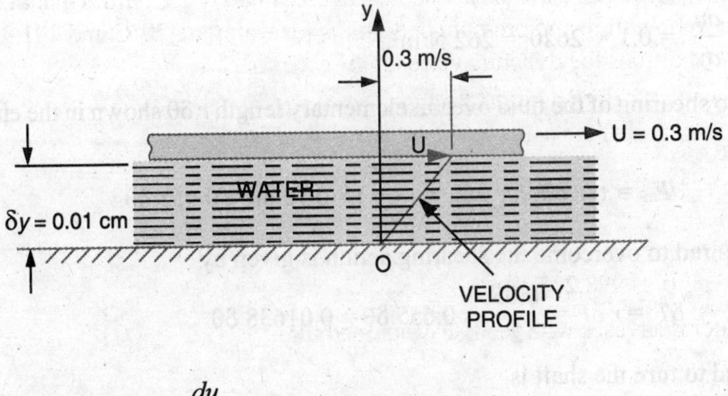

$$\text{Shear stress } \tau = \mu\frac{du}{dy} = 0.001 \times 3000 = 3 \text{ N/m}^2$$

Shear force, *i.e.*., the force required to maintain this condition

$$F = \tau.A = 3 \times 0.1 = 0.3 \text{ N}$$

Power required = $F.U$= .0.3 × 0.3 = 0.09 Nm/s = **0.09 W.**

Example 1.3. *Determine the torque and power required to turn a 10 cm long, 5 cm diameter shaft rotating at 500 revolutions per minute in a 5.1 cm diameter concentric bearing flooded with a lubricating oil of viscosity 100 centipoise.*

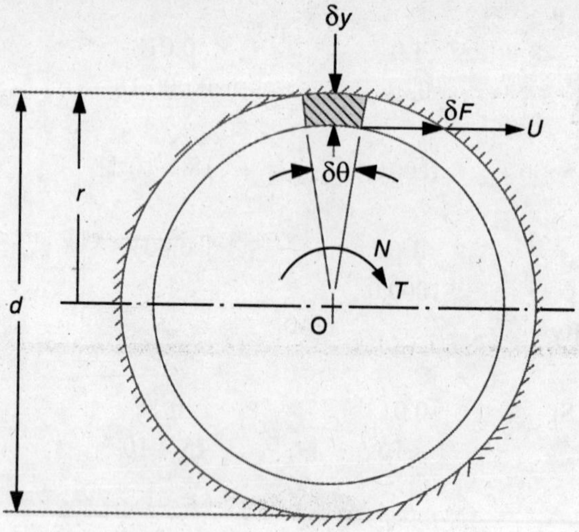

Solution.

Peripheral speed u of the shaft = $\frac{\pi d N}{60} = \frac{\pi \times 0.05 \times 500}{60}$ =1.31 m/s

Since the bearing is at rest,

$\delta u = 1.31 - 0 = 1.31$ m/s ; $\delta y = \frac{5.1-5}{2} = 0.05$ cm = 0.0005 m

$$\frac{du}{dy} = \frac{1.31}{0.0005} = 2620 \text{ rad /s}$$

μ = 100 centipoise = 1 poise = 0.1 N s/m^2

Shear stress $\tau = \mu \frac{du}{dy} = 0.1 \times 2620 = 262$ N/m^2.

The force δF due to shearing of the fluid over an elementary length $r\,\delta\theta$ shown in the end-view is given by

$$dF = \tau.\, r\,\delta\theta.\, l = 262 \times \frac{0.05}{2} \times 0.10\,\delta\theta = 0.655\,\delta\theta.$$

The torque δT required to overcome the shearing action is given by

$$\delta T = r\,\delta F = \frac{0.05}{2} \times 0.655\,\delta\theta = 0.01638\,\delta\theta$$

The torque required to turn the shaft is

$$T = \int dT = \int_0^{2\pi} 0.01638\, d\theta = \mathbf{0.1029\ Nm}$$

The power required to turn the shaft is given by

$$\text{Power} = T.\omega = 0.102 \times 1.31/0.025 = \mathbf{5.387\ W}$$

Example 1.4. *A thrust bearing consists of a 10 cm diameter pad rotating on another pad separated by an oil film μ = 80 cP, by 1.5 mm. Compute the power dissipated in the bearing if it rotates at 100 revolutions per minute.*

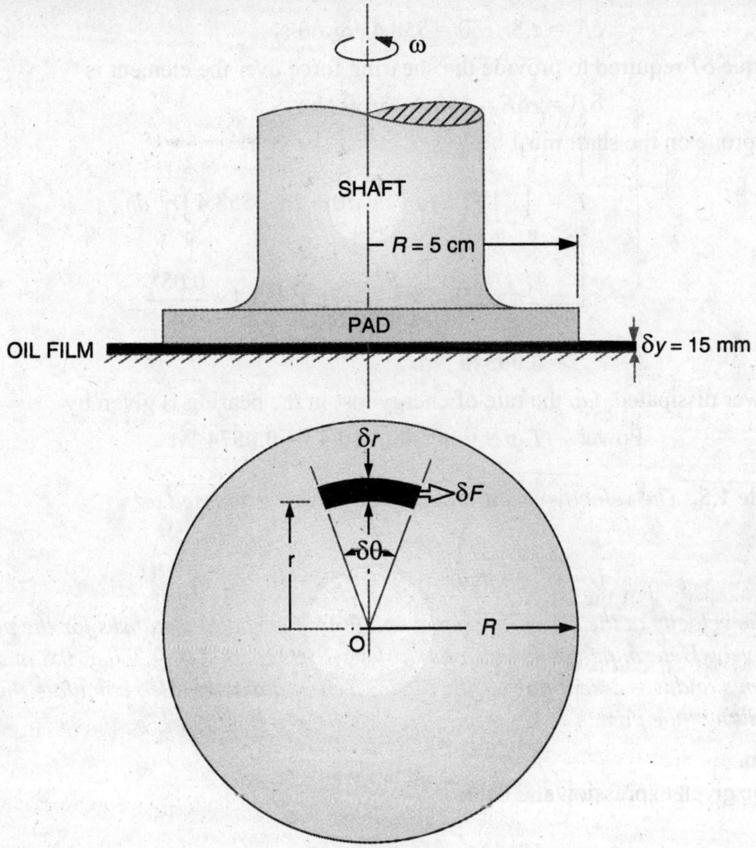

Thrust Bearing

Solution.

Consider a small element of width δr at a radius r subtending a small angle $\delta\theta$ at the centre of the pad as shown in the plan view. The area of the element is $\delta r \times r\delta\theta$.

It experiences a shear force δF due to the shearing action of the fluid contained between the pads.

The rotational speed of the pad is given by

$$\omega = 2\pi \times \frac{100}{60} = 10.47 \text{ rad/s}$$

The linear velocity u at a radial distance r must be

$$u = r\omega = 10.47r \text{ m/s}$$

Since the lower pad is at rest, the velocity gradient

$$\frac{du}{dy} = \frac{10.47r - 0}{0.0015} = 6980\, r \text{ rad/s}$$

From the data,

$$\mu = 80 \text{ cP} = 0.8\text{P} = 0.08 \text{ Ns/m}^2$$

The shearing stress at the location of the element must be

$$\tau = \mu \frac{du}{dy} = 0.08 \times 6980\, r = 558.4\, r \text{ N/m}^2$$

and the shearing force δF is given by

$$\delta F = \tau.\delta r.r\delta\theta = 558.4 r^2 \delta r\, \delta\theta \text{ N}$$

The torque δT required to provide the shearing force over the element is

$$\delta T = r\delta F = 558.4 r^3\, \delta r \delta\theta \text{ Nm}$$

and the total torque on the shaft must be

$$T = \int_0^R \int_0^{2\pi} 558.4\, r^3\, dr\, d\theta = 2\pi \times 558.4 \int_0^R r^3\, dr$$

$$= 2\pi \times 558.4 \times \frac{R^4}{4} = 2\pi \times 558.4 \times \frac{0.05^4}{4}$$

$$= \mathbf{0.00548\ Nm}$$

The power dissipated, *i.e.* the rate of energy lost in the bearing is given by

$$\text{Power} = T.\omega = 0.00548 \times 10.47 = \mathbf{0.0574\ W.}$$

Example 1.5. *The velocity variation in a pipe-flow is expressed as*

$$\frac{u}{U} = 1 - \left(\frac{r}{R}\right)^2$$

where U is the velocity at the centreline and r and R are the radial distances for the point and the wall respectively. Determine the shear stress τ for r/R ratios of 0.0, 0.2, 0.5, 0.8 and 1.0. Given that U = 10 m/s and μ = 2 centipoise. The pipe is 1 m in diameter. Also calculate the drag force per meter length of the pipe.

Solution.

From the given expression and data,

$$u = 10\left(1 - \frac{r}{R}\right)^2$$

and

$$\frac{du}{dr} = -20\left(\frac{r}{R}\right).\frac{1}{R} = -40\left(\frac{r}{R}\right)$$

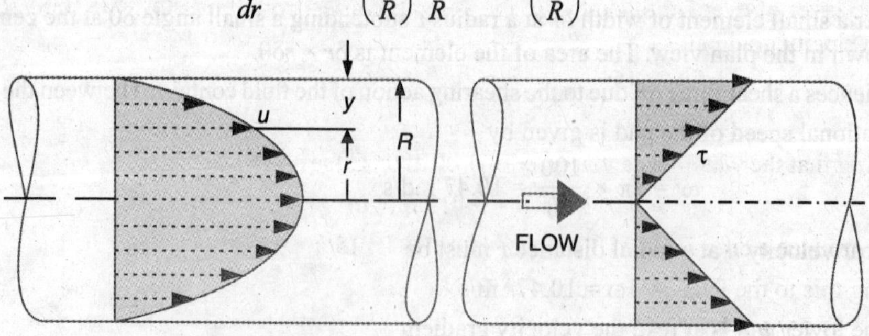

Laminar Flow through a Pipe

From the Newton's law of viscosity,

$$\tau = \mu \frac{du}{dy}$$

where y is the distance from the wall.

For a pipe of radius R,

$$y = R - r \text{ and } dy = -dr$$

Hence

$$\tau = -\mu \frac{du}{dr}$$

Substituting the given values,

$$\tau = -2 \times 0.001 \times (-40) \frac{r}{R} = 0.08 \frac{r}{R}$$

$\frac{r}{R}$	τ (N/m²)	Drag force $F = (2\pi rh)\tau$ (N)
0.0	0.000	0.0000
0.2	0.016	0.0503
0.5	0.040	0.1257
0.8	0.064	0.1885
1.0	0.080	0.2513

The given velocity distribution is parabolic and the corresponding shear stress variation is linear over the cross section. Such is the case for laminar flow in pipes.

Example 1.6. *A skater weighing 800 N skates at 54 km/hr on ice at 0° C. The average skating area supporting him is 10 cm² and the effective dynamic co-efficient of friction between the skates and the ice is 0.02. If there is actually a thin film of water between the skates and the ice, determine its average thickness.*

Solution.

The normal reaction by the ice on the skater equals the weight of the skater, i.e., 800 N.

The frictional reaction is, therefore, given by

$$F = f \times 800 = 0.02 \times 800 = 16 \text{ N}.$$

Over the layer of average thickness h, the velocity varies from 0 m/s to 54 × 1000/3600 or 15 m/s. The velocity gradient is

$$\frac{du}{dy} = \frac{15}{h} \text{ rad/s}$$

Assuming that the water has a viscosity of 1 cP at 0° C the shear stress

$$\tau = 0.001 \times 15/h = 0.015/h \text{ N/m}^2$$

which would produce a drag force of $0.015/h \times 10 \times 10^{-4} = 15/h \times 10^{-6}$ N

Equating this to the frictional reaction of 16 N.

$$\therefore \quad 15/h \times 10^{-6} = 16$$

$$\therefore \quad h = \mathbf{93.75 \times 10^{-6}\ m}$$

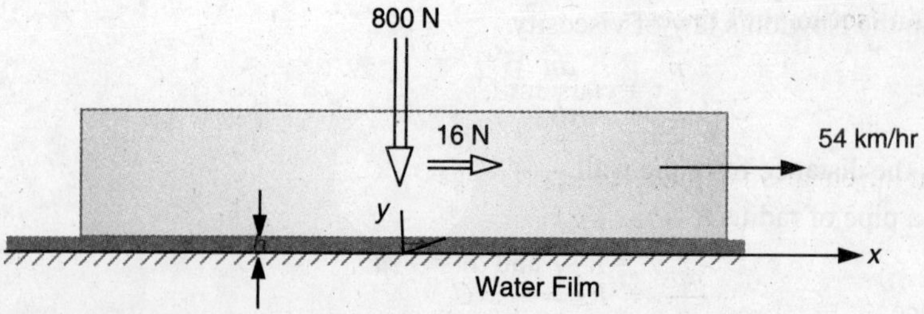

1.11.7 Compressibility

The measure of compressibility of a fluid is the bulk modulus of elasticity K defined as the compressive stress per unit volumetric strain. Alternatively, the coefficient of compressibility β is defined as volumetric strain per unit compressive stress, *i.e.,* the inverse of the bulk modulus K.

$$\beta = \frac{1}{K}$$

Consider the piston and cylinder arrangement shown in Fig. 1.12, let the volume and pressure at any instant of compression be v and p respectively. The compressibility curve drawn between the compressive stress p and the volumetric strain $\delta\, v/v$ is also shown. By definition,

$$K = \frac{dp}{dv/v} = \frac{dp}{d\rho/\rho} = \rho\frac{dp}{d\rho} = \frac{1}{\beta} \qquad ...(1.12)$$

because,

$$\frac{dv}{v} = \frac{d\rho}{\rho}$$

since $v \propto 1/\rho$

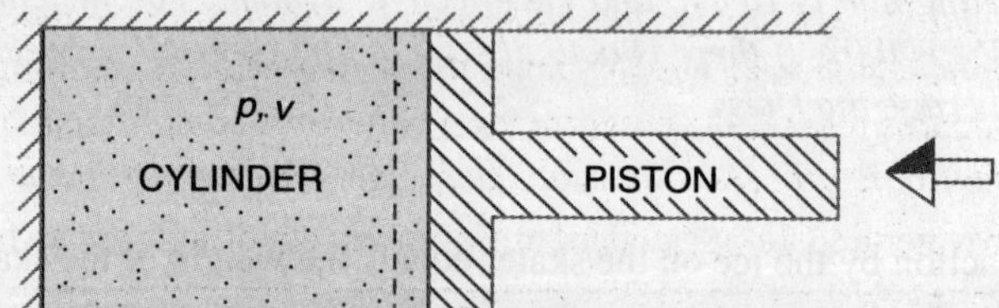

Fig. 1.12 (a)

Units of the bulk modulus of compressibility K can be derived from its definition. Since the units of compressive stress p are N/m^2 and the volumetric strain $\delta v/v$ is dimensionless, units of K must also be N/m^2. Values of K for different fluids are expressed in terms of kN/m^2 or MN/m^2 for the sake of convenience.

Let us now examine the relationships between the bulk modulus of elasticity and the local pressure for a perfect gas for two different processes of compression :

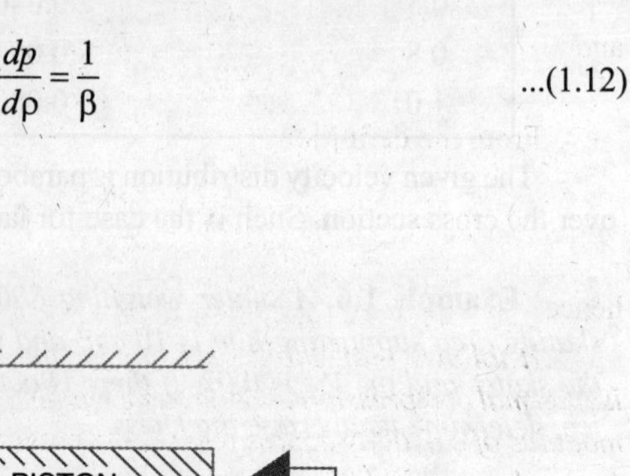

Fig. 1.12 (b): Compressibility of a Fluid

For an Isothermal Process,

$$\frac{p}{\rho} = \text{constant } C_1$$

From the equation of state for a perfect gas,

$$p = \rho RT$$

and $$\frac{dp}{d\rho} = RT = \frac{p}{\rho} = C_1$$

From the definition,

$$K = \frac{dp}{d\rho/\rho} = \rho\frac{dp}{d\rho} = \rho C_1 = p$$

hence, $$K = p \qquad \text{...(1.13)}$$

For an Isentropic Process,

$$\frac{p}{\rho^{\gamma}} = \text{constant } C_2$$

From the equation of state for a perfect gas,

$$p = \rho RT$$

and $$\frac{dp}{d\rho} = RT = \gamma\, \rho^{\gamma-1}.\, C_2$$

From the definition

$$K = \frac{dp}{d\rho/\rho} = \rho\frac{dp}{d\rho} = \gamma\, \rho^{\gamma}.\, C_2 = \gamma\, p$$

hence $$K = \gamma p \qquad \text{...(1.14)}$$

It follows that, for a perfect gas, the bulk modulus of elasticity equals its pressure for an isothermal compression and the same equals γ times the pressure for an isentropic process. In fact, the modulus of elasticity equals n times the pressure for a polytropic compression of index n ; the results for isothermal and isentropic process follow for the special cases of n = 1 and n = γ respectively.

Values of K for water and air at standard atmospheric conditions are as follows :

Fluid	Value of *K*, the Bulk Modulus of Elasticity
Water	2.07×10^6 kN/m^2
Air at STP (isothermal process)	100 kN/m^2
Air at STP (isentropic process)	140 kN/m^2

It may be noted that the property of compressibility plays no role when the fluid velocity is low even for seemingly compressible fluids like air. In the study of Fluid Mechanics, compressibility is regarded important only at high velocities, *i.e.,* nearing the local speed of sound in that medium ; otherwise the fluid flow is treated as incompressible.

The speed of sound in a fluid is given by

$$c = \sqrt{\frac{dp}{d\rho}} = \sqrt{\frac{\gamma p}{\rho}} = \sqrt{\gamma RT}$$

Example 1.7 *Assuming that the bulk modulus of elasticity of water is 2.07 x 10^6 kN/m² at standard atmospheric conditions determine the increase of pressure necessary to produce*

(a) *1% reduction in the volume at the same temperature and*

(b) *1% reduction in the volume of air undergoing an isentropic compression.*

Solution.

From the definition of the bulk modulus of elasticity

$$K = -\frac{dp}{dv / v}$$

Reduction of 1% of volume implies that

$$-\frac{dv}{v} = 0.01$$

(*a*) The increase of pressure for water is, therefore, given by

$$dp = 0.01 \times K = 0.01 \times 2.07 \times 10^6$$

$$= \mathbf{20,\ 700\ kN/m^2}$$

(*b*) For air, the bulk modulus of elasticity is given by

$$K = \gamma p$$

for an isentropic process.

At the standard atmospheric conditions,

$$K = 1.4 \times 101.325 = 141.86 \text{ kN/m}^2$$

The increase in pressure of air is, therefore, given by

$$dp = 0.01 \times 141.86 = \mathbf{1.42\ kN/m^2}$$

This example illustrates that the pressure required to bring about a reduction in the volume of water is approximately 15 000 times the pressure required for the same percentage reduction in the volume of air. This example demonstrates the fact that pipes and pressure vessels must be tested with water and not with air so that if there is a leak, a very small amount of water leakage will release the pressure. Testing with air could result in a blast!

1.11.8 Surface Tension σ

It is the property of the apparent tension effect which occurs at the interface of a liquid and a gas or at the interface of two immiscible liquids. The origin of surface tension is explained by considering the mechanism of cohesive forces within a liquid as in Fig. 1.13. A typical molecule A within the liquid is maintained in equilibrium by the action of cohesive forces exerted by the molecules surrounding it within a zone of action. The resultant force acting upon the molecule *A* is zero because the cohesive forces acting upon it cancel off mutually. Consideration of a molecule *B* located near the surface reveals that it may experience a net resultant force acting downwards. A typical molecule *C* located at the surface of the liquid does experience a net downward, force. While discussing the mechanism of cohesive forces, one should not forget that the zone of action of intermolecular cohesive forces is of the order of a millionth of a centimeter. This fact ensures that the molecules experiencing net downward forces are confined to a layer which defines the surface of the liquid. The surface of a liquid is, therefore, under a different state than the rest of the liquid ; the surface is in a state of tension akin to a stretched thin membrane, the surface tension decreases as the temperature of the liquid increases because the intermolecular cohesive forces decrease. At the critical temperature of a liquid, the surface tension becomes zero and the boundary between the liquid and its vapour vanishes !

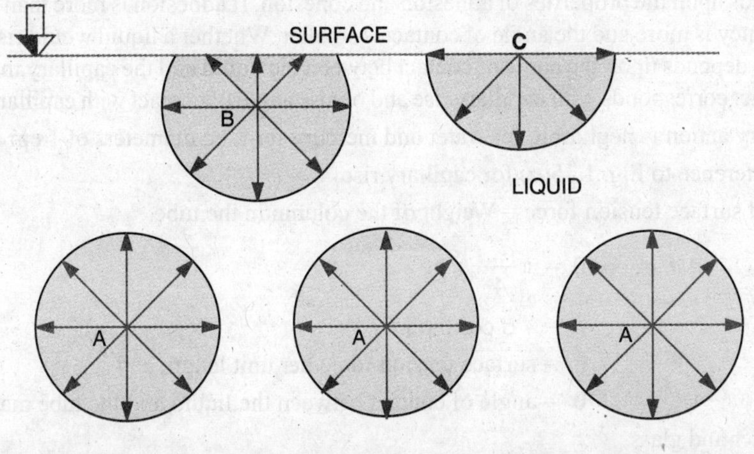

Fig. 1.13: Origin of Surface Tension

The force of surface tension is a line force ; it is expressed as force per unit length drawn on the free surface of a liquid. The phenomenon of surface tension is alternatively explained by the existence of free-surface energy at the interface of a liquid and another fluid. The surface tension is likewise expressed as free surface energy per unit length of the surface.

Water in contact with air has a surface tension of 0.073 N/m at 20°C. Mercury in contact with air has a surface tension of 0.5 N/m. The surface tension of water is considerably reduced by adding a small quantity of an organic solute such as soap or detergent. The surface tension of water increases if a salt such as sodium chloride is dissolved in it. Organic liquids such as alcohol have surface tension varying between 0.02 and 0.03 N/m.

1.11.9 Capillarity, Bubbles and Liquid-Jets

The force due to surface tension is generally negligible in comparison with the pressure and gravitational forces but the same assumes predominance for narrow liquid columns such as those in capillary tubes, in bubble mechanics ind liquid-jct studies and for flow through porous substances.

Capillarity implies the rise or depression of the level of a liquid in a capillary tube when it is held vertically or inclined in the liquid. The phenomenon is explained with reference to the angle of contact between the liquid and the capillary mâterial and the balance of force on the column of the liquid in them. The angle of contact is defined as the angle between the tangent to a drop at the solid boundary with the surface of the solid boundary on which it is placed as shown in Fig. 1.14.

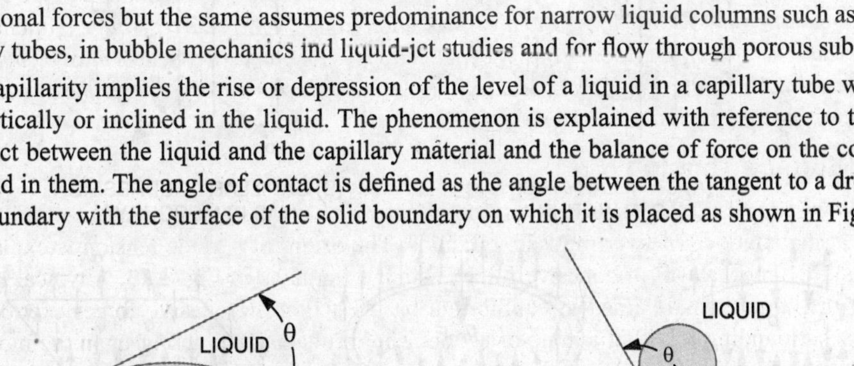

Fig. 1.14: Angle of Contact

If a liquid wets a surface and spreads out, the angle of contact between them is small but if a liquid does not do so, the angle of contact is large. The extent to which a liquid may wet a solid surface depends upon the properties of adhesion and cohesion. If adhesion is more than cohesion, the wetting tendency is more and the angle of contact is smaller. Whether a liquid would rise or drop in a capillary tube depends upon the angle of contact between the liquid and the capillary material. Acute angle of contact corresponds with capillary rise and obtuse angle of contact with capillary depression.

Capillary action is negligible for water and mercury for tube diameters of 1 cm or more.

With reference to Fig. 1.15(*a*) for capillary rise,

Upward surface tension force = Weight of the column in the tube

$$\pi\, d.\, \sigma.\, \cos\theta = \pi \frac{d^2}{4}.\, h.\, \rho g \qquad \text{... (1.15)}$$

whence $h = 4\,\sigma \cos\theta/\rho g d$

where σ = surface tension force per unit length and

θ = angle of contact between the liquid and the tube material.

For water and glass

$$\theta \approx 0$$

hence the capillary rise of water in a glass tube

$$h = 4\,\sigma/\rho g d$$

It may be noted that if a liquid surface supports another fluid of density ρ_b above the meniscus, then a balance of forces would result in

$$h = \frac{4\,\sigma\cos\theta}{(\rho - \rho_b)\, g d} \qquad \text{...(1.15 a)}$$

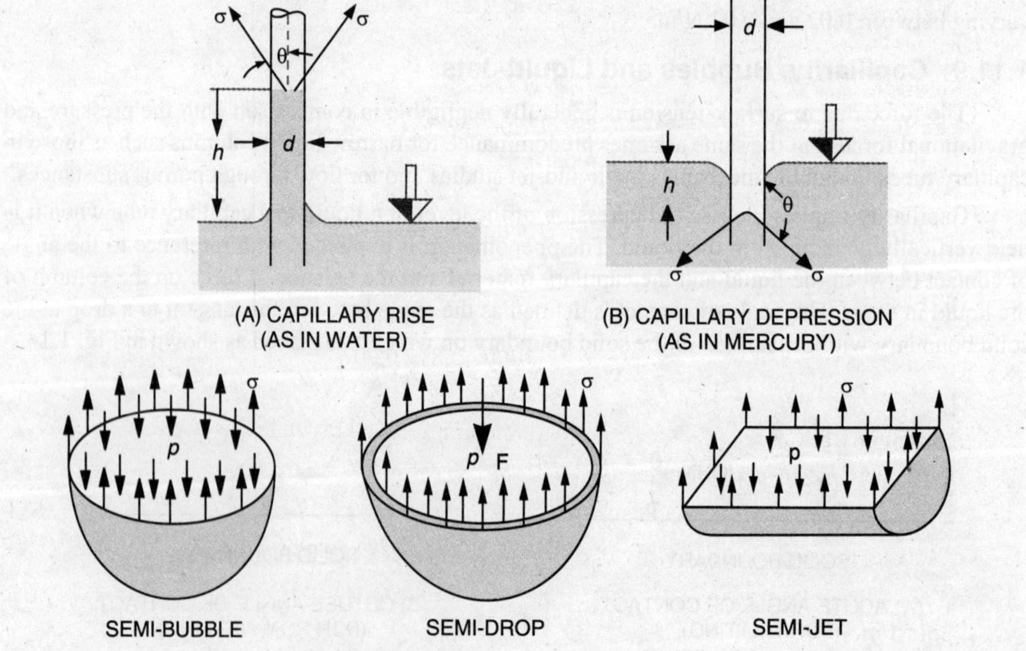

Fig. 1.15: Phenomena Explained by the Presence of Surface Tension

Referring now to Fig. 1.15(b) for capillary depression, the depression in given by the balance of downward force of surface tension and the upward hydrostatic force on the meniscus due to head h of the fluid outside the tube ;

$$\pi \,.\, d.\, \sigma \cos\theta = \rho.\, g.\, h.\, \pi\frac{d^2}{4}$$

or

$$h = 4\,\sigma\cos\theta/\rho g d.$$

A spherical liquid drop acted upon by the force of surface tension acquires a pressure of the liquid enclosed by it so as to equal the pressure force on it.

$$\pi d \,.\, \sigma = p.\, \pi d^2/4 \quad \text{... (1.16)}$$

or

$$p = 4\,\sigma/4$$

For a general curved surface with radii of curvature r_1 and r_2 at a point of interest,

$$p = \sigma\left(\frac{1}{r_1}+\frac{1}{r_2}\right)$$

Hollow bubbles such as soap bubbles have two surfaces on which surface tension can act. In such cases,

$$2\pi\, d.\, \sigma = p.\, \pi d^2/4$$

or

$$p = 8\sigma/d$$

Interestingly, soap solution has a high value of σ which makes it possible for the soap bubble to be large in diameter for small pressure of blowing !

A cylindrical liquid jet, of diameter d would acquire pressure p in the jet given by

$$2l.\, \sigma = p.\, ld$$

or

$$p = 2\,\sigma/d$$

Example 1.7. *Find the height through which water rises by capillary action in a glass tube of 2 mm bore if the surface tension at the prevailing temperature is 0.075 g/cm.*

Solution.

The angle of contact, θ between water and glass is 0°; cos θ = 1.

$$\sigma = 0.0.75 \text{ g/cm}$$

$$= 0.075\frac{\text{g}}{\text{cm}}\times\frac{1\text{ kgf}}{1000\text{g}}\times\frac{100\text{ cm}}{1\text{ m}}\times\frac{9.81\text{ N}}{1\text{ kgf}}$$

$$= 0.0736 \text{ N/m}$$

$$h = \frac{4\,\sigma\cos\theta}{\rho\, g\, d} = \frac{4\times 0.074\times 1}{1000\times 9.81\times 0.002} = 0.015 \text{ m} = \mathbf{1.5\ cm.}$$

Example 1.8. *Calculate the gauge pressure and the absolute pressure within*

(a) a droplet of water 0.4 cm in diameter and

(b) a jet of water 0.4 cm in diameter.

Assume the surface tension of water as 0.073 N/m and the atmospheric pressure as 101 300 N/m²

Solution.

Assuming the droplet to be spherical and the jet to be cylindrical, *for the droplet*

$$p_{gauge} = \frac{4\,\sigma}{d} = \frac{4 \times 0.073}{0.004}$$

$$= \mathbf{73\ N/m^2}$$

$$p_{abs} = 101300 + 73$$

$$= 101373\ \text{N/m}^2$$

$$= \mathbf{101.373\ kN/m^2}$$

for the jet

$$p_{gauge} = \frac{2\,\sigma}{d} = \frac{2 \times 0.073}{0.004}$$

$$= \mathbf{36.5\ N/m^2}$$

$$p_{abs} = 101\ 373 + 36.5$$

$$= 101\ 336.5\ \text{N/m}^2$$

$$= \mathbf{101.34\ kN/m^2}$$

Example 1.9. *A U-tube is made up of two capillaries, of diameters 1.0 mm and 1.5 mm 'respectively. The U-tube is kept vertically and partially filled with water of surface tension 0.0075 kg/m and zero contact angle. Calculate the difference in the level of the miniscii caused by the capillarity.*

Solution.

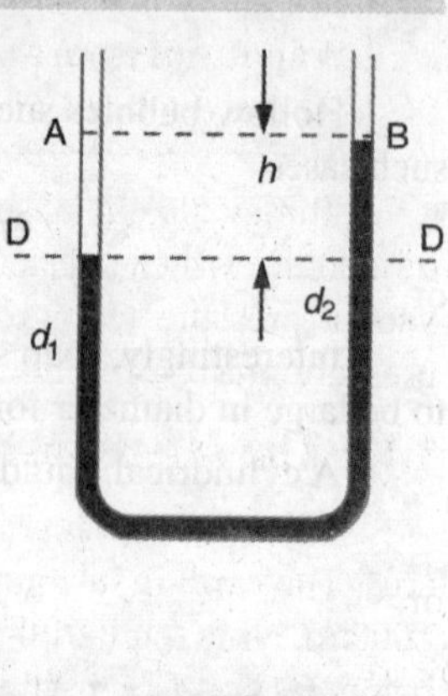

For the columns of water in the two vertical limbs of the U-tube referred'to datum.

Surface tension force at A = $\pi d_1 \sigma$ upwards and surface tension force at B = $\pi d_2 \sigma$ upwards.

The difference in the two surface tension forces supports the weight of the column of water in the smaller tube above the datum level.

$$\pi d_2\, \sigma - \pi\, d_1\, \sigma = \frac{\pi\, d_2^2}{4} h\rho g$$

whence,

$$h = \frac{4(d_2 - d_1)\,\sigma}{d_2^2\, \rho g}$$

$$= \frac{4(1.5 - 1.0) \times 0.0075}{(1.5/1000)^2 \times 1000 \times 9.81} = \mathbf{0.68\ mm}$$

Some observations are made from the preceding analysis.

* *If the diameters of the capillaries are equal, $d_2 = d_1$ then $h = 0$, i.e., no level difference will occur.*

* *If a capillary of diameter d_1 is attached to reservoir, $d_2 >> d_y$ then $h \to 4\sigma/\rho g d$ which is indeed the case for a single capillary dipped in a reservoir.*

* *If $d_2 < d_1$ then h becomes negative which implies that the level in the bigger capillary would rise, as expected.*

1.11 1.10 Vapour Pressure

Pressure exerted by the vapour formed at the free surface of a liquid is called vapour pressure. The phenomenon of vaporization is due to the surface molecules escaping the liquid by overcoming the molecular forces by virtue of their translational momentum. Vapour pressure of a liquid increases with its temperature due to increased molecular activity. The phenomena of vaporization and boiling are differentiated as follows :

* **Vaporization, if**

Vapour pressure < pressure above a liquid surface at that temperature.

* **Boiling, if**

Vapour pressure = pressure above a liquid surface at that temperature.

In other words, boiling point of a liquid is that temperature at which the vapour pressure equals the pressure above the liquid surface.

In a confined space, the partial vapour pressure exerted by the evaporating molecules to stay in equilibrium with the liquid is called the saturation pressure.

The vapour pressure consideration assumes importance where, in a liquid flow, the pressure falls equal to or below the vapour pressure. The liquid is rapidly vaporized giving rise to cavities or separated flow. The vapour cavities are carried along the flow and can prove harmful if subjected to an increasing pressure gradient.

Under high pressure, the cavities may collapse resulting in waves of high intensity. The short-lived high pressure-intensity sequence of cavities collapsing results in cavitation. The flow is then termed cavitating flow.

Temperature °C	10°	15°	20°	30°	40°	45°	50°	100°
Vapour pressure in meters head of water	0.12	0.17	0.25	0.44	0.76	0.97	1.26	10.33

It may also be observed that the saturation vapour pressure of mercury at 20°C is 1.3×10^{-6} m of mercury which speaks of its stability at vacuum in a barometer. On the other hand, the saturation vapour pressure for gasoline at 20°C is about 4.3 m of gasoline which means that much greater pressure must be maintained to keep gasoline in the liquid state.

1.12 The Atmosphere

The surface of earth consists of land to the extent of 29% and water over the remaining 71% of area. The earth is surrounded by the atmosphere which is predominantly air until an altitude of over 100 km. Nine tenth of the mass of the atmosphere is contained below an altitude of 20 km and 99% below 60 km. Life on the earth is mainly confined to biosphere which extends upto an altitude of 1 km. In the troposphere, the temperature, pressure and density of the air drop with increasing altitude. The rate of change of temperature with altitude is called lapse rate. The lapse rate for international standard atmopshere is taken as 6.5 K/km which means that, from the sea level upto the tropopause the temperature drops linearly at the rate of 6.5 degree kelvin per km altitude.

The standard atmospheric temperature, pressure and density referred to STP at the sea level are as follows :

Temperature	15°C or 288.15 K
Pressure	101.325 kN/m^2 or 10.333 m water gauge
Density	1.225 kg/m^3
Viscosity	1.79×10^{-3} Ns/m^2
Molecular Weight	28.966
Sonic Velocity in Air	340.29 m/s

It may be noted that STP in Chemistry is defined for 0°C temperature and atmospheric pressure as above. This is because the world of Chemistry is in the laboratory but the real world is in the open! The International Standard Atmosphere (ISA) for aircraft performance is the same as STP defined above at 0% humidity. Sometimes, the term NTP (Normal temperature and pressure) is used which refers to 20°C and the density of air 1.2 kg/m^3 at the atmospheric pressure.

Atmospheric and weather conditions are referred to 'bar' as the pressure:

$$1 \text{ bar} = 100 \text{ kPa} = 100 \text{ kN/m}^2$$

and it follows that 1 mbar = 0.1 kPa. Pressure above one bar is referred to as 1.2 bar, 2 bar and so on. Lines of constant pressure on an atmospheric pressure chart are called 'iso-bars'. Isobar contours therefore, connect he places with the same pressure. Isobars, typically, are drawn for 1.0, 1.1, 1.2, 0.9 etc. so that the regions of high and low pressures are distinguished and possibility of wind movement i.e., storms, from higher to lower pressures, are predicted.

1.13 Classification of Fluids and Regimes of Flow

Properties of viscosity and compressibility play a major role in determining the nature of fluid flow. As already stated, fluids are classified as Newtonian if they obey the Netwon's linear law of viscosity ; otherwise non-Newtonian. Fig. 1.6 and Tables 1.3 to 1.5 provide an insight into properties of fluids. It is interesting to appreciate that a solid does not deform continuously under the application of a shear stress and hence denoted along the stress axis. On the other hand, an inviscid fluid is one in which the viscosity is neglected. Hence it is represented as a line of zero shear stress. A fluid is said to be ideal if it is assumed both inviscid and incompressible. The words incompressible and compressible refer to the flow rather than to the fluid. At low velocities, a fluid flow is always considered to be incompressible. Approaching the sonic velocity, compressibility effects acquire importance rendering the flow compressible. The ratio of the velocity of a fluid to the sonic velocity in the same fluid is called the Mach number M of flow as already defined.

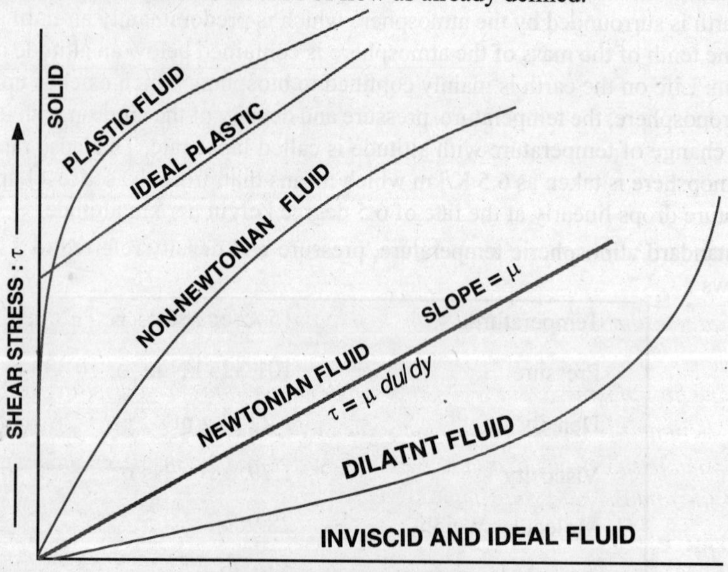

Fig. 1.16: Classification of Fluids

Types of fluids, on the basis of density and viscosity are listed in Tables 1.3 and 1.5 and type of flows on the basis of Mach numbers are listed in Table 1.4.

Table 1.3 **Classification of Fluids on the Basis of Density and Viscosity**

Type of Fluid	Density	Viscosity
1. Ideal Fluid	Constant	Zero
2. Incompressible Fluid	Constant	Non-Zero
3. Inviscid Fluid	Constant or Variable	Zero
4. Real Fluid	Variable	Non-Zero
5. Newtonian Fluid	Constant or Variable	$\tau = \mu \frac{du}{dy}$
6. Non-Newtonian Fluid	Constant or Variable	$\tau \neq \mu \frac{du}{dy}$
7. Compressible fluid	Variable	Zero
8. Perfect Gas	Variable $p = \rho RT$;	Zero

Table 1.4 **Classification of Flows on the Basis of Mach Number**

Type of Flow	Mach Number, M
Incompressible Flow	*M* less than 0.3
Compressible Subsonic Flow	*M* between 0.3 and 1.
Transonic Flow	*M* ranging between values less than 1 and more than 1.
Supersonic Flow	*M* greater than 1 but less than 7.
Hypersonic Flow	*M* greater than 7.

Example 1.10. *During the flow of a fluid following the law*

$$\tau = \mu\,(du/dy)^{1.3}$$

it is observed that the velocity distribution within the fluid film is given by

$$u/U_{max} = 2(y/h) - (y/h)^3/3$$

where h is the film thickness and U_{max} is the maximum velocity. The viscosity is 0.5 N s/m².

(a) Calculate the shear stress at the solid surface when U_{max} = 0.2 m/s and h = 1 cm.

(b) Estimate the viscosity of a Newtonian fluid to induce the same shear stress for the same velocity profile and the same maximum velocity.

Solution:

For the given non-Newtonian fluid,

$$u = U_{max} \times [(2/h)y - (1/3h^3) \times y^3]\ ;$$

$$du/dy = U_{max}\,(2/\,h - y^2/h^3)$$

at the wall, $y = 0.01$ cm, $h = 0.01$ cm.

With the given values $du/dy = 0.2(2/0.1 - 1/0.01) = 20\ s^{-1}$

The shear stress at the wall is given by $\tau = \mu(du/dy)^{1.3}$

$$= 0.5 \times 20^{13} = 0.5 \times 49.12 = 24.56\ N/m^2$$

For a Newtonian fluid, $\tau = \mu\, du/dy$; $24.56 = \mu \times 20$; $\mu = 1.23\ Ns/m^2$

Table 1.5 Rheological Classification of Fluids

Fluid behaviour	Time dependence	Type	Class
$\tau = \mu \dfrac{du}{dy}$ Air, water and many other engineering fluids behave as Newtonian fluids under normal circumstances.			**NEWTONIAN FLUIDS** $\tau = \mu \dfrac{du}{dy}$
Pseudoplastic Fluids $\tau = \mu \left(\dfrac{du}{dy}\right)^n, n < 1$ for example, fine particle suspensions.	Time-Independent	**Purely Viscous Fluids**	**NON-NEWTONIAN FLUIDS** $\tau \neq \mu \dfrac{du}{dy}$
Dilatant Fluids $\tau = \mu \left(\dfrac{du}{dy}\right)^n, n > 1$ for example, ultrafine irregular particle suspensions.	Time-Independent	**Purely Viscous Fluids**	**NON-NEWTONIAN FLUIDS**
Ideal Plastics or Bingham Fluids $\tau = \tau_0 + \mu \left(\dfrac{du}{dy}\right)^n$ for example, water suspensions of clay and fly ash.	Time-Independent	**Purely Viscous Fluids**	**NON-NEWTONIAN FLUIDS**
Thixotropic Fluids $\tau = \mu \left(\dfrac{du}{dy}\right)^n + f(t), f(t)$ decreasing, for example crude oils, bentonitic drilling fluid.	Time Dependent	**Purely Viscous Fluids**	**NON-NEWTONIAN FLUIDS**
Rheopectic Fluids $\tau = \mu \left(\dfrac{du}{dy}\right)^n + f(t), f(t)$ increasing, for example, rare liquid-solid suspensions	Time Dependent	**Purely Viscous Fluids**	**NON-NEWTONIAN FLUIDS**
Viscoelastic Fluids $\tau = \mu \dfrac{du}{dy} + \alpha E$ where E is the modulus of elasticity. For example, liquid-solid combinations in pipe flow and polymerised fluids with drag-reduction features are visco-elastic fluids.		**Visco-elastic Fluids**	**NON-NEWTONIAN FLUIDS**

SUPPLEMENTARY REVISION EXAMPLES

Example 1.11. *A shaft is rotating in a journal bearing. The clearance between the shaft and the bearing is filled with a viscous oil. Determine an expression for the power absorbed in overcoming the viscous resistance.*

Solution.

Let the diameter of the shaft be d, rotating at a speed of N revolutions per minute. Let the clearance between the shaft and the bearing be y, filled with a lubricant of viscosity μ.

Peripheral speed of the shaft

$$u = \frac{\pi d N}{60}.$$

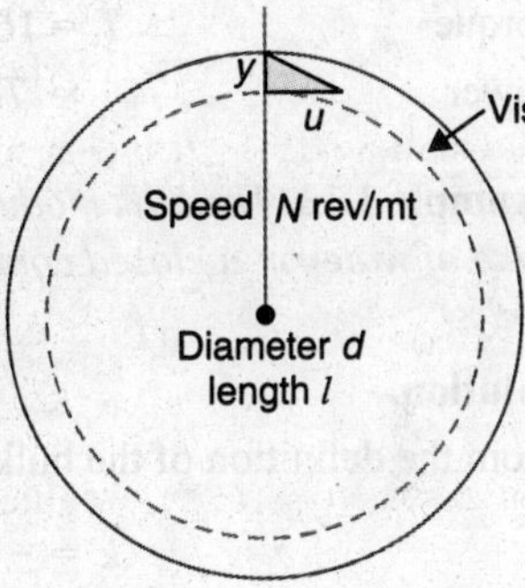

Velocity gradient in the gap

$$\frac{u}{y} = \frac{\pi d N}{60 y}.$$

Shear stress τ is given by

$$\tau = \mu \frac{u}{y} = \frac{\mu \pi d N}{60 y}.$$

Shear force over the periphery of the shaft of length l is

$$F = \tau . \pi d . l$$

$$= \mu (\pi d)^2 l N / 60 y$$

Torque = $F . d/2$ and power = $T . \omega$.

$$\text{Power} = \frac{\mu (\pi d)^2 . l . N}{60 y} . \frac{d}{2} . \frac{2 \pi N}{60}.$$

$$= \frac{\mu l N^2}{3600 y} . (\pi d)^3$$

Example1.12. *Lateral stability of a long shaft of 15 cm diameter is obtained by means of a 25 cm stationary bearing having an internal diameter of 15.05 cm. If the space between the shaft and the bearing is filled with a lubricant having a viscosity of 2.4 Pa.s, what power will be required to overcome the viscous resistance when the shaft is rotated at a constant rate of 180 revolutions per minute ?*

Solution.

From the data

$$\mu = 2.4 \text{ Pa.s or Ns/m}^2.$$

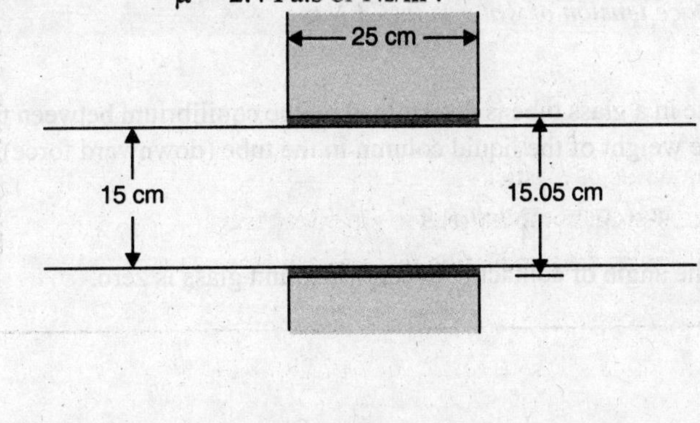

$$\text{Radial gap} = dr = (15.05 - 15)/(2 \times 100) = 0.00025 \text{ m}$$

$$\omega = 2\pi \times 180/60 = 6\pi = 18.85 \text{ rad/s}$$

Velocity of the shaft at its surface is given by

$$v = \frac{0.15}{2} \times 18.85 = 1.414 \text{ m/s.} = dv.$$

$$\tau = \mu \frac{dv}{dr} = 2.4 \times 1.414/0.00025 = 13{,}570 \text{ N/m}^2$$

Force is distributed around the peripheral area.

$$F = 13570 \times \pi \times 0.15 \times 0.25 = \mathbf{1600\ N}$$

Torque $T = 1600 \times 0.15/2 = \mathbf{120\ Nm}$

Power $= T\omega = 120 \times 18.85 = 2260 \text{ W} = \mathbf{2.26\ kW}$

Example 1.13. *The bulk modulus of water is 210 kN/cm². What pressure is required to reduce the volume of water in a closed container by 2%? Also show that the increase in mass density will be 2% only.*

Solution.

From the definition of the bulk modulus of elasticity,

$$k = -\frac{dp}{dv / v}$$

For 2% reduction in volume, $\frac{dv}{v} = 0.02$

Taking care of units,

$$dp = 210 \times 100 \times 100 \times 0.02 = 42{,}000 \text{ kN/m}^2.$$

$$= 42 \text{ MN/m}^2.$$

Now, by definition, mass density is the mass per unit volume i.e.,

$$\rho \propto \frac{1}{v}$$

$$\frac{d\rho}{\rho} = -\frac{dv}{v}$$

If $\frac{dv}{v}$ = 2% reduction in volume then $\frac{d\rho}{\rho}$ must be 2% increase in density.

Example 1.14. *What should be the minimum diameter of a glass tube to be selected to measure the water level (or to be used as a water manometer) if it is desired that the capillary correction be limited to 2 mm ? Surface tension of water = 0.073 N/m.*

Solution.

The capillary rise in a glass tube is determined by the equilibrium between the surface tension (upward force) and the weight of the liquid column in the tube (downward force).

$$\pi\, d.\, \sigma = \frac{\pi d^2}{4}.\, h.\, \rho\, g$$

Assuming that the angle of contact between water and glass is zero.

From the above,

$$h = \frac{4\sigma}{\rho g d}; \ d = \frac{4\sigma}{\rho g h}$$

For h to be limited to 0.2 mm.

$$d = \frac{4 \times 0.073 \times 1000}{1000 \times 9.81 \times 2} = 0.015 \text{ m}$$

$$= \mathbf{1.5 \text{ cm}}$$

Therefore, a tube of 1.5 cm bore (inside diameter) is the minimum size recommended.

Example 1.15. *If the surface tension at the air water interface is 0.073 N/m, estimate the pressure difference between inside and outside an air bubble of 0.01 mm.*

Solution.

Consider the semi-air bubble as a hollow hemisphere. It has two surfaces; inner and outer. The surface tension acts on both surfaces such that the surface tension force on it is $2\pi. d. \sigma$ outwards

This must equal to pressure force exerted by the other half of the bubble

$$F = p.\frac{\pi d^2}{4} \text{ inwards}$$

Hence, $$p.\frac{\pi d^2}{4} = 2. \pi d.\sigma$$

and $$p = \frac{8\sigma}{d}$$

For the given case

$$p = \frac{8 \times 0.073}{0.01/1000} = 58\ 400 \text{ N/m}^2$$

This is the pressure difference between inside and outside the bubble. The absolute pressure inside the bubble must be

$$p_{abc} = 101\ 300 + 58\ 400 = 159\ 700 \text{ N/m}^2$$

$$= \mathbf{159.7 \text{ kN/m}^2}.$$

Example 1.16. *If two vertical, parallel, clean glass plates are kept 2 mm apart partially submerged in water, how high will the water rise between them due to capillary action. Take the surface tension for water σ =0.075 N/m.*

Solution.

For a rise of h in level due to capillarity,

Upward surface tension force per unit width of the plate $= \frac{2}{1000} \times 1 \times 0.075$

$$= 0.150 \times 10^{-3} \text{ N}.$$

Downward acting force, *i.e.*, weight of the liquid column per unit width of the plate

$$= 1000 \times 9.81 \times h \times \frac{2}{1000}$$

$$= 19.62\ h$$

Equating the two for equilibrium,

$$19.62\ h = 0.150 \times 10^{-3}$$

$$h = 7.65 \times 10^{-6} \text{ m} = \mathbf{7.65 \times 10^{-3} \text{ mm}}.$$

This capillarity height is indeed very low compared to that in a tube of 2 mm diameter. The solved example 1.7 in the text shows that the capillarity height h in the tube would be 1.5 cm. A little thinking would show that the surface tension force is a great deal more and the weight of the water supported by it is very much less in a tube!

Example 1.17. *A thin flat plate of surface area 7.5 m² is pulled parallel to the upper and lower surfaces of a confinement of 12 mm gap which contains an oil of dynamic viscosity 0.12 Ns/m² on one side and another oil three times as viscous on the other side. If the force applied is 180 N, determine the velocity of the plate. Assume the plate to be in the middle of the confinement.*

Solution.

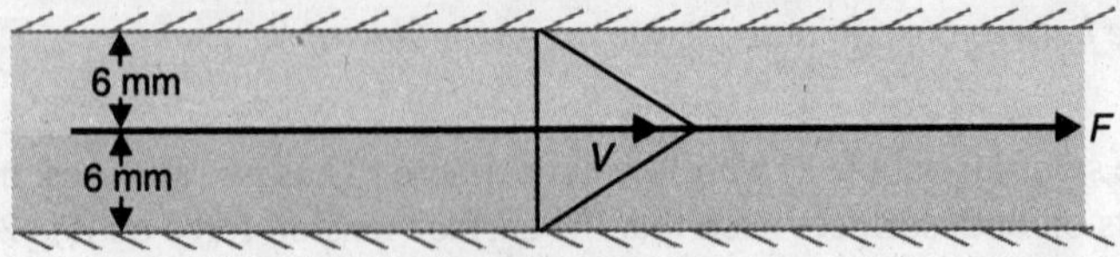

From the data,

$$F = 180 \text{ N} = \tau.A = 7.5\ \tau.$$

whence $\tau = 180/7.5 = 24$ Pa.

acting on both sides of the plate due to both layers of oil.

$$\tau = 0.12.\frac{V}{6/1000} + 0.12 \times 3.\frac{V}{6/1000}$$

$$= 20\ V + 60\ V = 80\ V. \text{ Pa}$$

if V is in m/s

Hence, $80\ V = 24;\ V = 0.3$ m/s.

Example 1.18. *The following data corresponds to an oil. Check whether the oil is Newtonian or non-Newtonian fluid. If yes, find its viscosity; if not what type of liquid is it ?*

Shear stress, Pa	*0*	*10*	*20*	*40*	*80*
Rate of angular deformation, s^{-1}	*0*	*1*	*2*	*3*	*4*

Solution.

One can plot the shear stress versus rate of strain graph and observe (even without plotting it) that the relationship is linear only until the shear stress is 20 *Pa*. Beyond that the shear stress rises faster and faster. So, we conclude that the fluid is non-Newtonian because, in a Newtonian fluid

$$\tau = \mu\frac{du}{dy}$$

i.e., linearity between shear stress and rate of strain should have been all along !

Recalling the types of fluids, from the text, it is a 'Dilatent' fluid in which shear stress-rate of strain relationship is linear at low values; shear stress rises faster with higher rates of strain.

Example 1.19. *Derive an expression for the capillary rise between two vertical parallel plates partially immersed in a liquid of relative density S in terms of the distance t between the plates, surface tension σ and the contact angle θ between the liquid and the plate.*

Solution.

When the liquid rises by a distance h, as shown the column of liquid between the plates is in equilibrium.

Upward surface tension force = $t.\ b.\ \cos\theta.\ \sigma$

Weight of the liquid column = $\rho.\ g.\ h.\ t.\ b.$

Equating the two for equilibrium,

$$\sigma.\ t.\ b.\ \cos\theta = \rho.\ g.\ h.\ t.\ b.$$

whence
$$h = \frac{\sigma \cos\theta}{\rho g}$$

In terms of the relative density S, which is

$$S = \frac{\rho}{\rho_{water}};\ \rho = S.\ \rho_{water}$$

we get
$$h = \frac{\sigma \cos\theta}{S.\ \rho_{water}.\ g}.$$

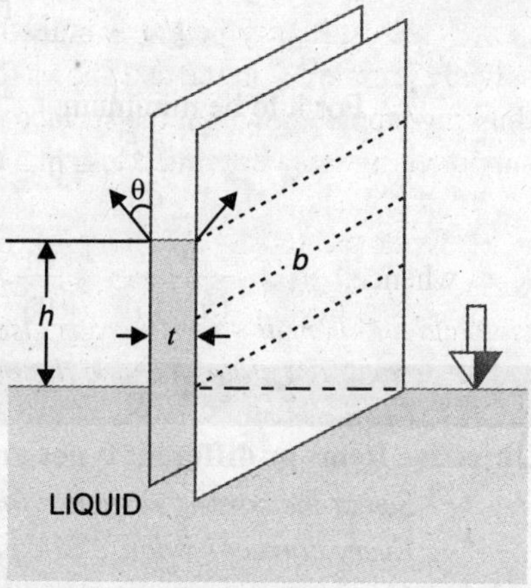

Example 1.20. *There is a pin hole of diameter 0.1 mm at the bottom of a glass vessel containing mercury. What depth of mercury can be put in the vessel without any leakage occurring?*

Take $\sigma = 0.55$ *N/m*

Solution.

The angle of contact θ shown as ABC of a mercury with glass is obtuse, which means that a capillary depression takes place when a thin glass tube is inserted in a jar of mercury. Likewise, if a thin glass passage on hole is created under mercury, it will resist the mercury from flowing out. For a diameter d, the surface tension force opposing the flow would be

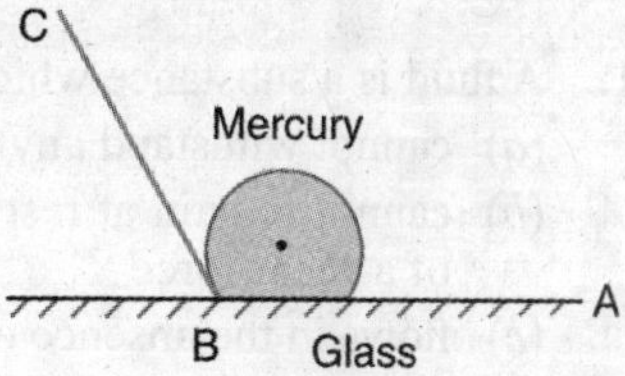

$$F_s = \pi\, d\, \sigma \cos\theta$$
$$= \pi \times 0.0001 \times 0.55 \cos\theta$$

This upward force in a pin hole can resist a hydrostatic force given by

$$F_h = \rho\, g\, h\, \pi\, d^2/r$$
$$= 13.6 \times 1000 \times 9.81 \times h \times \pi \times 0.0001^2/4$$
$$= 1.048 \times 10^{-3}\ h\ \mathrm{N}$$

Equating F_s to F_h,

$$h = \mathbf{0.165 \cos\theta\ m}$$

Example 1.21. *A thin plate of 'large extent' is pulled at a velocity v through a narrow gap of height h. There is oil of viscosity* μ_1 *on one side and another oil of viscosity* μ_2 *on the other side of the plate. Derive an expression for the position of the plate within the gap so that the pull required to drag the plate is the minimum.*

Solution.

Assuming the thin plate with negligible thickness, the oil layers on the two sides are as follows:

Oil of viscosity μ_1 with thickness y

and oil of viscosity μ_2 with thickness $h - y$.

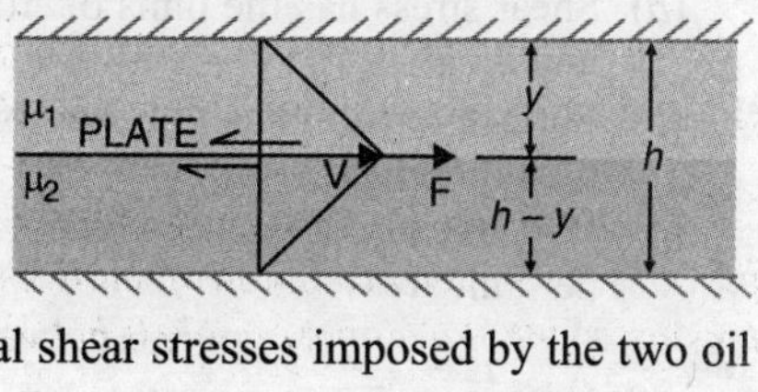

The force required to pull it is proportional to the total shear stresses imposed by the two oil layers:

$$F \propto \tau = \mu_1 \frac{V}{y} + \mu_2 \frac{V}{h-y} = \left[\mu_1 y^{-1} + \mu_2 (h-y^{-1})\right].V$$

For it to be minimum,

$$\frac{\partial F}{\partial y} = 0 = \frac{\partial \tau}{\partial y} = -1 \quad \mu_1 y^{-2} + (-1)\, \mu^2 (h-y)^{-2}$$

whence, $$\frac{y^2}{(h-y)^2} = \frac{\mu_1}{\mu_2}.$$

ITEM BANK

Objective Items or different types are given to review the concepts developed in the chapter.

* *Select the correct response for each item.*
* *Every correct response is expected in less than a minute and carries one mark.*
* *Compare your responses with the key answers given at the end.*

Multiple-Choice Items

Attempt the following Multiple-Choice Items by ticking the correct or the most appropriate response from among the alternatives given :

1. A fluid is a substance which
(*a*) cannot withstand any shear force
(*b*) cannot remain at rest under the action of a shear force
(*c*) flows in the absence of external forces
(*d*) is incompressible and inviscid

2. The Netwon's law of viscosity relates :
(*a*) the stress and strain in the fluid.
(*b*) the shear stress, pressure and velocity
(*c*) the shear stress and rate of strain.
(*d*) the viscosity and the shear stress

3. Select the correct statement:
(*a*) Viscosity of gas increases with temperature
(*b*) Density of a gas increases with temperature
(*c*) Surface tension of a liquid increases with temperature
(*d*) Bulk modulus of elasticity is independent of temperature

4. Point out the *incorrect* statement:
(*a*) Shear stress has the units of 'force per unit area'.
(*b*) Surface tension has the units of 'force per unit length'
(*c*) Dynamic viscosity is expressed as 'force per unit area'
(*d*) Bulk modulus is expressed as 'force per unit area'

5. Capillarity rise and depression
(*a*) depend solely upon the size of the tube.
(*b*) depend solely upon the surface tension of the liquid
(*c*) depend upon the surface tension of the liquid as well as the material of the tube
(*d*) depend upon the pressure difference between the liquid and the environment

6. The units of rate of strain in a fluid at a point are
(*a*) m (*b*) m/s (*c*) rad/s (*d*) nil

7. The velocity gradient in the transverse direction for a fluid flow equals
(*a*) the pressure gradient
(*b*) the rate of shear strain
(*c*) the stress at that point
(*d*) the strain at that point

Other Objective Item

8. Match the properties in block B with the types of fluids in block A :

Block A	Block B
(a) Ideal fluid	(k) Inviscid
(b) Newtonian fluid	(l) incompressible
	(m) viscous
(c) Perfect gas	(n) compressible

ANSWERS

1. (b), **2.** (c), **3.** (a), **4.** (c), **5.** (c), **6.** (c), **7.** (b), **8.** (akl, bm, ckn),

CHAPTER

2

Fluid Statics

LEARNING OBJECTIVES

On successful completion of all learning activities provided in the chapter, a student will be able to

- state and prove the Pascal's law for a static fluid ;
- derive the hydrostatic law in an incompressible and in a compressible fluid at rest ;
- determine the hydrostatic force and the point of application on a given plane or curved surface ;
- state the definition and importance of piezometric head ;
- determine the pressure at a desired point in a fluid by employing a single-tube, U-tube (upright or inverted) or a differential manometer ;
- compute the force of buoyancy on a partially or fully submerged body ;
- identify the states of equilibrium for a floating body and
- determine the surface profile of a liquid in a vessel subjected to a constant acceleration or a constant rotation.

2.1 The State of Rest ; Pascal's Law

A fluid is said to be at rest if there is no relative motion between adjacent layers. Fluid in a stationary container or at constant velocity or at constant acceleration are, therefore, in a state of 'rest'. The pressure at a point in a fluid is defined as the intensity of compressive force, *i.e.*, the normal compressive force per unit area at that point. Theoretically, an infinite number of planes, and hence an infinite number of normals to the planes may be drawn at a point. The question before us is to establish if the pressure is the same in all directions at that point. Yes, according to the Pascal's law, the pressure at a point in a static fluid is equal in all directions in space.

Let us prove this statement by considering the equilibrium of an infinitesimally small space element at a point O as shown in Fig. 2.1. Let us start by choosing the three coordinate axes *x, y* and *z* arbitrarily at that point. Let it be enclosed by a small tetrahedron OABC with its three faces OBC, AOC and AOB having areas δA_x, δA_y and δA_z normal to *x, y* and *z* axes respectively. Pressures on these faces are shown as p_x, p_y and p_z respectively. The fourth face, ABC has an area δA_n and the pressure normal to it is p_n along the normal, shown n. Since the tetrahedron is infinitesimally small and the axes are chosen arbitrarily, the weight of the fluid does not enter the analysis. Also, in the limit,

$$\delta A_x \to 0, \delta A_y \to 0, \delta A_z \to 0 \text{ and } \delta A_n \to 0$$

the pressures p_x, p_y, p_z and p_n refer to those at the point O.

The equilibrium of the fluid element under the action of pressure forces requires

$$\Sigma F_x = 0 \quad \Sigma F_y = 0 \text{ and } \Sigma F_z = 0$$

For x direction,

$$p_x \delta A_x - p_n \delta A_n (n.\, i) = 0$$

but $\delta A_n (n.\, i)$ is the projected area of δA_n normal to x-axis, which is δA_x.

Hence, $p_x\, \delta A_x - p_n \delta A_x = 0$

Cancelling δA_x from both sides, $p_x = p_n$

Similarly,

$$p_y = p_n \text{ and } p_z = p_n$$

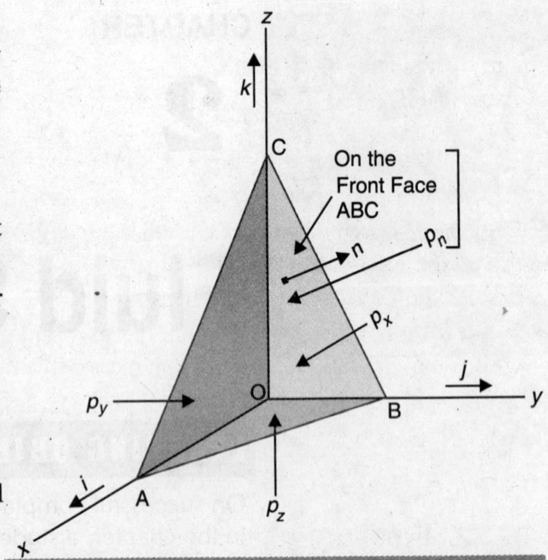

Fig. 2.1: Pressure at a point

for equilibrium along y and z directions respectively.

Hence, $p_x = p_y = p_z = p_n$ ....(2.1)

i.e., the pressure is the same on all faces of the tetrahedron. Since the choice of the axes is arbitrary *i.e.*, the orientation of the faces of the tetrahedron is arbitrary, it is concluded that the pressure at a point in a static fluid is the same in all directions in space.

A closer look at the proof of the Pascal's law shows that the implication of the fluid being at rest was the absence of shear stresses. But then, there are other cases of fluid flow where shear stresses do not exist. *Pascal's law of equal pressure in all directions in space is, therefore, valid*

(*a*) *for a fluid a rest, whether the fluid is viscous or inviscid, compressible or incompressible.*

(*b*) *for a liquid in solid-body movement, either at constant linear acceleration or at constant rotational velocity in a container; here again, there is no relative motion between different fluid layers and*

(*c*) *for the flow of an inviscid fluid and for ideal fluid flow where* $\mu = 0$ *corresponds to absence of shear stresses.*

Example 2.1. *A hydraulic jack consists of a handle cum lever of 30 cm length and an assembly of given dimensions. In order that a load of 20 kN be supported by the jack, what should be the force exerted on the handle ? The distance between the fulcrum of the lever and the point where the plunger is connected is 25 mm.*

Solution :

Area of cross section of the large piston

$$A = \pi \times (0.05)^2/4 = 0.00196 \text{ m}^2$$

Pressure of the hydraulic fluid under the piston should be such as to balance the force applied on the piston.

$$p = \frac{200 \times 1000}{0.00196} = 102 \times 10^6 \text{ N/m}^2$$

By Pascal's law, pressure is transmitted undiminished in all directions. The pressure at the bottom of the small piston must also be 10.2×10^6 N/m². The force exerted on the small piston must be

$$102 \times 10^6 \times \pi \times \frac{(0.015)^2}{4} = 18000 \text{ N}$$

The force desired to be exerted at the handle of the level should be

$$F = 18000 \times \frac{25}{300} = \mathbf{1500\ N}$$

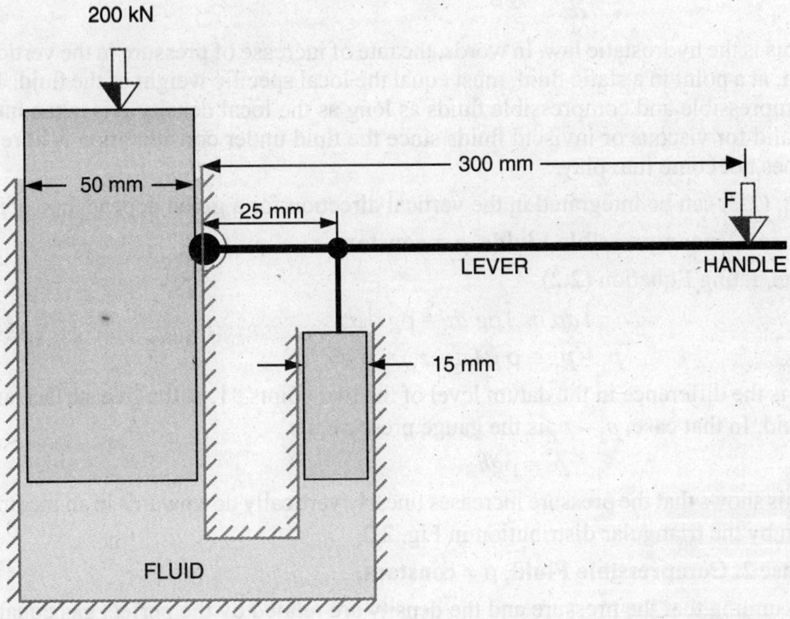

2. 2 Hydrostatic Law

In order to evaluate the pressure at a location O in a fluid at rest, consider an elementary vertical fluid prism of cross section area a and length δz enclosing O as shown in Fig. 2.2. For equilibrium in the z-direction, the sum of the vertical forces acting on it must vanish.

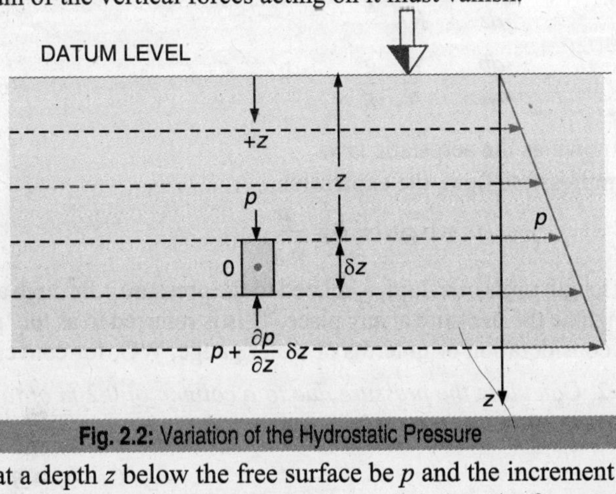

Fig. 2.2: Variation of the Hydrostatic Pressure

Let the pressure at a depth z below the free surface be p and the increment in pressure over a vertical depth δz be δp such that the pressure at a depth $z + \delta z$ below the free surface is $p + \delta p$. The net pressure force acting upwards is

$$(p + \delta p) . a - p . a = \delta p . a$$

The gravity force on the elementary prism equals the weight of the prism and acts down wards,

$$\text{Weight} = \rho\, g . (\delta z . a)$$

For equilibrium of the prism,

$$\delta p\, a = \rho\, g . (\delta z . a)$$

$$\delta p = \rho g\, \delta z$$

whence,

$$\frac{dp}{dz} = \rho g \qquad ....(2.2)$$

This is the hydrostatic law. In words, the rate of increase of pressure in the vertically downward direction, at a point in a static fluid, must equal the local specific weight of the fluid. The law is valid for incompressible and compressible fluids as long as the local density ρ is taken into account. The law is valid for viscous or inviscid fluids since the fluid under consideration is at rest and shearing effect does not come into play.

Eq. (2.2) can be integrated in the vertical direction for a given dependence *of p* on ρ.

Case 1. Incompressible Fluids, ρ = constant.

Integrating Equation (2.2)

$$\int dp = \int \rho g\, dz = \rho g \int dz$$

or

$$p_2 - p_1 = \rho\, g\, (z_2 - z_1) = \rho\, gh$$

where h is the difference in the datum level of the two points ; 1, at the free surface and 2, anywhere in the fluid. In that case, $p_2 - p_1$ is the gauge pressure, i.e.,

$$p = \rho gh \qquad ....(2.2\ a)$$

This shows that the pressure increases linearly vertically downwards in an incompressible fluid, as shown by the triangular distribution in Fig. 2.2.

Case 2. Compressible Fluid, ρ ≠ constant.

Assuming that the pressure and the density are related by the perfect gas equation,

$$p = \rho\, \mathrm{RT}$$

or

$$\rho = \frac{p}{RT}$$

equation (2.2) becomes

$$\frac{dp}{dz} = \frac{p}{RT} . g \qquad ... (2.2b)$$

or

$$\frac{dp}{p} = \frac{dz}{T} . \frac{g}{R}$$

This equation is known as the aerostatic law.

For an incompressible fluid, the expression

$$p = \rho\, gh \text{ or } h = \frac{p}{\rho\, g}$$

suggests that h, in length units of a fluid, is related to the pressure p through a constant (pg) and hence it can serve to designate the pressure at any place. This is referred to as the 'pressure head' h in terms of the fluid under consideration or in terms of water gauge, WG, for convenience.

Example 2.2. *Calculate the pressure due to a column of 0.2 m of (a) water (b) gasoline, sp. gr. 0.75 (c) mercury sp. gr. 13.6. Take ρ for water = 1000 kg/m³.*

Solution :

For incompressible fluids, the hydrostatic law states that

$$p = \rho\, gh$$

(*a*) For water, $p = 1000 \times 9.81 \times 0.2 = 1962$ N/m^2 = **1.962 kN/m²**

(*b*) For gasoline, $p = 0.75 \times 1000 \times 9.81 \times 0.2 = 1471.5$ N/m^2 = **1.4715 kN/m²**

(*c*) For mercury, $p = 13.6 \times 1000 \times 9.81 \times 0.2 = 26683.2$ N/m^2 = **26.7 kN/m²**

Example 2.3. *Convert the standard atmospheric pressure into metres of water gauge and metres of mercury.*

Solution :

The international standard atmospheric pressure is

$$p = 101\,325 \text{ N/m}^2$$

Equating it to ρ *gh* of the given fluids,

$$h = 101\,325/(1000 \times 9.81) = \mathbf{10.33 \text{ m WG}}$$

and $$h = 10.33/13.6 = \mathbf{0.759 \text{ m i.e., } 75.9 \text{ cm of mercury.}}$$

Example 2.4. (*a*) *In a chemical reactor 100 m tall, the density of the fluid mixture varies with the distance z in metres above the bottom of the reactor as*

$$\rho = 1000\left[1+\frac{z}{500}+\left(\frac{z}{1000}\right)^2\right]$$

Assuming the mixture to be effectively stationary, determine the pressure difference between the top and the bottom of the reactor.

Solution :

Considering the equilibrium of an elementary vertical prism of height *dz.* and cross sectional area *dA*.

$$pdA - (p + dp)dA = \text{weight of the prism}$$

$$= 1000\left(1+\frac{z}{500}\right)+\left(\frac{z}{1000}\right)^2 gdz\, dA$$

whence, $-dp = 1000\left[1+\frac{z}{500}+\left(\frac{z}{1000}\right)^2\right]$ and integrating between the limits 1 and 2

$$p_1 - p_2 = 1000 \times 9.81 \times \left(z+\frac{z^2}{1000}+\frac{z^3}{3\times10^6}\right)_0^{100} = 1000\times9.81\times110.3$$

$$= \mathbf{1.08 \times 10^6 \text{ N/m}^2}$$

(*b*) *Determine the density of sea water at a depth of 1 km, if the density at the surface is 1025 kg/m³ and the average bulk modulus of sea water is 2.35 × 10⁹ N/m²*

Solution:

$$\frac{\delta\rho}{\rho} = \frac{\delta p}{K} = \frac{\rho g h}{K} = \frac{1025\times9.81\times1000}{2.35\times10^9} = 4.28 \times 10^{-3}$$

$$\delta\rho = 4.28 \times 10^{-3} \times 1025 = 4.387 \text{ kg/m}^3$$

Density at 1 km depth = 1025 + 4.387 = **1029.387 kg/m³**

2.3. Hydrostatic Force on a Submerged Plane Surface

Let a plane surface of area A with its centroid at C be placed at an angle α with respect to the free surface of a liquid.

Consider an elementary shaded strip at a depth z as in Fig. 2.3 (a).

Pressure p on the strip = ρgz

Force on the strip, $\delta F = p\,\delta A = \rho gz\delta A = \rho\, gy \sin\alpha\, \delta A$

Force on the entire surface, $F = \int dF = \int p dA$

$$= \int \rho\, gy \sin\alpha\, dA = \rho\, g \sin\alpha \int y dA$$

$$= \rho\, g \sin\alpha\, \bar{y}\, A = \rho\, g\, \bar{z}\, A$$

or

$$F = \rho\, g\, \bar{z}.\, A \qquad \text{...(2.3)}$$

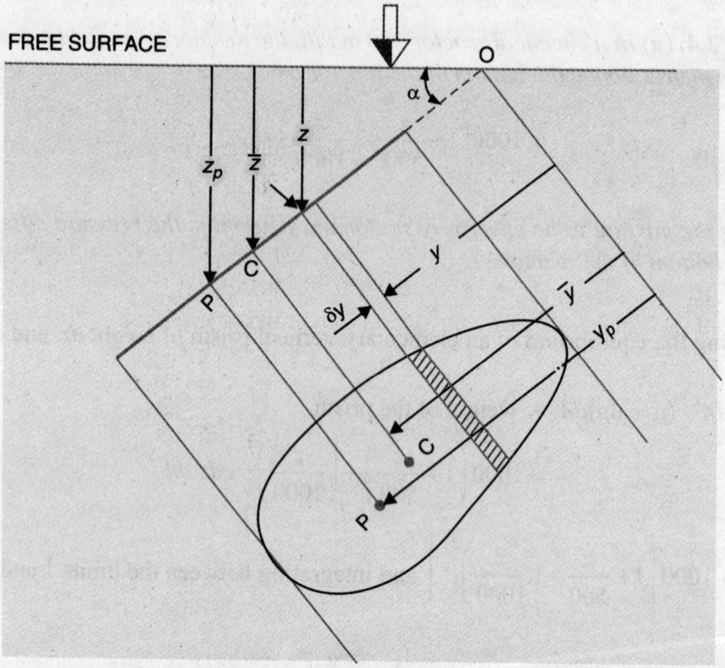

Fig. 2.3 (a): A Submerged Plane Surface

This equation may be interpreted by stating that the hydrostatic force on a plane surface equals the pressure $\rho\, g\, \bar{z}$ at the centroid of the area times the submerged area A of the surface and acts normal to it.

It is interesting to arrive at the same result by looking at the pressure distribution over the surface. The pressure at any point on the surface must act normal to the surface and the intensity of pressure is given by

$$p = \rho\, g\, z$$

where z is the depth of the liquid at that point. The pressure distribution is consequently linear, *i.e.*, trapezoidal as shown in a typical end view in Fig. 2.3 *(b)*. Volume under the pressure plot per unit width should represent hydrostatic force acting on the surface. Utilising the fact that the volume under the pressure plot must be given by

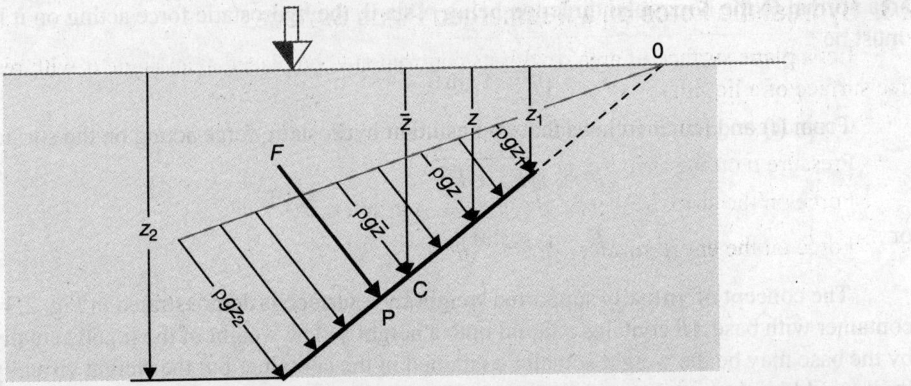

Fig. 2.3. (b) Pressure Distribution and Force on an Inclined Plane Surface Submerged in a Liquid

$$V = \int \rho gz.dA = \rho g \int zdA = \rho g\bar{z}.A$$

the force F on the surface is given by

$$F = \rho g\bar{z}.A \quad \text{....(2.3)}$$

An alternative method to determine the total hydrostatic force on a surface would be to consider as follows :

(*i*) The vertical component of the force acting on it is due to the weight of the liquid actually or virtually supported by the surface, which must be the weight of the volume of the liquid actually (or virtually) contained between the surface and the free level. Referring to Fig. 2.3 (c), it is observed that the volume of the liquid vertically supported by the surface must be

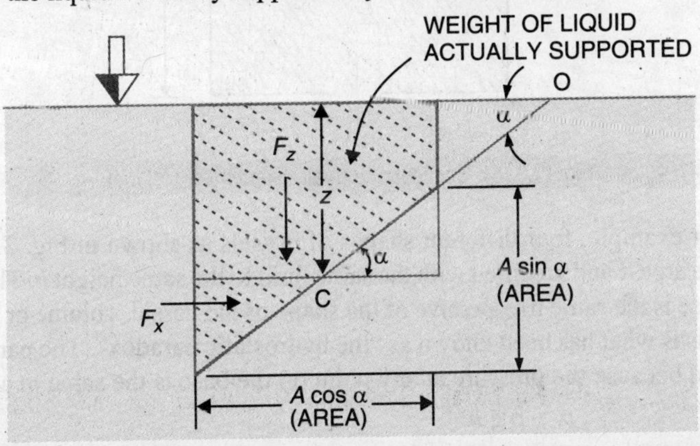

Fig. 2.3 (c): Vertical and Horizontal Components of Force Acting on an Inclined Plane Surface

$$\text{Volume} = \bar{z}\ .\ A\cos\alpha$$

and the weight of the same must be

$$\text{Weight} = \rho g\bar{z}.\ A\cos\alpha$$

This must also be the vertical component of the force acting on it.

$$F_z = \rho g\bar{z}\ .A\cos\alpha \quad \text{...}(i)$$

(*ii*) The horizontal component of the force acting on it arises from the fact that the surface has a projected area normal to the horizontal direction and a hydrostatic force acts on it. The projected area normal to the horizontal direction being A sin θ, the hydrostatic force acting on it horizontally must be

$$F_x = \rho g\bar{z} \,.A \sin\theta \qquad ...(ii)$$

From (*i*) and (*ii*), it follows that the resultant hydrostatic force acting on the surface must be

$$F = \sqrt{F_x^2 + F_z^2}$$

or

$$F = \rho\, g\, \bar{z}\, A \qquad .....(2.3)$$

The concept of virtually supported weight by a surface is demonstrated in Fig. 2.3 (*d*) where a container with base *AB* contains a liquid upto a height *h*. The weight of the liquid actually supported by the base may be the weight actually contained in the container but the weight virtually supported by it would be the weight as if the liquid were filled upto a height *h* as shown in dotted. The total hydrostatic force on the base *AB* is more than the weight of the fluid actually contained in the container but it equals the as if the total base is supporting a liquid of height *h*.

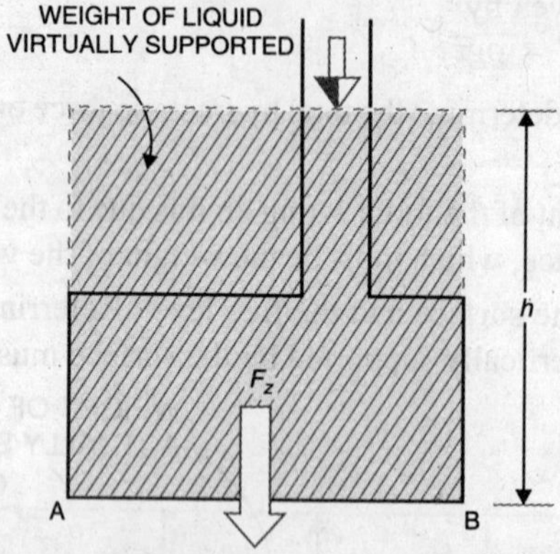

Fig. (2.3 (d): Concept of Virtually Supported Column

Consider, for example, four different shapes of vessels as shown in Fig. 2.3 (*e*) such that all have the same base area *A* and are filled with the same fluid to the same height *h*. The force exerted on the base in each case is the same irrespective of the shape of the vessel, volume or weight of the fluid contained in it. This is what has been known as "the hydrostatic paradox". The paradoxial behaviour is simply explained because the pressure at any point on the base is the same in all cases, namely.

$$p = \rho\, g\, h$$

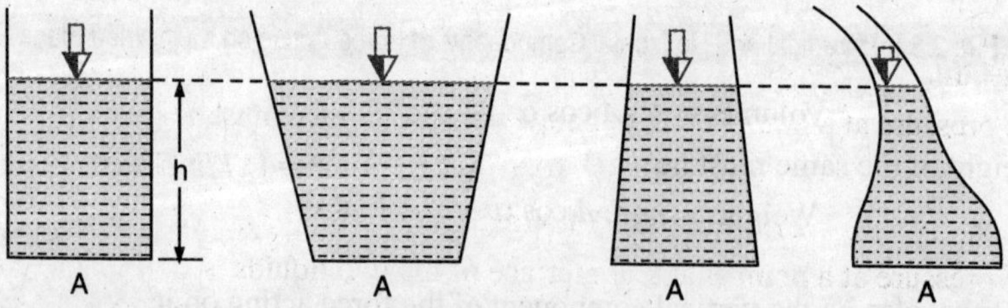

Fig. 2.3(e): The Hydrostatic Paradox

because the pressure depends only upon the head of the continuous fluid column in the vessel apart from the density and gravity. The force at the base of area *A* is given by

$$F = p \,.\, A = \rho\, g\, h \,.\, A$$

which is the same in each case.

In the case of a flat surface completely submerged by one liquid, on top at which a second liquid or gas is present, as shown in fig 2.3 (*f*). the force on the gate is calculated by finding the equivalent height of the lower liquid for the pressure exerted by the upper liquid. In the case of a flat surface partly submerged in one fluid and partly in another, as also shown in fig 2.3 (*f*), the equivalent height cannot be calculated as in the earlier case, as a proper pressure distribution cannot be obtained here on doing this.

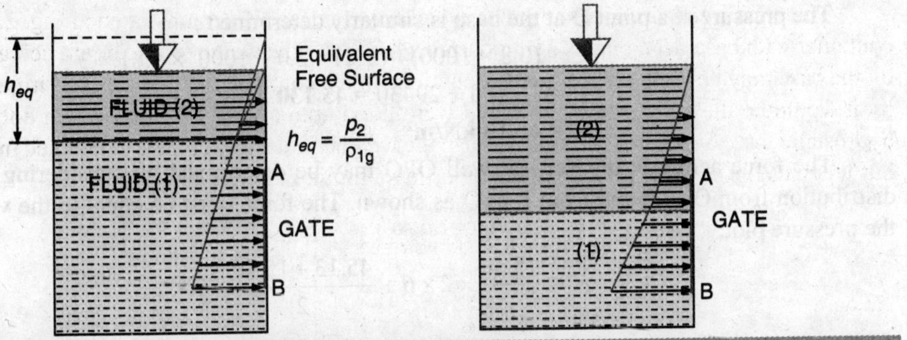

Fig 2.3(f): Equivalent Free Surface

Example 2.5. *A tank contains water of density 1000 kg/m³ upto a height of 3 m above the base. An immiscible liquid of specific gravity 0.8 is filled on top of that over 2 m depth. Calculate the pressures at a point 1.5 m below the free surface, at the interface and at another point 2.5 m below the free surface. Calculate also the force on a vertical wall, 6 m wide.*

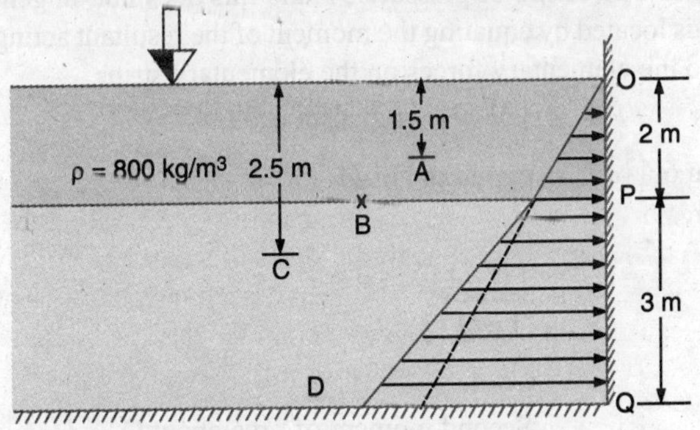

Solution.

The pressure at a point A, 1.5 m below the free surface must be given by

$$p_A = (0.8 \times 1000) \times 9.81 \times 1.5 = \mathbf{11770\ N/m^2}$$
$$= \mathbf{11.77\ kN/m^2}$$

The pressure at a point B at the interface of the two liquids is

$$p_B = (0.8 \times 1000) \times 9.81 \times 2 = \mathbf{15\,700\ N/m^2}$$
$$= \mathbf{15.70\ kN/m^2}$$

The pressure at a point C, 2.5 m below the free surface may be determined by either of the following considerations :

1. By considering that a liquid of specific gravity 0.8 acts over the 2.5 m depth and, in addition, liquid of differential specific gravity (10– 0.8), *i.e.*, 0.2 over the 0.5 m below the interface;

$$p_C = (0.8 \times 1000) \times 9.81 \times 2.5 + (0.2 \times 1000) \times 9.81 \times 0.5$$
$$= \mathbf{20\ 600\ N/m^2 = 20.6\ kN/m}$$

2. By determining the pressure at the interface and adding to it the pressure due to water for a further depth of 0.5 m;

$$p_C = (0.8 \times 1000) \times 9.81 \times 2.0 + 1000 \times 9.81 \times 0.5$$
$$= \mathbf{20600\ N/m^2 = 20.6\ kN/m^2}$$

The pressure at a point D at the base is similarly determined :

$$p_D = (0.8 \times 1000) \times 9.81 \times 2.0 + 1000 \times 9.81 \times 3$$
$$= 15\ 700 + 29430 = \mathbf{45\ 130\ N/m^2}$$
$$= \mathbf{45.13\ kN/m^2}$$

The force acting at the vertical wall OPQ may be determined by considering the pressure distribution from O to P and from P to Q as shown. The force must be equal to the volume under the pressure plot;

$$F = \frac{15.70}{2} \times 2 \times 6 + \frac{45.13 + 15.70}{2} \times 3 \times 6$$

$$= 94.2 + 547.47 = \mathbf{641.67\ kN}$$

2.4 Location of the Hydrostatic Force

Location of the line of action of the hydrostatic force and the point of action on a plane submerged surface is often necessary and this is now considered.

The point of action of the total hydrostatic force on the surface is called the centre of pressure distribution or simply as the 'centre of pressure', P and this does not, in general, coincide with the centroid. This point P is located by equating the moment of the resultant acting at P to the summation of the moments due to the elementary forces on the elementary strips,

$$F.y_p = \int \delta F.y = \int \rho g y^2 \sin\alpha\, \delta A$$

or $$y_p \rho g \sin\alpha \int y dA = \rho g \sin\alpha \int y^2 dA$$

whence

$$y_p = \frac{\int y^2 dA}{\int y dA}$$

In other words,

$$y_p = \frac{\text{Second moment of area about O}}{\text{First moment of area about O}} = \frac{I_0}{\bar{y}A}$$

By parallel-axis theorem,

$$I_0 = I_c + A\bar{y}^2$$

Hence,

$$y_p = \frac{I_c + A\bar{y}^2}{\bar{y}A} \text{ and in terms of } z \text{ and } \bar{z}_p \quad \text{... (2.4)}$$

$$\frac{\bar{z}_p}{\sin\alpha} = \frac{I_c}{\bar{z}\,A/\sin\alpha} + \frac{\bar{z}}{\sin\alpha}$$

$$\overline{z}_p = \overline{z} + I_c \frac{\sin^2 \alpha}{A\overline{z}} \qquad \text{... (2.5)}$$

This relation locates the centre of pressure P in relation to the centroid of the submerged surface. It may be observed that each of the quantities I_c, $\sin^2 \alpha$, A and z may either be positive or zero. It means that z_p must be more than or at most equal to z. In other words, the centre of pressure must be below C or at most coincide with C. This fact can be appreciated by observing the pressure distribution on a plane surface as well. Since the intensity of pressure increases as depth increases, the point of action of the total force due to the pressure must lie below the centroid of the area on which it acts.

Let us examine the action of hydrostatic force and the location of centre of pressure for the special cases of horizontal and vertical surfaces:

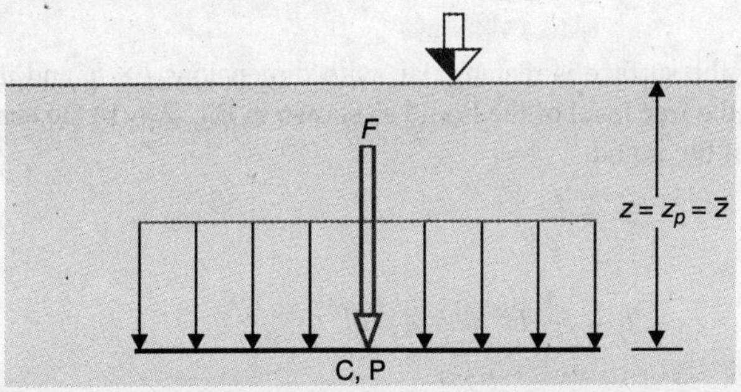

Fig. 2.4 (a): Pressure Distribution and Force on a Horizontal Submerged Plane Surface

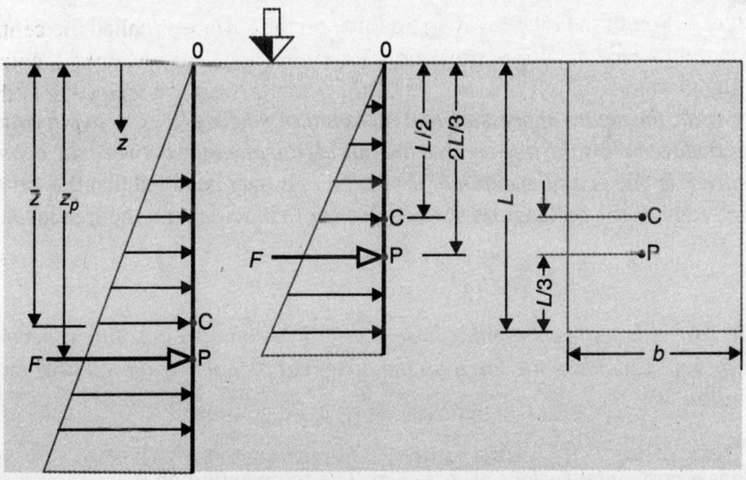

Fig. 2.4 (b): Pressure Distribution and Forces on Vertical Submerged Plane Surfaces

The pressure distribution on a **submerged horizontal surface** must be uniform over the entire area because every point on the surface is at a depth z below the free level of the liquid as shown in Fig. 2.4(a).

The hydrostatic force F on it must be

$$F = \rho g \overline{z} A$$

where $\overline{z} = z$, the depth of the centroid of the area A of the surface.

The centre of pressure P must be located on the surface at the centroid

$$z_p = z = \overline{z}$$

as is also seen by examining Eq. (2.5) by substituting $\sin \alpha = 0$ for $\alpha = 0$.

The pressure distribution on a **submerged vertical surface** must be such that, in the end view as shown in Fig. 2.4 (b), pressure increases linearly with the depth z, according to the relation

$$p = \rho g z$$

The hydrostatic force F on it must be

$$F = \rho g \bar{z} A$$

where $\bar{z}$ = depth of the centroid of the area A of the surface.

Since the pressure increases with the depth, the centre of pressure P must lie below the centroid of the area of the surface. From Eq. (2.5), for $\alpha = 90°$, $\sin \alpha = 1$,

$$z_p = \bar{z} + \frac{I_c}{A\bar{z}} \quad \text{...(2.6)}$$

In particular, if the surface is rectangular, with dimensions $l \times b$, and its one edge is placed immediately beneath the free level of the liquid as shown in Fig. 2.4 (b) the centroid is located at $l/2$ below the free level of the liquid,

$$\bar{z} = \frac{l}{2}$$

Also, $$I_C = \frac{1}{12} bl^3 \text{ and } A = bl$$

which yield $$z_p = \frac{l}{2} + \frac{l}{12}\frac{bl^3}{bl.l/2} = \frac{2}{3}l$$

and $$z_p - \bar{z} = \frac{l}{6} \quad \text{...(2.7)}$$

In other words, the centre of pressure of a rectangular surface lies at a depth equal to two-third the depth of the surface below the free level of the liquid; the distance between the centroid C and the centre of pressure P is one sixth the depth of the surface. It may be noted that this criterion does not hold good if one edge of the rectangular surface does not coincide with the free level. In that case,

$$\bar{z} < z_p \le \frac{2}{3}l$$

Example 2.6. *A box of rectangular base 2 m × 3 m contains gasoline (Specific gravity 0.7) up to a height of 5 m. Calculate the force on the base and on each of the vertical faces and locate their lines of action.*

Solution.

The base is a horizontal surface at $\bar{z} = 5$ m. The force acting on it

$$F = \rho g \bar{z} A = 1000 \times 0.7 \times 9.81 \times 5 \times (2 \times 3) = 206 \times 10^3 \text{ N} = 206 \text{ kN}$$

Since the pressure is uniform over the base, the resultant must pass through the centroid of the base and is directed vertically downwards.

Each vertical face has the centroid at depth

$$\bar{z} = \frac{5}{2} = 2.5 \text{ m}$$

Force on each 2 m × 5 m face is given by

$$F = 1000 \times 0.7 \times 9.81 \times 2.5 \times (2 \times 5) = 172 \times 10^3 \text{ N} = \mathbf{172\ kN}$$

and on a 3 m × 5 m face

$$F = 1000 \times 0.7 \times 9.81 \times 2.5 \times (3 \times 5) = 258 \times 10^3 \text{ N} = \mathbf{258\ kN}$$

The centre of pressure for each vertical face lies at a depth equal to 2/3 times the submerged height *i.e.*, at 5 × 2/3 or 3.33 m, below the level.

Example 2.7. *The gates of a lock are 5 m wide and 5 m high and, when closed, include an angle of 120°. Each gate is held on two hinges, one placed at the top and the other at the bottom of the gate.*

If the water levels are 4.5 and 3 m on the upstream and downstream sides respectively, determine the magnitude of the forces on the hinges due to the water pressure.

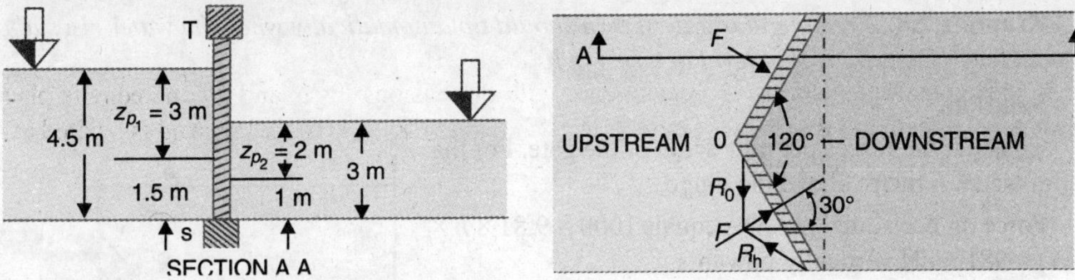

Solution.

With reference to the figure,

Upstream side	Downstream side	(Units)
$h_1 = 4.5$	$h_2 = 3$	(m)
$F_1 = 1000 \times 9.81 \times \frac{4.5}{2} \times (5 \times 4.5)$	$F_2 = 1000 \times 9.81 \times \frac{3}{2} \times (5 \times 3)$	
$= 497 \times 10^3$	$= 221 \times 10^3$	(N)
$= 497$	$= 221$	(kN)
$z_{p1} = \frac{3}{2} \times 4.5 = 3$	$z_{p2} = \frac{3}{2} \times 3 = 2$	(m)

The resultant force F acting on each gate acts normal to the gate as shown and, in magnitude,

$$F = F_1 - F_2 = (497 - 221) = \mathbf{276\ kN}$$

The line of action of F is h m from the base and can be determined by taking moments about the base of the gate

$$276 \times h = 497 \times (4.5 - 3) - 221 \times (3 - 2)$$

whence $h = 1.90$ m from the base.

The action force F on the gate must be balanced by the reaction at the hinges $\boldsymbol{R_h}$ and by the reaction $\boldsymbol{R_0}$ by the other gate at O.

$$F = R_h + R_0$$

For equilibrium R_0 being normal to the centreline, R_h must pass through the intersection of F and R_0. *Resolving the equilibrant forces along the direction of F.*

$$F = R_h \sin 30° + R_0 \sin 30° = (R_h + R_0) \sin 30°.$$

R_h and R_0 being equal in magnitude, by symmetry,

$$R_h = R_0 = \frac{1}{2}\frac{F}{\sin 30} = 276 \text{ kN}$$

Assuming that the total reaction by the hinges of a gate lies in the plane of F and R_0, i.e., located at $h = 1.90$ m from the base, the reactions R_T and R_B by the top and bottom hinges respectively can be evaluated by considering equilibrium under the action of forces and moments.

By equating the forces,

$$R_h = 276 = R_T + R_B \qquad \text{... (i)}$$

By equating the moments about the bottom hinge,

$$276 \times 1.9 = R_T \times 5 + R_B \times 0 \qquad \text{... (ii)}$$

From (*i*) and (*ii*)

$$\mathbf{R_T = 105 \text{ kN and } R_B = 171 \text{ kN}}$$

Example 2.8. *A rectangular gate shown should tip automatically when the water rises above a certain level. Determine that level in terms of h.*

Solution.

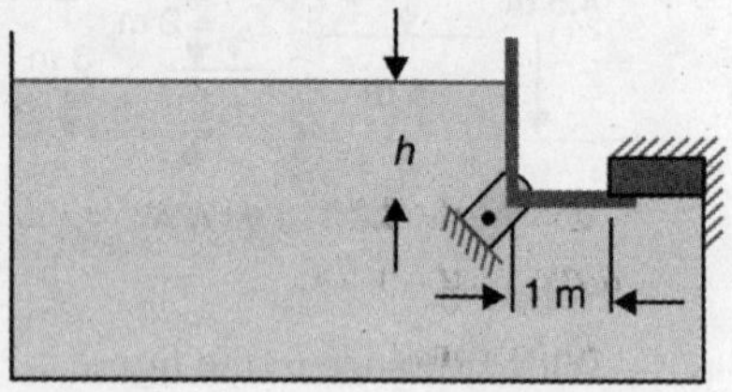

Consider the forces per unit width of the gate. For the level of water, h metre above the hinge,

Force on horizontal part OA equals $1000 \times 9.81 \times h \times (1 \times 1) = 9810\,h$ N vertically upwards

This force acts at a distance of 1/2 m from the hinge.

Force on the vertical part of the gate equals

$$1000 \times 9.81 \frac{h}{2} \times (h \times 1) = 9810 \frac{h^2}{2}$$

This force acts at $\frac{h}{3}$ m from the hinge.

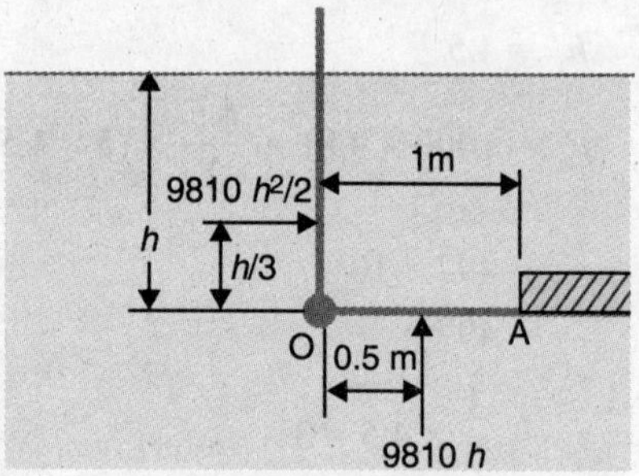

Tipping would occur when the overturning clockwise moment just exceeds the resisting anticlockwise moment about the hinge O.

$$9810 \frac{h^2}{2} \times \frac{h}{3} = 9810\, h \times \frac{1}{2}$$

or

$$h = \sqrt{3} = \mathbf{1.732 \text{ m}}$$

Example 2.9. *Determine and locate the total force F due to water pressure on the surface inclined at 30° to the horizontal, as shown. The surface is hinged on its top edge and is 2 m wide.*

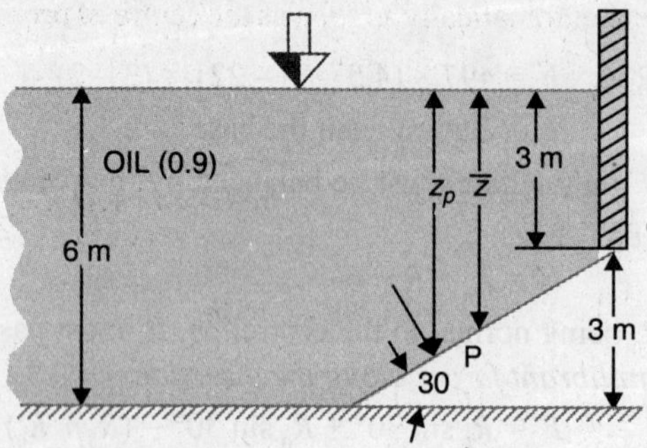

Solution.

$$\bar{z} = 3 + \frac{3}{2} = 4.5 \text{ m}$$

Total hydrostatic force on the surface

$$F = \rho g \bar{z} A = 0.9 \times 1000 \times 9.81 \times 4.5 \times \left(\frac{3}{\sin 30} \times 2\right)$$

$$= 477\,000 \text{ N} = \mathbf{477 \text{ kN}}$$

Depth of the centre of pressure below the liquid level,

$$z_h = \bar{z} + I_c \frac{\sin^2 \alpha}{A \bar{z}}$$

$$= 4.5 + \left(\frac{1}{12} \times 2 \times 6^3 \times 0.5^2\right) / (6 \times 2 \times 2 \times 4.5)$$

$$= \mathbf{4.67 \text{ m}}$$

Example 2.10. *The automatic tipper shown operates when the level reaches a certain height. Calculate that level h.*

Solution.

For a depth of water *h* and unit width of gate,

$$\bar{z} = \frac{h}{2}, \; A = \frac{h}{\sin 60°} \times 1 = 1.16\,h$$

$$I_c = \frac{h}{2} + 0.13h^3 \times (0.866)^2 / \left(1.16h \times \frac{h}{2}\right)$$

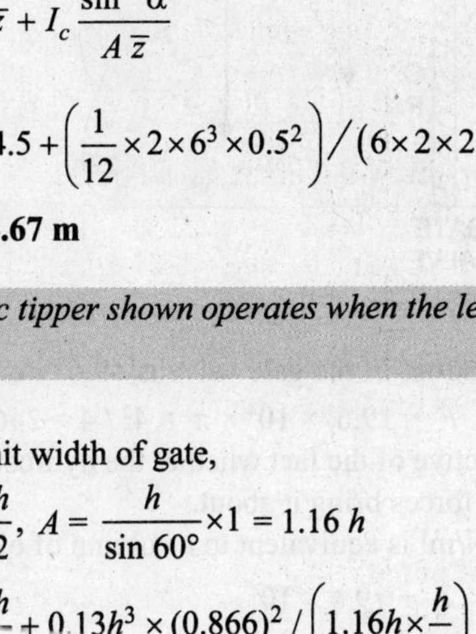

Tipping would occur automatically as soon as the centre of pressure just rises above the hinge O:

$$z_h = \bar{z} + I_c \frac{\sin^2 \theta}{A \bar{z}}$$

$$= \frac{h}{2} + 0.13h^3 \times (0.866)^2 / \left(1.16h \times \frac{h}{2}\right)$$

$$= \frac{h}{2} + 0.17\,h = 0.67\,h$$

$$= (h - 1)$$

from the given configuration.

Hence, $h = \mathbf{3.03 \text{ m}}$, the depth of water for tipping.

Example. 2.11. *A pipeline which is 4 m in diameter contains a gate valve. The pressure at the centreline of the pipe is 19.6 N/cm². If the pipe is filled with oil of specific gravity 0.87, find the force exerted by the oil upon the gate and the position of the centre of pressure.*

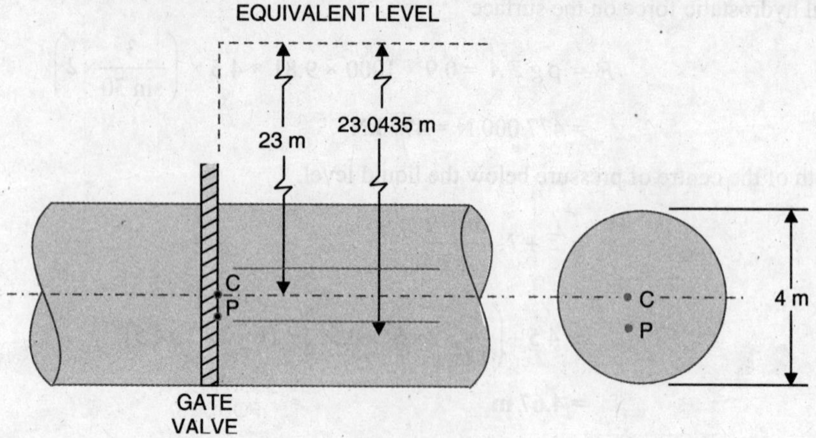

Solution.

The pressure at the centroid of the gate valve is 19.6 N/cm². The force on the gate is

$$F = 19.6 \times 10^4 \times \pi \times 4^2 / 4 = 2463008.6 \text{ N} = 2.463 \text{ MN}$$

This force acts irrespective of the fact whether the hydrostatic force alone or a combination of the hydrostatic and external forces bring it about.

The pressure of 19.6 N/m² is equivalent to a column of oil $\bar{z}$ high such that

$$(0.87 \times 1000) \times 9.81 \times \bar{z} = 19.6 \times 10^4$$

and $\bar{z} = 23$ m

The centre of pressure P is located by the expression

$$z_p = \bar{z} + I_c \frac{\sin^2 \alpha}{A\bar{z}}$$

so that $$z_p = 23 + \frac{\pi \times 4^4}{64} \times \frac{4}{\pi \times 4^2} \times \frac{1^2}{23}$$

$$= \mathbf{23.0435}$$

This distance is measured from the equivalent level surface of oil. It follows that this point lies at

23.0435 – 23 = **0.0435 m** below the centreline of the pipe

2.5. Force on a Submerged Curved Surface

On a curved surface, as shown in Fig. 2.5(*a*), the elemental force $p\,\delta A$ varies both in magnitude and in direction. The horizontal *i.e.*, *x* and *y* components of forces can be evaluated by the summation of elemental force components. For example, along the *x*-axis,

$$F_x = \int p\, \delta A \cos\theta \quad \text{...(2.8)}$$

Realising that $\delta A \cos\theta$ is the projection of the elemental area normal to the *x*-axis, the summation of the elemental projections amounts to the projected area of the whole curved surface. Hence

F_x = force on the projected area normal to the *x*-direction.

Similarly,

F_y = force on the projected area normal to the *y*-direction.

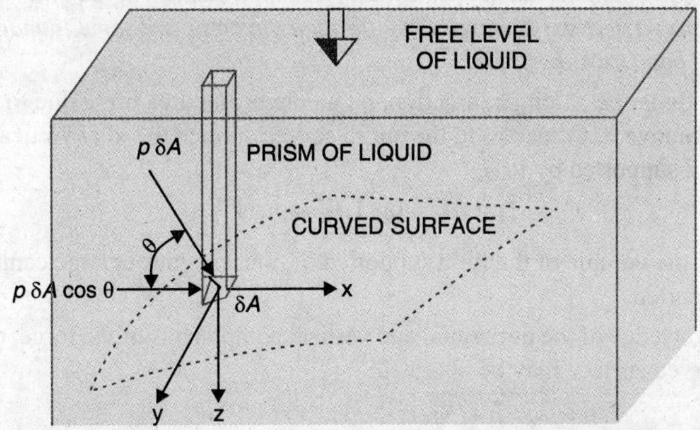

Fig. 2.5 (a) Pressure on a Curved Surface

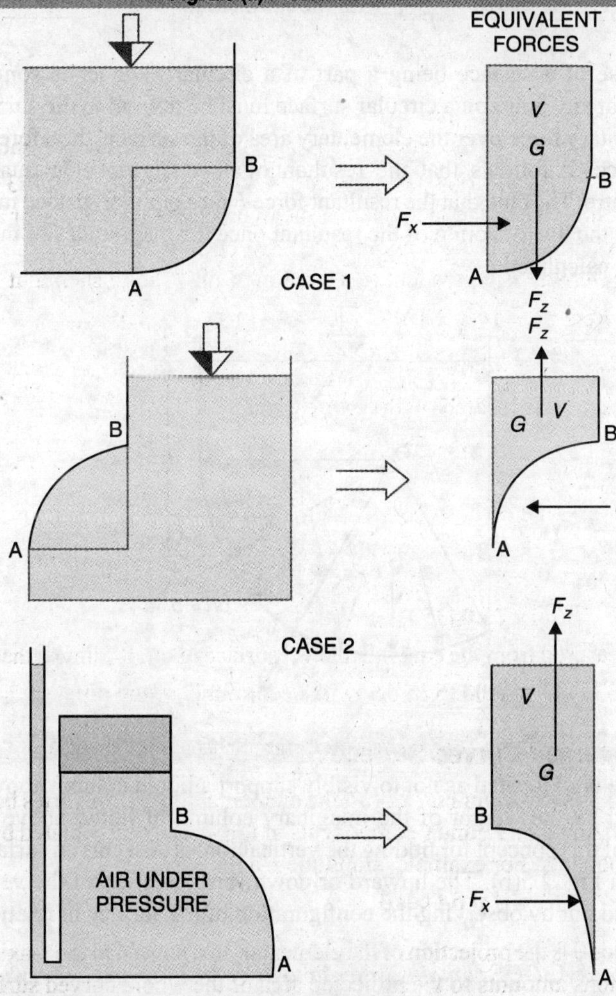

Fig 2.5(b) Forces on Curved Surfaces

It is intersting to note that the net horizontal force on a closed curved body must be zero since the projected areas on a vertical plane passing through the body, are same in magnitude due to the curved parts of the body on either side.

The vertical force, *i.e.*, along *z* direction, on an elemental area δA is due to the weight of the prism of the liquid above it. Considering the entire curved surface the total vertical force equals the weight of the liquid supported by it.

or $$F_x = \int \rho g \, dV = \rho g \int dV = \rho \, g \, V \qquad ...(2.9)$$

where V stands for the volume of the fluid supported. F_x must act through the centre of gravity G of the fluid mass supported.

From the knowledge of the horizontal and vertical components of the force, the resultant force F acting on the curved surface may be obtained.

Magnitude: $$F = \sqrt{F_x^2 + F_z^2}$$

Inclination: $$\alpha = \tan^{-1} \frac{F_z}{F_x}; \text{ with the horizontal.} \qquad ...(2.10)$$

The particular case of a surface being a part of a circular cylinder is somewhat easier to deal with. The pressure at any point on a circular surface must be normal to the surface as shown in Fig. 2.5 (c). Each elementary force over the elementary area of the surface, therefore, passes through the centre of the surface. It follows that the resultant of all concurrent elementary forces should also pass through the centre. The clue that the resultant force F on a circular surface must pass through its centre helps to locate the line of action of the resultant once the magnitudes of the horizontal and vertical components are calculated.

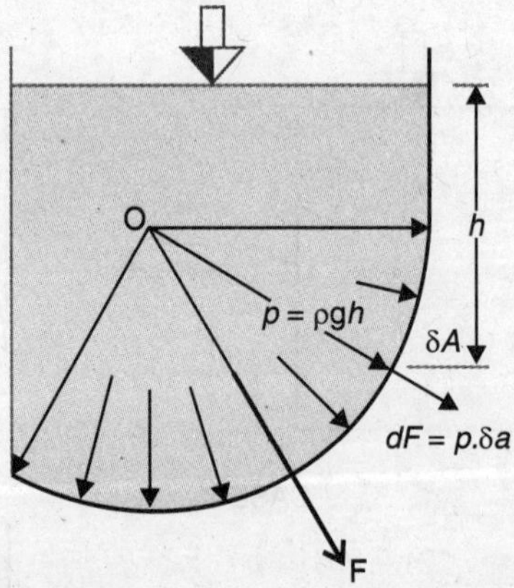

Fig 2.5(c): Force on a Circular Surface

If a curved surface is so located as not to visibly support a liquid column above it, the vertical force is still determined by the weight of the imaginary column of liquid above it upto the free surface. An illustration of this concept for finding the vertical forces on a curved surface AB in typical situations is provided in Fig. 2.5(b). The upward or downward direction of the vertical force on a surface should be figured out by observing the configuration of the surface in relation to the fluid.

Example 2.12. *A tank ABCDE contains water upto a depth of 1 m and is 2 m wide. The curve AB is defined by $z = x^2$ and curve CD is a quadrant of a circle of radius 0.3 m. Calculate the magnitude of the resultant forces on surfaces AB and CD. Find also the angle that the resultant force makes with the horizontal.*

Solution.

Force on surface CD:

Horizontal force along x-axis = 1000 × 9.81 × (0.7 + 0.3/2) (0.3 × 2) = 5000 N = **5.0 kN**

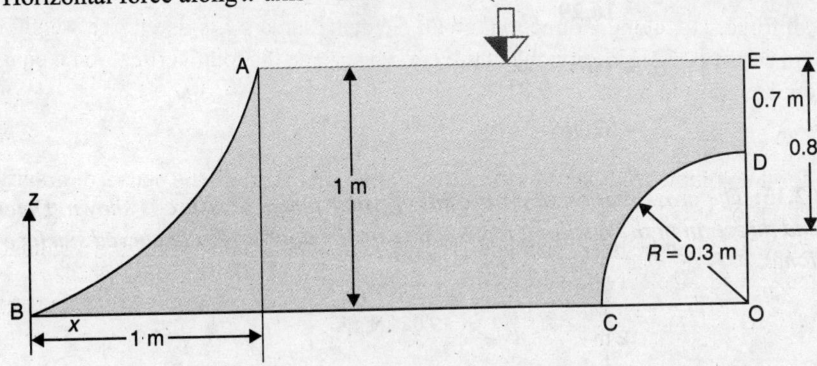

(*i*) Force on Surface CD

Vertical force along z-axis

= weight of the water actually supported by it

= weight of (prism OEFC – quadrant OCD) of water

$$= 1000 \times 9.81 \left(0.3 \times 1 \times 2 - \pi \times 0.3^2 \times \frac{2}{4}\right) = 4500 \text{ N}$$

= **4.5 kN** acting downwards

The resultant force on the surface CD is

$$F = \sqrt{5.0 + 4.5^2}$$

$$= \mathbf{7.0\ kN}$$

inclined at $\alpha = \tan^{-1}\dfrac{4.5}{5.0}$

= **42°** with the horizontal.

This force must pass through the centre O of the circle of which the surface CD is a part as shown.

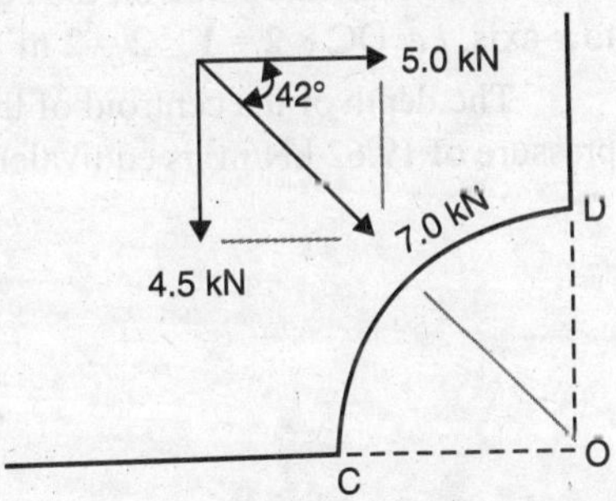

(i) Force on Surface CD

Force on Surface AB:

Horizontal force along x-axis

= 1000 × 9.81 × 0.5 × (1 × 2) = 9810 N = 9.81 kN

Vertical force on it acting upwards

= weight of the liquid AVB virtually supported by it

$$= \rho g[(1 \times 1 \times 2) - \int_0^1 | z\, dx \times 2 |]$$

$$= 2\rho g\left(1 - \int_0^1 x^2\, dx\right) = 2\rho g\left(1 - \left|\frac{x^3}{3}\right|_0^1\right)$$

= 2ρg (1 – 0.33) = 1.33 ρg

= 1.33 × 1000 × 9.81 = 13000 N = 13 kN

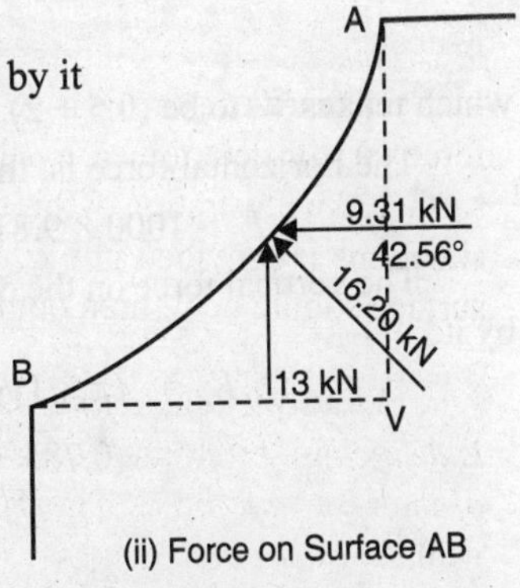

(ii) Force on Surface AB

The resultant force on the surface AB

$$F = \sqrt{(9.81)^2 + (13)^2}$$
$$= \mathbf{16.29}$$

inclined at
$$\alpha = \tan^{-1}\frac{13}{9.81}$$
$$= \mathbf{52.96°}$$

Example 2.13: *The cross section of a tank full of water under pressure is shown. Determine the magnitude and direction of the resultant force acting on the quarter-circle curved suirface BC if the tank is 2 m long.*

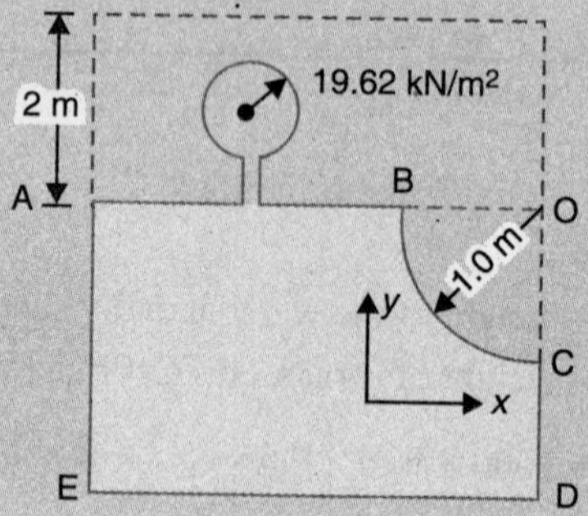

Solution.

The horizontal force on the curved surface BC equals the force on the projected area normal to *x*-axis, *i.e.* OC × 2 = 1 × 2 = 2 m²

The depth of the centroid of this projected area is 0.5 m below the top edge of water. But the pressure of 19.62 kN/m² is equivalent to a head of

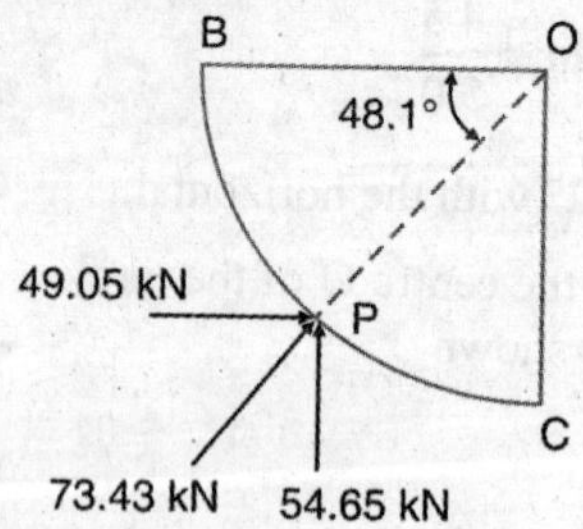

$$\frac{19.62 \times 1000}{1000 \times 9.81} = 2 \text{ m of water}$$

which makes $\bar{z}$ to be (0.5 + 2), *i.e.*, 2.5 m.

The horizontal force is, therefore,

$$F_x = 1000 \times 9.81 \times 2.5 \times 2 = 49050 \text{ N} = 49.05 \text{ kN}$$

The vertical force on the curved surface BC equals the weight of the liquid virtually supported by it.

$$F_z = (\pi \times 1^2/4 + 1 \times 2) \times 2 \times 1000 \times 9.81$$
$$= (0.785 + 2) \times 19620 = 54649.5 \text{ N}$$

$= 54.65$ kN upwards

The resultant force on the curved surface is

$$F = \sqrt{F_x^2 + F_z^2}$$

$$= \sqrt{49.05^2 + 54.65^2} = \mathbf{73.43\ kN}$$

inclined at $\alpha = \tan^{-1}(54.65/49.05)$

$= \tan^{-1}(1.114) = \mathbf{48.1°}$ with the horozontal

The resultant force must pass through the centre O since this is a circular curved surface.

Example 2.14. *A cylinder of 1 m diameter and 2 m length stays in equilibrium as shown. Calculate the specific gravity of the material of the cylinder.*

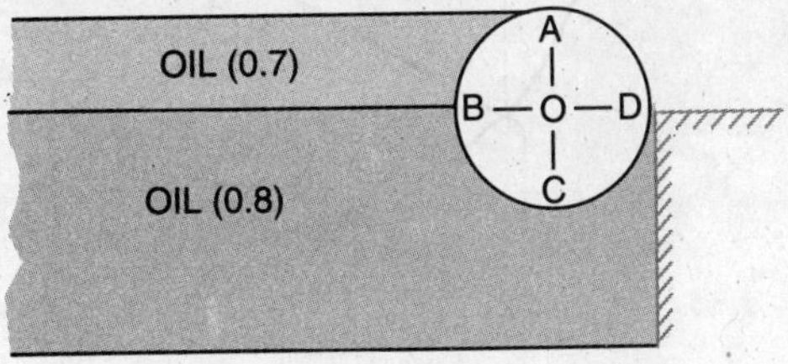

Solution.

In order to determine its specific gravity we are only concerned with the vertical force on the cylinder.

On surface AB, the vertical force equals the weight of the oil actually supported by it.

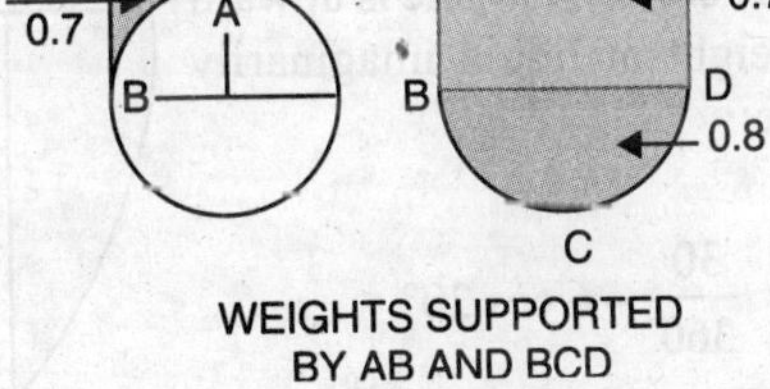

WEIGHTS SUPPORTED BY AB AND BCD

$$F_1 = \left(0.5 \times 0.5 - \pi \times 1 \times \frac{1}{16}\right) \times 2 \times \rho g$$

$= 0.054 \times 2 \times 1000 \times 0.7 \times 9.81 = 741.64$ N downwards

On surface BCD, the vertical force equals the virtual weight of oil supported by it

$$F_2 = \left(1 \times 0.5 \times 0.7 - \pi \times 1 \times \frac{1}{8} \times 0.8\right) \times 2 \times \rho_{water} \times g$$

$= 0.664 \times 2 \times 1000 \times 9.81 = 13027.7$ N

Net hydrstatic force on the cylinder

$F_2 - F_1 = 13027.7 - 741.64 = 12286$ N upwards

This must equal the weight of the cylinder

$$\rho_c \times 9.81 \times \pi \times 1^2 \times \frac{1}{4} \times 2 = 12236$$

whence, ρ_c = **797.3 kg/m³**

The specific gravity of the material should, therefore, be **0.797.**

Example 2.15. *A tainter gate is used as a control at the crest of a small spillway as shown. If the width of the gate is 3 m, find the magnitude of the force on the gate for the conditions shown. What angle will the resultant force make with the horizontay?*

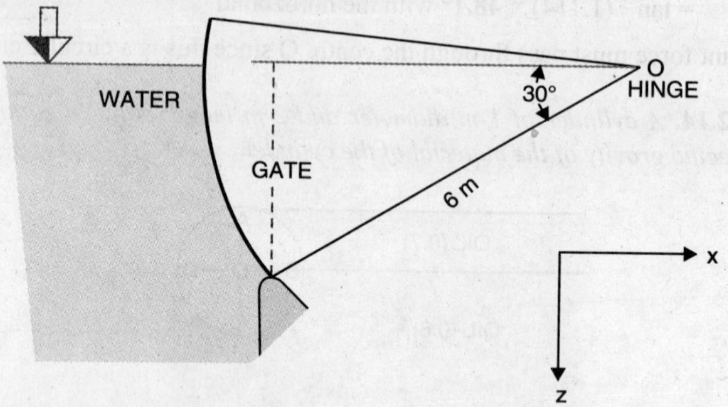

Solution.

The horizontal component of force acting on the gate is the horizontal force on the projected area normal to the x-axis;

$$F_x = 1000 \times 9.81 \times \frac{6 \sin 30°}{2} \times (6 \sin 30° \times 3)$$

$$= 132\,435 \text{ N} = 132.435 \text{ kN}$$

The vertical component of force on the gate is upward but numerically equals the weight of liquid imaginarily supported by the curved surface:

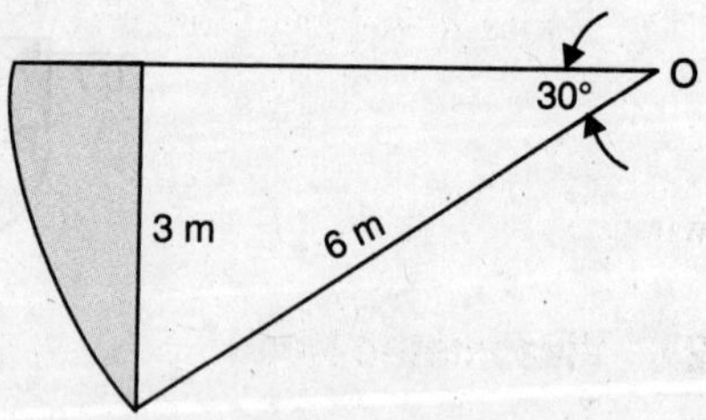

Area shown shade

$$= \pi \times 6^2 \times \frac{30}{360} - 6 \cos 30° \times \frac{3}{2}$$

$$= 9.424 - 7.794$$

$$= 1.63 \text{ m}^2$$

Weight of the water for the prism with this cross section is

$$1.63 \times 3 \times 1000 \times 9.81 = 47970 \text{ N}$$

The vertically upward force acting on the gate is, therefore, 47 970 N = 47.970 kN.

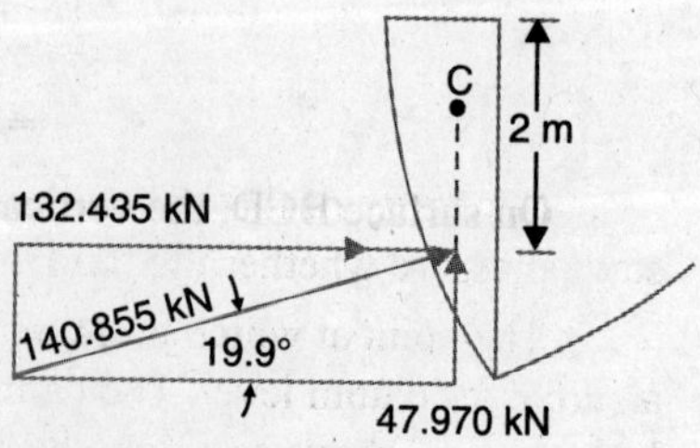

Resultant force = $\sqrt{132.435^2 + 47.970^2}$ = **140.855 kN**

The angle which the resultant force makes with the horizontal is

$$\tan^{-1}\frac{47.970}{132.435} = \tan^{-1} 0.362 = 19.9°$$

The resultant force and its inclination with the horizontal are shown acting on the gate. It may be observed that the vertical component of the force must pass through the centroid C whereas the horizontal component acts at two-third the vertical depth of the gate.

Example 2.16. *The gate OA shown is hinged at O and is in the form of a quadrant of a circle of radius 1 m. It supports water on one side as shown. If the width of the gate is 3 m, calculate the force required to hold the gate in position.*

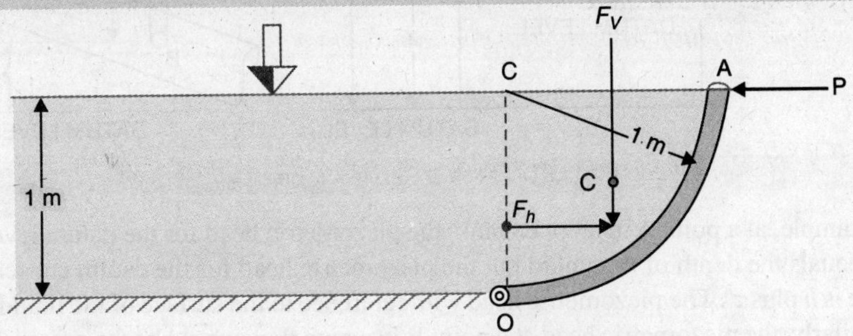

Solution.

$$F_h = \rho g \bar{z} A = 1000 \times 9.81 \times \frac{1}{2}(3 \times 1)$$

$$= 14\,715 \text{ N}$$

$$F_v = \frac{1}{4}\rho g \pi r^4 \times 3 = 1000 \times 9.81 \times \frac{\pi}{4} \times 1 \times 3$$

$$= 23114.3 \text{ N}$$

The horizontal force F_h acts at a height 1/3 m above the hinge O whereas the force F_v acts through the centroid of area OAC which is located at a distance 4 × 1/3 π m from OC.

Taking moments of forces about the hinge O,

$$P \times 1 = 14715 \times \frac{1}{3} + 23114.3 \times \frac{4 \times 1}{3\pi}$$

whence, $\quad P = \mathbf{14715\ N = 14.715\ kN.}$

2.6 Piezometric Head

The pressure at a point in a fluid is usually expressed as 'pressure head' or height of column of that fluid, h such that

$$h = \frac{p}{\rho g}$$

Similarity of this expression with the hydrostatic law suggests that the pressure referred to is static pressure whether the fluid in question is at rest or in motion.

The point at which pressure is being considered may have a 'datum head', z with reference to an arbitrary datum level. The datum head at a point may be positive or negative depending upon its location, *i.e.*, above or below the datum level.

Piezometric head at a point in a fluid at rest or in motion is defined as the sum of pressure head and datum head at that point with reference to a selected datum level,

$$\text{Piezometric head} = p/\rho g + z \qquad \text{...(2.11)}$$

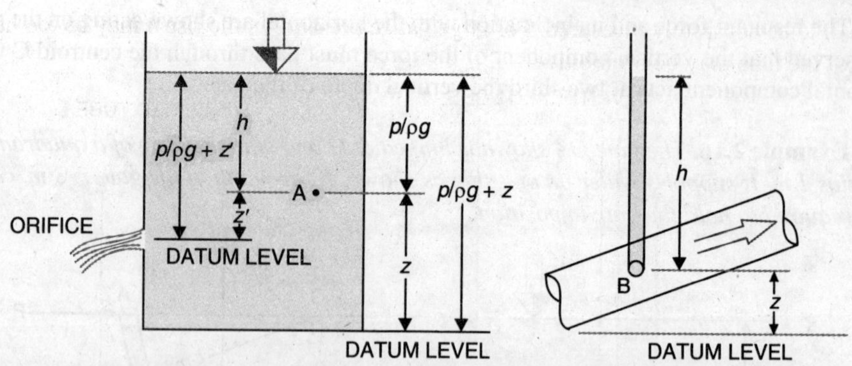

Fig. 2.6 Definition of Piezometric Head, h

For example, at a point A in an open tank, the piezometric head for the datum level at the base of the tank equals the depth of the liquid but the piezometric head for the datum chosen at the level of the orifice is h plus z'. The piezometric head with reference to the surface of the liquid itself would be zero. Similarly, the piezometric head at a point B in a pipe flow would be the pressure head h plus the datum head z above the datum level.

It is more convenient to deal with piezometric head than with pressure and datum heads individually because

(a) It is usually the sum of the pressure and datum heads that is of interest to us in calculations

(b) It is the sum of two that is easily measurable by employing simple 'piezometers', i.e., vertical tubes at different points in a fluid flow.

(c) The sum of the two remains constant at all points in a fluid at rest or in motion at constant velocity when referred to a datum level.

The technique of pressure measurement by employing the hydrostatic law, i.e.,

$$\frac{dp}{dz} = \rho g$$

is known as manometry and the devices used for measurement are called manometers. It is intended to introduce the principles of single tube and U-tube manometers, micro-manometers as well as differential manometers.

2.7. Single Tube Manometers

The pressure at a point in a liquid at rest or in motion may be measured by employing a single vertical tube, also referred to as a piezometer. The tube is attached to the point at its lower end and left open to the atmosphere at the upper end. If the pressure at the point is higher than the atmospheric pressure, the liquid will rise in the tube to a level which is proportional to the gauge pressure. Since the column of liquid in the tube is at rest, the gauge pressure at the lower end of the tube must be balanced by the hydrostatic pressure due to the column of liquid in it :

$$p = \rho g h \quad \text{... (2.12)}$$

Measurement of h on a scale attached to the piezometer would thus enable the gauge pressure p to be determined.

A single tube manometer may serve to measure the pressure if it is above the local atmospheric pressure, *i.e.*, if the gauge pressure p is positive because the height h of the liquid column in the tube will then be positive. The tube fails to measure vaccum, *i.e.*, sub-atmospheric pressure because it will then draw air from the atmosphere into the liquid. A single tube may also be inadequate if either the gauge pressure is very small or large. For every small gauge pressure p, the rise h in the tube may be too small to be accurately measured. For large gauge pressure p, the rise h may be too high and the length of the tube may be prohibitive.

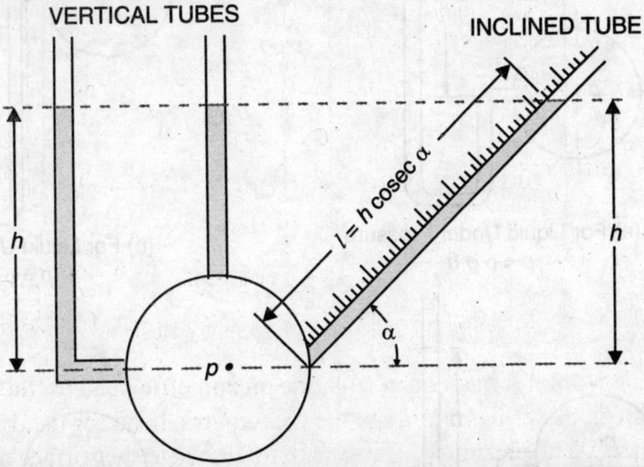

Fig. 2.7 Single Tube Manometer

For a small guage pressure p, it is better to use an inclined single tube. The reading l recorded on the scale of an inclined tube manometer is higher than that on a vertical tube manometer by a factor of cosec α where α is the angle of the inclined tube with the horizontal

$$l = h \operatorname{cosec} \alpha$$

This is clear from the fact that, for the same vertical h, the length of the liquid column in an inclined manometer is h cosec α. Smaller inclination α with the horizontal provide larger length of the liquid column but there is a limit to it. The limit is provided by the surface tension and capillarity effects which disallow accurate observation of the level of the liquid below a certain value of α. The lower limit for the water-in-glass tube manometers is around 20° inclination.

It may also be noted that a single vertical tube manometer can be used to measure the pressure in liquids only. It cannot be used to measure the pressure in a gas.

2.8. U-tube Manometers

A U-tube manometer consists of two tubes joined at one end to form a U-shaped tube. The two-tubes are said to constitute the limbs of the U-tube manometer.

A U-tube manometer can be employed to measure the pressure in any fluid; liquid or gas. Pressures above and below atmosphere may be measured with it equally comfortably. A U-tube may be used upright, inclined or inverted. Let us examine the use of a U-tube manometer in different situations as shown in Fig. 2.8.

For a liquid under moderate pressure, the U-tube may simply be connected as shown in Fig. 2.8(a). The level of the liquid rises in the open ended tube, above the point where the pressure is being measured, by a height h such that the pressure exerted due to the column h equals the pressure at the point;

$$p = \rho g h$$

For a liquid under vacuum, the U-tube should be connected such that it is mainly below the level of the point where the vaccum is measured. This is to allow for a depression h of the level of the liquid in the open ended tube such that the vacuum at the point equals the vaccum due to the depression h of the liquid;

$$p = -\rho g h$$

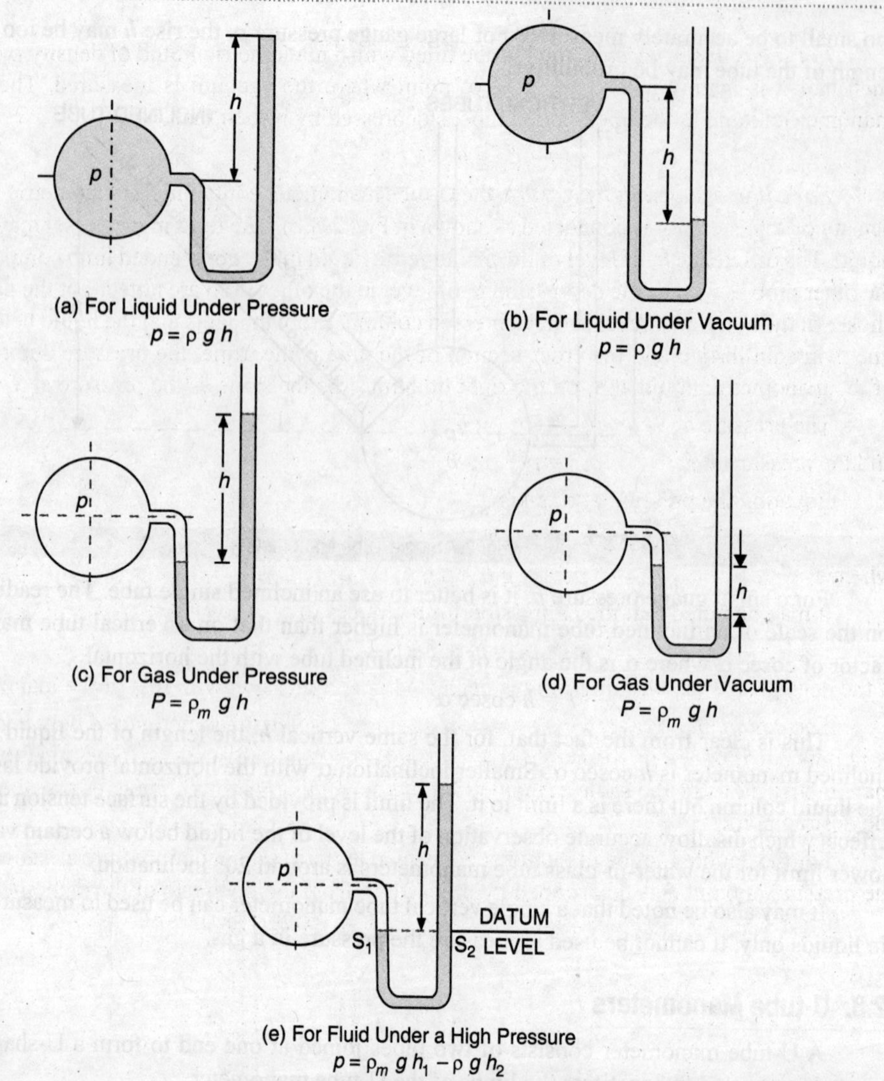

Fig. 2.8. Simple U-tube Manometers

For a gas under pressure, the manometric liquid is so selected that it does not dissolve the gas and that, for the range of pressure p, the rise h of the level of the liquid in the open end is measurable. For example, if the pressure of air in a vessel is to be measured, immiscible liquids such as alcohol, water and mercury qualify as manometric liquids. If the pressure of the air is of the order of 2 kN/m^2, water may be preferred to mercury because the height of the column would be of the order of 20 cm with water and only 1.5 cm with mercury. On the other hand, if the pressure is of the order of 0.2 kN/m^2, one may prefer to use alcohol and if the pressure if of the order of 20 kN/m^2, the obvious choice is in favour of mercury. The height h of the column in the open-ended tube in excess of that in the other tube in this case is related to the pressure p by

$$p = \rho_m g h \text{ where } \rho_m \text{ is the density of the manometric fluid.}$$

It may be noted that the density of the gas does not enter the relationship because the hydrostatic pressure due to the column of the gas is negligible in comparison with the hydrostatic pressure due to the liquid;

$$\rho << \rho_m$$

For a gas under vaccum, the U-tube filled with a manometric liquid of density ρ_m is connected such that it is mainly below the level of point where the vacuum is measured. The level of the manometric liquid in the open-ended tube is depressed by h such that,

$$p = -\rho_m g h$$

For a liquid under high pressure, the U-tube manometer containing a manometric liquid of high density ρ_m e.g. mercury is connected as shown in Fig. 2.8(*e*). The fluid in the vessel may be a gas or a liquid. The difference h_1 of level of the manometric liquid in the open-ended limb compared to that in the other limb as well as the depression h_2 of level in the other limb are noted. Let the datum level be chosen at the horizontal level in the depressed column. From the facts that the liquid below the datum line is in equilibrium and the cross-section of the tube is the same, the pressure due to the column of the manometric liquid at S_1 on the right limb must be the same as the pressure at S_2 on left limb.

The pressure at S_1 $\quad p_{S1} = p_{ms1}\, g\, h_1$

and the pressure at S_2 $\quad p_{S2} = p + \rho\, g\, h_2$

Equating the pressure at S_1 and S_2

$$p + \rho\, g\, h_2 = \rho_m g h_1$$

whence $\quad p = \rho_m g\, h - \rho\, g\, h_2$

It may be noted that the expression for p reduces to

$$p = \rho_m\, g\, h_1$$

if the density ρ of the working fluid is neglected in compression with that of the manometric fluid.

For a fluid under a very high pressure, a single U-tube manometer used may become too tall. A number of U-tubes of resonable size may then be used. Different U-tubes may have different manometric fluids or the same manometric fluid depending upon the range of pressure expected at the desired point.

For the double U-tube manometer connected to the vessel under pressure as shown in Fig. 2.9, the pressure at point P is determined by proceeding from the open end of the manometer.

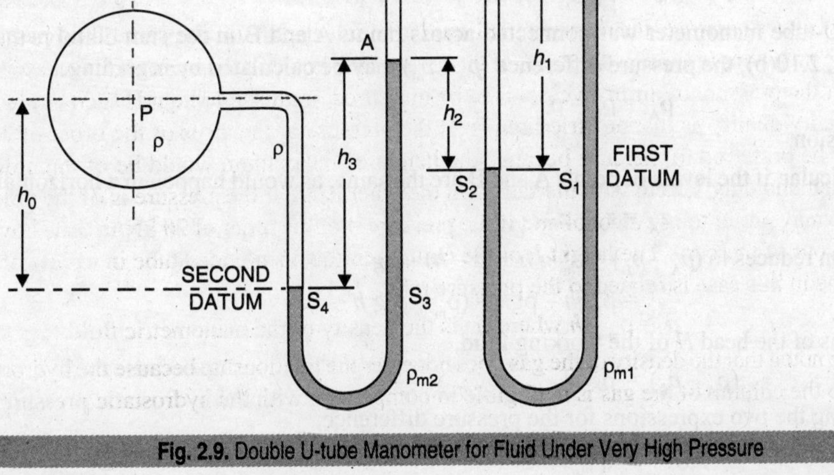

Fig. 2.9. Double U-tube Manometer for Fluid Under Very High Pressure

The pressure at point S_1 $\quad p_{S_1} = \rho_{m_1} gh_1$

and the pressure at point S_2 $\quad p_{S_2} = p_{s_1}$

The pressure at point A $\quad p_A = p_{S_2} - p_{m_2} gh_2$

$$= p_m gh_1 - \rho_{m_1} gh_2$$

The pressure at S_3 $\quad p_{S_3} = p_A + \rho_{m_2} g\, h_3$

$$= \rho_{m_1} gh_1 - \rho_{m_2} gh_2 + \rho_{m_2} gh_3$$

The pressure at point S_4 $\quad p_{S_4} = p_{S_3}$

and the pressure at the point p, the point of interest

$$p = p_{S_4} - \rho\, g\, h_0$$

giving the final result, $\quad p = \rho_{m_1} g\, h_1 - \rho_{m_2} g\, h_2 + \rho_{m_2} g\, h_3 - \rho\, g\, h_0$ ...(2.13)

This result could have been written straightaway by looking at the double U-tube configuration carefully.

2.9 Differential Manometers

A differential manometer is employed to measure the difference in pressure between two points. The two points , A and B, may either be located in the same vessel or in different vessels containing the same or different fluids. The two ends of the differential manometer are connected across the points A and B. The following devices for the measurement of differential pressure are studied:

(*a*) Upright U-tube manometer

(*b*) Well-type upright U-tube manometer

(*c*) Inverted U-tube manometer

(*d*) Inclined tube micromanometer.

(a) Upright U-tube Manometer. If an upright U-tube manometer is connected across points A and B in different fluid passages at different datum levels as shown in Fig 2.10(a), the pressure difference $(p_A - p_B)$ may be calculated by considering the equilibrium above the datum level S_1 S_2.

The pressure at S_1 $\quad p_{S_1} = p_A + \rho_A g z_A$

and the pressure at S_2 $\quad p_{S_2} = p_B + \rho_A g(z_B - h) + \rho m\, gh$

Equating p_{S_1} and p_{S_2}

$$p_A + \rho_A g z_A = p_B + \rho_B g(z_B - h) + \rho_m gh$$

whence $\quad (p_A - p_B) = \rho_m g\, h + \rho_B g(z_B - h) - \rho_A g z_A$ ...(2.14)

If the U-tube manometer was connected across points A and B in the same fluid passage as shown in Fig. 2.10(b), the pressure difference $(p_A - p_B)$ may be calculated by regarding.

$$\rho_A = \rho_B = \rho$$

in the expression.

In particular, if the levels of points A and B are the same, as would happen in a horizontal pipe,

$$z_A = z_B = z$$

the expression reduces to $(p_A - p_B) = \rho_m g\, h + (z - h) - \rho g$

$$= \rho_m gh - \rho gh = (\rho_m - \rho)g\, h$$

In terms of the head H of the working fluid,

$$(\rho_A - p_B) = \rho g\, H$$

Equating the two expressions for the pressure difference,

$$\rho\, g\, H = (\rho_m - \rho)g\, h$$

$$H = \left(\frac{\rho_m - \rho}{\rho}\right) h \qquad \text{... (2.15)}$$

Specifically, if the manometric fluid is mercury of specific gravity 13.6 and the working fluid is water

$$H = (13.6 - 1)h = 12.6\ h \qquad \text{... (2.15 a)}$$

If the manometric fluid is mercury of specific gravity 13.6 and the working fluid is an oil of specific gravity 0.8.

$$H = \left(\frac{13.6 - 0.8}{0.8}\right) h = 16\ h \qquad \text{... (2.15 b)}$$

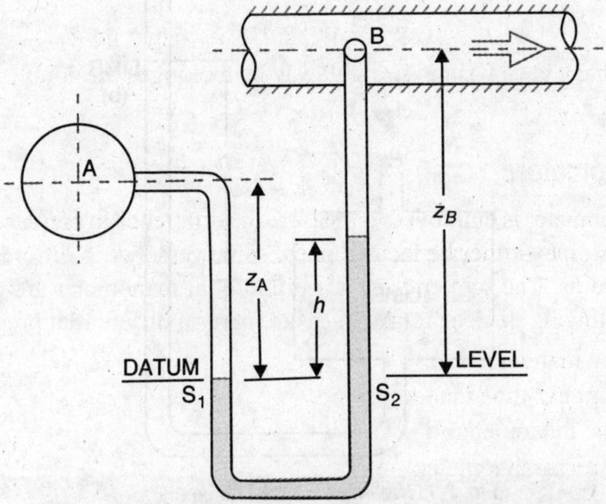

(a) U-TUBE MANOMETER FOR DIFFERENT FLUIDS

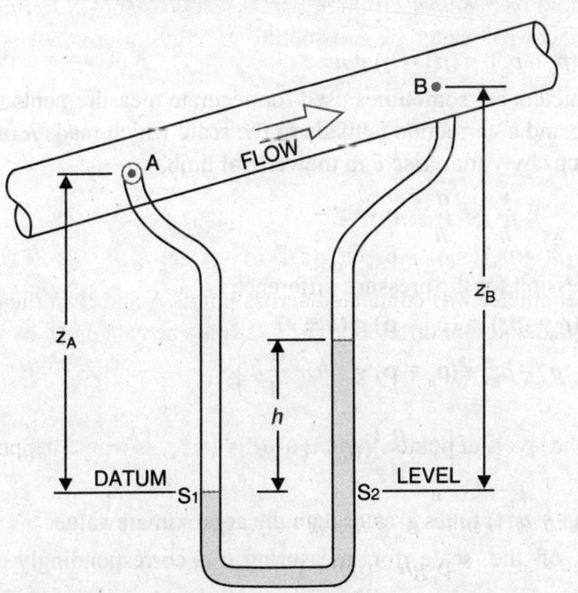

(b) U-TUBE MANOMETER FOR THE SAME FLUID

Fig. 2.10 Upright U-tube Manometer

(b) Well-Type U-tube Manometer. In the two-limb U-tube monometer, it is necessary to read the levels of the manometric liquid in both tubes and the difference in levels h is obtained by subtraction. In the well-type manometer, one of the limbs is replaced by a well of a large cross-section A in comparison with the cross section a of the tube.

$$A >> a$$

Fluctuations in the levels of the well may then be ignored i.e., a/A is assumed to be zero and the level in the well acts as a reference level for the other limb, as shown in Fig. 2.11.

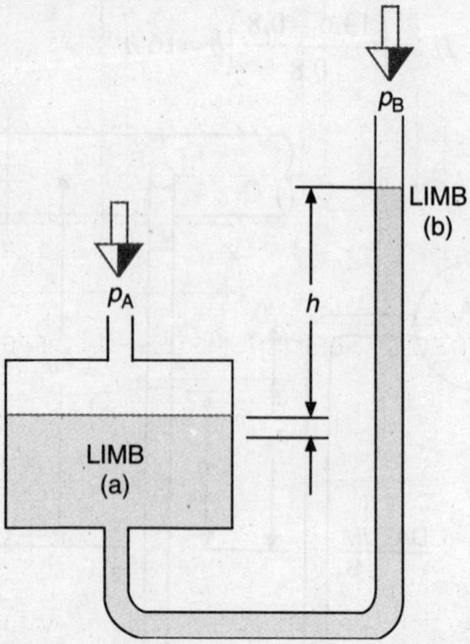

Fig. 2.11 Well-type U-tube Manometer

The pressure difference $(p_A - p_B)$ is related by the same expression as for a two limb U-tube manometer, *i.e.*,

$$(p_A - p_B) = (p_m - \rho)\, g\, h$$

Well-type manometers are sometimes used for accurate measurements where the ratio a/A is not assumed to be zero and a correction is made to the scale which reads h on the vertical limb. If the level in the well drops by y for a rise h in the vertical limb,

$$\frac{y}{h} = \frac{a}{A}$$

The correct expression for the pressure difference is

$$(p_A - p_B) = (\rho_m - \rho)\, g\, (h + y)$$

$$p_A - p_B = (\rho_m - \rho)\, g \left(h + \frac{a}{A} h \right)$$

$$= \left(1 + \frac{a}{A}\right)(\rho_m - \rho) gh \qquad \text{...(2.16)}$$

which shows that it is $(1 + a/A)$ times greater than the approximate value.

The calibration of the scale for measuring h is correspondingly done with a factor of $(1 + a/A)$.

(c) *Inverted U-tube Manometer.* Inverted U–tube manometers are employed to estimate the pressure difference between two points in the same conduit or in different conduits. The manometric fluid is lighter than the working fluids; consequently, the difference in two limbs is enhanced compared to that shown by a heavier manometric fluid in an upright U-tube manometer. This is, therefore, employed for accurate measurement of small pressure difference between any two points. For the configuration in Fig. 2.12 (a),

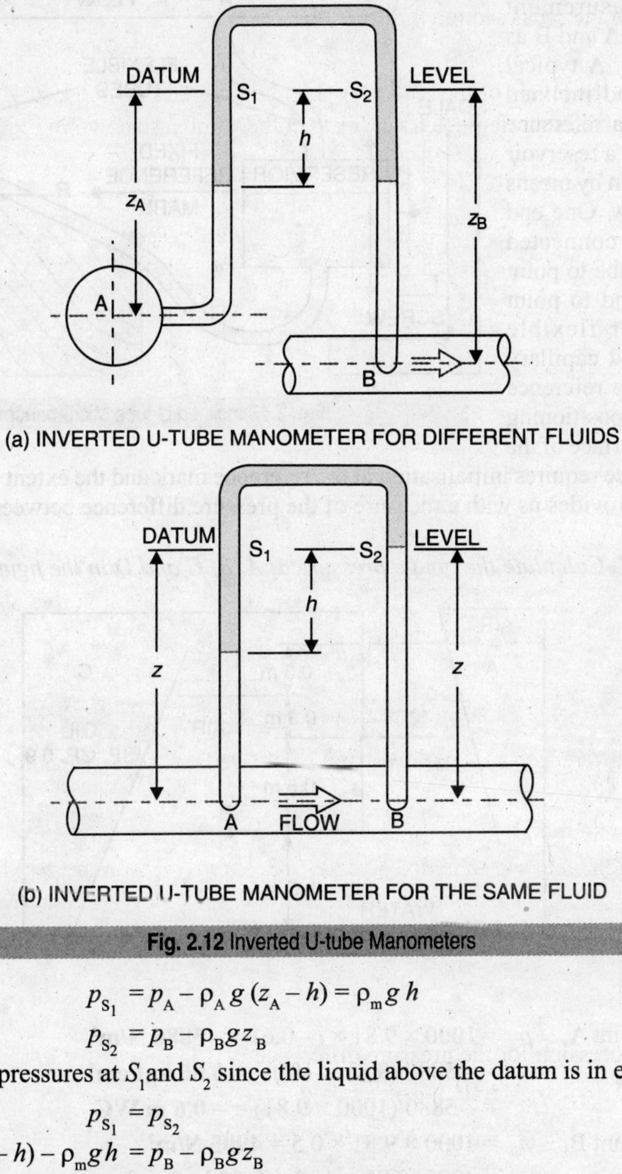

Fig. 2.12 Inverted U-tube Manometers

$$p_{S_1} = p_A - \rho_A g\,(z_A - h) = \rho_m g\,h$$

$$p_{S_2} = p_B - \rho_B g z_B$$

Equating the pressures at S_1 and S_2 since the liquid above the datum is in equilibrium.

$$p_{S_1} = p_{S_2}$$

$$p_A - \rho_A g(z_A - h) - \rho_m g h = p_B - \rho_B g z_B$$

whence $$(p_A - p_B) = \rho_m g h = \rho_A g\,(z_A - h) - \rho_B\,g\,z_B$$

If the points A and B were at the same level in the same fluid,

$$z_A = z_B = z$$

$$\rho_A = \rho_B = \rho$$

the pressure difference is given by

$$(p_A - p_B) = \rho_m g h + \rho g\,(z - h) - \rho g z$$

$$= \rho_m g h - \rho g h = (\rho_m - \rho)\,g h$$

(d) Inclined U-Tube Micromanometer. A micromanometer is a device for very accurate measurement between two points A and B as shown in Fig. 2.13. A typical device using a fixed inclined capillary tube with a reference mark *R* incorporates a reservoir movable up and down by means of micrometer screw. One end of the reservoir is connected through a flexible tube to point A and the other end to point B through another flexible tube and an inclined capillary tube provided with a reference mark and/or a fixed positioning telescope for the interface of the two fluids. The device requires initialisation at the reference mark and the extent of movement of the micrometer screw provides us with a measure of the pressure difference between points A and B.

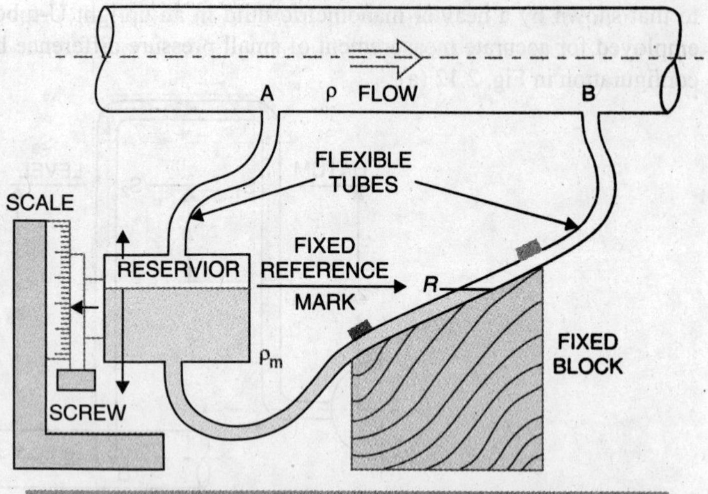

Fig. 2.13 Inclined U-tube Micromanometer

Example 2.17. *Calculate the gauge pressure at A, B, C and D in the figure in N/m², kgf/cm² and m of WG.*

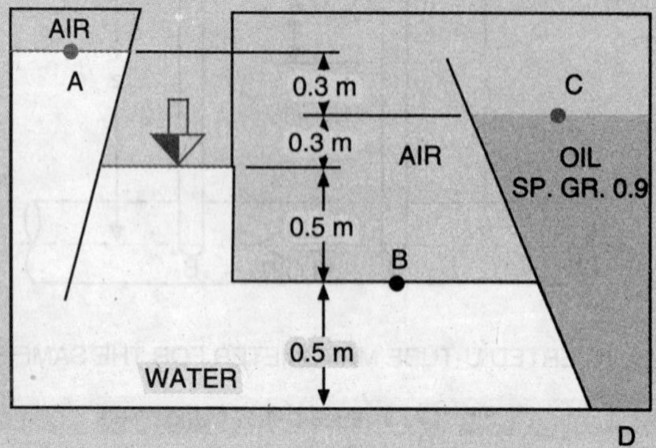

Solution.

At point A, $p_A = 1000 \times 9.81 \times (-0.6) = \mathbf{-5886\ N/m^2}$

$= -5880/(9.81 \times 10^4) = \mathbf{-0.06\ kgf/cm^2}$

$= -5880/(1000 \times 9.81) = \mathbf{-0.6\ m\ WG}$

At point B, $p_B = 1000 \times 9.81 \times 0.5 = \mathbf{4905\ N/m^2}$

$= 4905/(9.81 \times 10^4) = \mathbf{0.05\ kgf/cm^2}$

$= 4900/(1000 \times 9.81) = \mathbf{0.5\ m\ WG}$

At point C, p_C = pressure at B, air columns being too small to be accounted for in engineering practice

$= \mathbf{4905\ N/m^2 = 0.05\ kgf / cm^2 = 0.5\ m\ WG}$

At point D, $p_D = 4905 + 1000 \times 0.9 \times 9.871 \times 1.3$

$= \mathbf{16382.7\ N/m^2}$

$= 16382.7/(9.81 \times 10^4) = \mathbf{0.164\ kgf/cm^2}$

$= 16382.7/(1000 \times 9.81) = \mathbf{1.64\ m\ WG.}$

Example 2.18. *An inverted U-tube is connected across two pipes A and B carrying fluids as shown. Calculate the pressure difference between the centrelines of the pipes.*

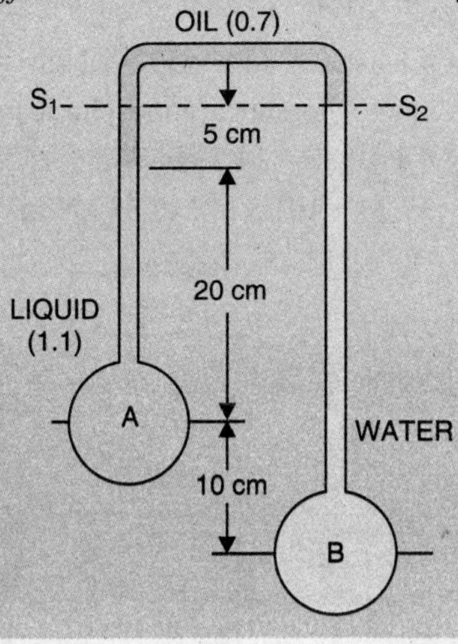

Solution.

The static columns of fluids in the two limbs are mutually in equilibrium, such that pressure at S_1 S_2 line is the same due to either column.

$$p_{S1} = p_A - 1000 \times 9.81(20 \times 1.1 + 5 \times 0.7)/100$$
$$= (p_A - 2502) \text{ N/m}^2 \text{ due to the left column.}$$
$$p_{S2} = p_B - 1000 \times 9.81(35 \times 1)/100$$
$$= (p_B - 3434) \text{ N/m}^2 \text{ due to the right column.}$$

Equating p_{S_1} and p_{S_2}

$$p_A - 2502 = p_B - 3434$$
$$p_A - p_B = \mathbf{-932 \text{ N/m}^2}$$

Example 2.19. *The pressure in a pipe at M is to be measured by the open manometer as shown. Fluid A is oil (Sp. gr. 0.8) and fluid B is mercury (Sp. gr. 13.6). Estimate the absolute pressure at M.*

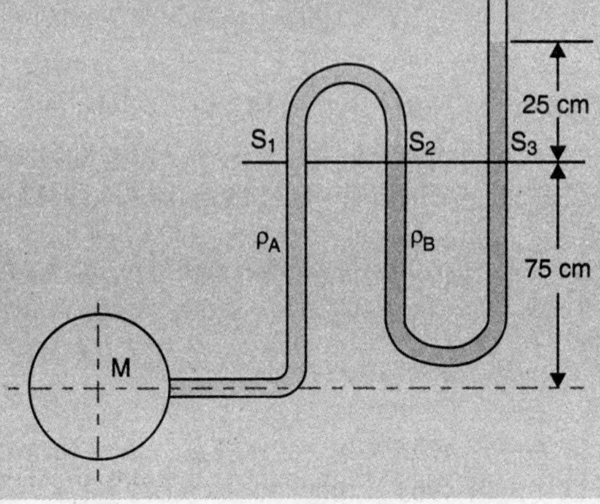

Solution.

$$p_{S_1} = 25 \text{ cm Hg} = \frac{25}{100} \times 13.6 = 3.4 \text{ m WG}$$
$$p_{S_1} = p_{S_2} = p_{S_3} = 3.4 \text{ m WG}$$

$$p_M = 3.4 + \frac{75}{100} \times 0.08 = 4 \text{ m WG.}$$

$$= 4 \times 1000 \times 9.81 = 39240 \text{ N/m}^2 = \mathbf{39.24 \text{ kN/m}^2}$$

Absolute pressure at M = gauge pressure + atmospheric pressure

$$= 4 + 10.33 = 14.33 \text{ m WG}$$

$$= 39.24 + 101.3 = \mathbf{140.54 \text{ kN/m}^2}$$

Example 2.20. *The figure shows a conical vessel having its outlet at A to which a U-tube manometer is connected. The reading of the manometer given in the figure is for the condition when the vessel is empty. Find the reading of the manometer when the vessel is completely filled with water.*

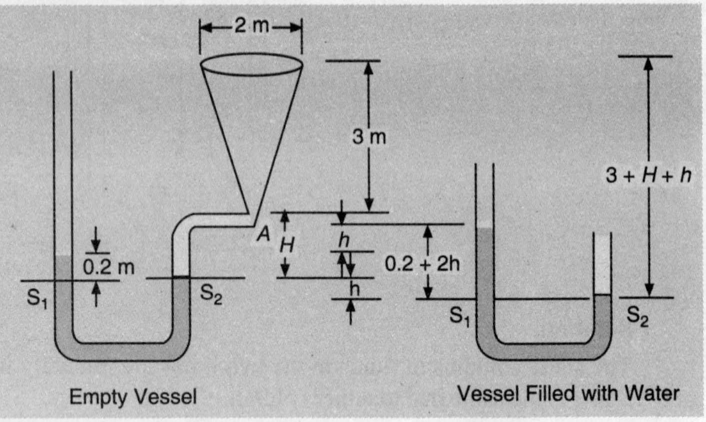

Solution.

Consider the equilibrium of the fluid columns in the manometer with reference to a datum level $S_1 S_2$.

When the vessel is empty,

$$p_{S_1} = (13.6 \times 1000) \times 9.81 \times 0.2 + p_{atm}$$
$$p_{S_2} = 1000 \times 9.81 \times H + p_{atm}$$

and equating these values $H = 2.72$ m

When the vessel is full of water, let the mercury level in the left limb rise by h and in the right limb drop by h. Again fixing the datum level $S_1 S_2$ such that the left column is mercury and the right column is water only.

$$p_{S_1} = (13.6 \times 1000) \times 9.81 \times (0.2 + 2h) + p_{atm}$$
$$p_{S_2} = 1000 \times 9.81 \times (2.72 + h + 3) + p_{atm}$$

and equating these values $h = 3/(27.2 - 1) = 0.1145$ m

Hence the manometer reading would be

$$0.20 + 2h = 0.20 + 2 \times 0.1145 = 0.429 \text{ m} = \mathbf{42.9 \text{ cm}}$$

Example 2.21. *Determine the differential reading h of an inverted U-tube manometer containing oil of specific gravity 0.8 as the manometric fluid when connected across pipes A and B conveying liquids of specific gravities 1.2 and 1.0 and immiscible with the oil. Pressure at A and B are equal and they are located at the same datum level.*

Solution.

Considering the static columns of fluid in the two limbs of the manometer, draw a horizontal line at a height of $(0.2 + h)$ metres above the pipes.

$$p_{S_1} = p_A - \rho_w g \times 1.2 \times 0.2 - \rho_w g \times 0.8 \times h$$
$$p_{S_2} = p_B - \rho_w g \times 1.0 \times (0.20 + h)$$

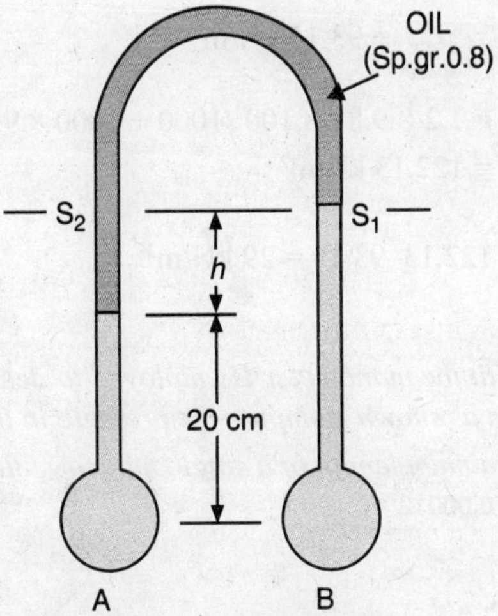

INVERTED U-TUBE

Equating p_{S1} and p_{S2} as well as p_A and p_B and cancelling $\rho_w g$,

$$-0.24 - 0.8\,h = -0.20 - h$$

whence $$h = 0.2 \text{ m} = 20 \text{ cm}$$

Example 2.22. *A closed tank A is filled with carbon tetrachloride (S = 2.25) and has a small air space at its top as shown. A manometer is connected between the A and a water tank as shown. If the pressure at point B in the water tank is 1.2 kg$_f$/cm^2 find the pressure in k N/m^2 in the closed air space at the top of tank A.*

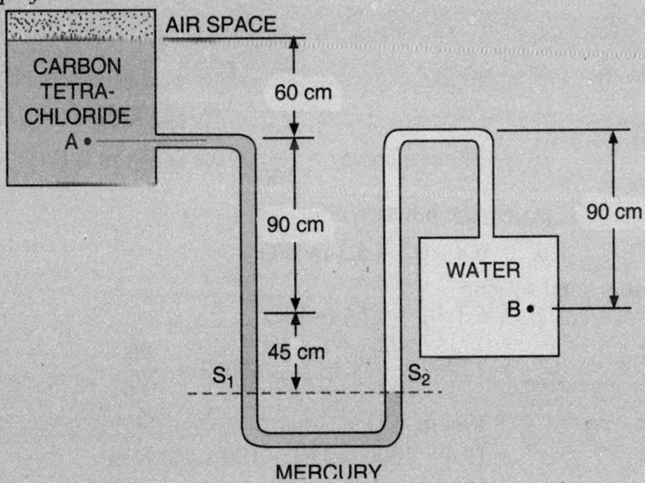

Solution.

Considering the datum reference $S_1\,S_2$, the pressure at S_1 must be equal to the pressure at S_2 for equilibrium.

$$p_{S_1} = p_{S_2}$$

From the data,

$$p_{S_1} = p_{\substack{air\\space}} + 2.25 \times 1000 \times 9.81(0.6 + 0.9)/1000$$

$$+ 13.6 \times 1000 \times 9.81 \times 0.45/1000$$

$$= p_{\substack{air\\space}} + 93.15 \text{ kN/m}^2$$

$$p_{S_2} = 1.2 \times 9.81 \times 100^2/1000 + 1000 \times 9.81 \times (-0.9 + 0.9) + 0.45)/1000$$

$$= 122.13 \text{ kN/m}^2$$

Equating p_{S_1} and p_{S_2}

$$p_{\substack{air\\space}} = 122.13–93.15 = 29 \text{ kN/m}^2$$

Example 2.23. *A multitube manometer is employed to determine the pressure in a pipe. For the levels in the manometers as shown, compute the pressure in the pipe.*

What would be the minimum length of a single mercury filled U-tube to record this pressure?

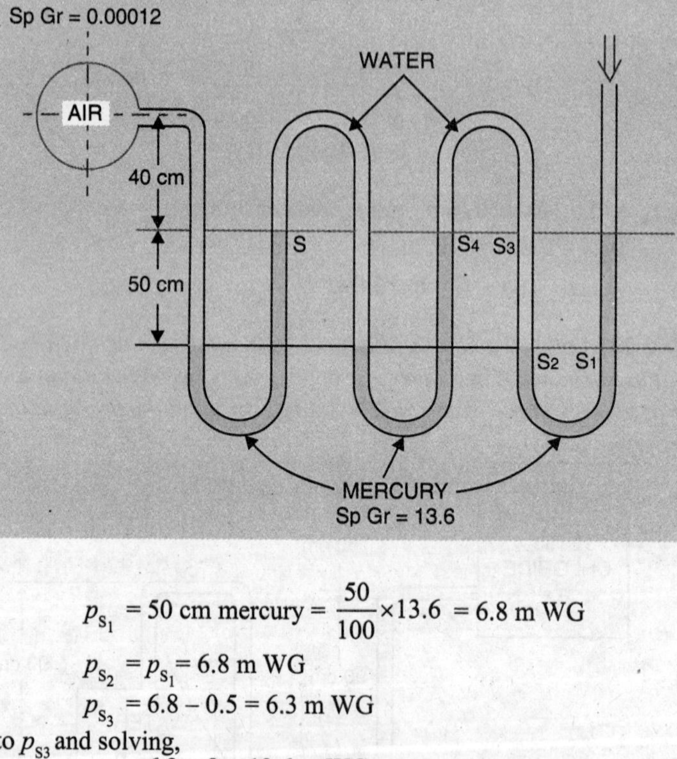

Solution.

$$p_{S_1} = 50 \text{ cm mercury} = \frac{50}{100} \times 13.6 = 6.8 \text{ m WG}$$

$$p_{S_2} = p_{S_1} = 6.8 \text{ m WG}$$

$$p_{S_3} = 6.8 - 0.5 = 6.3 \text{ m WG}$$

Equating p_{S4} to p_{S3} and solving,

$$p_S = 6.3 \times 2 = 12.6 \text{ m WG}$$

$$p_{\text{pipe}} = 12.6 + \frac{50}{100} \times 13.6 - 0.0012 \times \frac{90}{100}$$

$$= 19.4 \text{ m WG}$$

$$= 19.4 \times 1000 \times 9.81 = 190 \times 10^3 \text{ N/m}^2$$

$$= \mathbf{190 \text{ kN/m}^2 \text{ (gauge)}}$$

The pressure, 19.4 m WG equals 1.43 m mercury gauge. The length of the single - tube manometer would be at least 145 cm..

2.10 Buoyancy

If a body is submerged in a fluid, wholly or partially, the body experiences an upward force due to the fluid surrounding it. The phenomenon of upward force exerted by a fluid on the body is called buoyancy and the force is called 'buoyant force' or the force of buoyancy. *The body experiences buoyancy whether it floats or sinks under its own weight or due to other forces applied on it.*

The extent of the force of buoyancy experienced by a body may be estimated by employing the hydrostatic principle. Alternatively, the same may be computed with reference to the Archimedes principle of buoyancy.

The Archimedes principle of buoyancy states that a body immersed in a fluid experiences an upward buoyancy force equivalent to the weight of the fluid displaced by the body. It follows that a body floats partially immersed in a fluid to the extent that the weight of the fluid displaced equals the weight of the body.

A proof of this principle is given with reference to Fig. 2.14 (a). For an elemental prism of area δa, the net force is $(\rho g h_1 - \rho g h_u)\,\delta a$ upwards. For the entire body, the buoyancy force upwards,

$$F_B = \int(\rho g(h_1 - h_u)\,\delta a = \int \rho g h\,\delta a = \rho g \int h\delta a = \rho g V$$

which is the weight of the volume of fluid displaced and it must act through the centroid of the displaced fluid.

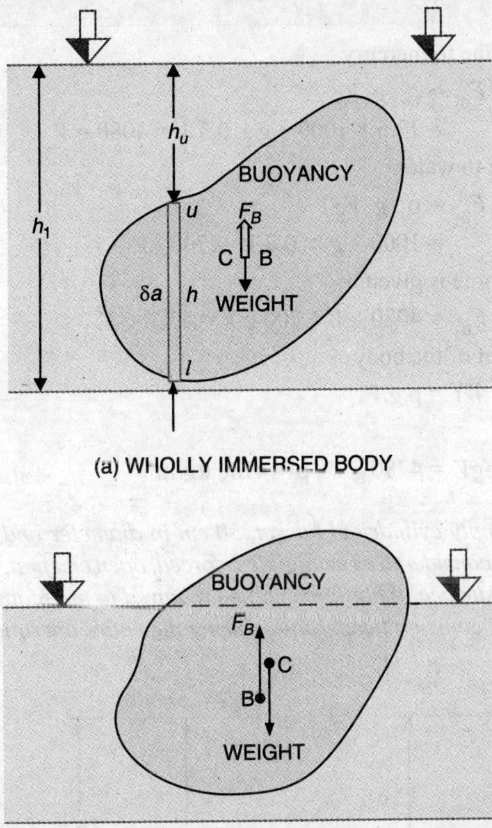

(a) WHOLLY IMMERSED BODY

(b) FLOATING BODY

Fig. 2.14. Buoyancy Effect

Example 2.24. *A metallic body floats at the interface of mercury (sp. gr. 13.6) and water such that 30% of its volume is submerged in mercury and 70% in water. Estimate the density of the metal.*

Solution.

The body floats in equilibrium under the action of buoyancy force, partly from water and partly from mercury, and under its own weight.

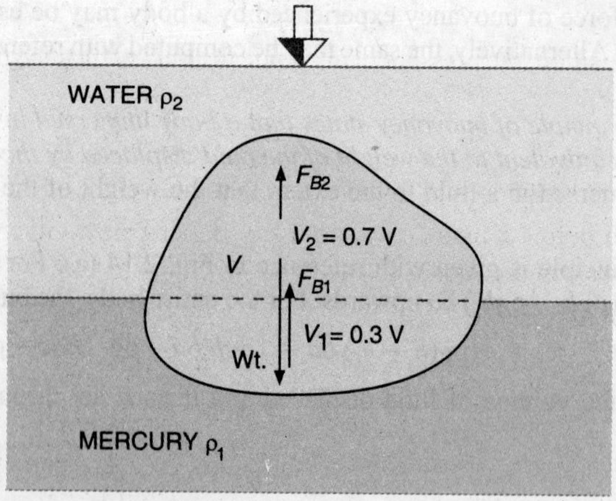

The buoyancy force due to mercury.

$$F_{B_1} = \rho_1 . g . V_1$$
$$= 13.6 \times 1000 \times g \times 0.3\, V = 4080\, g\, V$$

and the buoyancy force due to water,

$$F_{B_2} = \rho_2 . g . V_2$$
$$= 1000 \times g \times 0.7\, V = 700\, gV$$

The total buoyancy force is given by

$$F_{B_1} + F_{B_2} = 4080\, g\, V + 700\, g\, V = 4780\, g\, V$$

which must equal the weight of the body

$$Wt = \rho . g . V$$

Equating the two

$$\rho g V = 4780\, g\, V ; \rho = \mathbf{4780\ kg/m^3}$$

Example 2.25. *An empty cylindrical bucket, 30 cm in diameter and 50 cm long whose wall thickness and weight can be considered as negligible is forced, open end first, into water until its lower edge is 4 metres below the surface. What force will be required to maintain this position, assuming the trapped air to remain at constant temperature during the entire operation?*

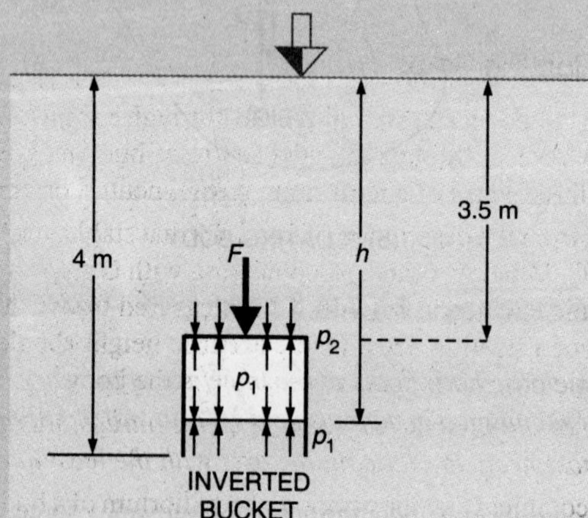

Solution.

In the immersed position, the pressure of air inside the bucket must be that of the liquid surface in contact with it.

$$p_1 = \rho g h + p_{atm} \quad ...(i)$$

Let $l = (h - 3.5)$, the length of air column.

The pressure of air before immersion was the atmospheric pressure and for a constant temperature compression,

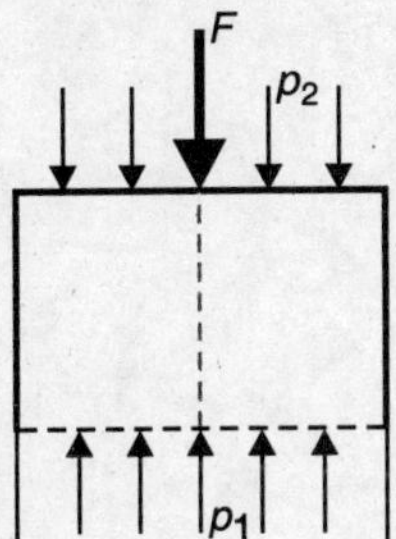

$$p_{atm} . v_{atm} = p_1 . v_1$$

or $$p_{atm} \times (0.50 \times \pi d^2/4) = p_1 \times (l \times \pi d^2/4).$$

and $$p_1 = p_{atm} \times 0.50/l \quad ...(ii)$$

Substituting (*ii*) in (*i*),

$$p_{atm} \times 0.50 / l = \rho g (3.50 + l) + p_{atm}$$

Assuming $p_{atm} = 101.3 \times 10^3$ N/m²

$\rho = 1000$ kg/m³ and $g = 9.81$ m/s²

$$l^2 + 13.83\, l - 5.16 = 0$$

whence $l = 0.36$ m

The force due to pressure difference above and below the base of the bucket is given by

$$F = (p_1 - p_2).\ \pi d^2/4 \quad ...(iii)$$

where p_2 = the pressure on the top surface

$$= 1000 \times 9.81 \times 3.5 + p_{atm} = 34\,335 + p_{atm}$$

$$p_1 = 1000 \times 9.81\ (3.50 + 0.36)\ p_{atm}$$

and $$= 37\,867 + p_{atm}$$

Hence $$F = (37\,867 - 34\,335) \times \pi \times \frac{0.3^3}{4} = \mathbf{250\ N.}$$

Alternatively, the force on the bucket is the force of buoyancy which must equal the weight of the volume of the water displaced,

$$F = \pi \times \frac{0.3^2}{4} \times 0.36 \times 1000 \times 9.81 = \mathbf{250\ N}$$

Metacentre M for a Floating Body

If a floating body is tilted by an external couple, through a small angle, the point of intersection of BC (produced) and a vertical through the new centre of buoyancy B, is called its metacentre. A floating body can be in three states of equilibrium; stable, neutral or unstable.

It can be observed from the description of Fig. 2.15 that stable, unstable and neutral equilibrium correspond to M above C, M below C and, M coinciding with C respectively. This criterion has long been used by the hydraulic engineers. Distance CM, measured from C to M is called the metacentric height. For the stability of a floating body, its metacentric height should be positive. It also follows that larger the metacentric positive height, more stable is the body.

A body is capable of tilting about any axis. A metacentre is, therefore, identified for tilt about that axis. The body is most likely to tilt about the axis with the least moment of inertia.

It is interesting to comment that the position of equilibrium of a body is the position of minimum potential energy corresponding to the equilibrium state. This is a general conclusion arrived at in the study of mechanics. The centre of gravity C should, therefore, be as low as possible. For a floating body, the condition of 'CM' to be positive is in conformity with this conclusion.

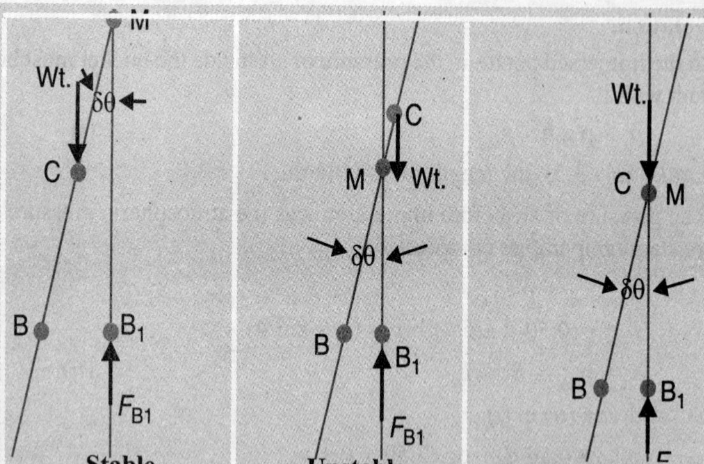

	Stable Equilibrium	Unstable Equilibrium	Neutral Equilibrium
Sense of tilt,	Clockwise,	Clockwise	Clockwise
Sense of external couple due to *Wt* and F_{B1}	Anticlockwise	Clockwise	Nil
Action of External, Couple	Restoring	Further Tilting	Nil
Behaviour of the body,	Returns to original position	Tilts further	Stays as it is
Condition of Equilibrium,	Stable	Unstable	Neutral
Sign of CM, the metacentric height	+ ve (M above C)	– ve (M below C)	Zero (M and C coincide)

Fig. 2.15. States of Equilibrium of a Floating Body

2.11. Determination of the Metacentric Height

Consider a floating body with a planform symmetrical about its fore-and-aft axis as shown in Fig. 2.16. In its upright position, it stays in equilibrium under the action of its weight W passing through its centre of gravity and the buoyancy force F_B passing through the centre of buoyancy. The two forces must be equal and opposite. The buoyancy force F_B is given by

$$F_B = \rho g V$$

where V is the volume of the fluid displaced and ρ is the density of the fluid.

When the floating body is tilted about its fore-and-aft axis, as shown in the figure, the centre of buoyancy shifts from B to a new position B_1, due to unsymmetrical submergence of the body in the liquid. The point of intersection of the vertical line drawn through B_1 and the line BC produced locates the metacentre M. For a small angle of tilt $d\theta$,

$$BB_1 = BM.\, d\theta$$

A small element of volume of the liquid $dx.dy.\, yd\theta$ at a distance y from the fore-and-aft axis, with weight $dx\, dy\, yd\theta\, \rho g$ exerts a moment about the fore-and-aft axis given by

$$dM = dx.\, dy.\, yd\theta\, \rho\, g.\, y$$

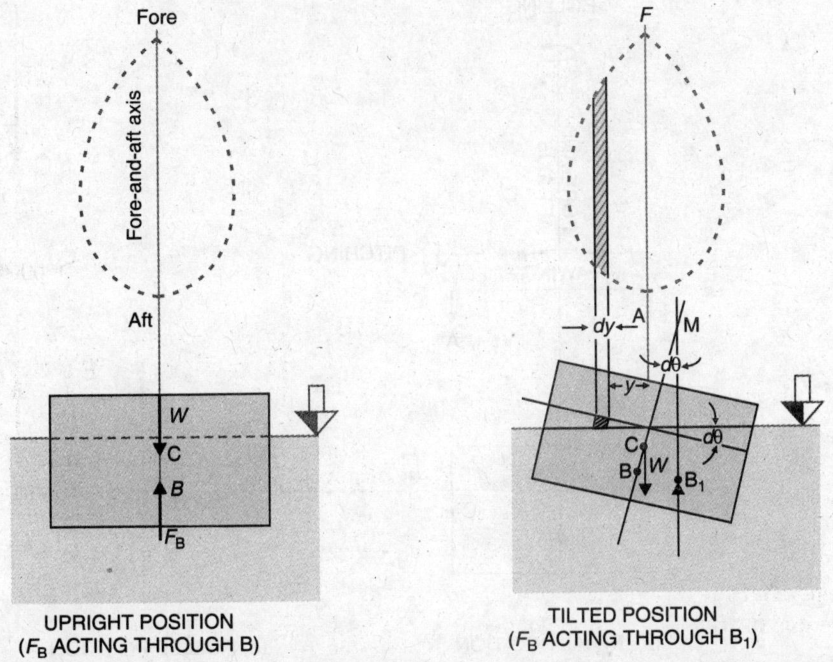

Fig. 2.16 Metacentric Height

The total moment by the liquid contained in the wedge shaped volume over the entire body is, therefore,

$$\rho g\, d\theta \int_{-y}^{+y}\int_{A}^{F} y^2 dx\, dy = \rho g\, d\theta \int_{A}^{F} y^2 dA = \rho g\, d\theta / I$$

This must equal the moment brought about by the horizontal shift BB_1 of the buoyancy force F_B,

$$F_B.\ BB_1 = \rho g\, d\theta . I$$

or $$\rho g V.\ BM.\ d\theta = \rho g\, d\theta .\ I$$

Cancelling $\rho g\, d\theta$ from both sides,

$$BM = I/V \qquad ...(2.17)$$

where I = Area moment of inertia of the planform at the liquid level about the fore-and-aft axis, i.e., axis of tilt.

V = Volume of the liquid displaced by the body.

Metacentric Height = CM = BM – BC = I/V – BC

The metacentric height of a floating body may also be determined experimentally.

2.12. Oscillation of a Floating Body: Time Period

A floating body has, in general six degrees of freedom; three translational and three rotational about the three mutually perpendicular axes. For fully and partially submerged bodies such as aircrafts and sea-vessels, these axes are recognised as the rolling, pitching and yawing axes as shown in Fig. 2.17. However, such bodies are constrained to move along the fore-and-aft axes. ***Of the three rotations, the rolling motion is the most likely to happen due to the least moment of inertia about the fore-and-aft axis. We, therefore, confine ourselves to determining the period of oscillation in rolling only.***

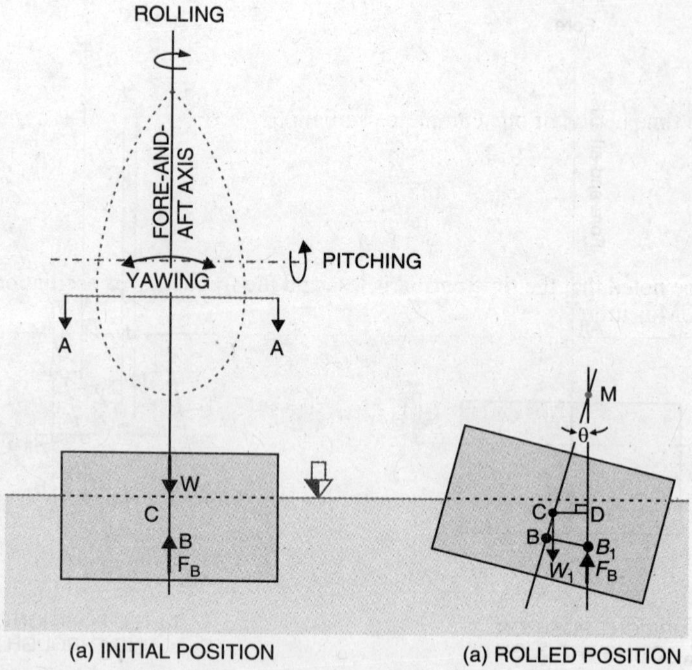

Fig. 2.17: Oscillation of a Floating Body

Consider a floating body in a rolled position, through an angle θ at an instant of time *t*. The body is acted upon by an external restoring moment due to its weight and buoyancy forces; acting parallel to each other CD apart. Since the weight and buoyancy force must be equal and distance CD equals CM sin θ . The external moment equals W.CM sin θ and acts so as to oppose the tilt θ.

Equating this to the product of moment of inertia I_{CC} about the rolling axis and the second temporal derivative of the angle of roll to obtain the equation of motion

$$-W.\text{CM}\sin\theta = I_{CC}\frac{d^2\theta}{dt^2}$$

Since $$I_{CC} = \frac{W}{g}k^2$$

where *k* is the radius of gyration about the rolling axis,

$$-W.\text{CM}\sin\theta = \frac{W}{g}k^2\frac{d^2\theta}{dt^2}$$

or $$\frac{d^2\theta}{dt^2}+\frac{g}{k^2}.\text{CM}\sin\theta = 0$$

For small tilt θ

$$\sin\theta \approx \theta$$

and $$\frac{d^2\theta}{dt^2}+\frac{g}{k^2}.\text{CM}\,\theta = 0$$

The solution of this second order linear differential equation appears in the form,

$$\theta = C_1\sin\sqrt{\frac{g\,\text{CM}}{k^2}}\,t+C_2\cos\sqrt{\frac{g\,\text{CM}}{k^2}}\,t$$

Using the boundary conditions,

$$\theta = 0 \text{ at } t = 0$$

and $$\theta = 0 \text{ at } t = \frac{T}{2}$$

where T is the time period of one complete oscillation,

$$C_2 = 0, C_1 \neq 0$$

and $$T = 2\pi\sqrt{\frac{k^2}{g\ \text{CM}}}$$

It may be noted that the time period is less and the frequency of oscillations is more if (*i*) k is less or if (*ii*) CM is more.

EXPERIMENT

Metacentric Height of a Floating Body

OBJECTIVE

To determine the metacentric height of a floating body experimentally.

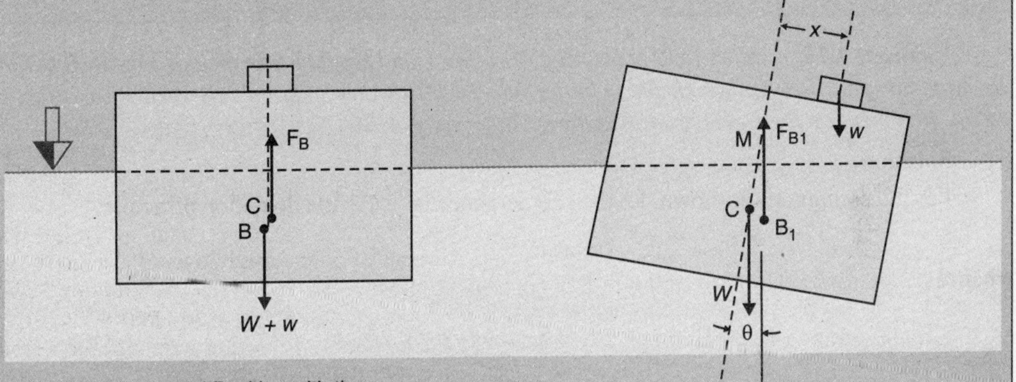

Initial Position with the Movable Weight in the centre

Titled Position by Displacing the Movable Weight

PRACTICAL RELEVANCE

Stability of a floating body depends on its metacentric height, CM. Greater the value of CM, greater is its degree of stability. Models of boats and ships are often made and tested before investments are made on building prototypes. More so, because the theoretical methods are difficult to apply due to irregular shapes of the floating vessels.

PROCEDURE

A movable weight w is placed initially at the centre of the deck of the floating vessel of weight W. The buoyancy force F_{B1} balances the total weight;

$$w + W = F_{B1}$$

When the movable weight w is displaced as shown, it acts at distance x along the deck, whereas the weight W of the ship still acts through C. The buoyant force, still the same in magnitude acts through a new point B1 in the tilted position.

Taking moments about the metacentre M; for equilibrium

$$wx \cos\theta = W.\text{CM} \sin\theta$$

whence $$\text{CM} = \frac{w}{W} \cdot \frac{x}{\tan\theta} \approx \frac{w}{W}\frac{x}{\theta}$$

for small angle of tilt θ.

RESULTS

It is noticed that the buoyancy force remains unaltered and

$$CM = \frac{w}{W} \frac{x}{\tan\theta} \approx \frac{w}{W} \frac{x}{\theta}$$

for small angle of tilt θ. The metacentric height is, therefore, determined.

POINTS FOR DISCUSSION

* *Does the metacentric height depend upon the initial submergence of the floating vessel?*
* *Should the movable weight be small compared to W? Why?*
* *What would happen if the metacentric height is negative?*

ANY OTHER OBSERVATION OR COMMENT?

Example 2.26. *A solid circular cylinder of radius r and height h has specific gravity 0.6. Find the minimum ratio r/h for which the cylinder will float in water with its axis vertical and just be stable. Comment on the stability with a change in its sp. gravity.*

Solution.

For the configuration shown, let the submergence be y. By the floatation principle,

$$\pi r^2 . h . \rho_{solid} = \pi r^2 . y . \rho_{liquid}$$

whence

$$y = 0.6\,h$$

$$BC = \frac{h}{2} - \frac{y}{2} = (h - 0.6h)/2 = 0.2h$$

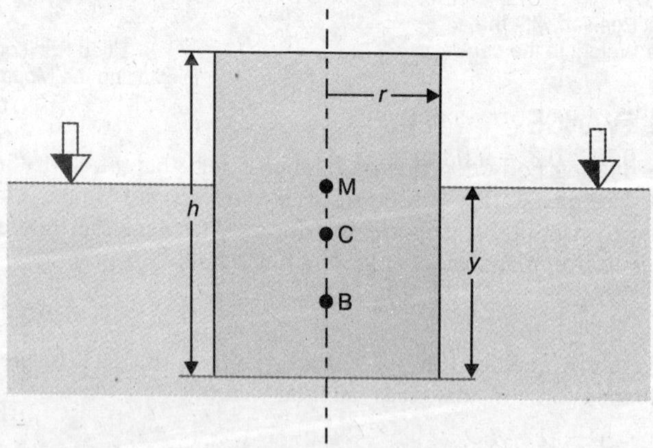

Since

$$BM = \frac{I}{V} : I = \frac{\pi}{4} r^4 ; V = 0.6\pi r^2 h$$

$$BM = \frac{r^2}{2.4h}$$

Also,

$$CM = BM - BC = \frac{r^2}{2.4h} - 0.2h$$

For stability, CM must be + ve, *i.e.*,

$$\frac{r^2}{2.4h} - 0.2h \geq 0$$

whence $\frac{r}{h} = \mathbf{0.693}$

For r/h greater than 0.693, CM is + ve assuring stability and for r/h less than 0.693, the cylinder is in unstable equilibrium. It can be seen that, for S greater than 0.6, r/h required for stability is less which means longer cylinders can float stable. For S less than 0.6, r/h required for stability is more which means that shorter cylinders can float stable.

Example 2.27. *An ocean liner 200 m long, 25 m wide displaces 50 MN of water. The second moment of inertia of the water plane about its fore-and-aft axis is 60% of the circumscribing rectangle. The position of the centre of buoyancy is 2.5 m below the centre of gravity. Locate the metacentre.*

Solution.

The volume displaced

$$V = \frac{50 \times 10^6}{10^3 \times 9.81} = 5096.84 \text{ m}^3$$

$$I = \frac{1}{12} \times 200 \times 25^3 \times 0.6 = 156\,250 \text{ m}^4$$

$$\text{BM} = \frac{I}{V} = \frac{156\,250}{5096.84} = 30.7 \text{ m}$$

$$\text{CM} = 30.7 - 2.5 = 28.21 \text{ m}$$

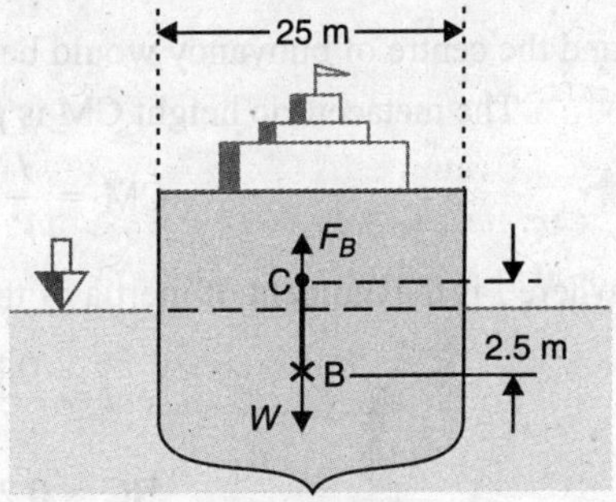

The metacentre is **28.2 m** above the centre of gravity; the liner is very stable!.

Example 2.28. *A block of wood, 0.2 m × 0.5 m in cross section and 0.8 m long has a mass of 64 kg. Can the block float with the 0.5 m side vertical?*

Solution.

The volume of the block of wood is

$$(0.2 \times 0.5) \times 0.8 = 0.08 \text{ m}^3$$

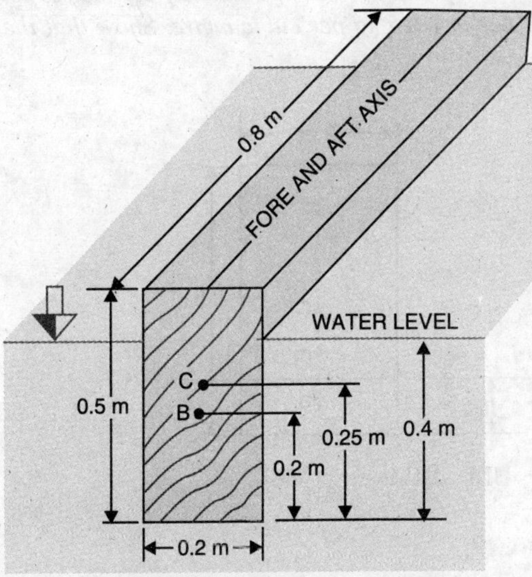

Block Floating with 0.5 m Side Upright

The density of the block is

$$\rho = \frac{64}{0.08} = 800 \text{ kg/m}^3$$

Since the density of the block of wood is less than 1000 kg /m^3, the density of water, the block must float in water. The weight of the volume of the water displaced would equal the weight of the block which is 64 kg. The volume of water displaced and hence the volume of the block submerged in water should be

$$V = \frac{64}{1000} = 0.064 \text{ m}^3$$

Let us examine its possibility of floating with the 0.5 m side vertical. If it did so, the height of submergence would be

$$h = \frac{0.064}{(0.2 \times 0.8)} = 0.4 \text{ m}$$

and the centre of buoyancy would be at 0.2 m from the base.

The metacentric height CM is given by

$$\text{CM} = \frac{I}{V} - \text{BC}$$

where I is the moment of inertia of the top view area about the fore-and-aft axis

$$I = \frac{0.8 \times (0.2)^3}{12} = 0.00053 \text{ m}^4$$

$$\text{BC} = 0.25 - 0.2 = 0.05 \text{ m}$$

$$\text{CM} = \text{BM} - \text{BC}$$

$$= 0.000\frac{53}{0.064} - 0.05$$

$$= \mathbf{-0.042 \text{ m}}$$

CM < 0 implies that the block is unstable while floating with its 0.5 m side vertical.

Example 2.29. *A cylindrical buoy is 2 metres in diameter, 2.5 metres long and weighs 2.2 metric tons. The density of sea water is 1025 kg per cubic metre. Show that the body cannot float with its axis vertical*

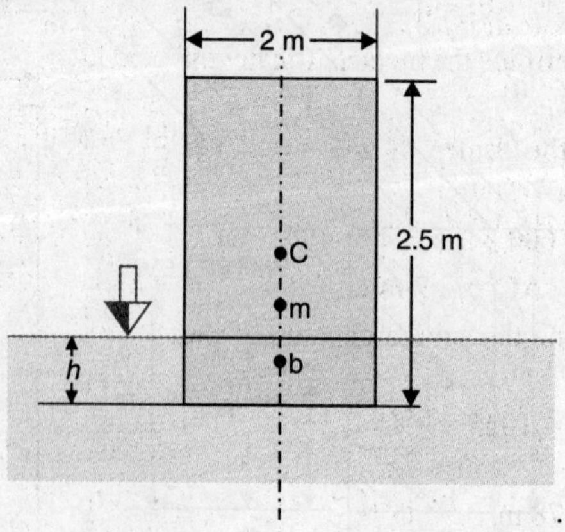

Solution.

Weight of the buoy $W = 2.2$ metric tons

$$= 2200 \times 9.81 = 21582 \text{ N}$$

For equilibrium in floatation,

$$F_B = W = 21\,582 \text{ N} = 21.6 \text{ kN}$$

where F_B = Weight of the volume of fluid displaced

and $F_B = 1025 \times 9.81 \times V$

Therefore, $$V = \frac{21\,582}{9.81 \times 1025} = 2.15 \text{ m}^2$$

Depth of immersion of the buoy

$$h = \frac{2.15}{(\pi \times 2/4)} = 0.68 \text{ m}$$

In order to investigate the stability of equilibrium, the metacentric height must be determined. The moment of inertia of the top view area at the sea level about the fore-and-aft axis is

$$I = \frac{\pi d^4}{64} = \frac{\pi \times 2^4}{64} = 0.785 \text{ m}^4$$

and $$\text{BM} = \frac{I}{V} = \frac{0.785}{2.15} = 0.365 \text{ m}$$

$$\text{CM} = \text{BM} - \text{BC} = \frac{I}{V} - \text{BC}$$

$$= 0.365 - (1.25 - 0.34) = 0.366 - 0.91 = -\mathbf{0.545\ m}$$

Negative value of CM implies instability of floatation with its axis vertical.

Example 2.30: *A rectangular pontoon floating in sea water (density 1025 kg/m³) is 8 m long, 7 m broad and 3 m deep. It weighs 600 kN. It carries on its upper deck an employ boiler 4 m diameter weighing 400 kN. The centre of gravity of the pontoon and boiler are at their geometric centres along a vertical line. Is the system stable?*

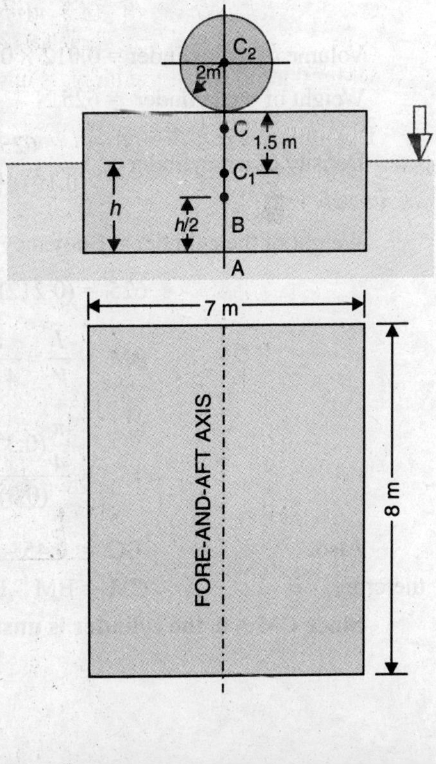

Solution.

It is necessary to determine the metacentric height for the given system.

Let us first locate the centre of gravity of the pontoon and boiler system.

$$(600 + 400) \times \text{AC} = 600 \times 1.5 + 400 \times (3 + 2)$$

whence $\text{AC} = 2.9$ m

Now, let us determine the submergence h of the system.

$$(600 + 400) \times 1000 = 1025 \times 9.81 \times h \times (8 \times 7)$$

whence $h = 1.78$ m

The centre of buoyancy must be $h/2$ above the base A.

$$AB = 0.89 \text{ m}$$

Then, applying the relationship, BM = I/V

For stability, the minimum distance BM corresponding to the minimum I should be examined. The area of moment of inertia about the fore-and-aft axis at the water line is

$$I = 8 \times 7^3/12 = 228.67 \text{ m}^4$$

Also $V = 1.78 \times 8 \times 7 = 99.68 \text{ m}^3$

and BM = 228.67/99.68 = 2.3 m

Now, BC = 2.9 – 0.89 = 2.01 m

CM = 2.3 – 2.01 = **0.29 m**

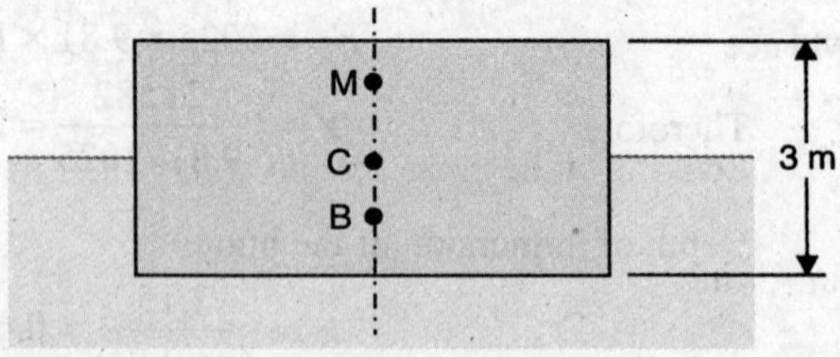

Since the metacentric height is positive, the system is in stable equilibrium.

Example 2.31. *A hollow cylinder open at both ends has internal diameter of 30 cm, wall thickness of 15 cm and a length of 90 cm. If it weighs 625 N, find whether the cylinder would be stable while floating in water with its axis vertical.*

Solution.

For the given hollow cylinder,

$$r_1 = 15 \text{ cm} = 0.15 \text{ m}$$
$$r_2 = 30 \text{ cm} = 0.30 \text{ m}$$
$$h = 90 \text{ cm} = 0.90 \text{ m}$$

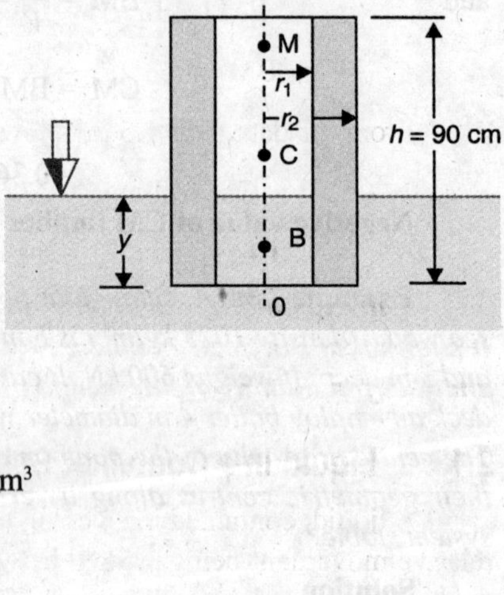

Area of cross section,

$$A = \pi(r_2^2 - r_1^2) = 0.212 \text{ m}^2$$

Volume of the cylinder = 0.012 × 0.9 = 0.191m³ .

Weight of the cylinder = 625 N

$$\text{Density of the cylinder} = \frac{625}{(0.191)(9.81)} = 333.56 \text{ kg/m}^3$$

Weight of the cylinder = Buoyancy force = $V.\ \rho_\omega\, g$

$$625 = (0.212)\,(y)\,(1000)\,(9.81)\ ;\ y = 0.3 \text{ m}$$

$$BM = \frac{I}{V} = \frac{\pi\,(r_2^4 - r_1^4)}{4\,(0.3)(0.212)}$$

$$= \frac{\frac{\pi}{4}(0.3^4 - 0.15^4)}{(0.3)(0.212)} = 0.094 \text{ m}$$

Also, BC = 0.45 – 0.15 = 0.3 m

therefore, CM = BM – BC = 0.094 – 0.3 = **– 0.206 m**

Since CM < 0, the cylinder is unstable.

Example 2.32. *A ship 12 m wide and 30 m long at the water line has a displacement of 6000 kN. The centre of gravity is 40 cm below the water line and the centre of buoyancy is 150 cm further below it. Determine the metacentric heights for rolling and pitching of the ship.*

Solution.

The displacement of 6000 kN must be the weight of the volume of the water displaced.

$$\rho g\, V = 6000 \times 1000 \text{ N}$$

whence
$$V = \frac{6000 \times 1000}{1000 \times 9.84} = 611.6 \text{ m}^3$$

Looking at the cross section at the water line;

for rolling
$$I_{xx} = \frac{30 \times 12^3}{12} = 4320 \text{ m}^4$$

for pitching,
$$I_{yy} = \frac{12 \times 30^3}{12} = 2700 \text{m}^4$$

Employing the relation

$$\text{BM} = \frac{I}{V}$$

$$\text{BM (rolling)} = I_{xx}/V = 4320/611.6 = 7.06 \text{ m}$$

$$\text{BM (pitching)} = I_{yy}/\ V = 27000/611.6 = 44.14\text{m}$$

From the data, B is 1.5 m below C. Therefore,

$$\text{CM (rolling)} = 7.06 - 1.5 = \mathbf{5.56\ m}$$

$$\text{CM (pitching)} = 44.15 - 1.5 = \mathbf{42.65\ m}$$

The example shows that the metacentre for pitching is much higher than that for rolling. It shows that the floating object is far more stable in pitching than in rolling. This is why most analyses for stability are limited to finding the metacentric height for rolling only.

2.13. Liquid in a Container Subjected to a Constant Linear Acceleration

A liquid, contained in a vessel, may be subjected to a constant linear acceleration without any relative movement being created between different elements of the liquid in the vessel. The liquid must, of course, orient itself once and for all to stay in that position for the given constant acceleration. The liquid is then said to be in a state of relative rest. Although the liquid mass is undergoing an acceleration, it is translated with the appearance of a rigid body. In the absence of relative motion between different fluid elements, there are no rates of strain and shear stresses in the liquid. The hydrostatic law is, therefore, applicable for the evaluation of hydrostatic forces etc. due to the liquid.

Consider a liquid contained in a vessel. While at rest, the free surface maintains a horizontal level as shown in Fig 2.18 but in the state of relative rest under a constant acceleration a, the liquid orients itself to maintain the free surface inclined at an angle with the horizontal. The inclination θ may be related to the magnitude and direction of the acceleration vector by considering the dynamics of an element of the liquid. The acceleration a is resolved into horizontal and vertical components a_x and a_z respectively and the attention is focussed in the $x - z$ plane. The element is chosen to have the dimensions δ_x , 1 and δ_z. It is subjected to a pressure force on its faces and the gravitational force acting downwards. The pressure may vary along x and z axes. Assuming that the pressure is p at the centre of mass of the element, the rates of variations of pressure along x and z directions are denoted as $\frac{\partial p}{\partial x}$ and $\frac{\partial p}{\partial z}$ respectively. The following information may be compiled by referring to Fig. 2.18.

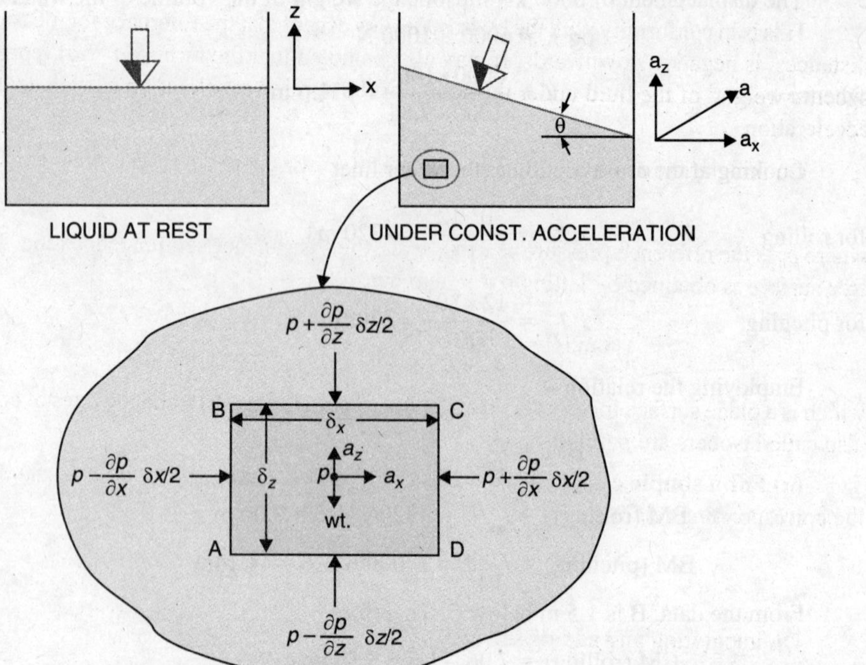

Fig. 2.18 Liquid under Constant Acceleration

Face	Area	Pressure	Force
AB	$\delta z \times 1$	$p - \frac{\partial p}{\partial x}.\frac{\delta x}{2}$	$\left(p - \frac{\partial p}{\partial x}.\frac{\delta x}{2}\right)\delta z$
CD	$\delta z \times 1$	$p + \frac{\partial p}{\partial x}\frac{\delta x}{2}$	$\left(p + \frac{\partial p}{\partial x}.\frac{\delta x}{2}\right)\delta z$
AD	$\delta x \times 1$	$p - \frac{\partial p}{\partial z}.\frac{\delta z}{2}$	$\left(p - \frac{\partial p}{\partial z}.\frac{\delta z}{2}\right)\delta x$
BC	$\delta x \times 1$	$p + \frac{\partial p}{\partial z}.\frac{\delta z}{2}$	$\left(p + \frac{\partial p}{\partial z}.\frac{\delta z}{2}\right)\delta x$

Writing the equation of motion in the x-direction,

$$\left(p - \frac{\partial p}{\partial x}.\frac{\delta x}{2}\right)\delta z - \left(p + \frac{\partial p}{\partial x}.\frac{\delta x}{2}\right)\delta z = \rho(\delta_x.\,\delta_z.\,1)a_x$$

whence

$$\frac{\partial p}{\partial x} = -\,\rho a_x \qquad \text{...(2.18}a\text{)}$$

Similarly, writing the equation of motion in the z-direction, accounting for the weight of the element,

$$\left(p-\frac{\partial p}{\partial z}.\frac{\delta z}{2}\right)\delta x-\left(p+\frac{\partial p}{\partial z}.\frac{\delta z}{2}\right)\delta x-\rho g(\delta x\,.\,\delta z\,.1) = \rho(\delta_x.\,\delta_z.\,1)a_z$$

whence $$\frac{\partial p}{\partial z} = -\rho(a_z + g) \qquad \text{... (2.18}b\text{)}$$

This is in conformity with the hydrostatic law except that the reference for measuring the vertical distances is negative downwards. It may also be noted that the term $\rho(a_z + g)$ represents the 'local specific weight' of the fluid under the action of the applied acceleration as well as the gravitational acceleration.

On integration of Eq. (2.18 *a*) and Eq. (2.18 *b*) w.r.t. x and z respectively, and adding,

$$p = -[\rho\, a_x x + \rho(a_z + g)\, z] + p_0 \qquad \text{... (2.19 a)}$$

where p_0 is the reference pressure at $x = 0, z = 0$, the free surface reference location. The shape of the free surface is obtained by letting $p = p_0$ and hence,

$$|\tan\theta| = \frac{a_x}{a_z + g} \qquad \text{...(2.20 a)}$$

which is a plane surface inclined at an angle θ as shown in Fig. 2.19 and the lines of constant pressure, also called isobars are parallel to it.

***(a)* For a simple case of acceleration only along *x* axis,** substitute $a_x = a$ and $a_z = 0$ to obtain the corresponding relations

$$\frac{\partial p}{\partial x} = -\rho\, a \text{ and } \frac{\partial p}{\partial z} = -\rho\, g$$

On integrating and adding the two,

$$p = -[\rho\, a\, x + \rho\, g\, z] + p_0 = -\rho(ax + gz) + p_0 \qquad \text{...(2.19 }b\text{)}$$

and the shape of the free surface, $p = p_0$ is given by

$$|\tan\theta| = \frac{a}{g} \qquad \text{...(2.20 }b\text{)}$$

a plane inclined down at θ along the direction of the acceleration.

***(b)* For another simple case of acceleration only along z-axis;** substituting $a_x = 0$ and $a_z = a$, we obtain,

$$\frac{\partial p}{\partial x} = 0$$

$$\frac{\partial p}{\partial z} = -\rho\, g - \rho\, a_z = -\rho\,(g + a_z)$$

On integrating and adding the two, $p = -\rho\,(g + a_z)\, z + p_0$

which shows that the liquid does not reorient itself but the pressure changes in the vertical direction.

It is interesting to observe the phenomenon for upward and downward acceleration.

(*i*) if a_z is upward, then the rate of change of pressure with z is more than that without acceleration. If a_z is of the order of g, (taking z +ve downwards)

$$p = 2\rho g z + p_0$$

which is double of that without acceleration.

(*ii*) if a_z is downward, then the rate of change of pressure with z is less than that without acceleration.

If a_z is downward, equal to g, (taking z + ve downwards)

$$p = \rho\,(g - g)\,z + p_0 = p_0$$

i.e., the pressure does not change with z.

It may be clarified that the shape of the container does not matter as long as the container provides a continuous connection in the liquid mass. It may as well be a U-tube.

The size of the container does not enter the analysis as far as the inclination θ of the surface and of the isobars is concerned but it is this parameter which decides the amount of spill-over from a small open container and the redistribution of the liquid in a closed container. This is shown in Fig. 2.19. Isobars in each of the three cases subjected to the same constant acceleration a are inclined at the same constant acceleration a are inclined the same angle θ. If the vessel is open and it is not large enough to contain the liquid with its surface inclined at θ, a part of the fluid is spilt which may be determined by difference between the quantity initially in it and that which remain finally. If the vessel is closed, the surface of the liquid must be located by considering the conservation of mas with due regard to its geometrical features.

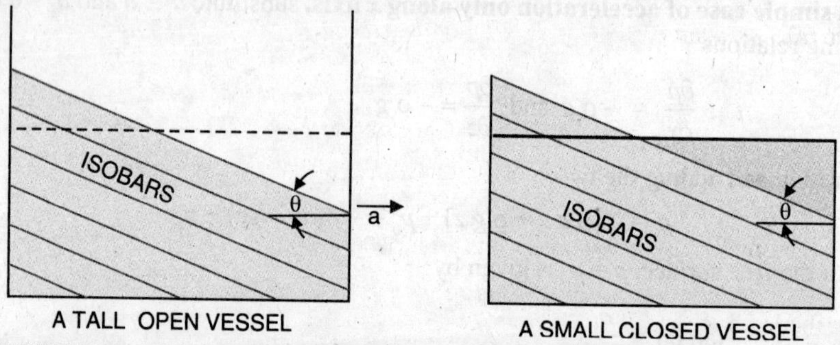

Fig. 2.19. Isobars in a Liquid under Constant Acceleration

It is interesting to evaluate the acceleration when the gauge pressure in the entire fluid vanishes. It is then subjected to the atmospheric pressure throughout.

From Eq. (2.19 a),

$$-\,[\rho\, a_x x + \rho(a_z + g)\, z] = 0$$

which is satisfied if the acceleration components are

$$\left.\begin{aligned} a_x &= 0 \\ a_z &= -\,g \end{aligned}\right] \text{ which is the case for a free fall !}$$

It implies that if a tank is lowered vertically at the gravitational acceleration g, e.g., just dropped as a free-fall then the liquid remains at the atmospheric pressure througout and there is no force on the base or on the walls of the tank. This is indeed expected during a free fall of the container as well as the liquid in it.

Example 2.33. *An open rectangular tank 5 m × 4m × 3 m high containing water upto a height of 2 m is accelerated at 3 m/s²*

(a) horizontally along the longer side,

(b) vertically upwards,

(c) vertically downwards and

(d) in a direction inclined at 30° to the horizontal along the longer side

Calculate, in each case, the shape of the free surface and the total force on the base of the tank as well as on the vertical faces of the container. At what acceleration will the force on each face be zero ?

Solution.

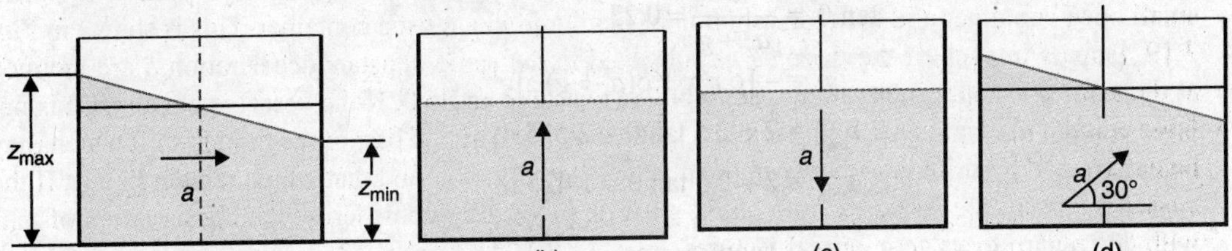

For Case (a), $\tan\theta = \dfrac{a_x}{g} = \dfrac{3}{9.81} = 0.306$

$$z_{max} = 2 + 2.5\tan\theta = 2.765 \text{ m}$$

$$z_{min} = 2 - 2.5\tan\theta = 1.235 \text{ m}$$

Force on the base is the same as that when at rest. Only the pressure distribution is changed, due to the slope of the free surface.

Force on the base = Weight of the water contained.

$$= 1000 \times 981 \times 5 \times 4 \times 2 = 392\,400 \text{ N} = 392.4 \text{ kN}$$

Force on the leading face

$$= 1000 \times 9.81 \times \frac{1.235}{2} \times \frac{4 \times 1.235}{1000} = 29.92 \text{ kN}$$

Force on the trailing face

$$= 1000 \times 9.81 \times \frac{2.765}{2} \times \frac{2.765 \times 4}{1000} = 150 \text{ kN}$$

Force on each of the side faces

$$= 1000 \times 9.81 \times \left(\frac{1.235 + 2.765}{4}\right) \times \frac{2 \times 5}{1000} = 98.1 \text{ kN}$$

For Case (b), $\tan\theta = 0$ as $a_x = 0$

The free surface remains horizontal

$$p = -\rho(a_z + g)z = -\rho(3 + 9.81)z$$

Force on the base $= 1000 \times 12.81 \times \dfrac{2 \times 5 \times 4}{1000} = 512.4$ kN

Force on each of the longer sides

$$= 1000 \times 12.81 \times \frac{5 \times 2 \times 1}{1000} = 128.1 \text{ kN}$$

Force on each of the shorter sides

$$= 1000 \times 12.81 \times \frac{4 \times 2 \times 1}{1000} = 102.48 \text{ kN}$$

For Case (c), This is similar to case (b), but here $a_z = -3$ m/s²
The free surface would remain horizontal

$$p = -\rho(-3 + 9.81)\,z$$

Force on the base $= 1000 \times 6.81 \times \dfrac{2\times5\times4}{1000} = 272.4$ kN

Force on the longer sides $= 1000 \times 6.81 \times \dfrac{5\times2\times1}{1000} = 68.1$ kN

Force on the shorter sides

$$= 1000 \times 6.81\times \frac{4\times2\times1}{1000} = 54.48 \text{ kN}$$

For Case (d), $a_x = 3 \cos 30° = 2.598$ m/s² $a_z = 3 \sin 30° = 1.5$ m/s²

$$\tan\theta = \frac{a_x}{a_z + g} = 0.23$$

$$p = -[\rho\, a_x x + \rho(a_z + g)\, z]$$

$$z_{max} = 2 + 2.5 \tan\theta = 2.575 \text{ m}$$

$$z_{min} = 2 - 2.5 \tan\theta = 1.425 \text{ m}$$

Force on the base $= 1000 \times 11.31 \times \dfrac{5\times4\times2}{1000} = 452.4$ kN

Force on the leading face

$$= 1000 \times 11.31 \times \frac{1.425}{2} \times \frac{1.425\times4}{1000} = 45.93 \text{ kN}$$

Force on the trailing face

$$= 1000 \times 11.31 \times \frac{2.575}{2} \times (4 \times 2.575) = 149.99 \text{ kN}$$

Force on each of the side faces

$$= 1000 \times 11.31 \times (1.425 + 2.575)/4 \times 2 \times 5 = 113.1 \text{ kN}$$

In order to determine the acceleration when the force on each face is zero, we examine the expression,

$$p = -\rho(a_z + g)z$$

The pressure will be zero when $a_z = -g$. In other words, the liquid mass will stay wholly at atmospheric pressure when accelerated downwards with a magnitude equal to that of g. This is identical to the case of free fall of a liquid mass where the pressure in the entire region adjusts itself to the local atmospheric value.

Example 2.34: *An open rectangular tank 5 m × 4 m × 3 m high containing water upto a height of 2 m is accelerated horizontally along the longer side.*

Determine the maximum acceleration that can be given without spilling the water and also calculate the percentage of water spilt over if this acceleration is increased by 20%.

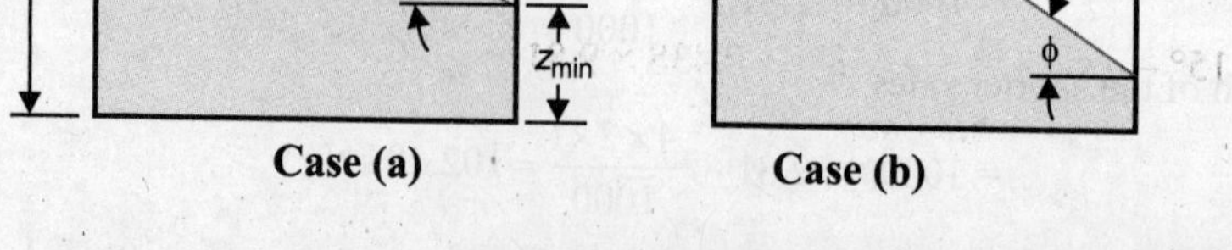

Case (a) Case (b)

Solution.

The water commences to spill over when z_{max} equals the height of the tank which is 3 m.

$$z_{max} = 2 + 2.5 \tan\theta = 3 \text{ m}$$

$$\tan\theta = 0.4 ; \theta = 22°$$

$$\frac{a_x}{g} = 0.4; a_x = 0.4 \times 9.81 = \mathbf{3.92\ m/s^2}$$

When it is accelerated at 20% excess acceleration,

$$a_x = 4.7 \text{ m/s}^2; \frac{a_x}{g} = 0.48 = \tan\phi, \phi = 25.6°$$

Again $z_{max} = 3$ m

and $z_{min} = 3 - 5 \tan\phi = 0.6$ m

Fraction of water spilt over

$$= \frac{(5\times4\times2) - \left[(\frac{1}{2}\times5\times2.4) + (5\times0.6)\right]\times4}{5\times4\times2} = 0.1$$

Percentage of water spilt over = **10%**

Example 2.35. *An open rectangular container 2.5 m × 1 m base and 2 m height, half full with water, is accelerated at 4 m/s² up a 15° incline along its length. (a) Will the water spill ? (b) At what acceleration would the water begin to spill?*

Solution.

(a) Let us compute the slope of the free surface of the water

$$\tan\theta = \frac{a_x}{g+a_z} = \frac{4\cos 15°}{9.81 + 4\sin 15°} = -0.35$$

whence $\theta = 19.8°$ backwards

The maximum slope which the surface can make with the horizontal; without siplling, is

$$\theta_{max} = \tan^{-1} 2/2.5 - 15° = 23.66° \text{ backwards}$$

The slope, 19.2°, made by the surface is less than this value; the water will not spill.

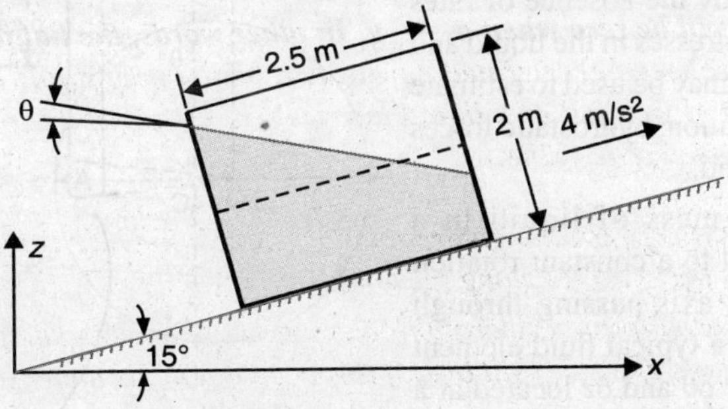

(b) Let the acceleration at which water would begin to spill be *a*,

then, $$\tan\theta_{max} = -\left(\frac{a\cos 15°}{g + a\sin 15°}\right) = \tan(-23.66)° = -0.438$$

whence $a \cos 15° - a \sin 15° \times 0.438 = +0.438 \times 9.81$

and $a = \mathbf{5.04\ m/s^2}$

Example 2.36. *An upright open U-tube, which can contain 0.5 m mercury in each limb has its limbs 20 cm apart. This is used as an accelerometer in a horizontal flight. What range of acceleration is within its scope?*

Solution.

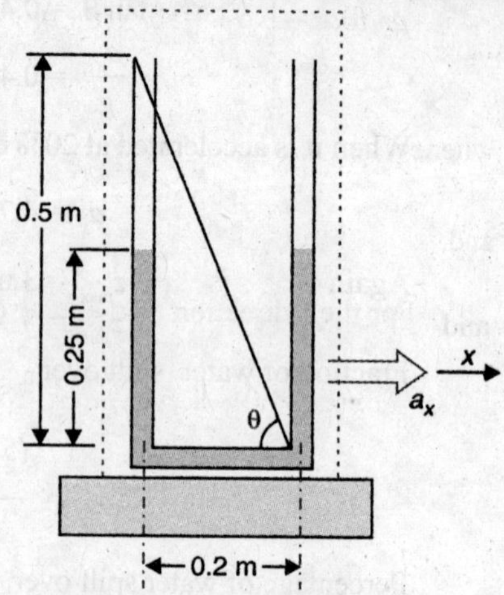

When subjected to horizontal acceleration, the fluid in one limb rises by the amount equal to the depressi on in the other limb. Thus, for maximum possible acceleration, i.e. for a maximum value of tan θ, the fluid should be initially filled to 50% height in the instrument i.e. 0.25 m in each limb, and should be at the verge of spilling over when accelerated.

$$z_{max} = 0.5 \text{ m} = 0.2 \tan\theta$$

whence $\tan\theta = 2.5$

But, $\tan\theta = \dfrac{a_x}{g} = 2.5$

Hence, $a_x = 2.5\, g = 24.5 \text{ m/s}^2$

The range of operation of this accelerometer is **0 to 24.5 m/s² in either direction.**

2.14. Liquid in a Container Subjected to a Constant Rotation

A liquid, contained in a vessel, may be rotated at a constant rotational speed ω without any relative movement being created between different elements of the liquid in the vessel. The liquid reorients itself once and for all to stay in that position with respect to the axis of rotation. The liquid mass is said to undergo a solid-body type of rotation, also known as a 'forced vortex' motion. This state of relative rest is also characterised by the absence of rates of strain and shear stresses in the liquid and the hydrostatic law may be used to estimate the pressure distribution, hydrostatic forces etc., due to the liquid.

Consider a mass of liquid in a container subjected to a constant rotation ω about a vertical axis passing through O. The velocity of a typical fluid element of dimensions δr, $r\,\delta\theta$ and δz located at a radial distance r from the axis of rotation must be

$$u = u_\theta = r\,\omega$$

The acceleration of the fluid element is given by $r\omega^2$ in a radially inward direction. Referring to the pressure acting upon the faces as shown in Fig. 2.20 the equations of motion may be written for the fluid element.

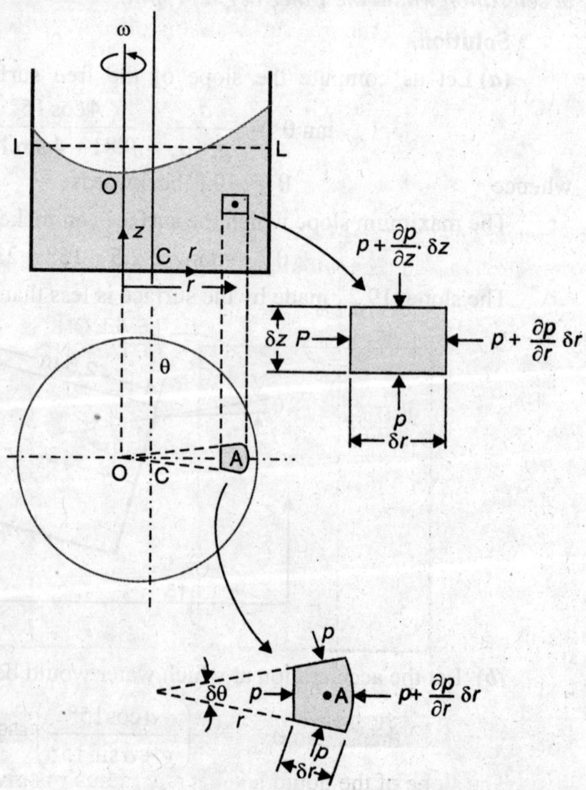

Fig. 2.20 Liquid in a Rotating Container

It may be noted that, owing to the axial symmetry of the motion, there is no pressure variation in the circumferential direction.

Considering the element as shown enlarged, for the v direction,

$$prd\theta dz - \left(p + \frac{\partial p}{\partial r}\delta r\right)(r+dr)(d\theta \, . \, dz) + 2pdr\frac{\delta\theta}{2}dz = -\,\omega^2 \rho\, r\, d\theta\, dz\, dr$$

whence $$-\frac{\partial p}{\partial r} r = -\,\omega^2 r^2 \rho$$

and $$\frac{\partial p}{\partial r} = \omega^2 r\, \rho \qquad \text{...(2.21a)}$$

For the z direction element, also shown enlarged,

$$p(\delta r \, . \, r\delta\theta) - \left(p + \frac{\partial p}{\partial z} . \, \delta z\right)(\delta r \, . \, r\delta\theta) - \rho g(\delta r \, . \, r\delta\theta . \, \delta z) = 0$$

$$\frac{\partial p}{\partial z} = -\,\rho g \qquad \text{...(2.21b)}$$

On integration of equations 2.21 a and b w.r.t. r and z respectively, and adding them,

$$p = -\rho g(z - z_0) + \frac{1}{2}\rho r^2 \omega^2 + p_0 \qquad \text{...(2.22)}$$

where p_0 is the reference pressure at $r = 0$, $z = z_0$ the free surface location in rotation.

The slope of the free surface is obtained by letting $p = p_0$ and hence it is given by;

$$0 = -\rho g(z - z_0) + \frac{1}{2}\rho r^2 \omega^2 \qquad \text{...(2.23)}$$

or $$z = \frac{r^2 \omega^2}{2g} + z_0$$

which is a paraboloid of revolution as shown in Fig. 2.21. Likewise, isobars, i.e., surfaces of constant pressure are also paraboloids of revolution.

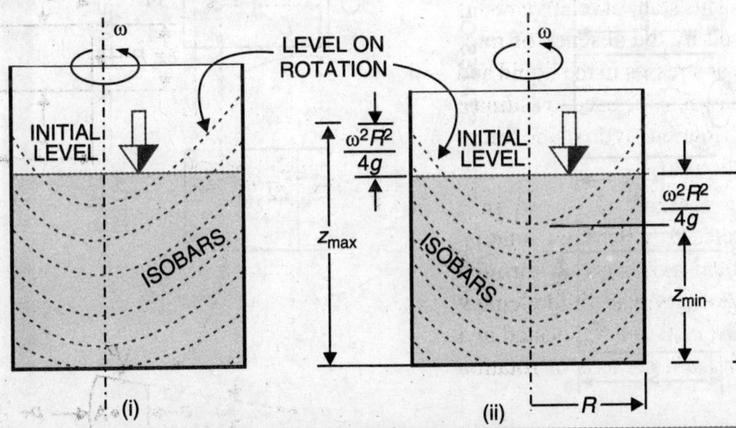

Fig. 2.21 Isobars in Rotating Liquid Masses

The slope of the liquid level at any radius r is given by

$$\frac{dz}{dr} = \frac{2r\omega^2}{2g} = \frac{r\omega^2}{g} = \tan\alpha$$

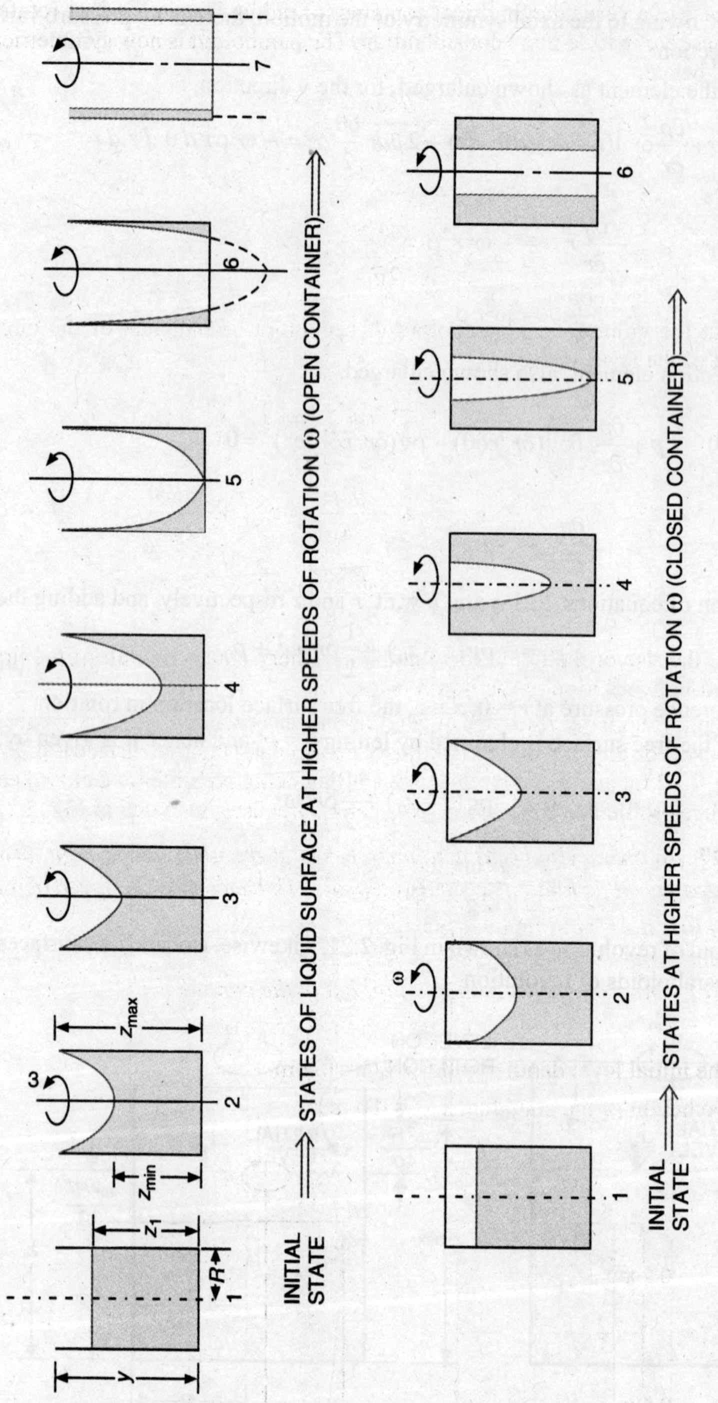

Fig 2.22. Rotating Fluid Mass at Various Speeds of Rotation

The special case of a vertical cylindrical container of radius R as in Fig 2.21 rotating about its axis of revolution deserves a little more consideration. The paraboloid is now symmetrical about the axis of rotation and hence,

at $\quad r = R, \quad z_{max} = \dfrac{R^2\omega^2}{2g} + z_0$

and at $\quad r = 0 \quad z_{min} = 0 + z_0$

Therefore, $\quad z = \dfrac{R^2\omega^2}{2g} + z_{min}$

Recalling that the volume of a paraboloid of revolution is half that of the circumscribing cylinder, the height of the level surface

$$z_1 = \frac{1}{2}(z_{max} + z_{min}) \qquad ...(2.25)$$

Finally, then

$$z_{max} = z_1 + \frac{R^2\omega^2}{4g} = z_{min} + \frac{R^2\omega^2}{2g} \qquad ...(2.26)$$

$$z_{min} = z_1 - \frac{R^2\,\omega^2}{4g} \qquad ...(2.27)$$

It is seen that the rise of the liquid level at the periphery equals the fall of the liquid level at the axis during solid-body rotation of the liquid mass.

The liquid begins to spill over the edge of the container at a speed for which z_{max} equals the height of the container. For speeds higher than this, spilling will occur and the free surface profile will change as illustrated in Fig. 2.22 for an open container. No spilling being possible for a closed container, the changes in the surface profile at highor speeds of rotation are also illustrated in Fig. 2.22.

Example 2.37. *An open cylindrical container, 0.3 m in diameter, 0.6 m high, two third filled with oil of specific gravity 0.8 is rotated about its vertical axis. Determine the speed of rotation when*

(a) the oil just starts spilling over the brim

(b) the point at the centre of the base is just exposed.

For case (b), find also the percentage of oil left in the container.

Solution.

Case (a) : The initial level depth $z_1 = 0.6 \times 2/3 = 0.4$ m

and the height of the container $z_{max} = 0.6$ m

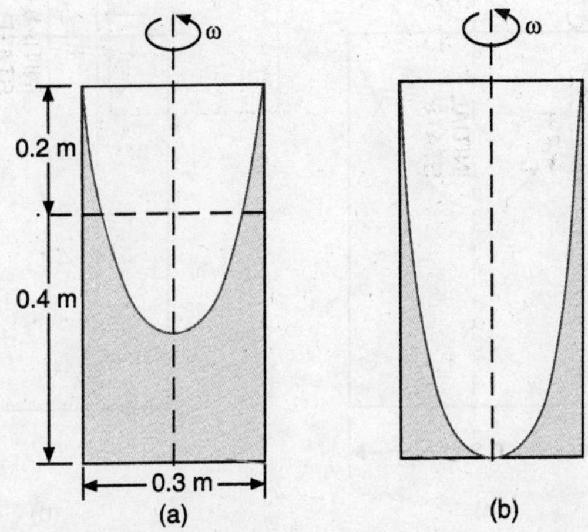

By Eq. (2.26),

$$z_{max} = z_1 + \frac{\omega^2 R^2}{4g}$$

$$0.6 = 0.4 + \frac{\omega^2 \times (0.3/2)^2}{4 \times 9.81}$$

whence $\omega = 18.68$ rad/s

Case (b) : Here again $z_{max} = 0.6$ m

but $z_{min} = 0$

$$z_{max} = \frac{R^2 \omega^2}{2g} + z_{min}$$

$$0.6 = \frac{0.3^2 \times \omega^2}{8 \times 9.81} + 0$$

and $\omega = \mathbf{22.9}$ **rad/s**

In this case, volume of oil left in the container

$$= \frac{1}{2} \times 0.6 \times A = 0.3\,A\,\text{m}^2$$

where A is cross-sectional area in m^2.

Comparing it with the initial volume, *i.e.*, 0.4 A m^2

the percentage of oil left $= \frac{0.3\,A}{0.4\,A} \times 100 = \mathbf{75\%}$

Example 2.38. *A closed cylindrical container, 0.3 m diameter and 0.6 m high, two-third filled with oil of specific gravity 0.8 is rotated about its vertical axis. Determine the speed of rotation when*

(a) the oil just starts touching the lid

(b) the point at the centre of the base is just clear of oil

Solution.

The problem is similar to Case (*a*) of Example 2.37.

Case (a) : Here, instead of spilling over, the oil starts touching the lid and continues to cover the lid from below.

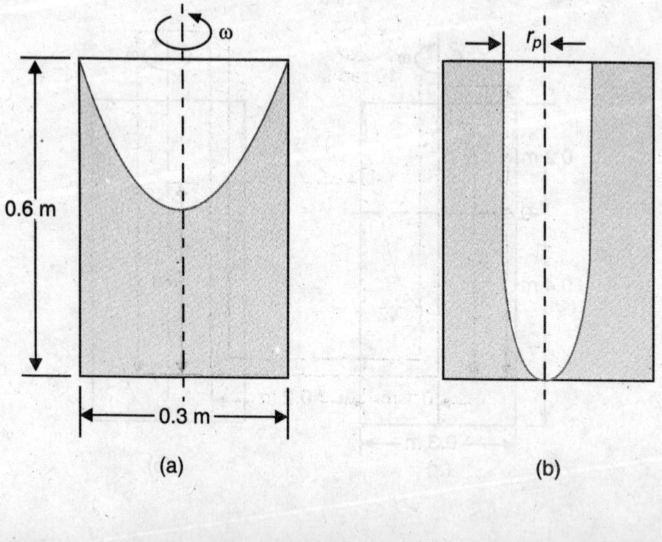

Case (b) : Now, the volume of the paraboloid must equal one third that of the container since the other two-third volume is occupied by the oil. From the consideration,

$$\frac{1}{2}\pi r_p^2 \times 0.6 = \frac{1}{3}\pi \times (0.3/2)^2 \times 0.6$$

$$r_p = 0.122 \text{ m}$$

By Eq. (2.26), $$z_{max} = 0.6 \text{ m} = \frac{0.122^2 \times \omega^2}{2 \times 9.81} + 0$$

whence, $$\omega = \mathbf{28.1\ rad/s}$$

Example 2.39. *An upright U-tube manometer with its limbs 0.6 m high and spaced 0.3 m apart contains a liquid to a height of 0.4 m in each limb. If the U-tube is rotated at 10 radians/second about a vertical axis at 0.1 m from one limb, what would the levels of the liquid be in the limbs?*

Solution.

Let z_{min} be the minimum reference level of the dotted parabola and z_1 and z_2 the liquid levels above the base.

$$z_1 = \frac{r_1^2\omega^2}{2g} + z_{min} = \frac{0.1^2 \times 10^2}{2 \times 9.81} + z_{min}$$

$$z_2 = \frac{r_2^2\omega^2}{2g} + z_{min} = \frac{0.2^2 \times 10^2}{2 \times 9.81} + z_{min}$$

But $$z_1 + z_2 = 2 \times 0.4 = 0.8 \text{ m}$$

Hence $$0.8 = \frac{0.2^2 \times 10^2 + 0.1 \times 10^2}{2 \times 9.81} + 2\, z_{min}$$

whence $$z_{min} = 0.273 \text{ m}$$

Consequently $$z_1 = 0.273 + \frac{0.1^2 \times 10^2}{2 \times 9.81} = 0.324 \text{ m}$$

and $$z_2 = 0.8 - 0.324 = \mathbf{0.476\ m}$$

Verifying *ab initio*, $$z_2 = 0.273 + \frac{0.2^2 \times 10^2}{2 \times 9.81} = \mathbf{0.476\ m}$$

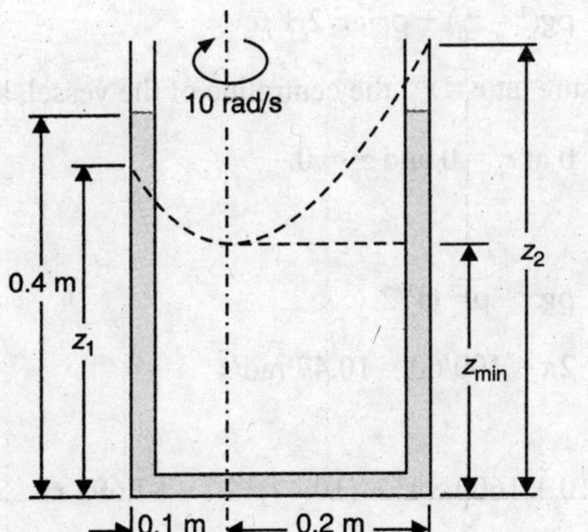

Example 2.40. *Find the speed of rotation of a 1 m diameter tank, initially full of water, such that water surface makes an angle of 45° with the horizontal at a radius of 30 cm. What is the slope of the surface at the wall of the tank.*

Solution.

The shape of the free surface must be a paraboloid of revolution given by

$$z = \frac{r^2\omega^2}{2g} + z_0$$

and the slope of the surface at any radius r is given by

$$\frac{dz}{dr} = \frac{2r\omega^2}{2g} = \frac{r\omega^2}{g} = \tan\alpha$$

At $\quad r = 0.3$ m; $\alpha = 45°$, $\tan\alpha = 1$

Hence $\quad 0.3\,\omega^2 = 9.81$

$$\omega = 5.72 \text{ rad/s}$$

$$N = 5.72 \times 60/2\pi = \frac{109.2}{2} = 54.62 \text{ r.p.m.}$$

The slope at the wall, $r = 1$ m, is given by

$$\tan\alpha = \frac{1 \times 5.72^2}{9.81} = 3.34 : \alpha = \mathbf{73.3°}$$

Example 2.41: *A closed cylindrical vessel of 0.5 m diameter and 2 m height is completely filled with glycerine of density 1600 kg/m³. It is rotated about its vertical axis at 100 revolutions per minute. Calculate the forces developed on the top, bottom and sides of the vessel.*

Solution.

Considering the equilibrium of a rotating fluid element,

$$\partial p/\partial z = \rho\, g \text{ in the vertical direction}$$

and

$$\partial p/\partial r = \rho\, r\, \omega^2 \text{ in the radial direction.}$$

On integration, $\quad p = \rho g(z - z_0) + \rho r^2\omega^2/2 + p_0$

where p_0 is the reference pressure at $r = r_0$, the centreline of the vessel, at $z = z_0$, at the top.

Let $\quad p_0 = 0$ at $r_0 = 0$ and $z_0 = 0$.

Then,

$$p = \rho g z + \rho r^2\, \omega^2/2$$

From the data, $\quad \omega = 2\pi \times 100/60 = 10.47$ rad/s

At the top of the vessel,

$$p = 0 + 1600 \times r^2 \times (10.47)^2/2 = 87.697\, r^2$$

which becomes

$$p = 87\ 697 \times (0.25)^2 = 5481\ \text{N/m}^2$$

at the periphery

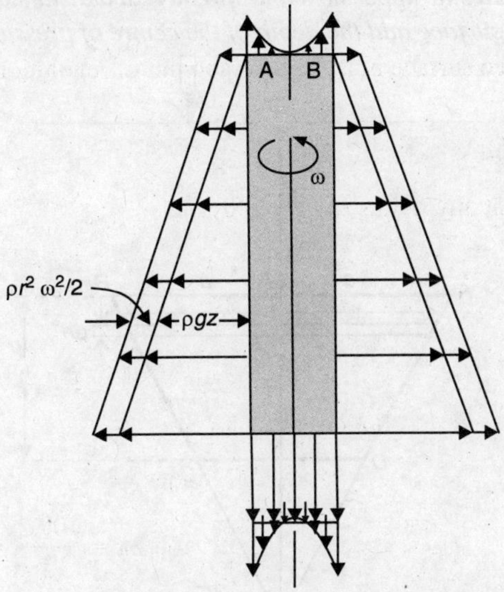

Pressure Variations in a Rotating Vessel

At the sides, the pressure is the sum of the hydrostatic pressure ρgz and the additional component of 5481 N/m²

At the bottom of the vessel

$$p = 1600 \times 9.81 \times 2 + 87\ 697\ r^2$$

which, at the periphery, becomes.

$$p = 31\ 392 + 5481 = 36\ 873\ \text{N/m}^2$$

The force on top end is

$$F = \int_0^{0.25} \rho r^2\omega^2 / 2.\, 2\pi r.dr$$

$$= \rho\pi\omega^2 \left|\frac{r^4}{4}\right|_0^{0.25} = \mathbf{538\ N}$$

At the bottom end,

$$F_b = 538 + 31\ 392 \times \pi \times (0.25)^2 = \mathbf{6702\ N}$$

At the sides, the pressure varies from 5481 N/m² at the top to 36 873 N/m² at the base. Since the pressure acts radially; equally in all directions, the net radial force must be zero.

Verification

The values of the pressures at the corner points are unique, e.g., pressure at the point A is 5481 N/m² normal to the top face on the one hand and normal to the side on the other. Further, the force on the top face should also be equal to the volume of the paraboloid of revolution generated by the pressure variation curve, i.e.

$$F = 5481 \times \pi \times (0.25)^2/2 = 538\ \text{N}$$

which is the same as before.

SUPPLEMENTARY REVISION EXAMPLES

Example 2.42. *A trapezoidal plate with parallel sides 2a and a at a distance h apart is immersed vertically in a liquid with its 2a side uppermost and spread it at a distance h below the surface of the liquid. Find the thrust on its surface and the depth of the centre of pressure.*

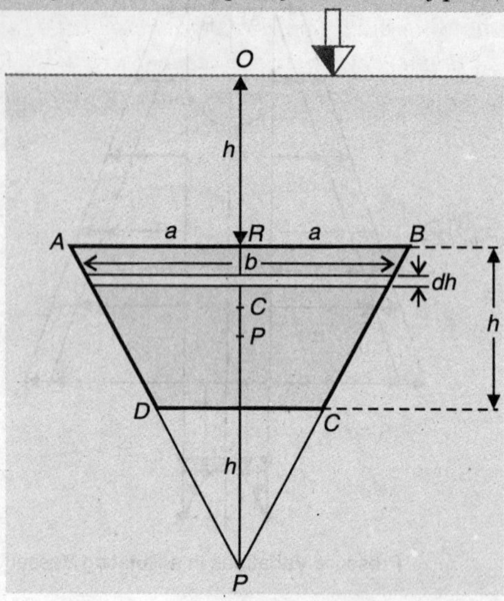

Solution.

The centroid C is located by taking first moments of the areas, considering that

ABCD = ABP – DCP

i.e., difference of two triangles.

Taking AB as reference,

$$\frac{2a \times 2h}{2} \times \frac{2h}{3} - \frac{ah}{2}\left(\frac{h}{3} + h\right) = \left(\frac{2a \times 2h}{2} - \frac{ah}{2}\right) \text{RC}$$

whence $\qquad \text{RC} = \frac{4}{9}h$

The thrust due to hydrostatic pressure is given by

$$F = \rho g \, \overline{z}. A$$

$$= \rho g \left(h + \frac{4}{9}h\right).\left(2ah - \frac{ah}{2}\right) = \frac{13}{6}\rho g \, ah^2.$$

The centre of pressure p is located by

$$z_p = \overline{z} + \frac{I_C}{A\overline{z}}$$

I_C for the trapezoidal area = $h^3 (2a + 3a)/12$

$$= \frac{5}{12}ah^3.$$

and $\qquad \overline{z} = \left(h + \frac{4}{9}h\right) = \frac{13}{9}h = 1.44\,h$

Hence, $$z_p = \frac{13}{9}h + \frac{5}{12}ah^3 \Big/ \left(\frac{3}{2}ah \times \frac{13}{9}h\right) = \frac{383}{234}h = \mathbf{1.64\ h.}$$

which means that p is below C by 0.2 h.

Example 2.43. *A U-tube contains a liquid of relative density 1.25 to a height of 25 cm in both the columns. It is rotated about a symmetrical vertical axis with 15 cm from both the limbs. Calculate the pressures at the bottom horizontal points and mid-point when the speed of rotation is 240 revolution per minute.*

Solution.

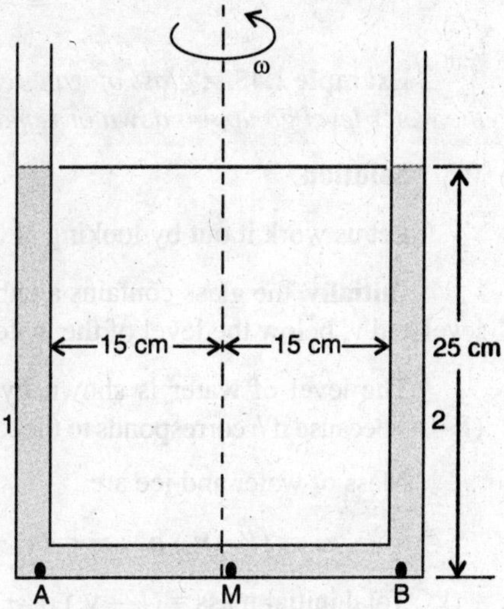

Since $$\omega = \frac{2\pi N}{60}$$

the angular speed of rotation is

$$\omega = \frac{2\pi \times 240}{60} = 25.13 \text{ rad/s}$$

Under constant rotation,

$$\frac{\partial p}{\partial r} = \rho r\omega^2 \,;\, p = \frac{1}{2}\rho r^2\omega^2 + p_0 \quad ...(i)$$

$$\frac{\partial p}{\partial z} = -\rho g \,;\, p = -\rho g\,(z - z_0) \quad ...(ii)$$

The pressure at A (and B) referred to p_0 is due to the columns of liquid supported and due to rotation *i.e.*,

$$p_A = p_B = 1.25 \times 1000 \times 9.81 \times 0.25 + \frac{1}{2}.\, 1.25 \times 1000 \times 0.15 \times 0.15 \times 25.13^2$$

$$= 3066 + 8881 = 11947 \text{ N/m}^2$$

$$= \mathbf{11.95\ kN/m^2}.$$

Also, $p_M = 3066 \text{ N/m}^2 = \mathbf{3.07\ kN/m^2}$.

Example 2.44. *A cylindrical vessel 15 cm in diameter and 40 cm high is completely filled with water and rotated about its vertical axis with a speed of 300 revolutions/minute. Find the quantity of water left in the vessel.*

Solution.

Diameter $D = 0.15$ m; $R = 0.075$ m

Height of the vessel = 0.40 m.

When it is rotated about a vertical axis, the liquid attains the shape of a paraboloid.

$$\omega = \frac{2\pi N}{60} = \frac{2 \times \pi \times 300}{60} = 10\pi \text{ rad/s.}$$

Liquid surface is given by $z = \dfrac{R^2\omega^2}{2g} + z_{min}$.

If z_1 is the mid-height $= \dfrac{1}{2}(z_{min} + z_{max})$

$$z_{max} = z_1 + \frac{\omega^2 R^2}{4g}; \; z_{min} = z_1 - \frac{\omega^2 R^2}{4g}$$

Now, $$0.4 = z_1 + \frac{(10\pi)^2 \; 0.075^2}{4 \times 9.81}; \; z_1 = 0.259$$

$$z_{min} = 0.259 - 0.1414 = 0.117 \text{ m}$$

Volume of water remaining in the vessel

$$= \frac{\pi \, 0.15^2}{4} \times 0.259 = \mathbf{4.58 \times 10^{-3} \; m^3}$$

Example 2.45. *A glass of water contains a single floating cube of ice. When the ice melts, does the water level go up, go down or remain the same ?*

Solution.

Let us work it out by looking at the initial and final conditions.

Initially, the glass contains a cube of ice of volume v with v_1 above the level and v_2 below the level of the water, such that $v_1 + v_2 = v$.

The level of water is shown by height h and the volume of water is $(V - v_2)$ because if h corresponds to the total volume V, then v_2 is occupied by ice.

Mass of water and ice are

$$m_w = (V - v_2)\,\rho_w \; ; \; m_i = (v_1 + v_2)\,\rho_i.$$

Total initial mass $= (V - v_2)\,\rho_w + (v_1 + v_2)\,\rho_i$

Now finally : When the ice melts the entire volume corresponding to a new height h' would be V' and the final mass would be $\rho_w V'$.

Now, by Archimides' principle of floatation, 'the magnitude of the buoyant force F_b must be equal to the weight of the fluid displaced.'

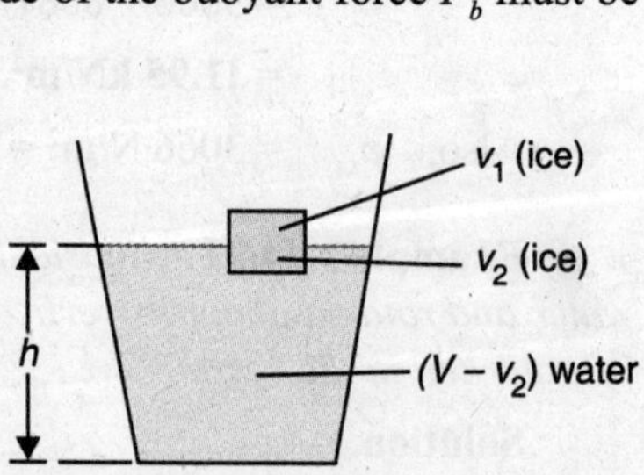

$$F_b = \rho_w . v_2 \, g; \; W = \rho_i \, (v_1 + v_2).\, g$$

$$F_b = W$$

$$\rho_w \, v_2 = \rho_i \, (v_1 + v_2)$$

whence $$\rho_i = \frac{v_2}{v_1 + v_2} . \rho_w$$

Conservation of mass requires

$$(V - v_2)\,\rho_w + (v_1 + v_2)\,\rho_i = V' . \rho_w$$

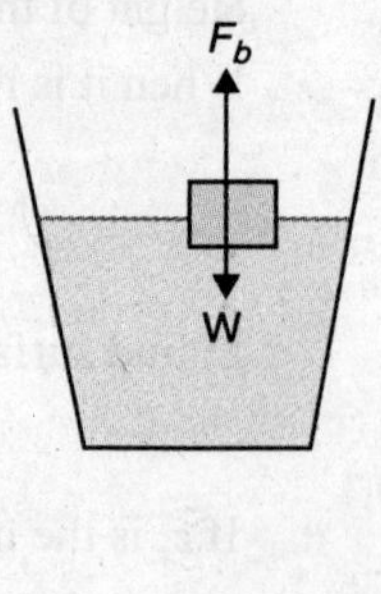

Using ρ_i, as above,

$$(V - v_2)\,\rho_w + (v_1 + v_2) . \frac{v_2}{v_1 + v_2} . \rho_w = V' . \rho_w$$

$$V \rho_w = V' \rho_w$$

which means $\boldsymbol{V' = V}$ **; level remains the same!**

One could have arrived at the same result by a simpler agrument as follows:

By Archimedes' principle.

$$F_b = \rho_w v_i g = W = \rho_i(v_1 + v_2).\ g$$

which shows that the total volume $(v_1 + v_2)$ of ice with density ρ_i results in a volume v_2 of water with density ρ_w. So, after melting, the volume v_2 (occupied inside the water by the ice) will be wholly occupied by the whole of the molten ice which had $(v_1 + v_2)$ volume. The level will remain the same !)

A very interesting corollary follows from the above analysis.

"If a glass of water, with ice floating above its surface, is full up to the brim, the level of water will remain the same, i.e., upto the brim during the melting of the ice and no water will be spilt out !"

We have already proved it. In this case, the level of water is up to the brim and the sum of the volume of all pieces of ice is v of which v_1 is above the level and v_2 below the level of water. The rest of the proof is the same !

Example 2.46. *A person in a row boat in a small pond throws an anchor overboard into the water, would the level of water in the pond rise, fall or remain the same ?*

Solution.

Initially, when the anchor is in the boat, it must be resulting in a water displacement such that the weight of the water displaced equals the total weight. Since the density of the anchor material is several times more then that of water, the volume of the water displaced is several times more than the volume of the anchor.

Now, when the anchor is thrown into the water, it would displace just the volume of water equal is its own volume. This is much less than before. Therefore, the level of water in the pond must go down!

Supposing the person threw an object with density less than that of water, like an inflated ball or a wooden item, then the volume of the water displaced would remain the same as it was initially; so no change in the level of water would have taken place!

Example 2.47. *Two tanks A and B, open to atmosphere, contaning oil and glycerine of sp. gravity 0.8 and 1.3 respectively are connected by a differential gauge. The manometric fluid shows an elevation of 4.5 m on the A side and 4 m on the B side. If the elevation of liquids in A and B are 28 m and 24 m respectively, find the specific gravity of the manometric fluid.*

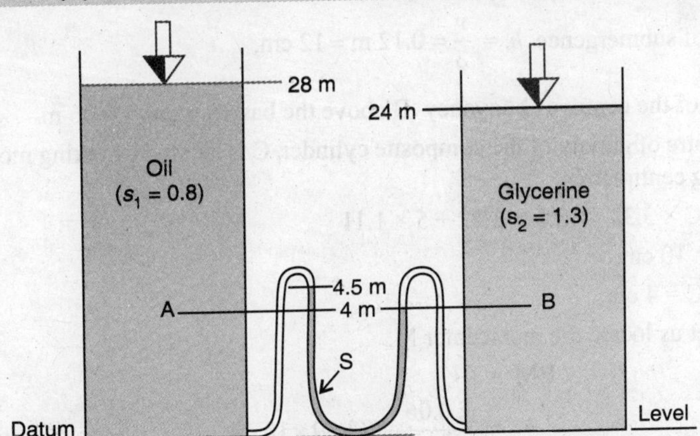

Solution.

Let us take the datum at the base from where all elevations are referred. Let the specific gravity of the manometric fluid be s. The system being in equilibrium the pressure at a point O, reckoned to the left and reckoned to the right must be the same.

Reckoned to the left of O,

the pressure at O = $0.8 \times 1000 \times g \times 28 + 4.5 \times s \times 1000 \times g$

Reckoned to the right of O,

the pressure at O = $1.3 \times 1000 \times g \times 24 + 4.0 \times s \times 1000 \times g$

Equating the two and cancelling $(1000 \times g)$ from all terms,

$$0.8 \times 28 + 0.5\,s = 1.3 \times 24$$

whence $\quad s = \mathbf{17.6}$

Example 2.48. *Two pistons of equal diameter, 6 cm, but different lengths, 5 cm, and 10 cm, with specific gravities 1.6 and 0.4 respectively are fastened together. Will they float in stable equilibrium with the heavier piston above the other ? Comment on the stability if the heavier piston is below the other.*

Solution.

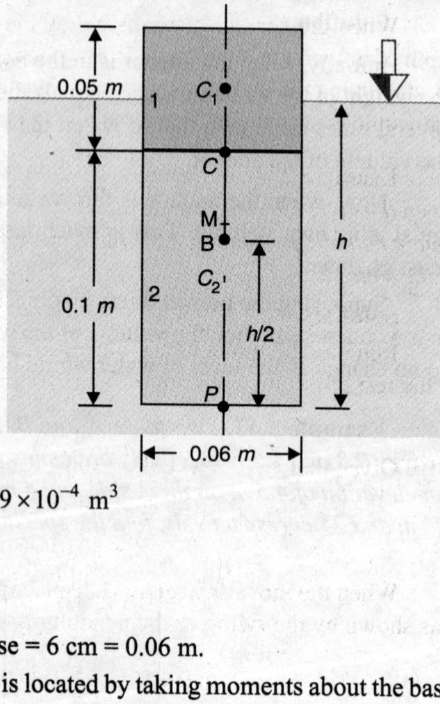

With 6 cm diameter

$$a = \frac{\pi \times 0.06^2}{4} = 2.83 \times 10^{-3}\ \text{m}^2$$

For piston 1,

$$v_1 = 0.05\,a = 1.41 \times 10^{-4}\ \text{m}^3$$

$$W_1 = 1.6 \times 9.81 \times 1000 \times v_1 = 2.22\ \text{N}$$

For piston 2,

$$v_2 = 0.1a = 2.83 \times 10^{-4}\ \text{m}^3$$

$$W_2 = 0.4 \times 9.81 \times 1000 \times v_2 = 1.11\ \text{N}$$

Total weight $W = W_1 + W_2 = 3.33$ N

Volume of water displaced, $v = \dfrac{3.33}{1000 \times 9.81} = 3.39 \times 10^{-4}\ \text{m}^3$

Depth of submergence $h = \dfrac{v}{a} = 0.12\ \text{m} = 12\ \text{cm}.$

Height of the centre of buoyancy B above the base = 6 cm = 0.06 m.

The centre of gravity of the composite cylinder, C is located by taking moments about the base point P. Using centimetres,

$$\text{PC} \times 3.33 = 12.5 \times 2.22 + 5 \times 1.11$$

whence, PC = 10 cm,

Also BC = 4 cm.

Now, let us locate the metacenter M.

$$\text{BM} = I/V$$

$$= \frac{\pi\, 0.06^4}{64} / (3.39 \times 10^{-4})$$

$$= 0.188 \times 10^{-3}\ \text{m} = 0.188\ \text{cm}$$

It shows that M is very close to B and the point C is far above it, making CM large and negative. The fact that CM is negative means that the equilibrium in this position is unstable.

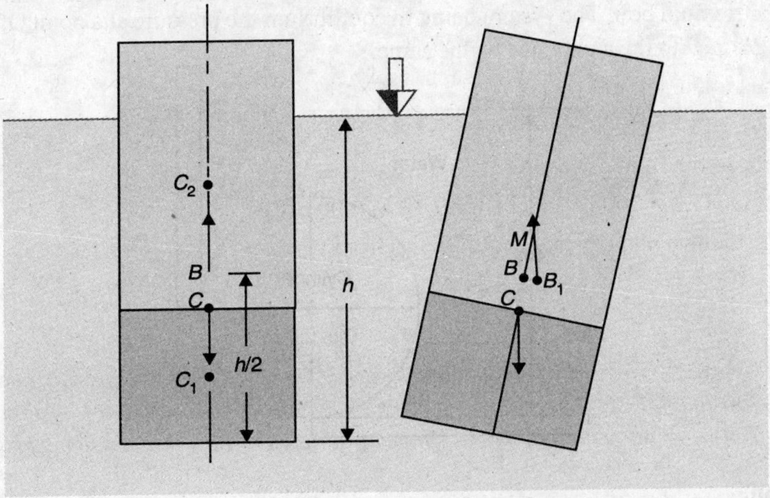

When the heavier piston is below the other, the location of C remains the same. So also the depth of immersion but then, B is now 1 cm above C. A small tilt will cause a resisting couple to set up, stabilizing the body. One does not have to locate M in a case where C is already below B because that would ensure the setting up of a stabilizing couple in the event of a small tilt.

Example 2.49. *A ship weighs 3000 tons, including a movable weight of 50 tons on its deck. When the movable weight is displaced along the lateral axis through a distance of 10 cm, a 5 m long simple pendulum in the ship swings through a distance of 30 cm. Determine the metacentric height of the ship.*

Solution.

Initially, the movable weight w, assumed to be at the centre of the deck, added to the weight of the rest of the ship W keep the ship upright as the centre of buoyancy B is just underneath them.

$$w + W = F_B$$

Given that $$w + W = 3000 \text{ Tons}$$

$$w = 50 \text{ Tons}$$

$$W = 3000 - 50 = 2950 \text{ Tons.}$$

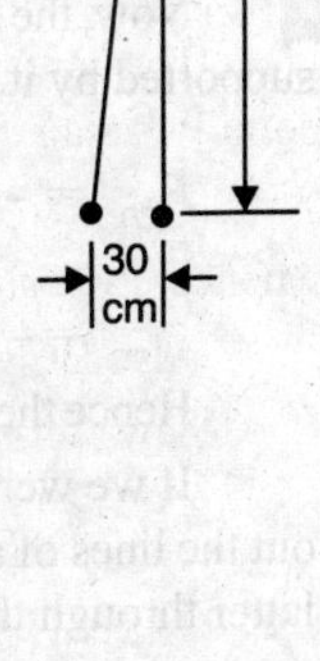

When the movable weight is displaced by $x = 0.1$ m, the ship tilts by an angle θ as shown by the swing of the pendulum,

$$\tan\theta \approx \theta = \frac{0.3}{5} = 0.06$$

Equating the moment due to the movable weight to the stabilising moment of the weight of the ship,

$$w\,x\cos\theta = W.\,\text{CM}\sin\theta$$

whence, $$\text{CM} = \frac{w}{W}\cdot\frac{x}{\tan\theta}$$

$$= \frac{50}{2950}\cdot\frac{0.1}{0.06} = 0.028 \text{ m} = \mathbf{2.8\ cm}$$

Example 2.50. *Determine the magnitude and direction of the net hydrostatic force acting on the curved surface of a cylinder of diameter 2.5 m, shown, per meter length of cylinder.*

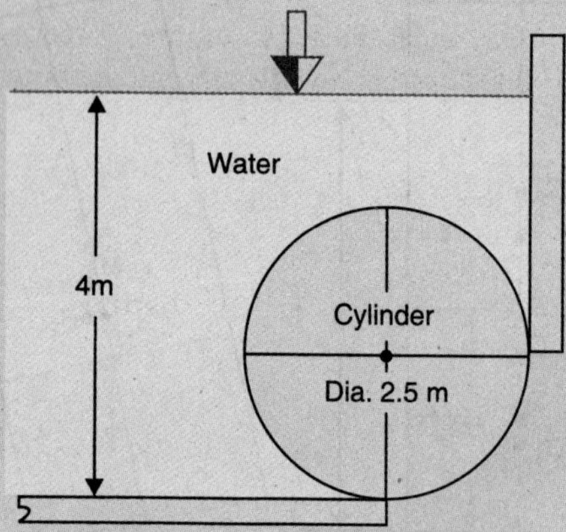

Solution.

Let us consider part ABC first. The hydrostatic force on it is due to the actual weight of water supported by it, as shown.

$$F_{ABC} = \left[1.5 \times 2.5 \times 1 + 2\left(1.25 \times 1.25 - \frac{1}{4}\pi 1.25^2\right) \times 1\right] \times \rho g$$

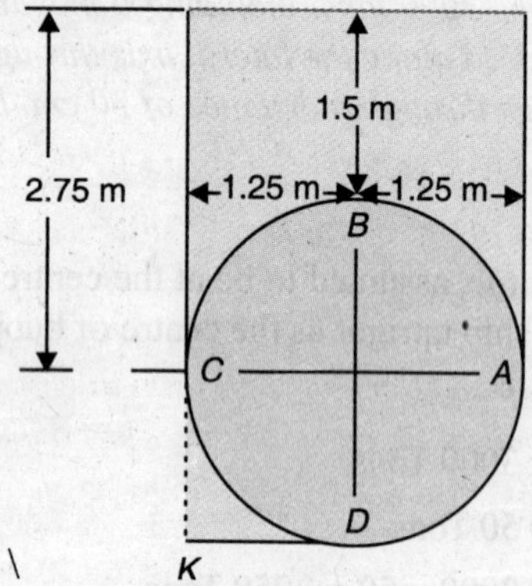

$= (3.75 + 0.67) \times 1000 \times 9.81 = 43{,}360$ N

$= 43.36$ kN downward

Now, the hydrostatic force on the curved part CD is determined by the weight of water virtually supported by it, as shown.

$$F_{CD} = -(1.25 \times 1.25 - \frac{1}{4}\pi\, 1.25^2) \times 1 \times 1000 \times 9.81$$

$= -32\ 86$ N $= -3.286$ kN upward

Hence the net hydrostatic force acting on the cylinder is $F = 43.36 - 3.2\ 86 = 40.07$ kN downward

If we were required to determine the line of action of the net force, we could do so by finding out the lines of action of the two components, former passing through the centre of the circle and the latter through the centroid of the part CKD.

Example 2.51. *A spherical balloon 1.5 m diameter is completely immersed in water and chained to the bottom. If the chain has a tension of 10 kN, find the weight of the balloon in air.*

Solution.

The balloon is in equilibrium under the action of three forces; weight W acting downwards, Pull in the chain 10 kN, *i.e.,* 10000 N acting downwards and the buoyancy acting upwards.

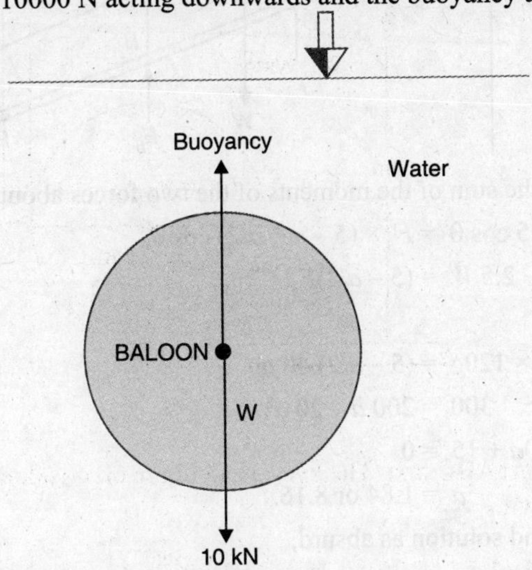

Buoyancy for the completely submerged balloon is given by the weight of the water displaced

Buoyancy $\quad F_b = \frac{4}{3}\pi(1.5/2)^3 \times 1000 \times 9.81$

$= W + 10000$

For equilibrium $\quad W = 17336 - 10000 = 7336$ N

whence $\quad W = \mathbf{7.336\ kN}$

Example 2.52. *A wooden beam 200 mm × 200 mm in cross section and 5 m long is hinged at one end and floats on water, as shown, in equilibrium. Find the inclination θ if the specific gravity of wood is 0.6.*

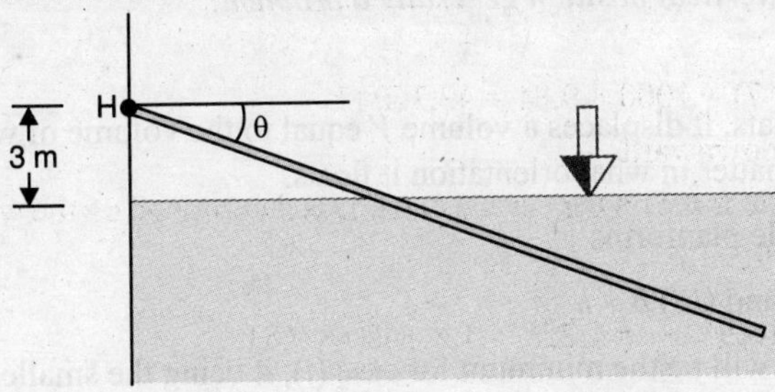

Solution.

Let us assume that a length 'a' is immersed in water and the remaining length $(5 - a)$ remains outside the water, as shown.

Weight of the beam

$$W = 5 \times 0.2 \times 0.2 \times 600 \times g = 120\, g \text{ N}.$$

Buoyancy on part DE equals the weight of the water displaced by it, *i.e.,*

$$F_b = a \times 0.2 \times 0.2 \times 1000\, g = 40\, ag \text{ N}.$$

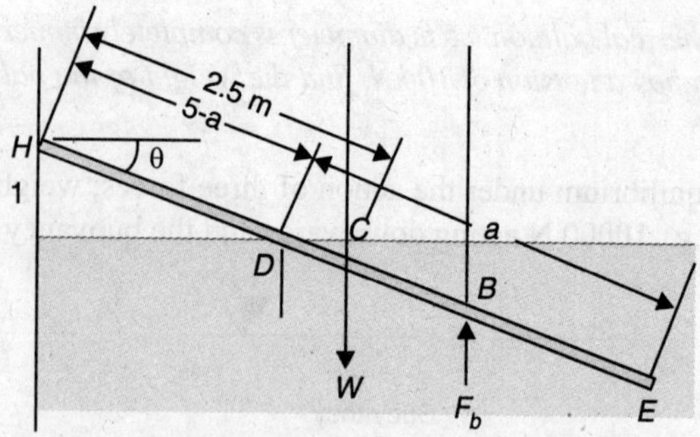

For equilibrium, the sum of the moments of the two forces about the hinge must vanish,

$$W \times 2.5 \cos\theta = F_b \times (5 - a + a/2) \cos\theta.$$

$$2.5\, W = (5 - a/2)F_b.$$

which means that

$$2.5 \times 120g = (5 - a/2)\, 40\, ag$$

$$300 = 200\, a - 20\, a^2$$

or $\quad a^2 - 10a + 15 = 0$

whence $\quad a = 1.84 \text{ or } 8.16.$

Ignoring the second solution as absurd,

$$a = 1.84 \text{ m}$$

$$5 - a = 3.16 \text{ m}$$

From the figure,

$$(5 - a) \sin\theta = 1$$

Hence $\quad \sin\theta = \dfrac{1}{3.16} = 0.316$

$$\theta = \mathbf{18.4°}$$

Example 2.53. *A block of wood measuring a × b × h (where h is the longest side), of specific gravity S, (S < 1), is made to float in water. Discuss the possible orientations and their pre-conditions. Why is its stability in the vertical position generally a problem?*

Solution.

When the block floats, it displaces a volume V equal to the volume of water equivalent to the weight of the block, no matter in what orientation it floats.

Of the three possible planforms

(*i*) $a \times b$, (*ii*) $a \times h$ and (*iii*) $b \times h$

the moment of inertia I_{CC} will be the minimum for case (*i*), it being the smallest area. I_{CC} will be the maximum for the larger of the two (*ii*) and (*iii*), depending upon a or b being larger side.

Now, $\quad BM = I/V.$

It shows that BM is the least for case (*i*) and greatest for either (*ii*) or (*iii*). In a floating body, the centroid C is always higher than the centre of buoyancy B. Therefore, the chance for CM to be positive, *i.e.*, M to be above C is great if BM is the greatest. The block acquires that position automatically.

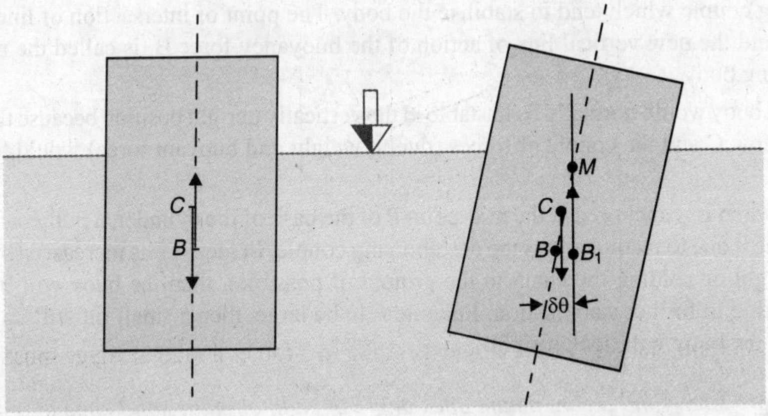

Initial Position with its Axis Vertical

Slightly Tilted Position to locate the Metacentre

(In this case, CM is positive, the couple of forces tend is restore the initial position; hence the equilibrium is stable)

The vertical position corresponds to planform (*i*) which has the least I_{CC}; hence problematic to ensure the metacentric height to be positive. BM should at least be equal to BC for C and M to coincide.

If M is above C, CM is said to be positive and this represents 'stable equilibrium'. On the other hand if M is below C, CM is negative, indicative of instability because the couple of forces tends to tilt the body further. The special case of C coinciding with M, *i.e.*, CM = 0 represents neutral equilibrium. In this case, the body stays as it is; it neither restores to original position, nor does it tilt any further.

Example 2.54. *Explain the concept of 'metacentre' of a cylindrical body floating upright with its axis vertical. State clearly how a light buoy can be made to float vertically by anchoring it with a chain and a suspended weight at the bottom of the buoy.*

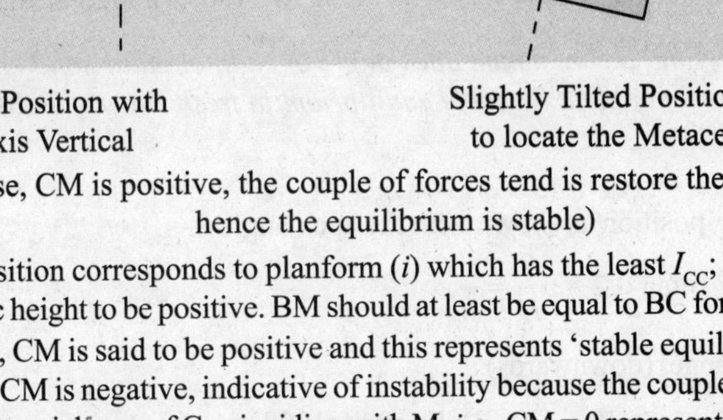

(Making a cylindrical buoy float stable in upright position by anchoring at the base.)

Solution.

A floating cylinder can be made to float and be in equilibrium in any position provided the centre of mass C and the centre of buoyancy B are vertically above each other, usually C above B. But the equilibrium may not be stable, *i.e.*, if it is slightly tilted from the equilibrium position it may not regain the equilibrium but further destabilise. In order to ensure its stability, the cylinder is tilted by a small angle $\delta\theta$ and the new centre of buoyancy B_1 is determined and a vertical line is drawn to show the new line of action of buoyancy force. The position of C remains the same even after tilting. If the new line of action of the buoyancy force intersects the original line of action (on which B and C lie) above C, then the weight acting through C and buoyancy force acting through B_1 result in a restoring couple which tend to stabilise the body. The point of intersection of line through BC (produced) and the new vertical line of action of the buoyancy force B_1 is called the metacentre *M* for the floating body.

A light buoy would normally be unstable in the vertically upright position because the metacentre would be below C and the couple of forces (due to weight and buoyant force) would tend is tilt the buoy further.

Now, when it is anchored at the mid-point P of the base of the cylinder, a pull *F* acts downward and the moment due to it counteracts the destabilising couple. In fact, if *F* is increased (by suspending a bigger weight or holding the chain to the ground, if possible), then the buoy will be submerged further resulting in further stabilisation. Imagine *F* to be large, then a small tilt will immediately be rectified and the buoy will float upright!

Example 2.55. *Find the minimum apex angle of right circular solid cone of a material with sp. gravity 0.8 so that it can float in stable equilibrium in fresh water with its axis vertical and the vertex downwards.*

Solution.

For the floating position as shown with the apex angle θ,

$$\tan \theta/2 = \frac{R}{H} = \frac{r}{h}.$$

Weight of the cone, (downwards)

$$W = \frac{1}{3}\pi R^2 H \times 0.8 \times 1000\, g$$

and the buoyancy force (upwards)

$$F_b = \frac{1}{3}\pi r^2 . h \times 1000\, g.$$

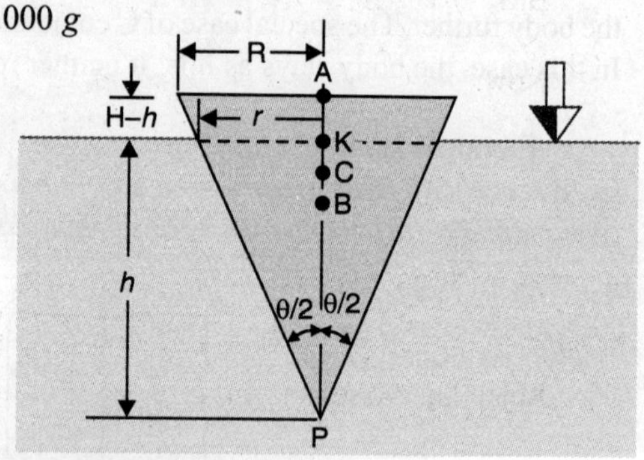

For equilibrium, $W = F_b$

$$0.8\, R^2 H = r^2 h$$

Eliminating *R* and *r*,

$$0.8\, H^3 = h^3;\; h = 0.928\, H.$$

The centroid of a cone is located at quarter height from the base.

$$AC = H/4 = 0.25\, H.$$

The centre of buoyancy is indeed the centroid of the immersed cone with height *h*, which gives

$$KB = h/4 = 0.232\, H$$

Since $AK = H - h = 0.072\, H,$

$$BC = (0.072 + 0.23 - 0.25)\, H = 0.054\, H.$$

The condition for stable equilibrium is that the metacentric height CM is positive; at least zero, which means that C and M coincide.

In that case, BM = BC.

Now, BM = 0.054 *H*

Using $r = h \tan \theta/2$

$$\frac{3h^2 (\tan \theta/2)^2}{4h} = 0.054\, \text{H}$$

$$(\tan \theta/2)^2 = \frac{0.054\, H \times 4}{3h}$$

$$= \frac{0.054\, H \times 4}{3 \times 0.928\, H} = 0.0776$$

$$\tan \theta/2 = 0.2785$$

$$\theta/2 = 15.56$$

$$\theta = \mathbf{31.13°}$$

This must be the minimum apex angle of the right circular solid cone for static equilibrium of floatation as required.

Example 2.56. *A cylindrical buoy 2 m diameter and 3.5 m height weight is 80 kN. Find the tension necessary in a vertical chain attached to the centre of the base of the buoy that will keep the buoy float vertically.*

Solution.

The purpose of tension in the chain, *i.e.,* downward force on the bouy is to immerse the cylindrical buoy to a depth such that the metacentric height is positive; at least zero!

In the limiting case of C coinciding with M,

BM = I/V = BC = AC – AB = 3.5/2 – h/2 = 1.75 – h/2

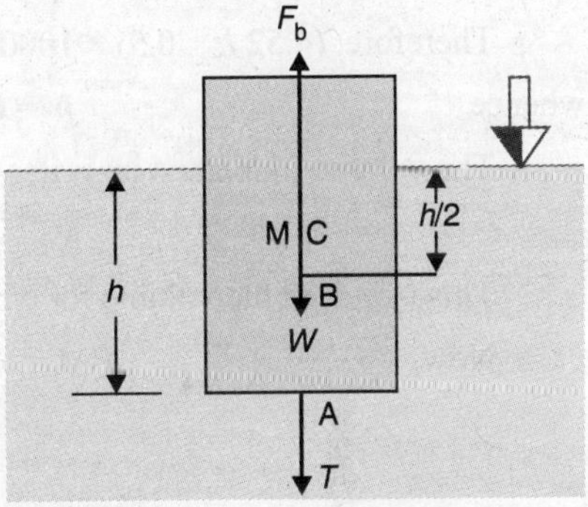

Now, $I = \frac{\pi}{64} d^4$ and $V = \frac{\pi d^2}{4} \times h$

$$\text{BM} = \frac{d^2}{16h} = \frac{2^2}{16h} = \frac{1}{4h} = 1.75 - \frac{h}{2}$$

or $\quad 2h^2 - 7h + 1 = 0$

whence $\quad h = 3.35$ m or 0.15 m

Rejecting the second value as too low and absurd,

$$h = 3.35 \text{ m.}$$

Buoyancy force $\quad F_b = \frac{\pi \times 2^2}{4} \times 3.35 \times 1000 \times 9.81$

$$= 10.3244 \text{ N} = 103.2 \text{ kN}$$

Weight of the buoy $\quad W = 80$ kN

For equilibrium, $T + W - F_b = 0$

whence $\quad T = \mathbf{103.2 - 80 = 23.2\ kN.}$

Example 2.57. *A buoy carrying a beacon has a cylindrical upper portion of 2.4 m diameter and 1.4 m depth. The curved lower portion has an external volume of 0.5 m³. The centre of buoyancy is 1.5 m below the top of the cylinder and the centre of gravity of the buoy and the beacon is 1.2 m below the top of the cylinder. The total weight is 2.7 tonnes. Taking the sp. gravity of the sea water as 1.02, calculate the metacentric height. Is the floating assembly in stable equilibrium ?*

Solution.

Assuming the depth of immersion of cylindrical part h, as shown, the total volume immersed in water is

$$V = h \times \frac{\pi}{4} \times 2.4^2 + 0.5$$

$$= 4.52\,h + 0.5 \text{ m}^3.$$

Buoyancy force equals the weight of the water displaced.

$$F_b = (4.52\,h + 0.5) \times 1000 \times g. \text{ N}$$

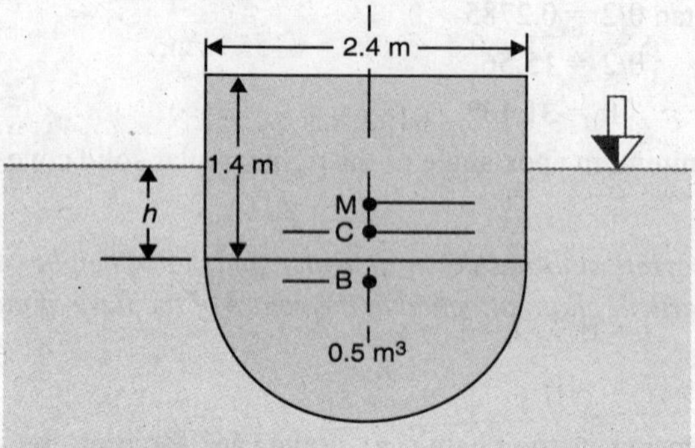

For equilibrium, it should be equal to the weight of the buoy, *i.e.*, 2.7 tonnes, *i.e.*

$$F_b = 2.7 \times 1000 \times \text{g N}$$

Therefore, $(4.52\,h + 0.5) \times 1000\text{ g} = 2.7 \times 1000\text{ g}$

whence $h = 0.487$ m

The volume immersed in water

$$V = 4.52 \times 0.487 + 0.5 = 2.7 \text{ m}^3.$$

This is indeed the volume for which the weight of water is 2.7 tonnes)

Now, $\text{BM} = I/V$

$$= \frac{\pi \times 2.4^4}{64} \frac{1}{2.7} = 0.603 \text{ m}$$

Given that

$$\text{BC} = 1.5 - 1.2 = 0.3 \text{ m}$$

We get $\text{CM} = 0.603 - 0.3 = \mathbf{0.303\ m}$

where M, the metacentre is above C. The metacentric height is 0.303, + ve. The buoy, therefore, floats in a stable equilibrium.

Example 2.58. *An open tank containing water to a depth of 1m is accelerated in a vertical direction at a rate of 2.5 m/s². Calculate the hydrostatic force per meter length on one of the side of the tank if the acceleration is*

(i) upwards and *(ii) downwards.*

Solution.

The pressure variation in the vertical direction for a liquid column subjected to a vertical acceleration a is given by

$$p = \rho(g + a_z)\,z + p_0$$

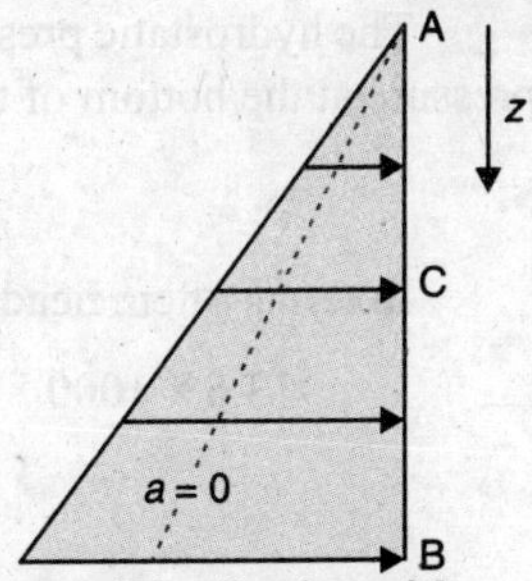

Pressure Distribution with Upward Acceleration

Since the tank is open, $p_0 = 0$, *i.e.*, atmospheric pressure.

$$p = \rho(g + a_z)\, z$$

(*i*) For upward acceleration, $a_z = 2.5$ m/s²

$$p = \rho(9.81 + 2.5)\, z = 12.31\, \rho z$$

$$p_c = 12.31 \times 0.5 \times 1000 = 6155 \text{ N/m}^2$$

$$F_{AB} = p_c \times 1 \times 1 = \mathbf{6155\ N = 6.155\ kN}$$

(*ii*) For downward acceleration,

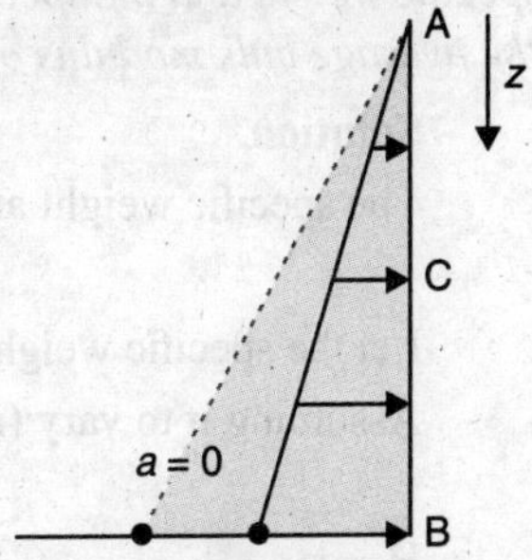

Pressure Distribution with Downward Acceleration

$$a_z = -\,2.5 \text{ m/s}^2.$$

$$p = \rho(9.81 - 2.5)\, z$$

$$= 7.31\, \rho z$$

$$p_c = 7.31 \times 0.5 \times 1000 = 3655 \text{ N/m}^2$$

$$F_{AB} = p_c \times 1 \times 1 = \mathbf{3655\ N = 3.655\ kN}$$

Note that, if downward acceleration was 9.81 m/s², pressure variation would be zero.

Example 2.59. *Find the intensity of pressure required to suck milk by a straw through a height of 200 mm from an open vessel. Sp. gravity of milk is 1.08.*

Solution.

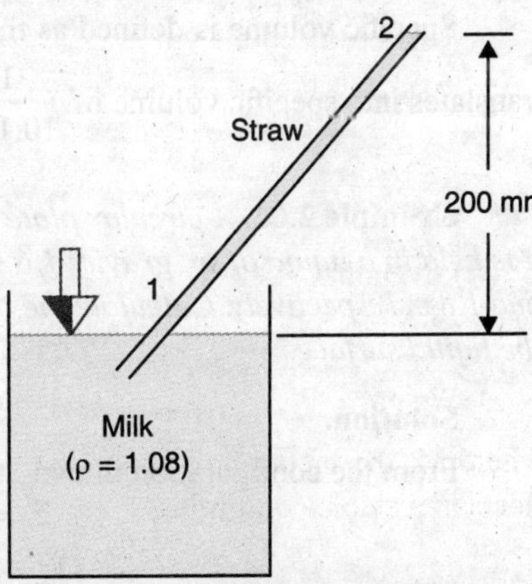

The pressure in the open vessel is atmospheric from where the milk is to be sucked.

The pressure at the other end of the straw should, therefore, be less than the atmospheric pressure.

Suction head, $h = 0.2$ m

Density of milk, $\rho = 1.08 \times 1000 = 1080$ kg/m³.

Hence $\Delta p = p_1 - p_2 = \rho g h$

$= 1080 \times 9.81 \times 0.2$

$= 2119$ Pa $= 2.12$ kPa

The pressure at point 2, should therefore be 2.12 kPa less than that at point 1 just to create suction.

Typically, atmospheric pressure is 1.013×10^5 N/m² or 101.3 kPa, therefore, the pressure at point 2 should be 2.12 kPa vacuum or (101.3 – 2.12), *i.e.*, **99.18 kPa absolute.**

Example 2.60. *A closed tank 8 m high is filled with bromine of sp. gravity 3.1 to a depth of 2 m and turpentine of sp. gravity 0.87 to a depth of 5 m. The upper part of the tank contains air at a pressure of 25 kPa (vacuum). Find the absolute pressure at the bottom of the tank in kPa and in cm of mercury. Take the atmospheric pressure as 101.3 kPa and neglect the density of air.*

Solution.

The absolute pressure of air at the top is

$$p_{air} = 101.3 - 25 = 76.3 \text{ kPa}.$$

The hydrostatic pressures due to bromine and turpentine are added to it to result in the absolute pressure at the bottom of the tank

$$p = 76.3 + (0.87 \times 1000 \times 5 + 3.1 \times 1000 \times 2) \times 9.81/1000$$
$$= \mathbf{179.8\ kPa.}$$

In terms of cm head of mercury of sp. gravity 13.6,

$$13.6 \times 1000 \times g \times \frac{h}{100} = 179.8 \times 1000$$
$$h = \mathbf{134.8\ cm\ of\ mercury}$$

Example 2.61. *The specific weight of water at the ocean surface is 10,000 N/m³. Calculate the specific weight and the specific volume at a depth of 12 km if the pressure there is 120.6 MN/m² and the average bulk modulus of sea water is 2.345 × 10⁶ kN/m².*

Solution.

The specific weight at the ocean surface

$$(\rho g)_s = 10{,}000\ \text{N/m}^3$$

Let the specific weight at the depth d be $(\rho g)_d$.

Assuming it to vary (increase) linearly, average specific weight is given by

$$(\rho g)_{av} = (10{,}000 + (\rho g)_d)/2\ \text{N/m}^3$$

Hence, $p = (\rho g)_{av} \times 12{,}000 = 120.6 \times 10^6$

$$(\rho g)_{av} = 10{,}050\ \text{N/m}^3$$

and $(\rho g)_d = 2 \times 10{,}050 - 10{,}000 = 10{,}100\ \text{N/m}^3$

Now, $K = \dfrac{-dp}{d\nu/\nu} = \dfrac{dp}{d\rho/\rho}$

Specific volume is defined as the volume per unit mass. The specific weight of 10,100 N/m³ translates into specific volume of $\dfrac{1}{10{,}100} \times 9.81$ m³/kg = **0.00097 m³/kg = 0.971 cm³/gram.**

Example 2.62. *A circular plate of 2 m diameter with a concentric hole of 0.5 m diameter in it is held in a liquid of sp. gravity 1.6 such that the least and greatest depths of submersion are 1 m and 1.8 m respectively. Calculate the total force and also the depth of the 'centres of pressures' from the liquid surface.*

Solution.

From the configuration drawn, the depth of the centroid of the area (on the upper surface)

$$\bar{z} = (1 + 1.8)/2 = 1.4\ \text{m}$$

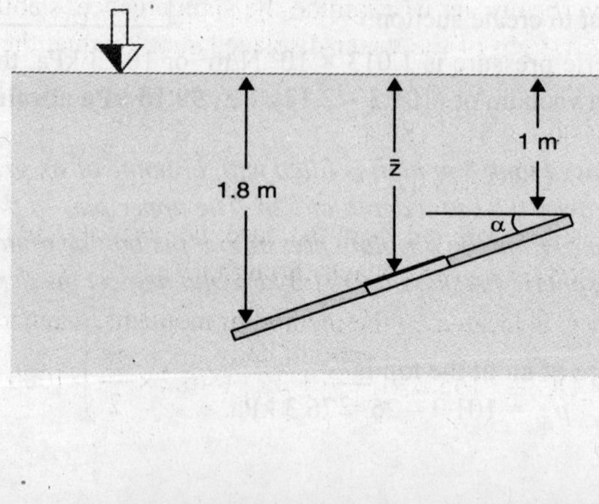

The hydrostatic force is given by

$$F = \rho.g.\bar{z}.A$$
$$= (1.6 \times 1000) \times 9.81 \times 1.4 \times \pi(2^2 - 0.5^2)/4$$
$$= 64720 \text{ N} = \mathbf{64.72 \text{ kN}}$$

The depth of the 'centre of pressure' is located by

$$\bar{z}_p = \bar{z} + I_C.\frac{\sin^2\alpha}{A\bar{z}}$$
$$= 1.4 + \left[\pi(2^4 - 0.5^4)/64 \times \left(\frac{0.8}{2}\right)^2 / (\pi(2^2 - 0.5^2)/(4 \times 1.4)\right]$$
$$= 1.4 + 0.03 = \mathbf{1.43 \text{ m.}}$$

Example 2.63. *A solid cylinder, 200 mm diameter and 800 mm long has its base 25 mm thick of a material of sp. gravity 6.0. The remaining part of the cylinder has a sp. gravity 0.6. State if it can float vertically in water and state the condition for critical stability.*

Solution.

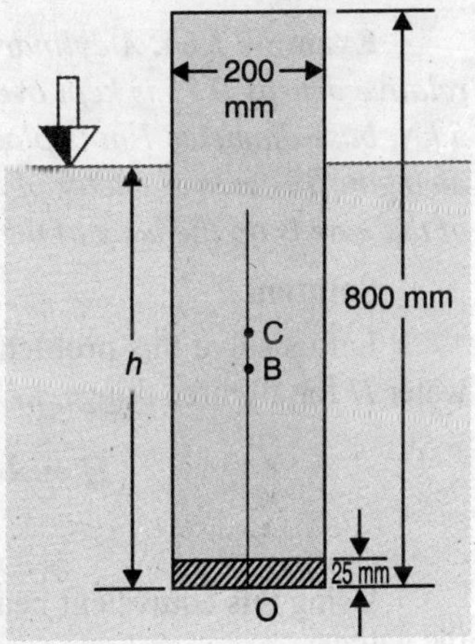

The volume of the entire cylinder is

$$V = \frac{\pi}{4} \times 0.2^2 \times 0.8$$
$$= 0.0314 \times 0.8$$
$$= 0.025 \text{ m}^3.$$

Weight of the cylindor

$$W = 0.0314 \times (0.8 - 0.025) \times 600 \text{ g}$$
$$+ 0.0314 \times 0.025 \times 6000 \text{ g}$$
$$= 143.2 + 46.2$$
$$= 189.4 \text{ N.}$$

Overall density of the cylinder

$$\rho = \frac{189.4}{0.025 \times 9.81} = 772 \text{ kg/m}^3.$$

which is less than that of water (of density 1000 kg/m³)

Since the cylinder is 'lighter' than water, it will float on water.

Now, for it to float vertically, let us examine the submergence, stability, etc. If the depth of submergence is h, then the weight of the water displaced should equal the weight of the cylinder which means that

$$h \times 0.0314 \times 1000 \times 9.81 = 189.4$$

whence, $$h = 0.615 \text{ m}$$

Centre of buoyancy is therefore $h/2$ from the base, *i.e.*, B is at 0.3075 m from the base. The buoyancy force of 189.4 × 9.81, *i.e.*, 1858 N acts through it.

The centre of gravity C is located by the method of moments, taken about the base.

$$189.4 \times OC = \left(143.2 \times \left(\frac{775}{2} + 25\right) + 46.2 \times \frac{25}{2}\right)/1000$$

$$= 143.2 \times \frac{417.5}{1000} + 46.2 \times \frac{12.5}{1000}$$

whence $\quad \text{OC} = 0.319 \text{ m}$

It is located just above the point B.

Now, to locate the metacentre M.

$$\text{BM} = I/V = \frac{\pi d^4}{64}/V$$

$$= \frac{\pi.\, 0.2^4}{64}/(0.615 \times 0.0314) = 0.0041 \text{ m}$$

It shows that M is 0.0041 m above B, 0.312 above the base. The point C is 0.0115 m above B. That means M is between B and C and CM is negative. Since the metacentric height CM is negative, the cylinder cannot float upright.

For it to be 'critically stable' in upright position, C and M should coincide which, means the bottom portion 'b' should be such that

$$189.4 \times 0.312 = \left(\frac{800 - b}{2} + b\right)(800 - b) \times 600g + \frac{b}{2} \times b \times 6000\, g.$$

Example 2.64. *A cylindrical vessel of 1 m diameter has water in it to a height of 2 m. Oil of relative density 0.75 is kept over the water column for another 1 m. Above the oil, a dead weight of 5 kN, base diameter 1 m is placed. Estimate the total hydrostatic force and the 'centre' of pressure on a gate 30 cm in diameter placed along the vertical surface of the vessel such that the lowest point of the gate is on the base of the vessel.*

Solution.

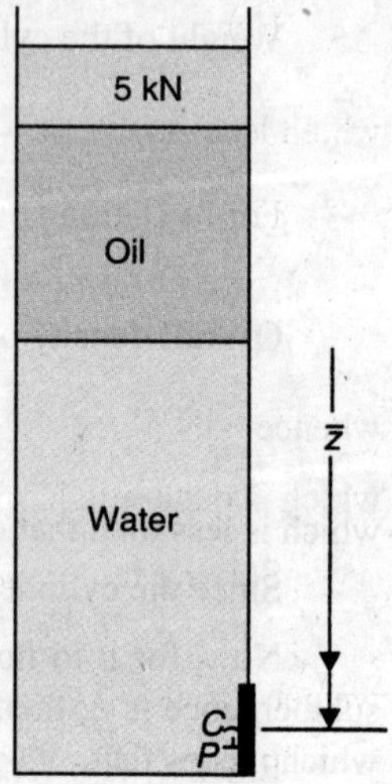

Let us solve this problem by considering an equivalent column of water H for all three things, *i.e.*, water, oil and the dead weight.

$$H = 2 + 1 \times 0.75 + \frac{5000}{\pi \times 1^2/4}./(1000 \times 9.81)$$

$$= 3.65 \text{ m of water}$$

Using this equivalent head of water, the depth of the centroid C of the gate is

$$\bar{z} = 3.65 - \frac{0.30}{2} = 3.5 \text{ m}$$

The total force on the gate must be

$$F = \rho\, g\, \bar{z} \times \pi d^2/4$$

$$= 1000 \times 9.81 \times 3.5 \times \pi \times 0.3^2/4$$

$$= \mathbf{2427\ N.}$$

We know that the equivalent depth of the centre of pressure P must be given by

$$z_p = \bar{z} + \frac{I_C}{A\bar{z}} \quad \text{for a vertical surface}$$

Hence, $\quad z_p = 3.5 + \dfrac{\pi d^4/64}{\pi d^2/4 \times 3.5} = 3.5 + \dfrac{d^2}{56}$

$$= 3.5 + (0.3)^2/56 = \mathbf{3.502\ m.}$$

which is 0.148 m from the base of the vessel.

Example 2.65. *A solid uniform cylinder of length 150 mm and diameter 75 mm is to float upright in water. Between what limits should its mass be?*

Solution.

Let the submergence in water be '*h*' for stable equilibrium in vertical position. Volume submerged

$$V = \frac{\pi d^2}{4} \times h$$

Weight of the water displaced must equal the weight of the cylinder with mass m.

$$mg = \frac{\pi d^2}{4} \times h \times 1000 \times g$$

whence $m/h = 4.42$

Now, $I = \frac{\pi d^4}{64}$,

$$\text{BM} = \frac{I}{V} = \frac{\pi d^4}{64} \cdot \frac{4}{\pi d^2 \times h} = \frac{d^2}{16h}$$

$$\text{CM} = \text{BM} - \text{BC} = \frac{d^2}{16h} - \text{BC}.$$

Depth of centroid of cylinder $= \frac{0.150}{2} = 0.075$ m

Depth of centre of buoyancy $= 0.15 - h/2$

Hence, $\text{BC} = 0.15 - h/2 - 0.075 = 0.075 - h/2$

which leads to $\text{CM} = \frac{d^2}{16h} - 0.075 + \frac{h}{2}$

For floatation in the vertical position CM must be +ve; at least zero.

$$\frac{d^2}{16h} - 0.075 + \frac{h}{2} = 0$$

whence $8h^2 - 1.2h + d^2 = 0$

which is quadratic in *h*.

Solving it, $h = \left(1.2 \pm \sqrt{1.44 - 32d^2}\right) / 16$

we get $d = 0.075$, $h = 0.145$ m or 0.0049 m.

Finally, $m = 4.42$ h, *i.e.*, **0.64 kg and 0.022 kg** are the limits for the mass.

Example 2.66. *A block of a material of specific gravity 1.6 floats at the interface of water and an 'unknown fluid' as shown. Find the density of the 'unknown fluid'.*

Solution.

Weight of the block $= (0.3 + 0.6) \times d \times b \times 1.6 \times 1000$

$= 1440\, d \times b$ N

where *b* is the dimension perpendicular to the plane of the paper.

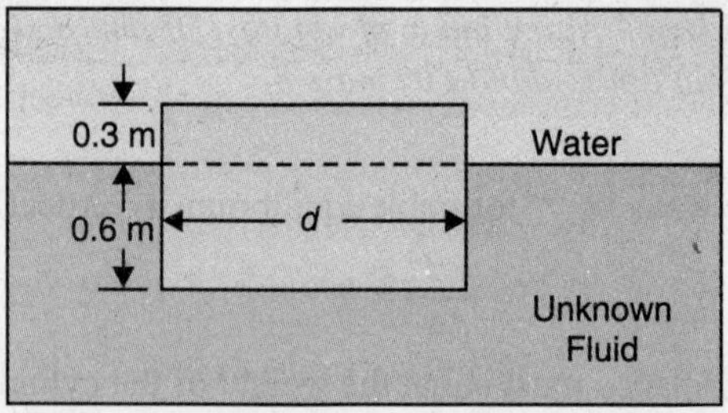

By Archimedes principle, the weight of the fluids displaced must be equal to the weight of the block. Hence,

$$0.3 \times d \times b \times 1000 + 0.6 \times d \times b \times \rho = 1440\, d \times b.$$

Cancelling $(d \times b)$ throughout

$$300 + 0.6\,\rho = 1440$$

and
$$\rho = (1440 - 300)/0.6$$
$$= \mathbf{1900\ kg/m^3}.$$

The specific gravity of the unknown fluid must be **1.9**.

Example 2.67. *Find the density of a metallic body which floats at the interface of mercury of sp. gravity 13.6 and water such that 40% of its volume is submerged in mercury and 60% in water.*

Solution.

$$\rho_{mercury} = 13600 \text{ kg/m}^3.$$
$$\rho_{water} = 1000 \text{ kg/m}^3.$$

Let the density of the metallic substance be ρ and its volume be V.

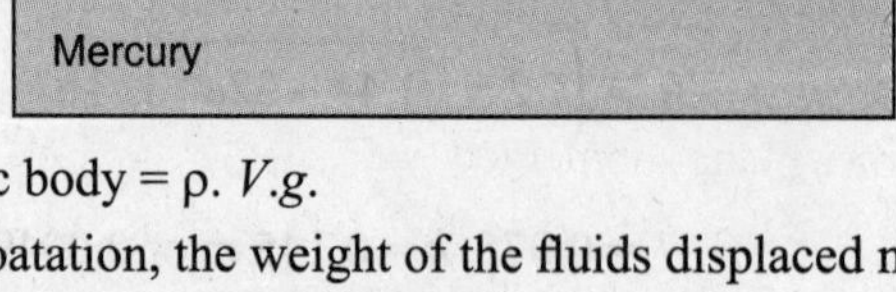

Weight of the metallic body = $\rho.\ V.g$.

By the principle of floatation, the weight of the fluids displaced must be equal to its weight.

Weight of the water displaced $= 0.6\, V \times 1000 \times g$
$= 600\, Vg$.

Weight of the mercury displaced $= 0.4\, V \times 13600 \times g$
$= 5440\, Vg$.

Hence, $600\, Vg + 5440\, Vg = \rho\, V g$.

Cancelling Vg throughout,

$$\rho = \mathbf{6040\ kg/m^3}.$$

This must be the density of the metallic body; its specific gravity is **6.04**.

ITEM BANK

Objective items of different types are given to review the concepts developed in the chapter.

* *Select the correct response for each item.*
* *Every correct response is expected in less than a minute and carries one mark.*
* *Compare your responses with the key answers given at the end.*

Multiple Choice Items

Attempt the following Multiple-Choice Items by ticking the correct or the most appropriate response from among the alternatives given:

1. The pressure at a point in a fluid is *not* equal in all directions if
(*a*) the fluid is at rest
(*b*) there are shear stresses
(*c*) the fluid is accelerated
(*d*) the fluid is rotated at a constant speed.

2. In the International Standard Atmosphere,
(*a*) the pressure stays constant with the altitude
(*b*) the lapse rate varies linearly with the altitude
(*c*) the rate of drop of pressure is over five times the rate of temperature.
(*d*) the temperature remains constant with the altitude

3. The centre of pressure of an inclined plane submerged surface
(*a*) should always coincide with the centroid of the surface
(*b*) should coincide with the centroid of the pressure prism
(*c*) may be above or below the centroid
(*d*) should be below the centroid of the surface

4. The hydrostatic force on a plane submerged surface
(*a*) remains the same no matter how the surface is turned as long as the centroid is unaltered
(*b*) acts vertically down no matter how the body is inclined
(*c*) changes with the shape of the area even if the centroid is unaltered
(*d*) depends upon the weight of the fluid displaced

5. The vertical component of the hydrostatic force on a submerged curved surface is
(*a*) the force on a vertical projection of the surface
(*b*) the product of pressure at centroid and the surface area
(*c*) the mass of the liquid vertically above it
(*d*) the weight of the liquid vertically above it

6. A U-tube differential manometer
(*a*) is used upright if the pressure difference is small
(*b*) is used inverted if the pressure difference is small
(*c*) cannot be inclined at any angle
(*d*) must be fitted with a well for the sake of accuracy

7. A solid body sinks in a fluid when
(*a*) the specific gravity of its material is greater than unity
(*b*) the buoyancy force does not pass through the metacentre
(*c*) the weight of the fluid displaced is less than the weight of the body
(*d*) the metacentre lies below the centre of gravity

8. A liquid mass undergoes a rigid-body type of motion when subjected to
(*a*) a constant rotational acceleration
(*b*) a constant velocity and a constant rotational speed
(*c*) a constant rotational speed
(*d*) variable linear velocity

9. The hydrostatic law
   (*a*) holds unaltered for a liquid in a vessel subjected to constant acceleration
   (*b*) holds unaltered for a liquid in a vessel subjected to constant rotation
   (*c*) does not hold for the atmosphere
   (*d*) holds for a liquid in equilibrium only

10. The buoyant force acting on a floating body passes through the
   (*a*) metacentre of the body
   (*b*) centre of gravity of the body
   (*c*) centroid of volume of the body
   (*d*) centroid of the displaced volume

11. An inverted U-tube manometer is more sensitive than an upright manometer because
   (*a*) the manometric fluid is lighter than the working liquids
   (*b*) the height of the levels is greater
   (*c*) the manometric fluid is heavier than the working liquids
   (*d*) it has no well on either side of it.

12. For stability of a floating body,
   (*a*) the metacentre M should lie between the centre of gravity C and the centre of buoyancy B.
   (*b*) M should lie above B and C
   (*c*) M should lie below B and C
   (*d*) M should coincide with B and C.

13. A cube of ice floating in a glass of water melts. After melting, the water level
   (*a*) rises
   (*b*) remains constant
   (*c*) falls
   (*d*) initially falls and then rises.

14. Piezometric head of a fluid is defined as
   (*a*) The sum of absolute pressure head and datum head.
   (*b*) the stagnation pressure head
   (*c*) the sum of stagnation pressure head and datum head
   (*d*) the sum of gauge pressure head and datum head.

Other Objective Items

15. Are the following statements true or false?
   (*a*) The centre of pressure coincides with the centroid for a horizontal surface. T/F
   (*b*) A body must be unstable if the metacentre is below the centre of gravity. T/F
   (*c*) The Pascal's law is valid for inviscid fluid flow. T/F
   (*d*) A U-tube manometer measures difference in pressure. T/F

16. Assertion A: Pressure is equal in all directions at a point in an ideal fluid flow.
   Reason R: Pascal's law is valid for all cases where shear stresses are zero.
   Which of the following options is right for the above assertion and reason?
   (*a*) A is true; R is true and it explains A.
   (*b*) A is true; R is true but it does not explain A.
   (*c*) A is true; R is false.
   (*d*) A is false; R is true.

17. Match the descriptions in block A with the consequences in block B.

Block A	**Block B**
(*a*) Fluids at rest	(*k*) no shear stresses
(*b*) Liquids at rest	(*l*) no normal stresses
(*c*) Fluids in solid body rotation	(*m*) constant shear stresses
(*d*) Laminar flow	

ANSWERS

1. (*b*), **2.** (*c*), **3.** (*d*), **4.** (*a*), **5.** (*d*), **6.** (*b*), **7.** (*c*), **8.** (*c*), **9.** (*b*), **10.** (*d*), **11.** (*a*), **12.** (*b*), **13.** (*b*), **14.** (*d*), **15.** (T, T, T, F), **16.** (*a*), **17.** (ak, bk, ck, dm)

QUESTION BANK

Questions on theory and numerical problems are given for tutorial work.

- * *Questions are arranged generally in order of the articles in the text.*
- * *Every question should take 15-20 minutes to solve..*
- * *Answers for the numerical problems are given alongside them.*

2.1. A hydraulic jack consists of a connected mass of oil with two pistons. The smaller piston 15 cm^2 in area is to take the force applied whereas the large piston 150 cm^2 in area carries the load. Assuming that a person can apply a force of 300 N, determine the maximum load that the jack can lift if

(*a*) the pistons are at the same level.

(*b*) the large piston is 1 m below the smaller, the liquid in the jack being oil with mass density 950 kg/m^3. **[306 kg, 320 kg]**

2.2. Starting with the hydrostatic law $dp = \rho g\, dz$, show that the pressure p at a point with respect to a reference pressure p_0 in constant temperature fluid is $\log_e (p/p_0)$ times the pressure in an incompressible fluid.

2.3. A gate ABC with a hinge at C, 2 m wide holds a 3 m high stationary column of water as shown in Fig. Problem 2.3. Calculate the tension in the string SA and the reaction at the hinge. **[17.66 kN, 70.63 kN]**

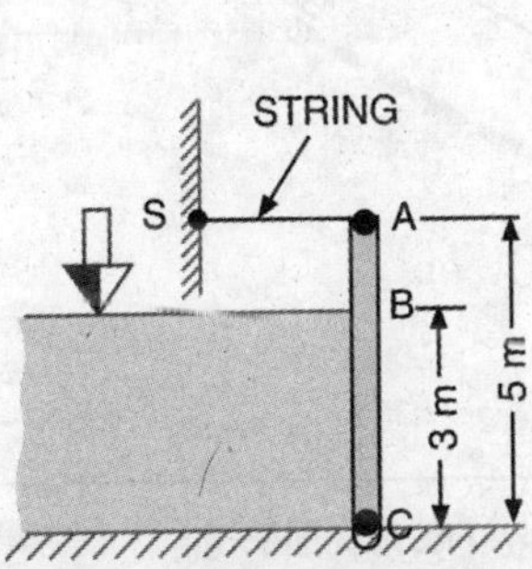

Fig. Problem 2.3

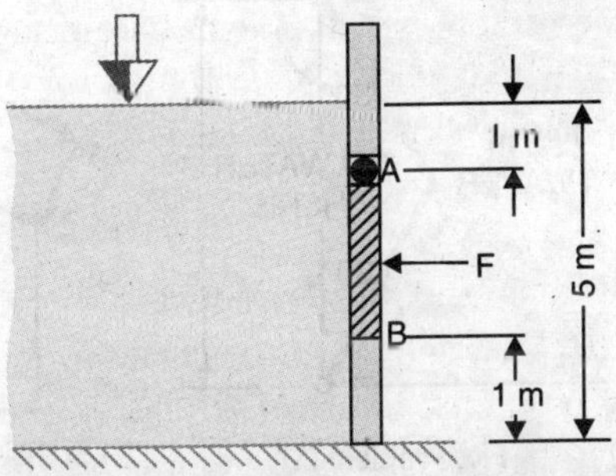

Fig. Problem 2.4

2.4. A 3 m diameter gate AB closes the side opening of a water tank as shown in Fig. Problem 2.4. Determine the magnitude and location of force F to hold the gate without causing a reaction at the hinge at A. **[173 kN, 2.72 m]**

2.5. Determine the height of liquid of specific gravity 0.8, to tip the rectangular flash board shown in Fig. Problem 2.5. Take the breadth of the board as 1 m. **[2.6 m]**

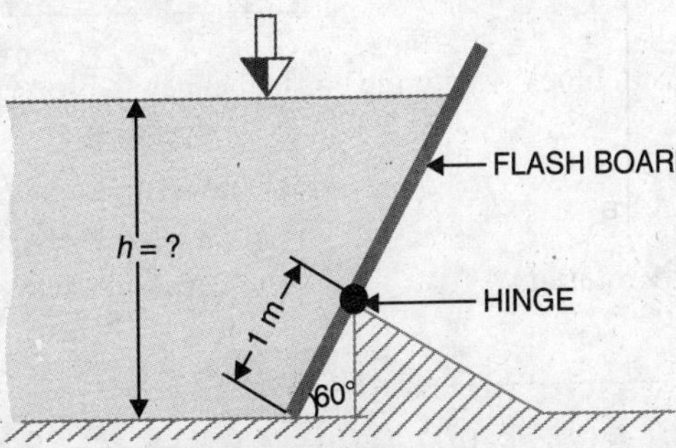

Fig. Problem 2.5

2.6. Determine the total force and its location on one face of the plates, shown in Fig. Problem 2.6, immersed in a liquid of specific gravity 0.9 ? **[132.44, 69.34 kN ; 3.16, 3.1 m]**

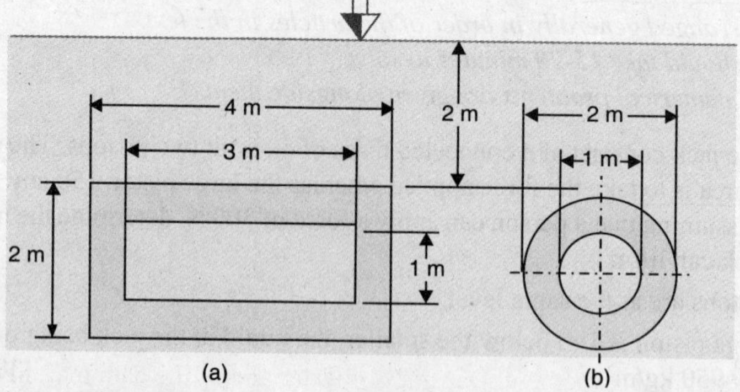

Fig. Problem 2.6

2.7. A pair of entrance lock-gates 6 m high is used to contain 5 m head of water on the upstream side, the downstream side being empty. If the gates are pivoted at the top and bottom, calculate the reactions of the pivots. **[206, 530 kN]**

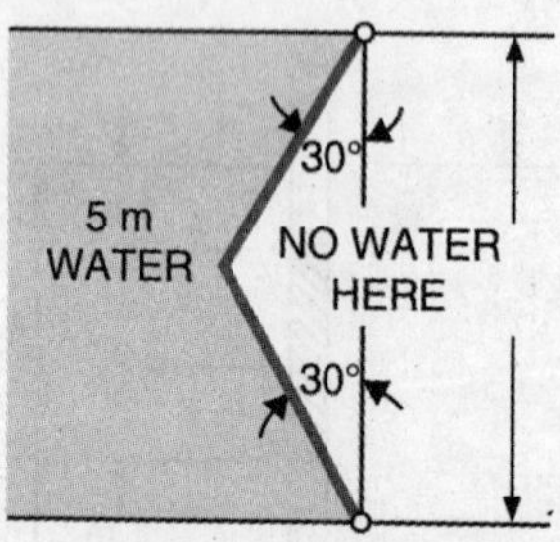

Fig. Problem 2.7

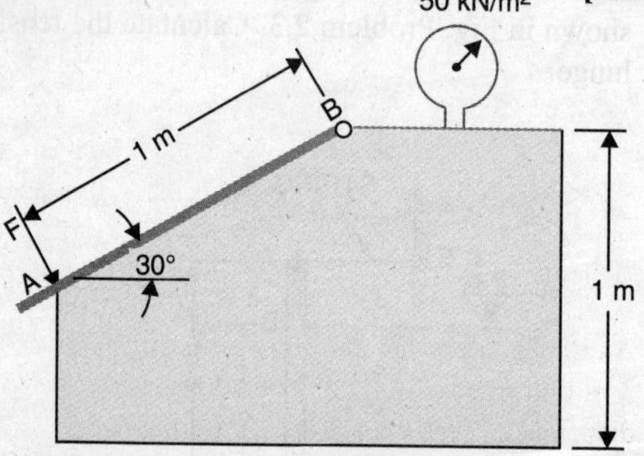

Fig. Problem 2.8

2.8. A 1 × 1 m cover AB on a container under pressure as shown in Fig. Problem 2.8 is held in position by a force F. Calculate the force and the reaction at the hinge B. **[26.24 kN]**

2.9. Calculate the magnitude of the horizontal components of the force experienced by each of the curved surfaces AB as shown in the figures. Calculate also the magnitude of the vertical components. Calculate also the magnitude of the vertical components of the force. The liquid is water. Consider unit width of the surface.

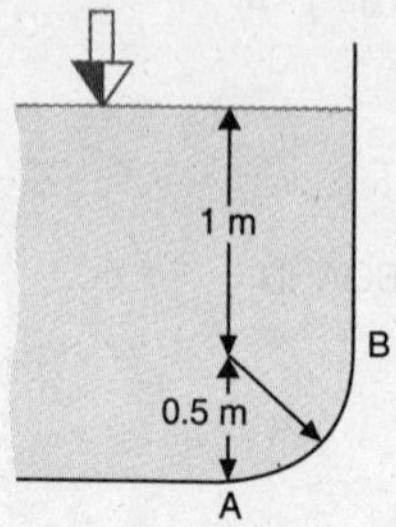

Fig. Problem 2.9 (a)

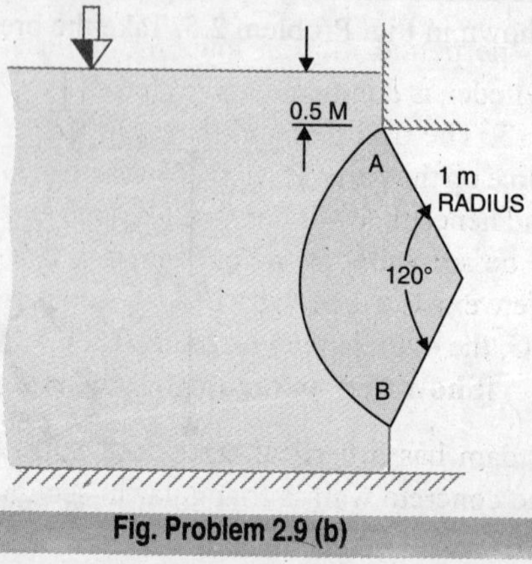

Fig. Problem 2.9 (b)

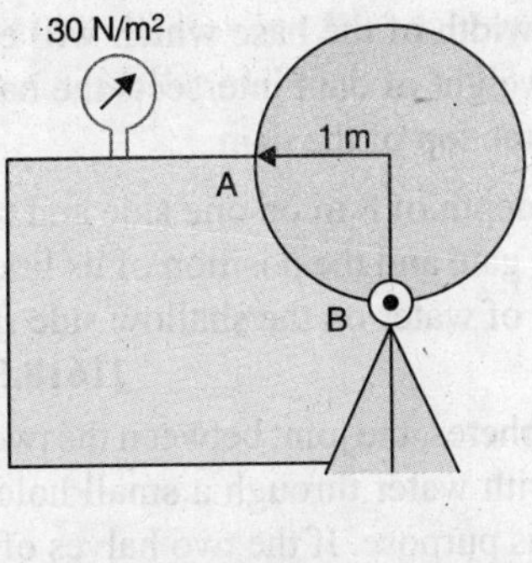

Fig. Problem 2.9 (c)

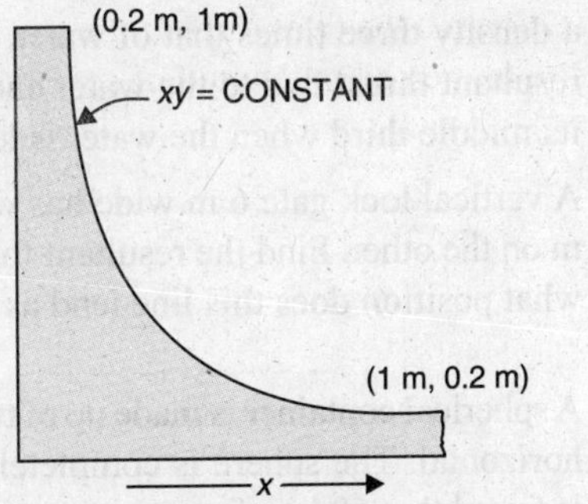

Fig. Problem 2.9 (d)

2.10. Specify the minimum B/H ratio at which the dam tipping moment relative to point O would be 50% of the holding moment. Densities of water and of the dam material are 1000 kg/m³ and 2250 kg/m³ respectively. **[B/H = 2/3]**

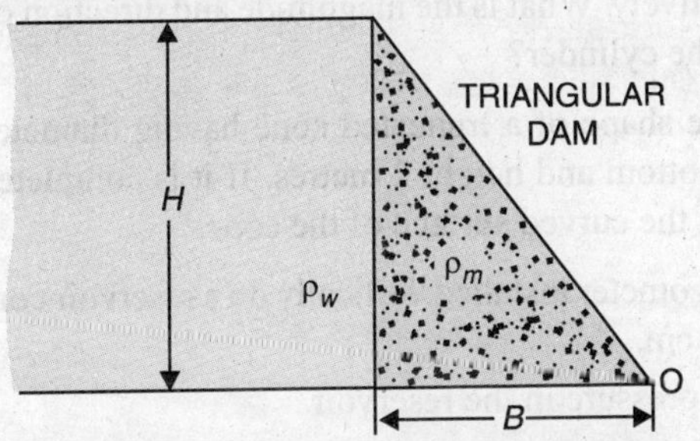

Fig. Problem 2.10

2.11. A circular opening, 3 m diameter, in the vertical side of water tank is closed by a disc of 3 m diameter which can rotate about a horizontal diameter. Calculate (i) the force on the disc, and (ii) the torque required to maintain the disc in equilibrium in the vertical position when the head of water above the horizontal diameter is 4 m. **[(i) 277.4 kN (ii) 38.8 kNm]**

2.12. A spherical vessel is filled completely with water of weight W. Show that the resultant fluid pressure on each of the halves into which it is divided by a vertical diametrical plane is $\frac{W\sqrt{13}}{4}$. If the diametrical plane is horizontal, show that the resultant fluid pressure on one half is 5 times that on the other.

[(a) Horizontal force 3/4 W, Vertical force 1/2 W (b) Vertical forces are 1/4 W and 5/4 W].

2.13. A 'perpetual motion machine' consisting of a cylinder is submerged in water as shown in Fig. 2.13. The designer surmises that the hydrostatic force on the part AB is greater than that on part BC and hence the cylinder should rotate and continue to do so. Show, by actual anaylsis, that although there exists a difference of forces on parts AB and BC; the cylinder cannot rotate.

[Show that no turning moment is possible]

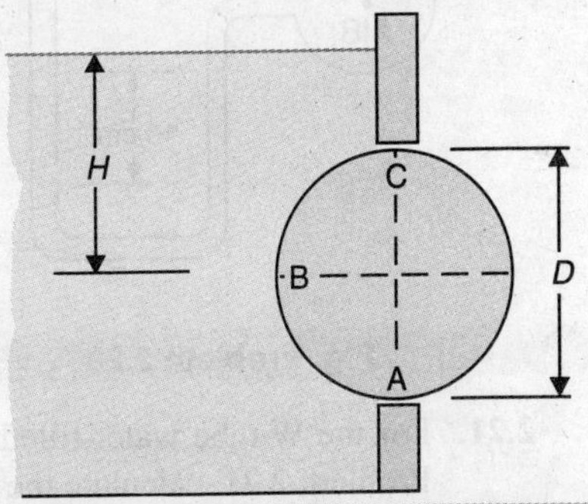

Fig. Problem 2.13

2.14. A dam has a vertical inner face 80 m high and the concrete wall is triangular in section and has a density three times that of water. Find the width of the base which will ensure that the resultant thrust due to the water and due to weight of dam intersects the base line within its middle third when the water is level with the top of the dam. **[46.2 m]**

2.15. A vertical lock gate 6 m wide has water to a depth of 8 m on one side and to a depth of 3 m on the other. Find the resultant force on the gate and the position of its line of action. To what position does this line tend as the depth of water on the shallow side rises to 8 m. **[1618.5 kN, 2.94 m]**

2.16. A spherical container is made up of two hemispheres, the joint between the two halves being horizontal. The sphere is completely filled with water through a small hole in the top. It is found that 50 kg of water is required for this purpose. If the two halves of the container are not secured together, what must be the mass of the upper hemisphere if it just fails to lift off the lower hemisphere. **[12.5 kg]**

2.17. A cylindrical drum 1.8 m diameter and 3 m wide is placed across a stream of the same width blocking it completely. The depth of water on the upstream and downstream sides are 1.8 and 0.9 m respectively. What is the magnitude and direction of the resultant pressure force of the water on the cylinder? **[66.58 kN 57.5°]**

2.18. A vessel is in the shape of a truncated cone having diameter of 1 metre at the top and 2 metres at the bottom and height 2 metres. If it is completely filled with water, find the upward thrust on the curved surface of the cone. **[25.6 kN]**

2.19. A single tube piezometer installed vertically on a reservoir containing water registers a rise in its level by 50 cm.

(*a*) Estimate the pressure in the reservoir.

(*b*) What would the piezometer reading be if the reservoir contained an oil of sp. gravity 0.8?

(*c*) How would you fit a water-filled manometer to a reservoir which contains air at the same pressure as in the original case and what would the manometer reading be ?

[4.905 kN/m² gauge; 62.5 cm; U-tube 50 cm]

2.20. Estimate the vacuum at A in an air pipe shown if a water manometer with the other end open to atmosphere reads a differential head of 10 cm. **[981 N/m² vacuum]**

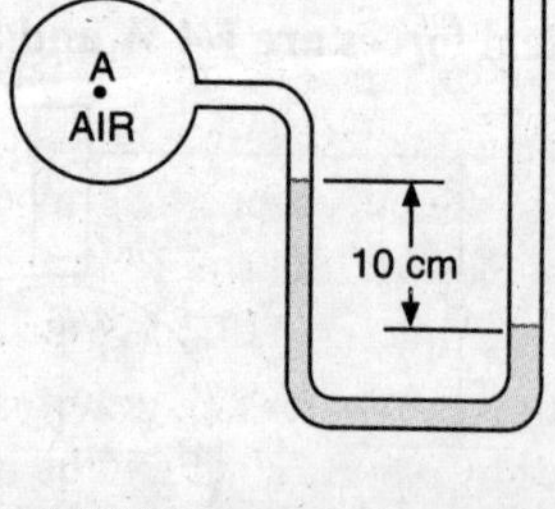

Fig. Problem 2.20

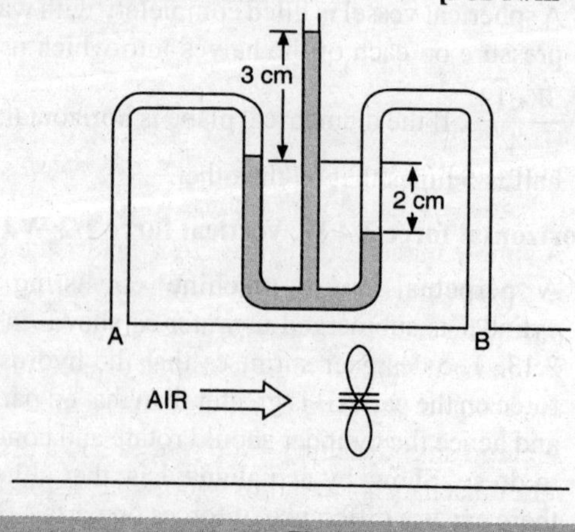

Fig. Problem 2.21

2.21. For the W-tube water-filled manometer configuration and the readings as shown in Fig. Problem 2.21, calculate the absolute pressures at A and B. Take the atmospheric pressure as 101.3 kN/m². **[101.594, 101.791 kN/m²]**

2.22. For the U-tube manometer connected as shown in Fig. 2.22, calculate h, the difference of mercury column in the two limbs. **[0.095 m]**

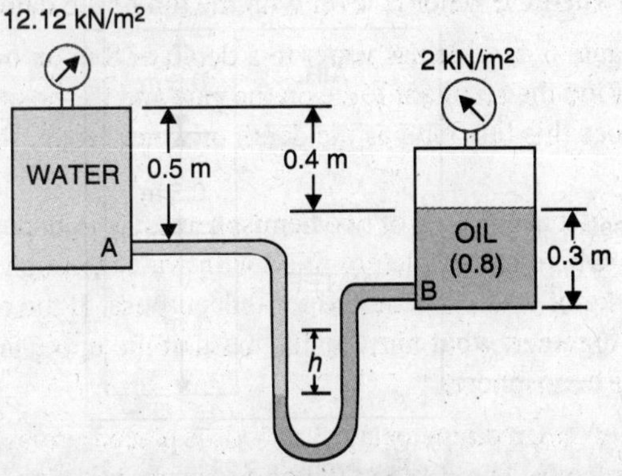

Fig. Problem 2.22

2.23. Two U-tube manometers, one upright and the other inverted type, are connected across a water line and on oil line as shown in Fig. 2.23. If h_1 = 5 cm, what shall h_2 be? **[0.197 cm]**

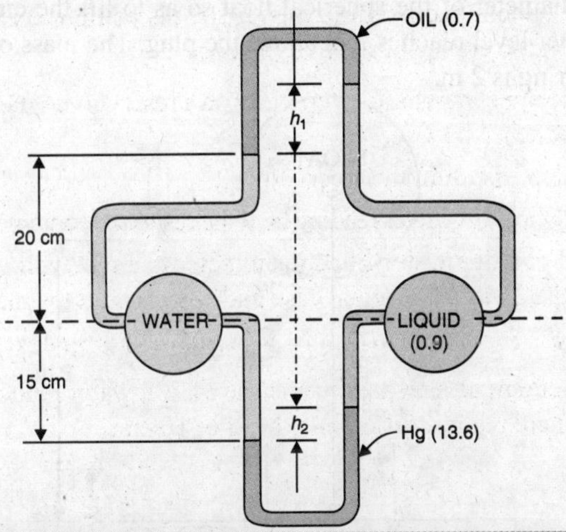

Fig. Problem 2.23

2.24. Show that in a micromanometer, the accuracy of recording depends on the closeness of densities of manometer liquids.

2.25. A simple manometer is used to measure the pressure of oil (specific gravity 0.8) flowing in a pipe line. It's right limb is open to the atmosphere and the left limb is connected to the pipe. The centre of the pipe is 90 cm below the level of mercury (specific gravity = 13.6) in the right limb. If the difference of mercury level in the two limbs is 15 cm, determine the absolute pressure of oil in the pipe in kN/m^2. **[127.2 kN/m^2 absolute]**

2.26. A cargo ship weighing 4000 tonnes has a draft of 7 m is sea water (sp. gravity = 1.035). After discharging cargo of 510 tonnes its draft reduces by 0.5 m. What will be its draft in a fresh water harbour after further discharging a cargo of 300 tonnes (assume no change in crosssectional area for depth under consideration). **[6.423 m]**

2.27. A rectangular pontoon floating in sea water is 21 m long, 7 m wide, 2.4 m deep and weighs 750 tonnes. It carries on its upper deck a normal vertical load of 500 tonnes. The centre of gravity of the load is 2.4 m above the deck and that of the pontoon is 1.2 m below the deck. Find the metacentric height. **[2.94 m]**

2.28. Determine the reading on the pressure gauge for the tank in Fig. prob. 2.28. Calculate also the pressure of the air entrapped. **[9810 N/m², 4905 N/m²]**

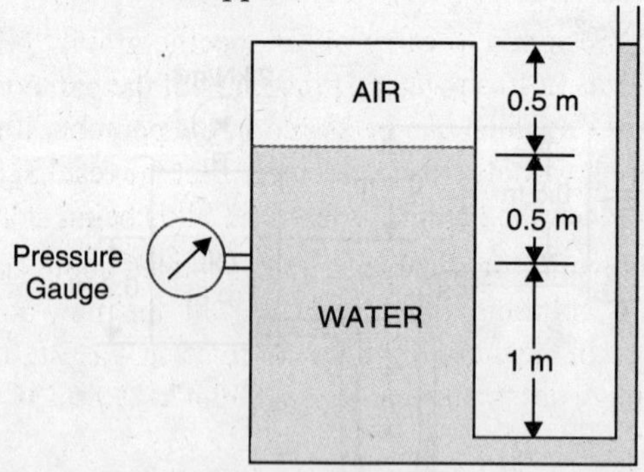

Fig. Problem 2.28

2.29. A right circular solid cone with apex angle 60° of relative density S w.r.t. the liquid in which it floats with its apex downwards should float stable. Show that the stability is ensured for $0.42 < S < 1$.

2.30. Calculate the diameter of the spherical float so as to lift the circular plug valve open as soon as the water level reaches 2 m above the plug. The mass of the plug is 5 kg and the length of the string is 2 m. **[0.43 m]**

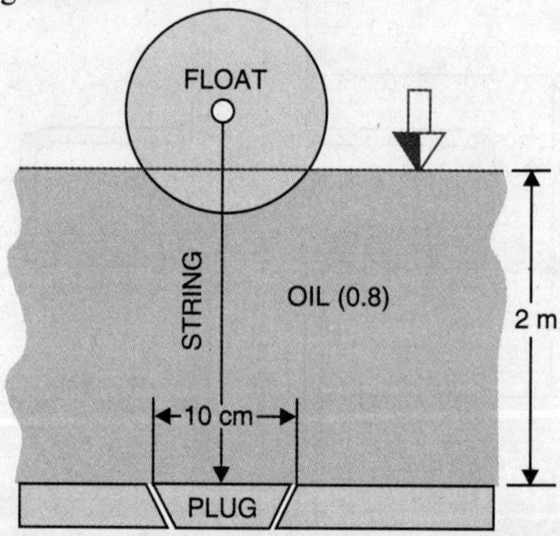

Fig. Problem 2.30

2.31. Calculate the minimum force F required to keep the timber floating horizontal. The timber cross-section is 10 cm × 10 cm and the specific gravity 0.6. **[73.6 N]**

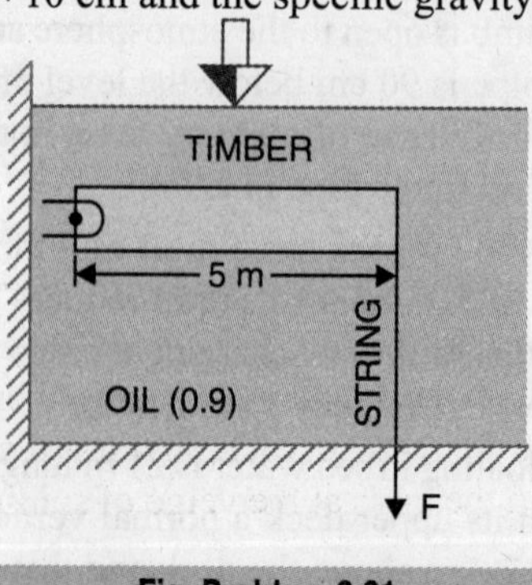

Fig. Problem 2.31

2.32. A solid cone of mass density 705 kg/m^3 floats in a liquid of density 950 kg/m^3. Determine the minimum apex angle if the cone is to float stable upside down with its axis vertical. **[61.05°]**

2.33. A rectangular pontoon, designed to carry oil of specific gravity 0.9 is 50 m long, 10 m broad, 5 m deep, and floats in fresh water. If the mass of the pontoon is 80 × 10^3 kg, what is the greatest volume of oil which can be stored in the pontoon, if the deck must be 1 m above water when floating with its axis horizontal? Plot the resultant pressure distribution on one side of the tank when the greatest volume of oil is being stored. **[2135 m^3]**

2.34. A W-tube system as shown in Fig. Problem 2.34 is rotated about an axis AA at 10 r.p.m. Determine the levels in columns C_1, C_2 and C_3 in the new position of equilibrium. Determine also the levels in the tubes if the axis of rotation was BB instead of AA. Assume the tubes to be long enough not to allow spillage. **[30.47, 29.06, 30.47 ; 29.07, 31.86 cm]**

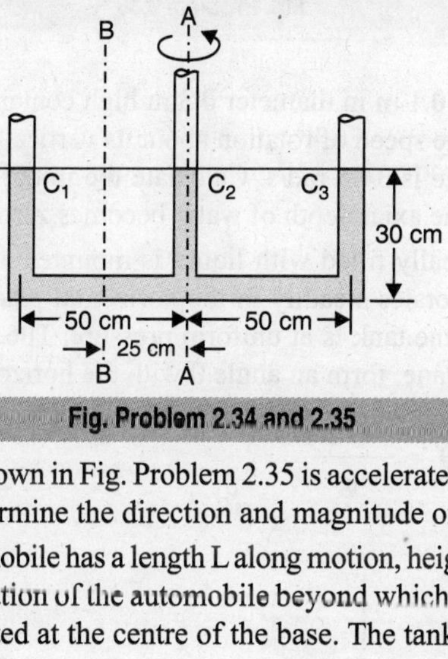

Fig. Problem 2.34 and 2.35

2.35. A W-tube system as shown in Fig. Problem 2.35 is accelerated rectilinearly such that column C_3 drops to zero. Determine the direction and magnitude of the acceleration. **[5.89 m/s^2]**

2.36. A fuel tank of an automobile has a length L along motion, height L/2 and width *W*. Determine the maximum acceleration of the automobile beyond which the fuel ceases to pass through the delivery hole located at the centre of the base. The tank at the instant is one-third full **[14.715 m/s^2]**

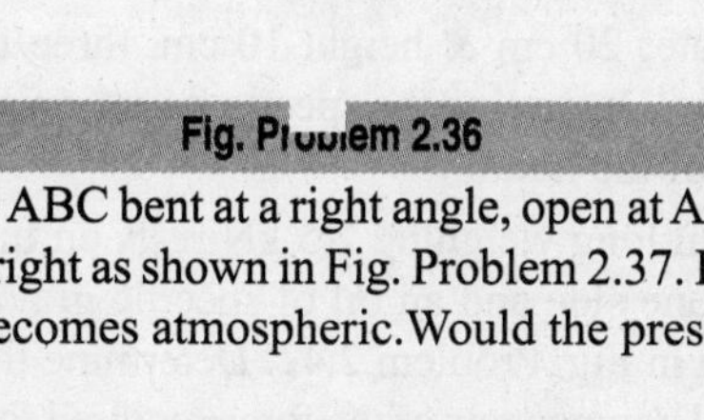

Fig. Problem 2.36

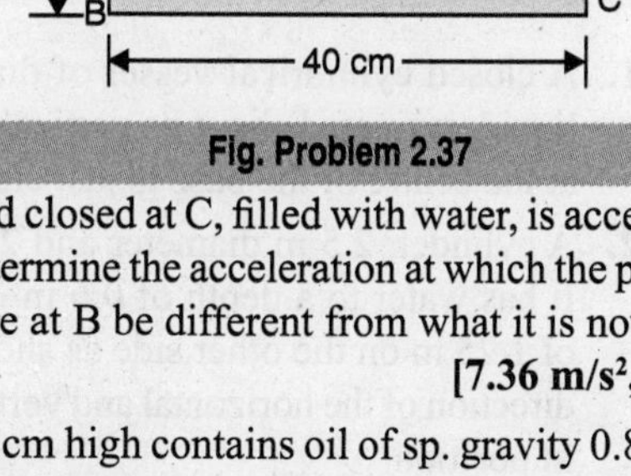

Fig. Problem 2.37

2.37. A tube ABC bent at a right angle, open at A and closed at C, filled with water, is accelerated to the right as shown in Fig. Problem 2.37. Determine the acceleration at which the pressure at C becomes atmospheric. Would the pressure at B be different from what it is now? **[7.36 m/s^2, same]**

2.38. A cylindrical vessel 20 cm in diameter and 30 cm high contains oil of sp. gravity 0.8 upto a height of 20 cm. Calculate the constant rotational speed ω about the vertical axis as shown in Fig. Problem 2.38 when the oil is at the verge of spilling over the brim. **[9.90 rad/s]**

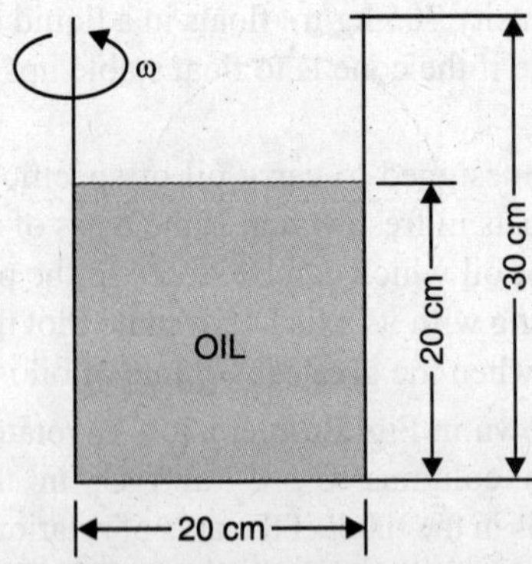

Fig. Problem 2.38

2.39. A cylindrical vessel 0.1 m in diameter 0.3 m high contains water upto a certain height. It was observed that the speed of rotation about its vertical axis when the water commences to spill over the edge is 34.5 rad/s. Calculate the initial depth of water in the vessel. At what speed would the axial depth of water becomes zero? **[22.4 cm, 48.5 rad/s]**

2.40. A small tank practically filled with liquid is mounted at the end of a long rotating arm of radius R which rotates steadily in the horizontal plane at an angular velocity ω. The remaining space in the tank is at uniform pressure. The free surface of the liquid, which may be taken as a plane, form an angle θ with the horizontal. Show that θ is given by

$$\tan\theta = -\frac{\omega^2 R}{g}$$

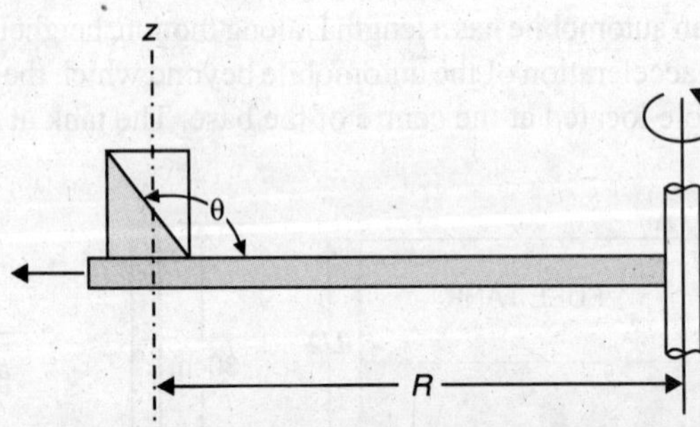

Fig. Problem 2.40

2.41. A closed cylindrical vessel of diameter 20 cm & height 10 cm, three fourth full with a liquid, is rotated about its vertical axis. Determine the speed of rotation at which the point at the centre of the base is just clear of the liquid. **[189.2 rpm]**

2.42. A cylinder, 2.5 m diameter and 2.0 m long weighing 2.5 kN rests on the floor of a tank. It has water to a depth of 0.6 m on one side and an oil of specific gravity 0.8 to a depth of 1.25 m on the other side as shown in Fig. Problem 2.42. Determine the magnitude and direction of the horizontal and vertical components of the force required to hold the cylinder in position. **[8.73 kN towards right, 25.65 kN upwards]**

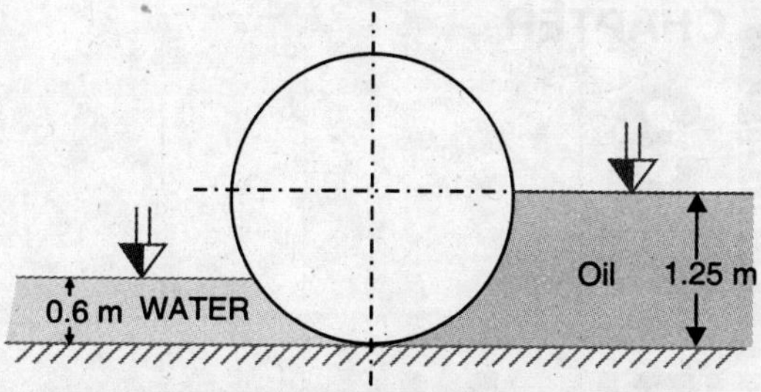

Fig. Problem 2.42

2.43. The hemispherical dome as shown in Fig. Problem 2.43, weighs 20 kg and is filled with water and attached to the floor by 6 equally spaced bolts. If the permissible force in each bolt is 57.7 kN find the required height of water in the neck. **[2.14 m]**

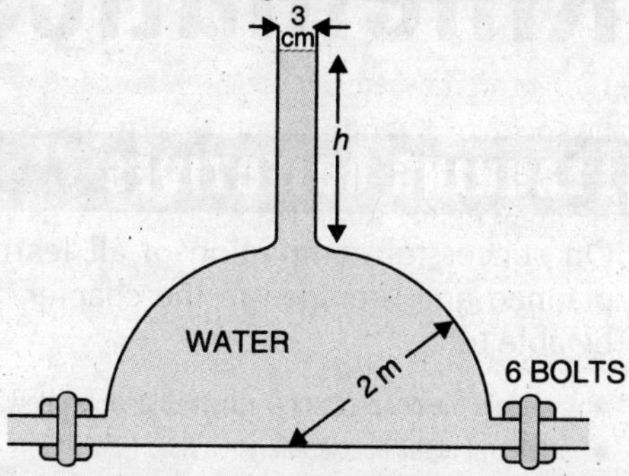

Fig. Problem 2.43

CHAPTER

3

Fluid Kinematics

LEARNING OBJECTIVES

On successful completion of all learning activities planned and provided in the chapter, a student will be able to

- explain the concepts of steady and uniform flow;
- determine the acceleration in a fluid flow;
- differentiate between streamlines, pathlines and streaklines;
- derive the continuity equation for an incompressible and a compressible flow;
- describe a real flow in terms of a stream function and
- explain the kinematics of a fluid element and establish the condition for irrotationality.

3.1. Description of Fluid Flow

Fluid kinematics refers to geometric and kinematic nature of a fluid in motion. Considerations of velocity, acceleration, rotation, irrotation, nature of flow and flow visualisation come under fluid kinematics.

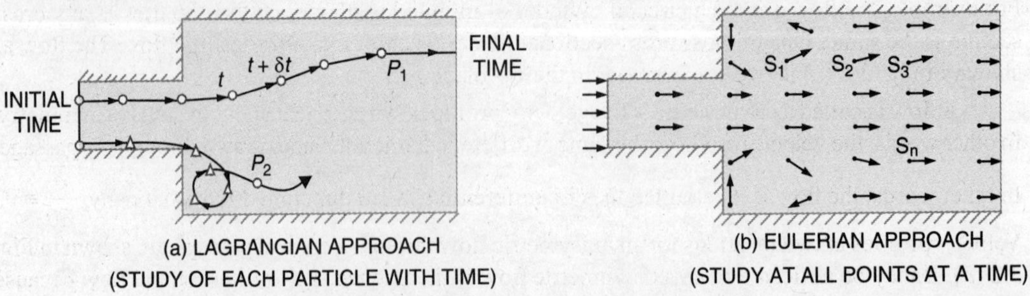

Fig. 3.1. Lagrangian and Eulerian Approaches

The terms 'velocity' and 'acceleration' apply 'at a point in a fluid'. In the rectangular coordinates system, the velocity components along the x, y and z directions are denoted by u, v and w respectively, and the velocity vector at a point is denoted by.

$$\boldsymbol{U} = ui + vj + wk$$

Similarly, the acceleration vector at a point is denoted by,

$$\boldsymbol{a} = a_x i + a_y j + a_z k$$

where a_x, a_y and a_z are the rectangular components of acceleration.

A fluid motion may be analysed by following one of the two alternative approaches; Lagrangian and Eulerian. In the Lagrangian approach, a particle or an element of fluid is identified and followed during the course of motion with time, as demonstrated in Fig. 3.1. The identified lump of fluid may change in shape, size and state as it moves. The laws of Mechanics must be applied to it at all times. Difficulty in tracing the lump of fluid rules out the possibility of applying the Lagrangian approach.

The alternative approach, called Eulerian approach, consists of observing the fluid by setting up fixed stations or points in the flow field. Motion of the fluid is specified by velocity components expressed as functions of space and time,

$$u = u\,(x, y, z, t)$$

$$v = v\,(x, y, z, t)$$

$$w = w\,(x, y, z, t)$$

In the Eulerian approach, the fluid motion at all points in the flow field is determined by applying the laws of Mechanics at all fixed stations. This is considerably easier than the Lagrangian approach and is followed in the study of Fluid Mechanics.

3.2. 'Dimensions' of Flow

A fluid flow is said to be one, two or three dimensional depending upon the number of independent space coordinates, *i.e.*, one, two or three respectively, required to describe the flow. When the dependent variables are functions of only one space coordinate, say x, it is one dimensional flow. The number of dependent variables does not matter. Examples of one dimensional flows are flow through pipes, channels and variable area ducts if the velocity distribution is considered constant at each cross section. Dependent variables such as velocity, pressure, density, temperature vary only with x, the only independent variable in one-dimentional flows as shown in Fig. 3.2*a*. The axis of the passage does not have to be a straight line for a flow to be one dimensional. For example, the flow shown in Fig. 3.2. (*b*) is one dimensional if s, the streamwise direction is chosen as the independent coordinate.

When the dependent variables in a fluid flow vary with only two space coordinates, the flow is said to be two dimensional. The flow does not vary along the third coordinate direction. What the assumption of two-dimensionality really amounts to, is that the flow is identical in parallel planes. For example, the flow around a circular cylinder of infinite length is such that the flow at any cross section is the same. One of those cross-sectional planes, is chosen to represent the flow. The flow as shown in Fig. 3.2 (*c*) is two-dimensional in the x-y plane.

A flow is called axisymmetric if the velocity profile is symmetrical about the axis of symmetry. In other words, the velocity profile is the same at different diametral planes drawn through the passage. In other words, the flow is invariant in the circumferential, *i.e.*, θ direction. Mathematically, $\dfrac{\partial}{\partial\theta} \equiv 0$. Velocity profiles at two locations for an axisymetric flow through a conical passage are shown in Fig. 3.2 (*d*). It may be noticed that an axisymmetric flow is, in essence, a two-dimensional flow because the only independent co-ordinates are x and y or r and x.

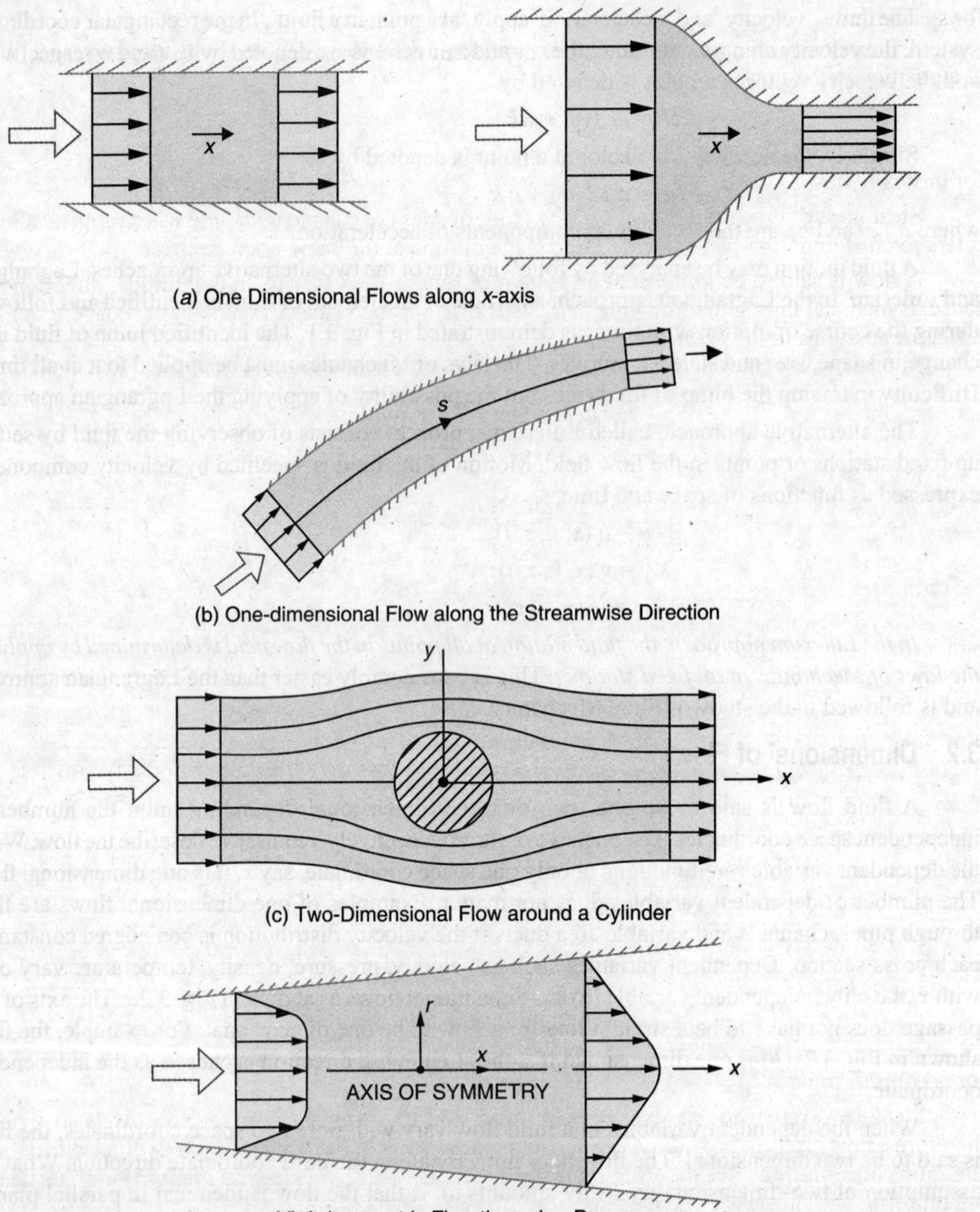

Fig. 3.2: One and Two Dimensional and Axisymmetric Flows

3.3. Steady Flow and Uniform Flow

A flow is considered to be *steady* if the dependent fluid variables at any point in the flow do not change with time. In other words, the time rate of the dependent variables at a position is zero.

$$\frac{\partial}{\partial t}\ \text{(dependent fluid variables)} = 0$$

It follows that

$$\frac{\partial u}{\partial t} = \frac{\partial v}{\partial t} = \frac{\partial w}{\partial t} = 0 = \frac{\partial p}{\partial t} = \frac{\partial \rho}{\partial t}, \text{ etc.} \qquad \text{... (3.1)}$$

for steady flow.

A flow is regarded as *unsteady* if the dependent fluid variables alter with the passage of time at a position in the flow,

$$\frac{\partial}{\partial t}\ (\quad) \neq 0$$

for unsteady flow.

Steadiness of flow implies that the flow-pattern does not change with time whereas unsteadiness refers to changing flow-pattern with the passage of time at the same point in space.

A flow is said to be *uniform* at an instant of time if the velocity, in magnitude, direction and sense, is identical throughout the flow field. This calls for the velocity components to be the same at different positions in the flow.

In other words, the space rate of change of velocity components at that time must vanish.

$$\frac{\partial}{\partial s}\ \text{(dependent fluid variables)} \equiv 0$$

It follows that

$$\frac{\partial u}{\partial x} = 0 = \frac{\partial u}{\partial y} = \frac{\partial u}{\partial z} = \frac{\partial v}{\partial x} = \frac{\partial v}{\partial y} = \frac{\partial w}{\partial y}, \text{ etc.} \qquad \text{... (3.2)}$$

for uniform flow.

It is conventional to define the term 'uniform flow' only in terms of the velocity components rather than in terms of all the dependent fluid variables.

Another relevant remark is that a flow may be considered uniform over the cross-sections although it may not be uniform longitudinally. For example, flow through a pipe with heat transfer is usually assumed to be uniform over the cross-sectional area although the velocity may vary longitudinally due to density variations. This would really imply a one-dimensional flow where

$$\frac{\partial u}{\partial r} = 0 = \frac{\partial u}{\partial \theta}$$

$$\frac{\partial u}{\partial z} \neq 0$$

If the velocity components at different locations are different at the same instant of time, the flow is said to be *non-uniform*.

$$\frac{\partial}{\partial s}\ (\,) \neq 0$$

for non-uniform flow.

Steadiness refers to 'no change with time' and uniformity refers to 'no change in space'; therefore a flow can be steady or unsteady quite independent of its being uniform or non-uniform. **All the four combinations are possible** i.e., 'steady and uniform', 'steady but non-uniform', unsteady but uniform' and 'unsteady and non-uniform' flows.

Consider, for example, the flow through the hose of a fire engine. When the pumping unit is started, the flow in the hose is unsteady; unsteady but uniform at the nozzle exit and unsteady and non-uniform between the hose and the nozzle. After the flow establishes, it is said to be steady; steady and uniform at the nozzle exit and steady but non-uniform between the hose and the nozzle.

It may also be noted that a flow can be steady or unsteady in one, two or three dimensional cases whereas uniformity of flow associated with invariance in space should ideally mean zero-dimensional flow.

3.4. Acceleration in Fluid Flow

Recognising the fact that the velocity components in a fluid flow are, in general, functions of space and time

$$u = u(x, y, z, t);\ v = v(x, y, z, t) \text{ and } w = w(x, y, z, t)$$

By differential calculus, an infinitesimal change in u is given by

$$\delta u = \frac{\partial u}{\partial x}\delta x + \frac{\partial u}{\partial y}\delta y + \frac{\partial u}{\partial z}\delta z + \frac{\partial u}{\partial t}\delta t$$

and the acceleration components are given by

$$a_x = \frac{Du}{Dt} = u\frac{\partial u}{\partial x} + v\frac{\partial u}{\partial y} + w\frac{\partial u}{\partial z} + \frac{\partial u}{\partial t} \quad \text{... (3.3 a)}$$

$$a_y = \frac{Dv}{Dt} = u\frac{\partial v}{\partial x} + v\frac{\partial v}{\partial y} + w\frac{\partial v}{\partial z} + \frac{\partial v}{\partial t} \quad \text{... (3.3 b)}$$

$$a_z = \frac{Dw}{Dt} = u\frac{\partial w}{\partial x} + v\frac{\partial w}{\partial y} + w\frac{\partial w}{\partial z} + \frac{\partial w}{\partial t} \quad \text{... (3.3 c)}$$

Total acceleration	=	***Convective acceleration***	***Local acceleration***

while left hand side terms of all equations are 'total acceleration', the first three right hand side terms are called 'convective acceleration' and the last term is said to be local acceleration for each of the three, i.e., x, y and z direction.

The total acceleration,

$$\frac{DU}{Dt} = (U.\nabla)\,U + \frac{\partial U}{\partial t} = \frac{Du}{Dt}i + \frac{Dv}{Dt}j + \frac{Dw}{Dt}k$$

$$= a_x i + a_y j + a_z k$$

represents the sum of convective acceleration, or the instantaneous space rate of change of velocity and the local acceleration or the local time rate of change of velocity. The two components may exist quite independent of each other. With reference to a differential control volume, the local acceleration refers to the net rate of change of velocity within the control volume whereas the convective acceleration is related to the net rate of effux of velocity across the control surface.

Expressions for the components of acceleration in cylindrical coordinates for an axisymmetric flow are given below:

$$a_r = u_r\frac{\partial u_r}{\partial r} + u_z\frac{\partial u_r}{\partial z} - \frac{u_\theta^2}{r} + \frac{\partial u_r}{\partial t}$$

$$a_\theta = u_r\frac{\partial u_\theta}{\partial r} + u_z\frac{\partial u_\theta}{\partial z} - \frac{u_r u_\theta}{r} + \frac{\partial u_\theta}{\partial t}$$

$$a_z = u_r\frac{\partial u_z}{\partial r} + u_z\frac{\partial u_z}{\partial_z} + \frac{\partial u_z}{\partial t} \quad \text{.... (3.4)}$$

- For steady flow,

$$\frac{\partial}{\partial t}\,(\,) \equiv 0$$

which makes only the local acceleration components zero.

For uniform flow,

$$\frac{\partial u}{\partial x} = 0 = \frac{\partial u_r}{\partial r}, \text{ etc.}$$

implying that only the convective acceleration components vanish. For steady uniform flow, the total acceleration also known as substantial, material or particle derivative becomes zero and the flow is non-accelerating in space and time.

Example 3.1. *A conical nozzle 2.0 m long converging from 0.2 m to 0.1 m diameter linearly is subjected to an inlet flow varying from 0 to 1 m/s in 5 seconds. Determine the total acceleration at the mid-length of the nozzle at the mid-time of variation. Assume uniform flow over each cross-section.*

Solution.

The inlet velocity is $U_1 = 0 + (1 - 0)/5t = 0.2\,t$ m/s

At the mid-length $U = 0.2\,t \times (0.2/0.15)^2 = 0.36\,t$ m/s

and $$\frac{\partial U}{\partial t} = 0.36 \text{ m/s}^2$$

At the mid-time of variation,

$$U = 0.36 \times 5/2 = 0.9 \text{ m/s}$$

Diameter at a distance x from inlet is

$$d = 0.2 + (0.1 - 0.2)/2.0x = 0.2 - 0.05x$$

Cross sectional area at that section is

$$a = \frac{\pi d^2}{4} = \frac{\pi}{4}(0.04 + 0.0025\,x^2 - 0.02x)$$

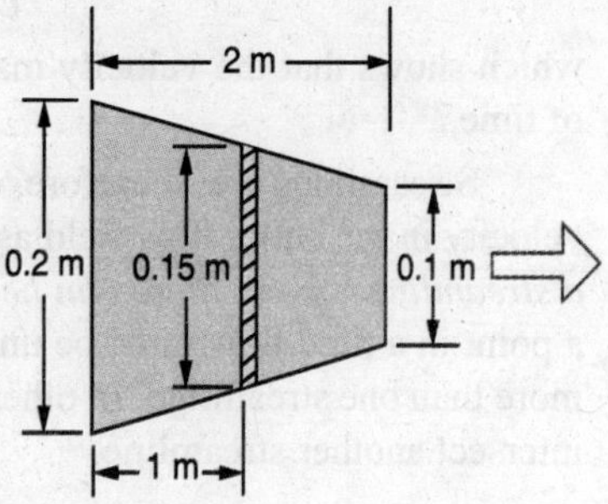

Since the flow rate is the same

$$Ua = \text{constant}$$

whence, $$U\frac{da}{dx} + a\frac{dU}{dx} = 0\,;\quad \frac{dU}{dx} = -\frac{U}{a}\frac{da}{dx}$$

Hence $$\frac{\partial U}{\partial x} = -\frac{0.9 \times 4}{\pi \times 0.15^2}\frac{\pi}{4}(0.005x - 0.02)$$

At the mid-plane, $$x = 1 \text{ m}$$

$$\frac{\partial U}{\partial x} = 0.6\,;\ U\frac{\partial U}{\partial x} = 0.9 \times 0.6 = 0.54 \text{ m/s}^2$$

Total acceleration at the mid-length at the mid-time is, therefore,

$$U\frac{\partial U}{\partial x} + \frac{\partial U}{\partial t} = 0.54 + 0.36 = \mathbf{0.9\ m/s^2}$$

3.5. Streamlines, Pathlines and Streaklines

A *streamline* in a fluid flow is a line tangent to which at any point is in the direction of velocity at that point at that instant as shown in Fig. 3.3.

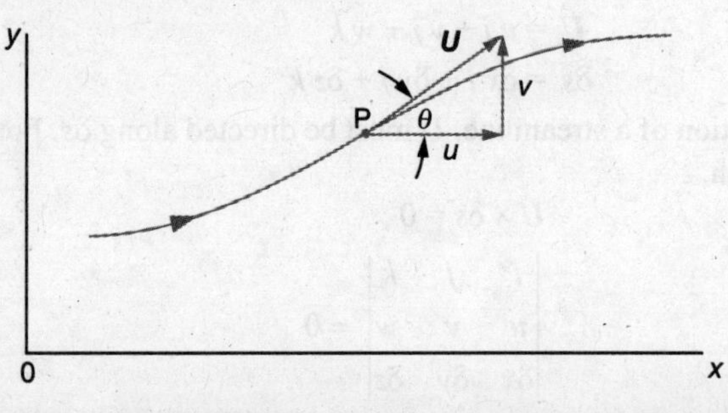

Fig. 3.3: A Streamline in the x-y Plane

Consider, a streamline in a plane flow in the *x-y* plane as shown in Fig. 3.3. By definition, the velocity vector $\boldsymbol{U}$ at a point P must be tangential to the streamline at that point. It follows that

$$\frac{dy}{dx} = \tan\theta = \frac{v}{u}$$

or

$$u\,dy - v\,dx = 0$$

where u and v are the velocity components along x and y directions respectively. It may also be noticed that the velocity vector is expressed as

$$\boldsymbol{U} = \boldsymbol{U}(s, t)$$

which shows that the velocity may vary along a streamline direction s as well as with the passage of time t.

Streamlines are, therefore, equivalent to instantaneous snap-shot indicating the directions of velocity in the entire flow field as in Fig. 3.4(*a*) and (*b*). *Since the component of velocity normal to a streamline is zero, there can be no flow across a streamline.* Further, the instantenous velocity at a point in a fluid flow must be unique in magnitude and direction, the same point cannot belong to more than one streamline. In other words, a streamline cannot intersect itself nor can any streamline intersect another streamline.

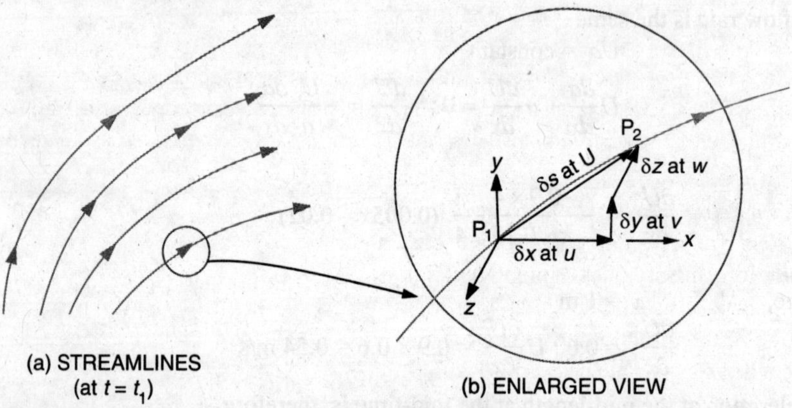

Fig. 3.4

In steady flow, the velocity components and hence the velocity vector cannot change with time. The same streamline-pattern should hold at all times. In unsteady flow, the streamline-pattern would change from instant to instant.

Consider next an elementary displacement element δs along a general streamline where the velocity is $\boldsymbol{U}$ such that

$$\boldsymbol{U} = u\,i + v\,j + w\,k$$

and

$$\boldsymbol{\delta s} = \delta x\,i + \delta y\,j + \delta z\,k$$

By the definition of a streamline, U must be directed along δs. For collinear vectors, the cross product must vanish.

Hence

$$\boldsymbol{U} \times \boldsymbol{\delta s} = 0$$

which means that

$$\begin{vmatrix} i & j & k \\ u & v & w \\ \delta x & \delta y & \delta z \end{vmatrix} = 0$$

or

$$(v\delta z - w\delta y)i - (u\delta z - w\delta x)j + (u\delta y - v\delta x)k = 0$$

Each component shall therefore, be zero.

$$\text{whence} \qquad \frac{\delta x}{u} = \frac{\delta y}{v} = \frac{\delta z}{w} \qquad \text{... (3.5)}$$

which is the equation of a streamline in a three-dimensional flow; steady or unsteady,uniform or non-uniform, viscous or inviscid, compressible or incompressible.

A *stream surface* is generated by a large number of closely spaced streamlines which pass through an arbitrary curve *AB* as in Fig. 3.4 (*c*). A stream surface can be plane, axisymmetric or spatial. If the arbitrary line constitutes a closed curve, *e.g.*, *ABCD* as in Fig. 3.4 (*d*), the closely spaced streamlines passing through the curve constitute a tabular stream surface called a *stream tube*. A stream tube with an elementary area δA of the cross section is termed as a *stream filament.*

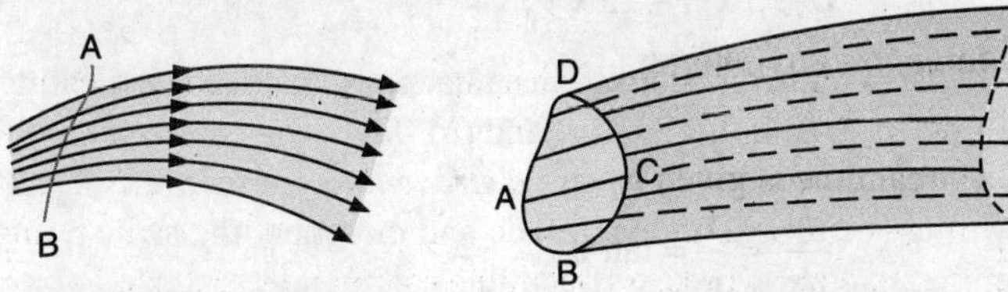

(c) STREAM SURFACE (at $t = t_1$) (d) STREAM TUBE (at $t = t_1$)

Fig. 3.4

From the fact that the velocity component normal to a streamline must be zero, no flow can penetrate a stream surface, a stream tube or a stream filament. Obviously, the impervious walls confining a fluid flow must also constitute stream surfaces, the confined passage reckoned as a stream tube.

A pathline in a fluid flow is the trajectory of a fluid particle, say P_1 as it advances with the passage of time, say from initial time t_i to final time t_f as in Fig. 3.4 (*e*). Pathlines are, therefore, history lines of individual fluid particles over a period of time. Tangent at a point on a pathline must be in the direction of velocity at that point at the time when the particle passes that point. A path line can, therefore, intersect itself at different times and in general pathlines may constitute a pattern of intersecting zig-zag lines. *In a steady flow*, the velocity vector at each point in the flow remains unaltered with time implying that the streamline pattern does not change when observed at different instance of time and that the individual fluid particles must follow the streamlines; *hence the streamlines and the pathlines coincide.*

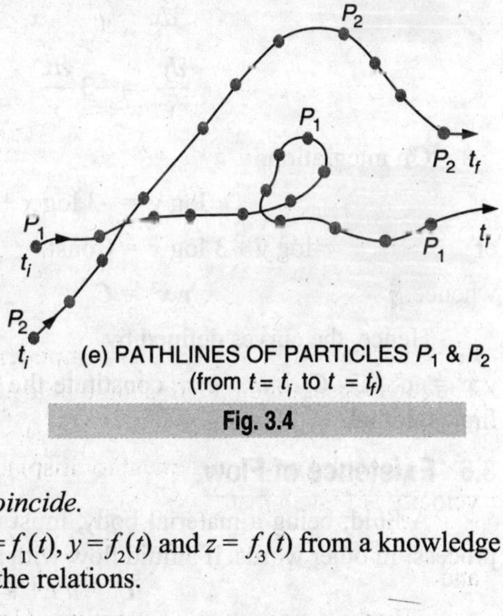

(e) PATHLINES OF PARTICLES P_1 & P_2 (from $t = t_i$ to $t = t_f$)

Fig. 3.4

The equation of a pathline may be obtained as $x = f_1(t)$, $y = f_2(t)$ and $z = f_3(t)$ from a knowledge of the velocity components *u*, *v* and *w* and employing the relations.

$$u = \frac{dx}{dt}, v = \frac{dy}{dt} \text{ and } w = \frac{dz}{dt}$$

$$\text{whence} \qquad x = \int u dt, y = \int v dt \text{ and } z = \int w dt$$

A *streakline* is the locus of locations, Fig. 3.4 (*f*), at an instant of time, of all the fluid particles which have passed through a fixed point in a flow field. A little consideration will show that the *streamlines, pathlines and streaklines are coincident in a steady flow.*

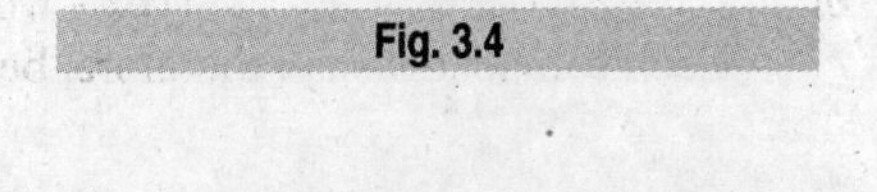

(f) STREAKLINE FOR P_1, P_2, P_3, ... (at $t = t_1$)

Fig. 3.4

It is instructive to associate these flows lines with the possible approaches to the study of fluid flow. Pathlines refer to the motion of identified fluid particles of elements and, therefore, constitute a feature of the Lagrangian approach. In fact, Fig. 3.1 (*a*) also represents a pathline. On the other hand, the Eulerian approach, demanding an instantaneous study at different locations, is concerned with the streamline-pattern of the flow. For steady flow, although the streamlines and the pathlines become identical, the Eulerian and the Lagrangian approaches coincide. This is clear from the fact that steadiness at a point (in the Eulerian approach) may not imply 'no change with time' of the velocity of a particle during its motion (in the Lagrangian approach).

Example 3.2. *Given the velocity field*

$$U = 5x^3 i - 15x^2 y j$$

Obtain the equation of the streamlines.

Solution.

The equation of a streamline is given by

$$\frac{dy}{dx} = \frac{v}{u} = \tan\theta$$

In words, the slope of a streamline at a point in *xy* plane equals the ratio of velocity components *v* and *u* at that point.

For the case in question,

$$\frac{v}{u} = \frac{-15x^2 y}{5x^3} = \frac{-3y}{x}$$

$$\frac{dy}{dx} = \frac{v}{u} = \frac{-3y}{x}$$

or $$\frac{dy}{y} = -3\frac{dx}{x}$$

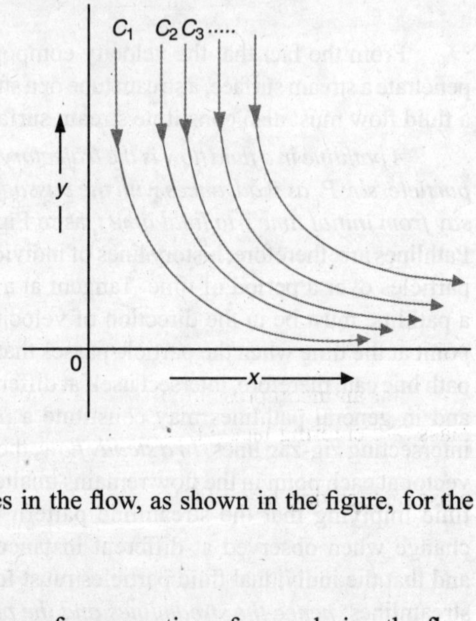

On integration,

$$\log y = -3 \log x + \text{Const.}$$

or $$\log y + 3 \log x = \text{Const.}$$

whence, $$y x^3 = C$$

Hence, the curves defined by

$y x^3 = C_1, C_2, C_3$ .... etc., constitute the streamlines in the flow, as shown in the figure, for the first quadrant.

3.6. Existence of Flow

A fluid, being a material body, must obey the law of conservation of mass during the flow process. In other words, if a fluid flow with a velocity field,

$$U = u i + v j + w k$$

were to exist, the velocity components must conform with the mass conservation principle. Velocity components in accordance with the mass conservation principle are also said to constitute a possible fluid flow. Violation of this principle would lead to non-existence or an impossible flow.

3.7. Continuity Equation: Differential Form

The law of conservation of mass for a differential control volume requires that the net mass rate leaving the control volume plus the rate of accumulation of mass within the control volume must equal zero.

In order to derive the continuity equation at a point in a fluid, the point is enclosed by an elementary control volume appropriate to the coordinate system and the efflux of mass across each surface as well as the rate of accumulation within the control volume is considered. A rectangular parallelopiped as shown in Fig. 3.5 (*a*) is the control volume in the rectangular coordinate system. Net efflux of mass along *x*-axis must be the excess outflow over inflow across faces normal to the *x*-axis. Assuming that the fluid enters the face ABCO with a velocity u and leaves the face EFGH with a velocity $u + \partial u / \partial x \,.\, \delta x$. If the fluid is compressible, the density at the face ABCO may be ρ and at the face EFGH, it will be $\rho + \partial \rho / \partial x \,.\, \delta x$. If the fluid is incompressible, its density ρ would be constant.

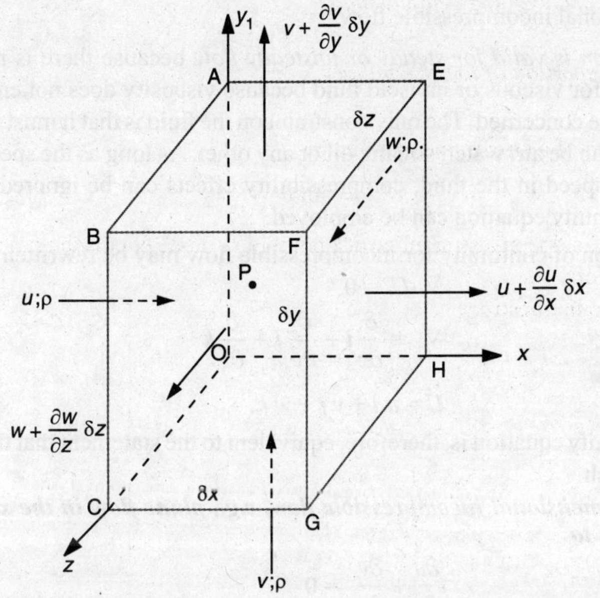

Fig. 3.5 (a) Control Volume in Rectangular Coordinates

For an incompressible flow of a fluid,

mass rate entering face ABCO

$$= \rho u \, \delta y \, \delta z$$

mass rate leaving face EFGH

$$= \rho \left(u + \frac{\partial u}{\partial x} \delta x \right) \delta y \, \delta z$$

$$= \rho u \, \delta y \, \delta z + \rho \frac{\partial u}{\partial x} \delta x \, \delta y \, \delta z$$

Hence, the net mass rate leaving in the *x*-direction, by difference,

$$= \rho \frac{\partial u}{\partial x} \delta x \, \delta y \, \delta z = \rho \frac{\partial u}{\partial x} \delta V$$

Similarly, net mass rates leaving in the *y* and *z* directions respectively are

$$\rho \frac{\partial v}{\partial y} \delta V \quad \text{and} \quad \rho \frac{\partial w}{\partial z} \delta V$$

The rate of accumulation of mass within the control volume is

$$\frac{\partial}{\partial t} (\rho \, \delta V) = \frac{\partial \rho}{\partial t} \delta V$$

Since ρ is constant for incompressible flow, this term vanishes showing that there can be no accumulation of mass with time within the control volume in an incompressible flow. By mass balance, the net rate of mass efflux across the constant volume should be zero.

$$\rho\frac{\partial u}{\partial x}\delta V + \rho\frac{\partial v}{\partial y}\delta V + \rho\frac{\partial w}{\partial z}\delta V = 0$$

Since $\delta V \neq 0$, however small the control volume may be, $\rho\ \delta\ V$ can be cancelled off, leaving the continuity equation as

$$\frac{\partial u}{\partial x} + \frac{\partial v}{\partial y} + \frac{\partial w}{\partial z} = 0 \qquad \text{... (3.6)}$$

for three dimensional incompressible flow.

The equation is valid for steady or unsteady flow because there is no $\partial / \partial t$ term in it. The equation is valid for viscous or inviscid fluid because viscosity does not enter the analysis as far as mass flow rates are concerned. The only constraint on the fluid is that it must undergo 'incompressible flow'. The fluid can be air, water, stream, oil or any other. As long as the speed of flow is appreciably below the sonic speed in the fluid, compressibility effects can be ignored and the incompressible form of the continuity equation can be employed.

The equation of continuity for incompressible flow may be rewritten in the form,

$$\nabla \cdot U = 0 \qquad \text{... (3.6 }a)$$

where
$$\nabla \equiv \frac{\partial}{\partial x}i + \frac{\partial}{\partial y}j + \frac{\partial}{\partial z}k$$

and
$$U = u\,i + v\,j + w\,k.$$

The continuity equation is, therefore, equivalent to the statement that the divergence of velocity vector must vanish.

For two dimensional incompressible flow, e.g., plane flow in the x-y plane, the continuity equation reduces to

$$\frac{\partial u}{\partial x} + \frac{\partial v}{\partial y} = 0 \qquad \text{...(3.6 }b)$$

because
$$\frac{\partial}{\partial z} \equiv 0.$$

Let us now obtain the continuity equation for a general three dimensional compresisible flow. The density variations across the faces of the control volume will now be taken into account.

The mass rate entering face ABCO

$$= \rho u\ \delta y\ \delta z$$

The mass rate leaving face EFGH

$$= \left(\rho + \frac{\partial \rho}{\partial x}\delta x\right)\left(u + \frac{\partial u}{\partial x}\delta x\right)\delta y\ \delta z$$

$$= \rho u\ \delta y\ \delta z + \rho\frac{\partial u}{\partial x}\delta x\ \delta y\ \delta z + u\frac{\partial \rho}{\partial x}\delta x\ \delta y\ \delta z + \text{smaller terms}$$

Hence, net mass rate leaving in the x-direction

$$= \rho\frac{\partial u}{\partial x}\delta x\ \delta y\ \delta z + u\frac{\partial \rho}{\partial x}\delta x\ \delta y\ \delta z$$

$$= \left(\rho\frac{\partial u}{\partial x} + u\frac{\partial \rho}{\partial x}\right)\delta x\ \delta y\ \delta z = \frac{\partial}{\partial x}(\rho u)\ \delta x\ \delta y\ \delta z$$

$$= \frac{\partial}{\partial x}(\rho u)\,.\,\delta V$$

Similarly, net mass rates leaving in the y and z directions are

$$\frac{\partial}{\partial y}(\rho v)\,\delta V \text{ and } \frac{\partial}{\partial z}(\rho w)\,\delta V \text{ respectively.}$$

The rate of accumulation of mass is given by

$$\frac{\partial}{\partial t}(\rho\,\delta x\,\delta y\,\delta z) = \frac{\partial \rho}{\partial t}\delta V$$

By mass balance,

$$\left[\frac{\partial}{\partial x}(\rho u) + \frac{\partial}{\partial y}(\rho v) + \frac{\partial}{\partial z}(\rho w) + \frac{\partial \rho}{\partial t}\right].\,\delta V = 0$$

Since $\delta V \neq 0$, however small the control volume be,

$$\frac{\partial}{\partial x}(\rho u) + \frac{\partial}{\partial y}(\rho v) + \frac{\partial}{\partial z}(\rho w) + \frac{\partial \rho}{\partial t} = 0 \qquad \text{... (3.7)}$$

In vector form, it may be written as

$$\nabla\,.\,(\rho \boldsymbol{U}) + \frac{\partial \rho}{\partial t} = 0 \qquad \text{... (3.7 a)}$$

This is the continuity equation

(*a*) *in three dimensions for a general fluid flow,*

(*b*) *for compressible or incompressible fluids;* ρ *variations taken into account,*

(*c*) *for viscous or inviscid flow; viscosity does not enter into this equation since viscosity implies the existence of stresses which are not related to mass,*

(*d*) *for steady or unsteady flow; temporal variations are taken into account,*

(*e*) *for uniform or non-uniform flow since u, v, w are the local values for the continuity at a point.*

It is interesting to examine how the continuity equation simplifies for some simpler cases of flow:

For incompressible 3-dimensional flow, ρ = constant,

whether steady or unsteady $\dfrac{\partial \rho}{\partial x} = 0 = \dfrac{\partial \rho}{\partial y} = \dfrac{\partial \rho}{\partial z} = \dfrac{\partial \rho}{\partial t}$

the continuity equation, then reduces to

$$\frac{\partial u}{\partial x} + \frac{\partial v}{\partial y} + \frac{\partial w}{\partial z} = 0$$

For incompressible 2-dimensional flow, $\dfrac{\partial}{\partial z} \equiv 0$

for flow in x-y plane; the continuity equation reduces to

$$\frac{\partial u}{\partial x} + \frac{\partial v}{\partial y} = 0$$

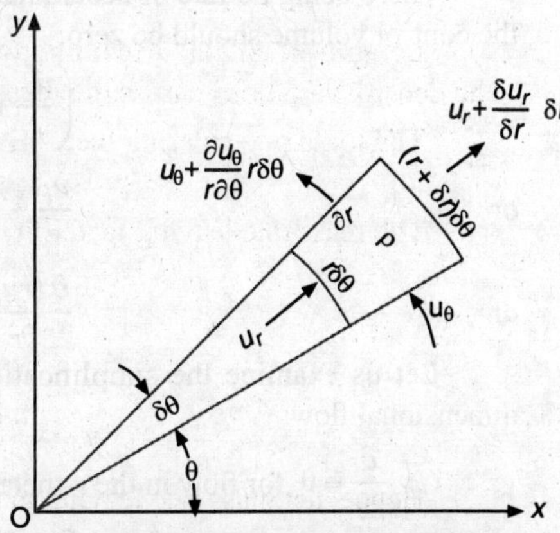

Fig. 3.5 (b) Control Volume in Cylindrical Coordinates

For compressible but steady flow, $\dfrac{\partial}{\partial t} \equiv 0$

the continuity equation reduces to

$$\frac{\partial(\rho u)}{\partial x} + \frac{\partial(\rho v)}{\partial y} + \frac{\partial(\rho w)}{\partial z} = 0 \qquad \text{... (3.8 a)}$$

For compressible plane and steady flow,

in addition, $\frac{\partial}{\partial z} \equiv 0$

for flow in *x-y* plane and the continuity equation reduces to

$$\frac{\partial(\rho u)}{\partial x} + \frac{\partial(\rho v)}{\partial y} = 0. \quad \text{... (3.8 b)}$$

Whether the fluid is inviscid or viscous, it makes no difference to the form of continuity at a point. In the cylindrical coordinates, the control volume enclosing the point is chosen as a slice of a cylinder as shown in Fig. 3.5(*b*). The depth of the slice perpendicular to the paper is unity and the dimensions in the plane of the paper are as shown. For incompressible flow, the density remains the same everywhere whereas for compressible flow, density variations must be considered. For incompressible flow,

Mass flow rate entering radially at radius $r = u_r\rho \,.\, r\delta\theta\ \delta z$

and the mass flow rate leaving radially at radius $(r + \delta r)$

$$= \left(u_r + \frac{\partial u_r}{\partial r}\delta r\right).\rho\ (r + \delta r)\ \delta\theta\ \delta z$$

$$= u_r\rho \,.\, r\delta\theta\ \delta z + u_r\ \rho \,.\, \delta r\ \delta\theta\ \delta z + \frac{\partial u_r}{\partial r}\delta r\ \rho\ r\delta\theta\ \delta z \,. + \text{smaller terms}$$

Hence, the net mass rate leaving in the radial direction

$$= u_r\rho \,.\, \delta r\delta\theta\ \delta z + \frac{\partial u_r}{\partial r}\rho \,.\, r\delta\theta\ \delta r\ \delta z$$

$$= \left(\frac{u_r}{r} + \frac{\partial u_r}{\partial r}\right)\rho \,.\, \delta r \,.\, r\delta\theta \,.\, \delta z = \left(\frac{u_r}{r} + \frac{\partial u_r}{\partial r}\right)\rho \,.\, \delta V$$

Similarly, the net mass flow rates leaving in the θ and *z* directions are given by

$$\left(u_\theta + \frac{\partial u_\theta}{r\partial\theta}.\, r\delta\theta\right)\rho \,.\, \delta r\delta z - u_\theta\rho\ \delta r\ \delta z = \frac{1}{r}\frac{\partial u_\theta}{\partial\theta}.\,\rho\delta V$$

and

$$\left(u_z + \frac{\partial u_z}{\partial z}.\,\delta z\right)\rho \,.\, \delta r \,.\, r\delta\theta - u_z\rho\ \delta r\ r\delta\theta = \frac{\partial u_z}{\partial z}.\,\rho\delta V$$

There being no rate of accumulation of fluid in an incompressible flow, the net efflux across the control volume should be zero:

$$\left(\frac{u_r}{r} + \frac{\partial u_r}{\partial r}\right).\,\rho\delta V + \frac{1}{r}\frac{\partial u_\theta}{\partial\theta}.\,\rho\delta V + \frac{\partial u_z}{\partial z}\rho\delta V = 0$$

or

$$\frac{u_r}{r} + \frac{\partial u_r}{\partial r} + \frac{1}{r}\frac{\partial u_\theta}{\partial\theta} + \frac{\partial u_z}{\partial z} = 0 \quad \text{... (3.9 a)}$$

or

$$\frac{\partial\,(ru_r)}{\partial r} + \frac{\partial u_\theta}{\partial\theta} + r\frac{\partial u_z}{\partial z} = 0 \quad \text{... (3.9 b)}$$

Let us examine the simplification in the continuity equation if the fluid undergoes two-dimensional flow;

(*i*) $\frac{\partial}{\partial z} \equiv 0$ for flow in the *r*-theta plane, the continuity equation becomes

$$\frac{\partial(ru_r)}{\partial r} + \frac{\partial u_\theta}{\partial\theta} = 0. \quad \text{... (3.9 c)}$$

(*ii*) $\frac{\partial}{\partial\theta} \equiv 0$ for axisymmetric flow, the continuity equation becomes

$$\frac{1}{r}\frac{\partial(ru_r)}{\partial r} + \frac{\partial u_z}{\partial z} = 0. \quad \text{... (3.9 d)}$$

Example 3.3. *Does a velocity field given by*

$$U = 5\ x^3 i - 15\ x^2 y j + t k$$

represent a possible fluid motion of an incompressible fluid ?

Solution.

In order to check for a physically possible or continuous fluid motion, we look for the compliance with the equation of continuity.

The continuity equation for three-dimensional incompressible flow states that,

$$\frac{\partial u}{\partial x} + \frac{\partial v}{\partial y} + \frac{\partial w}{\partial z} = 0$$

Now, for $u = 5x^3$, $v = -15\,x^2y$ and $w = t$

$$\frac{\partial u}{\partial x} = 15x^2$$

$$\frac{\partial v}{\partial y} = -15x^2$$

$$\frac{\partial w}{\partial z} = 0.$$

which, on substitution in the continuity equation, satisfy it for all x, y, z and t values because

$$15x^2 - 15x^2 \equiv 0.$$

This shows that the fluid motion represented by this velocity field is physically possible in the entire space at all times.

If the velocity components did not satisfy continuity equation, then they would not represent a possible flow.

Example 3.4. *Examine the following velocity components for a possible two-dimensional incompressible flow to be described.*

$$u_r = -\left(U \cos\theta - \frac{a^2U}{r^2}\right)$$

$$u_\theta = \left(U \sin\theta + \frac{a^2U}{r^2} \sin\theta\right)$$

Solution.

The physical possibility of the given velocity components can be checked with reference to the continuity equation,

$$\frac{\partial(ru_r)}{\partial r} + \frac{\partial u_\theta}{\partial \theta} = 0$$

From the given velocity components,

$$ru_r = -rU \cos\theta - \frac{a^2U}{r} \cos\theta$$

$$\frac{\partial(ru_r)}{\partial r} = -U \cos\theta - \frac{a^2U}{r^2} \cos\theta$$

$$\frac{\partial u_\theta}{\partial \theta} = U \cos\theta + \frac{a^2U}{r^2} \cos\theta$$

which, when substituted in the continuity equation, show that the continuity condition is identically satisfied. Hence, the given velocity components represent a possible incompressible flow in the $r - \theta$ plane.

Let us proceed, for the sake of interest, to see what physical situation is represented by the flow pattern:

At a radius, $r = a$

then $u_r = 0$

$u_\theta = 2U \sin\theta$

This is a purely circulatory flow, at a radius $r = a$

For r greater than a and

$$\theta = -\pi/2 \text{ to } \pi/2$$

$$u_r = -U\cos\theta\left(1 - \frac{a^2}{r^2}\right)$$

which is negative and for

$$\theta = \pi/2 \text{ to } 3\pi/2,$$

u_r is positive.

Similarly, for r greater than a,

$$u_\theta = U\sin\theta\left(1 + \frac{a^2}{r^2}\right)$$

which is positive for $\theta = 0$ to π and negative for $\theta = \pi$ to 2π.

Also noting that for $r >> a$, $\theta = 0$

$$u_r \approx -U\cos\theta = -U$$

$$u_\theta \approx U\sin\theta = 0$$

which implies a uniform free stream velocity. *We can plot the observed velocity components and conclude, by putting these facts together, that the flow in question is that of a uniform free stream with a circular cylinder of radius a in its passage.*

Example 3.5. *Two components of velocity in an incompressible fluid flow are given by*

$$u = x^3 - y^3 \text{ and } v = z^3 - y^3$$

Determine the third component, assuming that the origin is a stagnation point.

Solution.

The continuity equation requires that $\dfrac{\partial u}{\partial x} + \dfrac{\partial v}{\partial y} + \dfrac{\partial w}{\partial z} = 0$

From the given velocity components,

$$\frac{\partial u}{\partial x} = 3x^2 \text{ and } \frac{\partial v}{\partial y} = -3y^2$$

Hence, $$\frac{\partial w}{\partial z} = -3x^2 + 3y^3 = 3(y^2 - x^2)$$

Integrating it w.r.t. z,

$$w = 3(y^2 - x^2)z + f(x, y)$$

and employing the condition $w = 0$ at $(0, 0, 0)$; $f(x, y) = 0$

Hence, $\mathbf{w = 3(y^2 - x^2)\,z}$

Example 3.6. *Do the following velocity components represent a possible flow? Discuss the possibilities.*

$u = 2x + 3y$ and $v = 3y - 4x$

Solution.

From the given velocity components,

$$\frac{\partial u}{\partial x} = 2$$

and $$\frac{\partial v}{\partial y} = 3$$

These components do not satisfy the 2-dimensional continuity equation, $\frac{\partial u}{\partial x} + \frac{\partial v}{\partial y} = 0$

Hence the components cannot represent a 2-*d* incompressible flow. However, the flow could be 2-*d* unsteady compressible flow. Alternatively, it could be 3-*d* incompressible flow because the third component is not specified. For it to be a possible 3-d incompressible flow,

$$\frac{\partial w}{\partial z} = -5, w = \mathbf{5z + f(x, y)}$$

3.8. Stream Function for Two-Dimensional Incompressible Flow

Since no fluid can cross a streamline, the flow occurring between two streamlines must remain unchanged. For example, in Fig. 3.6(*a*), the mass flow between streamlines 1 and 2 taken across A_1A_2 must be the same as across B_1B_2 or A_1B_2. Density of fluid being constant, the volumentric flow rate between them must also be constant. A value of stream function ψ is assigned to each streamline such that the volumetric flow rate is given by $\psi_2 - \psi_1$.

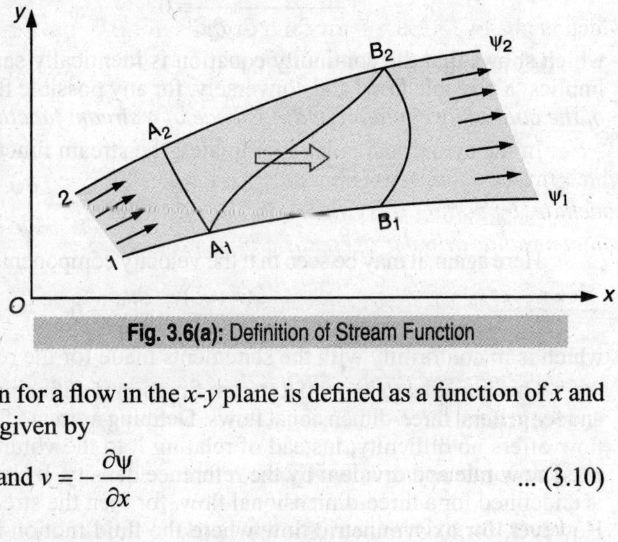

Fig. 3.6(a): Definition of Stream Function

Mathematically, the stream function for a flow in the *x*-*y* plane is defined as a function of *x* and *y* such that, the velocity components are given by

$$u = \frac{\partial \psi}{\partial y} \text{ and } v = -\frac{\partial \psi}{\partial x} \quad \text{... (3.10)}$$

The fact that this definition corresponds with the physical concept of stream function is demonstrated with reference to Fig. 3.6(b) where two neighbouring streamlines defined by stream functions ψ and $\psi + d\psi$ are shown.

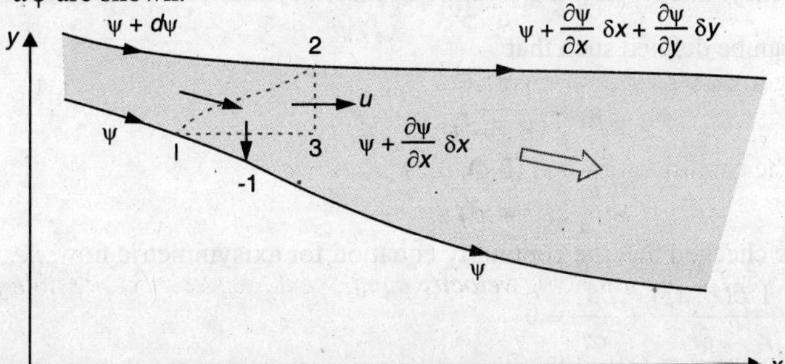

Fig. 3.6(b): Two Neighbouring Streamlines

Since $\psi = \psi(x, y)$

$$d\psi = \frac{\partial \psi}{\partial x}\delta x + \frac{\partial \psi}{\partial y}\delta y$$

Employing the definition of ψ

$$d\psi = -v\,\delta x + u\,\delta y$$

The r.h.s. of this equation represents the sum of flows across 1–3 and across 3–2. The sum must be equal to the flow across 1–2, which is denoted by $d\psi$, the difference in the value of stream functions assigned to the two streamlines.

Let us examine the continuity equation for two-dimensional incompressible flow,

$$\frac{\partial u}{\partial x} + \frac{\partial v}{\partial y} = 0 \quad \text{... (i)}$$

in the light of the definition of , the stream function, Eq. (3.10),

$$u = \frac{\partial \psi}{\partial y}, \quad v = -\frac{\partial \psi}{\partial x} \quad \text{... (ii)}$$

On substitution of (*ii*) in (*i*),

$$\frac{\partial^2 \psi}{\partial x\,\partial y} - \frac{\partial^2 \psi}{\partial x\,\partial y} \equiv 0$$

which shows that the continuity equation is identically satisfied. Hence 'existence of ψ' for a flow implies 'a possible flow' and conversely, for any possible flow, a function ψ must exist. *Compliance of the continuity equation and existence of a stream function are exactly the same thing.*

In the cylindrical polar coordinates, the stream function ψ is a function of r and θ such that

$$u_r = \frac{1}{r}\frac{\partial \psi}{\partial \theta} \text{ and } u_\theta = -\frac{\partial \psi}{\partial r} \quad \text{... (3.11)}$$

Here again, it may be seen that the velocity components identically satisfy the continuity equation

$$\frac{\partial(r \, . \, u_r)}{\partial r} + \frac{\partial u_\theta}{\partial \theta} \equiv 0$$

which is in conformity with the statements made for the rectangular coordinates.

One may like to know if a stream function is definable for two-dimensional compressible flows and for general three-dimensional flows. Defining a stream function for two-dimensional compressible flow offers no difficulty; instead of relating it to the volumetric flow rate, we would relate it to the mass flow rate and divide it by the reference density. It should be understood that a stream function is undefined for a three-dimensional flow, for then the stream surface are general arbitrary surfaces. However, for axisymmetric flow where the fluid motion is symmetrical about an axis, the stream surfaces can be generated by revolving arbitrary curves about the axis and the stream function ψ is definable.

In fact, axisymmetric flow is two-dimensional in cylindrical coordinate system because $\frac{\partial}{\partial \theta} = 0$ and hence ψ can be defined such that

$$u_z = \frac{1}{r}\frac{\partial \psi}{\partial r}$$

$$u_r = -\frac{1}{r}\frac{\partial \psi}{\partial z}$$

It can be checked that the continuity equation for axisymmetric flow, *i.e.*,

$$\frac{1}{r}\frac{\partial(r \, . \, u_r)}{\partial r} + \frac{\partial u_z}{\partial z} \equiv 0$$

is identically satisfied with the velocity components defined in this manner.

Example 3.7. *Does the stream function $\psi = x\,y$ represent a physically possible flow? If so, determine the velocity at a point (2, 3).*

Solution.

Existence of a stream function ψ implies that the equation of continuity is automatically satisfied because the velocity components derived from ψ satisfy the continuity equation identically. The stream function $\psi = xy$, therefore, must represent a physically possible flow. The streamlines are given by the equations $\psi = x\,y =$ constant which are rectangular hyperbolae.

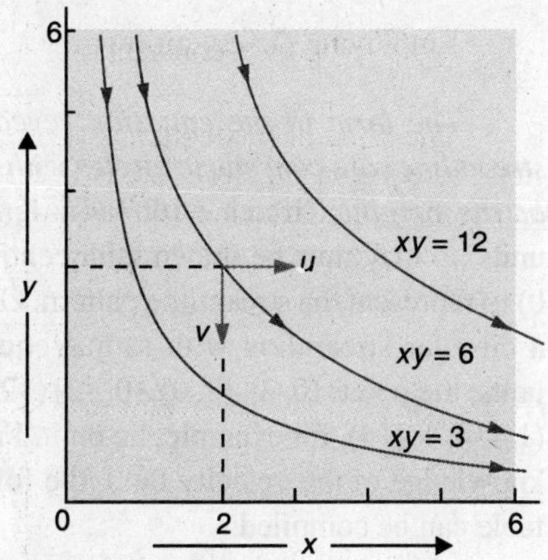

The velocity components are given by

$$u = \frac{\partial \psi}{\partial y} = x$$

$$v = -\frac{\partial \psi}{\partial x} = -y$$

In particular, at a point (2, 3), the velocity components must be

$$\mathbf{u = 2 \text{ and } v = -3 \text{ units.}}$$

Example 3.8. *Does a stream function exist for a flow field described by*

$$u = 2\,y \text{ and } v = -2\,x\,?$$

If so, determine it and plot a set of streamlines.

Solution.

Stream function for a flow field exists if the flow field is physically possible, *i.e.*, if the velocity components satisfy continuity equation,

$$\frac{\partial u}{\partial x} + \frac{\partial v}{\partial y} = 0 \qquad \text{... (i)}$$

for two-dimensional incompressible flow in the *x-y* plane.

Since

$$\frac{\partial u}{\partial x} = \frac{\partial (2y)}{\partial x} = 0$$

and

$$\frac{\partial v}{\partial y} = \frac{\partial (-2x)}{\partial y} = 0$$

the continuity equation is satisfied. A stream function, therefore, exists for the given flow field.

By definition,

$$\frac{\partial \psi}{\partial y} = u = 2\,y \qquad \text{... (ii)}$$

$$\frac{\partial \psi}{\partial x} = -v = 2\,x \qquad \text{... (iii)}$$

From (*ii*), by integration w.r.t. *y*,

$$\psi = y^2 + f(x)$$

and from (*iii*), by integration w.r.t. *x*,

$$\psi = x^2 + f(y)$$

which suggest that $\psi = x^2 + y^2$ must be the stream function for the flow.

A streamline is a line along which the stream function is constant. The equation of a streamline is, therefore,

$$x^2 + y^2 = \text{constant}$$

The form of the equation reveals that streamlines are concentric circles with centres at the origin. Circles with radii 1, 2, 3, 4 units..... may thus be drawn with centre at (0, 0) to represent the streamline pattern. Consider a circular streamline with radius equal to 2 units; the points (0, 2), (2, 0), (0, –2) (–2, 0) and (1.414, 1.414), for example, lie on it. From the knowledge of the velocity field, the following table can be compiled:

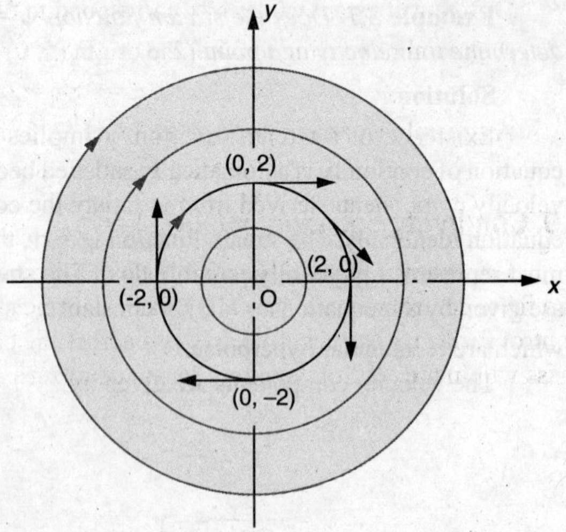

Point	*u*	*v*	$U = \sqrt{u^2 + v^2}$
(0, 2)	4	0	4
(2, 0)	0	–4	4
(0, –2)	– 4	0	4
(–2, 0)	0	4	4
(1.414, 1.414)	2.828	–2.828	4

It can be seen, by looking at the streamline with reference to the table, that the flow must be clockwise and that the speed of flow is 4 units all the way around that streamline.

Example 3.9. *The velocity in a flow field is given by*

$$u = 3 \text{ m/s} \qquad v = 6 \text{ m/s}$$

Determine the equation of the streamline passing through the origin and the one passing through a point (2 m, 3 m).

Solution.

The equation of a streamline is given by

$$\frac{dy}{dx} = \frac{v}{u} = \tan\theta$$

In this case, $\dfrac{dy}{dx} = 6/3 = 2$

whence, $y = 2x + C$

For it to pass through the origin (0, 0),

$$0 = 0 + C;\ C = 0;\ y = 2x.$$

For a streamline passing through (2, 3),

$$3 = 2 \times 2 + C;\ C = -1;\ y = 2x - 1$$

Alternatively, one may proceed to determine the stream function for the flow.

$$u = \frac{\partial \psi}{\partial y}, \qquad v = \frac{-\partial \psi}{\partial x} = -6$$

whence $\psi = 3y - 6x$

Different constant values of y correspond to different streamlines.

A streamline passing through the origin (0, 0) implies that

$$3y - 6x = 0\ ;\ \boldsymbol{y = 2x}$$

A streamline passing through the point (2, 3) implies that

$$3y - 6x = -3\ ;\ \boldsymbol{y = 2x - 1}$$

3.9. Continuity Equation: Integral Form

Let us now turn to the integral form of the continuity equation. The integral form is applicable to a finite control volume, virtually of any size and shape. If it has a volume V and is bounded by the control surface S, the net efflux of mass across the control surface S and the rate of accumulation of mass within the control volume V must add to zero.

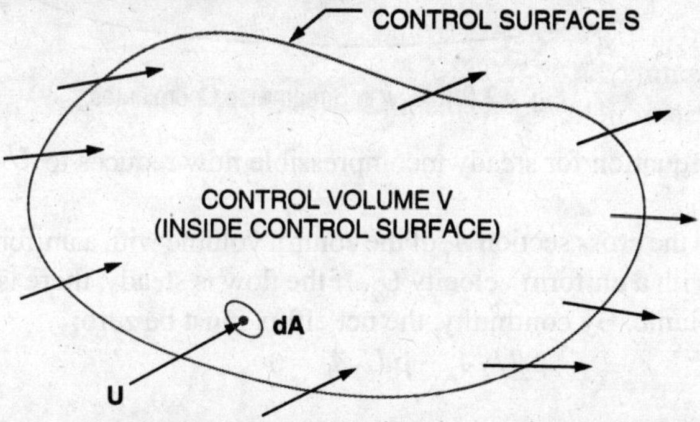

Fig. 3.7(a) Control Volume for the Continuity Equation : Integral Form

If the fluid crosses the elementary area dA with a velocity U then the net efflux across the control surface must be

$$\int_s \rho \boldsymbol{U} . d\boldsymbol{A}$$

The rate of mass accumulation within the control volume must be

$$\frac{\partial}{\partial t}\int \rho dV$$

The continuity equation, then, becomes

$$\int_s \rho U . dA + \frac{\partial}{\partial t}\int \rho dV = 0$$

The integral form of the continuity equation can be shown to be equivalent to the differential form as follows:

The first integral of Eq. (3.12) can be converted into a volume integral by the use of the divergence theorem. Since the control volume does not change with time, the order of differentiation and integration in the second term can be interchanged.

The operations result in

$$\int \nabla .(\rho U)\, dV + \int \frac{\partial \rho}{\partial t} dV = 0$$

or

$$\int \left\{ \nabla . (\rho U) + \frac{\partial \rho}{\partial t} \right\} dV = 0$$

Since the integral is zero for any arbitrary control volume, the integrand must vanish.

$$\nabla . (\rho U) + \frac{\partial \rho}{\partial t} = 0$$

$$\frac{\partial}{\partial x}(\rho u) + \frac{\partial}{\partial y}(\rho v) + \frac{\partial}{\partial z}(\rho w) + \frac{\partial \rho}{\partial t} = 0$$

Now, consider the mass balance for flow through a control volume such as that shown in Fig. 3.7 (*b*).

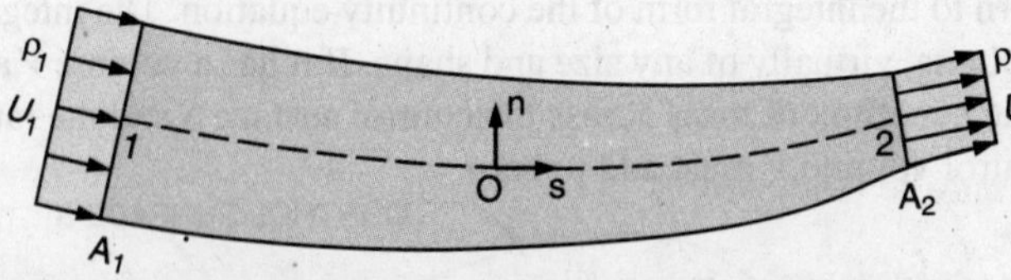

Fig. 3.7 (b): Flow in Streamwise Coordinates

The continuity equation for steady incompressible flow reduces to $U\dfrac{\partial \rho}{\partial S} + \rho\dfrac{\partial U}{\partial S} = 0$

The fluid enters the cross section A_1 of the control volume with a uniform velocity U_1 and leaves the cross section A_2 with a uniform velocity U_2. If the flow is steady, there is no rate of accumulation within the control volume. By continuity, the net efflux must be zero;

$$\rho_2 U_2 A_2 - \rho_1 U_1 A_1 = 0$$

or $$\rho_2 U_2 A_2 = \rho_1 U_1 A_1 \qquad \text{... (3.12 a)}$$

If the flow is incompressible, $\rho_2 = \rho_1$, the continuity equation is simply

$$U_2 A_2 = U_1 A_1 \qquad \text{... (3.12 b)}$$

Example 3.10. *The velocity of an incompressible fluid in a pipeline must be halved by providing a diverging piece only one diameter long. Determine the cone angle of the piece.*

Solution.

Applying the continuity equation between sections 1 and 2, assuming that the flow is uniform at the respective sections.

$$\rho_1 A_1 U_1 = \rho_2 A_2 U_2 = m$$

For incompressible flow $\rho_1 = \rho_2$

Given that

$$U_2 = \frac{U_1}{2}$$

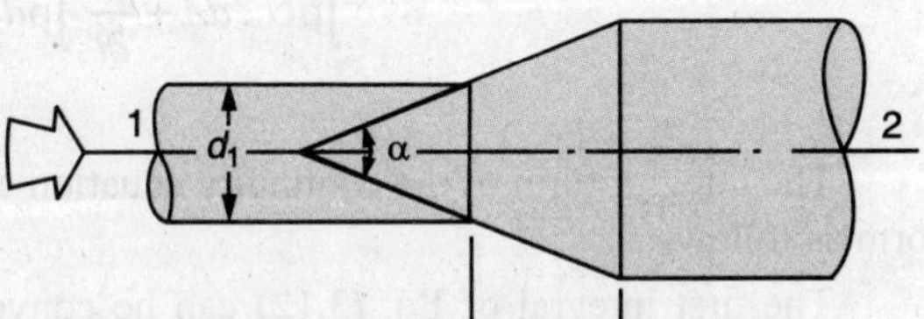

it follows that $A_2 = 2A_1$

or $d_2 = \sqrt{2}d_1\}$

Cone angle $\alpha = 2 \tan^{-1} \dfrac{(\sqrt{2}d_1 - d_1)/2}{d_1} = \mathbf{23.5°}$

Example 3.11. *A tank of volume 10 m³ contains gaseous methane at a constant temperature of 27° C. The methane is allowed to escape through a small hole of $10^{-7} m^2$ area. If the mass flow rate of methane is proportional to the pressure within the tank,*

$$m = 0.16 \times 10^{-9} . p \text{ kg/s}$$

Calculate the time required for the pressure in the tank to drop from 400 to 300 kPa.

Solution.

Let the control volume be comprised of the tank of volume

$$V = 10\ \text{m}^3$$

The principle of conservation of the mass requires that

$$\frac{\partial}{\partial t}\int_V \rho\, dV + \int_S \rho U \,.\, dA = 0$$

where the first term refers to the time rate of mass accummulation within the control volume and the second terms refers to the net outflow of mass through the control surface. The second term is given;

$$\dot{m} = \int_S \rho U \,.\, dA = 0.16 \times 10^{-9}\, p$$

The first term may be written as

$$\frac{d}{dt}\int \rho\, dV = V \,.\, \frac{d\rho}{dt} = 10\frac{d\rho}{dt}$$

By employing the perfect gas law, $p = \rho RT$

it is seen that $$dp = d\rho \,.\, RT$$

or $$d\rho = \frac{dp}{RT}$$

which, when substituted above, provides

$$10\frac{d\rho}{dt} + 0.16 \times 10^{-9}\, p = 0$$

or $$10\frac{dp}{dt} \,.\, \frac{1}{RT} + 0.16 \times 10^{-9}\, p = 0$$

and $$dt = \frac{-10}{R \times T \times 0.16 \times 10^{-9}} \frac{dp}{p}$$

Using $R = 518$ J/kg K and $T = 273 + 27 = 300$ K

$$t = \frac{-10}{518 \times 300 \times 0.16 \times 10^{-9}} \log_e\left(\frac{p_1}{p_2}\right)$$

Substituting $p_1 = 400$ kPa and $p_2 = 300$ kPa

$t = 115\ 700$ seconds = **32.14 hours**

Example 3.12. *A cubical air tank, 40 cm side, contains air at 900 kN/m² pressure and 20°C temperature. Air escapes through a 500 mm² area hole at a velocity of 50 m/s. Determine the rate of change of density at an instant when the density of the air in the tank is 7 kg/m³.*

Solution.

The continuity equation for the control volume (shown dotted) contained just within the tank is written down by visualising that the net mass efflux across it plus the rate of accumulation of mass within it must add to zero:

$$\int_S U\rho\, dA + \frac{\partial}{\partial t}\int \rho\, dV = 0$$

Simplifying the terms, with *SI* units

First term = $50 \times 7 \times 500 \times 10^{-6} = 0.175$

$$\text{Second term} = V\,.\frac{\partial \rho}{\partial t} = (40 \times 10^{-2})^3\,\frac{\partial \rho}{\partial t} = 0.064\,\frac{\partial \rho}{\partial t}$$

Therefore,

$$0.175 + 0.064\,\frac{\partial \rho}{\partial t} = 0$$

whence

$$\frac{\partial \rho}{\partial t} = -\,0.175/0.064 = \mathbf{-\,2.73\ kg/m^3 s}$$

3.10. Average Velocity

The concept of uniform flow given earlier is only theoretical and provides mathematical simplicity for analysis. A truly uniform flow cannot occur in nature due to the real-flow-effects such as viscous forces.However, in many engineering situations, flow in a passage may be nearly uniform or it must be so considered for the purpose of analysis. In such cases, care is taken to determine the average velocity in compliance with the law of conservation of mass.

Consider the flow normal to a cross section A of a duct. The mass flow rate across it is given by

$$\dot{m} = \int_A \rho u\, dA \qquad \text{... (3.13)}$$

which should be equated to ρUA where U is the average velocity. This yields

$$U = \frac{\int_A \rho u\, dA}{\rho A} = \frac{\int_A u\, dA}{A} \qquad \text{... (3.14)}$$

in an incompressible flow.

Example 3.13. *Air flows steadily through a long, straight round pipe of radius R. At a plane wall downstream the entrance, the velocity u varies with radius r according to the equation*

$$\frac{u}{U_{max}} = \left(1 - \frac{r}{R}\right)^{1/7}$$

where U_{max} is the maximum velocity at the centreline of the pipe. Evaluate the ratio U_{av}/U_{max}, if U_{av} is the average velocity in the pipe.

Solution.

The velocity profile at any diametrical plane looks as shown and hence the velocity through an elementary ring of radius r may be taken as u.

Area of the elementary ring = $2\pi r\, dr$

Mass flow through the ring = $u.\ \rho.\ 2\pi r\, dr$

where ρ is the mass density of the fluid.

Mass flow rate through the cross section of the pipe

$$\dot{m} = \int_0^R u\,.\,\rho\,.\,2\pi r\, dr$$

Also, the mass flow rate with the average velocity U_{av}

$$\dot{m} = U_{av}\,\rho\,.\,\pi R^2$$

Equating the two,

$$U_{av}\,\rho\,.\,\pi R^2 = \int_0^R u\,.\,\rho\,.\,2\pi r\, dr$$

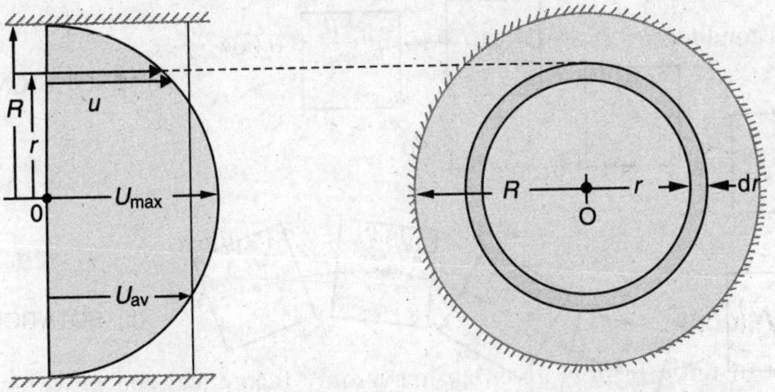

Fig. 3.8: Velocity Profile and Average Velocity.

whence
$$U_{av} = \frac{2\int_0^R u\, r\, dr}{R^2}$$

or
$$\frac{U_{av}}{U_{max}} = \frac{2\int_0^R \left(1-\frac{r}{R}\right)^{1/7} .\, r\, dr}{R^2}$$

Integrating the numerator and assigning the limits provide that

$$\frac{U_{av}}{U_{max}} = \frac{49}{60} = 0.816$$

which shows that **the average velocity is approximately 82% of the maximum velocity.**

Incidentally, the given velocity profile is the well known *one-seventh power law* which closely represents the velocity distribution in a fully developed turbulent flow through a pipe.

3.11 Kinematics of a Fluid Element

A fluid element may move in a flow and undergo

(a) pure or irrotational translation

(b) pure rotation or rotational translation

(c) pure distortion or deformation; angular or linear.

These are illustrated in Fig. 3.9 (*a*), (*b*) and (*c*) respectively by taking an upright cubical fluid element to start with in each case.

In actual practice, all the three phenomena may occur simultaneously ; a fluid element is translated, rotated and deformed as shown in Fig. 3.10.

The rate of shear or angular strain at a point in fluid is defined as half the difference of the rates of angular changes for two mutually perpendicular linear fluid elements enclosing that point. Referring to Fig. 3.10 the rate of shear at point *O* or *O′* or any point intermediate between them is given by the difference of the rates of angular changes for the line elements *OA* and *OB*.

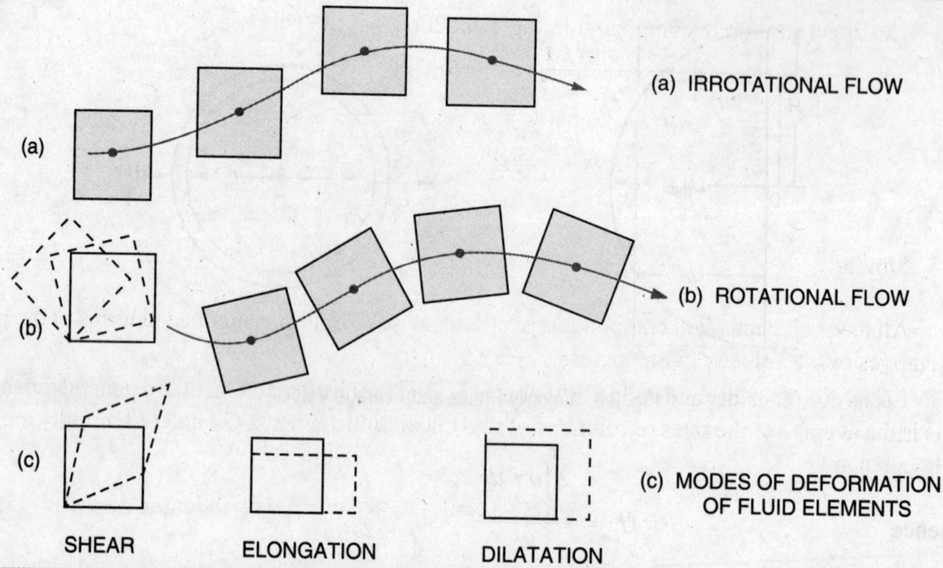

Fig. 3.9: Kinematics of a Fluid Element

$$\varepsilon_{xy} = \frac{1}{2}[\alpha - (-\beta)] / \delta t = \frac{1}{2}(\alpha + \beta) / \delta t \quad \text{... (3.15)}$$

Since α is caused by the variation of v along the x-axis,

$$\alpha = \frac{\partial v}{\partial x} \delta x \, \delta t / \delta x = \frac{\partial v}{\partial x} \delta t$$

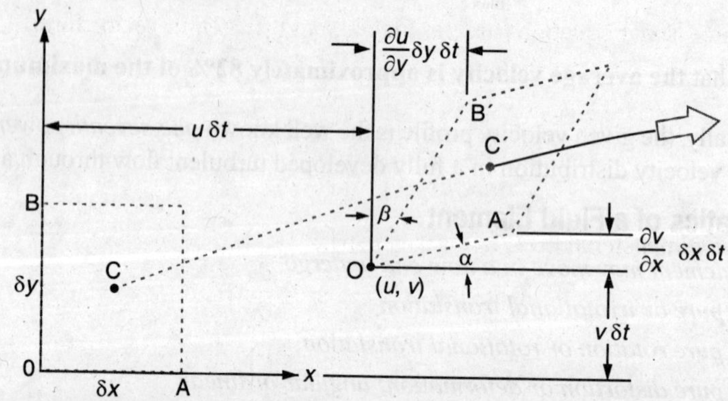

Fig. 3.10: Fluid Element Translated, Rotated and Sheared

and β is caused by the variation of u along the y-axis,

$$\beta = \frac{\partial u}{\partial y} \delta y \, \delta t / \delta y = \frac{\partial u}{\partial y} \delta t$$

Shear strain rate is given by

$$\varepsilon_{xy} = \frac{1}{2}\left(\frac{\partial v}{\partial x} + \frac{\partial u}{\partial y}\right) \text{ in the } (x - y) \text{ plane}$$

Similarly,
$$\varepsilon_{yz} = \frac{1}{2}\left(\frac{\partial v}{\partial z} + \frac{\partial w}{\partial y}\right) \text{ in the } (y - z) \text{ plane} \quad \text{... (3.16)}$$

and
$$\varepsilon_{xz} = \frac{1}{2}\left(\frac{\partial u}{\partial z} + \frac{\partial w}{\partial x}\right) \text{ in the } (x - z) \text{ plane}$$

The direct strain rate component in the x direction

$$\varepsilon_{xx} = \frac{O'A' - OA}{OA} \mid \delta t$$

or,
$$\varepsilon_{xx} = \frac{u + \dfrac{\partial u}{\partial x}\delta x - u}{\delta x} = \frac{\partial u}{\partial x}$$

Similarly,
$$\varepsilon_{yy} = \frac{\partial v}{\partial y} \text{ and } \varepsilon_{zz} = \frac{\partial w}{\partial z} \qquad \text{... (3.16 a)}$$

All the six independent components of strain rate have thus been defined in terms of the partial derivatives of the velocity components.

Let us now consider and define 'rotation' at a point for this fluid element in a general movement. This is the average of the rates of rotations of two linear fluid elements OA and OB initially mutually orthogonal at O.

$$\omega_z = \frac{1}{2}[(\alpha + (-\beta)] / \delta t = \frac{1}{2}(\alpha - \beta) / \delta t \text{ about the } z\text{-axis.} \qquad \text{... (3.17)}$$

Substituting for α and β as before,

$$\omega_z = \frac{1}{2}\left(\frac{\partial v}{\partial x} - \frac{\partial u}{\partial y}\right) \text{ about the } z\text{-axis}$$

Similarly,
$$\omega_x = \frac{1}{2}\left(\frac{\partial w}{\partial v} - \frac{\partial v}{\partial z}\right) \text{ about the } x\text{-axis} \qquad \text{... (3.18)}$$

and
$$\omega_y = \frac{1}{2}\left(\frac{\partial u}{\partial z} - \frac{\partial w}{\partial x}\right) \text{ about the } y\text{-axis.}$$

These three scalar equations are put together and written in vector form

$$\omega = \frac{1}{2}(\nabla \times U) = \frac{1}{2}\text{ curl } U = \frac{\Omega}{2} \qquad \text{... (3.19)}$$

Ω or curl U is called the vorticity;

hence,
$$\textit{rotation} = \frac{1}{2}\textit{ vorticity} \qquad \text{... (3.20)}$$

For a two dimensional flow, in the x-y plane,

$$\text{Rotation } \omega = \frac{1}{2}\left(\frac{\partial v}{\partial x} - \frac{\partial u}{\partial y}\right)$$

and,
$$\text{Vorticity } \Omega = \left(\frac{\partial v}{\partial x} - \frac{\partial u}{\partial y}\right).$$

It may be noted that the equations developed for the strain rate components and the rotation components are valid irrespective of the nature of fluid and the type of flow. This is so because the rate of strain and the rate of rotation originate from the kinematic behaviour of the particles. The spatial velocity gradients across the particles may have their origin in the viscosity of a fluid or the compressibility in flow. Eqs. (3.16), (3.19) and (3.20) have universal validity.

3.12 Irrotational Flow

For irrotational flow, each component of rotation or of vorticity must vanish, *i.e.*, $\omega_x = 0 = \omega_y = \omega_z$. In particular, for plane flow in $x - y$ plane,

$$\frac{\partial u}{\partial y} - \frac{\partial v}{\partial x} = 0 \qquad \text{... (3.21)}$$

is the condition of irrotationality.

A function $\phi = \phi(x, y)$ can be defined with an attempt to satisfy the condition of irrotationality identically.

For a three-dimensional flow,

$$\frac{\partial \phi}{\partial x} = u, \quad \frac{\partial \phi}{\partial y} = v \text{ and } \frac{\partial \phi}{\partial z} = w \text{ or } \nabla \phi = \boldsymbol{U}$$

satisfy the condition of irrotationality identically

$$\omega = \frac{1}{2}(\nabla \times \boldsymbol{U}) \equiv 0.$$

For a two dimensional flow in x-y plane,

$$\frac{\partial \phi}{\partial x} = u, \quad \frac{\partial \phi}{\partial y} = v \qquad \text{....(3.22)}$$

satisfy the condition of irrotationality identically

$$\frac{\partial v}{\partial x} - \frac{\partial u}{\partial y} = 0; \quad \frac{\partial^2 \phi}{\partial x \partial y} - \frac{\partial^2 \phi}{\partial y \partial x} \equiv 0$$

In general, the space derivatives of ϕ give the velocity components along the corresponding space coordinates. The function ϕ is called the velocity potential function.

It may be stressed that if a flow is irrotational, a velocity potential function ϕ must exist. Conversely, if a function ϕ is specified then the flow represented by the velocity components obtained from ϕ must be irrotational.

We may recall that the stream function ψ was also shown to provide the velocity components on partial differentiation. The two functions are related thus:

$$\frac{\partial \phi}{\partial x} = \frac{\partial \psi}{\partial y} \text{ and } \frac{\partial \phi}{\partial y} = -\frac{\partial \psi}{\partial x}$$

In polar coordinates, the rotation components are

$$\omega_z = \frac{1}{2r}\left(\frac{\partial}{\partial r}(ru_\theta) - \frac{\partial u_r}{\partial \theta}\right)$$

$$\omega_\theta = \frac{1}{2}\left(\frac{\partial u_\theta}{\partial z} - \frac{\partial u_2}{\partial r}\right) \qquad \text{... (3.23)}$$

$$\omega_r = \frac{1}{2}\left(\frac{1}{r}\frac{\partial u_z}{\partial \theta} - \frac{\partial u_\theta}{\partial z}\right)$$

and the condition of irrotationality for a plane flow in r–θ plane $\omega_z = 0$, reduces to

$$\frac{\partial}{\partial r}(ru_\theta) - \frac{\partial u_r}{\partial \theta} = 0 \qquad \text{... (3.24)}$$

A velocity potential function ϕ may be defined as follows to satisfy it.

$$\frac{\partial \phi}{\partial r} = u_r; \quad \frac{1}{r}\frac{\partial \phi}{\partial \theta} = u_\theta \qquad \text{... (3.25)}$$

A special case of purely circulatory flow ($u_r = 0$) is given by

$$\frac{\partial}{\partial r}(ru_\theta) = 0 \text{ or } ru_\theta = \text{constant.}$$

This is called a free vortex flow characterised by u_θ varying inversely as the radial distance from the origin as shown in Fig. 3.11.

Free vortex flow at the origin O is impossible because it demands an infinite velocity at this point. In practice, the flow near the origin must be rotational, *i.e.*, forced vortex so as to make the peripheral velocity distribution, shown dotted, dropping to zero at the origin.

Recapitulating the facts,

(*a*) Existence of $\psi \equiv$ compliance of continuity equation

$$u = \frac{\partial \psi}{\partial y}, v = -\frac{\partial \psi}{\partial x}$$

$$\frac{\partial u}{\partial x} + \frac{\partial v}{\partial y} = 0$$

(*b*) Existence of $\phi \equiv$ irrotationality of flow

$$u = \frac{\partial \phi}{\partial x}, v = \frac{\partial \phi}{\partial y}$$

$$\frac{\partial v}{\partial x} - \frac{\partial u}{\partial y} = 0$$

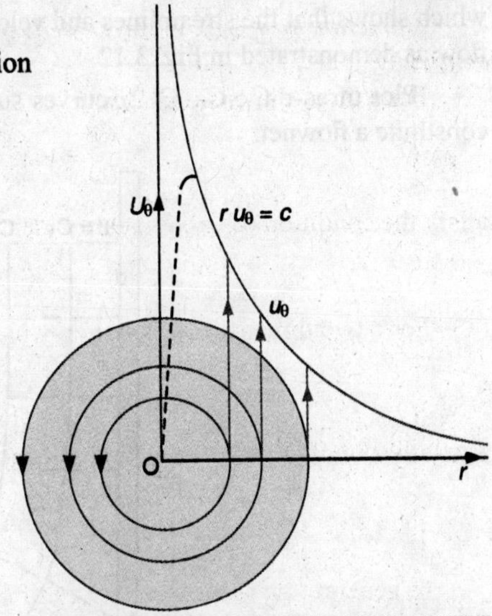

Fig. 3.11: Free vertex velocity Distribution

Substituting the definition of ψ in the irrotationality condition,

$$\frac{\partial^2 \psi}{\partial x^2} + \frac{\partial^2 \psi}{\partial y^2} = 0 \quad \text{... (3.26)}$$

and substituting definition of ϕ in the continuity equation,

$$\frac{\partial^2 \phi}{\partial x^2} + \frac{\partial^2 \phi}{\partial y^2} = 0 \quad \text{... (3.27)}$$

which show the ψ and ϕ are Laplacian functions of *x* and *y* respectively for a possible and irrotational flow.

3.13 Flownets

For a streamline defined by ψ = constant, *C*

$$d\psi = \frac{\partial \psi}{\partial x}\delta x + \frac{\partial \psi}{\partial y}\delta y = 0$$

the $\frac{dy}{dx}$ of a (ψ = constant) curve is

$$-\frac{\partial \psi}{\partial x} \Big/ \frac{\partial \psi}{\partial y} = \frac{v}{u}$$

For a velocity potential line defined by ϕ = constant, *K*

$$d\phi = \frac{\partial \phi}{\partial x}\delta x + \frac{\partial \phi}{\partial x}\delta y = 0$$

the $\frac{dy}{dx}$ of a (ϕ = constant) curve is

$$-\frac{\partial \phi}{\partial x} \Big/ \frac{\partial \phi}{\partial y} = -\frac{u}{v}$$

The product of the slopes of $\psi = C$ and $\phi = K$ curves at any place equals

$$\left(\frac{v}{u}\right) \times \left(-\frac{u}{v}\right) = -1 \quad \text{... (3.28)}$$

which shows that the streamlines and velocity potential lines form an orthogonal network in a fluid flow as demonstrated in Fig. 3.12.

Plot of $\psi = C_1, C_2, C_3$ .....curves superimposed upon the plot of $\phi = K_1, K_2, K_3$ ....curves constitute a flownet.

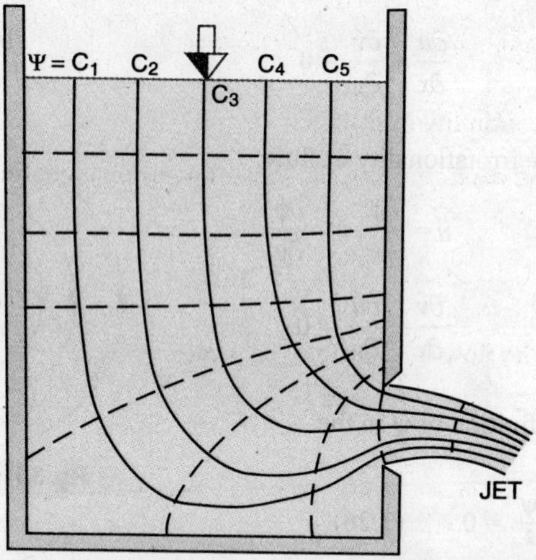

Fig. 3.12 : Orthogonal Network of Flowlines

Obviously, all flownets correspond to

(a) possible flow

(b) irrotational flow

(c) instantaneous conditions and

(d) one, two-dimensional or axisymmetric flows.

However, if a flow is not irrotational, ϕ is undefinable and hence $\phi = K$ lines cannot be drawn but ψ is defined and the streamline-pattern can still be plotted.

Example 3.14. *A fluid flow is described by the velocity vector*

$$U = 5x^2\, i - 15\, x^2\, yj$$

Test the flow for rotationality and shear strain rate.

Solution.

Since the given flow is in the *x-y* plane,

$$\omega_z = \frac{1}{2}\left(\frac{\partial v}{\partial x} - \frac{\partial u}{\partial y}\right) \text{ about the } z\text{-axis}$$

$$= \frac{1}{2}(-30xy - 0) = -15xy$$

showing that the condition of irrotationality is not satisfied. The flow is rotational except at the origin (0,0) where $\omega_z = 0$.

Shear strain rate

$$\varepsilon_{xy} = \frac{1}{2}\left(\frac{\partial v}{\partial x} + \frac{\partial u}{\partial y}\right)$$

in the $x - y$ plane which is also equal to $-15xy$ in this case.

Example 3.15. *What type of motion do the following velocity components constitute?*

$$u = a + by - cz$$
$$v = d - bx + ez$$
$$w = f + cx - ey$$

considering a, b, c, d, e and f as arbitrary constants.

Solution.

Checking for the continuity of flow;

$\frac{\partial u}{\partial x}, \frac{\partial v}{\partial y}$ and $\frac{\partial w}{\partial z}$ are each zero and the three-dimensional continuity equation

$$\frac{\partial u}{\partial x} + \frac{\partial v}{\partial y} + \frac{\partial w}{\partial z} = 0$$

is identically satisfied. The flow is, therefore, possible.

Further $\quad \omega_z = \frac{1}{2}(-b - b) = -b$

$$\omega_x = \frac{1}{2}(-e - e) = -e$$
$$\omega_y = \frac{1}{2}(-c - c) = -c$$

and $\quad \omega = -(e\,i + c\,j + bk)$

Shear strain-rate component,

$$\varepsilon_{xy} = \frac{1}{2}\left(\frac{\partial v}{\partial x} + \frac{\partial u}{\partial y}\right) = \frac{1}{2}(-b + b) = 0$$

Similarly, $\quad \varepsilon_{yz} = \varepsilon_{zx} = 0$

Normal strain-rate components

$$\varepsilon_{xx} = \frac{\partial u}{\partial x} = 0 = \varepsilon_{yy} = \varepsilon_{zz}$$

The fluid motion is devoid of strain rates in any plane but rotation components are finite and constant. These features suggest that the velocity field represents a rigid body rotation of fluid with

$$\omega = -(ei + cj + bk),$$

constant angular velocity. This is equivalent to pure rotation superimposed on pure translation.

Example 3.16. *A flow field is characterised by $\psi = x^3y$. Determine the velocity potential ϕ for the flow if the flow is irrotational.*

Solution.

For a given ψ the flow must be possible, but whether it is irrotational or not should be checked before attempting to evaluate ϕ. This is done by checking if the ψ is Laplacian, *i.e.*, if it satisfies,

$$\frac{\partial^2\psi}{\partial x^2} + \frac{\partial^2\psi}{\partial y^2} = 0$$

For the flow $\quad \psi = x^3y$

$$\frac{\partial\psi}{\partial x} = 3x^2y, \quad \frac{\partial^2\psi}{\partial x^2} = 6xy$$

$$\frac{\partial\psi}{\partial y} = x^3, \quad \frac{\partial^2\psi}{\partial y^2} = 0$$

and on substitution, we discover that ψ is not a Laplacian, *i.e.*, the flow exists but is not irrotational. Hence a **velocity potential function cannot be defined for this flow.**

Example 3.17. *Plot the flow field defined by $\phi = x^2 - y^2$ if it is a possible flow.*

Solution.

For a given ϕ the flow must be irrotational but whether this flow is possible should be checked either by seeing if ϕ is a Laplacian or by determining the velocity components from ϕ and then checking for the continuity equation.

$$\phi = x^2 - y^2$$

$$\frac{\partial \phi}{\partial x} = 2x, \quad \frac{\partial^2 \phi}{\partial x^2} = 2$$

$$\frac{\partial \phi}{\partial y} = -2y, \quad \frac{\partial^2 \phi}{\partial y^2} = -2$$

Hence $$\frac{\partial^2 \phi}{\partial x^2} + \frac{\partial^2 \phi}{\partial y^2} = 0$$

and the given ϕ depicts a possible irrotational flow

Alternatively, $$u = \frac{\partial \phi}{\partial x} = 2x$$

and $$v = \frac{\partial \phi}{\partial y} = -2y$$

For two-dimensional continuity equation,

$$\frac{\partial u}{\partial x} + \frac{\partial v}{\partial y} = 0; \; 2 - 2 \equiv 0$$

Compliance with the continuity equation also confirms the existence of a possible flow.

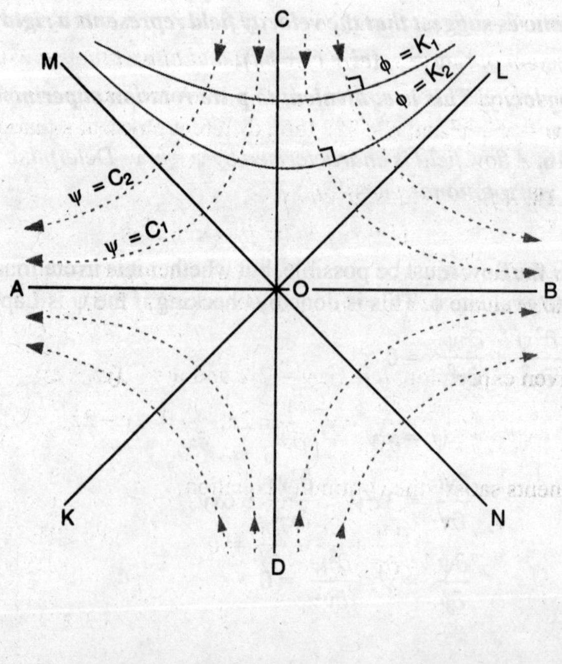

In order to plot the flow pattern, we need to determine the stream function ψ, too; from the definition of ψ,

$$\frac{\partial \psi}{\partial y} = u = 2x$$

Integrating w.r.t y alone

$$\psi = 2xy + f(x) + C$$

Differentiating partially w.r.t. x,

$$\frac{\partial \psi}{\partial x} = 2y + f'(x)$$

But $\frac{\partial \psi}{\partial x} = -v = 2y$

Hence $f'(x) = 0$

and $f(x) = K$

rendering $\psi = 2xy + \text{constant}$

Since $\psi = 0$ can be made to pass through (0, 0) the constant is meaningless.

Hence $\psi = 2xy$

is the stream function.

The streamlines $2xy = C$ lines are rectangular hyperbolae; so are $x^2 - y^2 = K$ lines.

Slope of a streamline is given by

$$\left(\frac{dy}{dx}\right)_s = -\frac{1}{2}Cx^{-2} = -\frac{y}{x}$$

and the slope of a potential line is given by

$$\left(\frac{dy}{dx}\right)_p = \frac{x}{y}$$

The product of the slopes results in '–1' which was expected as the condition for orthogonality.

The flownet sketched may represent a two-dimensional flow :

(a) near a right angled turn say COB,

(b) impingement of a jet from C over a flat plate AB and bifurcation into two streams, or

(c) impingement of two jets, from C and D and bifurcations towards A and B respectively.

It may be observed that $\psi = C$ and $\phi = K$ lines are interchangeable, *i.e.,* the flownet would appear similar, if $\psi = x^2 - y^2$ and $\phi = 2xy$ for a different physical situation.

Example 3.18. *A fluid field is given by*

$$U = xy\,i + 2yz\,j - (yz + z^2)\,k$$

Determine whether this is a possible steady incompressible fluid flow. If so, calculate the acceleration, angular velocity and velocity at the point (1, 2, 3).

Solution.

From the given expression, $u = xy$, $v = 2yz$ and $w = -(yz + z^2)$

and $$\frac{\partial u}{\partial x} = y,\ \frac{\partial v}{\partial y} = 2z,\ \frac{\partial w}{\partial z} = -y - 2z$$

The components satisfy the continuity equation,

$$\frac{\partial u}{\partial x} + \frac{\partial v}{\partial y} + \frac{\partial w}{\partial z} = 0$$

identically, showing the possibility of steady incompressible flow.

The acceleration along x-axis at the point of interest is given by

$$a_x = \frac{\partial u}{\partial t} + u\frac{\partial u}{\partial x} + v\frac{\partial u}{\partial y} + w\frac{\partial u}{\partial z}$$

$$= 0 + xy\,.y + 2yz.x + 0 = 16 \text{ units.}$$

Similarly, $a_y = 12$ units and $a_z = 84$ units

Hence $\boldsymbol{a} = \mathbf{16\,i + 12\,j + 84\,k}$

The angular velocity of rotation at the point of interest is given by

$$\omega = \frac{1}{2}(\nabla \times U) = \frac{1}{2}\begin{vmatrix} i & j & k \\ \partial/\partial x & \partial/\partial y & \partial/\partial z \\ xy & 2yz & -(yz+z^2) \end{vmatrix}$$

$$\omega = -\frac{\mathbf{1}}{\mathbf{2}}\,\mathbf{(7i + k) \text{ units}}$$

The velocity vector $\boldsymbol{U}$ at the point is obtained by substituting $x = 1$, $y = 2$ and $z = 3$ in the expression for velocity ;

or $\boldsymbol{U} = \mathbf{2i + 12j - 15k}$

Example 3.19. *If for a two-dimensional flow the stream function is given by $\psi = 2xy$, calculate the velocity at the point (3, 6).*

Show that the potential ϕ exists for this case and deduce it.

Solution. $\psi = 2xy$

$$u = \frac{\partial \psi}{\partial y} = 2x; \quad \frac{\partial^2 \psi}{\partial y^2} = 0$$

$$v = -\frac{\partial \psi}{\partial x} = -2y; \quad \frac{\partial^2 \psi}{\partial x^2} = 0$$

At the point $x = 3$ and $y = 6$

$u = 6$ units and $v = -12$ units.

The velocity is, therefore,

$\boldsymbol{U} = 6i - 12j$ units

Checking for the Laplacian equation,

$$\frac{\partial^2 \psi}{\partial x^2} + \frac{\partial^2 \psi}{\partial y^2} = 0$$

with $\frac{\partial^2 \psi}{\partial x^2} = 0$ and $\frac{\partial^2 \psi}{\partial y^2} = 0$, this is seen to be satisfied showing that the flow must be irrotational and that a velocity potential function must also exist.

Now, $$\frac{\partial \phi}{\partial x} = u = 2x \quad \text{... }(i)$$

$$\frac{\partial \phi}{\partial y} = v = -2y \quad \text{... }(ii)$$

From (*i*), by integration w.r.t. x, $\phi = x^2 + f(y)$ ... (*iii*)

and from (*ii*), by integration w.r.t. y, $\phi = -y^2 + f(x)$ ... (*iv*)

Comparing the results (*iii*) and (*iv*), $\phi = x^2 - y^2$

Example 3.20. *Given that* $\psi = 2xy$ *and* $\phi = x^2 - y^2$ *draw the streamlines corresponding to* $\psi = 100$ *and* $\psi = 300$*; also equipotential lines corresponding to* $\phi = 100$ *and* $\phi = 300$*. Six points on each line to represent the trend would suffice.*

Solution.

In order to plot the required streamlines, we make the following tabulated calculation of the coordinates in the first quadrant.

For $\psi = 100 = 2xy$; $xy = 50$							
x	0	1	2	5	10	25	$\to\infty$
y	$\to\infty$	50	25	10	5	2	0
For $\psi = 300 = 2xy$; $xy = 150$							
x	0	5	10	15	20	25	30
y	$\to\infty$	30	15	10	7.5	6	5
For $\phi = 100 = x^2 - y^2$							
x = 0 to 10	10	11	12	15	20	25	
y = non-existent	0	4.59	6.64	11.2	17.3	22.9	
For $\phi = 300 = x^2 - y^2$							
x = 0 to 17.3	17.3	18	20	22	25	30	
y = non-existent	0	4.9	10	13.6	18	24.5	

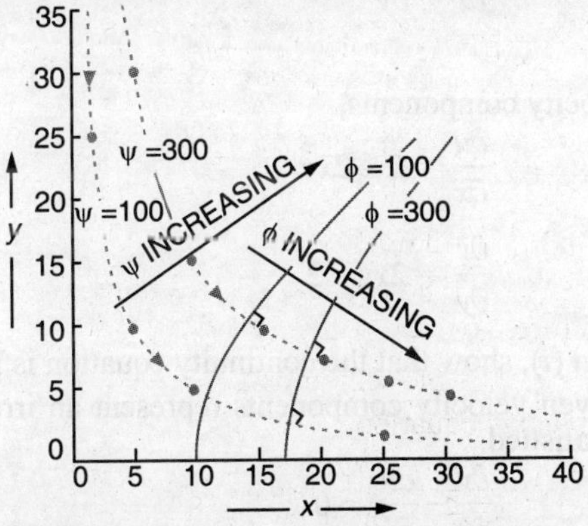

Example 3.21. *What is the irrotational velocity field associated with the potential*

$$\phi = 3x^2 - 3x + 3y^2 + 16t^2 + 12zt\ ?$$

Does the flow field satisfy the incompressible continuity equation?

Solution.

Starting with $\phi = 3x^2 - 3x + 3y^2 + 16t^2 + 12zt$

we get,

$$u = \frac{\partial \phi}{\partial x} = 6x - 3 \qquad \frac{\partial u}{\partial x} = 6$$

$$v = \frac{\partial \psi}{\partial y} = 6y \qquad \frac{\partial v}{\partial y} = 6$$

$$w = \frac{\partial \psi}{\partial z} = 12t \qquad \frac{\partial w}{\partial z} = 0.$$

The velocity field is thus

$$\boldsymbol{U} = (6x - 3)\,\boldsymbol{i} + 6y\,\boldsymbol{j} + 12t\,\boldsymbol{k}$$

The incompressible continuity equation

$$\frac{\partial u}{\partial x} + \frac{\partial v}{\partial y} + \frac{\partial w}{\partial z} = 0$$

reduces to $$6 + 6 + 0 = 0$$

which is untrue. The flow field, therefore, does not satisfy the equation and hence, it cannot exist as incompressible flow.

Example 3.22. *The velocity components in a two dimensional flow-field for an incompressible fluid are expressed as*

$$u = \frac{y^3}{3} + 2y - x^2 y$$

$$v = xy^2 - 2y - \frac{x^3}{3},$$

(*a*) *Show that these functions represents a possible case of an irrotational flow,*

(*b*) *obtain an expression for the stream function* ψ *and*

(*c*) *obtain an expression for the velocity potential* ϕ.

Solution.

(*a*) In order that the given velocity components represent a possible flow, the continuity equation

$$\frac{\partial u}{\partial x} + \frac{\partial v}{\partial y} = 0 \qquad \text{... (i)}$$

should be satisfied.

From the given velocity components,

$$\frac{\partial u}{\partial x} = 2 - 2xy$$

$$\frac{\partial v}{\partial y} = 2xy - 2$$

which, when substituted in (*i*), show that the continuity equation is identically satisfied.

In order that the given velocity components represent an irrotational flow, the condition of irrotationality should be satisfied.

$$\frac{\partial v}{\partial x} - \frac{\partial u}{\partial y} = 0 \qquad \text{... (ii)}$$

From the given velocity components,

$$\frac{\partial v}{\partial x} = y^2 - x^2$$

$$\frac{\partial u}{\partial y} = y^2 - x^2$$

which when substituted in (*ii*) show that the condition of irrotationality is identically satisfied.

The velocity components, therefore, represent a possible case of irrotational flow.

(*b*) From the definition of the stream function ψ,

$$u = \frac{\partial \psi}{\partial y},\; v = -\frac{\partial \psi}{\partial x}$$

Substituting the given functional forms of u and v and integrating w.r.t. and x and y respectively,

$$\psi = \frac{y^4}{12} + 2xy - \frac{x^2y^2}{2} + f(x) \qquad \text{... (iii)}$$

$$\psi = -\frac{x^2y^2}{2} + 2xy + \frac{x^4}{12} + g(y) \qquad \text{... (iv)}$$

Inspecting equations (*iii*) and (*iv*),

$$\psi = 2xy - \frac{x^2y^2}{2} + \frac{x^4}{12} + \frac{y^4}{12} \qquad \text{... (v)}$$

(*c*) From the definitions of the velocity potential function ψ,

$$u = \frac{\partial \phi}{\partial x}, \; v = \frac{\partial \phi}{\partial y}$$

Substituting the given functional forms of u and v and integrating w.r.t. x and y respectively,

$$\phi = \frac{y^2x}{3} + x^2 - \frac{x^3y}{3} + f(y) \qquad \text{... (vi)}$$

$$\phi = \frac{xy^2}{3} - y^2 - \frac{x^3y}{3} + g(y) \qquad \text{... (vii)}$$

Inspecting equation (*vi*) and (*vii*),

$$\phi = \frac{xy^3}{3} - \frac{yx^3}{3} + x^2 - y^2 \qquad \text{... (viii)}$$

Self Check. The stream function ψ and the velocity potential function ϕ should satisfy the Laplacian equation for a possible and irrotational flow. Let us check if it is so!

From (*v*),

$$\frac{\partial \psi}{\partial x} = 2y - xy^2 + x^3/3$$

$$\frac{\partial^2 \psi}{\partial x^2} = -y^2 + x^2$$

$$\frac{\partial \psi}{\partial y} = 2x - x^2y + y^2/3$$

$$\frac{\partial^2 \psi}{\partial y^2} = -x^2 + y^2$$

which show that

$$\frac{\partial^2 \psi}{\partial x^2} + \frac{\partial^2 \psi}{\partial y^2} \equiv 0$$

Similarly, from (*viii*),

$$\frac{\partial \phi}{\partial x} = y^2/3 - yx^2 + 2x$$

$$\frac{\partial^2 \phi}{\partial x^2} = -2xy + 2$$

$$\frac{\partial \phi}{\partial y} = xy^2 - x^3/3 - 2y$$

$$\frac{\partial^2 \phi}{\partial y^2} = 2xy - 2$$

which show that

$$\frac{\partial^2 \phi}{\partial x^2} + \frac{\partial^2 \phi}{\partial y^2} \equiv 0$$

Example 3.23. *It is desired to obtain the convective acceleration and the kinematics of a fluid element at the origin (0, 0) in a steady fluid flow from the record of velocity reproduced below.*

Point (x, y)	*Velocity*
(0, 0)	10 i + 15 j
(0.1, 0)	11 i + 15.5 j
(–0.1, 0)	9 i + 14.5 j
(0, 0.1)	9.4 i + 14 j
(0, –0.1)	11.6 i + 16 j

Solution.

From the data at closely spaced points in the *x-y* plane, let us evaluate the required partial derivatives by centre point formula,

$$\frac{\partial u}{\partial x} = \frac{u\,(0.1, 0) - u\,(-0.1, 0)}{0.1 - (-0.1)}$$

$$= \frac{11-9}{0.1-(-0.1)} = \frac{11-9}{0.2} = 10$$

Similarly, $$\frac{\partial u}{\partial y} = \frac{(9.4 - 11.6)}{0.2} = -11$$

$$\frac{\partial v}{\partial x} = \frac{(15.5 - 14.5)}{0.2} = 5$$

or $$\frac{\partial v}{\partial y} = \frac{(14-16)}{0.2} = -10$$

Checking for the continuity equation,

$$\frac{\partial u}{\partial y} + \frac{\partial v}{\partial y} = 10 - 10 \equiv 0$$

shows that it is identically satisfied which means that the observations are reliable.

Applying Eq. (3.3, *a*, *b*, *c*) for the components of acceleration at the origin (0, 0) where $u = 10$ and $v = 15$,

$$a_x = 10 \times 10 + 15\,(-11) = -65$$

$$a_y = 10 \times 5 + 15(-10) = -100$$

Hence the convective acceleration at the origin equals $-65\,i\ -100\,j$

Further, $$\omega_z = \frac{1}{2}\left(\frac{\partial v}{\partial x} - \frac{\partial u}{\partial y}\right) = \frac{1}{2}[5 - (-11)] = 8$$

$$\varepsilon_{xy} = \frac{1}{2}\left(\frac{\partial v}{\partial x} + \frac{\partial u}{\partial y}\right) = \frac{1}{2}(5 - 11) = -3$$

$$\varepsilon_{xx} = \frac{\partial u}{\partial x} = 10 \quad \varepsilon_{yy} = \frac{\partial v}{\partial y} = -10$$

which shows that the fluid element in translation at $(10\ i + 15\ j)$ units velocity suffers rotation (8 units), shear strain (–3 units) and dilatation or linear strain (10 units in *x*-direction and –10 units in *y*-direction) while undergoing an acceleration of $(-65\ i - 100\ j)$ units.

Example 3.24. *A concial pipe 10 cm inlet diameter, 20 cm outlet diameter and 100 cm long is placed horizontally. The velocity over any cross-section may be considered to be uniform. Determine the convective and local acceleration at a section where the diameter is 15 cm for the following two cases:*

(a) Constant inlet discharge 0.2 cm²/s

(b) Inlet discharge varying linearly from 0.2 m²/s to 0.4 m³/s over two seconds. The time of interest is when t = 1 second. Comment on the assumption of uniform flow over any cross-section in the light of the continuity equation.

Solution.

Diameter at x meter from inlet = 0.1 (1 + x) m

Cross-sectional area at a section X-X at a distance x from inlet

$$A = \frac{\pi}{4} \times (0.1) \times (1 + x)^2 = 0.00785\ (1 + x)^2\ \text{m}^2$$

In terms of the discharge Q,

the velocity u at X-X = $\dfrac{Q}{A} = \dfrac{Q}{0.00785\ (1 + x^2)} = 127.5\ (1 + x)^{-2}\, Q$

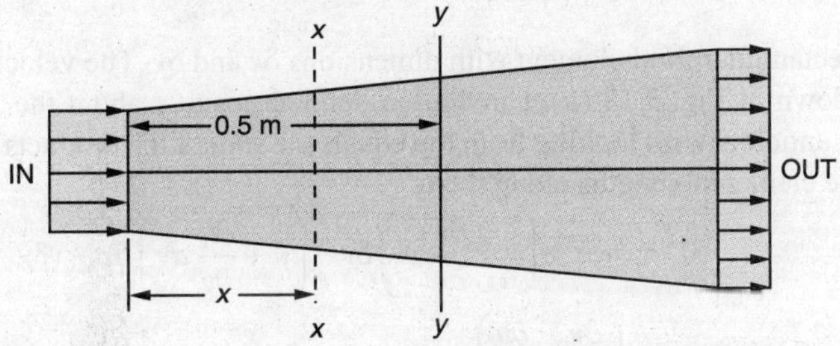

and the gradient $\dfrac{\partial u}{\partial x} = -255\ (1 + x)^{-3}\, Q$

The diameter of the concial pipe becomes 0.15 m at a distance

$$x = 0.5\ \text{m}$$

from the inlet, *i.e.*, at section Y-Y.

Case (a) The discharge is constant at 0.2 m²/s and the flow is steady. The local acceleration is $\dfrac{\partial u}{\partial t} = 0$ for steady state

The convective acceleration is

$$u\frac{\partial u}{\partial x} = 127.5\ (1 + 0.5)^{-2} \times 0.2 \times (-\,255)\ (1 + 0.5)^{-3} \times 0.2$$

$$= \mathbf{-171\ m/s^2\ directed\ longitudinally}$$

It may be noted that $\dfrac{\partial u}{\partial r}$ and $\dfrac{\partial u}{\partial \theta}$ are each zero and hence no other components of acceleration exist.

Case (*b*) The discharge varies from 0.2 m³/s to 0.4 m³/s over a period of 2 seconds.

At t = 1 second

the discharge $\quad Q = \dfrac{0.4 - 0.2}{2} \times 1 + 0.2 = 0.3\ \text{m}^3/\text{s}$

The convective acceleration is given by

$$u\frac{\partial u}{\partial x} = 127.5\ (1 + 0.5)^{-2} \times 0.3 + (-255)\ (1 + 0.5)^{-2} \times 0.3$$

and the local acceleration is given by

$$= -385 \text{ m/s}^2 \text{ directed longitudinally}$$

$$\frac{\partial u}{\partial t} = 127.5\,(1 + 0.5)^{-2}\,[0.3 - 0.2]/1$$

$$= 5.7 \text{ m/s}^2, \text{ also directed longitudinally}$$

The total acceleration at *Y-Y*, directed along the main flow, is

$$= -385 + 5.7 = \mathbf{-379.3\ m/s^2}$$

3.14 Circulation in a Flow

Along a closed contour in a flow field, the line integral of velocity is given the name circulation and is denoted by

$$\Gamma. = \oint U \,.\, ds \text{ , also written as } \int_C U \,.ds$$

For the three-dimensional flow,

$$\Gamma = \oint (udx + vdy + wdz)$$

and for a two-dimensional flow, $\Gamma = \oint (udx + vdy)$

$$\Gamma = \oint U \cos \alpha\, ds \qquad \text{... (3.29)}$$

Consider a rectangular fluid element with dimensions δx and δy. The velocities along the four sides are written down in Fig. 3.13 (*a*). Circulation defined positive about the *x*-axis by the right handed screw rule, anticlockwise looking from top equals the sum of the products of the components of velocity with the elemental lengths along them.

$$\delta\Gamma = u\delta x + \left(v + \frac{\partial v}{\partial x}\delta x\right)\delta y - \left(u + \frac{\partial u}{\partial y}dy\right)\delta x - v\delta y$$

$$= \left(\frac{\partial v}{\partial y} - \frac{\partial u}{\partial y}\right)\delta x\ \delta y = 2w_z\ \delta A$$

$$\delta\Gamma/\delta A = 2\ \omega_z = \Omega_z, \text{ the vorticity} \qquad \text{... (3.30)}$$

In words, *circulation per unit area equals the vorticity in flow.*

Let us now examine the circulation for a free vortex, $ru_\theta = C$

$$\Gamma = \int_c U \cos \alpha \,.\, r\delta\theta$$

For a circular path, as seen in Fig. 3.13 (*b-i*)

$$\alpha = 0 \text{ and } u_\theta = \frac{C}{r}$$

Hence $$\Gamma = \int_0^{2\pi} t\left(\frac{C}{r}\,.\,r\delta\theta\right) = 2\pi C \qquad \text{... (3.31)}$$

It may be understood that the integrand $\left(\frac{C}{r}\,.\,r\delta\theta\right) = C\delta\theta$ represents the elementary circulation $\delta\Gamma$ around an elementary contour *OMN* in the absence of radial velocity and the integration $\int_C d\Gamma$ represents the circulation over a circular streamline at radius *r*. Further, $\Gamma = 2\pi\ C$ shows that the circulation is not dependent upon the radius of the circular streamline.

In the first instance it may appear to be contradictory to describe a free vortex from the condition of irrotationality and then to get circulation around a contour as finite, $2\pi C$. However, it is not contradictory because, even there, it was pointed out that the origin is a singular point requiring infinite velocity. If therefore, circulation is determined along any closed contour excluding the origin, it should be zero. Let us take a closed contour *ABCDA* as shown in Fig. 3.13 (*b-ii*). The circulation on a contour is taken positive anticlockwise as stated earlier.

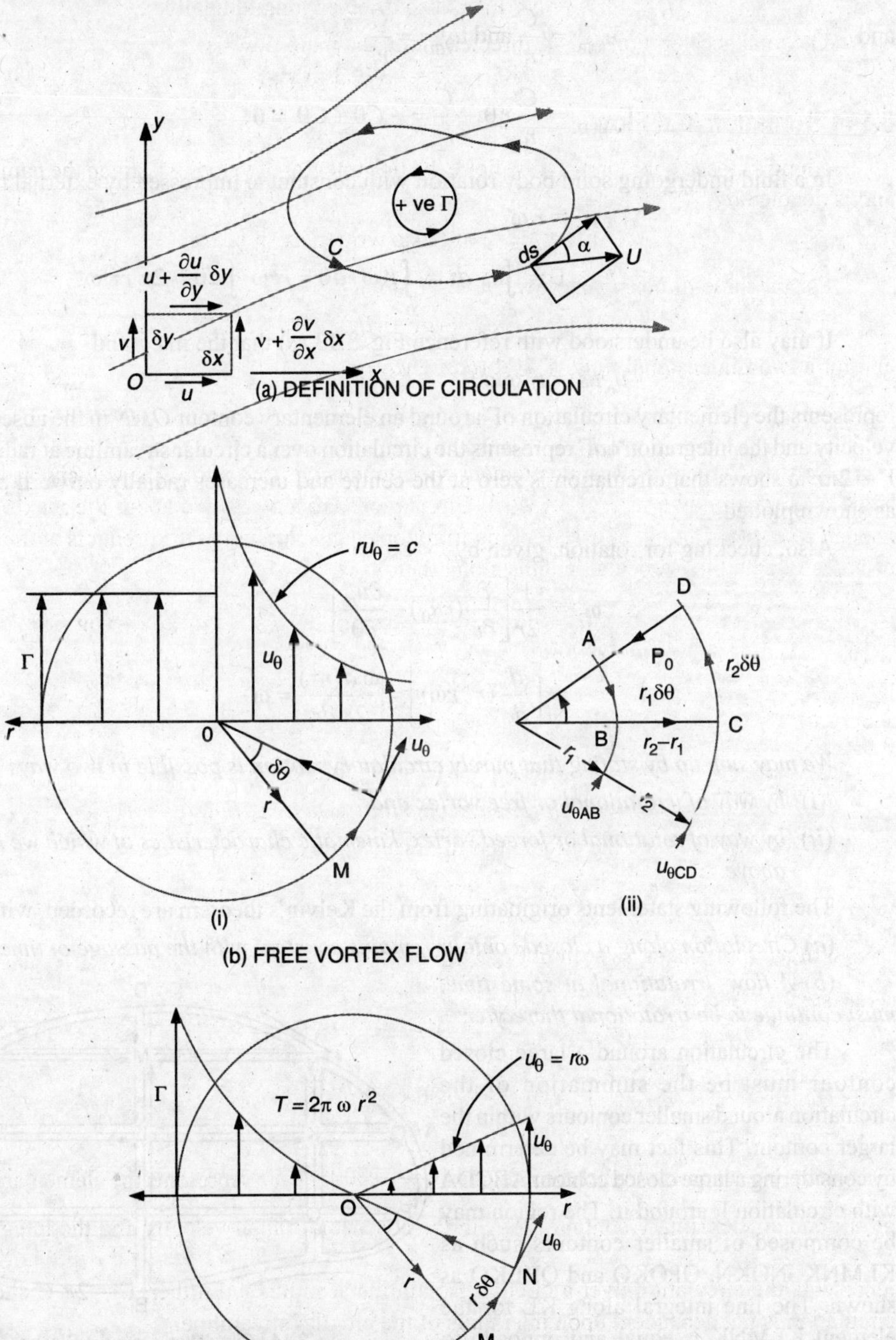

Fig. 3.13. Vortices and Circulation

$$\Gamma_{ABCDA} = -u_{\theta AB} \cdot r_1\, \theta \text{ along } AB - u_{rBC} \cdot (r_2 - r_1) \text{ along } BC$$
$$+ u_{\theta CD} \cdot r_2\, \theta \text{ along } CD - u_{rDA} \cdot (r_2 - r_1) \text{ along } DA$$

Since there is no radial flow. $U_{rBC} = 0 = u_{rDA}$

and $$u_{\theta AB} = \frac{C}{r_1} \text{ and } u_{\theta CD} = \frac{C}{r_2}$$

$$\Gamma_{ABCD} = \frac{C}{r_1} r_1 \theta + \frac{C}{r_2} = -C\theta + C\theta = 0 \quad \text{... (3.32)}$$

In a fluid undergoing solid body rotation with constant ω impressed by external means,

$$u_\theta = r\,\omega$$

$$\Gamma = \int_c u_\theta\, ds = \int_0^{2\pi} r\,\omega\, r d\theta = r^2 \omega \int_0^{2\pi} d\theta = 2\pi\, r^2\, \omega \quad \text{... (3.33)}$$

It may also be understood with reference Fig. 3.13 (*c*) that the integrand

$$u_\theta\, ds = r^2\, \omega\, d\theta$$

represents the elementary circulation $\delta\Gamma$ around an elementary contour *OMN* in the absence of radial velocity and the integration $\phi d\Gamma$ represents the circulation over a circular streamline at radius *r*. Further $\Gamma = 2\pi r^2 \omega$ shows that circulation is zero at the centre and increases radially outward parabolically as shown plotted.

Also, checking for rotation given by

$$\omega_z = \frac{1}{2r}\left[\frac{\partial}{\partial r}(r u_\theta) - \frac{\partial u_r}{\partial \theta}\right]$$

$$= \left[\frac{d}{dr}(r \cdot r\omega)\right] = \frac{\omega\, d(r^2)}{2r\, dr} = \omega$$

We may sum up by stating that purely circulatory motion is possible in two ways

(*i*) *by way of irrotational or free vortex and*

(*ii*) *by way of rotational or forced vortex, kinematic characteristics of which we have studied above.*

The following statements originating from the Kelvin's theorem are recorded, without proofs.

(*a*) *Circulation along a closed contour remains constant with the passage of time.*

(*b*) *A flow, irrotational at some time, must continue to be irrotational thereafter.*

The circulation around a large closed contour must be the summation of the circulation around smaller contours within the larger contour. This fact may be determined by considering a large closed contour ABCDA with circulation Γ around it. The region may be composed of smaller contours such as KLMNK, NOKN, OPQKO and QRLKQ as shown. The line integral along KL for the element KLMNK is equal and opposite to the line integral along LK for the neigbouring element QRLKQ. Similarly, line integrals along each of the internal boundaries such as NK, OK, PQ, KQ reduce to zero. Only the line integrals along the external boundaries such as NO and OP

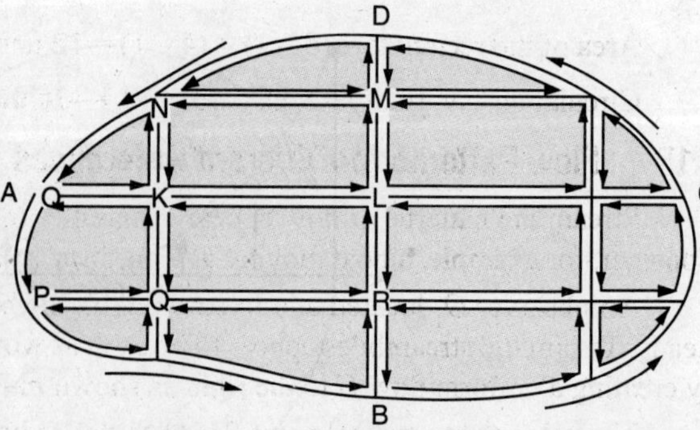

Fig. 3.14 : Circulation around a Closed Contour

remain non-zero, which indeed add up to the circulation Γ around the larger contour ABCDA, as shown in Fig. 3.14.

Example 3.25. *Determine the circulation Γ around a rectangle defined by*

$$x = 1,\ y = 1,\ x = 5 \text{ and } y = 4$$

for the velocity field $u = 2x + 3y$ *and* $v = -2y$.

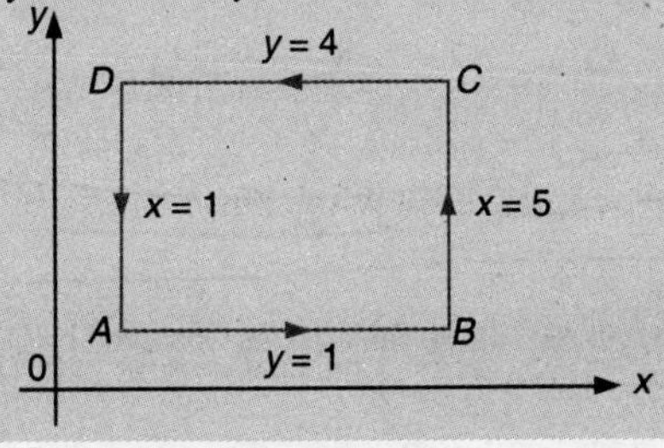

Solution.

Circulation is defined as the line integral of velocity around a closed contour:

$$\Gamma = \int_c u\,ds = \int_{c_{ABCD}} (u\,dx + v\,dy)$$

$$= \int_{ABCD} (2x + 3y)\,dx + \int_{BCDA} (-2y)\,dy$$

$$= \int_1 (2x + 3y)\,dx + \int_s^1 (2x + 3y)dx - \int_1^4 2y\,dy - \int_4^1 2\,y\,dy$$

Observing that $y = 1$ for the first integral and $y = 4$ for the second integral,

$$\Gamma = |\,x^2 + 3x\,|_1^5 + |\,x^2 + 12x\,|_5^1 - |\,y^2\,|_1^4 - |\,y^2\,|_4^1$$

$$= 36 - 72 = \mathbf{-36\ units.}$$

since the last two terms cancel mutually.

The result can be checked by employing the fact that circulation equals vorticity times the area.

$$\text{Vorticity} = 2\omega_z = \frac{\partial v}{\partial x} - \frac{\partial u}{\partial y} = 0 - 3 = -3$$

Area of the rectangle = (5 – 1) × (4 – 1) = 12 units

Circulation = vorticity × area = –3 × 12 = **–36 units.**

3.15 Flow Patterns for Different References

Streamline patterns of flow appear differently to observers on different frames of references. Consider, for example, a boat moving at a uniform velocity U to the left in an expanse of water.

An observer O_1 located at a fixed point on the boat observes that the relative flow pattern is steady. To him the streamlines appear to be such as would be formed by holding the boat at rest and by creating a uniform flow U to the right as shown in Fig. 3.15.

The velocity at a typical point P is observed to be U_p in a direction tangential to the streamline at that point.

An observer O_2 located at a pier on the other hand, notices the boat approaching him and then receding away from him. He observes that the absolute flow is unsteady. At a typical point P the absolute velocity U_a observed by the observer must be the sum of the velocity U_p observed by the observer on the boat and the velocity of the observer himself;

$$U_a = U_p + U$$

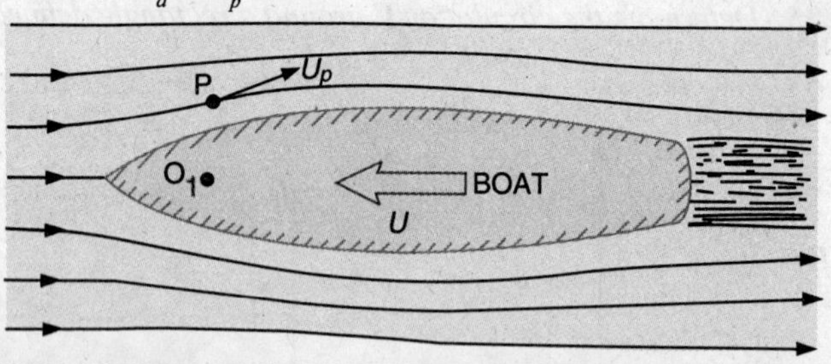

Fig. 3.15: Steady-flow Relative Streamlines Observed by O_1 on the Boat

The absolute streamline at the point P must, therefore, be tangential to U_a. The absolute streamline pattern is shown in Fig. 3.16.

The fact that the flow pattern past a body appears differently to observers in the inertial frames of reference may appear to be paradoxical in the first instance but it is explained as above. The steady flow picture corresponds to a control volume fixed with respect to the boat whereas the unsteady flow picture refers to the control volume fixed in space. One or the other is preferred for different objectives. When it is intended to analyse the flow on the basis of steady flow equations, the control volume fixed with respect to the boat is the obvious choice.

Opportunity is taken to differentiate between streamlines, pathlines and streaklines for the steady flow pattern shown in Fig. 3.16. Pathlines of particles 1, 2, 3......... which passed through a point Q are shown plotted. The locus of particles 1, 2, 3........at a time t constitutes a streakline at time t. Similarly, the locus of particles 1′, 2′, 3′ at a time t' constitutes another streakline at time t'.

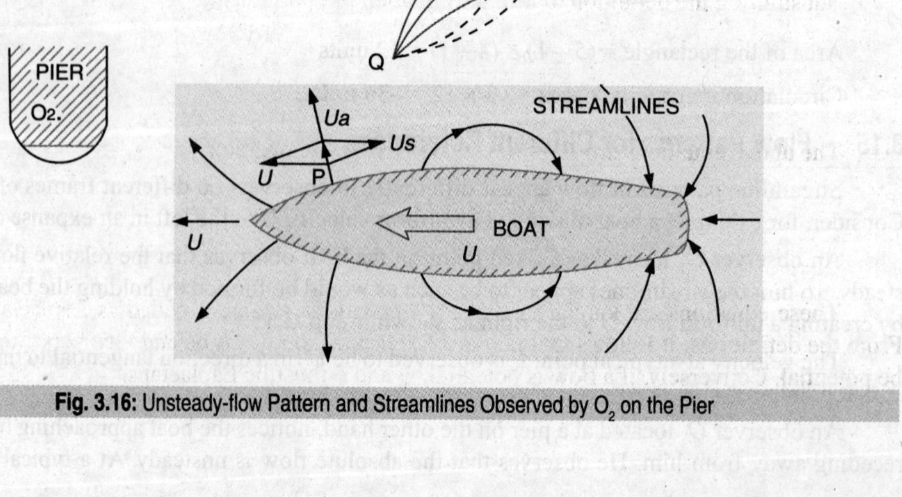

Fig. 3.16: Unsteady-flow Pattern and Streamlines Observed by O_2 on the Pier

Resume' of Equations for Fluid Kinematics

The equations which are applicable for the analysis of a two-dimensional, steady ideal flow in $x - y$ plane are as follows:

Continuity: Mass Conservation Equation: Existence of ψ

$$\rho\, UA = \text{constant} \quad \text{along streamline direction} \qquad \text{... (3.34)}$$

$$\left.\begin{array}{l} \dfrac{\partial u}{\partial x}+\dfrac{\partial v}{\partial y}=0 \\ \text{or} \\ u=\dfrac{\partial \psi}{\partial y},\ v=\dfrac{\partial \psi}{\partial x} \end{array}\right] \quad \text{in rectangular coordinates} \qquad \text{... (3.35)}$$

$$\left.\begin{array}{l} \dfrac{\partial (ru_r)}{\partial r}+\dfrac{\partial u_\theta}{\partial \theta}=0 \\ \text{or} \\ u_r=\dfrac{1}{r}\dfrac{\partial \psi}{\partial \theta},\ u_\theta=-\dfrac{\partial \psi}{\partial r} \end{array}\right] \quad \text{in cylindrical coordinates} \qquad \text{... (3.36)}$$

Irrotationality: Condition of Irrotationality: Existence of ϕ

$$\left.\begin{array}{l} \dfrac{\partial u}{\partial y}-\dfrac{\partial v}{\partial x}=0 \\ \text{or} \\ u=\dfrac{\partial \phi}{\partial x},\ v=\dfrac{\partial \phi}{\partial y} \end{array}\right] \quad \text{in rectangular coordinates} \qquad \text{... (3.37)}$$

$$\left.\begin{array}{l} \dfrac{\partial (ru_\theta)}{\partial r}-\dfrac{\partial u_r}{\partial \theta}=0 \\ \text{or} \\ u_r=\dfrac{\partial \phi}{\partial r},\ u_\theta=-\dfrac{1}{r}\dfrac{\partial \phi}{\partial \theta} \end{array}\right] \quad \text{in cylindrical coordinates} \qquad \text{... (3.38)}$$

For a fluid flow to be physically possible in two dimensions, the stream function ψ must exit and for a flow to be irrotational, the velocity potential ϕ must exist.

Substituting the definition of ψ in the condition of irrotationality,

$$\frac{d^2\,\psi}{dx^2}+\frac{d^2\psi}{dy^2}=0 \qquad \text{....(3.39)}$$

Substituting the definition of ϕ in the equation of continuity,

$$\frac{d^2\phi}{dx^2}+\frac{d^2\phi}{dy^2}=0 \qquad \text{... (3.40)}$$

The above equations are similar; ψ and ϕ satisfy the operator.

$$\boldsymbol{\frac{d^2(\,)}{dx^2}+\frac{d^2(\,)}{dy^2}=0}$$

These equations are known as Laplace's equations; ψ are ϕ called as Laplacian functions. From the definitions, it follows that if ψ is a Laplacian, so must ϕ be and vice versa; the flow must be potential. Conversely, if a flow is potential, ψ and ϕ must be Laplacians.

SUPPLEMENTARY REVISION EXAMPLES

Example 3.26. *For the following velocity vectors, determine the magnitude of the velocity at A(x = 2, y = 3, z = 1 and t = 2).*

(i) $V = (10t + xy)\, i + (-yz + 10t) j + (-yz + z^2/2)\, k.$

(ii) $V = 4xi + (-4y + 3t) j$

Solution.

Let us do the second question first as it appears simpler!

(ii) $$V = 4xi + (-4y + 3t)j$$

At the desired point A,

$$V = 8i + (-12 + 6)j = 8i - 6j$$

The magnitude of velocity is $\sqrt{8^2 + 6^2}$ = **10 units.**

(i) For the first question,

$$V = (20 + 6)i + (-3 + 20)j + (-3 + 1/2)\, k.$$
$$= 26i + 17j - 2.5k$$

The magnitude of velocity is

$$|V| = \sqrt{26^2 + 17^2 + 2.5^2} = \sqrt{676 + 289 + 6.25}$$

$$= \mathbf{31.2\ units.}$$

Example 3.27. *The x and y components in a 3- dimensional flow are given by $u = x^2 + z^2$ and $v = y^2 + z^2$. Find the simplest z component of the velocity.*

Solution.

Fluid flow must satisfy continuity equation for in compressible flow in 3 dimensions.

$$\frac{\partial u}{\partial x} + \frac{\partial v}{\partial y} + \frac{\partial w}{\partial z} = 0$$

From the data

$$\frac{\partial u}{\partial x} = 2x$$

$$\frac{\partial u}{\partial y} = 2y$$

Substituting in the continuity equation.

$$\frac{\partial w}{\partial z} = -2x - 2y$$

Integrating w.r.t. z,

$$w = -2xz - 2yz + f(x, y)$$

For the simplest case, let $f(x, y)$ be zero.

Hence, $$w = -2xz - 2yz$$

One can check the answer, whether it satisfies the continuity equation;

$$\frac{\partial w}{\partial z} = -2x - 2y, \quad \frac{\partial u}{\partial x} = 2x, \quad \frac{\partial u}{\partial y} = 2y$$

The sum is identically zero; showing that it is indeed true!

Example 3.28. *A conical pipe diverges uniformly from 100 to 250 mm diameter over a length of 1.2 m. If the rate of flow varies uniformly from 0.12 m³/s to 0.24 m³/s in 5 seconds, calculate at t = 3 seconds, the local and convective acceleration at the mid point.*

Solution.

At the mid point,

$$d = \frac{100 + 250}{2} = 175 \text{ mm} = 0.175 \text{ m},$$

$$a = \pi \times 0.175^2/4 = 0.024 \text{ m}^2.$$

At $t = 3$ seconds, at the mid point.

$$Q = 0.12 + (0.24 - 0.12) \times 3/5 = 0.192 \text{ m}^3/\text{s}$$

$$U = Q/a = 0.192/0.024 = 8 \text{ m/s}$$

In general, at time t, the velocity at mid point.

$$U_t = [0.12 + (0.24 - 0.12)\, t/5]/\, 0.024$$

$$= (5 + t) \text{ m/s}$$

Local acceleration, $\frac{\partial u}{\partial t} = \mathbf{1\ m/s^2}$.

Now, the diameter $d = 0.1 + (0.25 - 0.1)\, x/1.2$

$$= 0.1 + 0.125\, x$$

and area of cross section $a = \pi\, d^2/4$

$$= \frac{\pi}{4}(0.1 + 0.125x)^2$$

By continuity $U.a$ = constant

whence, $$U\frac{da}{dx} + a\frac{dU}{dx} = 0$$

where $$\frac{dU}{dx} = -\frac{U}{a}\frac{da}{dx}.$$

At mid length, $x = 0.6$ m,

$$\frac{da}{dx} = 2.\frac{\pi}{4}.(0.1 + 0.125 \times 0.6) \times 0.125 = 0.034 \text{ m}$$

Hence $$\frac{dU}{dx} = -\frac{8}{0.024} \times 0.034 = -11.4/s$$

Convective acceleration, $U\frac{dU}{dx} = -8 \times 11.4 = \mathbf{-91.2\ m/s^2}$.

Example 3.29. *The velocity components for the two dimensional flow of an incompressible fluid are stated as*

$$u = y^3/3 + 2x - x^2y$$

$$v = xy^2 - 2y - x^3/3.$$

Obtain an expression for the stream function ψ.

Solution.

It is wise to test if a flow can exist with the stated velocity components; then only would a stream function ψ exist!

So, continuity or existence of flow requires $\dfrac{\partial u}{\partial x} + \dfrac{\partial v}{\partial y} = 0$

For the given components,

$$\frac{\partial u}{\partial x} = 2 - 2xy \text{ and } \frac{\partial v}{\partial y} = 2xy - 2$$

which satisfy the continuity equation. So, a stream function ψ exists such that

$$\frac{\partial \psi}{\partial y} = 4 = y^3/3 + 2x - x^2y$$

Integrating it w.r.t. y,

$$\psi = y^4/12 + 2xy - x^2y^2/2 + f(x)$$

Likewise, $$\frac{\partial \psi}{\partial x} = -v = -xy^2 + 2y + x^3/3$$

whence $$\psi = -x^2y^2/2 + 2yx + x^4/12 + f(y)$$

Combining the two expressions,

$$\psi = 2xy - x^2y^2/2 + x^4/12 + y^4/12.$$

Example 3.30. *For the velocity field given by*
$V = 10xy\, i - 5y^2\, j$, find at point P(2, 3).
(a) absolute velocity
(b) absolute acceleration
(c) angular velocity and
(d) the value of the stream function if the streamline passing through the origin is assigned $\Psi = 0$.

Solution.

From the data

$u = 10\,xy$; at $P(2, 3)$ $u = 60$ units,

$v = -5y^2$; at $P(2, 3)$ $v = -45$ units,

(*a*) Absolute velocity at $P(2, 3)$ is

$$V = 60\,i - 45\,j$$

the magnitude being $\sqrt{60^2 + (-45)^2}$, ***i.e.*, 75 units**

making an angle θ of $\tan^{-1}\left(\dfrac{-45}{60}\right)$ *i.e.*, **–37.87° with *x* axis**

(*b*) Absolute acceleration $a = a_x\, i + a_y\, j$ where

$$a_x = \frac{\partial u}{\partial t} + u\frac{\partial u}{\partial x} + v\frac{\partial u}{\partial y} = 0 + 60 \times 10y - 45 \times 10x$$

$$= 1800 - 900 = 900 \text{ units}$$

$$a_y = \frac{\partial v}{\partial t} + u\frac{\partial v}{\partial x} + v\frac{\partial v}{\partial y} = 0 + 60 \times 0 - 45 \times (-10y)$$

$$= + 45 \times 10 \times 3 = 1350 \text{ units.}$$

Hence, $\quad a = \mathbf{900\, i + 1350\, j}$

the magnitude being $\sqrt{900^2 + 1350^2}$, *i.e.*, **1622 units**

making an angle θ of $\tan^{-1}\left(\frac{1350}{900}\right)$ *i.e.*, **56.3° with *x* axis**

(*c*) Angular velocity in a plane flow is about the axis normal to the plane, at point *P*(2, 3).

$$w_2 = \frac{1}{2}\left(\frac{\partial v}{\partial x} - \frac{\partial u}{\partial y}\right) = \frac{1}{2}(0 - 10x) = \mathbf{-10 \text{ units}}$$

(*d*) The value of the stream function can be determined for a flow that can exist, *i.e.*, which satisfied the continuity equation.

Let us check :

$$\frac{\partial u}{\partial x} = 10y$$

$$\frac{\delta v}{\partial y} = -10y$$

which means that $\frac{\partial u}{\partial x} + \frac{\partial v}{\partial y} \equiv 0$ and the flow can exist.

Now, by definition,

$$u = \frac{\partial \psi}{\partial y} \text{ and } v = -\frac{\partial \psi}{\partial x}$$

$$\frac{\partial \psi}{\partial y} = 10\, xy\, ; \frac{\partial \psi}{\partial x} = 5y^2$$

Integrating the first equation

$$\psi = 10x\, y^2/2 + f(x) = 5xy^2 + f(x)$$

and integrating the second equation,

$$\psi = 5y^2\, x + f(y)$$

The two expression lead us to conclude,

$$\boldsymbol{\psi = 5y^2\, x}$$

Since $f(x)$ or $f(y)$ must be zero if the $\psi = 0$ streamline has to pass through the origin (0, 0).

Example 3.31. *The stream function for a flow field is given by* $\psi = 3x^2y + (2 + t)y^2$ *Express the velocity field and determine the velocity at a point P with position vector.*

$$\boldsymbol{r} = \boldsymbol{i} + 2\boldsymbol{j} - 3\boldsymbol{k} \text{ at time } t = 2$$

Solution.

The velocity vector $\quad V = ui + vj$ is such that

$$u = \frac{\partial \psi}{\partial y}\, ; u = 3x^2 + 2\,(2 + t)\, y$$

$$v = \frac{-\partial \psi}{\partial x}; \ v = -6xy$$

Therefore, $V = [3x^2 + (4 + 2t)y]i - 6xy\,j$

At P(1, 2, –3), $t = 2$

$$u = 3 + 8 \times 2 = 19 \text{ units}$$

$$v = -6 \times 1 \times 2 = -12 \text{ units}$$

$$\boldsymbol{V} = \mathbf{19\,i - 12\,j}$$

Example 3.32. *Given the velocity vector in a flow,*

$$\boldsymbol{V} = -x\,\boldsymbol{i} + 2y\,\boldsymbol{j} + (3 - z)\boldsymbol{k}$$

Calculate the magnitudes of velocity, acceleration and the angular velocity at a point (1, 1, 2).

Solution.

From the given velocity vector, expressed as

$$\boldsymbol{V} = \boldsymbol{ui} + \boldsymbol{vj} + \boldsymbol{wk}$$

$$u = -x, \ v = 2y \text{ and } w = 3 - z$$

We may check for continuity of flow,

$$\frac{\partial u}{\partial x} + \frac{\partial v}{\partial y} + \frac{\partial w}{\partial z} = -1 + 2 - 1 = 0 \text{ at all points !}$$

The velocity at point (1, 1, 2) is

$$\boldsymbol{V} = -1\,\boldsymbol{i} + 2\,\boldsymbol{j} + 1\,\boldsymbol{k}$$

whence, the magnitude is $\sqrt{(-1)^2 + 2^2 + 1^2} = \sqrt{6}$.

The acceleration along the three axes is given by

$$a_x = \frac{\partial u}{\partial t} + u\frac{\partial u}{\partial x} + v\frac{\partial u}{\partial y} + w\frac{\partial u}{\partial z}$$

$$= 0 + (-x) \times (-1) + 0 + 0 = x$$

$$a_y = \frac{\partial v}{\partial t} + u\frac{\partial v}{\partial x} + v\frac{\partial u}{\partial y} + w\frac{\partial v}{\partial z}$$

$$= 0 + 0 + 2y.2 + 2 + 0 = 4y$$

$$a_z = 0 + 0 + 0 + (3 - z) \times (-1) = z - 3$$

Hence acceleration $= x\,\boldsymbol{i} + 4y\,\boldsymbol{j} + (z - 3)\,\boldsymbol{k}$

At point (1, 1, 2), $\boldsymbol{a} = \boldsymbol{i} + 4\boldsymbol{j} - \boldsymbol{k}$; $|a| = \sqrt{\mathbf{18}}$

Now, angular velocity for rotation is given by

$$\omega = \frac{1}{2}(\nabla \times \boldsymbol{U}) = 0$$

because the velocity components are only function of their respective coordinates *e.g.* $u = f(x), \neq (y, z)$, etc.

ITEM BANK

Objective items of different types are given to review the concepts developed in the Chapter.

- *Select correct response for each item.*
- *Every correct response is expected in less than a minute and carries one mark.*
- *Compare your responses with the key answers given at the end.*

Multiple-Choice Items

Attempt the following Multiple-Choice Items by ticking the correct or the most appropriate response from among the alternatives given:

1. A flow is said to be steady when

(*a*) conditions change steadily with time

(*b*) conditions do not change with time at any point

(*c*) conditions do not change steadily with time at any point

(*d*) the velocity does not change at all with time at any point.

2. Uniform flow occurs

(*a*) when the spatial rate of change of velocity is zero

(*b*) when the temporal rate of change of velocity is zero

(*c*) when the velocity changes steadily along the direction of flow

(*d*) only when the velocity vector at any point remains constant.

3. A streamline is a line

(*a*) drawn normal to the velocity vector at every point

(*b*) such that the streamlines divide the passage into equal number of parts

(*c*) which is along the path of a particle

(*d*) tangent to which is the direction of velocity at every point

4. The equation of continuity in fluid mechanics

(*a*) is a condition of equilibrium in the flow pattern

(*b*) is an embodiment of the law of thermodynamics

(*c*) relates work and energy

(*d*) is an embodiment of the law of conservation of mass.

5. A stream function

(*a*) is defined only for incompressible flow

(*b*) is defined for irrotational flow

(*c*) is defined when the flow is continuous

(*d*) is defined only for steady flow.

6. Irrotational flow is such that

(*a*) circulation is zero

(*b*) the particles go in a circular motion

(*c*) stream function exists

(*d*) vorticity may not be zero

7. In a converging steady flow, there is

(*a*) no acceleration

(*b*) no temporal acceleration

(*c*) only convective acceleration

(*d*) convective and temporal acceleration.

8. Dimensions of stream function are

(*a*) MLT^{-1} (*b*) LT^{-1}

(*c*) L^2T^{-1} (*d*) LT

9. Observation of a flow net enables us to

(*a*) determine the velocity at all points

(*b*) estimate the pressure at all points

(*c*) estimate the velocity variations

(*d*) determine the energy loss in flow.

Other Objective Items

10. Identify the following statements as true or false:

(*a*) Axisymmetric flows are two dimensional (T/F)

(*b*) One dimensional flow can take place in a curved passage (T/F)

(*c*) Local acceleration is zero in steady flow (T/F)

(*d*) Uniform flow can take place in a conical passage (T/F)

11. Are the following statements true or false?

(*a*) If y exist f must exist. (T/F)

(*b*) If f exist y must exist (T/F)

(*c*) If y is a Laplacian, f must exist (T/F)

(*d*) If f is a Laplacian, y must exist (T/F)

12. Are the following statements true or false?

(*a*) Irrotationality implies continuity of flow (T/F)

(*b*) Existence of stream function implies continuity (T/F)

(*c*) Irrotational flow implies zero vorticity (T/F)

(*d*) Intersection of two streamline implies a stagnation point (T/F)

13. Identify the following as true or false:

(*a*) Irrotational flow and inviscid flow are the same thing (T/F)

(*b*) For a possible irrotational flow, there must be a Laplacian (T/F)

(*c*) Circulation must be zero along a closed contour in an irrotational flow (T/F)

14. Match the flows in block A with the type of flow in block B.

Block A	**Block B**
(*a*) flow over a sphere	(*k*) two dimensional flow
(*b*) flow over a long circular cylinder	(*l*) three dimensional flow
(*c*) flow in a pipe bend	(*m*) axisymmetric flow
	(*n*) uniform flow

15. Match the conditions in block A with resulting facts in block B.

Block A	**Block B**
(*a*) existence of stream function	(*k*) irrotationality of flow
(*b*) existence of velocity potential function	(*l*) continuity of flow
(*c*) absence of temporal variations	(*m*) steady flow
(*d*) constant velocity vector	(*n*) uniform flow

16. Identify the following statements as true or false

(*a*) irrotational flow must be without shear (T/F)

(*b*) streamlines and velocity potential lines must constitute orthogonal net work (T/F)

ANSWERS

[1. *b*, **2.** *a*, **3.** *d*, **4.** *d*, **5.** *c*, **6.** *a*, **7.** *c*, **8.** *b*, **9.** *c*, **10.** (*a* T, *b* T, *c* T, *d* T), **11.** (*a* F, *b* F, *c* T, *d*T), **12.** (*a* F, *b* T, *c* T, *d* T), **13.** (*a* F, *b* T. *c* T), **14.** (*a m*, *b k*, *c l*), **15.** (*a l*, *b k*, *c m*, *d n*), **16.** (*a* F, *b* T)].

QUESTION BANK

Questions on theory and numerical problems are given for tutorial work.

* *The questions are arranged generally in order of articles in the text*
* *Every question should take 15-20 minutes for solution.*
* *Answers for the numerical problems are given alongside them.*

3.1. Do the following velocity components represent physically possible flows?

(a) $u = x^2 + z^2 + 5;\ v = y^2 + z^2;\ w = 4xyz$

(b) $u_r = \left(1 - \dfrac{A}{r^2}\right)\cos\theta;\ u_\theta = \left(1 + \dfrac{A}{r^2}\right)\sin\theta$

(c) $u_r = m/4\pi r;\ u_\theta = 0;\ u_z = 0$

(d) $U = 5x\, i + (3y + ty^2)\, j$ **[No, No, Yes, No]**

3.2. Examine whether the following velocity components can constitute two or three-dimensional incompressible flows. If three-dimensional, determine the third component

(a) $u = x^2 - y^2,\ v = y^2 + \log x$

(b) $u_r = r\sin\theta,\ u_\theta = 2r\cos\theta$

(c) $u_r = m/4\pi r^2,\ u_\theta = 0$

[(a) $w = -(2x + 2y)z + f(x, y)$
(b) Two-dimensional flow (c) $u_z = mz\,(4\pi r^3) + f(r, \theta)$]

3.3. The velocity profile for flow between two plates 2 m apart is given by

$$u = 10\,(1 - y^2)\ \text{m/s}$$

where u is the velocity at distance y metre from the plane lying midway between the surfaces.

(a) plot the velocity profile
(b) express the stream function ψ in terms of y
(c) plot the streamline pattern and
(d) relate the velocity profile with the streamline pattern qualitatively.

[$\psi = 10y\,(1 - y^2/3)$; Higher velocity corresponds to closer streamlines]

3.4. A conical nozzle 2.5 m long converging from 0.2 m to 0.1 m diameter linearly is subjected to an inlet flow varying from 0 to 1 m^3/s in 5 seconds. Determine the total acceleration at the mid-length of the nozzle at the mid-time of variation. Assume uniform flow over each cross-section.

3.5. Given a velocity field

$$U = (1 + 3t)\, i + yj$$

establish the streamlines passing through arbitrary points (x_1, y_1), (x_2, y_2), (x_3, y_3) etc. obtain also the parametric equations for the path of single particle passing through these points. Hence show that the streamline and pathline patterns coincide for steady flow.

3.6. Fluid velocity U and temperature T are known for a flow-field as

$$U = 6xy\, i + (xt - 3y^2)j$$

$$T = 20 + x^2 + yt$$

Evaluate the substantive rate of change of temperature in fluid flow at a point (2, 3) at time $t = 2$. **[101 units]**

3.7. Determine the stream function ψ as a direct consequence of the continuity equation, *i.e.*, function of x and y such that

$$\frac{\partial\psi}{\partial x} = u \text{ and } \frac{\partial\psi}{\partial y} = -v$$

Show that ψ is a constant along a streamline.

3.8. Obtain an expression for the component of rotation normal to $r-\theta$ plane in the cylindrical coordinate system; hence derive the condition of irrotationality for two dimensional flows.

3.9. Do the following velocity potentials represents possible flows? If so determine the stream functions and sketch the flow patterns.

(*a*) $\phi = y + x^2 - y^2$

(*b*) $\phi = U_r \cos\theta + \mu \cos\theta / r$

(*c*) $\phi = U\left(r + \dfrac{a^2}{r}\right)\cos\theta + \dfrac{\theta}{2\pi}$

[$x(2y-1)$, $Uv \sin\theta$, $-\mu \sin\theta/r$ $U(r - a^2/r)\sin\theta - \Gamma/2\pi \log r$]

3.10. For the flows represented by the following stream functions, determine

(*i*) $\psi = xy$ (*ii*) $\psi = \log r$ (*iii*) $\psi = 9r^2 \sin^2\theta$

(*a*) the velocity components

(*b*) the rotational character

(*c*) the shearing strain rates and

(*d*) the rates of dilatation.

[(*i*) $x, -y$; *zero*; 1, – 1 ; (*ii*) $y/(x^2+y^2), -x/(x^2/y^2)$; zero; $(x^2-y^2)/(x^2+y^2)^2$; $-2xy/(x^2-y^2)^2$; (*iii*) $18y$, 0; – 9; 9; 9]

3.11. Determine the stream function for a fluid flow if

$$u = 2x, \quad v = -2y$$

Plot the streamlines $\psi = 1$, 2, and 3.

Determine also the velocity potential function ϕ and plot a set of representative ϕ = constant lines in the same range of flow.

[$\psi = 2xy$, $\phi = x^2 - y^2$]

3.12. Evaluate the velocity and acceleration components at a point (1, 2, 3) at time $t = 1$ for a flow field, $U = 5x^3\,i - 15x^2\,yj + tk$

[$5\,i - 30j + k$; $75i + 150j + k$]

3.13. A two-dimensional flow is given by $u = y/b^2$ and $v = -x/a^2$. Determine the equation of the streamline passing through the point $(a, 0)$ Discuss the flow. **[$x^2/a^2 + y^2/b^2 = 11$]**

3.14. In a plane diffuser formed by an arrangement of stationary vanes, the flow patterns is approximately represented by the stream function

$$\psi = R\theta - 3\left[r + \frac{R}{3}e^{-3r/R}\right]$$

where r and θ are the coordinates of a point in the flow and R is the constant radial distance from the centre of the diffuser to the outer tip of each vane.

(*i*) Determine the radial and tangential components of velocity and sketch their variation in the region $0.5 \le r/R \le 1./0.$

(*ii*) Sketch the flow pattern in the diffuser and then plot a few points to show the likely shape of one vane in the region $0.5 \le r/R \le 1./0.$ **[$u_r = R/r$, $u_\theta = 3(1 - e^{-3r/R})$]**

3.15. If for a two-dimensional potential flow, the velocity potential is given by the expression $f = x(2y - 1)$.

(*i*) determine the velocity at the point P (4, 5).

(*ii*) What is the value of the stream function ψ at the point P ?

[(*i*) $u = 9$ and $v = 8$ units (*ii*) $y = y^2 - x^2 - y$, 4 units]

3.16. State if the flow represented by $u = 3x + 4y$ and $v = 2x - 3y$ is rotational or irrotational. Find the potential function if the flow is irrotational and vorticity if it is rotational.

[Rotational; vorticity = 2 units]

3.17. A stream function is given by

$$\psi = 2x^2 y + (x + t) y^2$$

Find the flow rates across the faces of the triangular prism OAB, shown in Fig. problem 3.17 having a thickness of 5 units in the z drection at time $t = 1$.

[OB, 20; OA, zero; AB, 20 units]

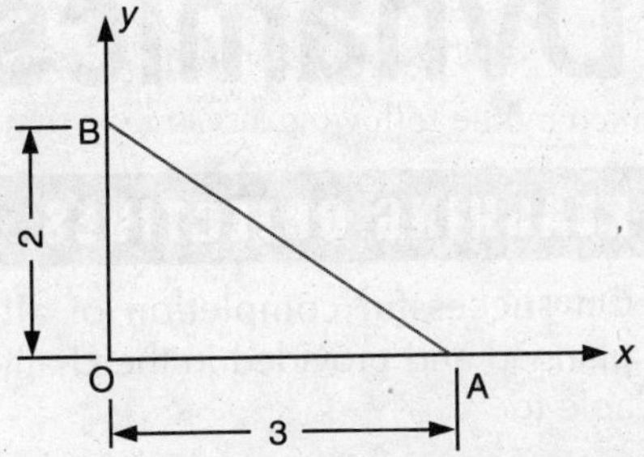

Fig. Problem: 3.17

CHAPTER

4

Fluid Dynamics

LEARNING OBJECTIVES

On successful completion of all learning activities planned and provided in the chapter, a student will be able to

- establish the Euler's equation of motion for a two-dimensional steady flow of an incompressible fluid and extend it for three-dimensional viscous flow;
- obtain the Bernoulli equation by integrating the Euler's equation under appropriate conditions;
- establish the steady flow energy equation and obtain the Bernoulli equation from it;
- state and apply the integral momentum equation and the angular momentum equation for simple engineering flows and
- determine the force exerted by a fluid flow on the passage confining it.

4.1. Statement of Laws for a System

'Dynamics' means the study of the state of motion of a body or a system under the influence of external forces and moments. We should, therefore, state and apply the accepted laws of motion and formulate the equations governing the fluids in motion. In Newtonian mechanics, a fluid system should obey the three laws of motion.

Newton's First Law of Motion: *Every body perseveres in its state of rest or of constant velocity unless it is compelled to change that state by some external action impressed thereon.* The term 'external action' is used to imply the external forces and moments acting on the system. The law provides the necessary conditions of equilibrium as

$$\Sigma F = 0 \text{ and } \Sigma M = 0 \qquad ...(4.1)$$

The above equations have been used in evaluating and locating the hydrostatic forces acting on a submerged body and in establishing the equilibrium and stability of a floating body.

Newton's Second Law of Motion: *The rate of change of momentum of a body is proportional to the impressed action.* The momentum implied may be linear or angular and the corresponding actions are force and moment respectively.

A large bulk of this chapter is devoted to the establishment of the equation of motion for a control volume, called the Euler's equation for *inviscid* flow.

Newton's Third Law of Motion. To every action, there is always an opposed and equal reaction. This law governs the action and reaction between two or more bodies. For fluid elements, the mutual actions upon each other are equal and oppositely directed. Between a fluid control volume and the confining passage, *e.g.*, in a jet engine, the force exerted by the fluid on the jet engine body must be equal and oppositely directed to the reaction force exerted by the jet engine body on the confined fluid. For a static column of a fluid in a container, the force exerted by the fluid on a part of the container must be equal and opposite to the reaction of that part of the container.

Reynolds Transport Theorem

The laws of motion can be stated either for a body or for an identified system. In a fluid motion, it is not possible to identify a 'lump of mass' as a body since it undergoes continuous deformation; it is easier to consider a 'control volume' instead. Reynolds transport theorem enables us to extend the laws to control volumes.

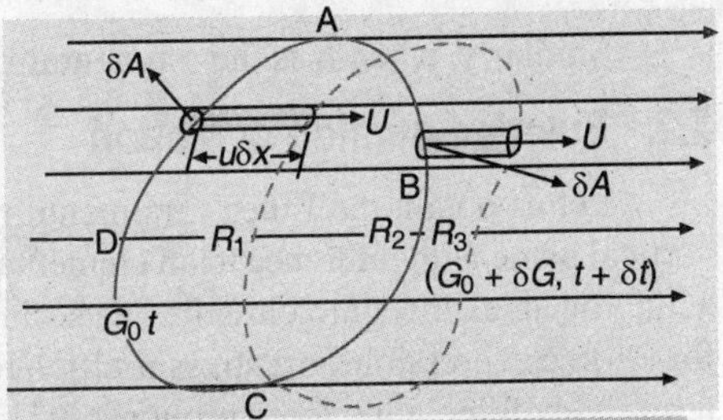

Fig. 4.1: Concept of 'Transport'

Consider a system *ABCD* bounded by a control surface *CS* at a time *t* as shown in Fig. 4.1. Let its volume define the control volume *CV*. Due to fluid motion, the system occupies a new position, shown dotted, at a time $t + \delta t$. Let *G* define the location of volume in the flow field. An extensive property *N* (dependent upon mass) is a function of intensive property *n* (independent of mass) such that

$$N = \int n\rho \, dV = N\,(G, t)$$

whence,
$$\left(\frac{dN}{dt}\right)_s = \frac{DN}{Dt} = \lim_{\Delta t \to 0} (N(G_0 + \delta G, t + \delta t) - N(G_0, t)) / \delta t$$

where G_0 and $G_0 + \delta G$ locate the volumes at *t* and $t + \delta t$ respectively.

Adding and subtracting $N(G_0, t + \delta t)$ in the term under the limit and rearranging them as the following two terms :

$$\lim_{\Delta t \to 0} (N(G_0, t + \delta t) - N(G_0, t)) / \delta t = \left(\frac{\partial N}{\partial t}\right)_{G_0} = \frac{\partial}{\partial t} \int_V n\rho \, dV$$

which means the local rate of change of *N* in the control volume and

$$N(G_0 + \delta G, t + \delta t) - N(G_0, t + \delta t) = N_3 - N_1$$

where N_1 and N_3 refer to the values of *N* for regions R_1 and R_3 respectively, the region R_2 being common to both positions.

Now,
$$N_3 = \int n\rho(dA.\,\boldsymbol{U}\delta t) = \int_{CDA} n\rho \, dt \, (\boldsymbol{U}.\,\boldsymbol{dA})$$

and
$$-N_1 = \int_{ABC} n\rho \, \delta t \, (\boldsymbol{U}.\,\boldsymbol{dA});$$

whence
$$N_3 - N_1 = \int_{ABCD} n\rho \, \delta t \, (\boldsymbol{U}.\,\boldsymbol{dA})$$

which reduces the second limit-term to $\int n(\rho \boldsymbol{U}.\,\boldsymbol{dA})$

Finally,
$$\boxed{\frac{DN}{Dt} = \int_{CS} n(\rho \boldsymbol{U}.\,\boldsymbol{dA}) + \frac{\partial}{\partial t} \int_{CV} n\rho dV} \quad ...(4.2)$$

This is the Reynolds transport equation. It shows that the rate of change of N (extensive property) for a system equals the sum of the efflux of N across the control surface and the rate of change of N within the control volume.

For example, when N is the mass, the l.h.s. terms is zero by the law of conservation of mass, then

$$\int_{CS} \rho U . dA + \frac{\partial}{\partial t} \int_{CV} \rho dV = 0$$

is the equation of continuity. It reduces to the familiar form

$$\nabla.(\rho U) + \frac{\partial \rho}{\partial t} = 0 >$$

Similarly, when N is the momentum, *i.e.*, $m\ U$, n becomes the velocity U.

4.2. Euler's Equation of Motion

Let us obtain the Euler's momentum equation for two-dimensional inviscid steady flow in a vertical plane with reference to an elementary fluid control volume in Fig. 4.2(*a*). The external forces acting on an element are classified as surface forces and body forces. We use the term surface force for a force expressible as a stress multiplied by the area of a surface *e.g.*, pressure force. We use the term 'body force' for a force proportional to the volume of the fluid element, *e.g.*, gravity force. Net surface force in the x-direction is

$$p\delta z\delta y - \left(p + \frac{\partial p}{\partial x}\right)\delta z.\delta y = \frac{\delta p}{\delta x}\delta x\delta y\delta z = \frac{\partial p}{\partial x}\delta V$$

Net surface force plus body force in the z-direction,

$$= -\frac{\partial u}{\partial z}\delta V - \rho g \delta V$$

Recalling the acceleration components for steady

flow in the x-direction, $a_x = u\dfrac{\partial u}{\partial x} + w\dfrac{\partial u}{\partial z}$

in the z-direction, $a_z = u\dfrac{\partial w}{\partial x} + w\dfrac{\partial w}{\partial z}$

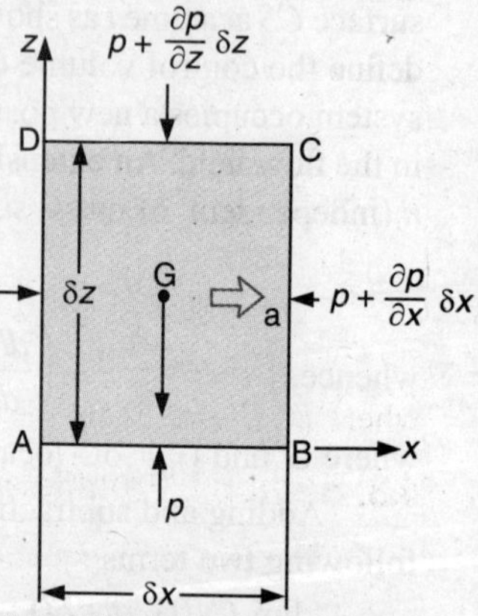

Fig. 4.2 (a): Force on an Element in the Rectangular Coordinates

The Newton's second law in the differential form can be written as

$$dF = dm\frac{DU}{Dt} = \text{mass} \times \text{total acceleration} \quad ...(4.3)$$

Therefore, $\left(u\dfrac{\partial u}{\partial x} + w\dfrac{\partial u}{\partial z}\right)\rho\delta V = -\dfrac{\partial p}{\partial x}\delta V$ along the x-axis.

and $\left(u\dfrac{\partial w}{\partial x} + w\dfrac{\partial w}{\partial z}\right)\rho\delta V = -\dfrac{\partial p}{\partial z}\delta V - \rho g\ \delta V$ along the z-axis.

Finally, for the two-dimensional steady flow of an inviscid fluid in a vertical plane, the Euler's momentum equations are

$$u\frac{\partial u}{\partial x} + w\frac{\partial u}{\partial z} = -\frac{1}{\rho}\frac{\partial p}{\partial x} \quad ...(4.4\ a)$$

and

$$u\frac{\partial w}{\partial x} + w\frac{\partial w}{\partial z} = -\frac{1}{\rho}\frac{\partial p}{\partial z} - g \quad ...(4.4\ b)$$

If the fluid flow occurs under the action of body forces (in addition to that due to gravitation), then B_x and B_z the external body force components per unit mass are written in general and Eqs. (4.4) becomes

$$u\frac{\partial u}{\partial x}+w\frac{\partial u}{\partial z} = B_x-\frac{1}{\rho}\frac{\partial p}{\partial x}$$

$$u\frac{\partial w}{\partial x}+w\frac{\partial w}{\partial z} = B_z-\frac{1}{\rho}\frac{\partial p}{\partial z}$$

The Euler's momentum equations for three-dimensional inviscid steady flow can be derived in a similar way and are:

$$u\frac{\partial u}{\partial x}+v\frac{\partial u}{\partial y}+w\frac{\partial u}{\partial z} = B_x-\frac{1}{\rho}\frac{\partial p}{\partial x}$$

$$u\frac{\partial v}{\partial x}+v\frac{\partial v}{\partial y}+w\frac{\partial v}{\partial z} = B_y-\frac{1}{\rho}\frac{\partial p}{\partial y} \qquad ...(4.5)$$

$$u\frac{\partial w}{\partial x}+v\frac{\partial w}{\partial y}+w\frac{\partial w}{\partial z} = B_z-\frac{1}{\rho}\frac{\partial p}{\partial z}$$

The Euler's equations in cylindrical coordinates for an axisymmetric steady flow can be derived with reference to the control volume in Fig. 4.2 (*b*) and appears as

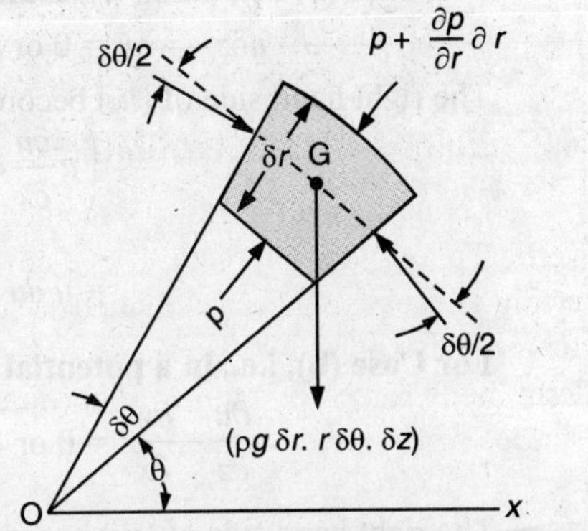

Fig. 4.2 (b): Forces on an Element in the Cylindrical Co-ordinates

$$u_r\frac{\partial u_r}{\partial r}+u_z\frac{\partial u_r}{\partial z}-\frac{u_\theta^2}{r} = B_r-\frac{1}{\rho}\frac{\partial p}{\partial r}$$

$$u_r\frac{\partial u_\theta}{\partial r}+u_z\frac{\partial u_\theta}{\partial z}+\frac{u_r u_\theta}{r} = B_\theta \qquad ...(4.6)$$

$$u_r\frac{\partial u_z}{\partial z}+u_z\frac{\partial u_z}{\partial r} = B_z-\frac{1}{\rho}\frac{\partial p}{\partial z}$$

where B_r, B_θ and B_z are the body forces per unit mass in r, θ and z directions respectively.

4.3. Bernoulli Equation: Derivation from the Euler's Equation

The Bernoulli equation, in its most general form, can be written as

$$\int\frac{dp}{\rho g}+\frac{U^2}{2g}+z = \text{constant} \qquad ...(4.7)$$

The interpretation of the terms and the scope of its application can be appreciated by deriving it from the Euler's equation. The very fact that it can be derived from the Euler's equation means that it is valid for inviscid fluid flow and it is contained in the Euler's equation!

Integration of the Euler's equation for two different cases results in the familiar form of the Bernoulli equation. The two cases are, therefore, considered in the process of derivation. These are also the cases where the equation is most frequently used.

Case (a) – Along a Streamline in an Inviscid and Steady Flow

Case (b) – Between Any Two Points in a Potential/Irrotational and Steady Flow.

For a two-dimensional steady flow in the presence of gravitational forces, the Euler's equations are

$$-\frac{1}{\rho}\frac{\partial p}{\partial x} = u\frac{\partial u}{\partial x}+w\frac{\partial u}{\partial z} \qquad ...(i)$$

and $$-g-\frac{1}{\rho}\frac{\partial p}{\partial z} = u\frac{\partial w}{\partial x}+w\frac{\partial w}{\partial z} \qquad ...(ii)$$

Multiplying each term of (*i*) by δx and each term of (*ii*) by δz and adding them,

$$-\frac{1}{\rho}\left(\frac{\partial p}{\partial x}\delta x+\frac{\partial p}{\partial z}\delta z\right)-g\delta z = u\frac{\partial u}{\partial x}\delta x + w\frac{\partial u}{\partial z}\delta x + u\frac{\partial w}{\partial x}\delta z + w\frac{\partial w}{\partial z}\delta z \qquad ...(iii)$$

The left hand side term is simply written as

$$-\frac{dp}{\rho} - gdz \qquad ...(iv)$$

The right hand side term can be simplified for the above two cases as follows:

For Case (*a*) i.e., along a streamline, invoke the equation of a streamline in the *x-z* plane.

$$u\delta z - w\delta x = 0 \text{ or } u\,\delta\, z = w\,\delta\, x$$

The right hand side of (*iii*) becomes

$$\left(u\frac{\partial u}{\partial x}\delta x + u\frac{\partial u}{\partial z}\delta z\right) + \left(w\frac{\partial w}{\partial x}\delta x + w\frac{\partial w}{\partial z}\delta z\right)$$

$$= u\,du + w\,dw = d\left(\frac{u^2+w^2}{2}\right) = d\left(\frac{U^2}{2}\right) \qquad ...(v)$$

For Case (b), i.e., in a potential flow, invoke the condition of irrotationality in the *x-z* plane,

$$\frac{\partial u}{\partial z} - \frac{\partial w}{\partial x} = 0 \text{ or } \frac{\partial u}{\partial z} = \frac{\partial w}{\partial x}$$

The right hand side of (*iii*) becomes

$$u\frac{\partial u}{\partial x}\delta x + w\frac{\partial w}{\partial x}\delta x + u\frac{\partial u}{\partial z}\delta z + w\frac{\partial w}{\partial z}\delta z$$

$$= \left(u\frac{\partial u}{\partial x}\delta x + u\frac{\partial u}{\partial z}\delta z\right) + \left(w\frac{\partial w}{\partial x}\delta x + w\frac{\partial w}{\partial z}\delta z\right)$$

$$= u\,du + w\,dw = d\left(\frac{u^2+w^2}{2}\right) = d\left(\frac{U^2}{2}\right) \qquad ...(vi)$$

which is precisely the same as for case (*a*).

Now, for either case (*a*) or case (*b*), it can be written as

$$-\frac{dp}{\rho} - gdz = d\left(\frac{U^2}{2}\right)$$

The assumption of two-dimensionality, as above, in Case (*a*) and in Case (*b*) is not necessary. It was made merely for the sake of reducing the number of terms for convenience. As a matter of fact, the Euler's equation for the steady flow of an inviscid fluid may be written, in vector notation, as

$$-\frac{\nabla p}{\rho} - g\nabla z = U\frac{\partial U}{\partial s}$$

Now, if *s* is the coordinate along a streamline or it is along any line in an irrotational flow, let us take the dot product of each term with an elementary displacement *ds* along *s*. Thus,

$$-\frac{\nabla p.ds}{\rho} - g\nabla z.ds = U\frac{\partial U}{\partial s}.ds$$

The terms $\nabla p.\, ds$ equals *dp*; the term $\nabla z.\, ds$ equals *dz* and the right hand side term simplifies to *UdU* or $d(U^2/2)$. Consequently it results in the same equation:

$$-\frac{dp}{\rho} - gz = d\left(\frac{U^2}{2}\right)$$

On rearrangement and integration,

$$\int \frac{dp}{\rho g} + \frac{U^2}{2g} + z = \text{constant} \quad ...(4.7\ a)$$

This is the general form of the Bernoulli equation, which may now be simplified for different cases.

(1) For an incompressible fluid flow, ρ = constant,

$$\frac{p}{\rho g} + \frac{U^2}{2g} + z = C \quad ...(4.7\ b)$$

(2) For a compressible fluid flow, undergoing an adiabatic process,

$$p = C\rho^\gamma$$

$$dp = \gamma C \rho^{\gamma-1} d\rho$$

and

$$\int \frac{dp}{\rho g} = \frac{\gamma}{g} C \int \rho^{\gamma-2} d\rho = \frac{\gamma}{g} \frac{p}{\rho^\gamma} \frac{\rho^{\gamma-1}}{(\gamma-1)} = \frac{\gamma}{\gamma-1} \frac{p}{\rho g}$$

The Bernoulli equation appears as

$$\frac{\gamma}{\gamma-1} \frac{p}{\rho} + \frac{U^2}{2} + gz = K \quad ...(4.8)$$

The assumptions made in the derivation of the Bernoulli equation may now be summarised as follows:

(i) inviscid flow; $\mu = 0$

(ii) steady flow; $\frac{\partial}{\partial t} = 0$

(iii) either the flow is along a streamline or the flow is irrotational

(iv) two-dimensional flow, in the presence of gravitational forces.

(v) for incompressible flow, the form is

$$\frac{p}{\rho g} + \frac{U^2}{2g} + z = C$$

and for compressible flow undergoing adiabatic process, the form is,

$$\frac{\gamma}{\gamma-1} \frac{p}{\rho g} + \frac{U^2}{2g} + z = K$$

The Bernoulli equation may, therefore, be applied with these assumptions.

It may be added to clarify that for an inviscid flow, not irrotational, the Bernoulli constants C_1, C_2, C_3 have different values along different streamlines whereas for irrotational flow, the Bernoulli constant C is the same in the entire flow field.

4.4. Flow through a Confined Passage

When a fluid flows through a confined passage such as a pipe, converging or diverging duct, a bend or a sudden expansion or a contraction, velocity of the fluid changes in accordance with the equation of continuity. Consequent to the change in velocity, the change in pressure is governed by the Bernoulli equation. If there is heat input/output or shaft work output/input during the flow, the steady flow energy equation is applied instead of the Bernoulli equation and the change in temperature is calculated.

Example 4.1. *Air flows steadily through a horizontal nozzle. At the nozzle inlet, the velocity is 6 m/s and the pressure is 100 kN/m². If the inlet area is 0.1 m² and the contraction ratio is 5, determine.*

(a) the velocity and the pressure at the exit.

(b) the discharge and mass flow rate through the nozzle.

Solution.

Consider a control volume bounded by a control surface as shown by dotted line in the figure.

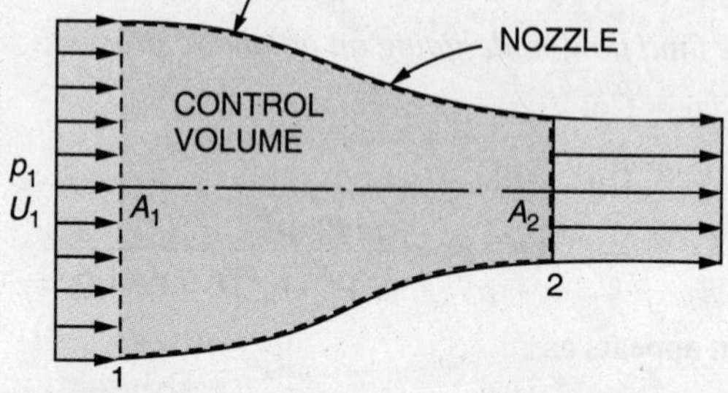

The air enters the inlet section 1 with a uniform velocity

$$U_1 = 6 \text{ m/s}^2$$

and at a uniform pressure

$$p_1 = 100 \text{ kN/m}^2$$

The areas of cross sections at entry and exit are

$$A_1 = 0.1 \text{ m}^2 \text{ and } A_2 = 0.1/5 = 0.02 \text{ m}^2.$$

(*a*)Assuming the flow of air as incompressible, the discharge at inlet should equal the discharge at exit.

$$U_1 A_1 = U_2 A_2$$

whence

$$U_2 = \frac{U_1 A_1}{A_2} = \frac{6 \times 0.1}{0.02} = \mathbf{30\ m/s}$$

It may be noted that the assumption of incompressibility of flow of air is well justified because the maximum velocity in the nozzle is only 30 m/s, which is well short of the speed of sound in air. (The speed of sound in air is 340 m/s at standard temperature and pressure conditions.)

Applying the Bernoulli equation between the inlet 1 and the exit 2,

$$\frac{p_1}{\rho g} + \frac{U_1^2}{2g} + z_1 = \frac{p_2}{\rho g} + \frac{U_2^2}{2g} + z_2$$

Since the axis of the nozzle is horizontal,

$$z_1 = z_2$$

Rearranging the terms, $p_2 = p_1 + \frac{\rho}{2}(U_1^2 - U_2^2)$

Taking the mass density of air as

$$\rho = 1.2 \text{ kg/m}^3$$

and the acceleration due to gravity

$$g = 9.81 \text{ m/s}^2$$

$$p_2 = 100000 + \frac{1.2}{2}(6^2 - 30^2)$$

$$= 99480 \text{ N/m}^2 = \mathbf{99.48\ kN/m^2}$$

(*b*) Now, $Q = U_1A_1$ or U_2A_2

$= 6 \times 0.1$ or 30×0.02

$= 0.6$ m^3/s

The mass flow rate is density times the velocity rate of discharge

$= 0.6 \times 1.2 =$ **0.72 kg/s**

Example 4.2. *A tube is used as a siphon to discharge an oil of specific gravity 0.8 from a large open vessel into drain at the atmospheric pressure. Calculate*

(a) the velocity of the oil through the siphon

(b) the pressure at the highest point B and

(c) the pressure at points A and C.

After calculating these values, comment on the following:

(i) Would the discharge be more, less or the same if the fluid were water ?

(ii) Is there a limit on the maximum height of B above the level in the vessel ?

(iii) Is there a limit on the vertical depth of the right limb of the siphon ?

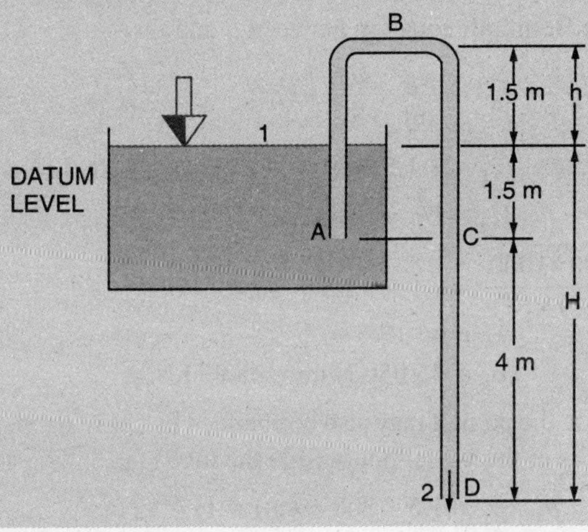

Solution.

Let us first state the assumptions :

1. The fluid is inviscid which means that there are no viscosity effects. It also means that the velocity is uniform across the tube.

2. The flow is incompressible which means that the density of the oil remains constant,

$$\rho = 0.8 \times 1000 = 800 \text{ kg/m}^3$$

3. The flow is steady which means that the instant of interest is neither the start nor the end of the process; there are no variations with the passage of time.

4. The vessel is large compared with the cross section of the tube:

$$A_1 >> A_2$$

Under these assumptions, the one-dimensional continuity equation and the Bernoulli equation may be applied between any two cross sections. Consider a cross section 1 of the large vessel and a cross section 2 at the exit of the siphon tube.

$$\rho U_1 A_1 = \rho U_2 A_2 \qquad ...(i)$$

$$\frac{p_1}{\rho g} + \frac{U_1^2}{2g} + z_1 = \frac{p_2}{\rho g} + \frac{U_2^2}{2g} + z_2 \qquad ...(ii)$$

Since $A_1 >> A_2$, it follows from (*i*) that

$$U_1 << U_2$$

i.e., U_1 is negligible in comparison with U_2, the velocity in the siphon. The velocity in the siphon will be designated as U.

(*a*) Taking the datum level at the free surface in the vessel

$$z_1 = 0 \text{ and } z_2 = -5.5 \text{ m}$$

and recognising that the atmospheric pressure acts at the oil surface in the tank as well as at the exit of siphon,

$$p_1 = p_{atm} = p_2 = p_D = 100 \text{ kN/m}^2$$

it follows from the Bernoulli equation that,

$$\frac{U^2}{2g} = z_1 - z_2 = 0 - (-5.5) = 5.5$$

$$U = \sqrt{2 \times 9.81 \times 5.50} = \mathbf{10.44 \text{ m/s}}$$

(*b*) Applying the Bernoulli equation between B and D.

$$\frac{p_B}{\rho g} + \frac{U_B^2}{2g} + z_B = \frac{p_D}{\rho g} + \frac{U_D^2}{2g} + z_D$$

and recognising that $z_B = +1.5$ m, $z_D = -5.5$ m, $p_D = 100$ kN/m^2

and $U_B = U_D = 10.44$ m/s

$$\frac{p_B - 100 \times 1000}{800 \times 9.81} = -5.5 - 1.5 = -7.0$$

$$p_B = 45\ 050 \text{ N/m}^2 = \mathbf{45.05 \text{ kN/m}^2}$$

(*c*) The pressure at the point A may also be obtained by applying the Bernoulli equation between 1 and A. If the point A is in the vessel but outside the tube then

$$p_A - p_1 = 1.5 \times 800 \times 9.81 = 11775 \text{ N/m}^2$$

which means that p_A is 11.75 kN/m^2 above the atmospheric pressure at 1,

$$p_A = 100 + 11.775 = \mathbf{111.775 \text{ kN/m}^2}$$

If the point A lies inside the siphon tube, the pressure there would be less.

$$\frac{p_A - p_1}{800 \times 9.81} = 1.5 - \frac{10.44^2}{2 \times 9.81} = -4.055$$

$$p_A = 100 \times 1000 - 31825$$

$$= 31.825 \text{ kN/m}^2 \text{ below atmospheric pressure}$$

$$= \mathbf{68.175 \text{ kN/m}^2 \text{ absolute.}}$$

Comments on the Siphon

(*i*) The velocity of the fluid in a siphon tube depends only upon the vertical height of the level surface above the exit of the siphon. The velocity of water would, therefore, be the same as that of oil, *i.e.*, 10.44 m/s. The volumetric discharge would also be the same.

(*ii*) It is seen that the pressure at B is sub-atmospheric. At a certain height, h, the pressure would be so low that the liquid would vaporise and the siphon action would break down. This may happen when the pressure falls below 3 m head, *i.e.*, 29.43 kN/m^2 in case of water.

The maximum height h of B should, therefore, be such that the local pressure does not fall below the vapour pressure of the liquid.

(*iii*) More the depth of the right limb of the siphon below the level surface in the vessel, more the velocity U through the siphon:

$$\frac{U^2}{2g} = z_1 - z_2 = H$$

The velocity U through the siphon increases as H increases. At large H, the velocity may tend to be so high as to render the pressure at B below vapour pressure although B may only be slightly above the level surface in the vessel.

4.5. Flow through Orifices and Mouthpieces

A fluid jet emerging from an orifice of area a tends to contract to form a vena contracta of area a_c at the atmospheric pressure beyond which the fluid-jet has a tendency to expand depending upon the resistance and entrainment of the surrounding fluid as shown in Fig. 4.3(*a*). The streamlines at vena contracta are substantially parallel to one another. A jet of water in air has significant tendency to expand whereas a jet of air in air or of water in water expands visibly.

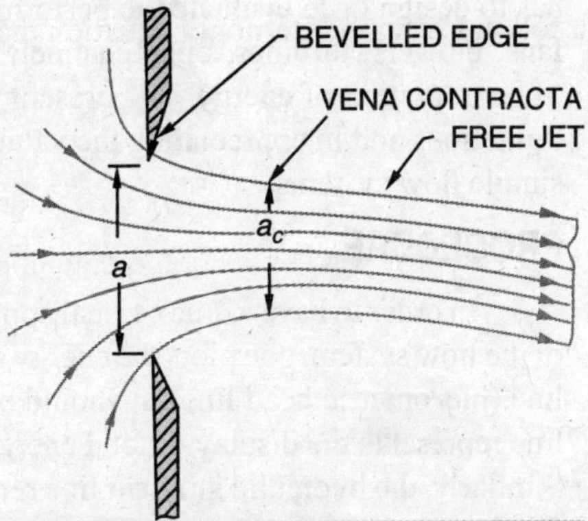

Fig. 4.3 (a): Flow through a Sharp Orifice

Coefficient of contraction

$$C_c = \frac{a_c}{u} \qquad ...(4.9)$$

The actual velocity at the vena contracta u_c is also less than the theoretical velocity u.

Coefficient of velocity $C_v = \dfrac{u_c}{u}$ ...(4.10)

The actual discharge $u_c \cdot u_c$ bears a ratio to the theoretical discharge $a.u$ called the coefficient of discharge.

Coefficient of discharge

$$C_d = \frac{a_c . u_c}{a . u} = C_c \times C_v \qquad ...(4.11)$$

Orifices may be of a variety of shapes and sizes and classified as small or large–depending upon whether or not the velocity through an orifice may be taken as uniform or not.

A mouthpiece, internal or external, straight or converging-diverging, has the basic effect of increasing the C_d through an increase of C_v by running full and limiting the extent of contraction of the jet.

Orifices and mouthpieces are usually employed to discharge the flow with a view to emptying the container. The time of emptying is determined by considering an elementary strip of the fluid in the container and writing for time δt to discharge it. For an arbitrary-sharp container, as in Fig. 4.3(*b*), let the level at an instant be at a height h above the orifice. For a fall δh in the level, the time taken is δt. If A is the local area of coss-section, the volume of liquid discharge over this time is

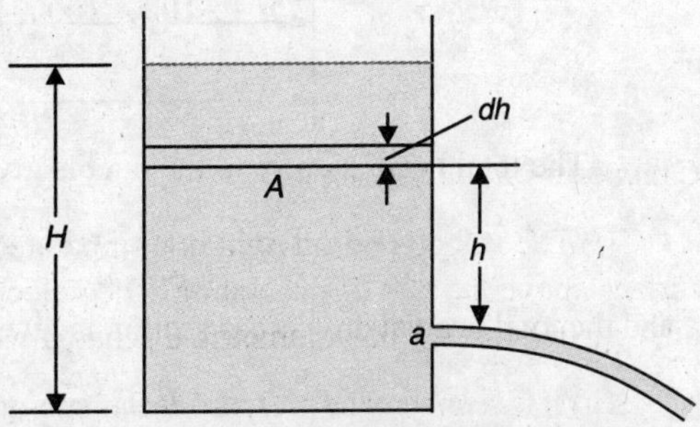

Fig. 4.3 (b): Discharge through a Sharp Orifice

$$\delta V = A \delta h$$

EXPERIMENT
Flow through a Converging-Diverging Duct

OBJECTIVE

To study the flow through a two-dimensional converging-diverging duct and thereby appreciate the scope and limitations of the Bernoulli equation as applied to incompressible flows.

PRACTICAL RELEVANCE

In the professional career of an engineer, he is often confronted with situations wherein he has to design or to evaluate the performance of fluid-flow devices such as pumps, compressors, fans, blowers, turbines, open channels etc. In this process he has often to apply the principle of conservation of energy. The present experiment helps the student in understanding the basic limitations and in appreciating the value of the principle of conservation of energy as applied to simple flow systems.

PROCEDURE

In order to have a quick visual appreciation of the energy levels at the various cross-sections of the flow system, one plots the energy gradient line (total energy line) and the hydraulic gradient line (piezometric head line). It should be apparent from the name itself that the energy gradient line represents the display of total energy h per unit weight of fluid as one moves along the flow. Similarly, the hydraulic gradient line represents the display of piezometric head ($p/\rho g + z$) along the flow.

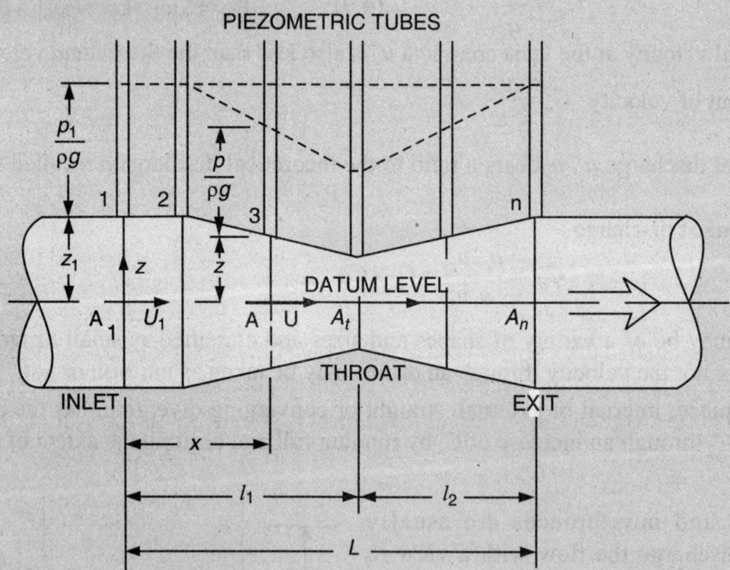

The total head at entry to the duct is given by

$$H_1 = \frac{p_1}{\rho g} + U_1^2 + z_1$$

and the total head at any cross section is given by

$$H = \frac{p}{\rho g} + \frac{U^2}{2g} + z$$

The drop or loss of energy is customarily represented by the loss coefficient

$$C_L = (H_1 - H)/H_1$$

* Measure the distance x of each pressure tap location from the pressure tap located at the entry section ahead of the converging-diverging duct. Also note the duct area A at each of these points.
* Adjust the overflow valve of the inlet tank such that a suitable supply head is maintained in the tank.
* Observe the nature of flow.
* Measure the piezometric head $p/\rho g + z$ at each of the pressure taps with reference to a datum.
* Determine the flow rate Q through the system (for law discharge rates, gravimetric method is preferable to the volumetric method due to higher accuracy).
* Repeat the test for sufficient number of values of the supply head.

RESULTS

Using the measured data for each inlet head H_1 calculate and tabulate:

1. the rate of flow Q by using the data of either volumetric or the gravimetric method used in the experiment.
2. the average velocity U at each cross-section where the pressure taps have been provided, using the equation, $Q = U.A$.
3. the velocity head $U^2/2g$ of the fluid flowing through the duct at each of the pressure tap location.
4. the total available head H at each pressure tap location by adding velocity and piezometric heads.
5. the loss coefficient, C_L for each location.
6. For each inlet supply head H_1 using the results of the above calculation draw the following non dimensional plots:

   (*i*) $(p/\rho g + z)/H_1$ *vs* x/L (*ii*) $(p/\rho g + z + U^2/2g)/H_1$ *vs* x/L (*iii*) C_L *vs* x/L

where L represents the total distance and x/L, the non-dimensional distance from the entry.

POINTS OF DISCUSSION

1. Is there any difference in the energy level of the supply reservoir and that at the first pressure tap location? If so, give reasons for the same.
2. You have calculated the velocity head at various sections along the conduit using the average velocity at that section. Is it correct to do so? Give your comments with reasons.
3. Did you notice fluctuations in the values of the pressure heads in any section of the duct? If so, in which sections of the duct were these fluctuations the largest and the smallest? Give reasons for the behaviour of the fluctuations.
4. Compare the loss of head per unit length in the converging portion of the duct with that in the diverging portion. Give reasons for any difference that occurs.
5. State giving reasons which of the assumptions made in the derivation of the Bernoulli equation are violated in the experiment performed by you and where?
6. Indicate giving reasons whether the results of your experimental plots will change if the duct is inclined in the vertical plane.

ANY OTHER OBSERVATION OR COMMENT?

The velocity with which the liquid escapes through the orifice is given by

$$u_c = C_v\sqrt{2gh} \qquad \text{...(4.12)}$$

and the area of the flow at the vena contracta is given by

$$a_c = C_c a$$

The discharge through the orifice is consequently given by

$$Q = C_d a\sqrt{2gh} \qquad \text{...(4.13)}$$

where C_d, the coefficient of discharge, is given by the product of coefficients of velocity and contraction,

$$C_d = C_v \times C_c$$

The volume of liquid passing through the orifice in time δt must be

$$\delta V = Q\delta t$$

With expressions for δV and Q and reckoning the fact that time increases as h decreases.

$$\delta t = \frac{-A\,\delta h}{C_d a\sqrt{2gh}} \qquad \text{...(4.14)}$$

On integration,

$$T = \int_0^T dt = -\int_{H_1}^{H_2} \frac{A}{C_d a\sqrt{2gh}}\,dh$$

$$= \frac{-1}{C_d a}\sqrt{2g}\int_{H_1}^{H_2} A h^{-1/2}\,dh$$

the time to empty it out from level H_1 to level H_2.

For different shapes of containers, the plane area A of the strip should be worked out as a function of the height h and substituted before integration for T.

Example 4.3. *An external mouthpiece converges from the inlet until the vena contracta to the shape of the jet and then it diverges gradually. The diameter at the vena contracta is 2 cm and the head over the centre of the mouthpiece is 1.44 metres. The head loss in the contraction may be taken as one per cent and that in the divergent portion five per cent of the total energy head before the inlet. What is the maximum discharge that can be drawn through the outlet and what should be the corresponding diameter at the outlet ?*

Assume that the pressure in the system may be permitted to fall to 8 metres below atmospheric, the liquid conveyed being water.

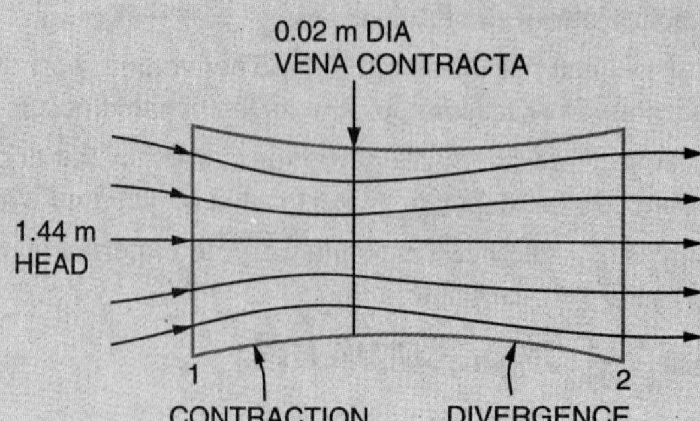

Solution.

In terms of metres of head

$$h_1 = 1.44 \text{ m}$$

The head loss in the contraction $= 0.01 \times 1.44 = 0.0144$ m

and the head loss in the divergence $= 0.05 \times 1.44 = 0.0720$ m

Head available at the vena contracta $= 1.44 - 0.0144 = 1.4256$ m

This must be equal to the sum of the pressure and the velocity head

$$\frac{p_c}{\rho g} + \frac{U_c^2}{2g} = 1.4256$$

For the maximum velocity U_c, the pressure p_c should be as low as practicable; the lower limit being specified as 8 m below atmosphere. Therefore,

$$-8 + \frac{U_c^2}{2g} = 1.4256$$

whence, $$U_c = 13.6 \text{ m/s}$$

The maximum possible discharge is thus

$$Q_{max} = 13.6 \times \pi \times 0.02^3/4 = \mathbf{0.004\ 27\ m^3/s}$$

Pressure at the exit, 2 is atmospheric. Conserving energy between the vena contracta and 2,

$$1.4256 = 0.0720 + \frac{U^2}{2g} + 0$$

whence, $$U_2 = 5.15 \text{ m/s}$$

Diameter d_2 at the exit is given by

$$\frac{\pi d_2^2}{4} \times 5.15 = 0.00427$$

and $$d_2 = 0.0325 \text{ m} = \mathbf{3.25\ cm}$$

Example 4.4. *For the configuration shown, calculate the minimum or just sufficient head H in the vessel and the corresponding discharge of water which can pass over the plate.*

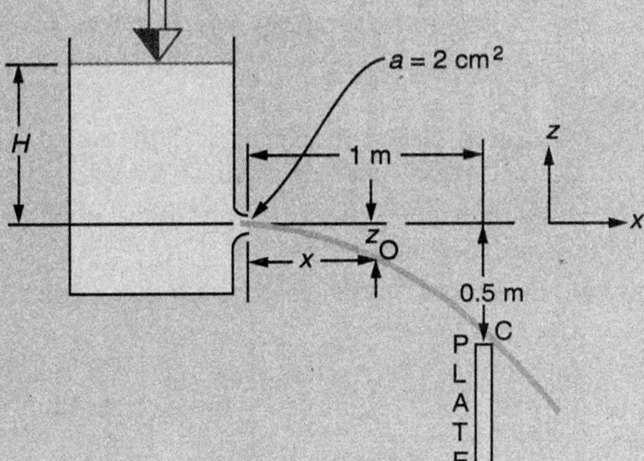

Solution.

For a point O on the trajectory,

$$x = u_1 t$$

$$z = -\frac{1}{2}gt^2$$

where u_1 is the velocity of efflux of the jet.

Eliminating t,

$$z = -\frac{g}{2}\frac{x^2}{u_1^2}$$

which shows that the trajectory must be parabolic.

Taking $\quad u_1 = \sqrt{2gh}, \text{ for } C_v \approx 1$

$$z = \frac{-x^2}{4H}, H = \frac{-x^2}{4z}$$

Hence, minimum head H to pass over C.

$$= -1^2/(4 \times -0.5) = \mathbf{0.5\ m}$$

The corresponding minimum discharge

$$Q = \sqrt{2gH} \times a \times C_d$$

Taking $\quad C_d = 0.8, Q = \sqrt{2 \times 9.81 \times 0.5} \times (2 \times 10^{-4}) \times 0.8 = \mathbf{0.000\ 5\ m^3/s}$

EXPERIMENT

Flow through Orifices and Nozzles

OBJECTIVE

To study the characteristics of free jets through orifices and mouth pieces, e.g., nozzles.

PRACTICAL RELEVANCE

Orifices and nozzles are devices which are used for discharging fluids into the atmosphere from tanks and conduits. In such case, it is usually required to know the actual flow rate and the time required for emptying a tank or a vessel through a nozzle or an orifice. Then, it is essential to know the characteristics of the free jet and constants of the device.This experiment illustrates a technique for determining the liquid jet characteristics and the constants of a nozzle and an orifice and their comparison.

PROCEDURE

1. Fit a sharp-edged small circular orifice of a given diameter to the side wall of the empty tank. Prior to fitting the orifice, measure its diameter, D.
2. Fix the position of the overflow pipe to get approximately the maximum head and start the flow.
3. Adjust the flow rate in and out of the system such that a constant head H is maintained in the tank.
4. Measure the flow rate through the system either gravimetrically or volumetrically. It may be noted that the gravimetric method is more accurate than the volumetric method for small flow rates.
5. Determine the x and z coordinates of the trajectory of the jet issuing out of the device up to the point that the jet disintegrates.
6. In order to determine the time of emptying, start the flow and the stop-watch simultaneously. Note the value of $\Delta H = H_1 - H_2$ and the time $T = (t_2 - t_1)$ required for the liquid level to all through the height ΔH.

7. Empty the tank and fit a nozzle of the same diameter as the orifice to the side wall of the tank noting its diameter.

RESULTS

Using the measured data for each inlet head *H*, tabulate:

1. The flow rate *Q by using the data of either the volumetric or the gravimetric method used in the experiment.*
2. The mean *z* coodinates of the jet corresponding to measured *x* coordinates. Also calculate the theoretical value of *z* coordinate by

$$z = -x^2/4H$$

and $$C_d = \frac{Q}{a.\sqrt{2gH}}$$

where *a* is the area of the device.

3. The coefficient of contraction, C_c as the ratio of area at vena contracta to that of the device.
4. Calculate λ as the ratio of the discharge coefficient of the nozzle to that of the orifice for the maximum inlet head.
5. Calculate the ratio of the time taken for the liquid level to fall by Δ*H* through the nozzle to that through the orifice. Compare this value with λ obtained in step 4.
6. Plot log *Q vs* log *H* for both devices; this is because we expect a straight-line plot in doing so.

POINTS FOR DISCUSSION

1. Why do you measure the coordinates of the jet boundaries instead of directly measuring the coordinates of the centre line of the liquid jet ?
2. Do you observe any difference between the experimental and the theoretical trajectories of the liquid jet for a particular device at a particular head *H* ?
3. Does the discrepancy between the experimental and the theoretical trajectories of the jet increase in the head *H* ? Explain why ?
4. Will the jet trajectory change if the axis of the orifice/nozzle is included at an angle α to the side wall of the tank ? If so, what will be its new trajectory ?
5. Why is the coefficient of discharge for a nozzle higher than that for an orifice?

ANY OTHER OBSERVATION OR COMMENT?

Example 4.5. *Two tanks of cross-sectional areas 6 m × 1.5 m and 1.5 m × 1.5 m are connected at their bottom by a short pipe 20 cm diameter. The difference of water levels in the tanks was 3 m when the flow started. Find the time to bring the difference down to 1.2 m. Take C_d = 0.8.*

Solution.

Cross-section areas of the tanks:

$$A_1 = 9 \text{ m}^2, A_2 = 2.25 \text{ m}^2$$

For the pipe, $$a = \pi/4 \times (0.2)^2 = 0.031 \text{m}^2$$

Consider the configuration when the level difference is *h* ; for a level drop of dh_1 in tank 1, rise in tank 2 is given by

$$= A_1/A_2\, dh_1 = dh_2$$

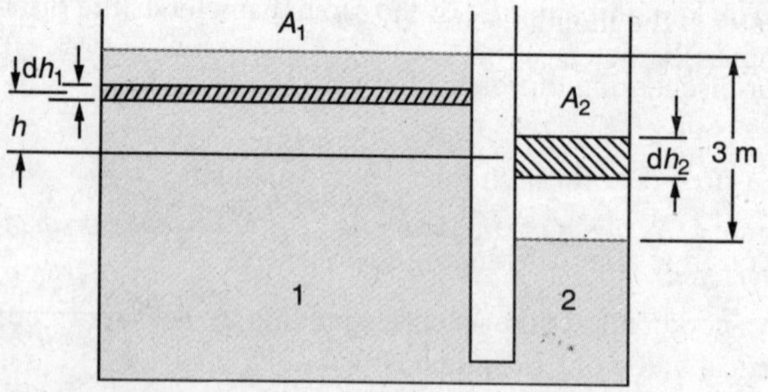

Change of level in time dt

$$dh = dh_1 + dh_2 = (A_1 + A_2)/A_2\, dh_1$$

because

$$dQ = A_1\, dh_1 = A_2\, dh_2 = a_c u_c\, dt$$

$$= C_d a\sqrt{2gh}\, dt$$

whence,

$$dt = -\frac{A_1 A_2}{(A_1 + A_2)C_d a\sqrt{2g}} h^{-1/2} dh$$

On integration,

$$T = \int_0^T dt = \left| \frac{2A_1A_2}{(A_1 + A_2)C_d a\sqrt{2g}} h^{1/2} \right|_{H_2}^{H_1}$$

$$= -\frac{2A_1A_2}{(A_1 + A_2)C_d a\sqrt{2g}}(\sqrt{H_1} - \sqrt{H_2})$$

Substituting the data,

$$T = -\frac{2\times 9\times 2.25}{(9+2.25)+0.8\times 0.031\times\sqrt{2}\times 9.81}(\sqrt{3} - \sqrt{1.2}) = \mathbf{20.59\ s}$$

Example 4.6. *A mouthpiece is fitted at the bottom of a boiler drum for emptying it. The drum is horizontal and half full of water. It is 10 m long and 2 m in diameter. Find the time required to empty the boiler if the diameter of mouthpiece is 5 cm and its C_d = 0.8. Sketch the boiler drum and derive the equation used, if any.*

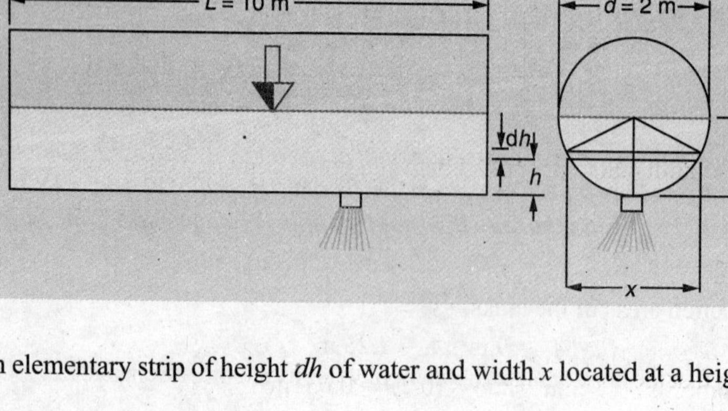

Solution.

Consider an elementary strip of height dh of water and width x located at a height h above the mouthpiece.

Area of the flat strip = $L\,x$

Volume of the water in this strip = $L\,x\,dh$

Velocity of water at the mouthpiece = $\sqrt{2gh}$

Time taken for discharging this elementary strip is given by

$$dt = -L\, x\, dh / (a\sqrt{2gh}\, Cd)$$

where $$x = 2\{r^2 - (r - h)^2\}^{0.5} = 2(hd - h^2)^{0.5}$$

Integrating w.r.t. time after substituting for x,

$$T = \int_0^T dt = -\int 2L\,(hd - h^2)^{0.5} / (a\sqrt{2g / h^{-0.5}}\, Cd)dh$$

$$T = \frac{-2L}{Cd\, a\sqrt{2g}} \int (hd - h^2)^{0.5} h^{-0.5} dh = \frac{-2L}{Cd\, a\sqrt{2g}} \int (d-h)^{0.5} dh$$

$$= \frac{2L}{1.5Cd\, a\sqrt{2g}} \left|(d-h)^{1.5}\right|_0^H = 2L\, 1.5Cd\, a\sqrt{2g}\,(d-H)^{1.5}$$

By straight forward substitution,

$$T = \frac{2\times 10}{1.5Cd \times \pi \times \frac{(0.05)^2}{4}\sqrt{(2\times 9.81)}}(2^{1.5} - 1^{1.5})$$

$$= \mathbf{3500\ s = 58.3\ minutes}$$

Example 4.7. *Trace the trajectory of a free jet issuing from a nozzle inclined at an angle α to the ground in a vertical plane. The discharge rate is Q and the area of nozzle cross section is a resulting in a velocity of U_1 at the nozzle.*

Solution.

Referring to the figure, take the origin at the nozzle exit, point 1, and consider the motion in the vertical $(x - z)$ plane.

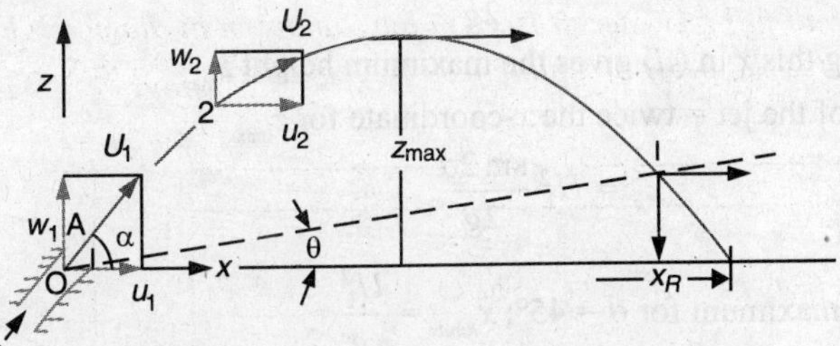

By the Bernoulli equation applied between point 1 and any other point 2,

$$\frac{U_1^2}{2} + 0 = \frac{U_2^2}{2} + gz$$

or $$U_2^2 = U_1^2 - 2gz \qquad ...(i)$$

Observing that the horizontal component of velocity should not change except due to air friction,

$$u_1 = u_2$$

and the vertical component would alter due to gravity effect

$$w_2 = w_1 - gt$$

If follows that, at any point.

$$x = u_1 t \text{ and } z = w_1 t - \frac{1}{2} g t^2$$

Eliminating t,

$$z = \frac{w_1}{u_1} x - \frac{g}{2} \frac{x^2}{u_1^2}$$

Observing that $\frac{w_1}{u_1} = \tan \alpha$

and $u_1 = U_1 \cos \alpha$

$$z = x \tan \alpha - \frac{gx^2}{2(U_1 \cos \alpha)^2} \quad \text{...(ii)}$$

Substituting for z from (*ii*) in (*i*)

$$U_2^2 = U_1^2 - 2g\left(x \tan \alpha - \frac{gx^2}{2(U_1 \cos \alpha)^2}\right) \quad \text{...(iii)}$$

Let us now inspect the equations to extract some useful information:

(*a*) The maximum attainable height z_{max} occurs when $w_2 = 0$ or $\frac{\partial z}{\partial x} = 0$

Differentiating (*ii*) w.r.t. x.,

$$\frac{dz}{dx} = \tan \alpha - \frac{gx}{U_1^2 \cos^2 \alpha} = 0$$

or $$\sin \alpha \cos \alpha = \frac{\sin 2\alpha}{2} = \frac{gx}{U_1^2}$$

whence $$\alpha = \frac{1}{2} \sin^{-1}\left(\frac{2gx}{U_1^2}\right) \quad \text{...(iv)}$$

and $$x = U_1^2 \frac{\sin 2\alpha}{2g} \quad \text{...(v)}$$

Substituting this x in (*ii*) gives the maximum height z_{max}.

(*b*) Range of the jet = twice the x-coordinate for z_{max}

$$x_R = U_1^2 \frac{\sin 2\alpha}{2g} \quad \text{...(vi)}$$

This is the maximum for $\alpha = 45°$; $x_{Rmax} = \frac{U_1^2}{g}$ ...(*vii*)

(*c*) Range of the jet on an inclined plane OI is determined by solving for (x, z), the point of intersection of the parabola given by equation (*ii*) and the inclined plane $z = x \tan \theta$ unless z is observable otherwise, and x calculated from equation (*ii*).

(*d*) Variation of z_{max} with α is observed from equation (*ii*) and (*v*) but a direct resort to $\frac{dz}{dx}$ from (*ii*) should be valuable.

(*e*) For a horizontal nozzle, $\alpha = 0$

$$z = -\frac{gx^2}{2U_1^2}$$

(*f*) For a vertical nozzle, $\alpha = 90°$, $\tan\alpha$ is infinite, the x-coordinate does not exist and the trajectory coincides with a vertical line from the nozzle. z_{max} for this case equals $U_1^2/2g$.

Example 4.8. *A fireman must reach a window 40 m above the ground with a water jet, from a nozzle 3 cm diameter discharging 30 kg/s. Assuming the nozzle height of 2 m, determine the greatest distance from the building where the fireman can stand and still reach the jet into the window.*

Solution.

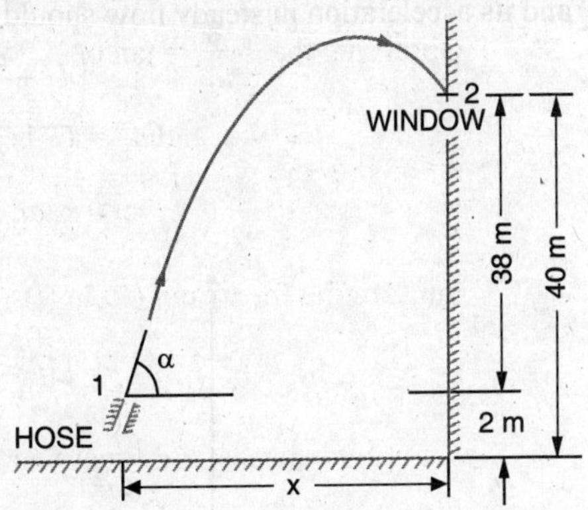

$$U_1 = \frac{m}{\rho A} = \frac{30}{1000 \times \pi / 4 \times (0.03)^2} = 42.4 \text{ m/s}$$

(Note that $U_1^2/2g$ should be more than the height of the window for the jet to reach at all; in this case $U_1^2/2g = 92 \text{ m} > 40 \text{ m}$)

The question requires an optimisation of α for the maximisation of x to achieve the desired height of z of 38 m.

$$38 = x\tan\alpha - \frac{9.81}{2 \times 42.5^2} x^2/\cos^2\alpha \quad ...(i)$$

Differentiating each term w.r.t. α

$$0 = x\sec^2\alpha + \tan\alpha\frac{dx}{d\alpha} - 0.0027x^2.\ 2\sec^2\alpha\tan\alpha - 0.0027/\cos^2\alpha.\ 2x\frac{dx}{d\alpha}$$

Maximum x for α demands that

$$\frac{dx}{d\alpha} = 0$$

hence
$$x\sec^2\alpha - 2 \times 0.0027x^2\sec^2\alpha\tan\alpha = 0$$

or
$$x\tan\alpha = \frac{1}{0.0054} = 185 \quad ...(ii)$$

Solving (*i*) and (*ii*) simultaneously

$$38 = 1.85 - 0.0027x^2/\cos^2\alpha$$

$$x^2/\cos^2\alpha = 54513$$

$$x/\cos\alpha = 233$$

But
$$x/\tan\alpha = x\sin\alpha/\cos\alpha = 185$$

so,
$$\sin\alpha = \frac{185}{233};\ \alpha = 52.4$$

and
$$\boldsymbol{x = 142 \text{ m.}}$$

4.6. Euler's Equation and Bernoulli Equation in the Streamwise Direction

Consider an element of an inviscid fluid in an incompressible flow enclosed between two stream surfaces such that the picture in an x-z plane appears as shown in Fig. 4.4. The streamwise direction is called s. Assuming s positive along the flow, the fluid velocity is U and the cross-sectional area is δa. The fluid velocity and pressure are given a positive increment along the flow. It may be noted that the sign would take care of itself.

The element experiences forces due to pressure on its surfaces and the gravitational force acts at its centre of gravity C.

Along the streamwise direction s, the net force acting on the element is

$$F = p\delta A - \left(p + \frac{dp}{ds}\delta s\right)\delta A - mg\sin\theta$$

$$= -\frac{dp}{ds}\delta s.\delta A - mg\sin\theta$$

The mass of the element must be

$$m = \rho\delta s\,\delta A$$

and its acceleration in steady flow should be

$$a = \frac{\left(U + \frac{dU}{ds}\delta s\right) - U}{\delta t} = \frac{ds}{dt}.\frac{dU}{ds} = U\frac{dU}{ds}$$

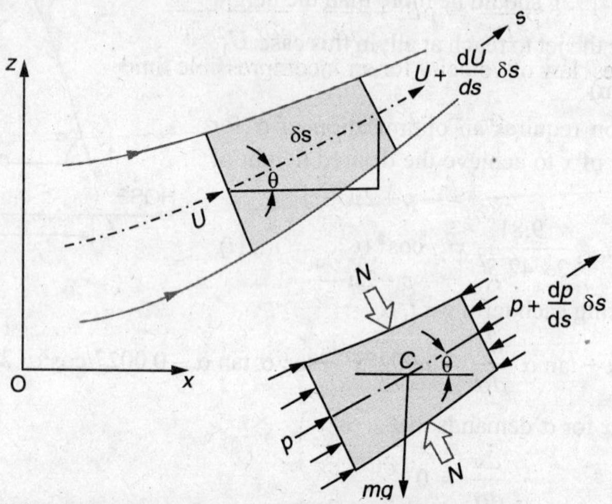

Fig. 4.4: Element of Fluid in the Streamline Coordinates

Applying the Newton's second law along the streamwise direction,

$$F = m\,a$$

$$-\frac{dp}{ds}\delta s\delta A - \rho\delta s\delta A g\sin\theta = \rho\delta s\delta A U\frac{dU}{ds} \quad \text{...(4.15)}$$

Dividing throughout by $\rho\delta s\,\delta A$ and recognising that

$$\sin\theta = \frac{\delta z}{\delta s} = \frac{dz}{ds}$$

the law reduces to

$$\frac{1}{\rho}\frac{dp}{ds} + g\frac{dz}{ds} + U\frac{dU}{ds} = 0 \quad \text{...(4.16)}$$

for the motion of an inviscid and steady flow along the streamwise direction.

This is the Euler's equation of motion along the streamwise direction.

Streamwise integration of the Euler's equation for the steady flow of an inviscid fluid results in the Bernoulli equation. Integrating each term and dividing by g,

$$\int\frac{dp}{\rho g} + \int\frac{UdU}{g} + \int dz = C$$

or,

$$\int\frac{dp}{\rho g} + \frac{U^2}{2g} + z = C$$

which is the Bernoulli equation.

4.7. Navier-Stokes Equations of Motion

The Navier-Stokes equation of motion is the general momentum equation for compressible or incompressible, viscous or inviscid flows. However, a rigorous derivation is not intended. When the shear stress on the faces of a control volume are included in the force-balance together with the normal stresses, the Newton's second law of motion provides

$$\rho\frac{Du}{Dt} = \rho B_x + \frac{\partial \sigma_{xx}}{\partial x} + \frac{\partial \tau_{yx}}{\partial y} + \frac{\partial \tau_{zx}}{\partial z}$$

$$\rho\frac{Dv}{Dt} = \rho B_y + \frac{\partial \sigma_{yy}}{\partial y} + \frac{\partial \tau_{xy}}{\partial x} + \frac{\partial \tau_{yz}}{\partial z} \quad \text{...(4.17)}$$

$$\rho\frac{Dw}{Dt} = \rho B_z + \frac{\partial \sigma_{zz}}{\partial z} + \frac{\partial \tau_{zx}}{\partial x} + \frac{\partial \tau_{zy}}{\partial y}$$

By the Stokes' law of velocity for an incompressible fluid,

$$\sigma_{xx} = -p + 2\mu\frac{\partial u}{\partial x}$$

$$\sigma_{yy} = -p + 2\mu\frac{\partial v}{\partial y}$$

$$\sigma_{zz} = -p + 2\mu\frac{\partial w}{\partial z} \quad \text{...(4.18)}$$

$$\tau_{xy} = \mu\left(\frac{\partial v}{\partial x} + \frac{\partial u}{\partial y}\right)$$

$$\tau_{zy} = \mu\left(\frac{\partial w}{\partial y} + \frac{\partial v}{\partial z}\right)$$

$$\tau_{zx} = \mu\left(\frac{\partial u}{\partial z} + \frac{\partial w}{\partial x}\right)$$

From the first three relations,

$$p = -\frac{\sigma_{xx} + \sigma_{yy} + \sigma_{zz}}{3} \quad \text{...(4.19)}$$

since, by continuity $\frac{\partial u}{\partial x} + \frac{\partial v}{\partial y} + \frac{\partial w}{\partial z} = 0$

On substitution, we obtain the Navier-Stokes momentum equation for a three-dimensional incompressible fluid flow as

$$\rho\frac{Du}{Dt} = \rho B_x - \frac{\partial p}{\partial x} + \mu\nabla^2 u$$

$$\rho\frac{Dv}{Dt} = \rho B_y - \frac{\partial p}{\partial y} + \mu\nabla^2 y \quad \text{... (4.20)}$$

$$\rho\frac{Dw}{Dt} = \rho B_z - \frac{\partial p}{\partial z} + \mu\nabla^2 w$$

The Navier-Stokes equations appear in the following form for axisymmetric incompressible fluid flow.

$$\rho\left(\frac{Du_r}{Dt} - \frac{u_\theta^2}{r}\right) = \rho B_r - \frac{\partial p}{\partial r} + \mu\left(\nabla^2 u_r - \frac{u_r}{r^2}\right)$$

$$\rho\left(\frac{Du_\theta}{Dt} + \frac{u_r u_\theta}{r}\right) = \rho B_\theta + \mu\left(\nabla^2 u_\theta - \frac{u_\theta}{r^2}\right) \quad \text{...(4.21)}$$

$$\rho\frac{Du_z}{Dt} = \rho B_z - \frac{\partial p}{\partial z} + \mu\nabla^2 u_z$$

Care should be taken to set $\frac{\partial}{\partial\theta} \equiv 0$ for axisymmetry and ρ = constant for incompressibility while expanding the D/Dt and ∇^2 terms. It may also be noticed that these equations reduce to the Euler's form for an inviscid fluid, *i.e.*, for $\mu = 0$. The Euler's equation is, therefore, a special case of the Navier-Stokes equation.

4.8. Integral Momentum Equation

Turning now to the integral form of the momentum equation for a given control volume, the summation of the external surface and body forces must equal the substantive rate of change of momentum which should equal the rate of efflux of momentum across the control surface plus the rate of change of momentum within the control volume.

$$\mathbf{F} = \mathbf{F}_S + \mathbf{F}_B = \frac{D(m\mathbf{U})}{Dt}$$

$$= \int_S \mathbf{U}(\rho\mathbf{U}.d\mathbf{A}) + \frac{\partial}{\partial t}\int_V \mathbf{U}\rho dV \qquad ...(4.22)$$

In other words,

Surface forces + body forces = net momentum efflux + momentum rate stored.

For steady flow, $\frac{\partial}{\partial t} \equiv 0$

and stipulating the selection of a control volume such that the incoming and the outgoing flows are normal to the respective control surfaces such as that shown in Fig 4.5

$$\mathbf{F} = \mathbf{F}_s + \mathbf{F}_B = \mathbf{U}_2(\rho_2\mathbf{U}_2\mathbf{A}_2) - \mathbf{U}_1(\rho_1\mathbf{U}_1\mathbf{A}_1)$$

or $\qquad \mathbf{F} = \dot{m}\,(\bar{\mathbf{U}}_2 - \bar{\mathbf{U}}_1)$

since, by the continuity equation,

$$\rho_1 U_1 A_1 = \rho_2 U_2 A_2 = \dot{m} \qquad ...(4.23)$$

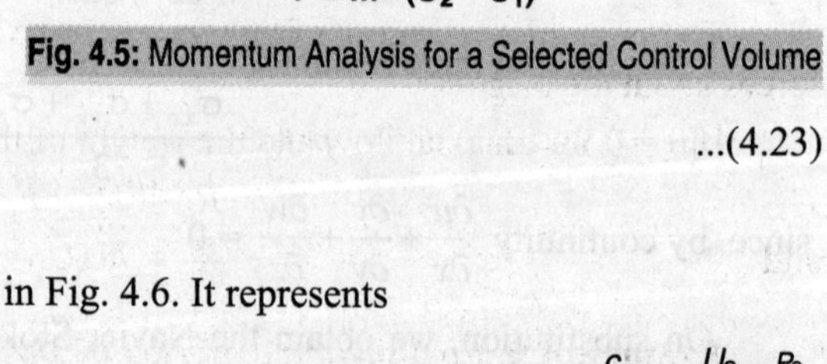

Fig. 4.5: Momentum Analysis for a Selected Control Volume

4.9. Forces on Confinements

Let us first consider a confinement shown in Fig. 4.6. It represents

(*i*) a reducing elbow ;

(*ii*) a pipe bend if $A_1 = A_2$; right angled if $\theta = 90°$; and

(*iii*) a conventional nozzle or a diffuser if the flow is uniaxial.

Let us take a control volume *abcde* such that surfaces *ab* and *cd* are normal to the inflow and outflow respectively, surface *ae* cuts through the support of the device and the surfaces *bc* and *de* enclose the confining passage arbitrarily.

The atmospheric pressure p_a acts all over the control surface normal to it and its net effect is zero. The fluid entering and leaving at p_1 and p_2 exerts gauge pressures

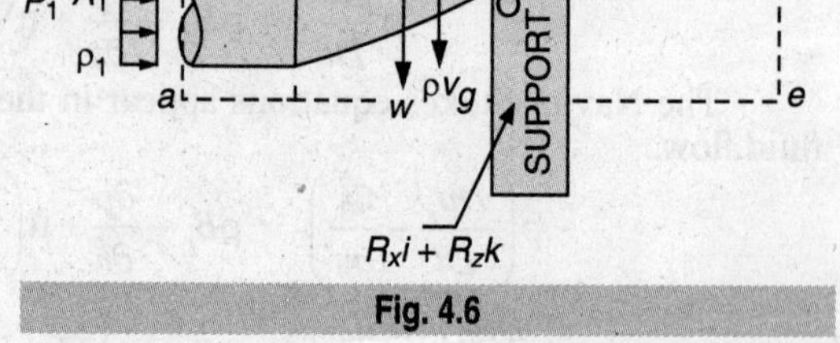

Fig. 4.6

$$(p_1 - p_a) = p_{g1} \text{ and } (p_2 - p_a) = p_{g2}$$

over areas A_1 and A_2 respectively.

Net surface force acting on it is,

$$p_{g1} A_1 - p_{g2} A_2 \cos\theta \text{ in the } x\text{-direction,}$$

and $$-p_{g2} A_2 \sin\theta \text{ in the } z\text{-direction.}$$

Net body force due to the weight of the fluid in the control volume as well as the weight of the device itself is

$$F_{net} = -(\rho Vg + W) \text{ in the } z\text{-direction}$$

where ρ = average mass density

Since the control surfaces cut through the support of the device, there must be a reaction R with components

$$R_x \text{ in the } x\text{-direction}$$
$$R_z \text{ in the } z\text{-direction}$$

Further, net momentum rate leaving the control volume is

$$(\rho_2 A_2 U_2) U_2 \cos\theta - (\rho_1 A_1 U_1) U_1 = \dot{m}(U_2 \cos\theta - U_1) \text{ in the } x\text{-direction}$$

and $$(\rho_2 U_2 A_2) U_2 \sin\theta = \dot{m}\ U_2 \sin\theta \text{ in the } z\text{-direction}$$

Applying the integral momentum equation,

$$(p_{g1} A_1 - p_{g2} A_2 \cos\theta) + R_x = \dot{m}(U_2 \cos\theta - U_1) \text{ in the } x\text{-direction} \quad ...(4.24)$$

$$(p_{g2} A_2 \sin\theta + w + \rho vg) + R_z = \dot{m}\ U_2 \sin\theta \text{ in the } z\text{-direction} \quad ...(4.25)$$

where $\dot{m}$ = mass flow rate $= \rho_1 A_1 U_1 = \rho_2 A_2 U_2$

Solution for R_x and R_z gives the resultant reaction by the support on the device as $(R_x i + R_z k)$ and the force F by the device on the support is equal and opposite to it.

$$F_x = -R_x \text{ and } F_z = -R_z$$

The point of application of this force can be located by taking moments of all the constituent forces about the origin O.

If $\theta = 0$ for uniaxial flow and the weight of the device and the fluid are neglected, then

$$R_z = 0 = -F_z$$

and $$R_x = \dot{m}(U_2 - U_1) - (\rho\, r_{g1} A_1 - p_{g2} A_2) = -F_x$$

Example 4.9. *Find the force exerted by the nozzle on the fireman for the configuration and data in the figure. Pressure of water entering the nozzle is 68.67N/cm².*

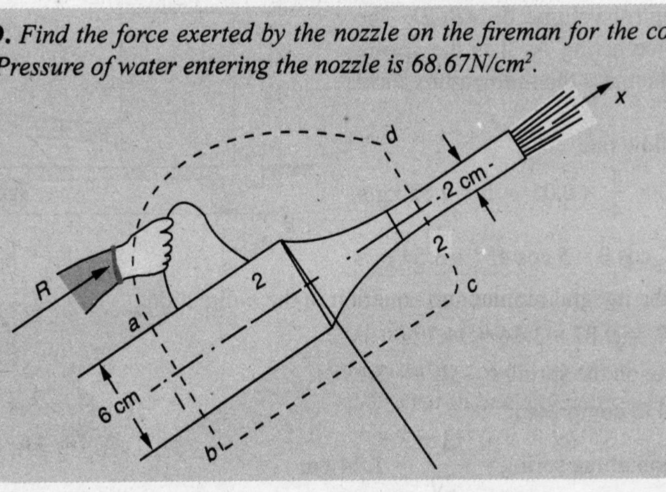

Solution.

Let the control volume be bounded by a surface *abcd* cutting through the hand of the fireman.

Take the *x*-axis along the flow and neglect the weight of the nozzle assembly and the water it contains.

Applying the Bernoulli equation between stations 1 and 2, neglecting their datum difference,

$$\frac{p_1}{\rho}+\frac{U_1^2}{2}=\frac{p_2}{\rho}+\frac{U_2^2}{2}$$

Since $(p_1 - p_2) = p_{1\text{gauge}} = 68.67 \times 100^2 = 686\,700 \text{ N/m}^2$

$$\frac{p_1-p_2}{\rho} = \frac{686\,700}{1000} = 686.7 \text{ m}^2/\text{s}^2$$

$$U_2^2-U_1^2 = 2 \times 686.7 = 1373.4 \text{ m}^2/\text{s}^2$$

But by continuity $\dfrac{U_2}{U_1} = \left(\dfrac{6}{2}\right)^2 = 9,$

Hence $(81-1)U_1^2 = 1373.4,\; U_1^2 = 17.17 \text{ m}^2/\text{s}^2$

$$U_1 = 4.14 \text{ m/s};\; U_2 = 37.26 \text{ m/s}$$

Mass flow rate $\dot{m} = 1000 \times \dfrac{\pi}{4} \times 0.06^2 \times 4.14 = 11.71 \text{ kg/s}$

Applying the integral momentum equation between stations 1 and 2, reaction R_x along the *x*-direction,

$$R_x = 11.71 \times (37.26 - 4.14) - 686700 \times \frac{\pi}{4} \times 0.06^2$$

$$= -1554 \text{ N}$$

The force F_x exerted by the nozzle on the fireman is equal and opposite to it.

$$\mathbf{F_x = 1554 \text{ N along the flow.}}$$

Example 4.10. *A tank and a trough are placed on a trolley as shown. Water issues from the tank through a 5 cm diameter nozzle at 5 m/s and strikes the trough which turns it by an angle 45° upwards. Determine the compression of the spring of stiffness 20 N/cm.*

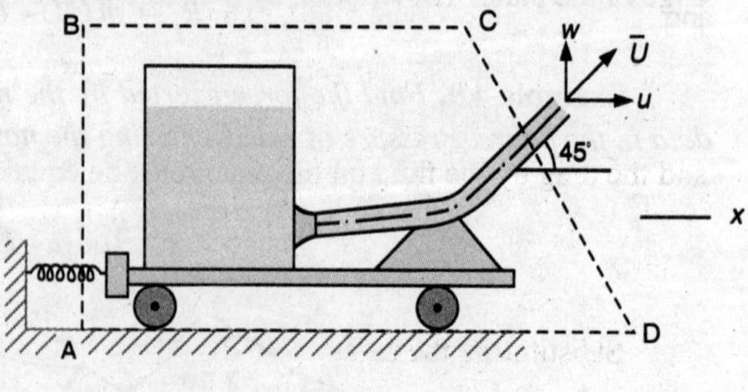

Solution.

Let the control volume be *ABCD*. The only external force acting on it is the spring reaction R_x and the mass the momentum leave normal to *CD*.

The mass flow rate

$$\dot{m} = 1000 \times \frac{\pi}{4} \times 0.05^2 \times 5 = 9.82 \text{ kg/s}$$

$$U_1 = 0,\; U_2 \cos\theta = 5 \cos 45° = 3.54 \text{ m/s}$$

Applying the integral momentum equation in the *x*-direction,

$$R_x = 9.82 \times 3.54 = 34.7 \text{ N}$$

Force acting on the spring $= -R_x = -34.7$ N

which could tend to compress it.

Compression of the spring $= \dfrac{34.7}{20} = \mathbf{1.74 \text{ cm}}$

Example 4.11. *A flat plate is towed longitudinally at 5 m/s in an otherwise stationary mass of water and it is observed that the velocity profile on either face changes from uniform at the leading edge to triangular up to 5 cm height at the trailing edge observed relative to the plate. Estimate the drag on the plate per unit width.*

Solution.

The flow over a flat plate is the same whether it is towed longitudinally in a stationary mass of fluid or it is held at rest and subjected to a uniform longitudinal flow. We, therefore, consider the plate at rest with uniform velocity U approaching at its leading edge. At the trailing edge, the velocity profile is triangular, such that $u = U y/\delta$ as shown in the figure where the flow on one side of the plate is sketched.

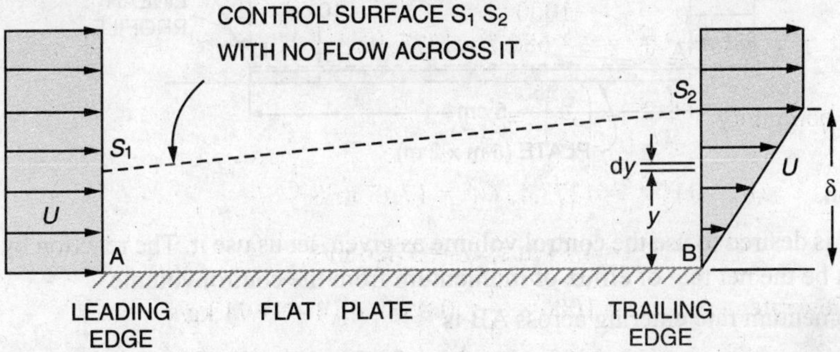

Consider a strip of height dy at a height of y above the plate at its trailing edge. The flow through the strip is $\rho u\, dy$ and the rate of momentum outflow is $\rho u^2\, dy$. Since the fluid approaches the plate with a velocity U, the rate of momentum in flow for this mass is $\rho u\, dy\, U$ such that the efflux of momentum is given by

$$\rho u(u - U)\, dy$$

The efflux of momentum over the height δ must therefore be

$$\int_0^\delta \rho u(u - U)dy$$

It is implied in this analysis that the control volume is bounded by a surface $S_1 S_2$ such that there is no flow across it. In other words, $S_1 S_2$ is a streamline passing through a distance δ at the trailing edge of the plate. The reaction by the plate on the fluid is given by

$$R = \int_0^\delta \rho u(u - U)dy$$

and the drag by the fluid on the plate must be equal and opposite to R

$$D = -R = -\int_0^\delta \rho u(u - U)dy$$

Substituting the data,

$$D = -\int_0^{0.05} 1000 \times 5 \times \frac{y}{0.05}\left(5\frac{y}{0.05} - 5\right)dy$$

$$= -500\,000 \int_0^{0.05}\left(\frac{y^2}{0.05} - y\right)dy = -500\,000\left|\frac{y^3}{0.15} - \frac{y^2}{2}\right|_0^{0.05}$$

$$= \mathbf{210\ N}$$

The total drag on both sides of the plate is twice this value, *i.e.*, **420 N.**

Example 4.12. *Consider the flow of water past a (5 m × 2 m) plate as shown. The velocity of water ahead of the plate is uniform and has a value of 2 m/s. At the trailing egde of the plate the velocity increases linearly from zero at the plate to 2 m/s at a distance of 10 cm from it, as shown in the figure. Using the linear momentum equation calculate the drag on the plate. Assume that the pressure is the same everywhere in the flow field.*

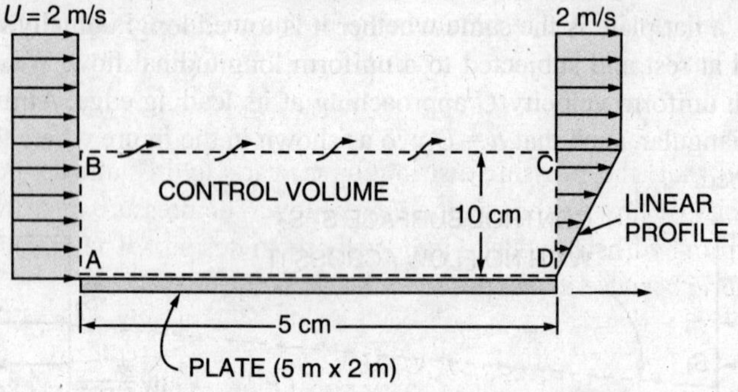

Solution.

Since it is desired to use the control volume as given, let us use it. The reaction by the plate on the fluid must be the net rate of efflux of momentum from the control volume.

The momentum rate entering across AB is

$$mU = 1000 \times 2 \times 0.1 \times 2 \times 2 = 400 \times 2 = 800 \text{ N}$$

and that leaving across *CD* is $1000 \times \int_0^{0.1} \frac{2}{0.1} y \times \frac{2}{0.1} y \times 2dy = 133 \times 2 = 266 \text{ N}$

Mass flow rate across *BC* must be the difference between that entering *AB* and that leaving *CD*

$$1000 \times (2 \times 0.1 \times 2 - 2/2 \times 0.1 \times 2) = 200 \text{ kg/s}$$

Momentum rate leaving $BC = 200 \times 2 = 400$ N

Hence $R_x = 400 + 266 - 800 = -134$ N

The drag force *D* exerted by the fluid on the plate is **134 N in the positive direction,** *i.e.*, along the motion.

Alternatively, a simpler control volume bounded by a streamline could have been chosen. In that case

$$D = -R = -\int_0^{\delta} \rho u(u - U)dy$$

per unit width.

Substituting the data,

$$D = -\int_0^{0.1} 1000 \times \frac{2}{0.1} y \times \left(\frac{2}{0.1} y - 2\right) dy$$

$$= -40,000 \int_0^{0.1} \left(\frac{y^2}{0.1} - y\right) dy$$

$$D = -40,000 \left| \frac{y^3}{0.3} - \frac{y^2}{2} \right|_0^{0.1}$$

$$= \textbf{67 N} \text{ per unit width of the plate.}$$

Since the plate is 2 m wide, **the drag on the plate is 134 N.**

4.10. Force on a Rectangular Sluice Gate

A gate placed across an open channel under which the water flows, as shown in Fig. 4.7, is called a sluice gate. A sluice gate is acted upon by a force which may be determined by employing the principles of hydrostatics and fluid dynamics.

A simple and direct method to determine the horizontal force acting on a vertical plate rectangular sluice gate is to record the pressure at a large number of points on the gate and to integrate the pressure over the surface of the gate,

$$F = \int_A p\, dA$$

It may be noticed that the pressure distribution is nearly hydrostatic on the upper part of the gate because the velocity of flow in that region is very low. The pressure distribution in the lower part of the gate is far from hydrostatic due to large velocity in this part. The pressure at the lower tip of the gate is atmospheric because it is exposed to the atmosphere.

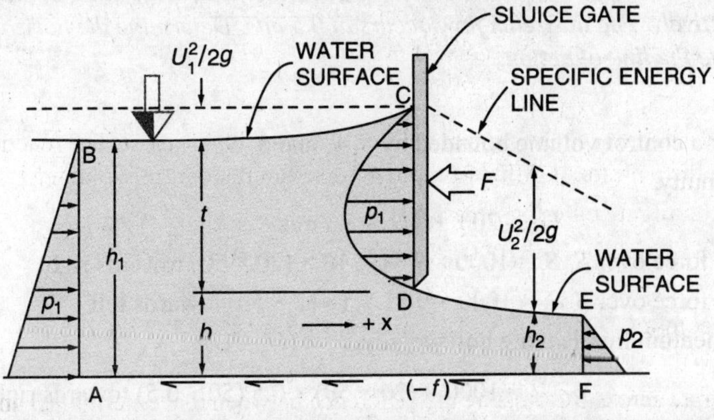

Fig. 4.7: Flow under a Sluice Gate

A simple and indirect method to compute the horizontal force acting on the gate is to apply the momentum equation to the liquid on a control volume so chosen as to just exclude the gate and so extended upstream and downstream in the channel where the inflow and outlow are steady and at uniform velocities U_1 and U_2 respectively. The analysis with reference to such a control volume *ABCDEFA* is as follows:

The control volume is acted upon by the following forces per unit width of the gate for a gate opening *b*.

(*i*) hydrostatic force due to the hydrostatic pressure acted by the liquid on the left of control surface *AB*, given by

$$\rho g \frac{h_1}{2} h_1 = \rho g \frac{h_1^2}{2}$$

(*ii*) hydrostatic force due to the hydrostatic pressure acted by the liquid on the right of the control surface *EF*, given by

$$-\rho g \frac{h_2}{2} h_2 = -\rho g \frac{h_2^2}{2}$$

(*iii*) reaction by the gate on the control surface *CD*. If the force exerted by the liquid on the gate is *F*, the reaction force by the gate must be – *F*.

(*iv*) frictional force by the base of the channel to the control surface *AF*, given by – *f*.

The net force acting on the control volume is, therefore,

$$\rho g \frac{h_1^2}{2} - \rho g \frac{h_2^2}{2} - F - f \qquad (i)$$

The net momentum rate leaving the control volume per unit width of the gate is

$$\rho h_2 U_2^2 - \rho h_1 U_1^2 \qquad (ii)$$

The integral momentum equation requires that the net force on the control volume should equal the net efflux of momentum. Equating (*i*) and (*ii*) and neglecting the friction force *f*,

$$F = \frac{\rho g}{2}\left(h_1^2 - h_2^2\right) + \rho q(U_1 - U_2) \qquad ...(4.26)$$

where *q* is the volumetric flow per unit width of the gate. The force *F* is exerted by the liquid on the gate per unit width of the gate.

Example 4.13. *Upstream and downstream elevations of water in a 50 m wide dam are 20 m and 2 m respectively. The upstream flow occurs at 0.5 m/s. Determine the horizontal force acting on the dam. Locate the line of action.*

Solution.

Consider a control volume bounded by $S_1 S_1$ and $S_2 S_2$.

By continuity,

$$U_2 = 20 \times 0.5/2 = 5.0 \text{ m/s}$$

Pressure force over $S_1 S_1$ = 1000 × 9.81 × 10 × (20 × 50) towards right

Pressure force over $S_2 S_2$ = 1000 × 9.81 × 1× (2 × 50) towards left

Net momentum flux leaving horizontally, = $\dot{m}(U_2 - U_1)$

= 1000 × (20 × 50) × 0.5 (50 – 0.5) towards right

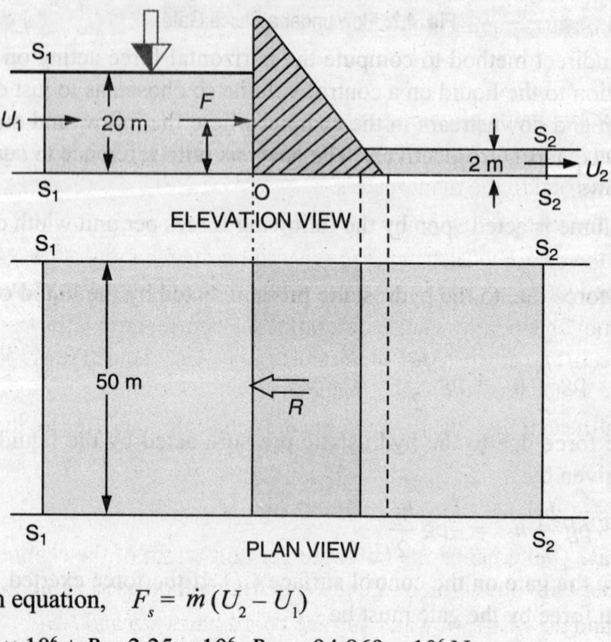

By momentum equation, $F_s = \dot{m}(U_2 - U_1)$

$(98.1 - 0.981) \times 10^6 + R = 2.25 \times 10^6;\ R = -94.869 \times 10^6$ N

This is the reaction by the dam on the water.

Horizontal force on the dam, **F = 94.869 × 10⁶ N to the right**

The line of action of this force can be located by locating the lines of action of the surface forces on S_1S_1 and S_2S_2 as well as of the momentum fluxes, and then taking the moments about a chosen origin, say O on the base line.

Here, the pressure force over S_1S_1 acts at 20/3 m from the base and the pressure force over S_2S_2 acts at 2/3 m from the base. For one-dimensional flow, the corresponding momentum fluxes are assumed to act at the centre-lines of the channels. Hence, taking moments about O,

$$98.1 \times 10^6 \times \frac{20}{3} - 0.981 \times 10^6 \times \frac{2}{3} - 1000 \times (2 \times 50) \times 5 \times 5 \times 1$$

$$+ 1000 \times (20 \times 50) \times 0.5 \times 0.5 \times 10 - 94.869 \times 10^6 \times h = 0$$

or $$654 - 0.654 - 2.5 + 2.5 - 94.869\, h = 0$$

whence, h the height from the base at which the resultant force acts is **6.887 m.**

This height appears to be reasonable from the consideration that the resultant force R is mainly due to the pressure force upstream which acts at 20/3 m from the base.

EXPERIMENT

Flow Under a Sluice Gate

OBJECTIVE

To examine the fundamental characteristics of flow under a sluice gate through the application of the continuity, momentum, energy and similarity relationships.

PRACTICAL RELEVANCE

Sluice gates are used in open channels as rate control devices and thus a basic understanding of the flow characteristics provides information for the hydraulic and structural design of systems with gates. Such systems are commonly found in irrigation canals and in water cooling systems of steel plants, aluminium plants, etc.

PROCEDURE

1. Select a suitable value of gate opening
2. Bleed (remove) the air from all the tubes connecting the piezometer openings on the gate and the flume bed to the manometers.
3. Determine the rate of flow Q through the flume.
4. Determine the piezometric heads h_i by recording the readings of the gate piezometers and the flume bed static pressures h_b by recording the readings of the flume bed piezometers.
5. Using the movable pointer gauge, determine the coordinates of the free water surface both at the upstream and downstream of the sluice gate. Also, note the upstream and downstream free surface water levels h_1 and h_2 respectively.
6. Place a total-head tube at the mid-point in the vena contracta. Record the total head reading.

RESULTS

Calculate the following :

1. The flow rate Q and hence the flow rate per unit width of the channel Q/w by using the data of either the volumetric or the gravimetric method used in the experiment.
2. Upstream and downstream water surface coordinates x/b and h/b.
3. Pressure coefficient on the face of the gate,

$$\frac{p_i}{\rho g b} = \frac{h_i}{b} \quad \text{and plot } p_i/(\rho g b) \text{ vs } h/b$$

4. Pressure coefficient on the base of the channel,

$$\frac{p_b}{\rho g b} = \frac{h_b}{b}$$

and plot $p_b/(\rho g b)$ *vs* x/b.

5. Force per unit width on the gate by using the momentum equation and the pressure distribution and also by using Eq. (3). In order to evaluate the integral make graphical integration of the pressure diagram. Compare the results of the two.
6. Determine the coefficients of contraction (C_c) and discharge (C_d) for various values of h/b ratios by using the following formulae:

$$C_c = \frac{\text{Depth of water at vena contracta}}{\text{Gate opening}} = h_2 / b$$

and

$$C_d = \frac{q}{q_{th}} = \frac{q(h_1 + h_2)}{\sqrt{2g}\, h_1 h_2}$$

POINTS FOR DISCUSSION

1. Discuss the experimental results, *i.e.*, force exerted, C_c and C_d for the flow.
2. Give reasons for the peculiar shape of the pressure distribution on the sluice gate. What type of pressure distribution would you expect across the flow under the sluice gate. Give reasons to justify your answer.
3. Can a sluice gate be used as a discharge measuring device ? If so, spell out the method of its use.
4. Discuss why the surface of the water rises as it approaches a sluice gate and why it continues to drop after it leaves the gate.

ANY OTHER OBSERVATION OR COMMENT?

4.11. Impact of a Jet of Fluid on a Vane

The momentum theorem has wide applications in numerous engineering systems. For example, it is readily applied to carry out a gross analysis of flow through turbines, pumps and other rotodynamic machines, as well as pipe bends and curved channels. It can also be applied in special engineering situations like hydraulic jump, hydraulic ram, etc. In particular, the impact of a fluid jet on a vane or a cascade of vanes has wide application in power generating rotodynamic machines.

When a jet of fluid strikes a solid surface at any angle it does not rebound from the surface. Instead, a stream of fluid is formed which moves over the surface. At the edges of the surface, the fluid stream will leave the surface tangentially provided the cross-sectional area of the jet is small in comparison with the area of the surface.

The striking jet possesses momentum which results in a force being applied on the surface as the momentum changes either because of a change in the magnitude of the velocity or because of change in the direction of velocity.

Consider the jet, shown in the Fig. 4.8 being deflected through an angle by a stationary vane. The two components of force on the vane can be obtained by applying the momentum principle, on a suitably chosen control volume. In applying the momentum principle, the assumptions made are the fluid is inviscid and compressible; the diameter of the jet is small in comparison with the surface area of the vane; surface tension forces are negligibly small; the flow is steady and that the velocity distribution across the cross-section is uniform.

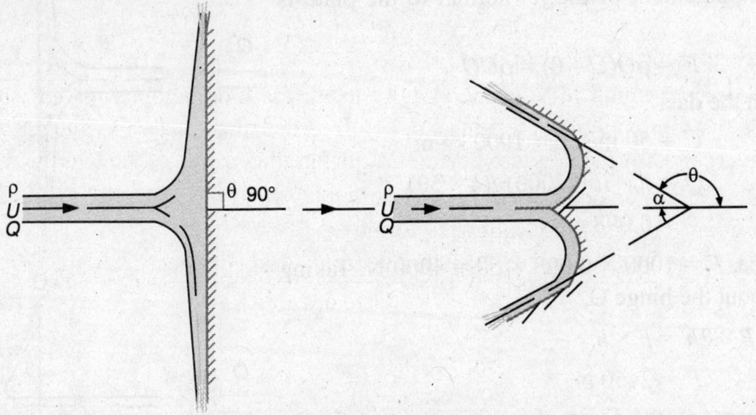

Fig. 4.8. Impact of a Jet on a Vane

Applying the momentum principle,

$$-F_x = \text{Rate of change of momentum in the } x\text{-direction}$$

$$= \int_A U\cos\theta(\rho U dA) - \int_A U(\rho U dA)$$

whence $$F_x = \rho QU(1-\cos\theta)$$

Also, $$F_z - \text{Rate of change of momentum in the } z\text{-direction}$$

$$= \int_A U\sin\theta(\rho U dA) - 0 = \rho QU\sin\theta$$

Often, the jet strikes a vane and results in an axisymmetrical flow with respect to the axis of the jet. Since the entire flow is turned by the same angle θ, the force exerted on the vane along the direction of the jet is given by

$$F_x = \rho QU(1-\cos\theta) \qquad \text{...(4.27)}$$

and there is no component of force normal to it due to axisymmetry.

For a jet striking a flat vane normal to it, $\theta = 90°$;

$$F_x = \rho QU$$

and for a jet striking a curved vane which turns the flow by an obtuse angle θ,

$$F_x = \rho QU(1-\cos\theta)$$

$$= \rho QU(1+\cos\alpha)$$

where $$\alpha = 180° - \theta.$$

The maximum force a jet can exert on a vane is $2\rho QU$ when $\alpha = 0°$, *i.e.*, when the flow is completely reversed in direction. In practice, however, the minimum value of α is limited to 20°.

Example 4.14. *A flat plate, hinged about its top edge as shown, is suspended vertically. It weights 8 kN. A jet of water, 50 mm in diameter and 50 m/s velocity, strikes normal to it at its mid-point. (a) Determine the angle of deflection where it stays in equilibrium. (b) Determine the horizontal force that should be applied to it at the bottom edge to keep it vertical.*

Solution.

It is easier to attempt the second part (*b*) first. The force of impingement of the jet normal to the plate is given by

$$F = \rho Q(U - 0) = \rho UQ$$

From the data,

$$U = 50 \text{ m/s}, \rho = 1000 \text{ kg/m}^3$$

$$Q = \pi \times (50/1000)^2/(4 \times 50)$$

$$= 0.098 \text{ m}^3/\text{s}$$

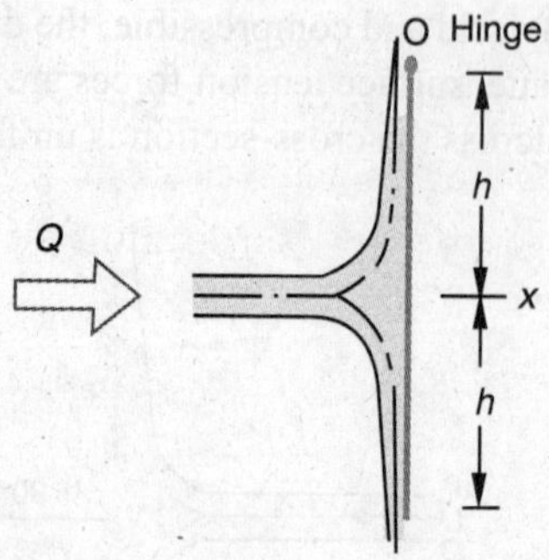

Hence, $F = 1000 \times 0.098 \times 50 = 4900$ N. Taking moment about the hinge O,

$$P \times 2h = F \times h$$

whence, $P = 2450$ N

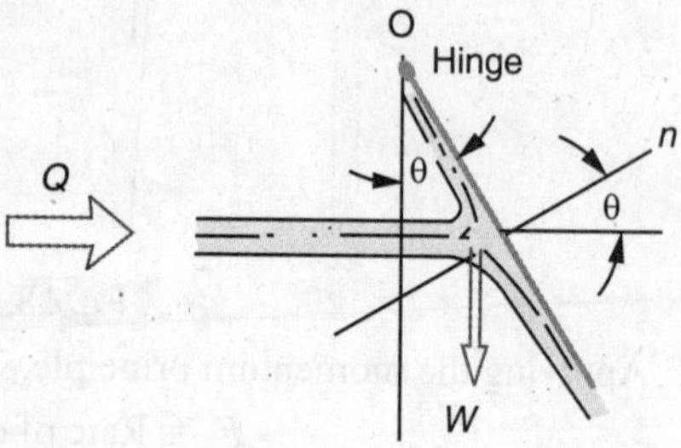

For part (*a*), the force of impingement of the jet normal to the plate is given by

$$F = \rho Q(- U \cos\theta + 0) = - \rho QU \cos\theta$$

Taking moments about the hinge O,

$$\rho QU \cos\theta\, h = Wh \sin\theta$$

whence $\tan\theta = \rho QU/W = 4900/8000 = 0.612$

and $\theta = \mathbf{31.47°}$

EXPERIMENT

Impact of a Jet on Vanes

OBJECTIVE

To study the application of the momentum principle for the impact of a jet on a vane.

PRACTICAL RELEVANCE

The impact of jets on vanes has a wide range of applications in rotodynamic machines. All the impulse driven machines such as Pelton's wheels, windmills, etc. are based on this principle. All such machines are used for the purpose of energy conversion and power generation. These facts make the study of impact of jets on vanes important from the engineering point of view.

PROCEDURE

1. Fix the flat vane or plate together with the appropriate deflector assembly on the apparatus.
2. Note the constants of the force and discharge measuring devices, the jet diameter *d* and the vane angle θ with respect to the jet axis.
3. Check to see that the gauges or meters on the force measuring equipment read zero for conditions of no flow.
4. Start the pump and adjust the jet flow to a very low value by manipulation of the control valve. At this low flow rate adjust the position of the vane so that the jet impinges and passes over the vane smoothly.
5. Adjust the inlet valve to get a suitable flow rate and measure the flow rate either gravimetrically or volumetrically (for low flow rates the gravimetric method is preferred because of the higher accuracies).

6. Note the values of the components (F_x and F_z) of the force acting on the vane as indicated by the force measuring device.
7. Stop the flow and recheck the zero of the force measuring device.
8. Repeat steps 4-7 for six different flow rates within the allowable limits of the equipment.

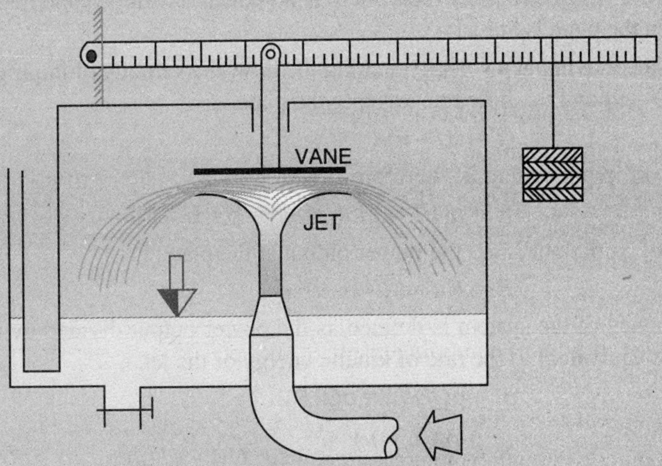

RESULTS

For each case, using the measured data, calculate and tabulate:

1. The flow rate Q
2. The jet velocity U
3. The Reynolds number ($Re = Ud\rho/\mu$) based on jet diameter d and jet velocity U.
4. The force components F_x and F_z respectively acting on the vane on the basis of the momentum principle.
5. The vane coefficient, K. In the case of the flat vane or plate the average value of the coefficient may be calculated.
6. Plot both the experimental and calculated values of the force components F_x and F_z as a function of the Reynolds number for the flat vane or plate. Distinguish between the force components as well as the experimental and calculated values.

POINTS FOR DISCUSSION

1. Give an explanation for the discrepancy, if any, between the measured and calculated force components acting on various vanes.
2. Discuss the dependence of the vane coefficient on the vane geometry.
3. Discuss the results presented in the graph.

ANY OTHER OBSERVATION OR COMMENT?

Example 4.15. *A jet of water with a velocity U and jet area A strikes a flat plate normal to it. Determine the force of impingement, power developed and efficiency when the plate is at rest and when the plate is permitted to move along the direction of the jet at a velocity u. Determine also the condition for maximum possible efficiency.*

What would happen if a series of plates were to face the jet in quick succession?

Solution.

For a flat plate at rest, the force of impingement equals the rate of change of momentum normal to the plate.

$$F = \rho Q(U - 0) = \rho UQ$$

where $Q = UA$, the volumetric flow rate. There is no power developed and hence zero efficiency of the system when the plate is at rest.

For the plate moving at a velocity u along the flow, the velocity of impingement is the relative velocity, *i.e.*, $(U - u)$ and the flow rate arriving the plate is

$$Q = (U - u)A$$

The force of impingement is, therefore,

$$F = \rho(U - u)A.\ (U - u) = \rho A(U - u)^2$$

The rate of work done, *i.e.*, the power output at the plate, is

$$P = F.u = A(U - u)^2.u$$

The efficiency of the system is defined as the power output divided by the power of the jet. The jet power is equivalent to the rate of kinetic energy of the jet.

$$\rho_j = \rho QU^2/2 = \rho QU^3/2$$

Hence $$\eta = \frac{\rho A(U-u)^2 \cdot u}{\rho AU^3/2} = 2u/U(1-u/U)^2$$

The efficiency would be maximum when

$$\frac{\partial \eta}{\partial u} = 0 \text{ or } \frac{\partial \eta}{\partial (u/U)} = 0$$

$$2u/U.(-2)(1-u/U) + (1-u/U).\ 2^2 = 0$$

or $$\boldsymbol{u/U = 1/3}$$

The value of the maximum efficiency is

$$\eta = 2 \times 1/3 \times (1 - 1/3)^2 = 8/27 \text{ or } \mathbf{30\%}$$

In case a series of plates arrive in quick succession, the entire flow rate of the jet, *i.e.*,

$$\rho Q = \rho AU$$

is utilised, but the velocity of impingement remaining the same, $(U - u)$, since every plate moves forward. In this case,

$$F = \rho AU(U - u)$$

and the power output $$P = \rho AU(U - u).\ u$$

The efficiency of the system improves as

$$\eta = \frac{\rho AU\ (U-u).u}{\rho AU^3/2} = 2u/U(1-u/U)$$

It is the maximum when $\partial n/\partial\left(\frac{U}{u}\right) = 0$

$$2u/U.\ (-1) + (1 - u/U).\ 2 = 0$$

or $$u/U = 1/2$$

and the maximum efficiency becomes,

$$\eta = 2.1/2(1 - 1/2) = 1/2\ ;\ \mathbf{50\%}$$

Example 4.16. *A jet of water strikes a flat plate inclined to it. Work out the distribution of discharge over the plate and the force of impact on it, if*

(a) the plate is stationary

(b) the plate moves in the same direction as the incoming flow.

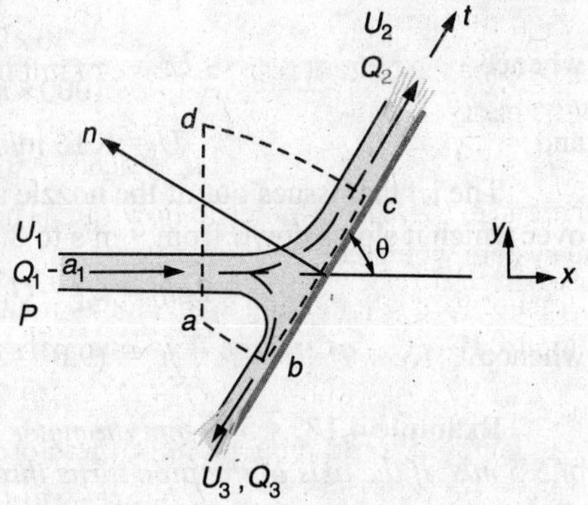

Solution.

The force exerted by a jet of incompressible fluid impinging on a plate can be estimated by the integral momentum analysis as follows:

Incoming fluid mass rate = $\rho Q_1 = \rho U_1 a_1$

Outgoing fluid mass rate = $\rho Q_2 + \rho Q_3$

By continuity, $\rho Q_1 = \rho Q_2 + \rho Q_3;\ Q_1 = Q_2 + Q_3$

(*a*) For the control volume *abcd* just excluding the plate,

$$F_t = (\rho Q_2 U_2 - \rho Q_3 U_3) - \rho Q_1 U_1 \cos\theta \text{ along the plate}$$

$$= \text{zero, for frictionless plate surface}$$

$$U_1 = U_2 = U_3 = U$$

by the Newton's first law, there being no external force in the *t*-direction.

Hence $$F_t = (Q_2 - Q_3 - Q_1)\cos\theta\,\rho U = 0$$

or $$Q_1 \cos\theta = Q_2 - Q_3$$

This equation and the continuity equation result in

$$\boldsymbol{Q_2 = \frac{Q_1}{2}(1+\cos\theta)} \quad \text{...(4.28)}$$

and $$\boldsymbol{Q_3 = \frac{Q_1}{2}(1-\cos\theta)} \quad \text{...(4.29)}$$

as the division of the incoming flow into the two streams.

By momentum balance normal to the plate, $F_n = \rho Q_1 U_1 \sin\theta$ normal to the plate

The force F_n is the force exerted by the plate on the control volume.

The force by impinging jet on the plate is, therefore, $\rho Q_1 U_1 \sin\theta$ acting in a direction normal to the plate along –ve axis. The components of this force along the incoming jet and perpendicular to it are $\rho Q_1 U_1 \sin^2\theta$ and $\rho Q_1 U_1 \sin\theta\cos\theta$ respectively.

(*b*) If the plate in question is made to move externally at a uniform velocity *u* in the same direction as the incoming flow, then the rate of fluid mass arriving the plate will be reduced in the ratio $(U_1 - u)/U_1$ as will be the velocity of the fluid striking the plate. The force of impact would then be

$$\rho(U_1 - u)a_1 \times (U_1 - u)\sin\theta = \boldsymbol{\rho a_1 (U_1 - u)^2 \sin\theta} \text{ normal to the plate} \quad \text{...(4.30)}$$

Example 4.17. *A vertically upward jet of water 7.5 cm in diameter issuing from a nozzle with a velocity of 9 m/s strikes normal to a flat circular plate of mass 30 kg and diameter 50 cm and supports it. Determine the vertical distance above the nozzle where the plate is held in equilibrium.*

Solution.

In the equilibrium position, the upward force of impingement of the jet must just balance the weight of the plate.

$$F = \rho A U^2 = 30 \times 9.81$$

whence $$U^2 = \frac{30\times 9.81\times 4}{1000\times\pi\times 0.075^2} = 66.62$$

and $$U = 8.16 \text{ m/s}$$

The jet that issues out of the nozzle at 9 m/s slows down as it goes up. The vertical distance h over which it slows down from 9 m/s to 8.16 m/s is determined by the Bernoulli principle

$$2gh = U_2^2 - U_1^2$$

whence $$h = (9.0^2 - 8.16^2)/2g = \mathbf{0.73\ m}$$

Example 4.18. *A 300 mm diameter pipe carries water under a head of 20 m with a velocity of 3.5 m/s. If the axis of the pipe turns through 45° in the horizontal plane, find the magnitude and direction of the force at the bend.*

Solution.

Let us consider the free-body diagram of a control volume shown dotted so that the reaction at the bend is assumed to $R_x\, i + R_y\, j$ with the x and y coordinates as shown.

For the pipe,

$$d = 0.3 \text{ m};\ a = \frac{\pi\times 0.3^2}{4} = 0.071 \text{ m}^2.$$

$$p = \rho gH = 1000 \times 9.81 \times 20 = 196200 \text{ N/m}^2.$$

The mass flow rate is given by = 3.5 × 0.071 × 1000 = 248.5 kg/s.

At section AA, $F_{xA} = 196200 \times 0.071 = 13930$ N, $F_{yA} = 0$

At section BB, $F_{xB} = -13930 \cos 45° = -9849$ N; $F_{yB} = -9849$ N.

Momentum rate entering AA = 248.5 × 3.5 = 870 N along x.

Momenum rate leaving BB, along x = 870 cos 45 = 615 N and along y = 615 N.

Rate of change of momentum in x direction = – 255 N.

and in y direction = + 615 N.

For equilibrium of the pipe bend,

$$R_x + 13930 - 9849 - 255 = 0;\ R_y - 9849 + 615 = 0$$

whence $$R_x = 3826 \text{ N},\ R_y = -9234 \text{ N}$$

$$R = \sqrt{3826^2 + 9234^2}$$

$$= 9995 \text{ N},\ \tan\theta = \frac{9234}{3826} = 2.41;\ \theta = 67.5°$$

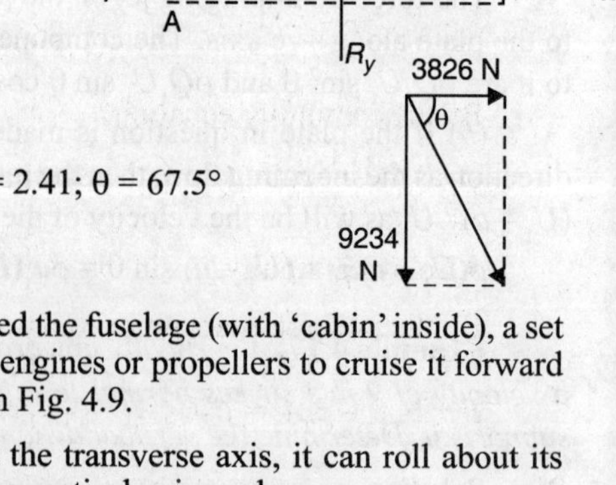

4.12. Aircraft, Jet Engine and Propeller Action

An aircraft is a system consisting of a main body called the fuselage (with 'cabin' inside), a set of main wings and tail wings to lift and steer it, a set of jet engines or propellers to cruise it forward and several other components, some of which are shown in Fig. 4.9.

An aircraft can pitch (with nose up or down) about the transverse axis, it can roll about its fore-and-aft axis and it can yaw (turn right or left) about the vertical axis as shown.

A jet engine consists of an air-intake system, a compressor, a diffuser, a combustion chamber, a turbine and a jet pipe, all arranged in one line for the high velocity air stream/combusting gases to proceed least interrupted. Issuance of the exhaust gases at high velocity through the jet pipe results in the propulsive force and hence, power.

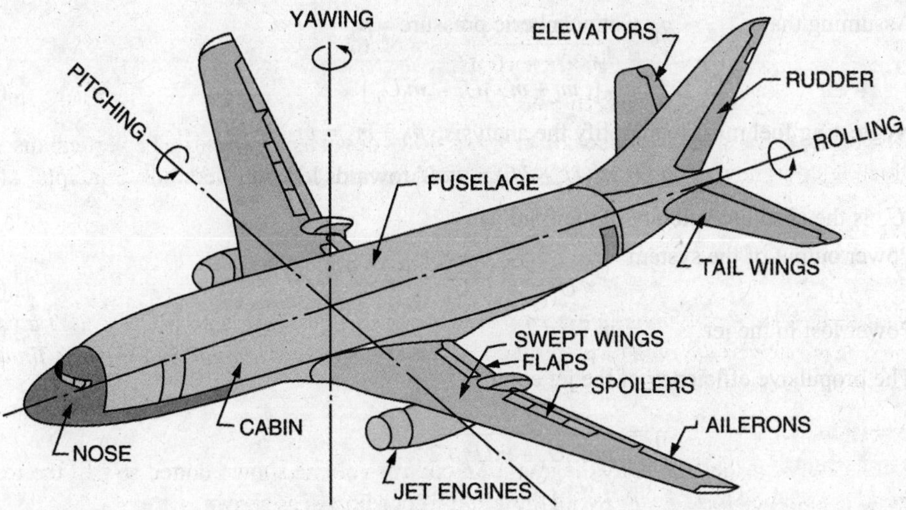

Fig 4.9: Components and Operation of an Aircraft

A propeller-system consists of similar basic facility except that the jet velocity is kept low so that the propeller, in front of the aircraft or behind a ship, provides the major proportion of the thrust. This is also called turboprop system.

A rocket motor, on the other hand, carries with it an oxidising agent to mix with its fuel within the system to develop a thrust independent of the medium through which it travels.

Jet Action. Consider a jet engine in flight at a constant velocity U_1 to the left. If we superimpose on the entire engine a velocity U_1 equal and opposite to the engine velocity, the engine is brought to rest, with inlet velocity U_1 and exit velocity U_2 with respect to engine as shown in Fig 4.10. The control volume encloses the fluid within the inlet and outlet. The jet engine body impresses upon the control volume a reaction force F assumed positive.

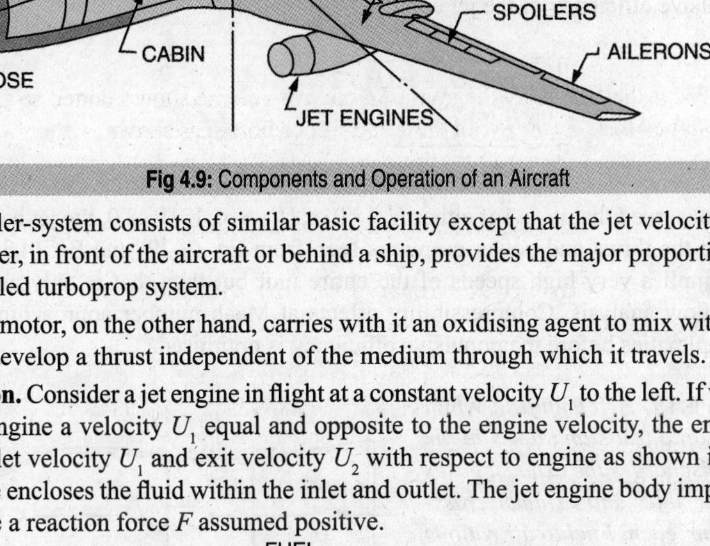

Fig. 4.10: Analysis of Jet Action

By the continuity equation,

$$\rho_1 U_1 A_1 + \frac{1}{N}\rho_1 U_1 A_1 = \rho_2 U_2 A_2 \text{ where } N \text{ is the fuel fraction}$$

Also by continuity, $\dot{m}_1 + \dot{m}_f = \dot{m}_2$

By the integral momentum equation,

$$F + p_1 A_1 - p_2 A_2 = \rho_2 U_2^2 A_2 - \rho_1 U_1^2 A_1 = \dot{m}_2 U_2 - \dot{m}_1 U_1$$

or

$$F = (\dot{m}_1 + \dot{m}_f)U_2 - \dot{m}_1 U_1 - (p_1 A_1 - p_2 A_2)$$

Action of the control volume on the jet engine body is the negative of this.

The thrust is given by $-F = -[(\dot{m}_1 + \dot{m}_f)U_2 - \dot{m}_1 U_1 - (p_1 A_1 - p_2 A_2)]$

Assuming that $p_1 \approx$ atmospheric pressure $\approx p_2$

and $A_1 \approx A_2$

$$-F = -[(\dot{m}_1 + \dot{m}_f)U_2 - \dot{m}_1 U_1] \qquad ...(4.28)$$

Neglecting fuel mass to simplify the analysis; $\dot{m}_1 + \dot{m}_2 = \dot{m}$

$$F = \dot{m}(U_1 - U_2) = \dot{m}\, U_j \text{ towards left,} \qquad ...(4.29)$$

where U_j is the absolute velocity of the fluid jet.

Power output of the system

$$= F \times U_1 = \dot{m}\, U_j U_1 \qquad ...(4.30)$$

Power lost in the jet $= \dot{m}\, U_j^2 / 2$ ...(4.31)

The propulsive efficiency of the jet engine,

$$\eta = \frac{\dot{m}U_jU_1}{\dot{m}U_jU_1 + \dot{m}U_j^2/2}$$

or

$$\eta = \frac{1}{1 + U_j / 2U_1} \qquad ...(4.32)$$

It may be observed that $\eta \to 1$ if either $U_j \to 0$ or $U_1 \to \infty$. If $U_j \to 0$, the jet has zero absolute velocity but then the thrust and power output as seen from Eq. (4.29) and Eq. (4.30) also tend to zero. $U_1 \to \infty$ implies very high speeds of the entire unit but then that is only according to the incompressible flow analysis. Compressibility effects at Mach number approaching unity reveal many more complexities before the populsive efficiency is optimised.

Example 4.19. *A jet engine when tested statically on a test stand takes in air at 200 m/s and discharges the exhaust gases at 1500 m/s. The inlet and exhaust cross-sections are 0.5 m² each. Fuel to air ratio is 1: 50. Determine the force expected to act on the engine. Assume the density of air at entry = 1.2 kg/m3.*

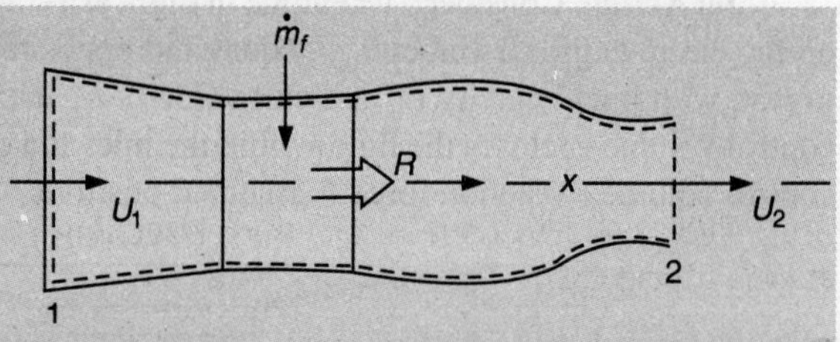

Solution.

Choosing the control volume comprising the fluid within the jet engine, the reaction R chosen to be positive, by the engine body is an external force on it.

Net external force $F_x = \dot{m}_2 U_2 - \dot{m}_1 U_1$

or $R + p_1A_1 - p_2A_2 = (\dot{m}_1 + \dot{m}_f)U_2 - \dot{m}_1 U_1$

Neglecting the initial momentum of the fuel along the x-axis, considering $p_1 = p_2$ in the absence of further information and

$$\dot{m}_1 = \rho_1 A_1 u_1 = 1.2 \times 0.5 \times 200 = 120 \text{ kg/s}$$

$$\dot{m}_f = 120/50 = 2.4 \text{ kg/s}$$

$$R = (120 + 2.4) \times 1500 - 120 \times 200$$

$$= 159\,600 \text{ N} = 159.6 \text{ kN}$$

Thus T acting on the engine = – **159.6 kN.** The negative sign implies that it acts towards the left, as expected, tending to move the engine forward towards the left.

Example 4.20. *A small jet engine in flight at 30 m/s at low altitude takes in air through a 0.4 m diameter circular inlet in the front and discharges the exhaust gases through a 0.1 m diameter nozzle at 100 m/s relative to the engine. Neglecting the fuel to air ratio, estimate the performance of the engine.*

Solution.

Superimposing a velocity field equal and opposite to the engine flight speed, the engine is conceived to be at rest with the flow velocities as

$$U_1 = 30 \text{ m/s}, U_2 = 100 \text{ m/s}$$

with respect to the engine.

$$U_j = \text{absolute velocity of the jet}$$

$$= 100 - 30 = 70 \text{ m/s}$$

$$\dot{m} = 1.2 \times 30 \times \frac{\pi}{4} \times 0.4^2 = 4.52 \text{ kg/s}$$

Thrust $F = \dot{m}(U_1 - U_2) = \dot{m}U_j$

$$= 4.52 \times 70 = \mathbf{316.67\ N}$$

Power loss in the jet $= \dot{m}U_j^2/2$

$$= \frac{4.52 \times 70^2}{2} = 11\,074 \text{ W} = \mathbf{11.1\ kW}$$

Power output of engine $= F.\, U_1$

$$= 316.67 \times 30 = 9500.18 \text{ W} = \mathbf{9.5\ kW}$$

Propulsive efficiency $\eta = \dfrac{1}{1 + U_j/2U_1} = \dfrac{1}{1 + 70/60} = 0.46 = \mathbf{46\%}$

The small jet engine is, therefore, operating at 46% efficiency.

Propeller Action

The operation of a propeller is explained by considering the flow pattern between a far upstream station 1 and a far downstream station 4 bounded by a slipstream just enclosing the tips of the propeller but leaving out the propeller between stations 2 and 3 shown in Fig. 4.11. Whether the propeller is travelling upstream at velocity U_1 or it is stationary and a flow U_1 from upstream is impressed upon it, the picture and hence the analysis is the same.

The mass flow rate at each section must be the same; areas at 2 and 3 being nearly equal, $U_2 = U_3 = U$.

Take the control volume as the fluid enclosed between sections 1 and 4 with the propeller thrust force F acting on it. By the integral momentum equation applied along the flow,

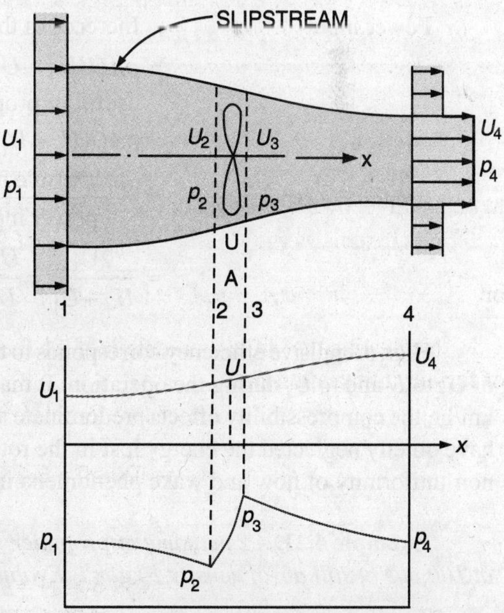

Fig. 4.11: Propeller Action

$$F = \rho AU(U_4 - U_1) = \dot{m}\,(U_4 - U_1)$$

and $$F = (p_3 - p_2)A = \text{net pressure force on the propeller.}$$

Hence $\rho U(U_4 - U_1) = (p_3 - p_2)$

By the Bernoulli equation,

between 1 and 2, $$p_1 + \frac{\rho U_1^2}{2} = p_2 + \frac{\rho U_2^2}{2}$$

between 3 and 4, $$p_3 + \frac{\rho U_3^2}{2} = p_4 + \frac{\rho U_4^2}{2}$$

Taking $$p_1 = p_4$$

$$(p_3 - p_2) = \frac{1}{2}\rho(U_4^2 - U_1^2)$$

From these equations,

$$\rho U(U_4 - U_1) = \frac{1}{2}\rho(U_4^2 - U_1^2) \quad \text{...(4.33)}$$

or $$U = \frac{U_1 + U_4}{2}$$

Hence, the velocity U through the propeller is the average of the far upstream and far downstream velocities.

In other words, the induced velocity u through the propeller equals half the difference in U_1 and U_4.

Velocity $$u = U - U_1 = U_4 - U = \frac{U_4 - U_1}{2} \quad \text{...(4.34)}$$

Propulsive power = Thrust × propulsive velocity

$= \rho AU(U_4 - U_1)U_1$

Power input = Increase in the rate of kinetic energy of the fluid from 1 to 4

$= \rho AU(U_4^2 - U_1^2)/2$

= useful or propulsive power plus power lost in the flow

$= \rho AU(U_4 - U_1)U_1 + \rho AU(U_4 - U_1)^2/2$

Propulsive efficiency $$= \frac{\text{propulsive power}}{\text{power input}}$$

$$= \frac{2U_1}{U_4 + U_1} = \frac{U_1}{U} \quad \text{...(4.35)}$$

High propulsive efficiency corresponds to the closeness of U to U_1, *i.e.*, requires minimal increase of U_1 to U and to U_4 during the operation. It may also be noted that at higher speeds, exceeding 700 km/hr, the compressibility effects predominate and the efficiency drops off drastically. Moreover, we have quietly neglected the energy lost in the rotational component of the flow, propeller blade-drag, non-uniformity of flow and wake phenomena in the analysis.

Example 4.21. *A 2 m diameter propeller of an aeroplane discharges 400 m³/s during a flight at 100 m/s in still air of density 1 kg/m³. Evaluate its performance in terms of thrust and efficiency.*

Solution.

Given that $U_1 = 100$ m/s

$$U = 400/\left(\frac{\pi}{4}\times 2^2\right) = 127 \text{ m/s}$$

$$= \frac{(U_1 + U_4)}{2}$$

$$U_4 = 154 \text{ m/s}$$

Thrust $\quad F = \dot{m}\ (U_4 - U_1)$

$$= (1 \times 400)(154 - 100)\ 21\ 600 \text{ N} = \mathbf{21.6\ kN}$$

Propulsive power $\quad = 21.600 \times 100 = 2.16 \times 10^6 \text{ W} = \mathbf{2.16\ MW}$

Power input $\quad = (1 \times 400)(154^2 - 100^2)/2$

$$= 2.74 \times 10^6 \text{ N} = 2.74 \text{ MN}$$

Propulsive efficiency $\quad = \dfrac{2.16\times10^6}{2.74\times10^6} = 0.788$ or 78.8%

This could have also been obtained by recalling Eq. (4.35),

whence, the propulsive efficiency $\eta = \dfrac{U_1}{U} = \dfrac{100}{127} = 0.788$ or **78.8 %**

Example 4.22. *A helicopter of mass 1000 kg has a rotor of diameter 10 m. It descends at a constant speed of 20 m/s. Calculate the induced velocity through the rotor.*

Solution.

Here, the propeller axis is vertical and the drag produced by the propeller must balance its weight for equilibrium.

The slipstream in this case is such that its areas of cross-section is more above and less below the rotor, the flow being upwards.

Drag on the rotor $= \rho AU\,(U_1 - U_4) = \rho AU\,(U_1 - 2U + U_1)$

Employing the fact $\quad u = U - U_1 = U_4 - U$

the drag is given by $\quad D = \rho A(U_1 - u) \times 2u$

$$= 1.2 \times \pi \times \frac{10^2}{4}(U_1 - u) \times 2u$$

where u, the induced velocity through the rotor, equals half the increment of the velocity from inlet to outlet taken far apart. Equating the drag to the weight of the helicopter,

$$D = W$$

$$188.8\,(20 - u)\,u = 1000 \times 9.81$$

$$u^2 - 20u + 52 = 0$$

solving it, $\quad u = \mathbf{3.1\ m/s}$

The other solution, $u = 16.9$ is meaningless because it would require U_4 to be reversed in relation to the direction of U_1.

Example 4.23. *A rocket moving forward at a speed of 600 m/s discharges its exhaust gases with an absolute velocity of 200 m/s. The exhaust gases, 5 kg/s, pass through an exit area of 0.06 m² where the absolute pressure is 100 kN/m². The atmospheric pressure at that alitude is 75 kN/m². Determine the thrust and the power of the rocket.*

Solution.

For the control volume attached to the rocket, enclosing the fluid within it until the exit, the velocity of the exhaust gases appears to be – 200 – 600, *i.e.*, – 800 m/s. Rate of change of momentum is, therefore,

$$4.5 \times (-800 - 0) = -3600 \text{ kg m/s}^2$$

This must equal the external force acting on it,

$$F + 0.06 \times (75 - 100) \times 1000 = -3600 \text{ N}$$

whence $F = -3600 + 1500 = -2100$ N

This is the backward force exerted by the rocket on the fluid in the control volume. The force, F, exerted by the fluid on the rocket is, therefore, 2100 N in the forward direction.

The power of the rocket is given by

$$p = F \times U = 2100 \times 600$$
$$= 1260\,000 \text{ Nm/s} = \mathbf{1260\ kW}$$

4.13. Angular Momentum Equation

The Newton's second law of motion expressed as 'net external moment on a system equals the rate of change of angular momentum' is also called the angular momentum equation. For an integral control volume, the substantive rate of change of angular momentum is due to the angular momentum crossing the control surface as well as due to the rate of accumulation within the control volume. Hence,

$$\mathbf{M} = \mathbf{M}_S + \mathbf{M}_B = \int_S (\mathbf{r} \times \mathbf{U})\rho U.dA\} + \frac{\partial}{\partial t}\int_V (\mathbf{r} \times \mathbf{U})\rho V$$

Let us consider practical application of this equation for simple problems in axisymmetric flows. This refers to circulatory flows in rotodynamic machines, garden sprinklers etc. The simplified angular momentum equation expressed in cylindrical polar coordinate is

$$T = \int_S rU_\theta(\rho U.dA) + \frac{\partial}{\partial t}\int_V rU_\theta \rho dV \qquad \text{...(4.36)}$$

where T is the torque about the axis of rotation.

Again $\frac{\partial}{\partial t} \equiv 0$ for steady flow and stipulating the incoming and outgoing flows to be normal to the local control surfaces,

$$T = r_2 U_{\theta 2}(\rho_2 U_2 A_2) - r_1 U_{\theta 1}(\rho_1 U_1 A_1)$$
$$T = (r_2 U_{\theta 2} - r_1 U_{\theta 1})\, \dot{m} \qquad \text{...(4.37)}$$

This is known as the Euler's principal equation.

In the absence of external momentum, $T = 0$

$$r_2 U_{\theta 2} = r_1 U_{\theta 1} = \text{constant} \qquad \text{...(4.38)}$$

where $U_{\theta 1}$ and $U_{\theta 2}$ are the absolute circumferential velocities of the fluid entering and leaving at radial locations r_1 and r_2 respectively. Examples of such torque-free motions are the free-vortex flows, free-spiral flows and garden water-sprinklers. Also, by continuity

$$\rho_1 U_1 A_1 = \rho_2 U_2 A_2 = \dot{m}$$

4.14. Energy Equation: Steady Flow Energy Equation

The energy per unit mass, e, of a fluid is the sum of

(*a*) kinetic energy, $\frac{U^2}{2}$
(*b*) potential energy, gz
(*c*) internal energy, i

} together called mechanical energy

which is stated as $e = U^2/2 + gz + i$

Invoking the first law of thermodynamics and using the fact the enthalpy $h = i + p/\rho$, it is possible to obtain the steady flow energy equation:

$$q - \dot{w}_s = \left| U^2/2 + gz + h \right|_1^2$$

where q is the heat added per unit mass and w_s is the shaft work done per unit mass.

4.15. Bernoulli Equation: Derivation From the Energy Equation

The steady flow energy equation states that

$$\left| \frac{U^2}{2} + gz + h \right|_1^2 = q - \dot{w}_s$$

In the absence of the heat transfer between the fluid and the surroundings, $q = 0$ and in the absence of any shaft work input or output, $w_s = 0$, the steady flow energy equation reduces to the form

$$\frac{U_2^2}{2} + gz_2 + \frac{p_2}{\rho} = \frac{U_1^2}{2} + gz_1 + \frac{p_1}{\rho}$$

or

$$\frac{U^2}{2} + \frac{p}{\rho} + gz = \text{constant}$$

which is the Bernoulli equation for an incompressible flow.

In order to obtain the Bernoulli equation for a compressible flow, let us rewrite the steady flow energy equation in the absence of heat transfer and shaft work,

$$\frac{U^2}{2} + gz + h = \text{constant}$$

Substituting

$$h = C_p T = C_p \frac{p}{\rho R} = \frac{C_p}{C_p - C_v} \frac{p}{\rho} = \frac{\gamma}{\gamma - 1} \frac{p}{\rho}$$

in the above equation, we get

$$\frac{U^2}{2} + \frac{\gamma}{\gamma - 1} \frac{p}{\rho} + gz = \text{constant}$$

which is the Bernoulli equation for a compressible flow.

4.16. Reflection on Bernoulli Equation

The Bernoulli equation is indeed a very important equation in fluid mechanics. One may well wonder why the Bernoulli equation can be obtained either from the law of motion or from the first law of Thermodynamics.

The Bernoulli equation can be derived either by the integration of the Euler's equation or by a degeneration of the steady flow energy equation. The answer is that the Bernoulli equation is arrived at after a number of assumptions in either case that it is applicable to a very narrow range of fluid flows. An examination of assumptions made in each case would reveal that the assumptions are not exactly the same but lead to similar interpretations.

The Bernoulli equation is basically an energy equation. In the form,

$$\frac{U^2}{2} + \frac{p}{\rho} + gz = C_1$$

each term represents energy per unit mass. Units of either term are

$$\frac{\text{m}^2}{\text{s}^2} \equiv \frac{\text{kg m}^2}{\text{kg s}^2} \equiv \frac{\text{kg m}^2}{\text{s}} /\text{kg} \equiv \text{J/kg}$$

It may also be noted that the units of each term in the Bernoulli equation could also be energy per unit weight, also called the head of the fluid.

$$\frac{U^2}{2g} + \frac{p}{\rho g} + z = C$$

Units of $p/\rho g$ are $\frac{\text{N}}{\text{m}^2}\frac{\text{m}^3}{\text{kg}}\frac{\text{s}^2}{\text{m}} = \frac{\text{Ns}^2}{\text{kg}} \equiv \frac{\text{kgm}}{\text{s}^2} \cdot \frac{\text{s}^2}{\text{kg}} = \text{m}$

Units of $U^2/2g$ are $\frac{\text{m}^2}{\text{s}^2}\frac{\text{s}^2}{\text{m}} = \text{m}$

Units of z are m head of the fluid

Units of Energy per unit weight are $\frac{\text{J}}{\text{N}} = \frac{\text{Nm}}{\text{N}} = \text{m}$

In the context of the discussion on the Bernoulli equation it may be noted that it reduces to the hydrostatic law when the fluid is at rest. In that case $U = 0$, and the

$$z + \frac{p}{\rho g} = \text{constant}$$

If the reference pressure is the atmospheric pressure p_0, at the position $z = 0$, the constant equals $p_0/\rho g$. Consequently,

$$p_0 - p = \rho g z$$

or

$$p_{\text{gauge}} = \rho g z$$

where z is taken positive downwards. *This is the hydrostatic law for an incompressible fluid at rest.*

Another case of interest is a forced vortex such as rotation of a liquid contained in a vessel. In that case, $U = r\,\omega$ and the Bernoulli equation reduces to

$$\frac{r^2\,\omega^2}{2g} + \frac{p}{\rho g} + z = C$$

$$\frac{1}{2}\rho r^2\,\omega^2 + p + \rho g z = K$$

which is the same as has been derived for a solid body rotation of a fluid.

4.17. Tips for Choosing an Appropriate Control Volume

Major simplification in the analysis of a fluid flow can be made by a thoughtful choice of the control volume. Some of the tips are:

(i) *The control volume should enclose all the fluid between the inlet and outlet, i.e., between the stations where the conditions are known.*

(ii) *The control surface should be drawn normal to inlet and exit flows so that calculations are simpler and shear stresses do not enter into the picture.*

(iii) *The control surface should cut across the supports or fixtures to include the external reactions on the control volume which can then be calculated.*

(iv) *The control surface should just surround the working fluid; the reaction of the container should enter the picture which can be evaluated.*

(v) *The control volume for which our analysis applies is inertial or non-accelerating; therefore, care should be taken to take it stationary or moving with constant velocity.*

4.18. Correction Factors for Non-Uniform Flows

It is usual in engineering analysis to assume that the flow through a passage takes place with an average velocity. The average velocity is calculated on the basis of mass conservation. The average velocity thus calculated does not correspond with the kinetic energy and momentum associated with the flow. It is, therefore, necessary to define and to compute correction factors to account for discrepancies in kinetic energy and momentum respectively.

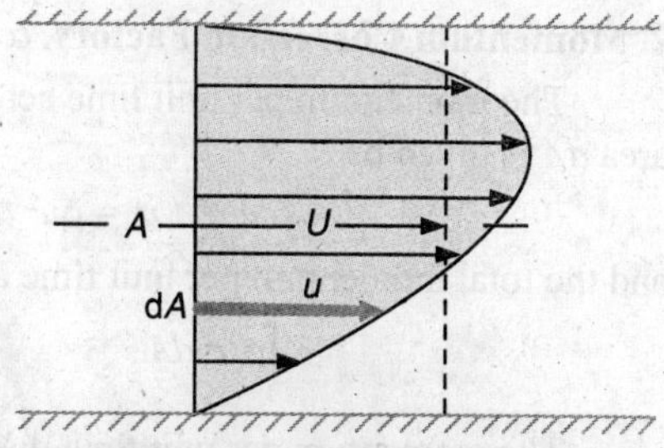

Fig. 4.12: Non-Uniform Velocity Profile in a Passage

Consider a passage of a cross-sectional area A through which the flow takes place with a given velocity distribution as shown in Fig. 4.12. Let the velocity of the fluid passing through an elemental area dA be u. Then, the mass flow rate passing through the elemental area would be given by

$$d\dot{m} = \rho u dA$$

and the total mass flow rate through the passage would be

$$\dot{m} = \int_A \rho u \, d A$$

the average velocity U should be such that it results in the same mass flow rate, *i.e.*,

$$\rho UA = \int_A \rho u \, d A$$

whence
$$U = \int_A u \, dA / A$$

1. Kinetic Energy Correction Factor, α

The kinetic energy per unity time actually associated with the fluid flowing through the elemental area dA is given by

$$\frac{1}{2}\rho u \, dA u^2 = \frac{1}{2}\rho u^3 \, dA$$

and the total kinetic energy per unit time actually passing through the entire cross-section must be

$$\frac{1}{2}\int \rho u^3 \, dA$$

The kinetic energy per unit time due to the average velocity U in the passage is

$$\frac{1}{2}\int \rho UAU^2 = \frac{1}{2}\rho AU^3$$

The kinetic energy correction factor α is defined as the ratio of the actual kinetic energy to that represented by the averaged flow:

$$\alpha = \frac{\frac{1}{2}\int \rho u^2 dA}{\frac{1}{2}\rho AU^3} = \frac{1}{A}\int \left(\frac{u}{U}\right)^3 dA \qquad ...(4.39)$$

The value of the kinetic energy correction factor α is unity when $u = U$, *i.e.*, when the flow is actually uniform. The value of α is greater than 1 for non-uniform velocity distributions. For laminar flow through a round pipe, $\alpha = 2$ and for turbulent flow through a pipe, $\alpha = 1.05$. It is conventional to use a value, $\alpha = 1$ for turbulent flows.

2. Momentum Correction Factors, α

The momentum per unit time actually associated with the fluid flowing through the elemental area dA is given by

$$\rho\, u\, dA\, u = \rho u^2\, dA$$

and the total momentum per unit time actually passing through the entire cross-section must be

$$\int \rho u^2 dA$$

The momentum per unit time due to the average velocity U in the passage is given by

$$\rho UAU = \rho U^2 A$$

The momentum correction factor, β is defined as the ratio of the actual momentum to that represented by the averaged flow,

$$\beta = \frac{\int \rho u^2 dA}{\rho U^2 A} = \frac{1}{A}\int \left(\frac{u}{U}\right)^2 dA \qquad \text{...(4.40)}$$

The value of the momentum correction factor β is unity when $u = U$, *i.e.*, when the flow is actually uniform. The value of α is greater than 1 for non-uniform velocity distributions. However, the value of β must be less than that of α. For laminar flow through a round pipe, the value of β is 1.33 and for turbulent flows, β is very close to unity, it being taken as 1 for all practical purposes.

Example 4.24. *A centrifugal pump has an impeller with inner and outer diameters 15 cm and 25 cm respectively. Effective outlet area is 180 cm² and discharge 0.03 m²/s. The blades are curved backwards with $\beta_2 = 40°$. Suction and delivery diameters are 5 cm each and the pressure heads at suction and delivery are –5 m and + 15 m respectively with respect to atmospheric pressure. The motor delivers 8 kW at 1450 r.p.m. Assume that the true outlet whirl velocity is 70% of ideal. Determine the manometric efficiency and the overall efficiency for medium as water.*

Solution.

With reference to the pump,

$$Q = \dot{m}/\rho = 0.03 = 2\pi U_2 \cos\theta_2 r_2 b_2 = \frac{180}{100^2} \times U_2 \cos\theta_2$$

whence, $$U_2 \cos\theta_2 = 1.67 \text{ m/s}$$

$$U_2 = \omega r_2 = \frac{\pi \times 1450}{60} \times \frac{25}{100} = 19 \text{ m/s}$$

$$W_2 = \frac{1.67}{\cos 50°} = 2.6 \text{ m/s}$$

$$U_{\theta 2} = (1912.6 \times \cos 40°) \times 0.70 = 11.9 \text{ m/s}$$

along the direction of rotation.

$$T = r_2 U_{\theta 2} \dot{m} = \frac{25}{100} \times 11.9 \times 0.03 \times 1000 = 44.6 \text{ Nm}$$

$$\omega = 2\pi \times \frac{1450}{60} = 152 \text{ rad/s}$$

Power input to water = 44.6 × 152 = 6780 W = 6.78 kW

Manometric head = head at outlet – head at inlet

= 15 + 5 = 20 m

since there is no difference in the velocity heads in the suction and delivery pipes.

Water power as displayed by the manometer readings

$$P = 20 \times 0.03 \times 1000 = 6000\ W$$

Manometric efficiency $\eta_m = \frac{6000}{6780} = \mathbf{0.89\ or\ 89\%}$

Overall efficiency $\dot{\eta}_0 = \frac{6000}{8000} = \mathbf{0.75\ or\ 75\%}$

Example 4.25. *A lawn sprinkler with two nozzles 0.5 cm diameter each at 20 cm and 15 cm radii is connected across a tap capable of 6 litres/minute discharge. The nozzles discharge water upwards and outwards from the plane of rotation. What torque will the sprinkler exert on the hand, if held stationary and at what angular velocity will it rotate free?*

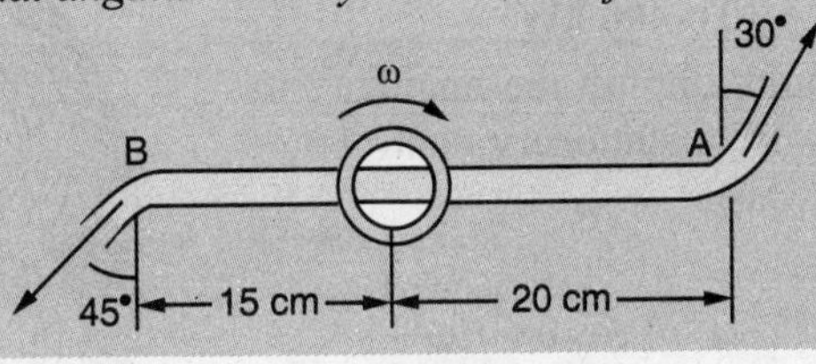

Solution.

Assume the discharge to be equally divided between the two nozzles.

$$Q_A = Q_B = 3 \times 10^{-3}/60 = 50 \times 10^{-6}\ m^3/s$$

$$U_A = U_B = 50 \times 10^{-6}(\pi 4 \times 0.005^2) = 2.54\ m/s$$

$$U_{\theta A} = 2.54 \cos 30° = 2.2\ m/s$$

$$U_{\theta B} = 2.54 \cos 45° = 1.8\ m/s$$

Assume that the water entering the sprinkler through a tap does not involve any angular momentum.

(*a*) When stationary, the torque due to nozzles discharging water,

for nozzle *A*, $T_A = 1000 \times 50 \times 10^{-6} \times 2.2 \times 0.20 = 0.0220$ Nm

for nozzle *B*, $T_B = 1000 \times 50 \times 10^{-6} \times 1.8 \times 1.5 = 0.0135$ Nm

Total torque due to nozzles *A* and *B*, T = **0.0355 Nm**

(*b*) When rotating free, let the angular velocity be ω.

Now the absolute velocities of the nozzle-discharge in the circumferential direction are

for nozzle *A*, $U_{\theta A} = (2.2 - 0.2\ \omega)$ m/s

for nozzle *B*, $U_{\theta B} = (1.8 - 0.15\ \omega)$ m/s

There being no external moment, the angular momentum will be conserved,

$$\rho Q_A\ (2.2 - 0.2\ \omega) \times (-0.2) - \rho Q_B(1.8 - 0.15\omega) \times 0.15 = 0$$

Cancelling ρ and using $Q_A = Q_B$,

$$\omega = \mathbf{9.72\ rad/s}$$

Example 4.26. *An impulse water turbine (Pelton turbine) having 22 vanes operates under a head of 50 m and 1 m³/s discharge. The mean vane speed is 10 m/s and the deflection angle is 150°. Calculate the output power. The vane radius is 1 m.*

Solution.

Consider a control volume fixed relative to a vane as shown in the figure.

$$U_1 = \sqrt{2gH} = (2 \times 9.81 \times 50)^{1/2} = 31.4 \text{ m/s}$$

Incoming velocity at $S_1S_1 = U_1 - r\omega = 31.4 - 10 = 21.4$ m/s

Outgoing velocity at S_2S_2 and S_3S_3 sections must also be 21.4 m/s in the absence of friction. Its component along the jet axis = – 21.4 cos 30° = 18.6 m/s.

By the integral momentum equation,

$$R_x = 1 \times 1000 \times (-18.6 - 21.4) = -40000 \text{ N}$$
$$= -40 \text{ kN}$$

The force on the vane is + 40 kN along the +ve x-axis.

Power output $= 40 \times 10 =$ **400 kW**

Alternatively, one could employ the angular momentum principle. Consider a stationary control volume enclosing the entire wheel.

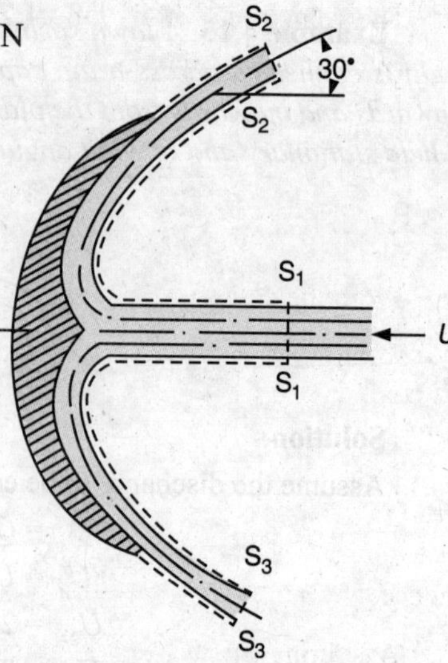

$$U_{\theta 1} = U_1$$
$$U_{\theta 2} = -(U_1 - r\omega)\cos\theta + r\omega$$
$$T = -r\,([U_1 - r\omega)\cos\theta + U_1 - r\omega]\,\dot{m}$$
$$= r\,([U_1 - r\omega)(1 + \cos\theta)\,\dot{m}$$

= torque by the shaft on the control volume

Torque by the fluid on the shaft is negative of this value.

Power output $=$ torque on the shaft $\times\ \omega$

$= 1 \times [21.4 \times (1 + .866)] \times 10 \times 1000$ W $=$ **400 kW**

SUPPLEMENTARY REVISION EXAMPLES

Example 4.27. *A nozzle at the end of a 80 mm hose pipe produces a jet 40 mm in diameter. Determine the force on the joint at the base of the nozzle when it is discharging 1200 litres of water per minute.*

Solution.

Given that

$$Q = 1200 \text{ litres/mt} = \frac{1200 \times 1000}{(100)^3}\,\frac{1}{60} = 0.02 \text{ m}^3/\text{s}$$

whence $\dot{m}$, the rate of mass flow is given by

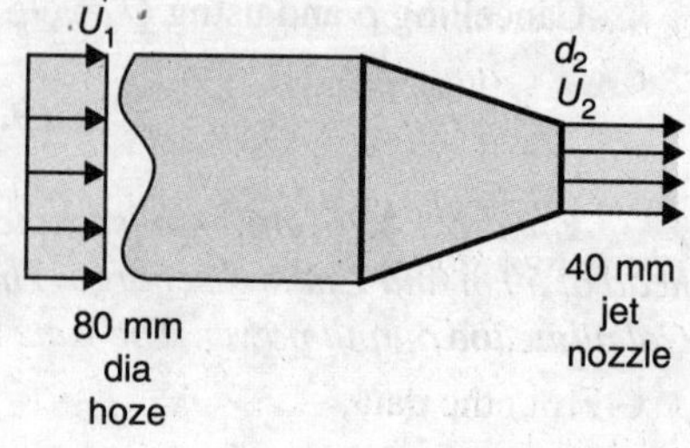

$$\dot{m} = \rho Q = 1000 \times 0.02 = 20 \text{ kg/s.}$$
$$d_1 = 80 \text{ mm} = 0.08 \text{ m}; \quad d_2 = 0.04 \text{ m}$$
$$a_1 = \frac{\pi d_1^2}{4} = 0.005 \text{ m}^2; \quad a_2 = 0.00125 \text{ m}^2$$
$$U_1 = 0.02/0.005 = 4 \text{ m/s}, \quad U_2 = 16 \text{ m/s}$$

Now,
$$F = \dot{m}\,(U_2 - U_1)$$
$$= 20\,(16 - 4) = 240 \text{ N.}$$

Example 4.28. *A rectangular duct of width 25 cm has an elbow made up of circular arcs of radii 40 cm and 65 cm for the inner and outer walls respectively. Calculate the discharge per unit width of the duct when the pressure difference between outer and inner walls is 30 kPa. Assume the flow to be irrotational.*

Solution.

Irrotation of flow implies $r\,U_\theta$ = constt.

$$0.4\,U_{\theta 1} = 0.65\,U_{\theta 2} = \text{c.}$$

The velocity is wholly tangential in the circular flow; higher at 1 than at 2.

By Bernoulli equation,

$$p_1 + \rho\frac{U_1^2}{2} = p_2 + \rho\frac{U_2^2}{2}$$

or
$$p_1 - p_2 = \rho/2\left(U_2^2 - u_1^2\right) = \frac{\rho}{2}(U_{\theta_2}^2 - u_{\theta_1}^2)$$

From the data,

$$30{,}000 = \frac{\rho}{2}\left(\frac{0.4^2\,U_{\theta_1}^2}{0.65^2} - U_{\theta_1}^2\right)$$

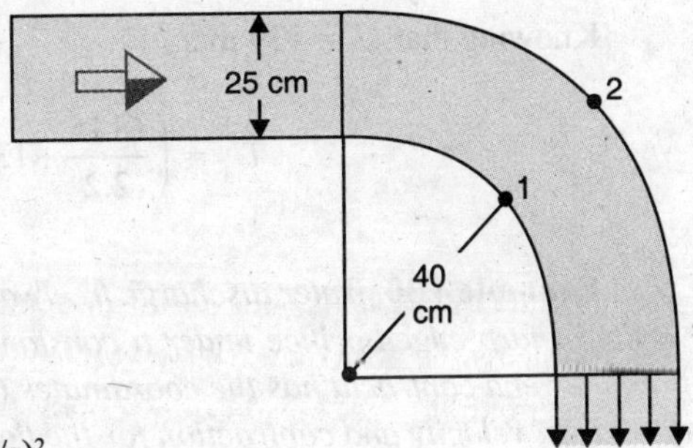

$$= \rho\frac{U_{\theta_1}^2}{2}\,0.62$$

whence, $\rho\,U_{\theta_1}^2 = 96774$

Assuming that water is flowing in the duct,

$$\rho = 1000\ \text{kg/m}^3$$

$$U_{\theta_1}^2 = 96774/1000 = 96.774\ (\text{m/s})^2$$

$$U_{\theta_1} = 9.84\ \text{m/s}$$

and
$$U_{\theta_2} = \frac{0.4}{0.65} \times 9.84 = 6.05\ \text{m/s}$$

At the mean radius (40 + 65)/2 cm i.e., 0.525 m

$$U_\theta = \frac{0.4}{0.525}\,u_{\theta 1} = 7.5\ \text{m/s}$$

With this velocity in the duct, the discharge per unit width is given by

$$Q = 7.5 \times 0.25 \times 1 = \mathbf{1.875\ m^3/s.}$$

Example 4.29. *A single engined plane flies at 500 km/hour in still air of density 1.1 kg/m³. If the disc area swept by the propeller blades is 4 m² and the power supplied to the propeller shaft is 1000 kW, estimate*

(a) the velocity of the jet far downstream

(b) the thrust exerted on the plane

if the conversion efficiency is 85%.

Solution.

From the data,

the speed of the plane is 500 km/hr; propulsive velocity

$$U_1 = \frac{500 \times 1000}{3600} = 139 \text{ m/s.}$$

Available Power = Thrust × propulsive velocity

1000 × 1000 × 0.85 = Thrust × 139

Hence, Thrust = 6115 N.

We need to determine the discharge in order to estimate the velocity of the jet

$$Q = \text{Average velocity (through the propeller)} \times A$$

$$= \frac{U_1 + U_4}{2} . 4 = 2(U_1 + U_4) \text{ m}^3\text{/s}$$

where U_1 & U_4 are the far-upstrean & far-downstream velocities.

$$\text{Thrust} = \text{mass flow rate } (U_4 - U_1) = \rho Q\,(U_4 - U_1)$$

$$= 1.1 \times 2\,(U_1 + U_4) \times (U_4 - U_1)$$

$$= 2.2\,(U_4^2 - U_1^2) = 6115 \text{ N}$$

Knowing that U_1 = 139 m/s

$$U_4 = \left(\frac{6115}{2.2} + 139^2\right)^{\frac{1}{2}} = \mathbf{149\ m/s}.$$

Example 4.30. *Water discharge freely at a rate of 98 litres/second through a 120 mm diameter vertical sharp edged orifice under a constant load of 10 m of water. A point on the jet measured from the vena contracta has the coordinates (+ 4.5 m, – 0.54 m). Find the hydraulic coefficients of discharge, velocity and contraction for the flow through the orifice.*

Solution.

Theoretically the velocity of the jet at the vena contracta is given by

$$v_{th} = \sqrt{2\,gh} = \sqrt{2 \times 9.81 \times 10} = 14.0 \text{ m/s}$$

In actual flow; the velocity v is such that

$$x = v\,t;\ z = -\frac{1}{2}gt^2$$

Eliminating t, $\quad z = -g/2\ x^2/v^2$

and substituting data, $\quad -0.54 = -9.81/2 \times 4.5^2/v^2$

$$v = 13.56 \text{ m/s}$$

Hence, the coefficient of velocity $C_v = \frac{13.56}{14.0} = \mathbf{0.97}$.

For the orifice, $\quad a = \pi \times 0.12^2/4 = 0.0113 \text{ m}^2$.

Theoretical discharge $\quad = 14 \times 0.0113 = 0.158 \text{ m}^3\text{/s}$

and the actual discharge $\quad$ = 98 litres/second = 0.098 m^3/s

Hence the coefficient of discharge $C_d = \frac{0.098}{0.158} = \mathbf{0.62}$

Now, $\quad C_d = C_v \times C_c$

which provides $\quad C_c = 0.62/0.97 = \mathbf{0.64}$.

Students are advised to view a video
MECHANICS: An Insight
by Prof. K. L. Kumar, I.I.T. Delhi. The video reviews some basic concepts, i.e., laws, principle and their applications.

ITEM BANK

Objective items of different types are given to review the concepts developed in the Chapter.

**Select correct response for each item.*

**Every correct response is expected in less than a minute and carries one mark.*

**Compare your responses with the key answers given at the end.*

Multiple Choice Items

Attempt the following multiple-choice items by ticking the correct response from among the alternatives given :

1. The Euler's equation for fluid flow refers to motion with
(*a*) constant velocity
(*b*) acceleration in general
(*c*) acceleration along the flow
(*d*) constant acceleration.

2. The assumption NOT made in the derivation of Bernoulli equation is
(*a*) inviscid flow
(*b*) steady flow
(*c*) two-dimensional flow
(*d*) uniform flow

3. A control surface is one which
(*a*) controls mass transfer across it
(*b*) controls momentum transfer across it
(*c*) surrounds a control volume
(*d*) surrounds a specific mass.

4. A control volume implies
(*a*) an isolated system
(*b*) a closed system
(*c*) a specified mass in a fluid flow
(*d*) a fixed region in space.

5. The Euler's equation of motion
(*a*) can be derived from the Navier-Stokes equation
(*b*) is a statement of energy balance
(*c*) is a preliminary step to derive the Bernoulli equation
(*d*) cannot be applied to a fluid at rest.

6. The first law of thermodynamics for a steady flow process
(*a*) is an alternative to the Bernoulli equation
(*b*) is an energy balance for a specified mass of fluid
(*c*) accounts for energy flux across a control volume
(*d*) is concerned with heat transfer.

7. The integral momentum equation requires the assumption that
(*a*) the flow is uniform
(*b*) the flow is uni-directional
(c) the fluid is incompressible
(*d*) the flow is steady.

Other Objective Items

8. Match the statements in Group A with those in Group B to obtain the maximum number of correct statements.

Group A	Group B
(*a*) The continuity equation	(*k*) must be satisfied for any fluid flow
(*b*) The Euler's equation	(*l*) is obeyed only for irrotational flow

(*c*) The Navier-Stokes equation
(*d*) The steady flow energy equation

(*m*) is not applicable to viscous flow
(*n*) is not necessary for incomp. flow
(*o*) is an alternative to the moment of momentum equation
(*p*) is a kinematic equation
(*q*) does not apply to compressible flows
(*r*) is a consequence of mass conservation
(*s*) is a consequence of the law of motion
(*t*) is a consequence of the first law of thermodynamics.

9. Select the minimum number of assumptions from group B for each of the applications listed in group A.

Group A	**Group B**
(*a*) Impact of a jet on vanes	(*k*) steady flow
(*b*) Propulsion by a jet-engine	(*l*) uniform flow
(*c*) Trajectory of a jet through an orifice	(*m*) frictionless passage
(*d*) Garden sprinkler	(*n*) irrotational flow field
	(*o*) conservation of mass
	(*p*) conservation of momentum
	(*q*) conservation of energy
	(*r*) cons. of moment of momentum

10. Identify the following statements as true or false :

(*a*) A control volume has a fixed amount of mass within it. (T/F)
(*b*) Euler's equation of motion is applicable for viscous flow. (T/F)
(*c*) Bernoulli equation is invalid for unsteady flow. (T/F)
(*d*) Some control surfaces of a control volume may have no flow across them. (T/F)
(*e*) The angular momentum equation refers to substantive rate of change of moment of momentum. (T/F)
(*f*) Bernoulli equation is an energy equation in the absence of heat transfer, work done and dissipation. (T/F)
(*g*) A garden water sprinkler operates on the principle of conservation of angular momentum (T/F)

ANSWERS

1. (*b*), **2.** (*d*), **3.** (*c*), **4.** (*d*), **5.** (*a*), **6.** (*c*), **7.** (*d*), **8.** (*a k, a r, a p, b m, b s, c k, c s, d k, d t*), **9.** (*a k, a l, a m, a o, b k, b o, c k, c l, c m, c o, c q, d k, d o, d r*), **10.** (*a*) F, (*b*) F, (*c*) T, (*d*) T, (*e*) T, (*f*) T, (g) T).

Questions on theory and numerical problems are given for tutorial work.

* *The questions are arranged generally in order of articles in the text*
* *Every question should take 15-20 minutes to solve.*
* *Answer for the numericál problesm are given alongside them.*

4.1 Air flows through a duct, and the pitot-static tube measuring the velocity is attached to a differential gauge containing water. If the deflection of the gauge is 100 mm, calcuate the air velocity. Assume density of air is constant at 1.22 kg/m³ and that the coefficient of the tube is 0.98. **[39.3 m/s]**

4.2 The figure shows a liquid flowing through a convergent-divergent section set into a horizontal pipe of uniform diameter. The pressure at section A is p_A and the pressure at any downstream section, x from A is p. Sketch on the same axes the graph of $(p - p_A)$ versus x, assuming no wall friction.

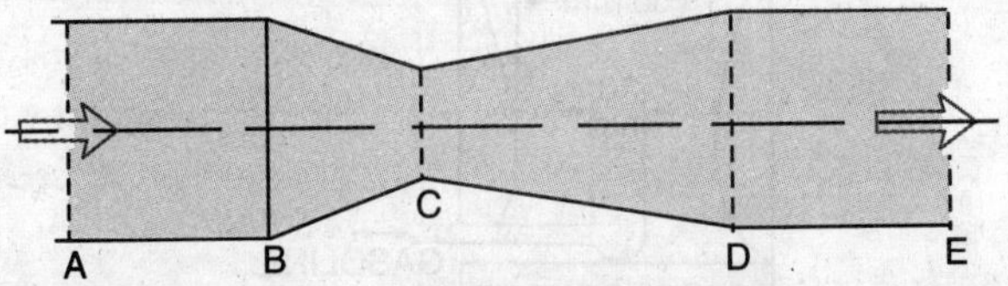

Fig. Problem 4.2

4.3 A 'perpetual motion' machine consisting of a cylinder is submerged in water as shown. The designer surmises that the hydrostatic force on the part AB is greater than that on part BC and hence the cylinder should rotate and continue to do so. Show, by actual anaylsis, that although there exists a difference of forces on parts AB and BC, the cylinder cannot rotate.

[No turning moment possible]

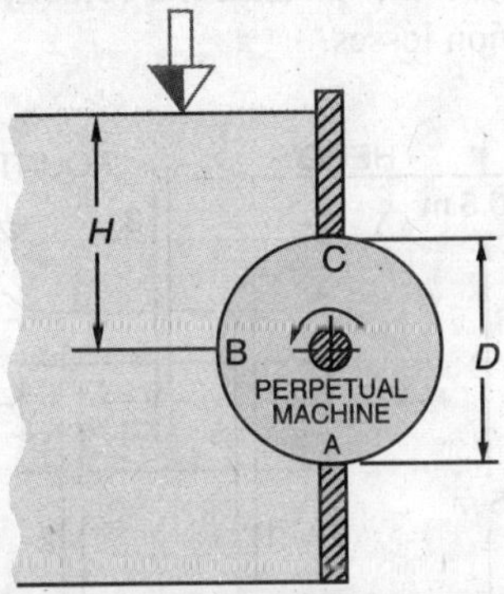

Fig. Problem 4.3

4.4 Consider two identical reservoirs discharging water through equal openings at the base; one directly into the atmosphere and the other by installing a vertical frictionless pipe flowing full as shown in the figure. Express the velocities and pressures at points 1, 2, and 3 in both cases in terms of h and H.

[$p_1 = p_2 = p_3$, *atmospheric pressure;* $U_1 = \sqrt{2gh}$, $U_2 = \sqrt{2g(h + H/2}$ $U_3 = \sqrt{2g(h+H}$ $p'_3 =$ *atmospheric pressure;* $U'_1 = U'_2 = U'_3 = 2g(h + H)$ $p'_1 = -\rho gH, p'_2 = -\rho gH/2$

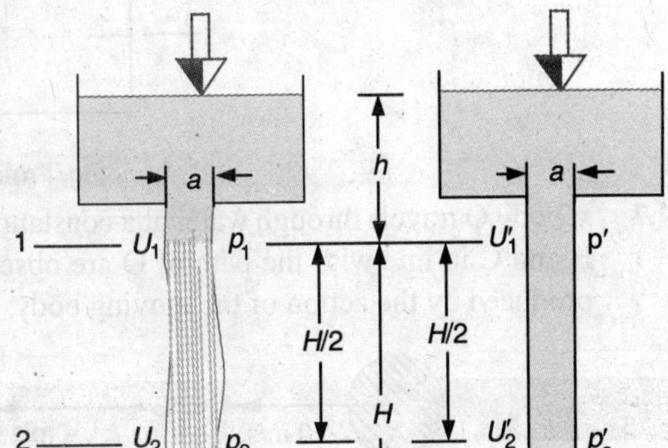

Fig. Problem 4.4

4.5 The carburettor of an engine sucks air directly from the atmosphere at a rate of 0.1 m³/s. The gasoline is atomised through an atomiser of outer diameter 10 mm at a location where the venturi diameter is 50 mm. Taking the density of air as 1.25 kg/m², calculate the vacuum in the location of the atomiser. **[1.76 kN/m² vacuum]**

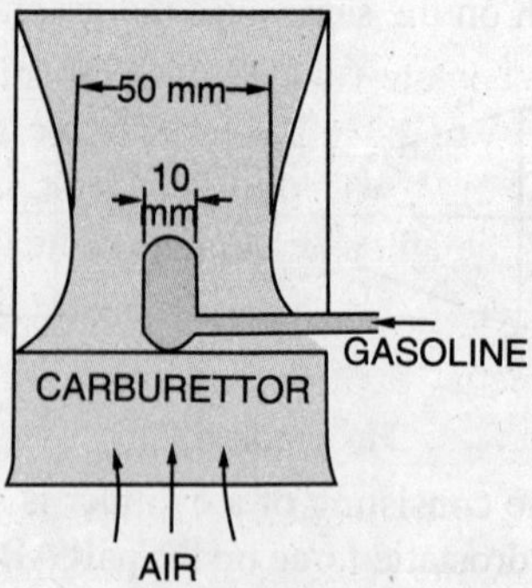

Fig. Problem 4.5

4.6 A system of three reservoirs may be arranged to comprise the Hero's fountain as shown in the figure. For the configuration shown, calculate the velocity of water fountain issuing from the nozzle. Neglect the pipe friction losses. **[19.56 m/s]**

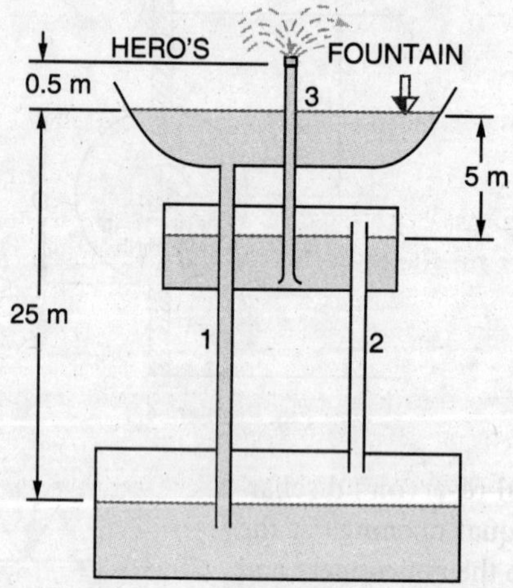

Fig. Problem 4.6

4.7 A body O travels through water at a constant speed of 10 m/s, as indicated. Velocities at points B and C in line with the path of O are observed to be 4 m/s and 2 m/s. These velocities are produced by the action of the moving body. What is $(p_B - p_C)$?

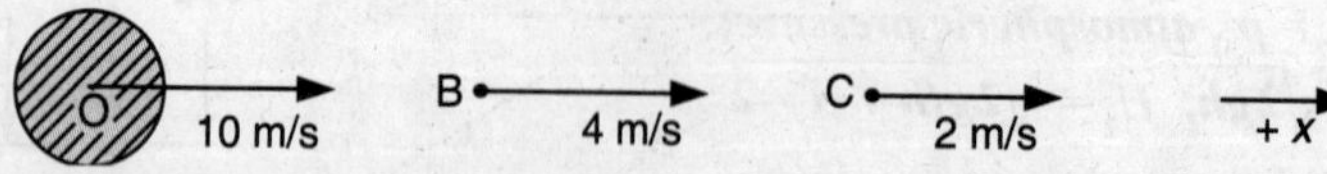

Fig. Problem 4.7

(**Hint.** Bring the body to 'rest' by superimposing a flow at 10 m/s in the – ve *x* direction) **[14 kN/m²].**

4.8 A U-tube differential manometer connects two points *P* and *Q* on a pipeline carrying liquid of specific gravity 2.14. The pressure at point *P* is 15 kN/m^2 and *Q* is 21 kN/m^2. The point *P* is located 3 m above the point *Q*. Determine the manometer 'reading' if the manometric liquid is mercury. **[0.507 m; higher in limb of *Q*]**

4.9 A jet propelled boat with an absolute velocity of 8.7 m/s is moving upstream in a river. The stream is flowing with a velocity of 2.3 m/s. A jet of water is ejected astern at a relative veocity of 18 m/s. If the flow rate is 1.4 m^3/s, what thrust is developed on the boat? Find also the power required to drive the boat and the efficiency of the propulsion device. **[4.6 m/s]**

4.10 A ship is propelled by hydraulic jet propulsion and maintains a constant velocity of 1 km/minute due North. The total drag force is 15000 N. Assuming the exit jet diameter of 1 m, calculate the absolute velocity of the jet. **[17.8 m/s]**

4.11 The rotor of a helicopter is 12 m in diameter. The weight of the helicopter together with its occupants is 19.6 kN. Calculate the velocity of air passing through the motor in the hovering condition, *i.e.*, when the helicopter is standstill in flight. Take the density of air as 1.25 kg/m^3. **[21.5 m/s]**

4.12 A container is moved horizontally by a constant tractive force of 5 k N. It has a velocity of 20 m/s when its own weight plus the contents equal 1 kN. At a certain time, the contents begin to drop out vertically from the base of the container at a uniform rate of 100 N/s. Calculate the change in the velocity of the container with the passage of time, i.e. after 1 second and 5 seconds. **[76.7 m/s, 580 m/s]**

4.13 A jet of water issues from a nozzle with a velocity of 40 m/s at the rate of 0.15 m^3/s. What is the power of the jet? **[120 kW]**

4.14 A flat plate is struck normally by a jet of water 50 mm in diameter with a velocity of 18 m/s. Calculate (*a*) the force on the plate when it is stationary, (*b*) the force on the plate when it moves in the same direction as the jet with a velocity of 6 m/s, (*c*) the work done per second and the efficiency in case (*b*). **[636.2 N, 282.7 N, 16.965 W, 29.6 %]**

4.15 Water flows steadily through a Y-junction in horizontal plane as shown in the figure. The velocities in the downstream limbs are equal and all velocity profiles are uniform. Calculate the magnitude and direction of force required to hold the junction stationary in the *x*-*y* plane. **[$(-1.45\,i + 0.165\,j)$ kN]**

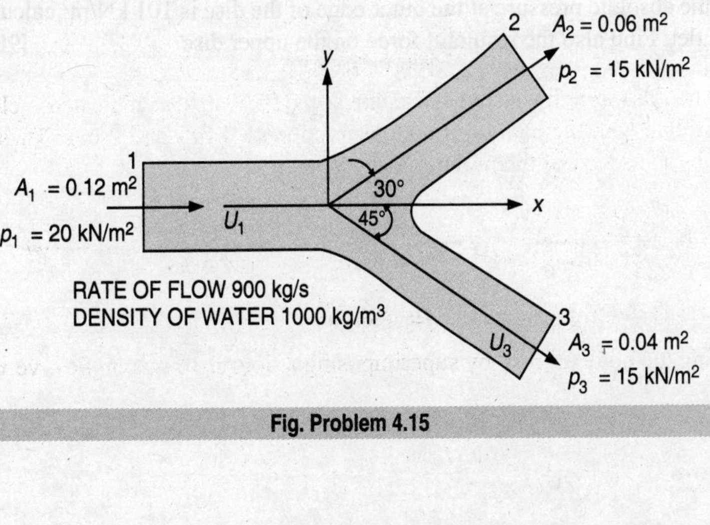

Fig. Problem 4.15

4.16 A stream of water issues from a nozzle of area 2580 mm^2 with a velocity of 30 m/s. The water is deflected by a curved vane through an angle of 2.62 rad when the vane is stationary. If the vane moves with a velocity of 12 m/s in the initial direction of the jet, find (*a*) the force on the vane in the direction of motion of the vane, (*b*) the force perpendicular to that direction and (*c*) the work done per second on the vane. **[1580 N, 424 N, 19.3 kW]**

4.17 A reducing bend is incorporated in a pipeline so that the direction of flow is turned through 60° in the horizontal plane and the pipe diameter is reduced from 0.25 m to 0.15 m. The velocity and pressure at the entry to bend are 1.5 m/s and 300 kN/m^2 gauge respectively and at the exit the pressure is 287.2 kN/m^2 gauge. Find the magnitude and direction of the reaction force in the bend in the horizontal plane due to the flowing water. **[13 kN at 21.4°]**

4.18 Show that the propulsive efficiency of a propeller-engine is given

$$\eta = \frac{U_1 + U_4}{2U_4} = \frac{2U_1}{U_1 + U_4}$$

where U_1 and U_4 are the relative airstream speeds far upstream and downstream respectively.

4.19 A lawn sprinkler consists of two nozzles placed at 10 cm radii capable of rotating in a horizontal plane. The nozzles are inclined up with the plane of rotation at 30° as shown in the figure. For a total discharge of 15 litres per minute and nozzle cross sections of 0.5 cm^2 each, determine the rotational velocity if left free and also the torque acting on the rotor if allowed to rotate at half the free rotational speed. Assume the incoming water to enter axially upwards. **[21.5 rad/s, 0.027 Nm]**

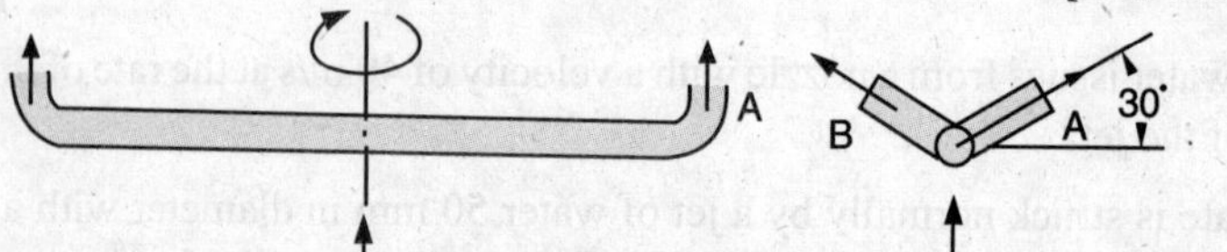

Fig. Problem 4.19

4.20. A pump discharges 2 m^3/s of water through a pipeline. If the pressure difference at the inlet and the outlet of the pump is equivalent to 10 m of water, what power is being transmitted to the water from the pump? **[196.2 kW]**

4.21 Inward radial flow occurs between two horizontal discs 0.6 m in diameter and 75 mm apart, the water leaving through a central pipe 150 mm in diameter in the lower disc at the rate of 0.17 m^3/s. If the absolute pressure at the outer edge of the disc is 101 kN/m^2 calculate the pressure at the outlet. Find also the resultant force on the upper disc. **[90 kN/m^2, 567 N]**

CHAPTER

5

Flow Measurement

LEARNING OBJECTIVES

On successful completion of all learning activities provided in the chapter, a student will be able to

- discuss different techniques of flow measurement;
- determine the velocity profile for flows through an open channel by employing a V-notch or a rectangular notch;
- obtain the discharge through an open channel by employing a V-notch or a rectangular notch ;
- obtain the discharge through a pipe by employing a venturimeter, a flow nozzle or an orificemeter;
- describe the technique of hot-wire anemometry;
- state the advantages of laser anemometry; and
- outline some techniques of flow visualisation.

5.1. Introduction

It is often necessary to measure the pressure and velocity at any point in a flow field as also the discharge through a passage. For example, we must determine the velocity profile at a cross section in a pipe flow in order to determine the nature of flow in it. Discharge through open field channels must be estimated for the purpose of distribution of water in the fields. Discharge through oil pipelines and gas conduits is likewise necessary in order to calculate the patterns of consumption.

Various methods of flow measurement have come into existence. An attempt has been made to study some simple and more commonly used techniques based upon the principles of fluid mechanics.

5.2. Measurement of Static, Stagnation and Dynamic Pressures and Velocity

The thermodynamic pressure, frequently referred to as pressure p is the *static pressure* in a fluid flow. The static pressure is a parameter to describe the state of a fluid irrespective of its velocity. If the fluid flows in a passage, a simple flushed hole in the wall normal to the flow called a 'wall tap' is made to sense the static pressure as shown in Fig. 5.1. The wall tap is connected to a suitable manometer. The fact that a wall tap indeed senses the static pressure can be appreciated by noticing that there is no component of velocity along the axis of the hole. The wall tap measures the static pressure at the wall. In most circumstances, the static pressure at a cross-section remains the same. The wall tap may then be assumed to register the pressure at that cross-section. In practice, a single wall tap is not relied upon. A number of holes, say three, are drilled through the wall normal to it and they are mutually connected by flexible tubes in order to register the most likely static pressure at the cross section as shown in Fig. 5.2. In this arrangement, the reading on the manometer would be reliable even if all but one hole are blocked.

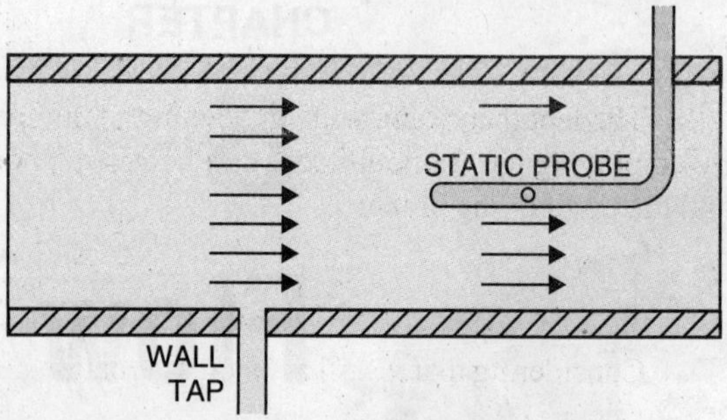

Fig. 5.1: Measurement of Static Pressure

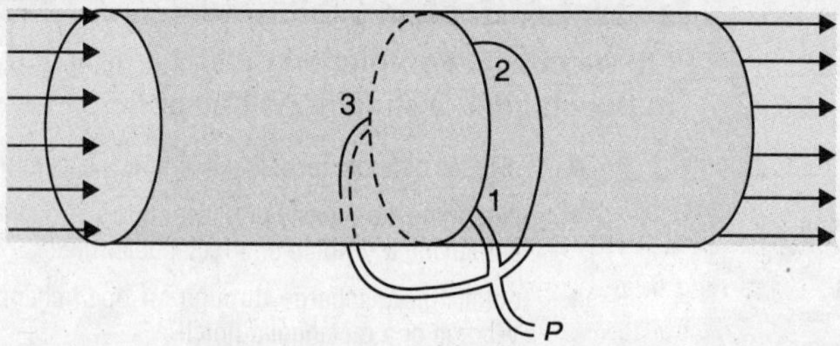

Fig 5.2: Static Pressure Measurement by connecting three tapped holes at a cross-section

The static pressure in a fluid flow at a cross section may alternatively be measured by a static probe inserted in the fluid as also shown in Fig. 5.1. The static pressure hole in the probe is such that the axis of the hole is normal to the velocity of the fluid. Here again, there could be three or more holes on the periphery of the probe and the same, when connected to the manometer, provide the most likely static pressure at that cross section.

The *stagnation pressure* at a point in a fluid flow is the total pressure which would result if the fluid were brought to rest isentropically. In actual practice, a stagnation point is created by bringing the fluid to rest at the desired point and the pressure at the point corresponds to the stagnation pressure as shown in Fig. 5.3. A simple device to measure the stagnation pressure is a tube with a hole in the front inserted in the flow such that the velocity is normal to the plane of the hole, *i.e.*, along the axis of the hole.

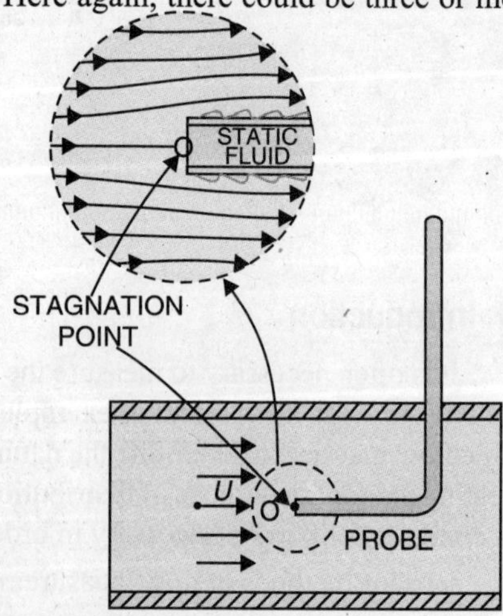

Fig. 5.3: Stagnation Pressure Probe in Pipe Flow

It may be noted that the stagnation pressure is in excess of the static pressure by an amount equivalent to the dynamic pressure.

Stagnation pressure = static pressure + dynamic pressure

i.e., $$p_0 = p + \frac{\rho u^2}{2}$$

The fact that a tube with an opening in the front senses the stagnation pressure may be shown by employing the Bernoulli equation between a point in the free stream just upstream of the probe and the tip O of the probe:

$$\frac{p}{\rho g} + \frac{u^2}{2g} + z = \frac{p_0}{\rho g} + \frac{u_0{}^2}{2g} + z_0$$

Considering that $u_0 = 0$ at O, the stagnation point as also shown in the enlarged view and $z = z_0$,

$$p_0 = p + \frac{\rho u^2}{2} \quad \text{... (5.1)}$$

which shows that the pressure at O is indeed the stagnation pressure.

If the stagnation pressure and the static pressure at a point are known, the velocity at that point may be determined by making use of the above equation,

whence $$p_0 - p = \frac{\rho u^2}{2}$$

and $$u = \sqrt{\frac{2(p_0 - p)}{\rho}}$$

In an open channel or a flume, the liquid surface is at the atmospheric pressure and the pressure at any depth d below the free surface must be

$$p = \rho g d \quad \text{... (5.2)}$$

If the rise of the liquid in the stagnation tube is h above the free surface, the pressure p_0 at stagnation point O must be as is clear in Fig. 5.4

$$p_0 = \rho g\,(h + d) \quad \text{... (5.2 a)}$$

The velocity of the liquid is, therefore, given by

$$u = \sqrt{\frac{2(\rho g(h + d) - \rho g d)}{\rho}} = \sqrt{2gh} \quad \text{... (5.3)}$$

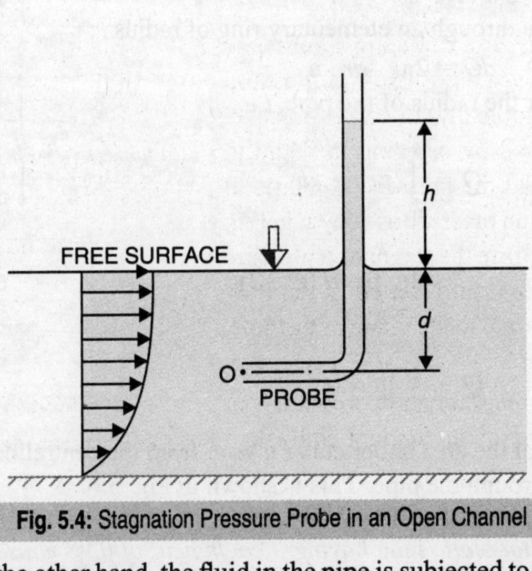

Fig. 5.4: Stagnation Pressure Probe in an Open Channel

In a pipe flow, on the other hand, the fluid in the pipe is subjected to a pressure p which should be taken into account.

This is achieved, in practice, by employing a Prandtl probe or a pitot-static probe which has provisions to register both the stagnation pressure p_0 and the static pressure p as shown in Fig. 5.5

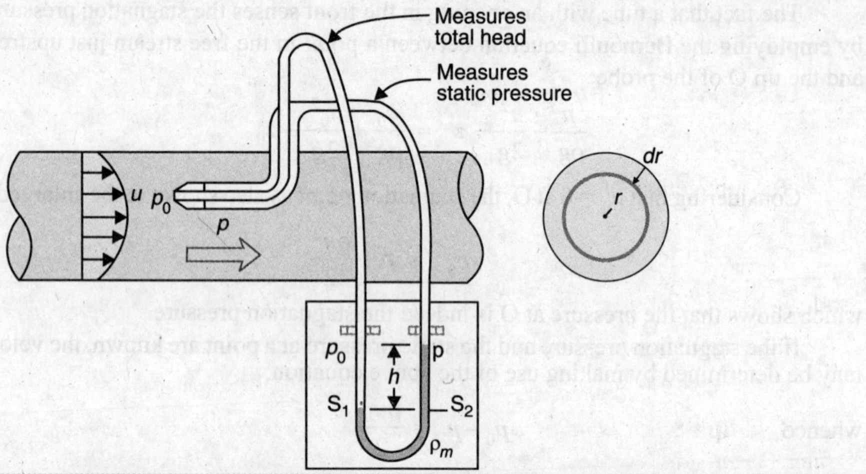

Fig. 5.5: Measurement of Velocity by a Prandtl Probe and a U-Tube Manometer

The difference between the two pressures is recorded on a U tube differential manometer. In the position of equilibrium, the static pressures at S_1 and S_2 must be equal,

$$p_{s1} = p_{s2}$$

or
$$p_0 + \rho g h = p + \rho_m g h$$

whence
$$p_0 - p = (\rho_m - \rho)\, g h$$

The expression for velocity at the point of interest, therefore, reduces to

$$u = \sqrt{2\frac{(\rho_m - \rho)}{\rho} g h} \quad \text{... (5.4)}$$

in terms of the differential head h registered on the manometer with a manometric fluid of density ρ_m.

The discharge through a pipe is computed by way of considering the discharge through an elementary ring of radius r;

$$dQ = 2\pi r \,.\, dr \,.\, u$$

and by integrating it over the radius of the pipe, *i.e.*,

$$Q = \int_0^R 2\pi r \,.\, u \,.\, dr$$

$$= 2\pi \int_0^R u \,.\, d\,(r^2/2)$$

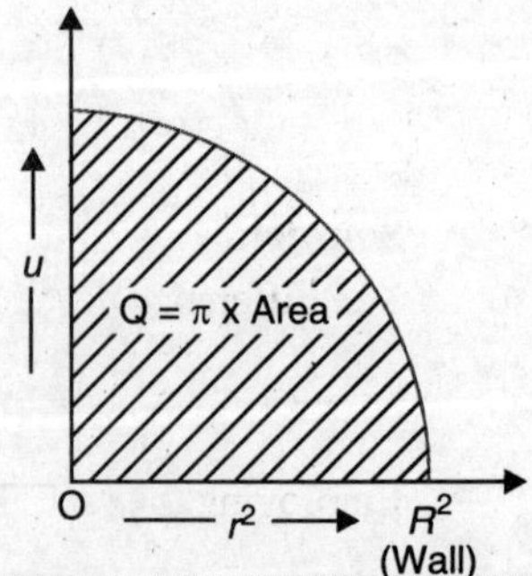

Fig. 5.6

or
$$Q = \pi \int_0^R u \; d\,(r^2) \quad \text{... (5.5)}$$

The expression shows that the area under curve u vs r^2 from the centreline to the wall multiplied by π equals the discharge through the pipe. This is shown in Fig. 5.6.

Example 5.1. *A pitot-static tube having a coefficient of 0.98 is used to measure the velocity of water in a pipe. The stagnation pressure recorded is 3 m and static pressure 2 m. What velocity does it indicate?*

Solution.

By the Bernoulli equation, the stagnation and static pressure heads are expressed as

$$\frac{p}{\rho g} + \frac{u^2}{2g} = 3 \quad \text{... } (i)$$

$$\frac{p}{\rho g} = 2 \quad \text{... } (ii)$$

By subtraction,

$$\frac{u^2}{2g} = 1$$

and $u = \sqrt{2 \times 9.81 \times 1} = 4.429$ m/s

The figure is multiplied by the coefficient of the pitot-static tube to obtain

$$u_{\text{actual}} = 4.429 \times 0.98 = \mathbf{4.341\ m/s}$$

Example 5.2. *A submarine moves horizontally in sea and has its axis much below the surface of water. A pitot-static tube properly placed just in front of the submarine and along its axis is connected to two limbs of a U-tube manometer containing mercury. The difference of mercury levels is found to be 17 cm. Find the speed of the submarine knowing that the density of mercury is 13.6 and that of sea water 1.026 with respect to fresh water.*

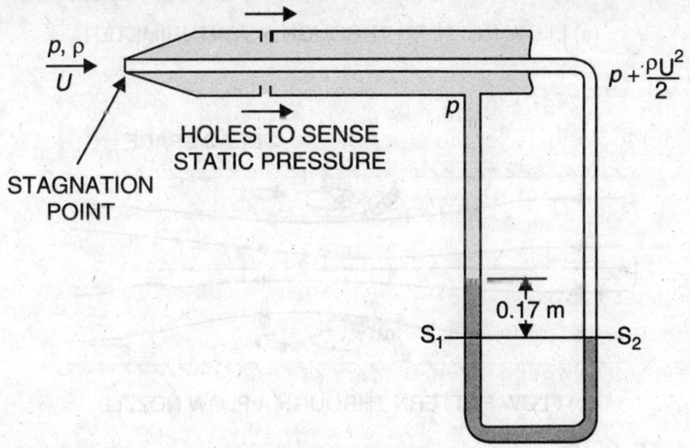

Solution.

The difference of mercury levels is equivalent to a differential head of the sea water equal to

$$h = \frac{0.17 \times (13.6 - 1.026)}{1.026} = 2.083 \text{ m of sea water}$$

Employing the fact,

$$h = U^2/2g$$

$$U = \sqrt{2.083 \times 2 \times 9.81} = \mathbf{6.39\ m/s}$$

5.3. Measurement of Discharge Through a Pipe by a Venturimeter, a Flow Nozzle and an Orificemeter

Discharge through a pipe is usually measured by providing a coaxial area contraction within the pipe and by recording the pressure drop across the contraction.

Three different flow meters operate on this principle. A venturimeter is a device consisting of a short length of gradual convergence and a longer length of gradual divergence. The semi-angle of convergence is 8 to 10° and the semi-angle of divergence is 3 to 5°. A pressure tapping 1 is provided at a location before the convergence commences and another pressure tapping 2 is provided at the throat section of the venturimeter. The pressure difference $(p_1 - p_2)$ between the two tappings is measured by means of a U-tube manometer. The manometer may contain water or mercury as the manometric fluid depending upon the pressure difference estimated. A flow-nozzle is a device through which the contraction of area is brought about by nozzle. One of the pressure tappings is provided at a distance of one diameter upstream the nozzle plate and the other at the nozzle exit. An orificemeter is a circular plate with an orifice in its centre. The flow adjusts itself such that it contracts until a point downstream the orifice plate and then expands to fill the passage. One of the pressure tappings is provided at a distance of one diameter upstream the orifice plate and the other at a distance of half a diameter downstream the plate. The pressure difference in each case is measured by means of a U-tube manometer connected across the two pressure tappings.

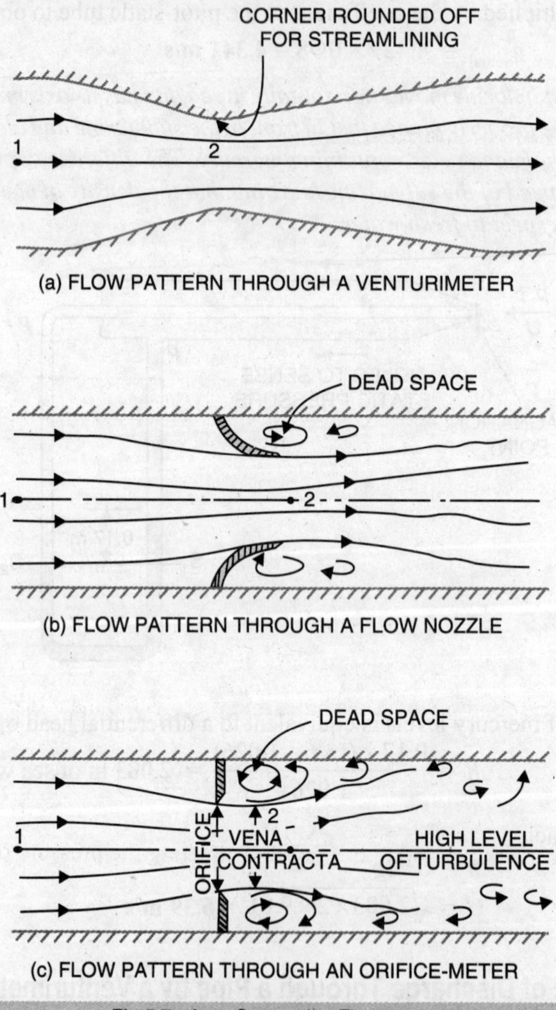

Fig 5.7 : Area Contraction Flowmeters

In each of these devices shown in Fig. 5.7 the flow accelerates until the location of the vena contracta where the cross-section of the fluid stream is the minimum and decelerates to fill the passage thereafter. The pressure tapping 1 corresponds to the free-stream inlet to the device and the pressure tapping 2 is designed to lie at the vena contracta. In the case of the venturimeter, the minimum cross-section of the flow stream occurs at the throat because the angle of convergence is kept low. In the case of flow nozzle, the minimum cross-section of the fluid stream occurs at the exit of the nozzle and in the case of the orificemeter, it occurs at the vena contracta approximately half a pipe diameter downstream the edge of the orifice plate.

Assuming the flow to be incompressible and inviscid between the inlet section 1 and the minimum cross-section 2 and assuming the flow to be one-dimensional, the continuity equation and the Bernoulli equation may be applied.

By continuity,

$$Q = U_1 A_1 = U_2 A_2 \quad \text{... (i)}$$

By the Bernoulli equation,

$$\frac{p_1}{\rho g} + \frac{U_1^2}{2g} + z_1 = \frac{p_2}{\rho g} + \frac{U_2^2}{2g} + z_2 \quad \text{... (ii)}$$

From (*i*)

$$U_1 = \frac{A_2}{A_1} U_2$$

Substituting the value of U_1 in (*ii*) and rearranging the terms,

$$\frac{U_2^2 - \left(\frac{A_2}{A_1} U_2\right)^2}{2g} = \frac{p_1 - p_2}{\rho g} + (z_1 - z_2)$$

or

$$U_2^2 = \frac{A_1^2}{A_1^2 - A_2^2} \cdot 2g \left[\frac{p_1 - p_2}{\rho g} + (z_1 - z_2)\right]$$

whence

$$U_2 = \frac{A_1}{\sqrt{A_1^2 - A_2^2}} \sqrt{2g \left[\frac{p_1 - p_2}{\rho g} + (z_1 - z_2)\right]}$$

The discharge through the venturimeter is given by Q,

$$Q = U_1 A_1 = U_2 A_2$$

The specific gravities S and S_m of the working fluid and the manometric fluid respectively are given by

$$S = \frac{\rho}{\rho_{water}}; \; S_m = \frac{\rho_m}{\rho_{water}} \quad \text{... (5.7)}$$

where ρ and ρ_m are the corresponding mass densities and ρ_{water} is the mass density of water at 4°C.

In terms of S and S_m,

$$h = h_m \left(\frac{\rho_m}{\rho} - 1\right) = h_m \left(\frac{S_m}{S} - 1\right)$$

This is affirmed by considering a datum $S_1 S_2$ and equating the pressure on either limb above $S_1 S_2$ level.

For the left limb,

$$\frac{p_1}{\rho} + g z_1$$

and for the right limb,

$$\frac{p_2}{\rho} + g h_m \frac{\rho_m}{\rho} + (z_2 - h_m)$$

Equating these pressures and rearranging the terms,

$$\frac{p_1 - p_2}{\rho} + g(z_1 - z_2) = g h_m \left(\frac{\rho_m}{\rho} - 1\right) = g h_m \left(\frac{S_m}{S} - 1\right) \quad \text{.... (5.8)}$$

The actual discharge through the flowmeter is less than that given by the theoretical expression derived on the basis of incompressible and inviscid one-dimensional flow. In actual practice, the flow is not one-dimensional; the velocity across the section does not remain constant. This is because of the real-flow effects, *e.g.*, boundary-layer formation near the wall and eddy formation in the vicinity of the vena contracta, particularly so in the flow nozzle and the orifice-meter. In the case of orifice meter, the area of the vena contracta is appreciably less than the area A_2 provided by the orifice. In order to take into account the wall friction and area-contraction effects, a multiplying factor called 'the coefficient of discharge' C_d is taken. The value of C_d is determined by the ratio of the actually measured discharge to the theoretical value.

$$C_d = \frac{\text{Actual discharge}}{\text{Theoretical discharge}} \quad \text{... (5.9)}$$

The expression for the actual discharge, therefore, becomes

$$Q = C_d \cdot \frac{A_1 A_2}{\sqrt{A_1^2 - A_2^2}} \cdot \sqrt{2g\left[\frac{p_1 - p_2}{\rho g} + (z_1 - z_2)\right]}$$

$$= C_d \cdot \frac{A_1 A_2}{\sqrt{A_1^2 - A_2^2}} \cdot \sqrt{2g\, h_m\left(\frac{\rho_m}{\rho} - 1\right)}$$

$$= C_d \cdot \frac{A_1 A_2}{\sqrt{A_1^2 - A_2^2}} \cdot \sqrt{2g\, h_m\left(\frac{S_m}{S} - 1\right)} \quad \text{... (5.9a)}$$

$$= C_d \cdot \frac{A_1 A_2}{\sqrt{A_1^2 - A_2^2}} \cdot \sqrt{2g\, h} \quad \text{... (5.9b)}$$

It may be noted that the areas of cross-sections are related to the corresponding diameters:

$$A_1 = \frac{\pi}{4} D_1^2 \text{ and } A_2 = \frac{\pi}{4} D_2^2$$

and
$$\frac{A_1 A_2}{\sqrt{A_1^2 - A_2^2}} = \frac{\pi}{4} \cdot \frac{D_1^2 D_2^2}{\sqrt{D_1^4 - D_2^4}}$$

In some situations the diameter D_2 is much smaller than D_1. Consequently,

$$D_2^2 << D_1^2 \text{ and } D_2^4 <<< D_1^4$$

or
$$A_2 << A_1 \text{ and } A_2^2 <<< A_1^2$$

such that

$$\frac{A_1 A_2}{\sqrt{A_1^2 - A_2^2}} = \frac{A_2}{\sqrt{1 - \left(\frac{A_2}{A_1}\right)^2}} \approx A_2 = \frac{\pi}{4} D_2^2$$

The mass flow rate through the pipe is given by multiplying the volumetric flow rate with the mass density of the fluid ;

Mass flow rate $\dot{m} = \rho Q$

TABLE 5.1 COMPARISON OF VENTURIMETER, FLOW NOZZLE AND ORFICE-METER

Type of Flowmeter	*Diagram*	*Value of Cd*	*Head loss across it*	*Length*	*Cost*	*Accuracy*	*Typical Installations*
Venturi meter	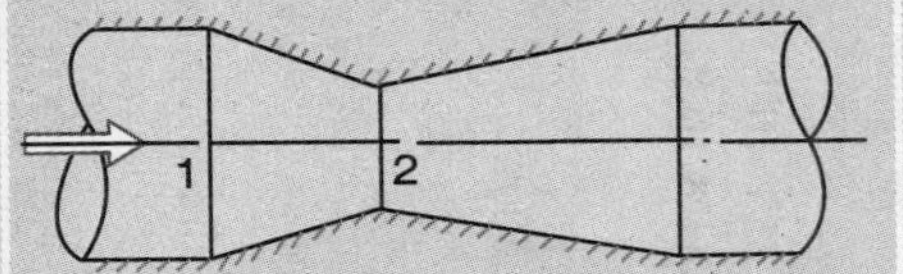	0.95 to 0.98	Low	Long	High	Very Good	In a pipe of small diameter
Flow Nozzle	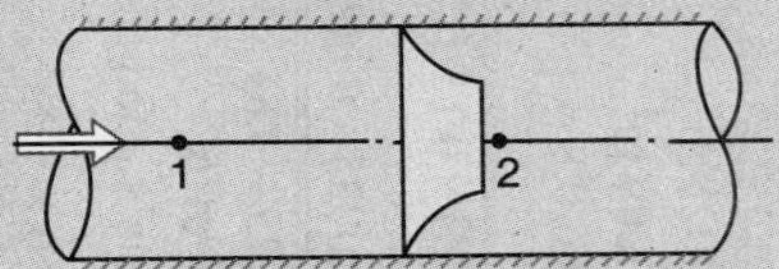	0.70 to 0.80	Intermediate between a Venturi meter and an orifice meter				At entry to a plenum chamber
Orifice meter	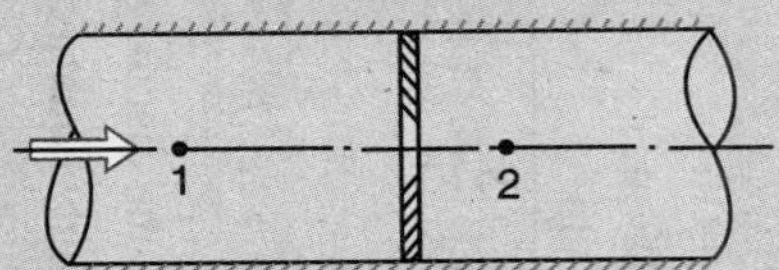	0.60 to 0.65	High	Short	Low	Good	In a pipe of large diameter

Example 5.3. *Predict the flow rate through a venturimeter, placed at 30° to the horizontal, carrying gasoline of specific gravity 0.80 as shown. If the venturimeter is turned into horizontal or vertical position, what will the differential manometer read?*

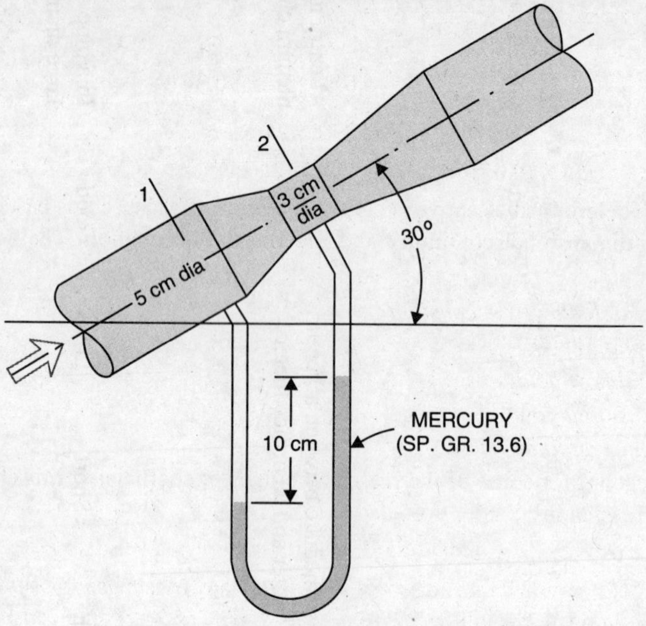

Solution.

$$Q = \frac{A_1 A_2}{\sqrt{A_1^2 - A_2^2}} \sqrt{2gh_m\left(\frac{S_m}{S} - 1\right)}$$

$$= \frac{\pi}{4} \frac{D_1^2 D_2^2}{\sqrt{D_1^4 - D_2^4}} \sqrt{2gh_m\left(\frac{S_m}{S} - 1\right)}$$

From the data ;

$D_1 = 0.05$ m $\quad D_1^2 = 0.0025$ m^2 $\quad D_1^4 = 6.25 \times 10^{-6}$

$D_2 = 0.03$ m $\quad D_2^2 = 0.0009$ m^2 $\quad D_2^4 = 0.81 \times 10^{-6}$

$h_m = 0.1$ m $\quad S_m/S = 13.6/0.8 = 17$

$$\sqrt{2g\,h_m\left(\frac{S_m}{S} - 1\right)} = \sqrt{(2 \times 9.81 \times 0.1 \times (17 - 1))} = 5.60 \text{ m}$$

Hence $$Q = \frac{\pi}{4} \times \frac{0.0025 \times 0.0009}{\sqrt{6.25 - 0.81} \times 10^{-6}} \times 5.60$$

$$= \mathbf{0.004\ 26\ m^3/s}$$

Alternatively, the discharge could be obtained starting from the first principles :

By the continuity equation,

$$Q = A_1 U_1 = A_2 U_2$$

$$U_2 = \frac{25}{9} \times U_1 = 2.78\, U_1$$

By the Bernoulli equation,

or $$\frac{p_1}{\rho g} + \frac{U_1^2}{2g} + z_1 = \frac{p_2}{\rho g} + \frac{U_2^2}{2g} + z_2$$

or
$$\frac{(U_2^2 - U_1^2)}{2g} = \frac{p_1 - p_2}{\rho g} + (z_1 - z_2)$$

$$\frac{(7.7-1)}{2 \times 9.81} U_1^2 = h_m \left(\frac{S_m}{S} - 1 \right)$$

or
$$U_1^2 = 0.1\ (17-1)/0.341$$

whence
$$U_1 = 2.16 \text{ m/s}$$

and
$$\pi/4 \times (0.05)^2 \times 2.16 = \mathbf{0.004\ 26\ m^3/s}$$

The discharge determined as above is called 'theoretical' because it is based upon the perfect validity of the one-dimensional continuity and the Bernoulli equation. The assumptions in the background are :

(a) *incompressible flow,* ρ = *constant*
(b) *inviscid fluid,* $\mu = 0$
(c) *irrotational flow field*
(d) *one-dimensional flow*
(e) *steady state conditions* .

In order to account for some of the real-flow effects, a coefficient of discharge $C_d = 0.96$ is introducted so that discharge actually expected

$$Q = 0.00426 \times 0.96 = \mathbf{0.0041\ m^3/s.}$$

If the venturimeter is turned into horizontal, vertical or any other positions for the same discharge rate Q, an inspection of the terms in Eq. 5.9 (a) reveals that A_1 and A_2 being constant, the expression under the square root sign must not alter. Also S and S_m being constant, h_m should stay unaltered. This is physically explained by the fact that the pressure and datum head at 1 and 2 rearrange between themselves so as to render $\left[\frac{p_1 - p_2}{\rho g} + (z_1 - z_2)\right]$ constant, which equals $h_m \left(\frac{S_m}{S} - 1\right)$ and, therefore **h_m remains unaltered.**

Example 5.4. *Calculate the maximum allowable discharge of water through a venturimeter, throat 5 cm, fitted in a 10 cm diameter line with its inlet at an open channel. Assume $C_d = 0.95$.*

Solution.

It is well known that as the local pressure in a confined passage drops below a certain value, the liquid starts evaporating 'very fast' at the existing temperature. In order to avoid the problems of vapour lock and discontinuities, the pressure is not allowed to drop below a certain limit. For water, under atmospheric temperature the absolute pressure is desired to remain above 3 m of water.

Closeness to an open channel suggests that, neglecting the datum head difference, the inlet head, due to atmospheric pressure.

$$\frac{p_1}{\rho g} + \frac{U_1^2}{2g} = 10.33 + 0 = 10.33 \text{ m}$$

and from the consideration of the vapour pressure, the minimum value of $\frac{p_1}{\rho g}$ is 3 m

By the Bernoulli equation,

$$\frac{p_1}{\rho g} + \frac{U_1^2}{2g} = \frac{p_2}{\rho g} + \frac{U_2^2}{2g}$$

$$10.33 = 3 + \frac{U_2^2}{2 \times 9.81}$$

whence the maximum velocity at the throat,

$$U_2 = 12 \text{ m/s}$$

and the maximum actual discharge $Q = C_d A_2 U_2$

$$Q = 0.95 \times \frac{\pi}{4}(0.05)^2 \times 12 = 0.0224 \text{ m}^3/\text{s}$$

The maximum mass flow rate is given by

$$\dot{m} = \rho Q = 1000 \times 0.0224 = \mathbf{22.4 \text{ kg/s}}$$

Example 5.5. *A subsonic wind tunnel has a test section 0.5 m × 0.5 m and a settling chamber 1 m × 1 m. A U-tube manometer installed between the two locations register 0.1 m difference between the vertical columns. The manometric fluid is alcohol of sp. gr. 0.8. Estimate the test section velocity.*

Solution.

A subsonic wind tunnel is employed to test the lift and drag forces on models placed in the test section. A relative wind is created to simulate the model in motion. An outline of a closed circuit wind tunnel is shown.

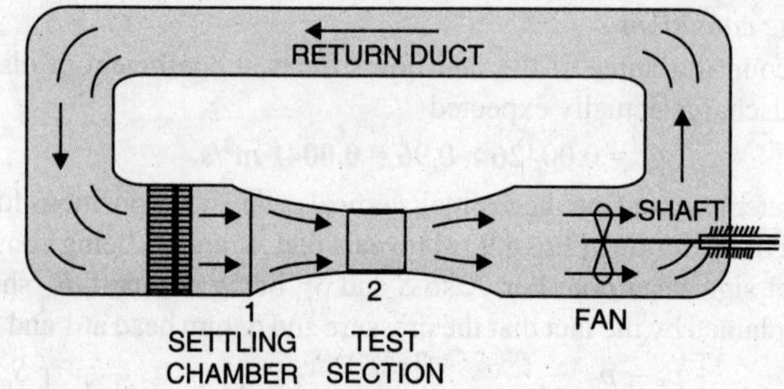

For the problem in hand

$$A_1 = 1 \times 1 = 1 \text{ m}^2$$

$$A_2 = 0.5 \times 0.5 = 0.25 \text{ m}^2$$

By the equation of continuity,

$$U_1 A_1 = U_2 A_2$$

and $U_1 = 0.25\, U_2$

By the Beronulli equation for incompressible flow,

$$\frac{p_1}{\rho g} + \frac{U_1^2}{2g} + z_1 = \frac{p_1}{\rho g} + \frac{U_2^2}{2g} + z_2$$

For a horizontal wind tunnel, $z_1 = z_2$

which yields, $$\frac{U_2^2 - U_1^2}{2g} = \frac{p_1 - p_2}{\rho g} = h_m\left(\frac{\rho_m}{\rho} - 1\right)$$

Substituting the values of

$$U_1 = 0.25\, U_2$$

$$\rho_m = 0.08 \times 1000 = 800 \text{ kg/m}^3\ ;\ \rho = 1.2 \text{ kg/m}^3$$

$$U_2^2 - (0.25\, U_2)^2 = 2 \times 9.81 \times 0.1\left(\frac{800}{1.2} - 1\right)$$

whence $$U_2 = \mathbf{37.32 \text{ m/s}}$$

The assumption of incompressibility of flow is justified because the maximum velocity in the test section is only 41.73 m/s which is substantially less than the velocity of sound in air.

Rotameter

Another device for measuring the discharge in a vertical segment of a pipe line is a rotameter shown in Fig. 5.8 (*a*). It consists of an accurately ground glass tube diverging upwards, *i.e.*, in the direction of flow with a float in the tube. For a particular discharge, the position of the float is established by a balance of the drag force and buoyancy acting in the upward direction and the weight force acting in the downward direction, as shown in Fig. 5.8(*b*). Thus the position of the float is related to the discharge. In practice, spiral slits are cut on a part of the float to set it rotating for stability against tilting.

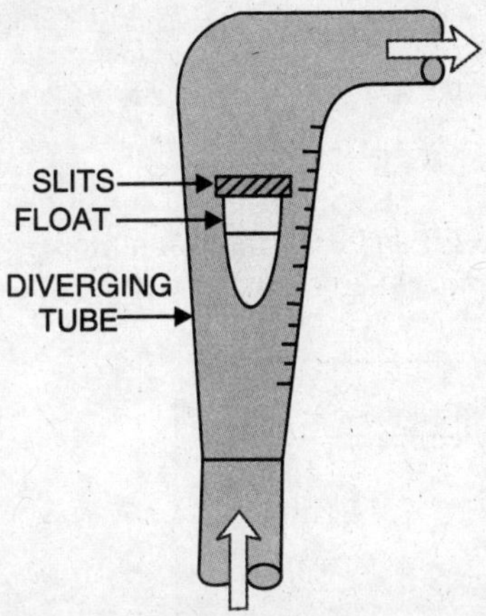

Fig 5.8 (a) : Rotameter

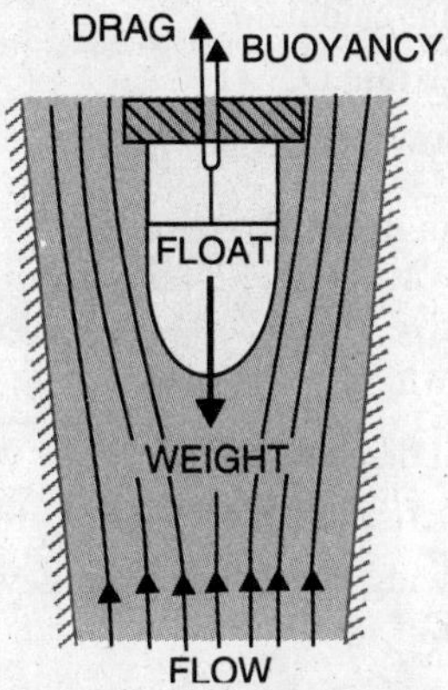

Fig 5.8 (b) : Balance of Forces on a Float

EXPERIMENT

Study of Flow Rate Measuring Devices

OBJECTIVE

Two calibrate a flow rate measuring device, namely, an orifice-meter, a venturimeter or a flow nozzle.

PRACTICAL RELEVANCE

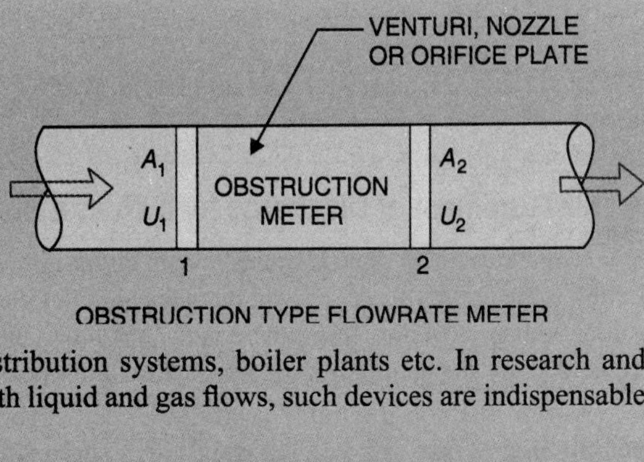

OBSTRUCTION TYPE FLOWRATE METER

Flow rate measurement is a fundamental necessity in almost all flow situations of engineering importance. For confined flows the main devices used are a class of meters called obstruction meters. Among these, the orifice meter, the flow nozzle and the venturimeter are the most commonly used devices. Specifically, such meters will be found in chemical processing plants, water distribution systems, boiler plants etc. In research and development laboratories dealing both with liquid and gas flows, such devices are indispensable tools for getting quantitative results.

PROCEDURE

1. Note the inlet pipe diameter d_1 and the constriction/throat diameter d_2 of the device. Note the density of the manometer fluid ρ_m and that of the flowing fluid ρ.

2. Start the pump and adjust the control valve in the pipeline for maximum discharge.
3. Measure the pressure difference h_m across the meter.
4. Measure the flow rate through the piping system using either the volumetric or the gravimetric technique.
5. Decrease the flow rate through the system by regulating the control valve.
6. Repeat steps 3–5 for at least ten different settings of the control valve.

RESULTS

At each setting, calculate

1. the actual flow rate Q.
2. the average flow velocity U_1
3. the Reynolds number $R_e = \dfrac{\rho U_1 d_1}{\mu_d}$

Also calculate

1. the theoretical discharge Q_{th}
2. the discharge coefficient $C_d = Q/Q_{th}$ for each device.

Plot the following curves for each device:

1. $\log Q$ vs $\log h_m$
2. C_d vs R_e

Superimpose the standard results for C_d vs R_e for the individual meters given in text on the corresponding experimental curves obtained by you.

POINTS FOR DISCUSSION

1. Comment on the results presented in your graphs.
2. Find the sensitivity of the meters.
3. Can the same calibration be used if the meters are installed in inclined pipes?
4. Compare the value of C_d obtained by you with the corresponding expected values for the flow meters.
5. Give reasons for the values of C_d for the venturimeter being higher than that for the nozzle meter and that for the nozzle meter being higher than for orificemeter.

ANY OTHER OBSERVATION OR COMMENT?

5.4. Measurement of Discharge by a Weir, a Notch or a Venturi flume

Flow through an open channel occurs under atmospheric pressure. The motive force to create the flow is the gravity force derived from the slope of the bed and the kinetic energy possessed by the liquid. Mechanics of flow through open channels will be presented in a separate chapter; the purpose here is to study the common methods of discharge measurement.

A weir is an obstruction placed in an open channel over which the flow occurs. The weir is generally in the form of a vertical wall with a sharp edge at the top, running all the way across the open channel. When the liquid flows over the weir, the height of the liquid above the tip of the sharp edge bears a relationship with the discharge across it.

A notch is a sharp-edged device which permits the liquid to go through it, the liquid being exposed to the atmospheric pressure. Notches may be rectangular, triangular, circular or trapezoidal in shape. A triangular notch, also called V-notch, consists of a V-cut sharp-edged passage through which the liquid passes. The only difference between a weir and a rectangular notch is that a weir runs all the way across the channel whereas a rectangular notch may be as wide as required, say over a part of the channel.

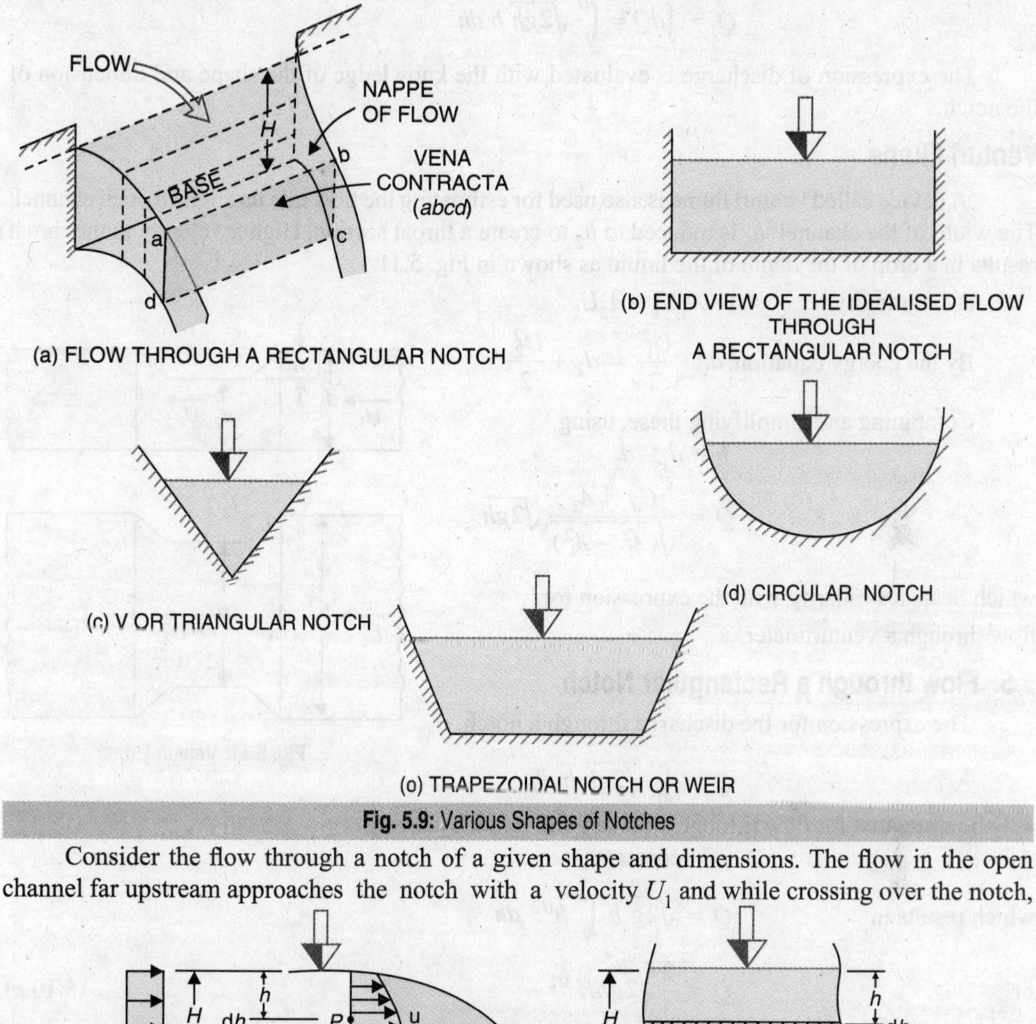

Fig. 5.9: Various Shapes of Notches

Consider the flow through a notch of a given shape and dimensions. The flow in the open channel far upstream approaches the notch with a velocity U_1 and while crossing over the notch,

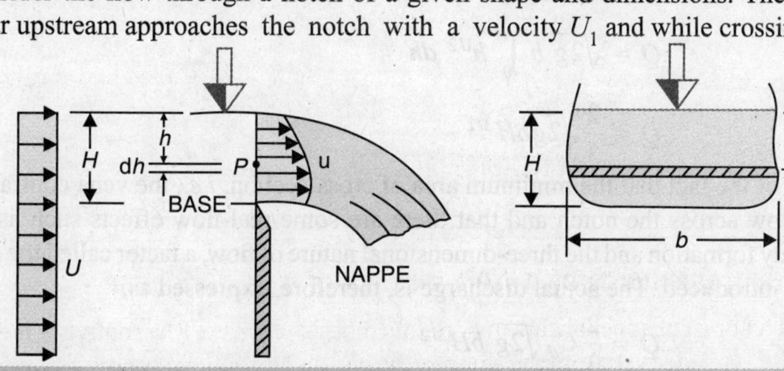

Fig 5.10: Flow through a Notch

the total height H and the velocity distribution must be in accordance with the continuity principle and the Bernoulli equation. Consider a horizontal strip of depth dh located at a depth h below the liquid level. If the width of the strip is b, the area is $b\ dh$. The velocity of the liquid at a point on the strip must be given by

$$u = \sqrt{2gh}$$

The flow rate across the strip is

$$dQ = u\,dA = \sqrt{2gh}\; b\,dh$$

The total discharge across the notch must be given by

$$Q = \int dQ = \int_0^H \sqrt{2gh}\; b\,dh$$

The expression of discharge is evaluated with the knowledge of the shape and dimension of the notch.

Venturi Flume

A device called venturi flume is also used for estimating the flow rate through an open channel. The width of the channel b_1 is reduced to b_2 to create a throat section. Higher velocity at the throat results in a drop of the depth of the liquid as shown in Fig. 5.11.

By continuity, $A_1U_1 = A_2U_2$

By the energy equation, $d_1 + \dfrac{U_1^2}{2} = d_2 + \dfrac{U_2^2}{2}$

Combining and simplifying these, using

$$h = d_1 - d_2$$

$$Q = \frac{C_d\, A_1 A_2}{\sqrt{(A_1^2 - A_1^2)}}\sqrt{2gh}$$

which bears a similarity with the expression for flow through a venturimeter.

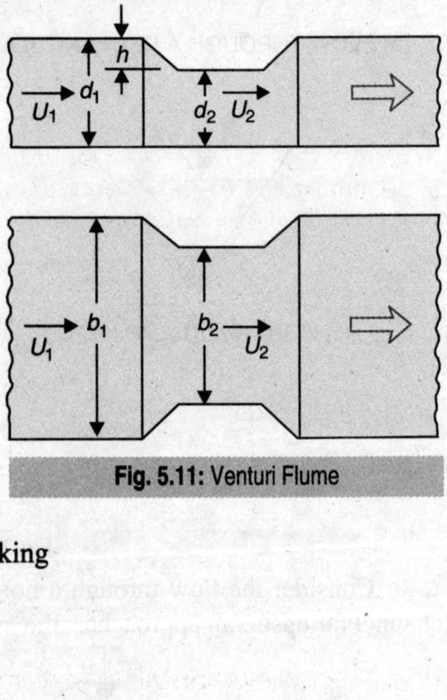

Fig. 5.11: Venturi Flume

5.5. Flow through a Rectangular Notch

The expression for the discharge through a notch,

$$Q = \int_0^H \sqrt{2gh}\; b\,dh$$

may be evaluated for flow through a rectangular notch by taking

$$b = \text{constant}$$

which results in $\quad Q = \sqrt{2g}\; b \int_0^H h^{1/2}\,dh$

or $\quad Q = \dfrac{2}{3}\sqrt{2g}bH^{3/2}$ ... (5.10 *a*)

Taking note of the fact that the minimum area of cross-section, *i.e.*, the vena contracta, is less than the area of flow across the notch and that there are some real-flow effects such as side-wall boundaries and eddy formation and the three-dimensional nature of flow, a factor called the coefficient of discharge C_d is introduced. The actual discharge is, therefore, expressed as,

$$Q = \frac{2}{3} C_d \sqrt{2g}\; bH^{3/2} \qquad \text{... (5.10 } b\text{)}$$

This expression takes no note of the fact that the liquid approaches the notch with an initial velocity U_1 and that the liquid possesses knietic energy per unit weight given by $U_1^2/2g$. In order to take the velocity of approach into account, the limits of integration in the expression for Q should be

$$\frac{U_1^2}{2g} \text{ to } \left(\frac{U_1^2}{2g} + H\right)$$

instead of 0 to H. With this correction, the improved expression for the discharge through a rectangular notch becomes,

$$Q = \frac{2}{3} C_d \sqrt{2g}\, b \left[\left(H + \frac{U_1^2}{2g} \right)^{3/2} - \left(\frac{U_1^2}{2g} \right)^{3/2} \right] \quad \text{... (5.10 c)}$$

When the flow occurs over a rectangular notch, the nappe of flow is contracted at the ends and an area of minimum cross-section akin to vena contracta occurs at a short distance from the notch as shown in Fig. 5.9(*a*) The effect of the end contraction is that the effective width of flow is reduced. According to Francis, each end contraction is of the order of 10 per cent of head over the notch, *i.e.*, 0.1 *H*. For a rectangular notch with two end contractions,

$$\text{Effective width} = b - 0.2\,H$$

and the expression for the discharge becomes

$$Q = \frac{2}{3} C_d \sqrt{2g}\, (b - 0.2\,H)\, H^{3/2} \quad \text{... (5.10 d)}$$

which implies that the end contractions reduce the discharge by

$$q = \frac{2}{3} C_d \sqrt{2g}\, (0.2\,H)\, H^{3/2}$$

$$= \frac{2}{15} C_d \sqrt{2g} H^{5/2}$$

Comparison of this expression with the expression for discharge for a triangular notch (derived later) shows that, for an equivalent triangular weir,

$$\frac{2}{15} C_d \sqrt{2g} H^{5/2} = \frac{8}{15} C_d \sqrt{2g} \tan\theta/2\, H^{5/2};\ \tan\theta/2 = \frac{1}{4} \quad \text{... (5.11)}$$

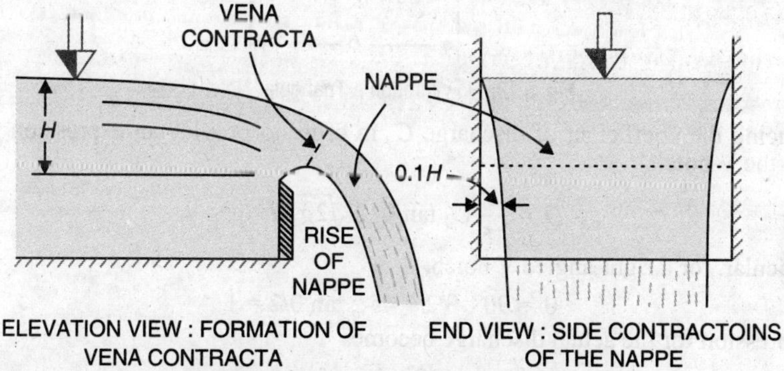

Fig. 5.12: Flow over a Rectangular Weir

If the rectangular weir is enlarged to become a trapezoidal weir with the side slopes '1 horizontal to 4 vertical', it would give the discharge that an uncontracted rectangular weir would have given. This was first discovered by Cippoleti and the compensated weir is called *Cippoleti weir.*

It may also be noted that the lower surface of the nappe rises by 10% of the head over the notch. This however does not change the width of flow. Effect of the rise of the nappe is included in the coefficient of discharge C_d for the notch.

5.6. Flow through a Triangular Notch

Consider a triangular or V-notch with an included angle θ and semi-angle θ/2 . Let the liquid flow through it with the level *H* above the base point. Consider the flow through an elementary strip of depth *dh* located at a depth *h* below the level of the liquid. The discharge through it must be

$$dQ = \sqrt{2gh}\, b\,dh$$

and the discharge through the entire notch should be

$$Q = \int dQ = \int_0^H \sqrt{2gh}\, b\, dh$$

For a triangular notch,

$$b = 2\,(H - h)\tan\theta/2 \qquad \text{... (5.12)}$$

Substituting this relation in the expression for Q

$$Q = 2\tan\theta/2\,\sqrt{2g}\int_0^H (H-h)\,h^{3/2}\,dh$$

or

$$Q = 2\tan\theta/2\,\sqrt{2g}\left[\frac{2}{3}H\,h^{3/2} - \frac{2}{5}h^{5/2}\right]_0^H$$

$$= \frac{8}{15}\tan\theta/2\,\sqrt{2g}\,H^{5/2} \qquad \text{... (5.13 a)}$$

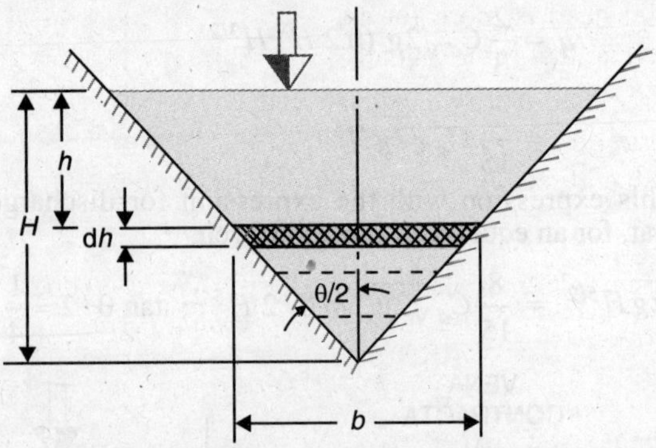

Fig. 5.13: Flow through a Triangular Notch

Introducing the coefficient of discharge C_d in order to provide an expression for the actual flow through the V-notch,

$$Q = \frac{8}{15}C_d\tan\theta/2\,\sqrt{2g}\,H^{5/2} \qquad \text{... (5.13 b)}$$

In particular, for a right-angled V notch,

$$\theta = 90°.\ \theta/2 = 45°;\ \tan\theta/2 = 1$$

the expression for the actual discharge becomes

$$Q = \frac{8}{15}C_d\sqrt{2g}\,H^{5/2} \qquad \text{... (5.13 c)}$$

A word of comparison between the accuracy of measurement by a rectangular and for a V-notch is worthwhile.

For a rectangular notch, $Q = K H^{3/2}$

and for V-notch, $Q = C H^{5/2}$

where K and C are appropriate constants.

The accuracy of measurement of the discharge depends upon the accuracy of measurement of the head H. A small error dH in H would correspond to an error dQ in Q.

For a rectangular notch,

$$dQ = \frac{3}{2}K\,H^{1/2}\,dH$$

and

$$\frac{dQ}{Q} = \frac{3}{2}\frac{dH}{H} = 1.5\frac{dH}{H} \qquad \text{... (5.14 a)}$$

Similarly, for a V-notch

$$dQ = \frac{5}{2} CH^{3/2}\ dH$$

and
$$\frac{dQ}{Q} = \frac{5}{2}\frac{dH}{H} = 2.5\frac{dH}{H} \quad \text{... (5.14 b)}$$

It is seen that a 1% error in the measurement of the head results in 1.5% error in discharge measured by a rectangular notch and 2.5% error in discharge measured by a V-notch.

There is one more point to it. For low discharge, the reading 'H' for a V-notch is greater than the reading 'H' for a rectangular notch for the same discharge. This is obvious because the width of a rectangular notch is b everywhere whereas the width at the base of a V notch is zero. A V-notch is, therefore, preferred for low discharge. For high discharge, a rectangular notch is preferred because the incremental head is more for an increment in discharge than that for a V-notch.

If it is required to have a single notch suitable for a large range of discharge, a combination of 'V-notch and retangular notch' may be used.

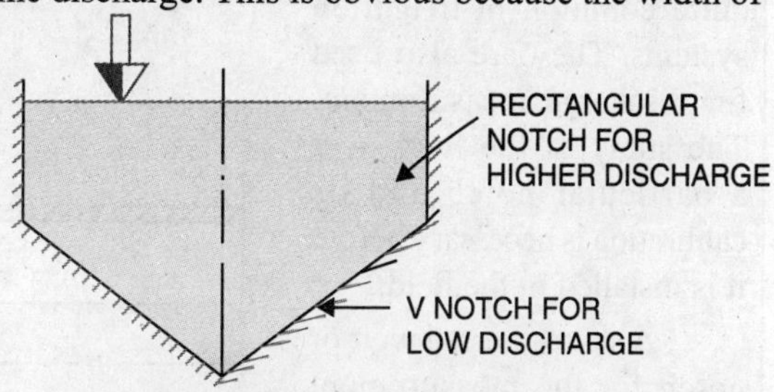

Fig 5.14: A Combination Notch

Example 5.6. *A reservoir having a surface area of 500 m² is emptied by a 0.5 m wide rectangular weir. How long should it take to empty the reservoir from a height 3.2 m to 0.1 m above the sill ? Take $C_d = 0.65$*

Solution.

With reference to a rectangular weir, $A\ dh = -\ Q\ dt$

The negative sign appears because h decreases as t increases.

For a rectangular weir, $\quad Q - \frac{2}{3}\sqrt{2g}\ h^{3/2}\ C_d$

Hence
$$dt = \frac{-A\ h^{-3/2} dh}{\frac{2}{3}\sqrt{2g}\ bC_d}$$

$$T = \int_0^T dt = \frac{3A}{\sqrt{2gb}\ C_d}\left(\frac{1}{\sqrt{H_2}} - \frac{1}{\sqrt{H_1}}\right)$$

$$= 3\times\frac{500}{(4.43\times 0.5\times 0.65)}\times\left(\frac{1}{\sqrt{0.1}} - \frac{1}{\sqrt{0.2}}\right)$$

= 965 s = **16 minutes and 5 seconds.**

EXPERIMENT

Flow over a Notch or a Weir

OBJECTIVE

To study the flow over a notch or a weir and to calibrate it for the discharge measurement in a free surface flow.

PRACTICAL RELEVANCE

A weir is an obstruction placed across a free-surface flow such that the flow takes place over it. A weir with a sharp edge is commonly referred to as a notch. Generally, notches are openings cut in metallic plates and installed in flumes or small channels. Installation of a notch is exclusively for the purpose of measuring the discharge in the stream. Other types of weir shaped like dams whose primary purpose is to harness the flow are also quite common in irrigation systems. They are also used for discharge measurement. The study of the flow over a particular notch and its calibration is necessary before it is installed in the field.

A sharp-crested weir or notch for the measurement of discharge should have a regular geometrical shape, *i.e.*, rectangular, triangular, circular, etc.

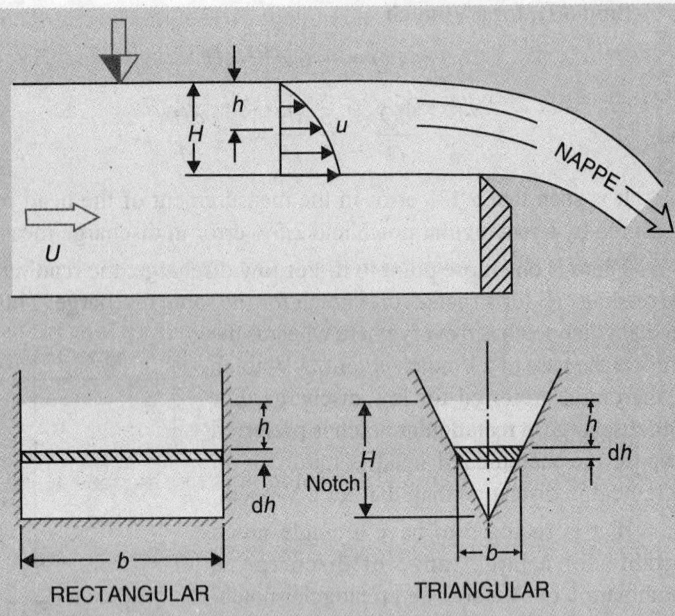

Flow over the Notch

PROCEDURE

1. Note the geometrical aspects of the notch.
2. Measure the level of the crest h_1 of the notch relative to the bed of the flume.
3. Start the water supply and allow the flow to take place over the notch. After steady state conditions have been attained, measure the level h_2 of the free liquid surface relative to the bed of the flume at a point sufficiently upstream of the notch. For the above flow conditions, measure the discharge Q over the notch by a standardised gravimetric or volumetric method.
4. Repeat step 3 for different flow rates until the entire range of flow rates has been investigated.

   Qualitatively observe the mechanism of the nappe formation, the side contraction effects and other features of the flow for the various flow rates.

RESULTS

1. Plot log H *vs* log Q
2. Determine the constants k and n for the calibration curve in the form

$$Q = k H^n$$

Compare the experimentally determined value of k with the appropriate theoretical value for the notch tested.

POINTS FOR DISCUSSION

1. Comment on the values of the calibration constants.
2. Why should the geometrical shape of a notch be well-defined and regular ?

3. What is meant by end contractions ? Do they occur in all types of notches ?
4. Are there any means of increasing the value of k or of C_d for a given notch ?

ANY OTHER OBSERVATION OR COMMENT?

Example 5.7. *A venturimeter with 7 cm throat diameter is fitted in a 15 cm diameter vertical pipeline with upward discharge. Tlie absolute pressure at a point 6 m below the throat is 5 atmospheres which the pressure at the throat is such that it supports a column of 20 cm water gauge.*

(a) Estimate the discharge through the pipe.

(b) Find the pressure at a point 6 m above the throat.

Solution.

(*a*)Applying the Bernoulli equation between sections 1 and 2, with pressure expressed as absolute heads of water,

$$5 \times 10.33 + U_1^2/2g + 0 = (10.33 + 0.20) + U_2^2/2g + 6$$

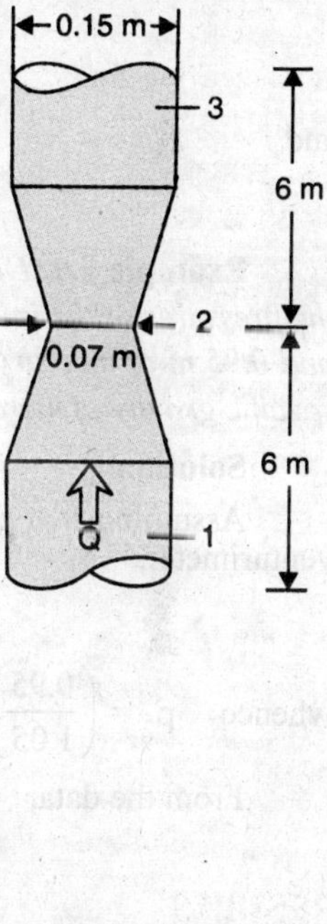

whence $U_2^2 - U_1^2 = 35.12 \times 2 \times 9.81 = 689$

From the data

$$A_2 = \frac{\pi \times 0.07^2}{4} = 0.00385\,\text{m}^2$$

$$A_1 = \frac{\pi \times 0.15^2}{4} = 0.0177\,\text{m}^2$$

$$U_1 = Q/A_1 = 56.6\,Q$$

$$U_2 = Q/A_2 = 260\,Q$$

where $(260^2 - 56.6^2)\,Q^2 = 689$

and $Q = \mathbf{0.1034\ m^3/s}$

whence $U_1 = 56.6 \times 0.1034 = 5.85$ m/s

and $U_2 = 260 \times 0.1034 = 26.88$ m/s

(*b*) Applying the Bernoulli equation between sections 2 and 3,

$$(10.33 + 0.20) + U_2^2/2g = h_3 + U_3^2/2g + 6$$

Since $U_2 = 26.88$ m/s and $U_3 = U_1 = 5.85$ m/s

$$h_3 = 10.53 - 6 + 26.88^2/2g - 5.85^2/2g$$

$$= 39.61 \text{ m absolute head of water}$$

Of course, the pressure at 3 could be obtained directly by applying the Bernoulli equation between sections 1 and 3.

$$5 \times 10.33 + U_1^2/2g + 0 = h_3 + U_3^2/2g + 12$$

Since $U_1 = U_3$

$$h_3 = 5 \times 10.33 - 12 = 39.65 \text{ m}$$

which is the same as that calculated above.

This pressure may also be expressed in terms of atmospheres as

$$p_3 = 39.65/10.33 = \mathbf{3.84\ atmospheres\ absolute.}$$

Example 5.8. *A vertical water pipe 1.5 m long tapers from 75 mm diameter at the bottom to 150 mm diameter at the top when the rate of flow is 50 litres/second upwards. Determine the pressure at the top if the pressure at the bottom end is 150 kN/m².*

Solution.

For the given configuration,

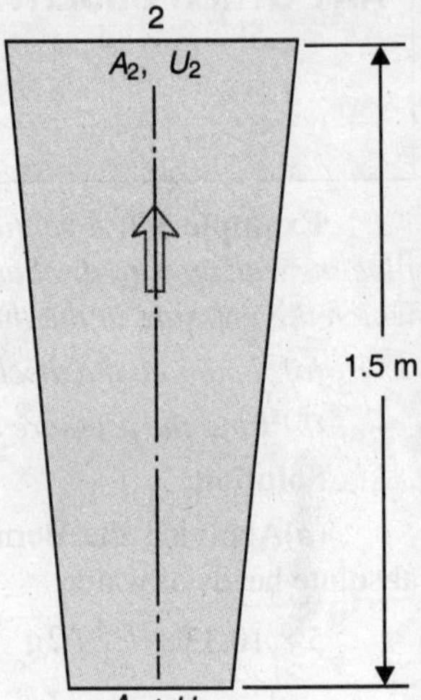

$$A_1 = \pi \times (75/1000)^2/4 = 0.00442 \text{ m}^2$$
$$A_2 = (150/75)^2 \times 0.00442 = 0.0177 \text{ m}^2$$
$$Q = 50 \text{ litres/s} = 0.05 \text{ m}^3/\text{s}$$
$$U_1 = 0.05/0.00442 = 11.31 \text{ m/s}$$
$$U_2 = 0.05/0.0177 = 2.83 \text{ m/s}$$

Applying the Bernoulli equation between the two points 1 and 2,

$$\frac{p_1}{\rho g} + \frac{U_1^2}{2g} + z_2 = \frac{p_2}{\rho g} + \frac{U_2^2}{2g} + z_2$$

whence
$$\frac{p_2}{\rho g} = \frac{150 \times 1000}{1000 \times 9.81} + \frac{(11.31^2 - 2.83^2)}{2 \times 9.81} - 1.5$$
$$= 15.29 + 6.11 - 1.5 = 19.90 \text{ m}$$

and
$$p_2 = 19.90 \times 1000 \times 9.81 \text{ N/m}^2$$
$$= \mathbf{195.23 \text{ kN/m}^2}$$

Example 5.9. *Determine the flow rate of a gas through a 75 mm diameter horizontal pipe if the pressures at the inlet and the 25 mm diameter throat of a venturimeter installed in it are 1.05 m and 0.95 m of mercury absolute. Assume γ = 1.40, density of the gas in the pipe as 1.6 kg/m³ and specific gravity of mercury as 13.6.*

Solution.

Assuming that the flow takes place isentropically from the inlet, 1 to the throat, 2 of the venturimeter,

$$p_1 / \rho_1^\gamma = p_2 / \rho_2^\gamma$$

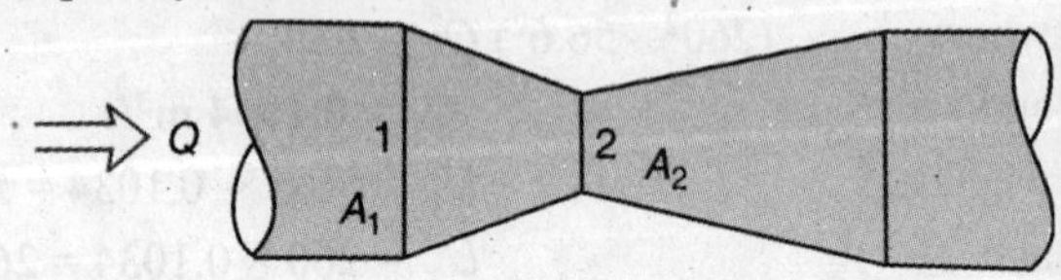

whence
$$\rho_2 = \left(\frac{0.95}{1.05}\right)^{1/1.4} \times 1.6 = 1.49 \text{ kg/m}^3$$

From the data,

$$A_1 = \frac{\pi}{4}\left(\frac{75}{1000}\right)^2 = 0.0044 \text{ m}^2$$

and
$$A_2 = \frac{\pi}{4}\left(\frac{25}{1000}\right)^2 = 0.000491 \text{ m}^2$$

By continuity,

$$\rho_1 A_1 U_1 = \rho_2 A_2 U_2$$

whence
$$U_2/nU_1 = \frac{1.6 \times 0.0044}{1.49 \times 0.000491} = 9.625$$

Applying the Bernoulli equation,

$$\frac{\gamma}{\gamma - 1}\frac{p_1}{\rho_1} + \frac{U_1^2}{2} = \frac{\gamma}{\gamma - 1}\frac{p_2}{\rho_2} + \frac{U_2^2}{2}$$

whence $\qquad \dfrac{U_2^2 - U_1^2}{2} = \left(\dfrac{1.05}{1.6} - \dfrac{0.95}{1.49}\right) \times \dfrac{13.6 \times 101\,325}{10.33} \times \dfrac{1.4}{1.4-1}$

$$= (0.656 - 0.638) \times 466956 = 8405.21$$

Substituting $\qquad U_2 = 9.625\, U_1$

$$(9.625^2 - 1)\, U_1^2 + 2 \times 8405.21 = 16810.42$$

whence $\qquad U_1 = 13.54$ m/s.

The volumetric flow rate is given by

$$Q = A_1 U_1 = 0.0044 \times 13.54 = 0.0595 \text{ m}^3/\text{s}$$

The mass flow rate is given by

$$\dot{m} = \rho_1 A_1 U_1 = \rho_1 Q = 1.6 \times 0.06 = \mathbf{0.095\ kg/s.}$$

Example 5.10. *A closed tank partially filled with water to a depth of 1 m has an orifice of diameter 1.5 cm at its bottom. The pressure of air above the water surface is raised until the discharge becomes 2 litres/second through the orifice of $C_d = 0.6$. Calculate the pressure of the air.*

Solution.

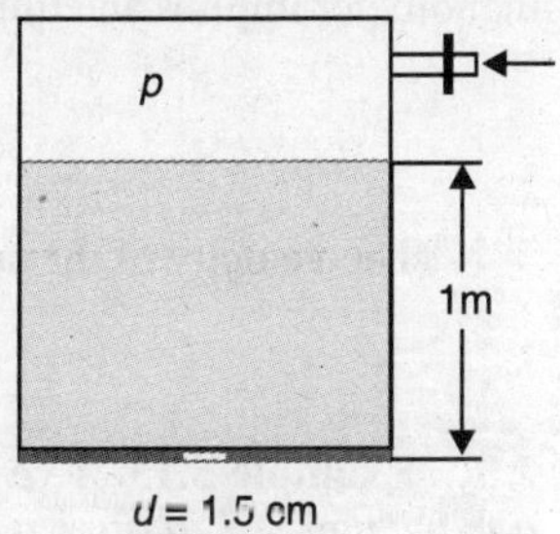

The discharge required is

$$Q = 2 \text{ litres/second} = 0.002 \text{ m}^3/\text{s}$$

which should be given by

$$Q = \sqrt{2gH} \times \pi d^2 / 4 \times C_d$$

whence $\qquad H = \left(\dfrac{0.002 \times 4 \times 100 \times 100}{\pi \times (1.5)^2 \times 0.6}\right)^2 / 2g$

$$= 18.13 \text{ m}$$

The depth of water is only 1 m. The gauge pressure on top of the water column should, therefore, be

$$p = \rho g \times (18.13 - 1)$$

$$= 1000 \times 9.81 \times 17.13$$

$$= 168045 \text{ N/m}^2 = \mathbf{168.045\ kN/m^2}$$

Example 5.11. *A pitot tube inserted in a pipe of 30 cm diameter records the stagnation pressure at the centre line as 1 m head of water gauge. A wall tap measures the static pressure in the pipe as 10 cm of mercury (vacuum). Calculate the rate of flow in the pipe assuming that the average velocity is 85% of the maximum velocity and the coefficient of the velocity for the probe is 0.98.*

Solution.

From the data,

$$d_1 = 0.3 \text{ m};\ A_1 = \pi \times 0.3^2/4 = 0.071 \text{ m}^2$$

$$\frac{p_o}{\rho g} = 1.0 \text{ m head of water gauge}$$

$$\frac{p}{\rho g} = -0.1 \times 13.6 = -1.36 \text{ m head of water gauge}$$

Both p_o and p refer to the atmospheric pressure as reference. Now,

$$\frac{p_o}{\rho g} = \frac{p}{\rho g} + \frac{u^2}{2g} = -1.36 + \frac{U^2}{2g} = 1.0$$

whence $\qquad U = 6.805$ m/s

Since the coefficient of velocity is 0.98, the actual velocity at the pipe centre line is

$$U_{act} = 6.805 \times 0.98 = 6.67 \text{ m/s}$$

The average velocity in the pipe is

$$U_{av} = 6.67 \times 0.85 = 5.67 \text{ m/s}$$

and the discharge is given by

$$Q = 5.67 \times 0.071 = \mathbf{0.40\ m^3/s}$$

Example 5.12. *A submarine moves horizontally in sea water (density 1026 kg/m³) with its axis 20 m below the surface of water. A pitot tube placed in front of the submarine along its axis is connected to a differential mercury manometer showing a deflection of 20 cm. Determine the speed of the submarine.*

Solution.

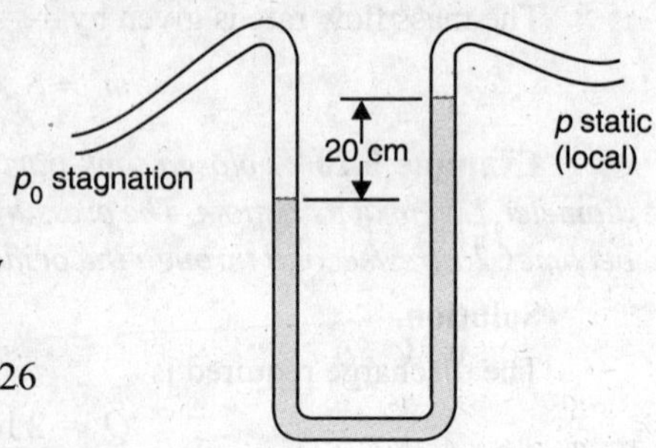

The pitot probe in front registers the stagnation pressure and the other limb of the mercury-filled differential manometer must be subjected the local static pressure.

The differential head of 20 cm in a U-tube mercury manometer implies an equivalent column of sea water as

$$h = 20 \times (13.6 - 1.026)/1.026$$

$$= 245.1 \text{ cm} = 2.451 \text{ m}$$

The velocity of the submarine is

$$U = \sqrt{2gh} = \sqrt{(2 \times 9.81 \times 2.451)} = \mathbf{6.93\ m/s}$$

Example 5.13. *A venturimeter is to be introduced in a 25 cm horizontal pipe line conveying a maximum of 150 litres per second. Tlie pressure in the pipe at the point of interest is 6 m head of water gauge. Determine the minimum diameter of the throat for the venturimeter to be selected so that a negative head is formed in it.*

Calculate the discharge when a U-tube differential mercury manometer reads a deflection of 20 cm.

Solution.

Applying the Bernoulli equation between the entry and the throat of the venturimeter,

$$\frac{p_1}{\rho g} + \frac{U_1^2}{2g} + z_1 = \frac{p_2}{\rho g} + \frac{U_2^2}{2g} + z_2$$

For a horizontal venturimeter, $z_1 = z_2$; from the data,

$$A_1 = \pi \times 0.25^2/4 = 0.0491 \text{ m}^2$$

The pressure at the entry to the venturimeter is 6 m water gauge which is equivalent to an absolute pressure head;

$$\frac{p_1}{\rho g} = 6 + 10.33 = 16.33 \text{ m}$$

assuming the local atmospheric pressure as 10.33 m head of water.

For the maximum discharge of 0.15 m³/s

$$U_1 = 0.15/0.0491 = 3.06 \text{ m/s}$$

At this discharge

$$\frac{p_2}{\rho g} = \text{atmospheric pressure head} = 10.33 \text{ m (absolute)}$$

Finally,

$$16.33 + 3.06^2/2g = 10.33 + U_2^2/2g$$

whence $U_2 = 11.56$ m/s

and $A_2 = 0.15/11.56 = 0.013$ m²

$d_2 = 0.13$ m = 13 cm

The minimum diameter at the throat is, therefore, **13 cm.**

Now, the U-tube differential manometer reading of 20 cm of mercury is equivalent to a water head

$$h = 0.20\,(13.6/1 - 1) = 2.52 \text{ m}$$

The discharge through a venturimeter is given by the expression

$$Q = \frac{A_1^2\, A_2^2}{\sqrt{A_1^2 - A_2^2}}\sqrt{2gh}$$

$$= \frac{0.0491 \times 0.0133}{(0.0491^2 - 0.0133^2)^{1/2}} \times \sqrt{2 \times 9.81 \times 2.52}$$

$$= 0.097 \text{ m}^3\text{/s} = \mathbf{97 \text{ litres/s}}$$

Example 5.14. *A rectangular weir, 1 m wide and a right angled V notch are both installed in an open channel. If the water flows over the weir with a head of 15 cm, determine the head of water over the V-notch. Assume the coefficient of discharge for the rectangular weir and the V-notch as 0.65 and 0.60 respectively.*

Solution.

The discharge over a rectangular weir is given by

$$Q = \frac{2}{3} C_d \sqrt{2g}\,.\, b\,.H^{3/2}$$

With the given data,

$$Q = \frac{2}{3} \times 0.65 \times \sqrt{2g} \times 1 \times 0.15^{3/2} = 0.025\sqrt{2g} \quad \text{... }(i)$$

The discharge over a right angled V-notch is given by

$$Q = \frac{8}{15} \times 0.60 \times \sqrt{2g}\, H^{5/2} = 0.32\sqrt{2g}\, H^{5/2} \quad \text{... }(ii)$$

Equating (*i*) and (*ii*) for the same discharge,

$$H^{5/2} = \frac{0.025}{0.32} = 0.078$$

and $H = 0.36$ m = **36 cm**

5.7. Hot-Wire Anemometry

The technique of using hot-wire anemometer in air has acquired popularity in recent years for the measurement of

(*a*) *fluid velocity in magnitude and direction;*

(*b*) *level of turbulence; and*

(*c*) *shear stresses in fluid flow.*

The principle of measurement is based on the convective heat loss from an electrically heated thin wire by the flow of fluid over it. Cooling of the wire results in a decrease of resistance which causes an unbalance of the bridge across which the hot-wire probe is connected as shown in Fig. 5.15. The error-voltage undergoes amplification and a feedback bridge current is supplied to bring the resistance and hence the temperature of the hot-wire to its initial balanced position. The hot-wire will thus tend to operate at a constant temperature.

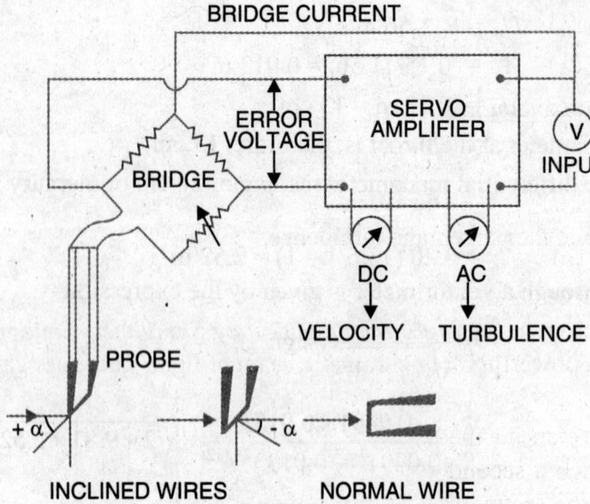

(a) CONSTANT TEMPERATURES SYSTEM

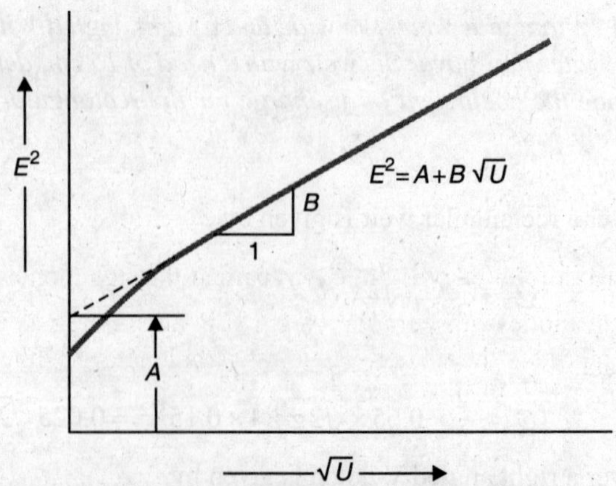

Fig. 5.15: (b) Calibration of Hot-Wire

The mean velocity of flow is measured by the bridge *DC* voltmeter.

By the King's law, (stated without proof)

$$E^2 = A + B\sqrt{U} \quad \text{... (5.15)}$$

A linear calibration curve E^2 *vs* $\sqrt{U}$ is thus plotted to evaluate the constants *A* and *B*. The level of turbulence and the shear stresses in turbulent flows are estimated with reference to the bridge *ACRMS* voltmeter

$$\text{Percentage turbulence} = \frac{\sqrt{\overline{u'^2}}}{U} \times 100$$

Starting with the King's law,

$$E^2 = A + B\sqrt{U}$$

$$2\,E\,\delta E = \frac{1}{2} B\,U^{1/2}\,\delta U$$

Denoting

$$\delta E = e', \text{ a small change in voltage}$$

$$\delta U = u', \text{ a small change in velocity}$$

$$2Ee' = \frac{1}{2} BU^{-1/2}u'$$

or $$e' = \frac{1}{4}\frac{E^2 - A}{E}\frac{u'}{U} \qquad \text{... (5.16)}$$

Squaring and time-averaging,

$$\overline{e'^2} = \frac{1}{16}\left(\frac{E^2 - A}{E}\right)^2 \frac{\sqrt{\overline{u'^2}}}{U^2}$$

which is used to evaluate the percentage turbulence.

5.8. LASER Anemometry

LASER is an acronym for *Light Amplification by Stimulated Emission of Radiation.* It is a device for producing a powerful monochromatic beam of light. There are gaseous, liquid and solid-state lasers.

Doppler effect refers to the shift in frequency which a second person observes after the scattered light reaches it. LASER Doppler Anemometry (LDA) is concerned with the measurement of fluid velocity, turbulence and other parameters where a LASER beam is focussed into a moving zone which has tiny particles introduced artificially. The beam is scattered by the particles and the shift in frequency is directly recorded through a photomultiplier.

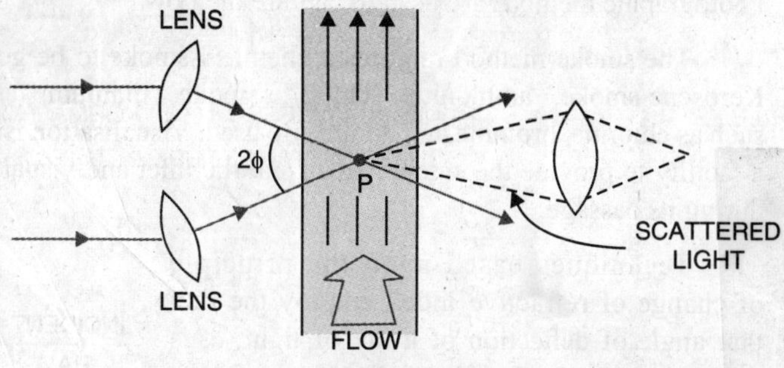

Fig 5.16

There are different modes of operation of LASER anemometres. One such arrangement is shown in Fig. 5.16. It measures the velocity, level of turbulence and other parameters at a point *P* where the two incident LASER beams intersect.

Advantages of LDA are listed as follows :

* *It is a non-obtrusive method where the flow is not disturbed physically.*
* *It has high spatial and temporal resolutions.*
* *It can be used in highly turbulent and eddying flows.*
* *It is combustion-compatible, i.e., it can be used in combusting flows.*
* *It needs no calibration and the measurements are absolute.*

There are, however, some limitations of the *LDA*. It results in intermittent signal with low signal-to-noise ratio, it requires artificial seeding of same-density particles as the fluid and it is an expensive optical apparatus. Even so, *LDA* is being increasingly employed in complex flow situations, two-phase flows and in boundary layer regions.

5.9. Flow Visualisation Techniques

For water and liquids	For air and gases
1. Coloured fluid or Powder injection	1. Smoke insertion Methods
2. Hydrogen Bubble Technique	2. Change of refractive index
3. Ultramicroscopy based upon the scattering of light by fine impurities in liquids.	(*a*) The Shadowgraph (*b*) The Schlieren method (*c*) The Interferometer

Visualisation of the flow of a liquid is best achieved by injecting another coloured fluid into it. Red ink and potassium permanganate dye are commonly employed. Surface flow studies are carried out by spraying aluminium or lycopodium powder. Addition of small droplets of oil which shine when illuminated is also employed frequently.

The hydrogen bubble technique consists of placing a fine wire in water and passing D.C. through it to electrolyse local water. The hydrogen bubbles thus generated are illuminated and observed for flow studies.

The ultramicroscope method utilises the presence of very fine impurities invariably present in the liquids. Under intense illumination very powerful microscopes are used to observe the motion of these subparticles. These subparticles may be smaller than the wavelength of light but are observable. Photographic methods are used to capture the flow.

The smoke method requires a chemical smoke to be generated or introduced into fhe fluid. Kerosene smoke, ammonium chloride smoke, titanium tetrachloride smoke or coloured gases such as chlorine, bromine and iodine are used. Visualisation is best achieved in a smoke tunnel with a facility to provide the smoke streams at the inlet and visual or photographic study of the smoke during its passage.

Techniques based upon the principle of change of refractive index employ the fact that angle of deflection of a ray of light, δ, is proportional to the density gradient in the flow,

$$\delta = K\left(\frac{d\rho}{dy}\right)_{y=y_1} . l$$

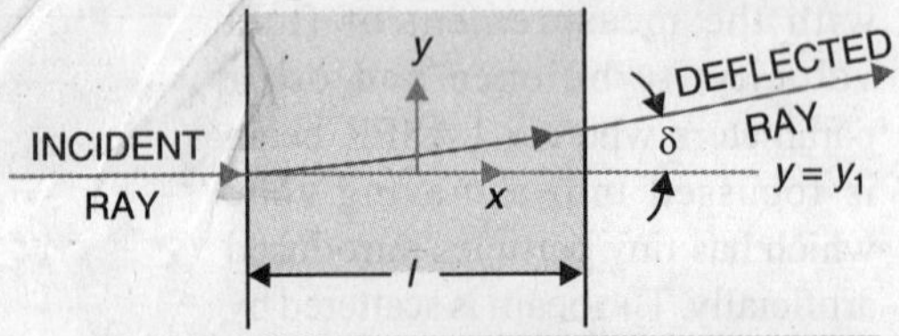

Fig. 5.17

where K is a constant for the fluid.

Shadowgraph	Schlieren	Interferometer
Direct viewing of the deflected ray pattern on a screen. The illumination depends upon the relative deflection of the light rays, *i.e.*, $d\delta/dy$ or $d^2\rho/dy^2$, the second derivative of the density. Simple optical tool. Good for qualitative and visual studies in turbulent and compressible flows with any types of shocks.	A device which indicates the density gradient $d\rho/dy$. The illumination or contrast on a screen is dependent upon the focal length of a lens, adjustment of a knife edge and the density gradient in fluid flow. Accurate equipment. No aberration. Universally used for supersonic-to-subsonic shock flow visualisation.	The most precise visualisation method based upon the principle of interference of light waves at a phase angle. The number of fringes are in relation to the change in density $(\rho - \rho_0)$ w.r.t. the zero fringe condition. High quality and precise visualisation and quantisation of density changes. Useful in the entire flow regime.
$C = K\frac{d^2\rho}{dy^2} . l$	$C = \frac{f}{y_1} K\left(\frac{d\rho}{dy}\right) . l$	$N = K(\rho - \rho_o) . l$

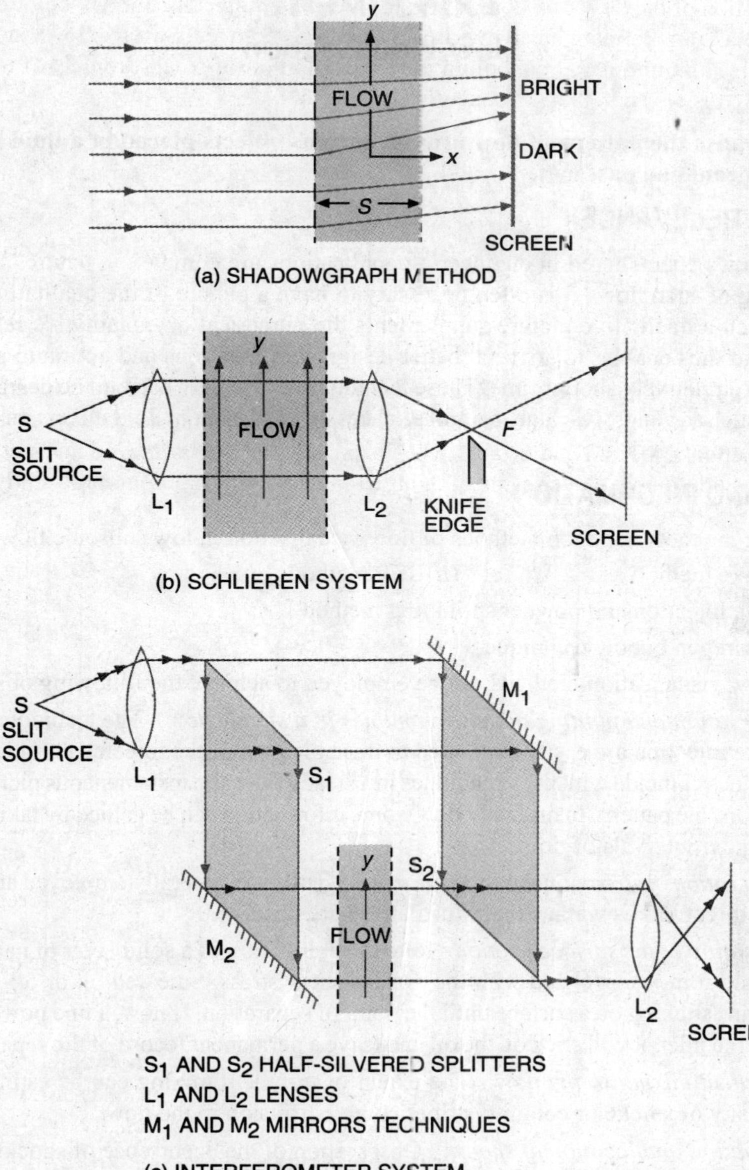

Fig. 5.18: Optical Flow Visualisation

Students are advised to view a video programme entitled

FLOW VISUALISATION

by Prof. K. L. Kumar

Indian Institute of Technology, Delhi.

The video demonstrates different techniques of flow visualisation and some very interesting flow patterns such as flow over a cylinder, an aerofoil, a bank of cylinders, etc.

EXPERIMENT

Flow Visualisation

OBJECTIVE

To visualise the pattern of flow around various objects placed in a fluid stream and flow through confined passages.

PRACTICAL RELEVANCE

Most flows encountered in engineering applications are complex in nature. To enable an understanding of such flows it is often necessary to have a picture of the qualitative pattern of the flow . Such a qualitative picture supplements the numerical or quantitative results for the given flow and thus enables to arrive at better design configurations and optimum performance parameters in a relatively shorter time. These techniques are widely used in the design of aircraft components such as wings, fuselage, control surfaces; in the design of fluid mixers, in determining the shape of bridge piers, etc.

BACKGROUND INFORMATION

Among the most common methods of flow visualisation in low subsonic flow are:

(*i*) Smoke method ; (*ii*) Tuft method ;

(*iii*) Dye injection and powder sprinkling method ;

(*iv*) Hydrogen bubble technique.

The flow visualisation methods can be employed to achieve the following objectives :

1. *Observation and plotting of the streamlines in a steady flow :* The technique of dye and smoke injection are easily employed to trace the path of the injected particles. Since the pathlines coincide with the streamlines in a steady flow, the instantaneous picture provides a streamline pattern. In unsteady flow, some information can be gained by taking the snaps at regular intervals of time.
2. *Observation of streamlines in a surface flow :* Tufts are generally employed at the surface; their directions as well as their fluctuations are observed.
3. *Estimation of the surface friction :* Rate of sublimation of a solid layer of naphthalene or a liquid film is correlated with the surface shear stress. Smeared oil or oil-dye mixture remains sticking on a surface until the point of separation. If now, a fine powder is blown from the inlet, it will stick on the oil and leave a permanent record of the separation point.
4. *Turbulent mixing in the flow :* The extent of turbulent mixing can be estimated by the intensity of smoke or colour distribution and diffusion in the flow.
5. *Formation and location of shocks :* Observation of the occurrence of shockwave and its intensity is very important in passages such as nozzles and in external flows such as over aerofoils.

RESULTS

The observed flow-patterns should be sketched on laboratory record sheets. For every sketch, the shape and size of the model, angle of incidence, Reynolds number and speed of flow must be recorded.

POINTS FOR DISCUSSION

1. Discuss the flow-patterns obtained ; locate the regions of velocity equal to, less than and more than the free-stream value.

2. What would happen to the pattern if the size of the body, the angle of incidence and the Reynolds number are changed ?
3. Define streamlines, path lines and streak lines or filament lines and state which lines are observed by the dye injection method.
4. Can an idea be made of the speed of flow by taking photograph at regular intervals of time ?
5. Comment on the type of body studied by you. Is it a streamlined body or a bluff body ?

ANY OTHER OBSERVATION OR COMMENT?

SUPPLEMENTARY REVISION EXAMPLES

Example 5.15. *An orifice meter is to be fitted into a horizontal pipe 20 cm dia, carrying oil of sp. gravity 0.85 to measure the flow rate. The differential head is to be indicated by a U-tube manometer containing mercury (sp. gr. 13.6). If the manometric reading is not to exceed 0.2 m when the flow is 15 kg/s, what should be the diameter of the orifice? Assume $C_d = 0.62$.*

Solution.

The expression for the actual volumetric discharge through an orifice meter is

$$Q = C_d . \frac{A_1 A_2}{\sqrt{A_1^2 - A_2^2}} . \sqrt{2g\, h_m \left(\frac{S_m}{S} - 1\right)}$$

with usual notation.

Mass flow rate $= \rho . Q$.

From the data, h_m (maximum) $= 0.2$ m

$S_m = 13.6;\ S = 0.85$

$C_d = 0.62$

$Q = 15/(0.85 \times 1000) = 0.0176\ m^3/s$

$A_1 = \pi\, (0.2)^2/4 = 0.0314\ m^2$

$A_1^2 = 0.000987\ m^4$.

Substituting these values in the expression for discharge,

$$0.0176 = 0.62 \frac{0.0314 A_2}{\sqrt{0.000987 - A_2^2}} \times 2 \times 9.81 \times 0.2 \left(\frac{13.6}{0.85} - 1\right)$$

$$0.0176 = \frac{0.0195 A_2}{\sqrt{0.000987 - A_2^2}} \times 58.9 = \frac{1.15 A_2}{\sqrt{0.000987 - A_2^2}}$$

or $$0.000987 - A_2^2 = 65.2\, A_2^2$$

$$A_2^2 = 0.00015\ m^4$$

whence $$A_2 = 0.0122\ m^2$$

and $$d_2 = \sqrt{\frac{0.0122 \times 4}{\pi}} = 0.125\ m = \mathbf{12.5\ cm.}$$

Example 5.16. *A venturi meter has its axis vertical, the inlet and throat diameter being 150 mm and 75 mm respectively. The throat is 225 mm above the inlet and k = 0.96. Petrol of specifice gravity 0.78 flows up through the meter at a rate of 0.029 m³/s. Find the pressure difference between the inlet and the throat.*

Solution.

From the data

$$d_1 = 0.15 \text{ m}, d_2 = 0.075 \text{ m}$$
$$\rho = 0.78 \times 1000 = 780 \text{ kg/m}^3.$$
$$Q = 0.029 \text{ m}^3/\text{s}$$
$$k = 0.96 = C_d$$

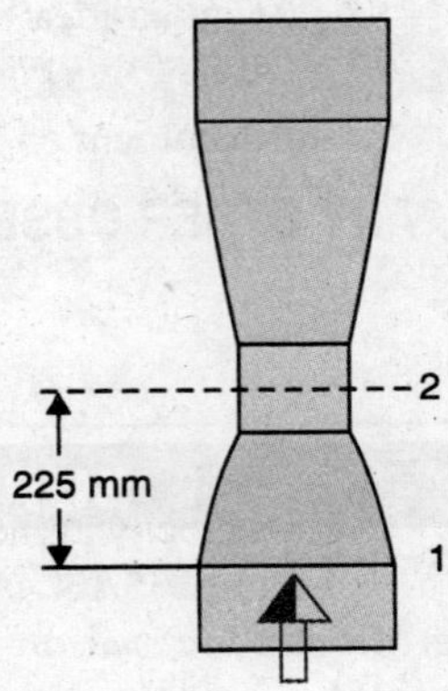

From the continuity and Bernoulli principle,

$$Q = C_d \cdot \frac{A_1 A_2}{\sqrt{A_1^2 - A_2^2}} \cdot \sqrt{2g\left[\frac{(p_1 - p_2)}{\rho g} + (z_1 - z_2)\right]}$$

Since $A = \pi d^2/4$ at any cross-section.

$$Q = C_d \cdot \frac{d_1^2 d_2^2}{\sqrt{d_1^4 - d_2^4}} \cdot \sqrt{2g\left[\frac{(p_1 - p_2)}{\rho g} + (z_1 - z_2)\right]}$$

With the given values,

$$0.029 = 0.96 \frac{0.15^2 \times 0.075^2}{\sqrt{0.15^4 - 0.075^4}} \cdot \sqrt{2 \times 9.81\left[\frac{(p_1 - p_2)}{780 \times 9.81} + (-0.225)\right]}$$

whence $(p_1 - p_2)$, the pressure difference can be calculated!

Example 5.17. *A sharp-crested rectangular notch of 0.8 m width and a sharp 90° V-notch are to be used alternatively for measuring an expected flow of 45 litres/second of a liquid. Find the percentage error in discharge that would result in the two cases if an error of 1 mm is made in the head measurement. Assume C_d = 0.6 for both the notches.*

Solution.

Firstly, we need to find out the expected heads H_1 and H_2 for the two cases: For the rectangular notch,

$$Q = \frac{2}{3} . C_d \sqrt{2g} . b . H_1^{3/2} = \frac{2}{3} \times 0.6 \times \sqrt{2 \times 9.81} \times 0.8 . H_1^{3/2} = 0.045 \text{ m}^3/\text{s}$$

$$H_1^{3/2} = 0.0317; H_1 = 0.1 \text{ m}$$

For the *V*- notch,

$$Q = \frac{8}{15} \tan 45° . \sqrt{2g} . H_2^{5/2}; H_2 = 0.071 \text{ m}$$

For the rectangular notch,

$$\frac{dQ}{Q} = \frac{3}{2}\frac{dH}{H}$$

$$\% \text{ Error in discharge} = \frac{3}{2} \times \frac{0.001}{0.1} \times 100 = \mathbf{1.5\%}$$

For the *V*-notch,

$$\frac{dQ}{Q} = 2.5 \frac{dH}{H}$$

$$\% \text{ Error in discharge} = \frac{5}{2} \times \frac{0.001}{0.071} \times 100 = \mathbf{3.5\%}$$

Example 5.18. *A pitot static tube placed in the centre of 20 cm diameter pipe has an orifice pointing upstream and the other perpendicular to it. If the pressure difference between the two orifices is 5 cm of water when the discharge through the pipe is 25.5 litres per second, calculate the coefficient of the meter. Take the mean velocity in the pipe to be 0.83 lines the central velocity.*

Solution.

From the data,

$$h = \frac{p_1 - p_2}{\rho g} = 0.05 \text{ m (of water)}$$

$$Q = 25.5 \text{ litres/second} = \frac{25.5}{1000} \text{ m}^3/\text{s} = 0.0255 \text{ m}^3/\text{s}.$$

The velocity at the centre of the pipe, U_c is given by

$$U_c = C_v \times \sqrt{2\,gh}$$

$$= C_v \times \sqrt{2 \times 9.81 \times 0.05} = 0.99\, C_v \text{ m/s}$$

and the discharge Q must be

$$Q = 0.83 \times 0.99\, C_v \times \pi \frac{0.2^2}{4} = 0.0258\, C_v \text{ m}^3/\text{s}$$

$$= 0.0255 \text{ m}^3/\text{s}$$

whence $C_v = 0.0255/0.0258 = \mathbf{0.988}$

Example 5.19. *A rectangular orifice 1.5 m wide and 1 m deep is discharging water from a large tank. If the water level in the tank is 3 cm above the top edge of the orifice, find the discharge through it. Take the coefficient of discharge of the orifice as 0.6.*

Solution.

The dimensions of the orifice (in meters) are large and comparable to the depth of water above it. It should, therefore be treated as a 'large' orifice and the discharge should be determined by considering an elementary discharge dQ through an elementary area dA of height dh, as shown, and integrating it over the height of the orifice.

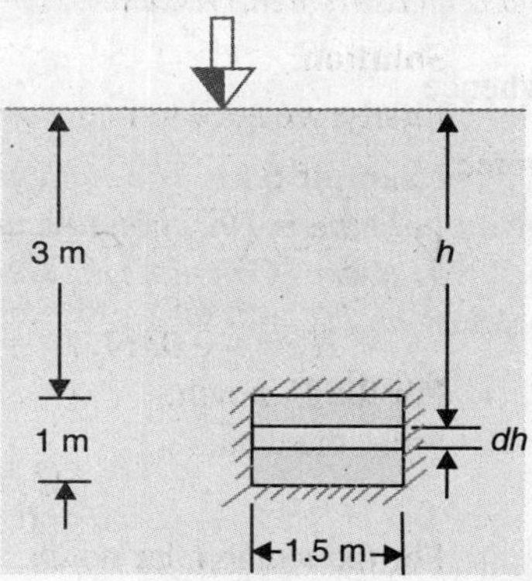

Considering an elementary area with height dh at a depth of h from the water level,

$$dQ = 0.6 \times \sqrt{2gh} \times (1.5 \times dh)$$

$$= 4\sqrt{h}\, dh.$$

By integration over the orifice,

$$Q = 4 \left| \frac{h^{3/2}}{3/2} \right|_3^4 = 2.67\,(4^{1.5} - 3^{1.5})$$

$$= 2.67\,(8 - 5.2) = \mathbf{7.46\ m^3/s.}$$

If we calculated the discharge by considering the mean velocity through the orifice at its mid-height, there would be an error involved in doing so.

Example 5.20. *The pressure inside a 4 cm diameter hose is 700 kPa (gauge). Find the velocity of the free jet coming out of a 2 cm nozzle fitted at the exit of the hose.*

Solution.

Applying the Bernoulli equation between a section 1 in the hose and a section 2 just outside the nozzle,

$$\frac{p_1}{\rho g} + \frac{U_1^2}{2g} + z_1 = \frac{p_2}{\rho g} + \frac{U_2^2}{2g} + z_2$$

Considering the centreline as the datum,

$$z_1 = z_2 = 0$$

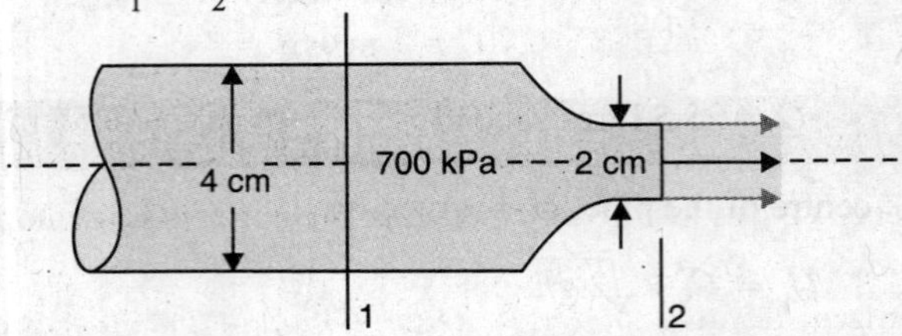

The pressure at 1 is 700 kPa (gauge), which means that $(p_1 - p_2) = 700$ kPa because the pressure at section 2 is atmospheric.

$$\frac{U_2^2 - U_1^2}{2g} = \frac{700 \times 1000}{\rho g}$$

By continuity, $\quad U_1 A_1 = U_2 A_2 \,;\; U_1 = U_2/4$

Cancelling g on both sides,

$$U_2^2 - \frac{1}{16} U_2^2 = \frac{2 \times 700 \times 1000}{1000} = 1400$$

$$15/16\, U_2^2 = 1400$$

whence $\quad U_2 = \mathbf{38.6\ m/s}$

Example 5.21. *The static pressure in a 30 cm diameter pipe flowing full with water measured by a pitot tube is 10 cm of mercury (vacuum). The stagnation pressure at the centreline is known to be 0.981 N/cm². Calculate the rate of flow, assuming that the average velocity is 85% of the maximum velocity and $C_v = 0.98$.*

Solution.

From the data,

$$d = 0.3 \text{ m};\ a = \frac{\pi d^2}{4} = 0.0707 \text{ m}^2.$$

The static pressure p = 10 cm of mercury (vacuum)

= 0.1 × 13.6 = 1.36 m of water (vacuum)

= – 1.36 m of water

The stagnation pressure p_0 at the centreline is given by

$$p_0 = 0.981 \text{ N/cm}^2 = 0.981 \times 10^4 \text{ N/m}^2.$$

In terms of head of water,

$$p_0 = \frac{0.981 \times 10^4}{\rho g} = \frac{0.981 \times 10^4}{1000 \times 9.81} = 1 \text{ m.}$$

$$h = (p_0 - p) \text{ in terms of head of water} = 1 - (-1.36) = 2.36 \text{ m.}$$

Velocity at the centre $= C_v \times \sqrt{2gh}$

$= 0.98 \times \sqrt{2 \times 9.81 \times 2.36}$

$= 6.67$ m/s.

and the average velocity $= 0.85 \times 6.67 = 5.67$ m/s

Rate of flow $Q = 5.67 \times 0.707 = \mathbf{4\ m^3/s}$

ITEM BANK

Objective items of different types are given to review the concepts developed in the Chapter.

* *Select correct response for each item.*
* *Every correct response is expected in less than a minute and carries one mark.*
* *Compare your responses with the key answers given at the end.*

Multiple-Choice Items

Attempt the following Multiple-Choice Items by ticking the correct or the most appropriate response from among the alternatives given :

1. Velocity at a point in a pipe flow may be measured by installing
(*a*) a pitot probe at the point
(*b*) a wall tap
(*c*) a stagnation pressure probe at that point
(*d*) a Prandtl probe at that point

2. Of the various methods of measuring discharge through a pipe line, the one with the least loss of energy and direct reading is
(*a*) by means of a venturimeter
(*b*) by means of an orificemeter
(*c*) by means of a flow nozzle
(*d*) by traversing a pitot-static probe

3. A horizontal pipe line conveys a constant rate of flow which is measured by a venturimeter installed in it. When the pipe is inclined upward in the .direction of flow, the reading of level difference on a differential manometer
(*a*) will remain the same
(*b*) will increase
(*c*) will decrease
(*d*) may fluctuate with time

4. Coefficients of velocity, contraction and discharge arranged in increasing order of value are
(*a*) C_v, C_C, C_d (*b*) C_d, C_c, C_v
(*c*) C_v, C_d, C_c (*d*) C_c, C_d, C_v

5. A triangular notch is a more accurate measuring device than a rectangular notch
(*a*) for low flow rates
(*b*) for high flow rates
(*c*) for medium flow rates
(*d*) for all flow rates

6. Venturimeter (*V*), flow nozzle (*N*) and orificemeter (*O*) arranged in increasing order of coefficient of discharge are
(*a*) *V, N, O* (*b*) *N, O, V*
(*c*) *O, N, V* (*d*) *O, V, N*

7. Measurement of velocity without any obstruction to the flow in a pipe may be made by
(*a*) pitot-static probe
(*b*) hot-wire anemometer
(*c*) hot-film anemometer
(*d*) laser-doppler anemometer

8. According to the King's law for a constant temperature hot-wire anemometer,
(a) $E \propto U$ (b) $E^2 \propto U$
(c) $E^2 \propto \sqrt{U}$ (d) $E \propto U^2$

9. Schlieren flow visualisation technique operates by employing
(a) density variations in flow
(b) density gradients in flow
(c) second derivatives of density
(d) higher derivatives of density

True-False Objective Items

10. Identify the following statements as true or false :
(a) The converging part of a venturimeter is shorter than the diverging part. (T/F)
(b) The 'dead space' in a flow nozzle is just downstream of the nozzle. (T/F)
(c) Rotation of the float in rotameter enables its stability in vertical position. (T/F)
(d) There are no end contractions in a trapezoidal notch. (T/F)
(e) The trajectory of a jet issuing from a nozzle is parabolic. (T/F)
(f) Introduction of a mouthpiece of the same diameter as the jet issuing from an orifice results in an increase in the discharge. (T/F)

ANSWERS

1. (d), **2.** (a), **3.** (a), **4.** (b), **5.** (a), **6.** (c), **7.** (d), **8.** (c), **9.** (b), **10.** (a)T, (b)T, (c)T, (d)F, (e)T, (f)T).

Questions on theory and numerical problems are given for tutorial work.

* *The questions are arranged generally in order of the articles in the text.*
* *Every question should take 15-20 minutes to solve.*
* *Answers for the numerical problems are given alongside them.*

5.1. A bellmouth entry in front of an air-compressor is to be calibrated for discharge through it in terms of height h of water in a single tube manometer as shown. If $h = 0.2$ m and the density of air is 1.2 kg/m³, estimate the discharge of air through the compressor. **[1.8 m³/s]**

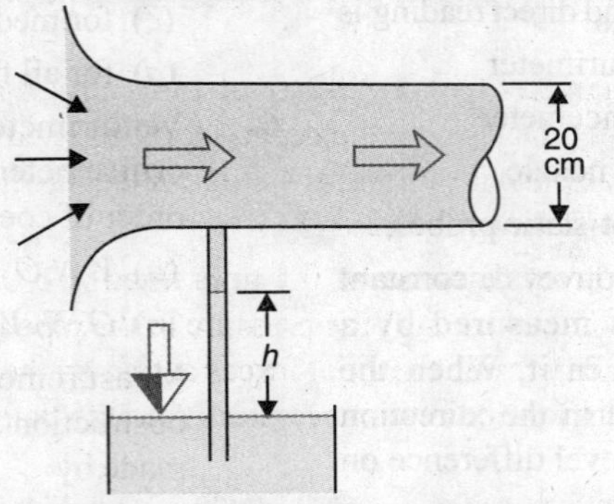

Fig. Problem 5.1

5.2. A vertical nozzle 2.5 cm in diameter discharges water at a velocity of 12 m/s. Calculate the diameter of the jet at 5 m height above the nozzle. State the assumption made. **[3.33 cm]**

5.3. Calculate the discharge of air through a pipe if an inlet orifice plate installed as shown records 10 cm of water difference in the U-tube manometer. **[0.846 m³/s]**

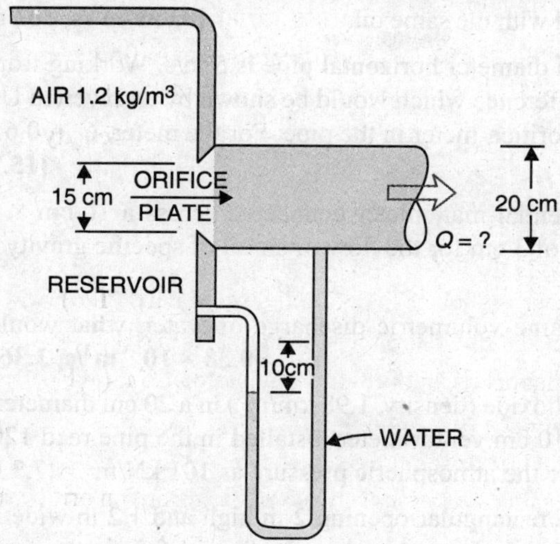

Fig. Problem 5.3

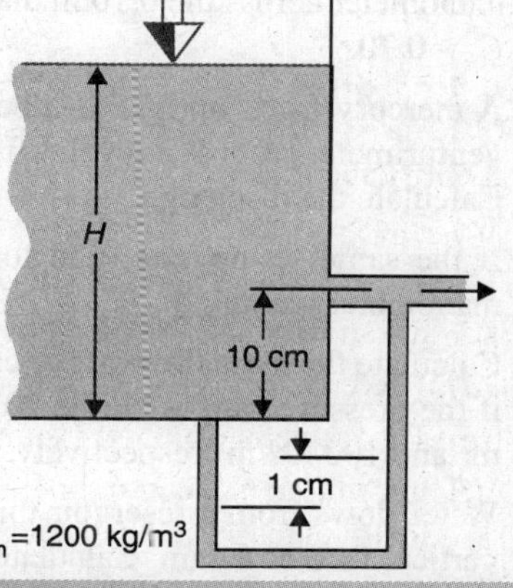

Fig. Problem 5.4

5.4. Calculate the height of level surface H of liquid of sp. gr. 0.7 from the U-tube manometer reading as shown. **[0.107 m]**

5.5. A pitot static probe is used to measure the flow of water in a 5 cm diameter pipe. If the mean velocity is 5 m/s and pitot static probe is connected across a mercury filled differential manometer, what should be the level difference of the mercury columns ? **[10.1 cm]**

5.6. Describe each of the following statements as either true or false :

(*a*) A streamline can never form a closed loop.

(*b*) In a moving fluid, the stagnation pressure is always greater than the static pressure at the same point. **[(a) False, circles for a vortex flow (b) True]**

5.7. In a 100 mm diameter horizontal pipe a venturimeter of 0.5 contraction ratio has been filled. The head of water on the meter when there is no flow is 3 m gauge. Find the rate of flow for which the throat pressure will be 2 m of water absolute. The coefficient of the meter is 0.97. **[293×10^{-3} m³/s]**

5.8. A jet of water 50 mm diameter with a velocity of 1 m/s strikes a flat plate inclined at an angle of 25° to the axis of the jet. Determine the normal force exerted on the plate (*a*) when the plate is stationary, (*b*) when the plate is moving at 4.5 m/s in the direction of the jet and (*c*) determine the work done and the efficiency for case (*b*). **[269 N, 151.5 N, 5%]**

5.9. A venturimeter with a throat diameter of 0.1 m is fitted in a 0.2 m diameter horizontal pipe carrying water at a rate of 0.04 m³/s. If the pressure in the pipe is 2.5 bar, gauge, and all losses are assumed negligible, determine the axial forces acting on the throat of the venturimeter and determine whether it is in tension or compression. Assume that the pipe does not provide any resisting force. **[5.83 kN compression]**

5.10. A venturimeter with a throat of 100 mm is fitted in a vertical pipeline of 200 mm diameter with oil of specific gravity 0.88 flowing upwards at a rate of 0.06 m³/s. The venturimeter coefficient is 0.96. Two pressure gauges calibrated in kN/m² are fitted at tapping points, one at the throat and the other in the inlet pipe 320 mm below the throat. The difference between the two gauge pressure readings is 28 kN/m².

Working from the Bernoulli equation determine (*a*) the volume flow rate of oil through the pipe, (*b*) the difference of level in the two limbs of a mercury manometer if it is connected to the tapping points and the connecting pipes are filled with the same oil. **[0.059 m³/s, 202 mm]**

5.11. The mean velocity of water in a 0.15 m diameter horizontal pipe is 5 m/s. Working from the first principles calculate the pressure difference which would be shown by an inverted U-tube manometer across the 0.10 m diameter orifice-meter in the pipe. For the meter, $C_d = 0.61$ and $C_c = 0.70$. **[15.7 m]**

5.12. A mercury filled upright U-tube differential manometer connected across a 10 cm × 5 cm venturimeter records a level difference of 3 cm for the flow of an oil of specific gravity 0.90. Calculate the discharge.

If the same set-up was used for the same volumetric discharge of water, what would the manometer reading be ? **[0.58×10^{-3} m³/s, 3.36 cm]**

5.13. Calculate the mass flow rate of carbon dioxide (density, 1.98 km/m³) in a 20 cm diameter pipe if the pressure gauges on the 20 cm × 10 cm venturimeter installed in the pipe read 120 kN/m² and 100 kN/m² respectively. Assume the atmospheric pressure as 101 kN/m². **[7.3 kg/s]**

5.14. Water flows from a reservoir through a rectangular opening 2 m high and 1.2 m wide in the vertical face of a dam. Calculate the discharge in m³/s when the free surface in the reservoir is 0.5 m above the top of the opening assuming a coefficient of discharge of 0.64. **[8.16 m³/s]**

5.15. A vertical triangular orifice in the wall of a reservoir has a base 0.9 m long, 0.6 m below its vertex and 1.2 m below the water surface. Determine the theoretical discharge. **[1.19 m³/s]**

5.16. The discharge over a rectangular notch is to be 0.14 m³/s when the water level is 23 cm above the sill. Assuming a coefficient of discharge as 0.6 and deducing the formula for theoretical discharge from first principles, calculate the width of notch required. **[0.72 m]**

5.17. In an experiment on a 90° V-notch the flow is collected in a 0.9 m diameter vertical cylindrical tank. It is found that the depth of water increases by 0.685 m in 16.8 s when the head over the notch is 0.2 m. Determine the coefficient of discharge of the notch. **[0.615]**

5.18. A sharp-edged notch is in the form of a symmetrical trapezium. The horizontal base is 10 cm wide, the top is 50 cm wide and the depth is 30 cm. Assuming the coefficient of discharge to be 0.6 and that the velocity of approach is negligible, calculate the height of the water level above the base of the notch if the discharge is 0.043 m³/s. **[22.9 cm]**

5.19. A vertical pipe 1 m diameter conveys water from an overhead tank which is barely 0.2 m filled as shown in Fig. Prob. 5.19. Assuming that the water flows as a sheet over the periphery rather than filling up the pipe, calculate the discharge. **[0.293 m³/s]**

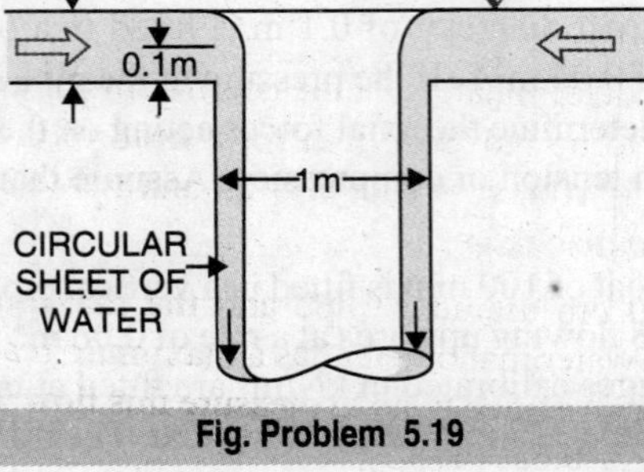

Fig. Problem 5.19

5.20. A right-angled triangular weir discharges water into a tank provided with an orifice at the base as shown in Fig. Prob. 5.20. Relate h_1 and h_2 for given weir and orifice sizes.

[$32\ h_1^{5/2} = 15\ \pi\ h_2^{1/2}\ d^2$]

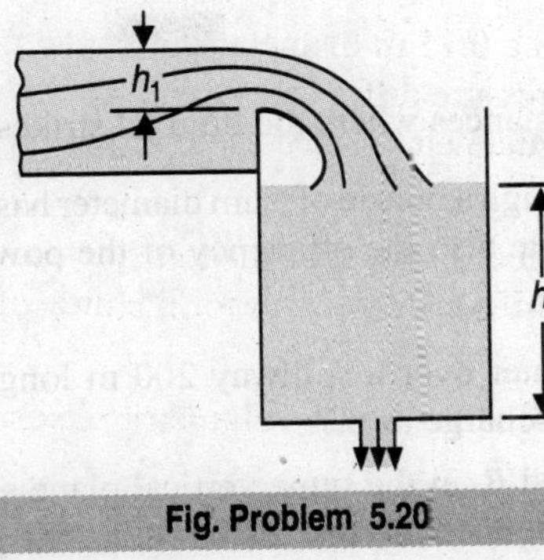

Fig. Problem 5.20

5.21. Calculate the increase in discharge recorded by a V-notch if the initial and final h are 0.3 m and 0.8 m respectively in the configuration shown in Fig. Prob. 5.21. Assume $C_d = 0.62$.

[0.917 m³/s]

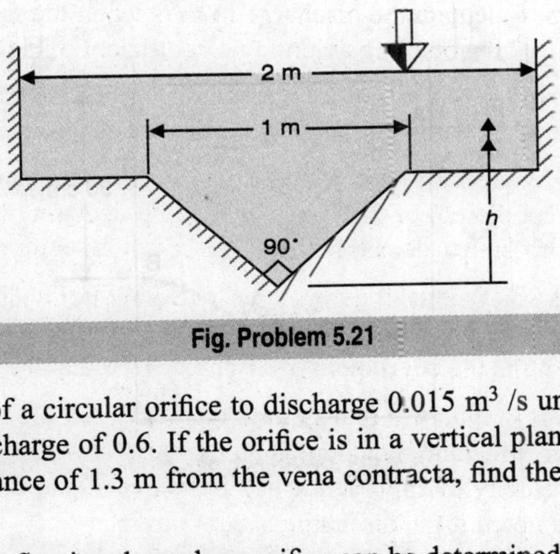

Fig. Problem 5.21

5.22. Find the diameter of a circular orifice to discharge 0.015 m³ /s under a head of 2.4 m using a coefficient of discharge of 0.6. If the orifice is in a vertical plane and the jet drops 0.25 m in a horizontal distance of 1.3 m from the vena contracta, find the value of the coefficient of contraction. **[6.82 cm, 0.715]**

5.23. The velocity of a jet flowing through an orifice can be determined from the position of three points A_0, A_1 and A_2, in the path of the jet. If the position of A_1 and A_2 are x_1 and x_2 to the right and y_1 and y_2 below A_0 respectively show that the velocity,

$$U = \sqrt{\frac{gx_1\ x_2\ (x_2 - x_1)}{2(x_1 y_2 - y_1\ x_2)}}$$

In an experiment on the determination of the coefficients of an orifice 30 mm in diameter, the rate of flow was found to be 0.078 m³/min for a head of 0.435 m. The distances measured in millimetres were : x_1 = 100, x_2 = 200, y_1 = 25 and y_2 = 62.5. Calculate the coefficients of velocity, contraction and discharge. **[0.96, 0.656, 0.63]**

5.24. Air flows through a 20 cm diameter pipe and the maximum expected rate of flow is 3 m³/min. If the available water manometer has a maximum reading of 200 mm, what would be the suitable diameter for an orifice plate to measure this flow ? Assume the density of air to be 1.2 kg/m³, C_v = 0.97, C_c = 0.65. **[4 cm]**

5.25. A fluid jet issues from a nozzle held at a certain inclination to the surface of the earth. Develop an equation for the jet trajectory and obtain the condition of maximum height acquired by the fluid if the inclination of the nozzle is

(*a*) fixed

(*b*) adjustable

Locate the corresponding distances where the fluid jet strikes a level surface.

5.26. A jet of water discharged through a nozzle of 5 cm diameter has a velocity of 22.5 m/s. Calculate the power in the jet. Calculate also the efficiency of the power transmission if the source of water is at 32 m head. **[11.2 kW, 80.6%]**

5.27. Estimate the flow rate of water over a spillway 200 m long, flowing under a head of 1 m. Assume the coefficient of discharge as 0.7. **[413 m³/s]**

5.28. Two tanks with orifices *A* and *B*, in the same vertical plane are placed on the same level and it is desired to adjust the distance between them so that the jets intersect in the plane of the bases. Calculate the distance *x* if the depth of water in each tank is 40 cm.
Is the point *P* equidistant from the two tanks ? What would happen if the orifices *A* and *B* are located differently ? **[69.28 cm; Yes, always equidistant)**

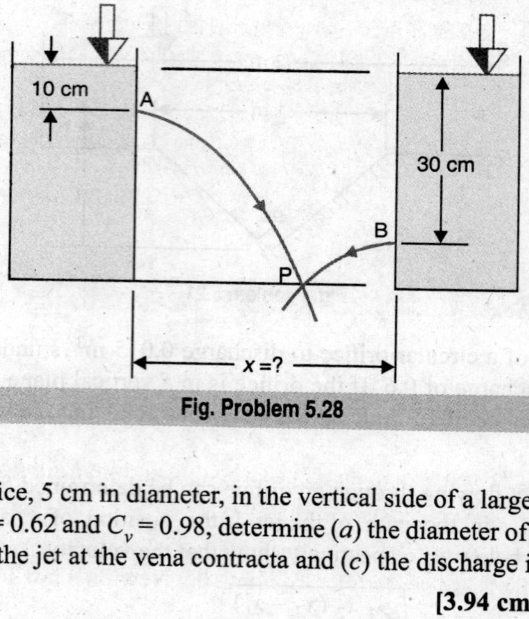

Fig. Problem 5.28

5.29. A sharp-edged orifice, 5 cm in diameter, in the vertical side of a large tank discharges a under head of 5 m. If $C_c = 0.62$ and $C_v = 0.98$, determine (*a*) the diameter of the jet at vena contracta, (*b*) the velocity of the jet at the vena contracta and (*c*) the discharge in m³/s. **[3.94 cm, 9.71 m/s, 0.0119 m³/s]**

5.30. A rectangular orifice, 1.5 m wide and 1.0 m deep is discharging water from a tank. If the water level in the tank is 3.0 m above the top edge of the orifice, find the discharge through the orifice. Take the coefficient of discharge for the orifice as 0.6. **[7.45 m³/s]**

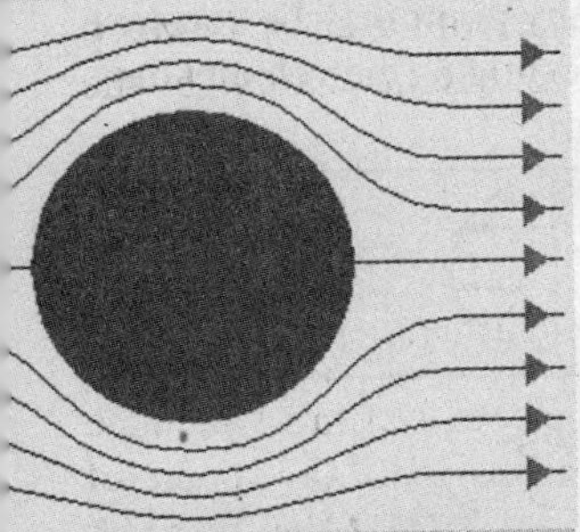

CHAPTER

6

Ideal Fluid Flow

LEARNING OBJECTIVES

On successful completion of all learning activities planned and provided in the chapter, a student will be able to

- state the importance of ideal fluid flow analysis;
- describe flownets for simple ideal flows, i.e., uniform flow, source, sink and free vortex;
- describe simple superpositions of basic ideal flows; and
- simulate the flow over a cliff and the flow over a circular cylinder by the superposition of basic ideal flows.

6.1. Introduction

The term ideal fluid flow refers to an incompressible and inviscid flow, *i.e.*, when

$$\rho = \text{constant and } \mu = 0$$

An ideal fluid flow is, therefore, an approximation of flow where density is assumed to be constant and viscosity effects are negligible. Invoking the Newton's law of viscosity,

$$\tau = \mu \frac{du}{dy} \quad \text{...(6.1)}$$

it follows that, in the absence of viscosity, there cannot be any shear stresses in an ideal flow. Also, there cannot be any shear forces between the ideal fluid and the confining boundaries whether the velocity of the fluid in the immediate vicinity of the boundary is zero or non-zero.

Absence of shear stresses throughout the ideal fluid flow means that

(*a*) *the only stress at a point in an ideal fluid flow must be the pressure and*

(*b*) *the only stress the fluid may act on the solid boundary must be normal to it at that point.*

An ideal fluid flow may be steady or unsteady, *i.e.*, the velocity at any point may be constant or variable with time. Also, an ideal fluid flow may be uniform or non-uniform, *i.e.*, the velocity may be constant or variable across the cross section of flow as shown in Fig. 6.1.

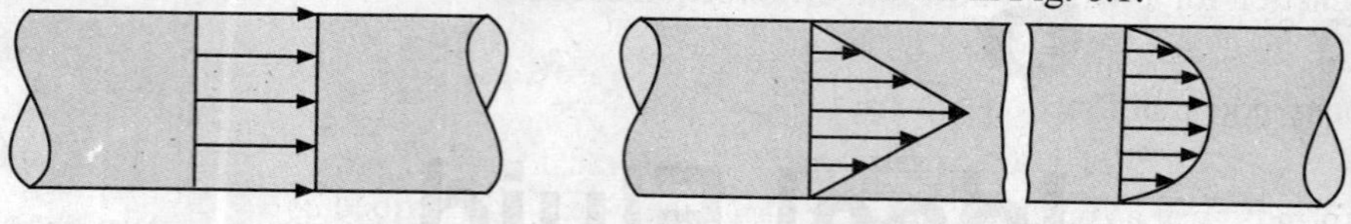

Fig. 6.1: Examples of Ideal Fluid Flow

6.2. Importance of Ideal Fluid Flow

Engineers deal with flows of various types in different situations, *e.g.*, flow through confined passages, fluid machines, control systems, etc.. In reality, the flow may be complex due to such phenomena as boundary layers, separation, recirculation etc. and an exact analysis of flow may be difficult and often not even required for engineering estimation where 'factor of safety' is applied in design calculations.

Even if one needs to study a complex fluid motion through computational technique, it is necessary to proceed by iteration, starting with a good first guess. The first approximation with ideal fluid flow analysis is often adequate. It may also be noted that the flow away from the solid boundaries, *i.e.*, where shear stresses are negligible, is more like ideal fluid flow.

Flow around some bodies have been studied through conformal transformation of simple ideal flows. For example, the flow and consequent lift for an asymmetric aerofoil may be studied by the conformal transformation of flow around a rotating circular cylinder which indeed is obtained by the superposition of uniform stream over a doublet and a free vortex. It is for these reasons that the simple technique of ideal fluid flow analysis assumes considerable importance in engineering.

In the following sections we are mainly concerned with two-dimensional flow. A two-dimensional flow, by definition, is such in which the velocity components and other dependent variables vary only with two space coordinates. A two-dimensional flow must, therefore, be identical in parallel planes; one such plane is chosen to represent the flow and the flow is unaltered in a direction normal to that plane.

6.3. The Uniform Flow

A uniform flow, also called a free-stream, is a flow with a constant velocity as shown in Fig. 6.2. Thus, a uniform flow at a constant velocity U along the x-axis is described by

$$u = U = \frac{\partial \Psi}{\partial y} = \frac{\partial \phi}{\partial x}$$

and

$$v = 0 = -\frac{\partial \Psi}{\partial x} = \frac{\partial \phi}{\partial y}$$

which provide that $\Psi = Uy$ and $\phi = Ux$

Streamlines are given by

$$\Psi = Uy = \text{constant} \quad ...(6.2)$$

which are straight lines parallel to the x-axis and velocity potential lines are given by

$$\phi = Ux = \text{constant} \quad ...(6.3)$$

which are straight lines parallel to the y-axis.

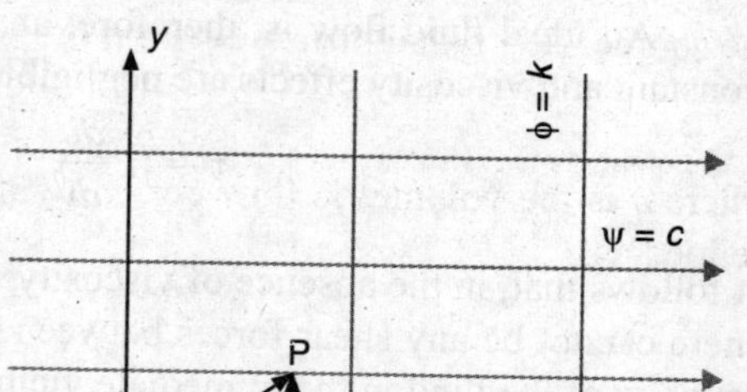

Fig. 6.2. Flownet for a Uniform Flow

It is sometimes necessary to represent a uniform flow in terms of cylindrical coordinates.

Invoking the relations,

$$x = r\cos\theta \text{ and } y = r\sin\theta$$

the stream function for a uniform flow in the x direction becomes

$$\Psi = Uy = Ur\sin\theta \qquad \text{...(6.4)}$$

and the velocity potential function becomes

$$\phi = Ux = Ur\cos\theta \qquad \text{...(6.5)}$$

A uniform flow at a constant velocity V along the y-axis is described by

$$u = 0 = \frac{\partial\Psi}{\partial y} = \frac{\partial\phi}{\partial x} \qquad \text{...(6.6)}$$

$$u = V = -\frac{\partial\Psi}{\partial x} = \frac{\partial\phi}{\partial y} \qquad \text{...(6.7)}$$

which provide that $\qquad \Psi = -Vx \text{ and } \phi = Vy \qquad$ ...(6.8)

Streamlines, Ψ = constant lines, are parallel to the y axis and velocity potential lines. ϕ = constant lines are lines parallel to the x-axis.

A uniform flow at a constant velocity Q inclined at an angle α to the x-axis may be considered to be made up of a uniform flow at a constant velocity U along the x-axis and a uniform flow at a constant velocity V along the y-axis as shown in Fig. 6.3. The stream function for such a flow should be given by

$$\Psi = Uy - Vx = Qr\sin(\theta - \alpha) \qquad \text{...(6.9)}$$

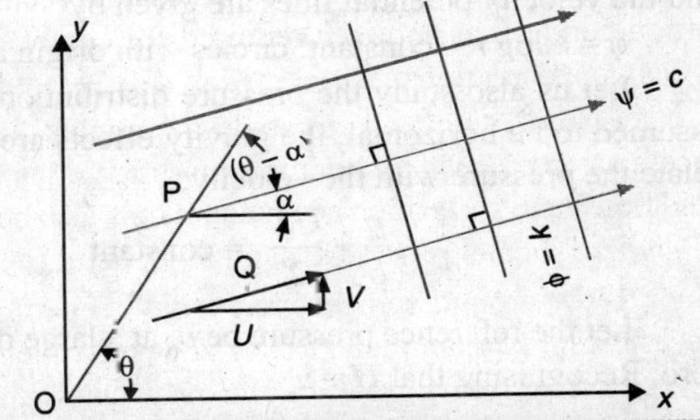

Fig 6.3: Flow Pattern for a Uniform Flow

and the velocity potential function should be given by

$$\phi = Ux + Vy = Qr\cos(\theta - \alpha) \qquad \text{... (6.10)}$$

6.4. The Source Flow

A plane source is a point in the plane from where fluid emerges at constant rate in all directions in that plane. A plane source is, in fact, a straight line source (normal to the plane) from which fluid is emitted in all directions at a constant rate per unit length of the line. A cross-section of a line source is the plane point source.

The point source is taken as the origin O and the fluid velocity is directed radially away from the source. At a radial distance r from the source, the radial velocity, u, is given by

$$u_r = \frac{q}{2\pi r} = \frac{k}{r} \qquad \text{...(6.11)}$$

where q is the volumetric flow per unit length emitted by the source and k, the source strength, is defined as

$$k = \frac{q}{2\pi} \qquad \text{...(6.12)}$$

In other words, the radial velocity at a unit radius from the point source is a measure of the source strength.

Recalling the definition of Ψ,

$$\frac{1}{r}\frac{\partial\Psi}{\partial\theta} = u_r = \frac{k}{r} = \frac{q}{2\pi r} \qquad \text{...(6.13)}$$

and $$-\frac{\partial\Psi}{\partial r} = u_\theta = 0 \qquad \text{...(6.14)}$$

whence $\Psi = k\theta = \frac{q}{2\pi}\theta$

Similarly, from the definition of ϕ,

$$\frac{\partial\phi}{\partial r} = u_r = \frac{k}{r} = \frac{q}{2\pi r} \quad ...(6.15)$$

and $$\frac{1}{r}\frac{\partial\phi}{\partial\theta} = u_\theta = 0 \quad ...(6.16)$$

whence $$\phi = k \log r = \frac{q}{2\pi} \log r \quad ...(6.17)$$

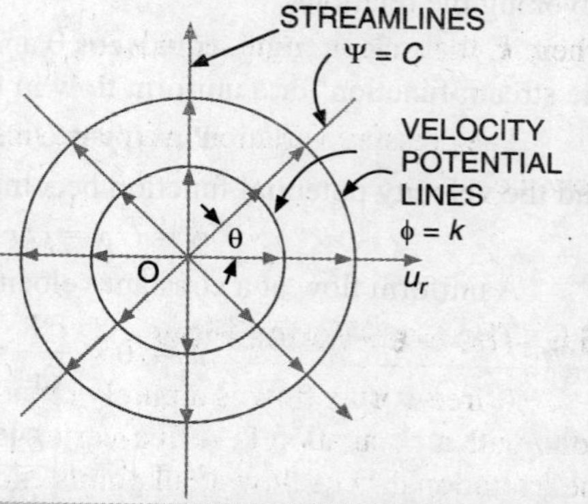

Fig. 6.4: Flow for a Plane Source

Streamlines are given by

$\Psi = k\,\theta$ = constant; radial lines

and the velocity potential lines are given by

$\phi = k \log r$ = constant; circles with origin at the source as shown in Fig. 6.4

Let us also study the pressure distribution in a plane source flow. If the plane of the flow is assumed to be horizontal, the gravity effects are absent and the Bernoulli equation may be used to relate the pressure with the velocity;

$$\frac{p}{\rho} + \frac{U^2}{2} = \text{constant} \quad ...(6.18)$$

Let the reference pressure be p_0 at a large distance away from the source where the velocity is zero. Recognising that $U = u_r$

$$\frac{p}{\rho} + \frac{u_r^2}{2} = \frac{p_0}{\rho}$$

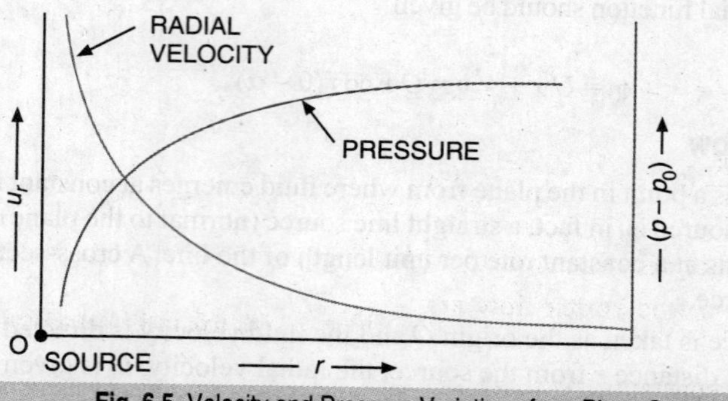

Fig. 6.5. Velocity and Pressure Variations for a Plane Source

$$(p - p_0) = -\frac{\rho\, u_r^2}{2} = -\frac{\rho\, k^2}{2r^2} \quad ..(6.19)$$

It is thus seen that, in a plane source flow, the velocity decreases inversely as the radial distance from the source whereas the pressure increases inversely as the square of the radial distance from the source as shown in Fig. 6.5.

6.5. The Sink Flow

A sink is a source with a negative strength. In a plane point-sink, fluid moves radially inwards in a plane towards a point where it disappears at a constant rate. For a plane sink flow,

$$u_r = -\frac{q}{2\pi r} = \frac{k}{r} \quad ...(6.20)$$

where k, the sink strength, equals $-\dfrac{q}{2\pi}$.

The pressure variation in a plane sink flow is given by the same expression as in a plane source flow, *i.e.*,

$$(p - p_0) = -\frac{\rho k^2}{2r^2} \quad ...(6.21)$$

6.6. The Free-Vortex Flow

A free-vortex flow is a purely circulatory flow such that the centre of the vortex is a singular point with a circulation Γ. A free-vortex is also called irrotational or potential in order to emphacise the irrotationality of flow at all points except at the singular point. At any radius r from the centre, the tangential velocity u_θ is given by

$$\Gamma = 2\pi\, r\, u_\theta$$

or

$$u_\theta = \frac{\Gamma}{2\pi r} = \frac{C}{r} \quad ...(6.22)$$

where C, the vortex strength, is defined by

$$C = \frac{\Gamma}{2\pi} \quad ...(6.23)$$

Observing that

$$u_\theta = -\frac{\partial \Psi}{\partial r} = \frac{1}{r}\frac{\partial \phi}{\partial \theta}$$

and

$$u_r = \frac{1}{r}\frac{\partial \Psi}{\partial \theta} = \frac{\partial \phi}{\partial r} = 0$$

we deduce that

$$\Psi = -C \log r = \frac{-\Gamma}{2\pi} \log r \quad ...(6.24)$$

$$\phi = C\theta = \frac{\Gamma}{2\pi}\theta \quad ...(6.25)$$

for a positive or anticlockwise free vortex.

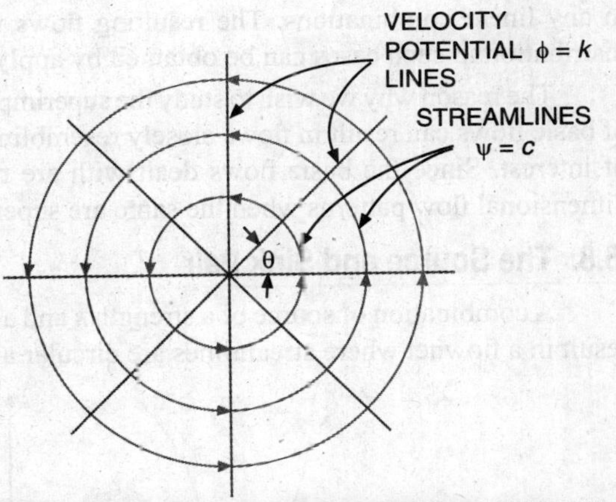

Fig. 6.6: Free Vortex Flow Pattern

Streamlines for a free-vortex flow are given by

$$\Psi = -C \log r = \text{constant} \quad ...(6.26)$$

which shows that streamlines are lines of constant r, *i.e.*, concentric circles as shown in Fig. 6.6.

Velocity potential lines are given by

$$\phi = C\theta \quad ...(6.27)$$

which are radial lines drawn from the centre.

The pressure variation in a plane free-vortex flow is governed by the Bernoulli equation,

$$\frac{p}{\rho} + \frac{U^2}{2} = \text{constant}$$

Assuming the reference pressure p_o at a large distance from the centre where the velocity is zero and recognising that $U = u_\theta$,

$$\frac{p}{\rho} + \frac{u_\theta^2}{2} = \frac{p_0}{\rho}$$

whence
$$(p - p_0) = -\frac{\rho u_\theta^2}{2} = -\frac{\rho C^2}{2r^2} \qquad ...(6.28)$$

which shows that the pressure increases inversely as the square of the radial distance from the centre of the free-vortex as shown in Fig. 6.7.

It is easy to observe that, for a negative or anticlockwise free vortex, circulation Γ should be negative. The streamlines will become concentric circles in clockwise direction; the tangential velocity and pressure variations will remain the same.

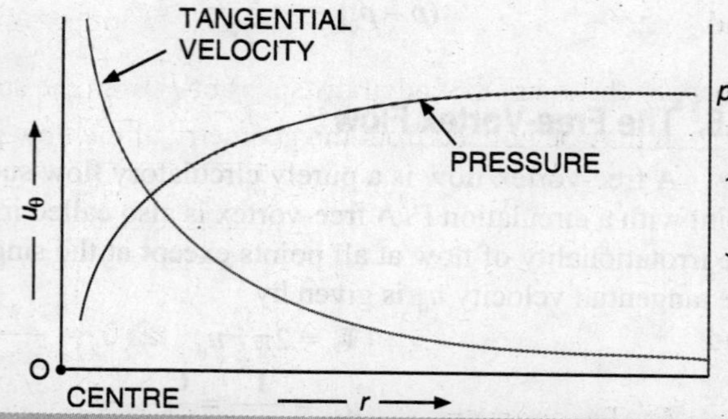

Fig. 6.7: Velocity and Pressure Variation in a Free-vortex Flow

6.7. Superimposed Flow Patterns

Basic ideal flow patterns due to a free stream, a source and a free-vortex may be suprimposed in any linear combinations. The resulting flows will still be irrotational and ideal. The pressure distribution in such cases can be obtained by applying Bernoulli's equation.

The reason why we wish to study the superimposition of ideal flows is that a suitable combination of basic flows can result in flows closely resembling flows around bodies and flows within passages of interest. Since the basic flows dealt with are two-dimensional, we expect to obtain only two-dimensional flow patterns when the same are superimposed.

6.8. The Source and Sink Pair

A combination of source of a strength k and a sink of strength $(-k)$ placed at a distance $2s$ apart result in a flownet where streamlines are circular arcs, as shown in Fig. 6.8.

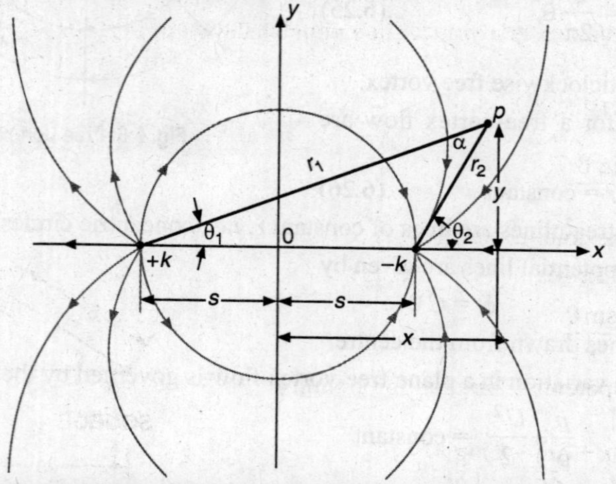

Fig. 6.8: Source and Sink Pair

If the x-axis is chosen along the join of the source and the sink, the stream function and the velocity potential function at a point P are given as follows:

$$\Psi = k\,\theta_1 - k\,\theta_2 = k\,(\theta_1 - \theta_2) = -\,k\,\alpha \qquad \text{...(6.29)}$$

where $\alpha = (\theta_2 - \theta_1)$

the angle subtended at P by the join of the source and the sink

and $$\theta = k \log r_1 - k \log r_2 = k \log \frac{r_1}{r_2} \qquad \text{...(6.30)}$$

where r_1 and r_2 are the radial distances of P from the source and the sink respectively.

It may be noticed from the geometry of the flow pattern that

$$r_1 = \sqrt{y^2 + (x+s)^2}$$

$$r_2 = \sqrt{y^2 + (x-s)^2}$$

and $$\tan\theta_1 = \frac{y}{x+s}; \quad \tan\theta_2 = \frac{y}{x-s}$$

By Trigonometry,

$$\tan(\theta_2 - \theta_1) = \frac{\tan\theta_2 - \tan\theta_1}{1 + \tan\theta_1 \tan\theta_2}$$

It, therefore, follows that

$$\tan\alpha = \frac{2ys}{x^2 + y^2 - s^2} \qquad \text{...(6.31)}$$

and $$\alpha = \tan^{-1}\left(\frac{2ys}{x^2 + y^2 - s^2}\right) \qquad \text{...(6.32)}$$

6.9. The Doublet

A doublet is a special case of a source and sink pair when the two approach each other under the limiting case of

(*i*) the distance δs between them tends to zero and

(*ii*) the product $\delta s.q$, also called the doublet strength μ, remains constant.

The stream function at a point P in a doublet shown in Fig. 6.9 is given by θ and dimension lines (See).

$$\Psi = k\,\theta - k\,(\theta + \delta\theta)$$

$$= -\,k\delta\theta$$

$$= -\frac{\delta s \sin\theta}{r}$$

In terms of the doublet strength,

$$\mu = \delta s.q = 2\pi\, k\delta s \qquad \text{...(6.33)}$$

$$\Psi = -\frac{\mu}{2\pi}\frac{\sin\theta}{r} \qquad \text{...(6.34)}$$

The velocity potential function for a point P is given by

$$\phi = k \log(r + \delta r) - k \log r$$

$$= k \log\left(1 + \frac{\delta r}{r}\right) - k\,\frac{\delta r}{r} + \text{smaller terms}$$

$$= \frac{q}{2\pi}\frac{\delta s \cos\theta}{r}$$

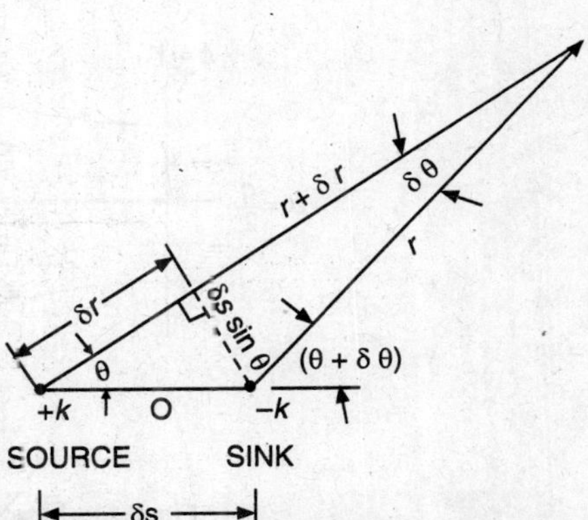

Fig. 6.9: Analysis of a Doublet

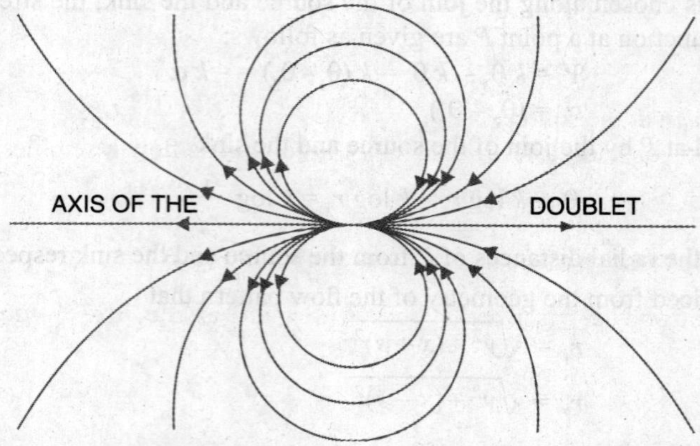

Fig. 6.10: Streamlines for a Doublet

Streamline pattern for a doublet, as above, is shown in Fig. 6.10.

In terms of the doublet strength μ

$$\phi = \frac{\mu}{2\pi}\frac{\cos\theta}{r} \quad ...(6.35)$$

It may be remarked that the signs of expression for Ψ and φ are changed if the source and sink are interchanged.

6.10. A Plane Source in a Uniform Flow

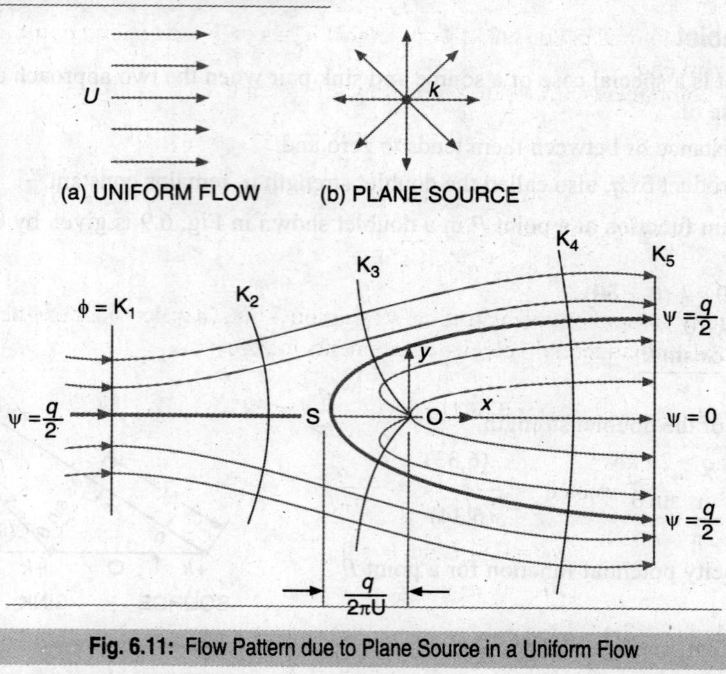

Fig. 6.11: Flow Pattern due to Plane Source in a Uniform Flow

The flow pattern resulting from the combination of a uniform flow and a plane source would be such that the stream function is given by

$$\Psi = Uy + k\,\theta = U\,r\sin\theta + k\,\theta$$

where $$k = \frac{q}{k\pi}$$

and the origin is taken at the source.

A plot of (Ψ = constant) lines reveals that the composite flow resembles (*i*) flow over a plane half-body, outside $\Psi = \frac{q}{2}$ and (*ii*) source flow within a plane half-body as shown in Fig. 6.11.

The plane half-body is described by the dividing streamline, $\Psi = \frac{q}{2}$ which is

$$Uy + \frac{q\,\theta}{2\pi} = \frac{q}{2}$$

or $$y = r\sin\theta = \frac{q\left(1-\frac{\theta}{\pi}\right)}{2U} \quad \text{...(6.36)}$$

At $\theta = 0$, $$y_{max} = \frac{q}{2U}, \text{ the maximum ordinate} \quad \text{... (6.37}a\text{)}$$

and at $\theta = \frac{\pi}{2}$, $$y = \frac{q}{4U}, \text{ the ordinate above the origin.} \quad \text{...(6.37}b\text{)}$$

At $\theta = \pi$, $y = 0$, the leading point of the half-body. ...(6.37 *c*)

and at $\theta = \frac{3\pi}{2}$, $y = \frac{-q}{4U}$, the ordinate below the origin. ...(6.37 *d*)

The reason why it is called a half-body is that it has only the leading point; it trails to infinity at the downstream end.

Velocity components anywhere in the flow are given by

$$u_r = \frac{1}{r}\frac{\partial\Psi}{\partial\theta} = U\cos\theta + \frac{q}{2\pi r} \quad \text{...(6.38a)}$$

$$u_\theta = -\,U\sin\theta \quad \text{...(6.38}b\text{)}$$

and the resultant velocity is obtained as

$$u_{res} = \sqrt{\left(u_r^2 + u_\theta^2\right)} \quad \text{...(6.39)}$$

The velocity components vanish at the stagnation point. In order to locate the stagnation point, equate the radial and tangential velocity components to zero,

$$u_r = 0 \text{ or } U\cos\theta + \frac{q}{2\pi r} = 0$$

whence $$r\cos\theta = \frac{-q}{2\pi U};\; x = -\frac{q}{2\pi U} \quad \text{...(6.40)}$$

Also, $$u_\theta = 0 \text{ or } U\sin\theta = 0$$

whence $$\theta = 0, \pi;\; y = 0$$

The stagnation point is, therefore, at $\left(-\frac{q}{2\pi U}, 0\right)$ which is the leading points of the half-body.

It may also be verified that the position of the maximum velocity on the half-body is given by

$$\theta = +\,67° \text{ and } \theta = -\,67° \quad \text{...(6.41)}$$

obtained by substituting r from Eq. (6.36) in Eq. (6.39) and setting $\partial u_{res} / \partial\theta$ to zero.

Example 6.1. *Obtain the equation of the dividing streamline for the flow resulting from a superposition of a free-stream at 20 m/s on a two-dimensional source with a strength of 10 m²/s. Sketch the flow pattern.*

Solution.

For a free-stream and source combination,

$$\Psi = Ur \sin \theta + k\, \theta$$

where $U = 20$ m/s

and $k = \dfrac{q}{2\pi} = 10$ m²/s

whence $q = 2\pi \times 10 = 62.83$ m²/s

Hence $\Psi = 20\, r \sin \theta + 10\, \theta$

represents the flow pattern.

The dividing streamline is defined by

$$\Psi = q/2 = 62.83/2 = 31.42 \text{ m}^2/\text{s}$$

which is represented by

$$20\, r \sin \theta + 10\, \theta = 31.42$$

or $\mathbf{20\, y + 10\, \theta = 31.42}$

Tabulating the values,

θ	y in metres	Comments
0	1.571	the maximum +ve ordinate.
π/2	0.786	the + ve ordinate above the origin.
π	0.0	the leading point.
3π/2	– 0.786	the –ve ordinate below the origin.
2π	– 1.571	the maximum –ve ordinate.

The stagnation point can be located by equating the velocity components to zero:

$$u_r = \frac{1}{r}\frac{\partial \Psi}{\partial \theta} = U \cos \theta + k/r = 0$$

$$u_\theta = -\frac{\partial \Psi}{\partial r} = -\, U \sin \theta = 0$$

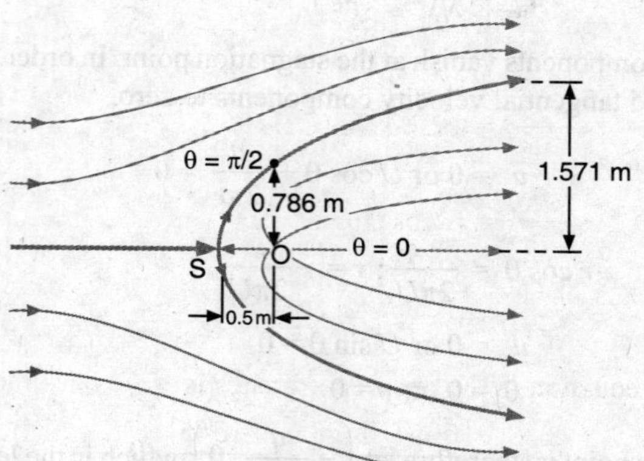

Solving the equations,

$$\theta = 0 \text{ or } \pi; \text{ only } \pi \text{ being a practical solution}$$

$$r = x = -\, k/U = -\, q(2\pi U) = -\, 0.5 \text{ m}$$

The stagnation point is, therefore, at S where $r = -\, 0.5$ m, $\theta = \pi$

Example 6.2. *Flow over a plane half-body is studied by utilising a free-stream at 5 m/s superimposed on a source at the origin. The body has a maximum width 2 m. Calculate*

(a) the coordinates of the stagnation point, (b) the width of the body at the origin and (c) the velocity at a point (0.5, π/2).

Solution.

The maximum ordinate of the body is given by

$$y_{max} = \frac{q}{2U} = \frac{2}{2} = 1\text{m}$$

whence $q = 2 \times 5 = 10 \text{ m}^2/\text{s}$

(*a*) For the stagnation point,

$$x = \frac{-q}{2\pi U} = -\frac{10}{2\pi \times 5} = \mathbf{-0.32\ m;\ y = 0}$$

(*b*) At the origin $\theta = \frac{\pi}{2};\ y = \frac{q}{4U} = \frac{10}{4 \times 5} = 0.5 \text{ m}$

The width of the body is 0.5 × 2 = **1 m**

(*c*) The velocity components are given by

$$u_r = U\cos\theta + \frac{q}{2\pi r} = 5 \times \cos\frac{\pi}{2} + \frac{10}{2 \times \pi \times 0.5} = 3.18 \text{ m/s}$$

$$u_\theta = -U\sin\theta = -5 \times \sin\frac{\pi}{2} = -5 \text{ m/s}$$

$$U_{res} = \sqrt{3.18^2 + (-5)^2} = \mathbf{5.93\ m/s}$$

Example 6.3. *Is the flow expressed by the velocity potential,*

$$\phi = Ux + K\log r$$

possible for all U and K ? Sketch the flow if U = 10 m/s and K = 5 m²/s.

Solution.

Starting with the given expression for the velocity potential function,

$$\phi = Ur\cos\theta + K\log r$$

we get, $$\frac{\partial\phi}{\partial\theta} = -Ur\sin\theta$$

and $$\frac{\partial^2\phi}{\partial\theta^2} = -Ur\cos\theta$$

Similarly, $$\frac{\partial\phi}{\partial r} = U\cos\theta + \frac{K}{r}$$

and $$\frac{\partial^2\phi}{\partial r^2} = -\frac{K}{r^2}$$

The Laplace's equation in the polar coordinates is

$$\nabla^2\phi = \frac{\partial^2\phi}{\partial r^2} + \frac{1}{r}\frac{\partial\phi}{\partial r} + \frac{1}{r^2}\frac{\partial^2\phi}{\partial\theta^2} = 0$$

Substituting the values of the derivatives

$$\nabla^2\phi = \frac{-K}{r^2} + \frac{U\cos\theta}{r} + \frac{K}{r^2} - \frac{U\cos\theta}{r} \equiv 0$$

which shows that the flow is possible and irrotational for all U and K. This is, of course, understandable because a source of arbitrary strength K can be placed in any free stream of uniform flow U.

For the given case, $U = 10$ m/s, $K = 5$ m²/s per metre length

The equation of a general streamline in the flow is given by

$$\Psi = 10\,y + 5\,\theta$$

which can be plotted by assigning different values to Ψ.

In particular, when $\Psi = \dfrac{q}{2} = \pi K$

the dividing streamline is given by the following :

$$10\,y + 5\theta = \pi \times 5 = 15.7$$

or

$$y = \frac{15.7 - 5\theta}{10} = 1.57 - 0.5\,\theta$$

The salient points are given by the following :

$$y = y_{max} = 1.57\text{m} \quad \text{at } \theta = 0$$

$$y = y_0 = 0.785\text{m} \quad \text{at } \theta = \pi/2$$

$$y = 0 \quad \text{at } \theta = \pi$$

The stagnation point is located by

$$x = \frac{-q}{2\pi U} = -\frac{K}{U} = -\frac{5}{10} = \mathbf{-\,0.5\ m}$$

and

$$\mathbf{y = 0}$$

Example 6.4. *Does the velocity potential*

$$\phi = \frac{-A\cos\theta}{2\pi r}$$

represent a possible irrotational flow ? If so, determine the stream function for the flow.

Solution.

For a plane irrotational flow,

$$u_r = \frac{\partial\phi}{\partial r} \text{ and } u_\theta = \frac{1}{r}\frac{\partial\phi}{\partial\theta}$$

For the given ϕ,

$$u_r = \frac{A\cos\theta}{2\pi r^2} \text{ and } u_\theta = \frac{A\sin\theta}{2\pi r^2}$$

The continuity equation for the 2-d incompressible flow is

$$\frac{\partial(ru_r)}{\partial r} + \frac{\partial u_\theta}{\partial\theta} = 0$$

Substituting the above velocity components,

$$\frac{\partial}{\partial r}\left(\frac{A\cos\theta}{2\pi r}\right) + \frac{\partial}{\partial\theta}\left(\frac{A\sin\theta}{2\pi r^2}\right)$$

$$= -\frac{A\cos\theta}{2\pi r^2} + \frac{A\cos\theta}{2\pi r^2} \equiv 0$$

Hence, the given potential represents a possible 2-d irrotational flow.

One could have checked the same by examining whether the given ϕ satisfied the Laplace's equation, *i.e.*,

$$\frac{1}{r}\frac{\partial\phi}{\partial r}+\frac{\partial^2\phi}{\partial r^2}+\frac{1}{r^2}\frac{\partial^2\phi}{\partial\theta^2}=0$$

In this case, $$\frac{\partial\phi}{\partial r}=\frac{A\cos\theta}{2\pi r^2};\quad \frac{\partial^2\phi}{\partial r^2}=\frac{-A\cos\theta}{\pi r^3}$$

$$\frac{\partial\phi}{\partial\theta}=\frac{A\sin\theta}{2\pi r}\text{ and }\frac{\partial^2\phi}{\partial\theta^2}=\frac{-A\cos\theta}{2\pi r}$$

Substituting these values in the Laplace's equation,

$$\frac{A\cos\theta}{2\pi r^3}-\frac{A\cos\theta}{\pi r^3}+\frac{A\cos\theta}{2\pi r^3}=0$$

Since ϕ satisfies the Laplace's equation identically, it represents a possible flow.

6.11 A Source and Sink Pair in a Uniform Flow

The combination of flows due to a source and sink pair and a uniform stream results in a flow pattern in which the dividing streamline forms the profile of a fully streamlined body in an ideal flow as shown in Fig. 6.12. The flow outside it corresponds to the flow over the streamlined body which is of considerable importance in engineering.

A source and sink pair in a uniform flow is described by stream function

$$\Psi=Uy-\frac{q}{2\pi}\,.\,\tan^{-1}\frac{2ys}{x^2+y^2-s^2}$$

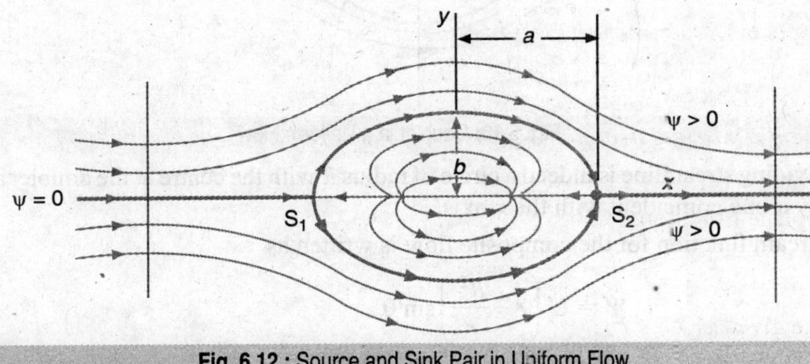

Fig. 6.12 : Source and Sink Pair in Uniform Flow

The streamline represented by $\Psi = 0$ is the dividing streamline representing a Rankine oval.

$$y=\frac{q}{2\pi U}\,.\,\tan^{-1}\frac{2ys}{x^2+y^2-s^2} \qquad ...(6.42)$$

with its semi-major and semi-minor axes a and b located by $y = 0$ and $x = 0$ respectively.

It is interesting to note the range of shapes a 'streamlined body for an ideal flow may have. The general shape is an oval. It may become a straight line along the flow when the source and sink are each of zero strength. It may become a circle when the source and sink are brought closer in the form of a doublet.

6.12 A Doublet in a Uniform Flow

A doublet placed in a uniform flow constitutes a very important flow pattern past a circular cylinder. The limiting case of a doublet from a source and sink pair placed in a uniform flow results in a Rankine oval of equal axes, *i.e.*, a circle.

Adding the stream functions for a uniform flow and a doublet,

$$\Psi = Uy - \frac{\mu}{2\pi r} \sin\theta$$

$$= \left(Ur - \frac{2}{2\pi r}\right) \sin\theta$$

For the dividing streamline,

$\Psi = 0$ line,

$$r^2 = \frac{\mu}{2\pi U} = \text{constant} = R^2 \qquad ...(6.43)$$

which is a circle of radius $R = \sqrt{\frac{\mu}{2\pi U}}$

and $\sin\theta = 0$ or $\theta = 0$ which corresponds to the x-axis as shown in Fig. 6.13.

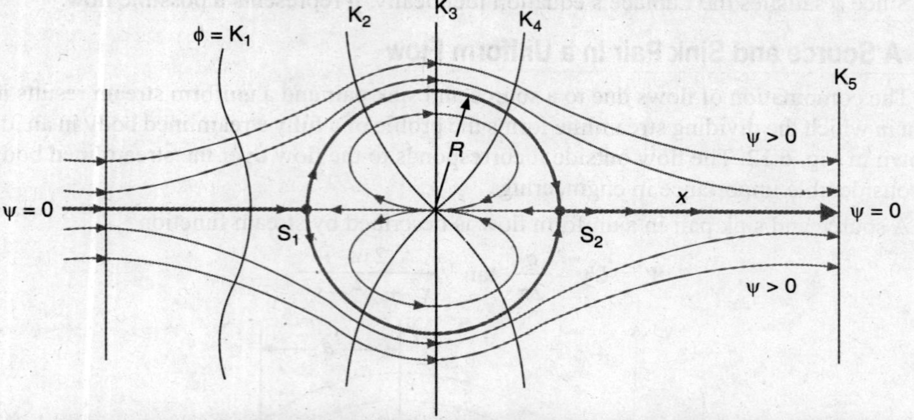

Fig. 6.13: Doublet in a Uniform Flow

The dividing streamline is indeed a circle of radius R with the centre at the doublet and extended both ways by a line coincident with the x-axis.

The stream function for the composite flow is written as

$$\Psi = U\left(r - \frac{R^2}{r}\right) \sin\theta \qquad ...(6.44)$$

Similarly, the velocity potential function is written as

$$\phi = Ux + \frac{\mu}{2\pi r} \cos\theta$$

$$= U\left(r + \frac{\mu}{2\pi r U}\right) \cos\theta$$

$$= U\left(r + \frac{R^2}{r}\right) \cos\theta \qquad ...(6.45)$$

The velocity components anywhere in the flow field are given by

$$u_r = \frac{1}{r}\frac{\partial \Psi}{\partial \theta} = U\left(1 - \frac{R^2}{r^2}\right) \cos\theta \qquad ...(6.46)$$

$$u_\theta = -\frac{\partial \Psi}{\partial r} = -U\left(1 + \frac{R^2}{r^2}\right) \sin\theta \qquad ...(6.47)$$

On the surface of the cylinder, $r = R$

$$u_r = 0$$

as expected because the flow will be wholly tangential.

$$U_\theta = -2U \sin\theta \qquad \text{...(6.48)}$$

Stagnation points in the flow correspond to $u_r = 0$ and $u_\theta = 0$

whence $\qquad r = R$ and $\theta = 0, \pi$

as the two stagnation points S_1 and S_2.

Pressure distribution for flow over a circular cylinder may be determined by employing the Bernoulli equation.

$$\frac{p}{\rho} + \frac{u^2}{2} = \text{constant}$$

The reference point is taken at a location far upstream in the uniform flow where the velocity is U. where the pressure is assumed as p_s.

The velocity, u, at any point on the surface of the cylinder is given by

$$u_r = 0; \quad u_\theta = -2U \sin\theta$$

The pressure p at that point must be given by

$$\frac{p}{\rho} + \frac{(-2U\sin\theta)^2}{2} = \frac{p_s}{\rho} + \frac{U^2}{2}$$

whence $$\frac{p - p_s}{\rho} = (1 - 4\sin^2\theta)\frac{U^2}{2}$$

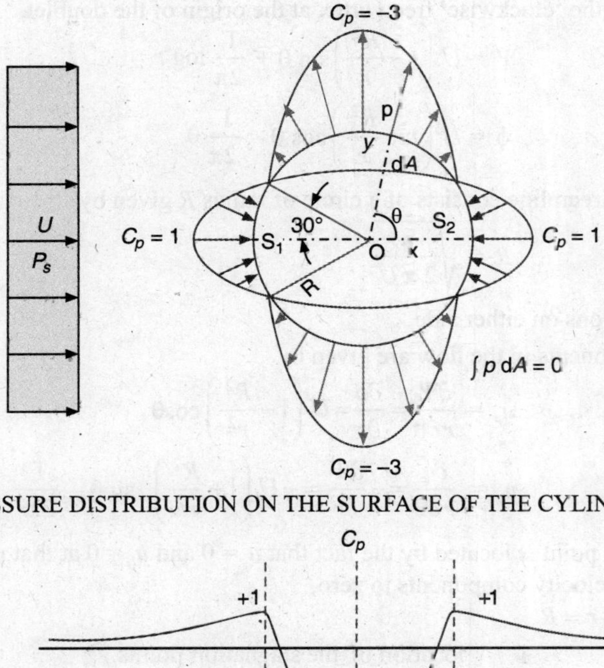

PRESSURE DISTRIBUTION ON THE SURFACE OF THE CYLINDER

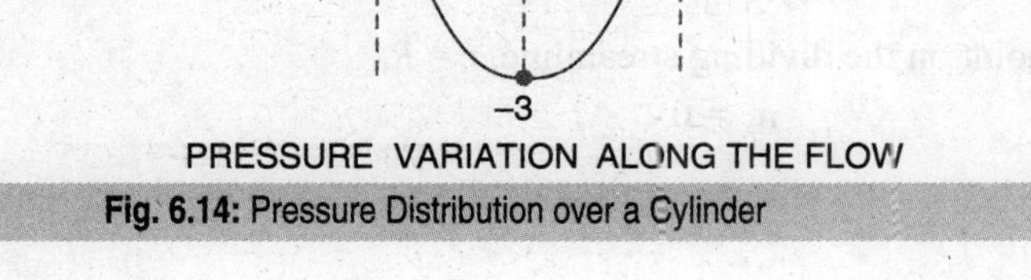

PRESSURE VARIATION ALONG THE FLOW

Fig. 6.14: Pressure Distribution over a Cylinder

and the dimensionless pressure coefficient is given by

$$C_p = \frac{p - p_s}{1/2\rho U^2} = 1 - 4\sin^2\theta$$

Values of the pressure coefficient at some circumferential positions is tabulated as follows and shown plotted in Fig. 6.14.

TABLE 6.1 : Pressure Distribution Over a Circular Cylinder

Position	Pressure Coefficients	Comment
0	1	Stagnation Point
30°	0	Zero Pressure
90°	–3	Least Pressure
150°	0	Zero Pressure
180°	1	Stagnation Point

6.13. A Doublet and a Free Vortex in a Uniform Flow

A doublet and a free-vortex at a common origin, placed in a uniform flow, constitutes by far the most important flow pattern. Its importance is due to the fact that the resulting flow pattern is asymmetrical about the x-axis. The extent of asymmetry is related to the transverse force acting on the dividing streamline.

Adding the stream functions and the velocity potential functions respectively for the 'doublet in a free stream' and the 'clockwise' free vortex at the origin of the doublet,

$$\Psi = U\left(r - \frac{R^2}{r}\right)\sin\theta + \frac{\Gamma}{2\pi}\log r \quad \text{... (6.49)}$$

$$\phi = U\left(r + \frac{R^2}{r}\right)\cos\theta - \frac{\Gamma}{2\pi}\theta \quad \text{... (6.50)}$$

The dividing streamline consists of a circle of radius R given by

$$R = \sqrt{\frac{\mu}{2\pi U}}$$

together with extensions on either side.

Velocity components in the flow are given by

$$u_r = \frac{\partial\Psi}{r\,\partial\theta} = \frac{\partial\phi}{\partial r} = U\left(1 - \frac{R^2}{r^2}\right)\cos\theta \quad \text{... (6.51)}$$

and

$$u_\theta = \frac{\partial\phi}{r\,\partial\theta} = -\frac{\partial\Psi}{\partial r} = -U\left(1 + \frac{R^2}{r^2}\right)\sin\theta - \frac{\Gamma}{2\pi r} \quad \text{... (6.52)}$$

The stagnation point is located by the fact that $u_r = 0$ and $u_\theta = 0$ at that point.

Equating the velocity components to zero,

$$\left.\begin{array}{ll} u_r = 0, & r = R \\ u_\theta = 0, & \sin\theta = -\dfrac{\Gamma}{4\pi RU} \end{array}\right] \text{location of the stagnation points.}$$

At any point on the dividing streamline, $r = R$,

$$u_r = 0$$

$$u_\theta = -\left(2U\sin\theta + \frac{\Gamma}{2\pi R}\right)$$

Since the dividing streamline is a circle, it is indeed expected that the radial velocity should be zero because no flow can cross a streamline. The tangential component of velocity is a measure of the flow along the circular streamline.

It is interesting to see that the location of the stagnation points given by

$$r = R \text{ and } \sin\theta = -\frac{\Gamma}{4\pi RU}$$

varies with the ratio of Γ to $4\pi RU$. The ratio

$$\left|\frac{\Gamma}{4\pi RU}\right|$$

may be anything from 0 to a value of 1 or even more than 1.

Let us examine four possible cases as shown in Fig. 6.15:

Case (a). $\Gamma = 0$, a case when the free-vortex is absent. In this case, $\sin\theta = 0$ which corresponds to the stagnation points being located at $\theta = 0$ and $\theta = \pi$.

Case (b). Γ less than $4\pi RU$, a case when the strength of the free-vortex is small. In this case, $\sin\theta$ lies between 0 and –1 and the stagnation points are located at S_1 and S_2. The former corresponds to θ between 0 and $-\pi/2$ and the latter corresponds to θ between π and $3\pi/2$.

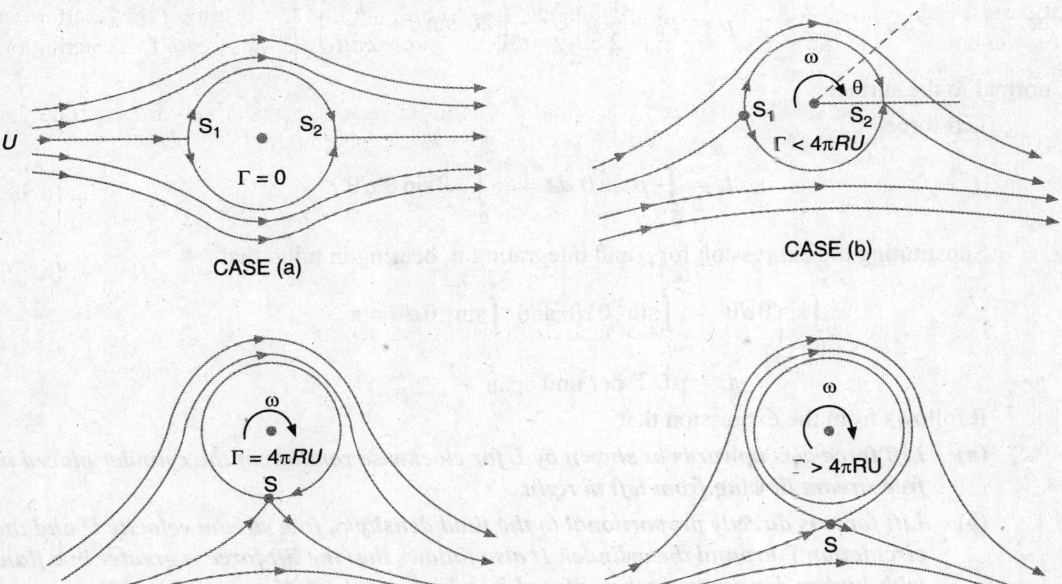

Fig. 6.15: A Doublet and a Free Vortex in a Free Stream.

Case (c). $\Gamma = 4\pi RU$, a case when the circulation due to free-vortex just equals the critical product $4\pi RU$. In this case, $\sin\theta = -1$ which corresponds to a single stagnation point at $\theta = -\pi/2$ or $3\pi/2$, *i.e.*, at the bottom of the dividing streamline.

Case (d). Γ more than $4\pi RU$, a case when the strength of the free-vortex is large. In this case, the numerical value of $\sin\theta$ exceeds 1 which makes θ imaginary ! The stagnation point is, consequently, not on the circular streamline but it is located away from it, as shown.

Combined flow pattern for a doublet and a free vortex placed in a free stream closely resembles the flow around a circular cylinder when the infinitely long cylinder is rotated at different speeds in a uniform flow.

Case (*a*) corresponds to 'no rotation' of the cylinder, Case (*b*) to a small rotational speed. Case (*c*) to a critical rotational speed and Case (*d*) to a 'high rotational' speed of the cylinder.

Let us estimate the 'lift' force over the rotating cylinder subjected to a free stream. The velocity at any point on the surface of a cylinder placed in a free stream with circulation Γ associated with its rotation is given by

$$u_\theta = -\left(2U\sin\theta + \frac{\Gamma}{2\pi R}\right)$$

$$u_r = 0$$

representing a purely circulatory motion.

Applying the Bernoulli equation,

$$p_0 + \frac{1}{2}\rho U^2 = p + \frac{1}{2}\rho u_\theta^2$$

or
$$p = p_0 + \frac{1}{2}\rho\left[U^2 - \left(2U\sin\theta + \frac{\Gamma}{2\pi R}\right)^2\right]$$

normal to the surface

Lift force

$$L = \int_S -p\sin\theta\, dA = -\int_0^{2\pi} pR\sin\theta\, d\theta$$

Substituting the expression for p and integrating it, bearing in mind that

$$\int_0^{2\pi}\sin\theta\,d\theta = \int_0^{2\pi}\sin^3\theta\,d\theta \text{ and } \int_0^{2\pi}\sin^2\theta\,d\theta = \pi$$

$$L = \rho U\Gamma \text{ per unit span}$$

It follows from the expression that

(*a*) ***Lift force acts upwards as shown by L for clockwise rotation of the cylinder placed in free stream flowing from left to right.***

(*b*) ***Lift force is directly proportional to the fluid density ρ, free stream velocity U and the circulation Γ around the cylinder. It also follows that the lift force is greater in a fluid with higher density and when placed in a higher velocity free stream and of course, when rotated at higher speeds!***

The ultimate significance of the expression $L = \rho U\Gamma$ is due to the fact that this is valid for flow over an object of any cross-section as long as Γ is produced around it. In general, Γ is not zero for most of the unsymmetrical bodies such as an aerofoil placed at an angle of attack by a free stream. The lift force is given by the same expression irrespective of the origin of circulation and the shape of the body.

Example 6.5. *A circular cylinder 0.5 m diameter rotates at 600 revolutions per minute in a uniform stream of 15 m/s.*

(*a*) *Locate the stagnation points.*

(*b*) *Calculate the minimum rotational speed for detached stagnation points.*

Solution.

For (*a*), from data, $u_\theta = \dfrac{\pi \times 0.5 \times 600}{60} = 15.7$ m/s

$$\Gamma = 2\pi R u_\theta = 2\pi \times 0.25 \times 15.7 = 24.6 \text{ m}^2/\text{s}$$

For the stagnation ponts,

$$\sin\theta = -\frac{\Gamma}{4\pi RU} = -\frac{24.6}{4\pi \times 0.25 \times 15} = -0.522$$

whence θ = **– 31.5° and – 148.5°.**

on the periphery of the cylinder for S_2 and S_1 respectively.

(*b*) The stagnation points are at the verge of detachment when they coincide;

$$\sin\theta = -\frac{\Gamma}{4\pi RU} = -1 \text{ at } \theta = -\frac{\pi}{2} \text{ or } \frac{3\pi}{2}$$

From the relation $\Gamma = 4\pi RU$

$$2\pi R u_\theta = 4\pi RU$$

$$u_\theta = 2U = 2 \times 15 = \mathbf{30 \text{ m/s}}$$

6.14. Vortex-and-Sink Flow

The flow obtained by superimposing a free vortex on a sink flow bears an interesting similarity with a stationary tornado. It explains the nature of a tornado which, while appearing as a circulatory flow, sucks in material bodies towards its centre.

The stream function for a vortex-and-smk flow is given by

$$\Psi = k\theta - C \log r$$

where $k = -q/2\pi$, the sink strength

and $C = \Gamma/2\pi$, the vortex strength.

Similarly, the velocity potential function ϕ is given by

$$\phi = k \log r + C\theta$$

The velocity components at any point (r, θ) are given by

$$u_r = \frac{1}{r}\frac{\delta\psi}{\delta\theta} = \frac{k}{r}$$

and

$$u_\theta = -\frac{\partial\Psi}{\partial r} = \frac{C}{r}$$

Applying the Bernoulli equation between the stagnation point and a desired point,

$$\frac{p_0}{\rho} = \frac{p}{\rho} + \left(\frac{u_\theta^2 + u_r^2}{2}\right)$$

whence

$$\frac{p - p_0}{\rho} = -\frac{C^2 + k^2}{2r^2}$$

It may be observed that the pressure variation across the flow is steeper than that for either the vortex or the sink flow alone, as shown in Fig. 6.15.

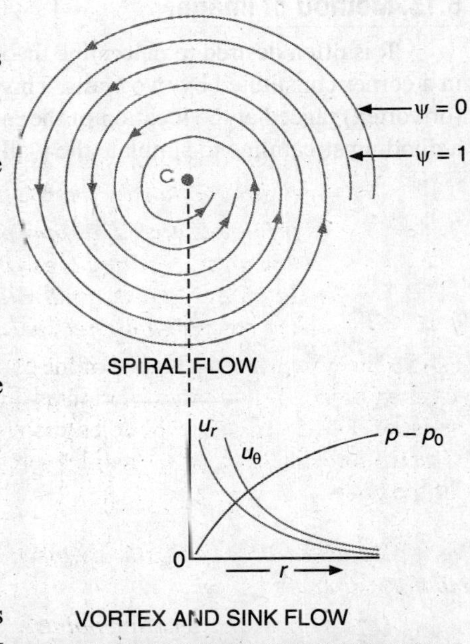

Fig. 6.16: Vortex-and-Sink Flow

Example 6.6. *Calculate the vortex and sink strengths to simulate a tornado in which the vacuum is 800 Pa at a radial distance of 200 m from the centreline and the influx of air is 20,000 m³/s per metre height.*

Solution.

For a vortex-and-sink flow,

$$\Psi = k\,\theta - C \log r$$

where $k = -\dfrac{q}{2\pi}$, the sink strength

and $C = \dfrac{\Gamma}{2\pi}$, the vortex strength.

The velocity components being $u_\theta = \dfrac{C}{r}$ and $u_r = \dfrac{k}{r}$

and the pressure distribution is given by

$$\frac{p - p_0}{\rho} = -\frac{C^2 + k^2}{2r^2}$$

From the data, $p - p_0 = -800$ Pa, $\rho = 1.2$ kg/m³ for $r = 200$ m,

Therefore, $C^2 + k^2 = \dfrac{800}{1.2} \times 2 \times 200^2 = 53.3 \times 10^6$.

Since q, the inflow per metre height, is 20,000 m³/s,

$$k = \frac{-20,000}{2\times 66} = -\mathbf{3183\ m^2/s}$$

and, hence $C^2 = 53.3 \times 10^6 - (-3183)^2$

whence $C = \mathbf{6572\ m^2/s.}$

6.15. Method of Images

It is often desired to determine the ideal flow due to a source or a vortex placed near a wall or in a corner constituted by two walls. This is made possible by realising that a corresponding source (or vortex) placed at the location of the mirror image of the given source (or vortex) would result in a dividing streamline to simulate the wall.

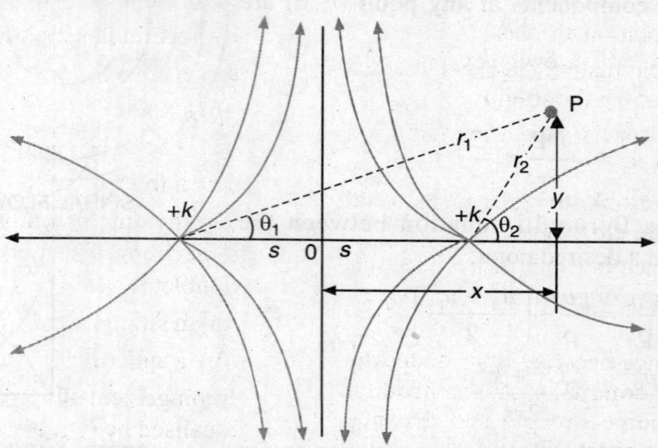

Fig. 6.17: Method of Images: Illustration

Let us illustrate this method by plotting the flow pattern due to a source of strength k placed in the vicinity of a wall.

For two sources of equal strength k placed a distance 2 s apart, on the x-axis,

$$\Psi = k(\theta_1 + \theta_2)$$
$$\phi = k \log r_1 + k \log r_2 = k \log r_1 r_2$$

From the geometry of the flow pattern,

$$r_1 = \sqrt{y^2 + (x+s)^2} \; ; \; r_2 = \sqrt{y^2 + (x-s)^2}$$

and $$\tan\theta_1 = \frac{y}{x+s} \quad \tan\theta_2 = \frac{y}{x-s}$$

ITEM BANK

Objective items of different types are given to review the concepts developed in the Chapter.

* *Select the correct response for each item.*
* *Every correct response is expected in less than a minute and carries one mark.*
* *Compare your responses with the key answers given at the end.*

Multiple-Choice Items

Attempt the following Multiple-Choice Items by ticking the correct or the most appropriate responses from among the alternatives given :

1. A source in two-dimensional flow is a
(*a*) point from which fluid issues
(*b*) line from which fluid issues
(*c*) straight line from which fluid issues
(*d*) straight line from which fluid issues steadily

2. A two-dimensional vortex has
(*a*) helical streamlines
(*b*) vorticity about any circular streamline including or excluding the origin
(*c*) zero circulation around it
(*d*) constant circulation around any closed path including the origin

3. A doublet is a special case of a source and sink pair when
(*a*) source and sink strengths are zero and the two are close together
(*b*) the distance between them becomes zero for given source and sink strengths
(*c*) the distance becomes zero under the limiting condition of the product of the source strength and distance remaining constant
(*d*) the source and sink coincide, while moving along the x-axis as a pair

4. The two stagnation points for an ideal flow over a circular cylinder coincide when
(*a*) the free stream is uniform and steady
(*b*) the circulation about the cylinder is very small
(*c*) the circulation about the cylinder is a critical value
(*d*) the circulation about the cylinder exceeds a critical value

5. Flow over an elliptical body is idealised by superimposing a free stream on
(*a*) a source
(*b*) a doublet
(*c*) a source and sink pair
(*d*) a free vortex

6. The dividing streamline for a uniform flow superimposed over a two-dimensional doublet is
(*a*) a straight line (*b*) a circle
(*c*) a sphere (*d*) an ellipse

7. Impingement of a jet on a flat plate may be idealised by
(*a*) $\Psi = xy$ (*b*) $\Psi = x^2 - y^2$
(*c*) $\Psi = x^2 + y^2$ (*d*) $\Psi = x / y$

8. On the surface of the rotating liquid contained in a beaker,

(*a*) velocity at the centre is zero

(*b*) pressure at the centre is zero

(*c*) velocity at the centre is maximum

(*d*) pressure at the centre is non-zero

Other Objective Items

9. Identify the following statements as true or false :

(*a*) All ideal flow patterns are superimposable. (T/F)

(*b*) In a source flow, pressure increases along the flow. (T/F)

(*c*) In a free vortex, the velocity potential lines are radial. (T/F)

(*d*) In a source and sink pair, the streamlines are arcs of circles. (T/F)

(*e*) In a doublet, there are arcs of circles. (T/F)

10. Match the types of flows in Block 'A' with the basic ideal flows in Block B.

Block A	Block B
(*a*) Flow over a stationary cylinder	(*k*) source
(*b*) Flow over a half body	(*l*) free stream
(*c*) Flow over a rotating cylinder	(*m*) sink
	(*n*) free vortex
	(*o*) doublet

ANSWERS

1. (*c*), **2.** (*d*), **3.** (*c*), **4.** (*c*), **5.** (*c*), **6.** (*b*), **7.** (*a*), **8.** (*d*), **9.** (*a T, b F, d T, e T*), **10.** (*alo, blk, clon*)].

QUESTION BANK

Questions on theory and numerical problems are given for tutorial work.

* *The questions are arranged in generally order of the articles in the text.*
* *Every question should take 15-20 minutes to solve.*
* *Answers for the numerical problems are given alongside them.*

6.1. Determine the equation of the dividing streamline for the flow constituted by superimposing a free stream of 10 m/s on a two-dimensional source of strength 5 m^2/s. Locate the stagnation point. How would the position of the stagnation point change if the source strength is increased or decreased? **[$2y + \theta = 2\pi$]**

6.2 The stream function for the two-dimensional flow of a liquid is given by $\Psi = 2xy$. In the range of values of x and y between 0 and 5, plot the streamlines and equipotential lines passing through the coordinates (1, 1), (1, 2), (2, 2). Also determine the velocity in magnitude and direction at point (1, 2). **[4.47 m/s, 63.4°]**

6.3 A source of strength 30 m^2/s is located at the origin and another source of strength 20 m^2/s is located at (1, 0). Find the velocity components u and v at (– 1,0) and (1, 1). Also, if the dynamic pressure at infinity is zero for $\rho = 2.0$ kg/m^3, calculate the dynamic pressure at the above points.

[u = 0 and 22.15 m/s; v = 21.33 m/s and – 25.23 m/s; 40.5 N/m^2 and 39.6 N/m^2]

6.4. A source of strength m at the origin and a uniform flow of 15 m/s are combined in a two-dimensional flow so that a stagnation point occurs at (1, 0). Obtain the velocity potential and the stream function. **[$\Psi = -15 \tan^{-1} (y/x) + 15\, y$, $\phi = -7.5 \ln (x^2 + y^2) + 15x$]**

6.5. A source discharging 20 m^3/s is located at (–1,0) and a sink of twice the strength is located at (2, 0). For the pressure 100 N/m^2 at the origin and density 1.8 kg/m^3, find the velocity and pressure at points (0, 1) and (1, 1). **[u = 27.69 m/s, 22.89 m/s; v = –6.15 m/s, –29.24 m/s]**

6.6. How does the pressure vary in the radial direction (*a*) in a source flow and (*b*) in a free-vortex flow ?

Justify your answers by relating the pressure distributions with the corresponding velocity distributions.

6.7. Determine the pressure variation across a tornedo constituted by a vortex of strength 5000 m^2/s and a sink of strength 3000 m^2/s in air. Compute the pressure at a point 100 m from the centre. **[51.67 Pa Vacuum]**

6.8. Determine the flow pattern by superimposing a uniform flow on a vortex pair placed across the flow. Assume that the free stream velocity is U and that the vortices of strength C, *i.e.*, ($\Gamma/2\pi$) are located a distance 2 s across the flow as shown in Fig. Prob. 6.8.

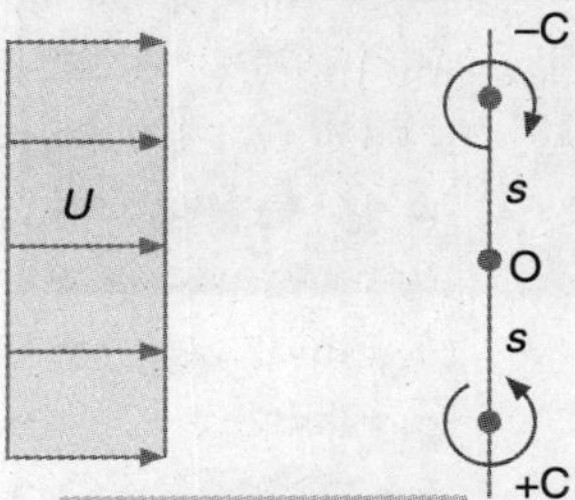

Fig. Prob. 6.8

6.9. Explain the non-linear path followed by a table tennis ball hit by a bat and imparted a back-spin.

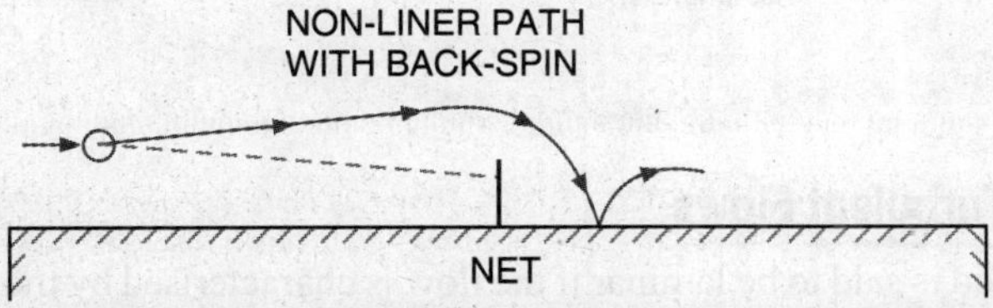

Fig. Prob. 6.9

6.10. A source of strength k is located at a distance d from a wall. Employing the method of images, *i.e.*, assuming the existence of another source of the same strength located at its mirror image, determine the stream function of the flow. Plot the streamlines Estimate the pressure distribution at the wall.

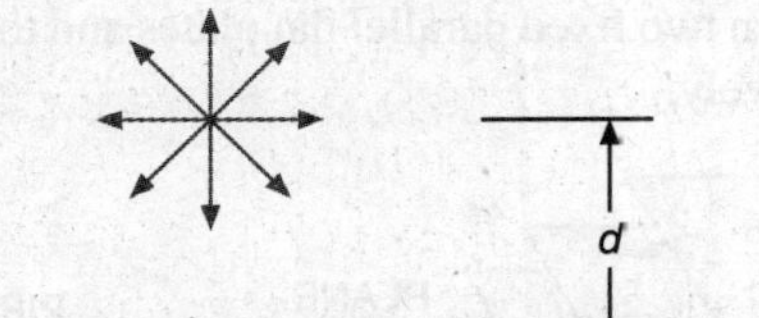

Fig. Prob. 6.10

CHAPTER

7

Laminar Flow

LEARNING OBJECTIVES

On successful completion of all learning activities planned and provided in the chapter, a student will be able to

- describe the conditions under which the flow is laminar or turbulent ;
- establish and apply the Hagen-Poiseuille equation for laminar flow in a pipe;
- solve simple laminar flow problems ; and
- determine the viscosity of a liquid by employing viscometers.

7.1. Laminar and Turbulent Flows

The flow of a fluid is said to be laminar if the flow is characterised by the fluid particles staying in laminae, *i.e.*, thin sheets. Each lamina may be considered to be a stream surface. The shape of a lamina depends upon the shape of the boundaries of the passage. For example, if the flow takes place between two parallel flat plates, the laminae must be plane sheets parallel to each other. If the flow takes place through a round pipe, the laminae must be concentric cylindrical sheets. If the flow takes place in a varying-area passage, the laminae should also be such that they constitute stream surfaces in the flow. Laminar flow between two fixed parallel flat plates and through a round tube are sketched in Figs. 7.1. (*a*) and (*b*) respectively.

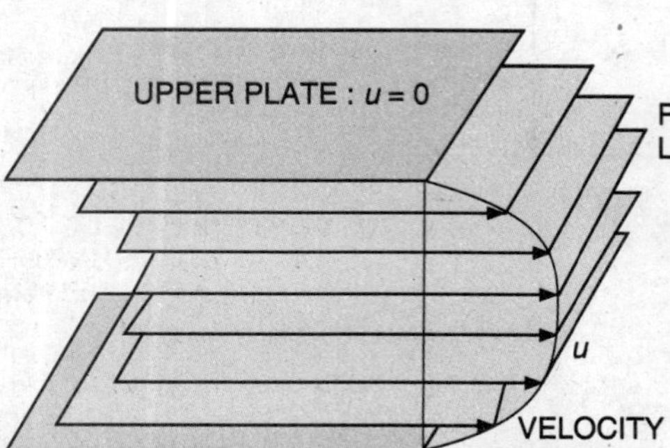

Fig. 7.1 (a) LAMINAR FLOW BETWEEN TWO PARALLEL PLATES

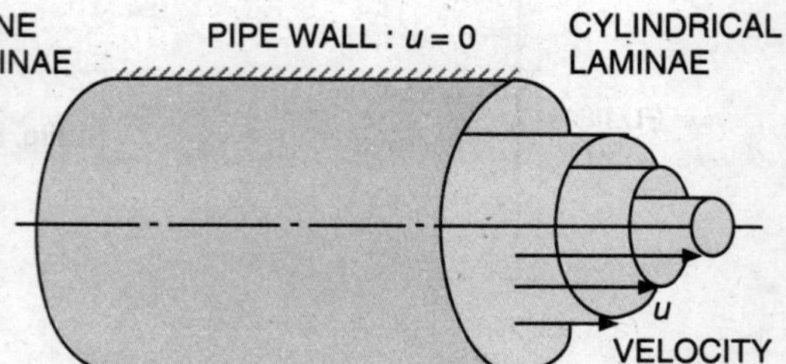

Fig. 7.1 (b): LAMINAR FLOW THROUGH A ROUND TUBE

It is interesting to discuss the factors which decide whether the flow of a fluid is laminar or not. At low velocities, a fluid flow may be laminar but at high velocities, the tendency of fluid particles is to mix between different layers and to become turbulent. Highly viscous fluids tend to flow laminar because the shear forces due to viscosity tend to oppose motion and inhibit free mixing whereas low-viscosity fluids tend to flow turbulent. The dimension across the passage also has an influence on the nature of flow. If the width of a passage is small or if the diameter of a pipe is small, the fluid is likely to flow laminar because of the influence of the boundaries on the entire fluid. If the width of passage is large or the diameter of a pipe is large, the flow may still be laminar in the vicinity of the wall but in the core region the flow is likely to be turbulent. Lastly, it is observed that low-density fluids are more likely to flow laminar than are denser fluids. This is because the exchange of momentum and hence the effectiveness of mixing is less for low-density fluids than for denser fluids. Finally to sum up, **a flow is more likely to be laminar if the velocity U is low ; the diameter d is small ; the density ρ is low; and the viscosity μ is high.**

These four variables are grouped in the form of a non-dimensional paramater called Reynolds number,

$$Re = \frac{Ud\rho}{\mu} \qquad \text{....(7.1)}$$

The Reynolds number actually represents the ratio of inertial forces to viscous forces in the fluid. Thus when U and d are large and μ is small, inertial forces predominate over viscous forces and the flow becomes turbulent.

From the foregoing discussion, it follows that the flow of a fluid is more likely to be laminar at low Reynolds number and more likely to be turbulent at high Reynolds number. Flow through a pipe is laminar at Reynolds number less than 2000 and it is turbulent at Reynolds number more than 3000. The flow is said to be in transition stage at Reynolds numbers between 2000 and 3000. The Reynolds number beyond which the flow changes from laminar to turbulent is called the critical number. The velocity of the fluid corresponding to the critical Reynolds number is called the critical velocity. The critical Reynolds numbers in different situations may be different depending upon the choice of the characteristic length l, *e.g.*, the diameter, d is chosen for a pipeflow.

In order to study the effect of Reynolds number on the nature of flow in a pipe, a simple experiment may be performed as follows :

Apparatus

A constant-head water tank, a transparent pipe, a control valve at the downstream end, a dye injection system and a suitable dye, *e.g.*, potassium permanganate solution as shown in Fig. 7.2 (*a*) and (*b*).

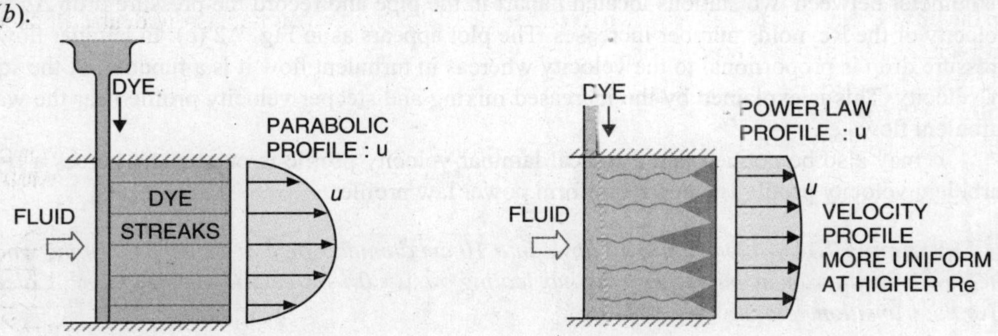

(a) LAMINAR FLOW : Re < 2000 (b) TURBULENT FLOW : Re > 3000

Fig. 7.2: Characteristics of Laminar and Turbulent Flows

Observations and Conclusions

Start the experimental observations by opening the control valve slightly and injecting very little dye. The dye streaks appear straight.

Slight increase of flow rate makes no significant difference to the streak form but after a certain stage, the dye streaks appear wavy and disturbed. Further opening of the valve results in complete dispersion of the dye in the fluid. This is the process of change-over from laminar to turbulent flow by changing u, i.e., Re since d, ρ and μ are kept constant.

Under carefully controlled conditions the laminar flow can be maintained up to $Re = 40{,}000$ or so but under normal circumstances, the transition occurs between Re of 2000 and 3000 in the process of changing from laminar to turbulent flow. Below 2000, laminar flow persists whereas above 3000, laminar flow may change to turbulent flow in the presence of the slightest disturbance. The Reynolds number up to which laminar flow may be maintained is said to be the *upper critical Reynolds number.* For a round pipe of given diameter d and the fluid properties ρ and μ, the corresponding velocity is said to be *upper critical velocity.*

When the velocity is gradually reduced to change the nature of flow from turbulent to laminar, the Reynolds number is always required to be reduced below 2000. The Reynolds number to which the flow should be reduced to change from turbulent to laminar flow is called the *lower critical Reynolds number* and the corresponding velocity of flow for a given pipe and fluid is called the *lower critical velocity.*

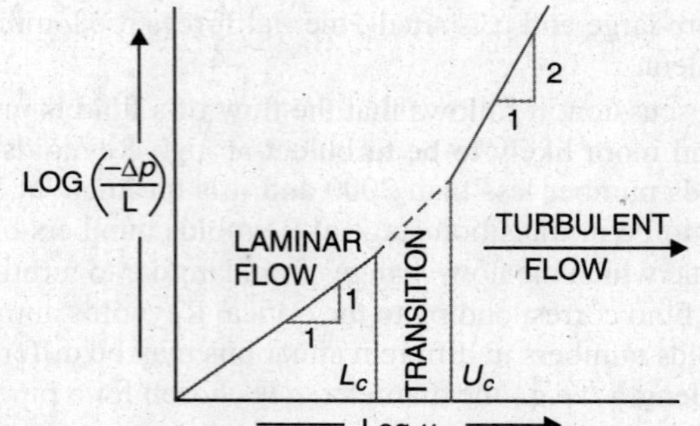

(c) LOWER (L_C) AND UPPER (U_C) CRITICAL VELOCITIES

Fig. 7.2: Characteristics of Laminar and Turbulent Flows (Contd.)

In order to appreciate the drastic change in the nature of flow, we instal a differential pressure manometer between two stations located l apart in the pipe and record the pressure drop Δp as the velocity or the Reynolds number increases. The plot appears as in Fig. 7.2 (*c*). In laminar flow, the pressure drop is proportional to the velocity whereas in turbulent flow it is a function of the square of velocity. This is explained by the increased mixing and steeper velocity profile near the wall in turbulent flow.

It may also be noticed that a typical laminar velocity profile is parabolic whereas a typical turbulent velocity profile is a nearly uniform power law profile.

Example 7.1. *A lubricating oil flows in a 10 cm diameter pipe at 1 m/s. Determine whether the flow is laminar or turbulent. For the lubricating oil,* $\mu = 0.1$ *Ns/m*^{2} *and* $\rho = 930$ *kg/m*^{3}*. Calculate also the transition velocity.*

Solution.

The Reynolds number is given by

$$Re = \frac{\mu d \rho}{\mu} = \frac{1 \times 0.1 \times 930}{0.1} = 930$$

Since this is appreciably less than 2000, the flow must be **laminar.**

Transition may occur if $Re > 2000$

i.e., $$\frac{ud\rho}{\mu} > 2000$$

or $$u > \frac{2000 \times 0.1}{930 \times 0.1}$$

or $$u > \mathbf{2.15\ m/s.}$$

7.2 Laminar Flow through a Round Pipe

Consider a laminar flow through a horizontal round pipe as shown in Fig. 7.3. This is also referred as Hagen-Poiseuille flow. The flow proceeds in cylindrical laminae, as shown earlier. The equilibrium of a cylinder of radius r contained in a lamina of radius $r + \delta r$ is considered. For Equilibrium, the net pressure force on cylindrical element due to pressures p and $p + \delta p$ on its ends must equal the shear force on its periphery due to the presence of a cylindrical lamina over it.

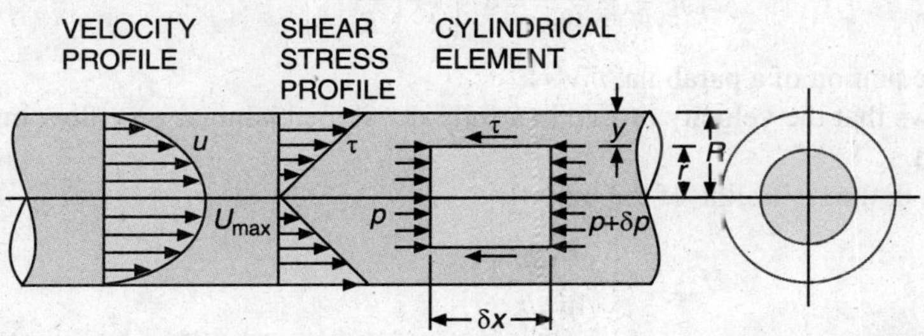

Fig. 7.3. Laminar Flow through a Pipe

$$p\pi r^2 - (p + \delta p)\pi r^2 = \tau.\ 2\pi\ r\ \delta x$$

or $$-\ \delta p\pi r^2 = \tau.\ 2\pi\ r\delta x$$

whence $$\tau = -\frac{dp}{dx}\frac{r}{2} \qquad ...(7.2)$$

If the laminar flow in a pipe is fully developed, *i.e.,* the velocity profile does not change in the longitudinal direction, the velocity u varies only with r and the pressure p remains constant over the cross section. Under such conditions, longitudinal pressure gradient dp/dx remains constant. Hence τ is proportional to r, i.e., shear stress varies linearly with radius.

The maximum shear stress τ_{max} occurs at the pipe wall, *i.e.,* at $r = R$.

$$\tau_{max} = -\frac{dp}{dx}\frac{R}{2} \qquad ...(7.3)$$

The velocity distribution in the pipe may be obtained by invoking the relationship

$$\tau = \mu\frac{du}{dy}$$

where y is measured from the wall.

Since $$y = R - r$$

and $$dy = -\ dr$$

it follows that $$\tau = -\mu \frac{du}{dr} \quad ...(7.4)$$

Substituting this relationship in Eq. (7.2)

$$\frac{du}{dr} = \frac{1}{2\mu}\frac{dp}{dx}r$$

On integration,

$$u = \frac{1}{4\mu}\frac{dp}{dx}r^2 + K \quad ...(7.5)$$

At $r = R$, $u = 0$, hence $K = -\frac{1}{4\mu}\frac{dp}{dx}R^2$

and $$u = -\frac{1}{4\mu}\frac{dp}{dx}(R^2 - r^2)$$

or $$u = -\frac{1}{4\mu}\frac{dp}{dx}R^2\left(1-\left(\frac{r}{R}\right)^2\right) \quad ...(7.6)$$

which is an equation of a parabola, u v/s r.

It shows that the velocity profile in a fully developed laminar pipe flow must be a paraboloid of revolution.

At $r = 0$, the centreline of the pipe,

$$U_{max} = -\frac{1}{4\mu}\frac{dp}{dx}R^2 \quad ...(7.7)$$

Then $$u = U_{max}\left(1-\left(\frac{r}{R}\right)^2\right) \quad ....(7.8)$$

Total discharge through the pipe Q.

$$Q = \int dq = \int_0^R 2\pi r dr . U_{max}\left(1-\frac{r^2}{R^2}\right)$$

$$= \int_0^R 2\pi U_{max}\left(1-\frac{r^2}{R^2}\right) rdr.$$

$$= 2\pi\, U_{max}\left(\frac{R^2}{4}\right)$$

Substituting the value of U_{max} from Eq. (7.7).

$$Q = \frac{\pi}{8\mu}\left(\frac{dp}{dx}\right)R^4$$

or $$Q = -\frac{\pi}{128\mu}D^4\left(\frac{dp}{dx}\right) \quad ...(7.9)$$

in terms of diameter D, Eq. (7.9) is also known as the Hagen-Poiseuille equation for laminar flow through a round pipe.

The average velocity U is given by

$$U = \frac{Q}{\pi R^2} = -\frac{1}{8\mu}\left(\frac{dp}{dx}\right)R^2 \quad ...(7.10)$$

The power required to maintain the laminar flow in a pipe may be determined by employing the relationship

$$\text{Power} = F\,U$$

where F is the pressure force across the length of the pipe and U is the average velocity over the cross section. In terms of pressures,

$$\text{Power} = (p_1 - p_2)\, \pi R^2 U = (p_1 - p_2)\, Q \quad \text{....(7.11)}$$

which is the product of pressure difference across the length of the pipe and the discharge through it.

From Eq. (7.9), the relationship for the pressure gradient in the pipe may be expressed as

$$\frac{dp}{dx} = -\frac{8\,\mu\, U}{R^2} \quad \text{...(7.12)}$$

This equation shows that the pressure gradient must be negative, *i.e.,* the pressure must drop along the direction of flow. The drop in pressure over a finite length l is given by

$$\Delta p = \frac{8\,\mu\, U}{R^2} l \quad \text{....(7.13)}$$

The total drag force on the pipe due to the laminar flow through it can be calculated by considering that the wall shear stress τ_{max} acts over the peripheral area $2\pi R\, l$ for a length l of the pipe,

$$\text{Drag} = \tau_{max} . 2\pi\, R\, l \quad \text{....(7.14)}$$

A force equal and opposite to the drag force must be required to hold the pipe in position. The drag coefficient or skin-friction coefficient C_f must be given by

$$\tau_{max} = C_f \rho \frac{U^2}{2}$$

or

$$C_f = \frac{\tau_{max}}{\rho U^2 / 2} \quad \text{....(7.15)}$$

Employing Eqs. (7.3) and (7.9),

$$C_f = \frac{16\mu}{UD\,\rho} = \frac{16}{Re} \quad \text{...(7.16)}$$

The analysis of laminar flow through a pipe could have, alternatively, been initiated from the Navier-Stokes equations. Since u_r and u_θ vanish, the first two equations are identically satisfied and the third equation for the axial direction reduces to the following:

$$0 = 0 - \frac{dp}{dx} + \left(\frac{\partial^2}{\partial r^2} u + \frac{1}{r}\frac{\partial}{\partial r} u + 0 + 0 \right) \mu \quad \text{...(7.17)}$$

or

$$\frac{d^2u}{dr^2} + \frac{1}{r}\frac{du}{dr} = \frac{1}{r}\frac{d}{dr}\left(r\frac{du}{dr} \right) = \frac{1}{\mu}\frac{dp}{dx} \quad \text{...(7.18)}$$

or

$$\frac{d}{dr}\left(r\frac{du}{dr} \right) = \frac{1}{\mu}\frac{dp}{dx} r \quad \text{...(7.18a)}$$

Integrating w.r.t. r,

$$r\frac{du}{dr} = \frac{1}{\mu}\frac{dp}{dx}\frac{r^2}{2} + C \quad \text{...(7.19)}$$

Using the condition $r = 0;\ \dfrac{du}{dr} = 0$

since the velocity is the maximum at the centreline, we get $C = 0$

Integrating once more w.r.t. r:

$$u = \frac{1}{4\mu}\frac{dp}{dx} r^2 + K$$

We have again arrived at Eq. (7.5) ; further analysis is the same as already done.

Example 7.2. *A 20 cm diameter pipe 30 km long transports oil from a tanker to the shore at 0.01 m^3/s. Calculate the power required to maintain the flow. Take $\mu = 0.1\ Nm/s^2$, $\rho = 900\ kg/m^3$ for the oil.*

Solution.

From the data, $$U = \frac{0.01 \times 4}{\pi \times 0.2^2} = 0.318 \text{ m/s}$$

$$Re = \frac{0.318 \times 0.2 \times 900}{0.1} = 572.4$$

This being less than 2000, the flow is laminar.

From Eq. (7.13),

$$\Delta p = \frac{8\,\mu U}{R^2}.l = \frac{8 \times 0.1 \times 0.318}{0.1^2} \times 30 \times 1000$$

$$= 763,200 \text{ N/m}^2$$

Power loss = $\Delta p.\ Q$ = 763, 200 × 0.01 = 7632 W = 7.63 kW.

The power required to maintain the flow is **7.63 kW.**

Example 7.3. *In the injection moulding of a certain plastic, the heated compound is forced under a pressure difference of 7 kg_f/cm^2 through a tube 0.5 cm diameter and 25 cm length. If the mean velocity of flow is 40 cm/s, calculate the viscosity of the compound, assuming laminar flow. If the specific gravity of the compound is 1.5, calculate the Reynolds number for the flow and hence show that the flow is actually laminar.*

Solution.

$$\Delta p = -7 \text{ kg}_f/\text{cm}^2 = -7 \times 9.81 \times 10^4 = -686.7 \times 10^3 \text{ N/m}^2$$

$$U = 0.4 \text{ m/s}$$

$$\rho = 1.5 \times 1000 = 1500 \text{ kg/m}^3$$

$$d = 0.005 \text{ m};\ \Delta l = 0.25 \text{ m}$$

For laminar flow,

$$U = \frac{-1}{8\mu}\frac{dp}{dx}R^2$$

Hence, $$\mu = \frac{1}{8 \times 0.4} \times \frac{686.7 \times 10^3}{0.25} \times (0.0025)^2$$

$$= \mathbf{5.365\ Ns/m^2}$$

It should be checked that the nature of the flow is laminar.

$$Re = \frac{Ud\rho}{\mu} = \frac{0.4 \times 0.005 \times 1500}{5.365}$$

$$= \mathbf{0.559 \text{ which is} << 2000}$$

showing that the flow is **indeed laminar.**

Example 7.4. *An oil having a viscosity of 0.0098 kg_f second per square metre and a specific gravity of 1.59 flows through a horizontal pipe of 5 centimetre diameter with a pressure drop of 0.06 kg per square centimetre per metre length of the pipe. Determine (i) the rate of flow in kilograms per minute, (ii) the shear stress at the pipe wall, (iii) the total drag for 100-metre length of pipe and (iv) the power required for the 100-metre length of the pipe to maintain the flow.*

Solution.

Viscosity of the oil μ = 0.0098 $kg_f\ s/m^2$ = 0.0098 × 9.81 = 0.096 Ns/m^2

Specific gravity $S = 1.59$

Diameter $D = 5 \text{ cm} = 0.05 \text{ m}$

Pressure drop $\Delta p = 0.06 \text{ kg}_f/\text{cm}^2$ per metre length

$= 0.06 \times 9.81 \times 10000 = 5886 \text{ N/m}^2$ per metre length

$$\frac{dp}{dl} = -5886 \text{ N/m}^2$$

(*i*) From the Hagen-Poiseuille equation,

$$Q = \frac{-\pi D^4}{128\,\mu}\frac{dp}{dx}$$

$$= \frac{\pi \times 0.05^4}{128 \times 0.096} \times 5886 = 0.0094 \text{ m}^3/\text{s}$$

$$= 0.0094 \times 60 \times (100 \times 1.59)$$

$$= \mathbf{896.75 \text{ kg/minute}}$$

(*ii*) The shear stress is given by

$$\tau = -\frac{dp}{dx}\frac{r}{2}$$

At the pipe wall,

$$r = 0.025 \text{ m}$$

$$\tau_{max} = \frac{1}{2} \times 5886 \times 0.025 = \mathbf{73.58 \text{ N/m}^2}$$

(*iii*) Shearing area at the pipe wall for 100 m length of the pipe is

$$A = 100 \times \pi \times 0.05 = 15.71 \text{ m}^2$$

Drag on 100 m long pipe = shear stress × shearing area

$= 73.58 \times 15.71 = 1155.94 = \mathbf{1.156 \text{ kN}}$

(*iv*) Power required to maintain the flow through 100 m length of the pipe

$$\text{Power} = Q \times \Delta p$$

$$= 0.0094 \times 5886 \times 100 \text{ W}$$

$$= \mathbf{5.53 \text{ kW}}$$

The power required could also be worked out from a knowledge of total shearing force and the average velocity through the pipe

$$U = \frac{-1}{8\mu}\frac{dp}{dx}.R^2 = \frac{1 \times 5886}{8 \times 0.096}(0.05)^2 = 4.79 \text{ m/s}$$

$$F = 1.15 \text{ kN}$$

$$\text{Power} = F.\,U = 1.156 \times 4.79 = \mathbf{5.53 \text{ kW}}$$

Example 7.5. *The velocity distribution for laminar flow in a round pipe of radius R is given by the equation*

$$\frac{u}{U_{max}} = 1 - \left(\frac{r}{R}\right)^2$$

where u is the velocity at a radius r and U_{max} is the maximum velocity at the centreline. Determine the values of the kinetic energy and momentum correction factors if the average velocity is used to represent the flow.

Solution.

Consider an elementary area dA in the form of a ring at a radius r such that

$$dA = 2\pi r\, dr$$

The average velocity U in the pipe should be given by

$$U = \frac{1}{A}\int u\, dA$$

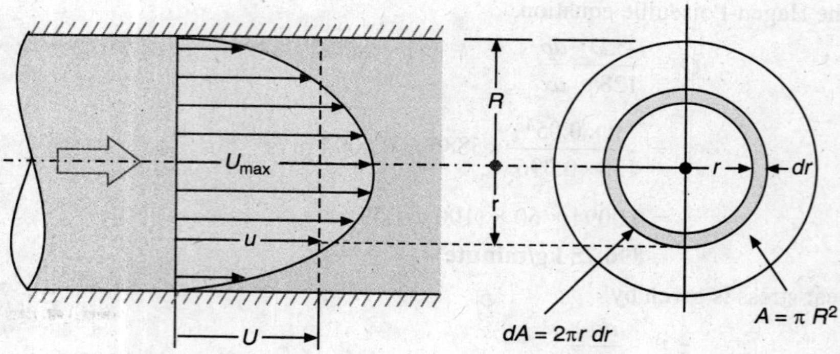

Laminar Flow through a Pipe

or

$$U = \frac{1}{\pi R^2}\int_0^R U_{max}\left(1-\frac{r^2}{R^2}\right)2\pi r dr = \frac{2U_{max}}{R^2}\int_0^R\left(r-\frac{r^2}{R^2}\right)dr$$

$$= \frac{2U_{max}}{R^2}\left(\frac{R^2}{2}-\frac{R^4}{4R^2}\right) = \frac{U_{max}}{2}$$

The kinetic energy correction factor a is given by

$$\alpha = \frac{1}{A}\int\left(\frac{u}{U}\right)^3 dA = \frac{1}{\pi R^2}\int_0^R \frac{u^3}{U^3}\,2\pi r\, dr$$

$$= \frac{16}{R^2U^3_{max}}\int_0^R u^3 r dr = \frac{16}{R^3}\int_0^R\left[1-\left(\frac{r}{R}\right)^2\right]^3 r\, dr = \mathbf{2.0}$$

The momentum correction factor β is given by

$$\beta = \frac{1}{A}\int\left(\frac{u}{U}\right)^2 dA = \frac{1}{\pi R^2}\int_0^R \frac{u^2}{U^2}\,2\pi r\, dr$$

$$= \frac{8}{R^2}\int_0^R\left[1-\left(\frac{r}{R}\right)^2\right]^2 r\, dr = \frac{4}{3} = \mathbf{1.33}$$

The implication of these correction factors is that the kinetic energy calculated on the basis of the average velocity must be doubled to yield the true kinetic energy of the flow and that the momentum calculated on the basis of the average velocity must be increased by 33% to yield the true momentum of the laminar flow through a pipe.

EXPERIMENT

Development of Flow

OBJECTIVE

To study the development of flow through a round duct.

PRACTICAL RELEVANCE

The study of boundary layer development over a flat plate is historically significant because it has led to the bridging of the gap between the empirical field of hydraulics and the mathematical science of hydrodynamics. The study of this simple case is also meaningful because it provides reasonable predictions for flows over more complicated shapes such as aerofoils and solid cylinders. The study of developing flow through a round duct is meaningful in determining the critical entry length beyond which the flow becomes fully developed. This information is relevant in the design of pipe-systems and in the estimation of power requirements for conveying a given throughput by a pipe.

BACKGROUND INFORMATION

A long horizontal smooth pipe having a flared entry, with a suction fan installed at its exit end is the basic facility needed for conducting this experiment. A number of pitot static probes suitably mounted at different distances from the pipe inlet enables the measurement of the velocity profiles at the various cross-sections.

The velocity distribution at the entry cross-section is uniform. As the flow proceeds along the duct, a boundary layer grows at the wall. Eventually the growing boundary layers on the periphery of the duct merge at the centreline. At this stage, the uniform entrance flow has developed into a well-defined pattern depending upon the entry Reynolds number. This development process from the uniform entry flow to the fully developed pattern is a gradual process.

Entry Reynolds Number	Developed State	Developed Velocity Profile
< 2000	laminar	parabolic
> 2500	turbulent	logarithmic

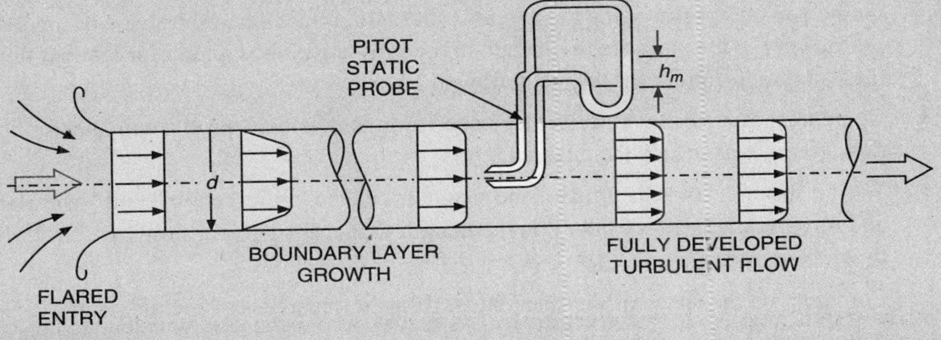

PROCEDURE

1. Set the flow such that the magnitude of the Reynolds number, $Re = Ud\rho/\mu$, is below the lower critical limit of 2000.
2. After steady-state conditions are attained, conduct pitot surveys at the desired cross-sections. It may be noted that steady-state conditions are indicated by the constancy of the manometer readings with the passage of time. One should note that the development of

flow takes place at a faster rate near the entry zone than at the later stages. In fact, beyond approximately 50 diameters from the entry cross-section, no further change in the velocity profile is expected to occur.

3. Note the static pressure readings at all the piezometer taps.
4. Repeat the experiment for inlet $Re > 2500$ for turbulent flow.

RESULTS

1. Calculate the local velocity ratio U/U_{max} as a function of the relative radial distance r/r_0 for the velocity profiles measured for the laminar. Plot the profiles for this case on one graph sheet.
2. Make a plot of the Hagen-Poiseuille velocity distribution law for fully-developed laminar flow through a circular duct (with the same volume rate of flow as for your experiments). Superimpose the data points for the laminar fully-developed case on the graph.
3. Make a logarithmic plot of the mean velocity distribution in the core region of the pipe for turbulent flow, *i.e.*, plot $(U_{max} - U)/U$ against In $(1 - r/r_0)$ and superimpose the data points for the turbulent, fully-developed case on this graph. The term $U = \sqrt{\tau_0/\rho}$ is the shear velocity. For fully-developed flow, τ_0 can be evaluated by carrying out a momentum balance between two sections over a longitudinal distance dx. The result of such an analysis will give $\tau_0 = -r_0/2 \cdot dp/dx$.
4. From the measurements, plot the piezometric head line, along the length of the duct. Determine the 'critical entry length'.

POINTS FOR DISCUSSION

1. Discuss the nature of various plots :

   (*a*) development of the velocity profiles for laminar and turbulent flows ;

   (*b*) 'theoretical' and experimental profiles for laminar flow ;

   (*c*) effect of the development of the boundary layer on the longitudinal pressure gradient.
2. What is meant by the expression 'fully-developed flow' as applied to duct flow ? How do the wall shear stress, the longitudinal pressure gradient and the shape of the velocity profile vary in the entrance region and in the fully-developed region for laminar flow ? Is the variation different for turbulent flow ?
3. Compare the shape of your fully-developed turbulent velocity profile with that for laminar flow giving reasons for the difference.
4. What is the effect of a disturbance and roughness on the 'critical entry length' in the laminar and turbulent regimes of flow ? Is this length longer for laminar flow than for turbulent flow ? If so, why ?
5. Comment on the shape of the shear stress distribution across a pipe diameter.
6. Any other observation or comment?

ANY OTHER OBSERVATION OR COMMENT?

7.3 Laminar Flow through an Annulus

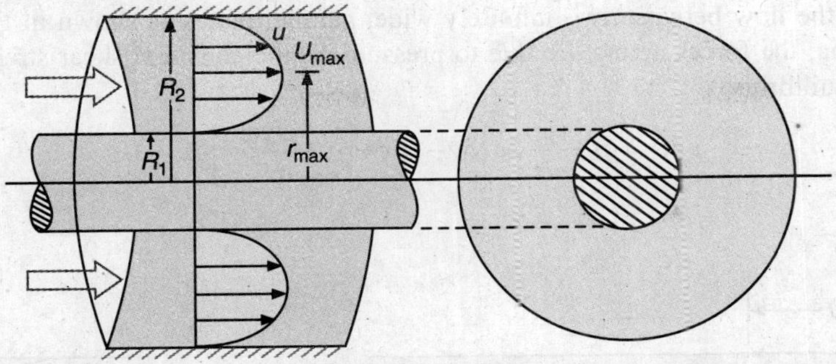

Fig. 7.4: Flow through an Annulus

In order to consider the laminar flow through an annulus, as in Fig. 7.4, between two co-axial cylinders, it is best to start from Eq. (7.19) and integrate it twice w.r.t. r for the new boundary conditions,

$$r\frac{du}{dr} = \frac{1}{\mu}\frac{dp}{dx}\frac{r^2}{2} + A \qquad \text{...(7.20)}$$

or

$$\frac{du}{dr} = \frac{1}{2\mu}\frac{dp}{dx}r + \frac{A}{r}$$

and by integrating,

$$u = \frac{1}{4\mu}\frac{dp}{dx}r^2 + A\log_e r + B \qquad \text{...(7.21)}$$

For this case, $u = 0$ at $r = R_1$ and also at $r = R_2$

$$0 = \frac{1}{4\mu}\frac{dp}{dx}R_1^2 + A\log_e R_1 + B$$

$$0 = \frac{1}{4\mu}\frac{dp}{dx}R_2^2 + A\log_e R_2 + B$$

From these equations,

$$A = -\frac{1}{4\mu}\frac{dp}{dx}(R_2^2 - R_1^2)/\log_e\frac{R_2}{R_1} \qquad \text{...(7.22)}$$

In order to locate r_{max} for the position of maximum velocity,

$$\frac{du}{dr} = 0;\ r^2_{max} = -2A\mu/(dp/dx)$$

whence

$$r_{max} = \sqrt{\left(\frac{R_2^2 - R_1^2}{2\log_e R_2/R_1}\right)} \qquad \text{...(7.23)}$$

In terms of r_{max}, expression for velocity u becomes

$$u = \frac{1}{2\mu}\frac{dp}{dx}\left(\frac{r^2 - R_1^2}{2} - r^2_{max}\log_e\frac{r}{R_1}\right) \qquad \text{...(7.24)}$$

7.4 Laminar Flow between Parallel Plates

Consider the flow between two infinitely wide, parallel plates as shown in Fig. 7.5. For an elemental lamina, the forces acting are due to pressure on the ends and shear stresses above and below it. For equilibrium,

$$p\delta y - \left(p + \frac{dp}{dx}\delta x\right)\delta y - \tau\delta x + \left(\tau + \frac{d\tau}{dy}\delta y\right)\delta x = 0$$

whence

$$\frac{dp}{dx} = \frac{d\tau}{dy} \qquad ...(7.25)$$

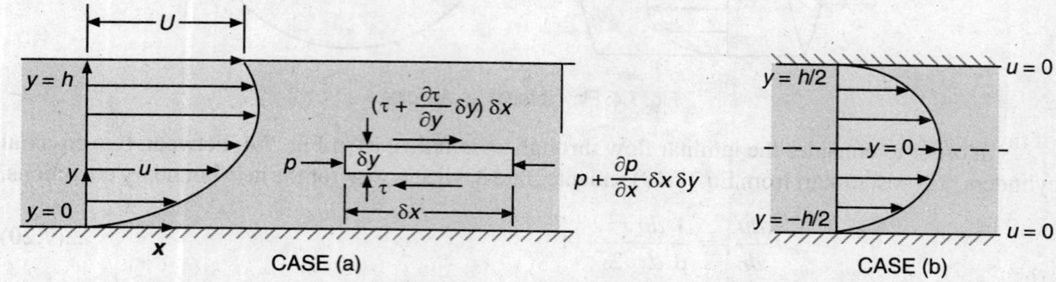

Fig. 7.5: Flow between Parallel Plates

This equation states that the axial pressure gradient and the transverse gradients of shear stress must be equal for laminar flow between flat plates. It does not matter whether one or both plates are stationary or one or both moving while being maintained parallel to each other.

By the Newton's law of viscosity, $\tau = \mu\dfrac{du}{dy}$,

therefore,

$$\frac{d}{dy}\left(\mu\frac{du}{dy}\right) = \frac{dp}{dx}$$

or

$$\frac{d^2u}{dy^2} = \frac{1}{\mu}\frac{dp}{dx} \qquad ...(7.26)$$

On integration,

$$\frac{du}{dy} = \frac{1}{\mu}\frac{dp}{dx}y + A \qquad ...(7.27)$$

and

$$u = \frac{1}{\mu}\frac{dp}{dx}y^2/2 + Ay + B \qquad ...(7.28)$$

The equation 7.26 can also be obtained directly from the Navier-Stokes. By applying Navier-Stokes equation with $v = 0$ we get

$$u\frac{\partial u}{\partial x} + 0 = \frac{-1}{\rho}\frac{\partial p}{\partial x} + \frac{\mu}{\rho}\left(\frac{\partial^2 u}{\partial x^2} + \frac{\partial^2 u}{\partial y^2}\right)$$

The equation in y – direction vanishes identically as $v = 0$. Also by continuity equation

$$\frac{\partial u}{\partial x} = 0 \qquad \frac{\partial^2 u}{\partial x^2} = 0$$

Therefore the above equation takes the form

$$\mu\left(\frac{\partial^2 u}{\partial y^2}\right) = \frac{\partial p}{\partial x}$$

which is the same as equation (7.26.) !

The constants of integration A and B are evaluated with regard to the boundary conditions of the flow regime. Let us consider two cases :

Case (*a*). Couette Flow. *The upper plate moving at U to the right and lower plate kept stationary as is shown in Fig. 7.5 (a).*

$$u = 0 \text{ at } y = 0;\ u = U \text{ at } y = h$$

$$B = 0,\ A = \frac{U}{h} - \frac{1}{\mu}\frac{dp}{dx}\frac{h}{2}$$

$$u = \frac{U}{h}y - \frac{1}{2\mu}\left(\frac{dp}{dx}\right)yh\left[1 - \frac{y}{h}\right]$$

It may be noticed from this expression that the variation of u w.r.t. y is parabolic. The location of the maximum velocity U_{max} may be found by evaluating $\frac{du}{dy}$ and setting it to nought.

$$\frac{du}{dy} = \frac{U}{h} - \frac{h}{2\mu}\frac{dp}{dx}\left(1 - \frac{2y}{h}\right) = 0$$

whence y_{max} the transverse distance from the wall where the velocity is the maximum is given by

$$y_{max} = \frac{h}{2} - \frac{\mu U}{h\dfrac{dp}{dx}} \qquad \text{...(7.30 a)}$$

and the maximum velocity is given by

$$U_{max} = -\frac{U}{2} - \frac{h^2}{8\mu}\frac{dp}{dx} \qquad \text{...(7.30 b)}$$

Case (b.) Plane Poiseuille Flow. *Both plates stationary as shown in Fig. 7.5 (b)*

$$u = 0 \text{ at } y = \frac{h}{2};\ u = 0 \text{ at } y = -\frac{h}{2}$$

$$A = 0,\ B = \frac{h^2}{8\mu}\frac{dp}{dx}$$

$$u = -\frac{h^2}{8\mu}\frac{dp}{px}\left(1 - 4\left(\frac{y}{h}\right)^2\right) \qquad \text{...(7.31)}$$

which implies that the velocity profile should be parabolic.

$$u = U_{max} \text{ at } y = 0, \text{ hence } U_{max} = -\frac{h^2}{8\mu}\frac{dp}{dx}$$

The average velocity $U_{av} = \dfrac{Q}{h} = \displaystyle\int \frac{dQ}{h} = \int_{-h/2}^{h/2} \frac{u\,dy}{h} = \frac{2}{3}U_{max}$

which is also so expected by virtue of the fact that the area under a parabolic curve is two-third of the area of the circumscribing rectangle. In terms of U_{av}

$$\frac{dp}{dx} = -12\,\mu\frac{U_{av}}{h^2} \qquad \text{...(7.32)}$$

The shear stress at a point is given by

$$\tau = \mu\frac{du}{dy} = \mu\frac{h^2}{\mu}\frac{dp}{dx}\frac{y}{h}\frac{1}{h} = y\frac{dp}{dx} \qquad \text{...(7.33)}$$

which shows that the shearing stress varies linearly across the section from 0 value at $y = 0$ to $\frac{h}{2}\frac{dp}{dx}$ at either wall.

It may be added that this analysis of the laminar flow is valid for horizontal plates. If the plates were inclined to the ground datum, the pressure p must be replaced by the piezometric head $H = \frac{p}{\rho g} + z$. Hence, instead of $\frac{dp}{dx}$ we write $\frac{\rho g\, dH}{dx}$ and obtain for cases (*a*) and (*b*).

$$u = U\frac{y}{h} - \frac{h^2}{2\mu}\frac{\rho g\, dH}{dx}\frac{y}{h}\left(1 - \frac{y}{h}\right) \qquad ...(7.34)$$

and

$$u = \frac{h^2}{2\mu}\frac{\rho g\, dH}{dx}\left(2 - 4\left(\frac{y}{h}\right)^2\right) \qquad ...(7.35)$$

The Reynolds number Re for the flow is given by $\frac{U_{av} h \rho}{\mu}$. It can be shown by an experiment similar to that for a pipe flow that until $Re = 2000$, laminar flow persists ; beyond 2000, laminar flow becomes unstable and changes to turbulent flow with the slightest disturbance anywhere.

Example 7.6. *A flat plate, 1 m x 1 m is pulled over a flat surface up a 30° incline at 1 m/s as shown. The oil between the flat surfaces has μ = 0.06Ns/m² and p = 900kg/m³.*

Calculate (a) the force and power required to maintain the speed against a pressure difference of 10 N/cm² and (b) the discharge rate of oil.

Solution.

For an inclined plate,

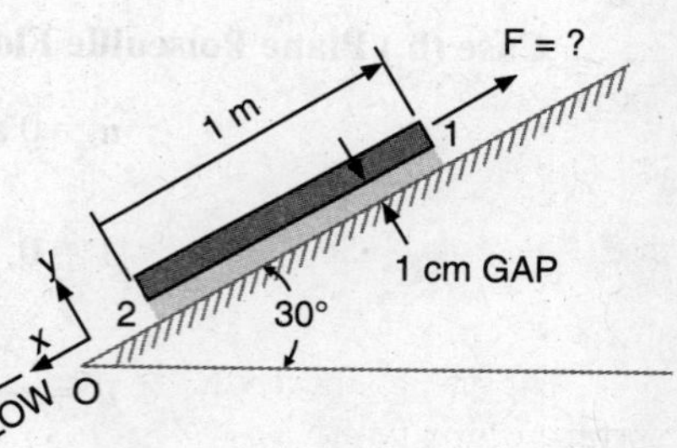

$$\rho g\frac{dH}{dx} = \frac{dp}{dx} + \rho g\frac{dz}{dx}$$

$$= -\frac{10\times10^4}{1} - 900\times9.81\times0.5$$

$$= -104,414.5 \text{ N/m}^3 \text{ along the flow}$$

By Eq. (7.34),

$$u = 1\times\frac{y}{0.01} + \frac{104\,414.5\times0.01^2}{2\times0.06}\times\frac{y}{0.01}\left(1 - \frac{y}{0.01}\right)$$

$$= 8801.21\, y - 870,120.83\, y^2$$

and

$$\frac{\partial u}{\partial y} = 8801.21 - 1740,241.67\, y$$

$$= 8801.21 - 1740,241.67(0.01) = -8601.21$$

shear stress $\tau = \mu\frac{\partial u}{\partial y} = 0.06\times8601.21 = -516.07 \text{ N/m}^2$

(*a*) Shear force $F = -516.07 \times 1 \times 1 =$ **– 516.07 N**

opposing the motion of the plate.

Power required $P = 516.07 \times 1 =$ **516.07 W**

(*b*) $Q = \int_0^h u\,dy = \left| 8801.21\frac{y^2}{2} - 870,120.83\frac{y^3}{3} \right|_0^{0.01}$

$= \mathbf{0.15\ m^3/s}$ **per unit width of the plate.**

Example 7.7. *A belt conveyer consists of a flat belt 0.5 m wide which slides at a velocity of 4 m/s parallel to a surface separated by a 6 cm thick layer of oil of viscosity 0.25 Ns/m. Determine*

(*a*) *the pressure gradient required to cause no shear stress at the belt surface.*

(*b*) *the average velocity and the discharge of oil to be maintained for the above.*

Solution.

This is a case of Couette flow where the upper surface moves parallel to the lower stationary surface.

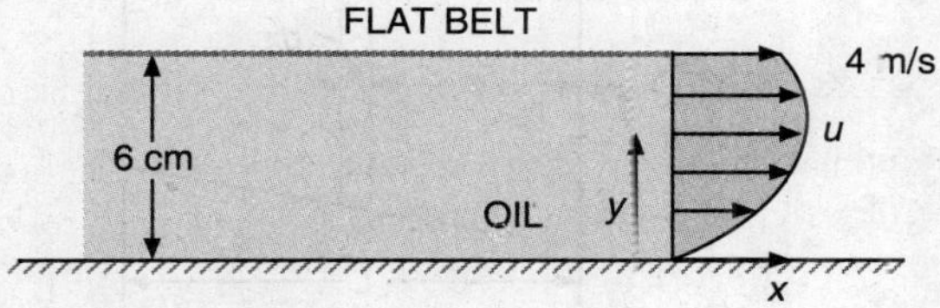

From the expression of velocity for laminar flow,

$$u = \frac{1}{\mu} \times \frac{dp}{dx} y^2/2 + Ay + B$$

with the boundary conditions, $u = 0$ at $y = 0$

and $u = U = 4$ m/s at $y = h = 0.06$ m

Therefore, $u = Uy/h - \frac{h}{2\mu}\frac{dp}{dx} y(1 - y/h)$

and $\frac{du}{dy} = U/h - \frac{h}{2\mu}\frac{dp}{dx}(1 - 2y/h)$

(*a*) For no shear, the velocity gradient must be zero, *i.e.*, the maximum velocity occurs at the surface of the belt:

$$\frac{du}{dy} = 0 \text{ at } y = h$$

Then, $U/h. \frac{h}{2\mu}\frac{dp}{dx}(1 - 2h/h) = 0$

and $\frac{dp}{dx} = \frac{-2\mu U}{h^2} = -\frac{2 \times 0.25 \times 4}{0.06^2}$

$= \mathbf{-\ 555.55\ N/m^3}$

(*b*) The average velocity is given by

$$U_{av} = \int_0^h u\,dy/h$$

$$= U/h^2 \left| y^2/2 \right|_0^h - \frac{1}{2\mu}(-555.55) \left| y^2/2 - y^3/3h \right|_0^h$$

$$= 4/2 + \frac{555.55}{2 \times 0.25} \times (0.06)^2 / 6 = \mathbf{2.67 \ m/s}$$

This result could have been obtained by using the fact that the area under a parabolic curve is two third of the area of the circumscribing rectangle. Since the maximum velocity on the parabolic profile is 4 m/s, the average must be two third of 4, *i.e.*, **2.67 m/s.**

The discharge of the oil required to maintain this condition is

$$Q = 2.67 \times (h \times 0.5) = 2.67 \times (0.06 \times 0.5) = \mathbf{0.08 \ m^3/s}$$

Example 7.8. *A conical contraction is used to reduce the diameter from D_1 to D_2 over a length L in a laminar flow. Determine the pressure loss across it.*

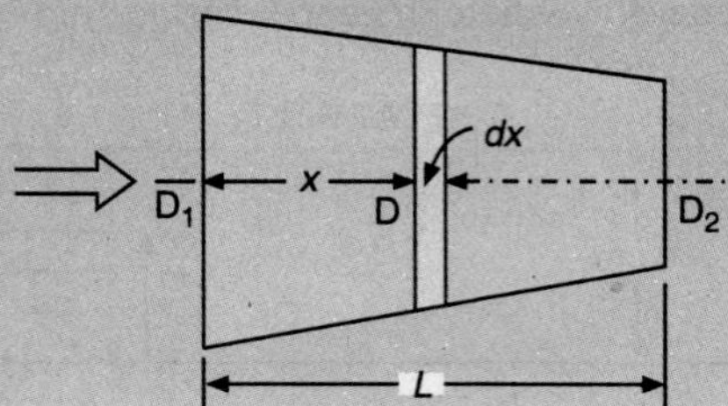

Solution.

The diameter of the conical contraction at a distance x from the inlet is given by

$$D = D_1 - (D_1 - D_2)\, x/L \qquad ...(i)$$

The Hagen-Poiseuille equation for laminar flow through a round pipe of diameter D is expressed as

$$Q = \frac{\pi D^4}{-128\,\mu}\frac{dp}{dx}$$

For a differential length dx, the differential pressure drop is expressed as

$$d(\Delta p) = \frac{128\,\mu\, Q}{\pi D^4}dx \qquad ...(ii)$$

From (*i*) $\quad dD = -(D_1 - D_2)\, dx/L$

whence $\quad dx = -(LdD)/(D_1 - D_2)$

Substituting it in (*ii*),

$$d(\Delta p) = \frac{128\,\mu\, Q}{\pi}\frac{-L}{(D_1 - D_2)}\frac{1}{D^4}dD$$

Integrating both sides to determine the pressure drop across the conical contraction,

$$\Delta p = \frac{128\,\mu\, Q}{3\pi}\frac{-L}{(D_1 - D_2)}\left|\frac{1}{D^3}\right|_{D_1}^{D_2}$$

$$= \frac{128\,\mu\, Q}{3\pi}\frac{-L}{(D_1 - D_2)}\left(\frac{1}{D_2^3} - \frac{1}{D_1^3}\right)$$

Finally $\quad \Delta p = -128\,\mu \frac{Q}{3\pi}\frac{(D_1^2 + D_1.D_2 + D_2^2)}{D_1^3 D_2^3}$

7.5. Laminar Flow between Co-axial Rotating Cylinders

Consider the purely circulatory flow of a fluid contained between two long co-axial rotating cylinders. Let the cylinders be of radii R_1 and R_2 rotating at angular velocities ω_1 and ω_2 respectively. The only component of velocity which may exist is u_θ and it may vary only in the radial direction.

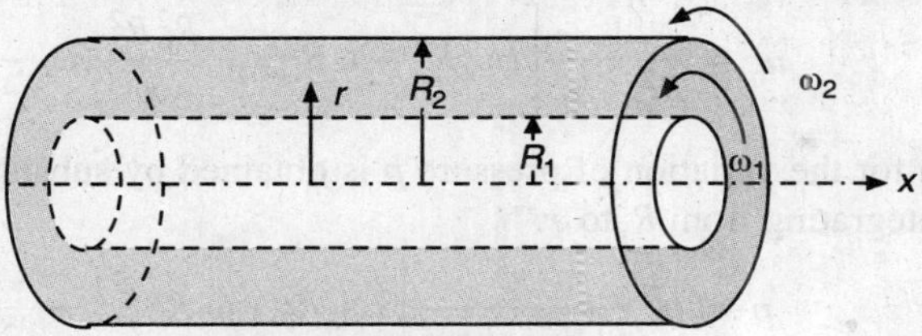

(A) PICTORIAL REPRESENTATION

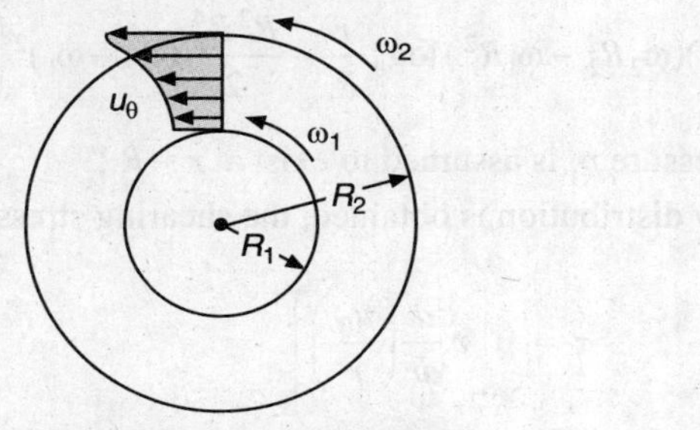

(B) VELOCITY PROFILE

Fig. 7.6: Laminar Flow between Co-axial Rotating Cylinders

The radial momentum equation provides that

$$\rho\frac{u_\theta^2}{r} = \frac{dp}{dr} \qquad ...(7.36)$$

and the circumferential momentum equation reduces to form

$$\frac{d^2u_\theta}{dr^2}+\frac{d}{dr}\left(\frac{u_\theta}{r}\right) = 0 \qquad ...(7.37)$$

Since the velocity varies only in the radial direction, Eq. (7.36) ensures that the pressure may also vary only in the radial direction.

Integrating each term in Eq. (7.37) w.r.t. r,

$$\frac{du_\theta}{dr}+\frac{u_\theta}{r} = C$$

$$\frac{1}{r}\frac{d}{dr}(ru_\theta) = C \qquad ...(7.38)$$

Integrating Eq. (7.38) w.r.t. r,

$$u_\theta = \frac{C_1 r}{2}+\frac{C_2}{r} \qquad ...(7.39)$$

where C_1 and C_2 are constants which may be determined from the boundary conditions. For the case in hand,

At $\qquad r = R_1;\ u_{\theta 1} = R_1\omega_1$

and $\qquad r = R_2;\ u_{\theta 2} = R_2\omega_2$

The values of the constants for these conditions are

$$C_1 = \omega_1+\frac{R_2^2}{R_2^2-R_1^2}(\omega_2-\omega_1) \qquad ...(7.40\ a)$$

$$C_2 = -\frac{R_1^2\ R_2^2}{R_2^2 - R_1^2}(\omega_2 - \omega_1) \quad \text{...(7.40 b)}$$

The expression for velocity μ_θ becomes

$$u_\theta = \frac{1}{R_2^2 - R_1^2}\left[(\omega_2 R_2^2 - \omega_1 R_1^2)r - \frac{R_1^2 R_2^2}{r}(\omega_2 - \omega_1)\right] \quad \text{...(7.41)}$$

The expression for the variation of pressure p is obtained by substituting u_θ from Eq. (7.41) into Eq. (7.36) and integrating from R_1 to r.

$$p = p_1 + \frac{\rho}{(R_2^2 - R_1^2)^2}\left[\left(\omega_2 R_2^2 - \omega_1 R_1^2\right)^2\left(\frac{r^2 - R_1^2}{2}\right)\right.$$

$$\left. -2R_1^2 R_2^2(\omega_2 - \omega_1)(\omega_2 R_2^2 - \omega_1 R_1^2)\log_e \frac{r}{R_1} - \frac{R_1^2 R_2^4}{2}(\omega_2 - \omega_1)^2\left(\frac{1}{r^2} - \frac{1}{R_1^2}\right)\right] \quad \text{...(7.42)}$$

where the reference pressure p_1 is assumed to exist at $r = R_1$;

Once the velocity distribution is obtained, the shearing stress may be evaluated by employing the definition

$$\tau = \mu\left[r\frac{d}{dr}\left(\frac{u_\theta}{r}\right)\right] \quad \text{...(7.43)}$$

and using the expression for u_θ from Eq. 7.41. It results in the following expressions for

$$\tau = \frac{2\mu}{R_2^2 - R_1^2}\frac{R_1^2\ R_2^2}{r^2}(\omega_2 - \omega_1) \quad \text{...(7.44)}$$

The shearing stress at the walls of the inner and outer cylinders may be evaluated by substituting $r = R_1$ and $r = R_2$ respectively.

$$\tau_1 = \frac{2\mu}{R_2^2 - R_1^2}R_2^2(\omega_2 - \omega_1) \quad \text{...(7.45)}$$

$$\tau_2 = \frac{2\mu}{R_2^2 - R_1^2}R_1^2(\omega_2 - \omega_1) \quad \text{...(7.46)}$$

It is interesting to observe that the torque on the inner cylinder would be

$$T_1 = \tau_1 2\,\pi R_1 .\ l R_1$$

$$= \frac{4\mu\pi}{R_2^2 - R_1^2}R_1^2 R_2^2(\omega_2 - \omega_1) \quad \text{...(7.47)}$$

and the torque on the outer cylinder would be

$$T_2 = \tau_2 2\,\pi R_2 .\ l R_2$$

$$= \frac{4\mu\pi}{R_2^2 - R_1^2}R_1^2 R_2^2(\omega_2 - \omega_1) \quad \text{...(7.48)}$$

The torques T_1 and T_2 are equal. This is understandable because the torque on one acts as action on the other and the reaction by that is acted upon the former, the two being equal and opposite by the Newton's third law of motion.

Let us next consider some special cases.

Case (a): When the two cylinders rotate with the same angular velocity. In that case $\omega_1 = \omega_2 = \omega$ From Eq.(7.41),

$$u_\theta = r\,\omega \qquad ...(7.49)$$

and from Eq. (7.42),

$$p = p_0 + \rho r^2\omega^2/2 \qquad ...(7.50)$$

Also from Eq. (7.44), it is noticed that the shear stress is zero everywhere. In this case, the two cylinders together with the fluid undergo solid-body rotation, *i.e.*, $\tau = 0$, it is a case of forced or rotational vortex in which there is no shearing between any two layers in the fluid. The velocity distribution is shown sketched in Fig. 7.7(a).

(a) COAXIAL CYLINDERS ROTATING AT THE SAME ANGULAR VELOCITY

(b) FLOW INSIDE A ROTATING PIPE

Fig. 7.7: Special Cases of Rotation

Case (b): If the inner cylinder does not exist, it reduces to the case of rotational flow in a pipe as shown in Fig. 7.7 (b). For this case

$$R_1 = 0;\ R_2 = R$$

and

$$\omega_1 = 0;\ \omega_2 = \omega$$

Under these conditions, the expression for u_θ reduces to

$$u_\theta = r\,\omega$$

and the expression for p reduces to

$$p = p_0 + \rho r^2\omega^2/2$$

where p_0 is the reference pressure at the centreline; also the expression for τ reduces to

$$\tau = 0$$

which means that the fluid undergoes solid-body rotation.

Case (c): If the outer cylinder does not exist, it reduces to the case of laminar flow around a long rotating cylinder in an expanse of fluid as shown in Fig. 7.7(c). For this case,

$$R_1 = R;\ \omega_1 = \omega$$

and

$$R_2 \to \infty;\ \omega_2 = 0$$

Under these conditions, the expression for u_θ reduces to

$$u_\theta = R^2\omega/r \qquad ...(7.51)$$

and the expression for p reduces to

$$p = p_0 - \rho R^4\omega^2/2r^2 \qquad ...(7.52)$$

where p_0 is the reference pressure at the surface of the cylinder.

The shear stress variation reduces to

$$\tau = -2\mu R^2/r\ \omega$$

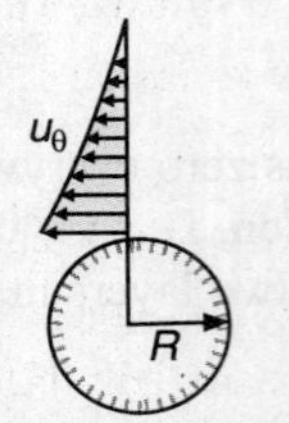

(c) AROUND A ROTATING CYLINDER

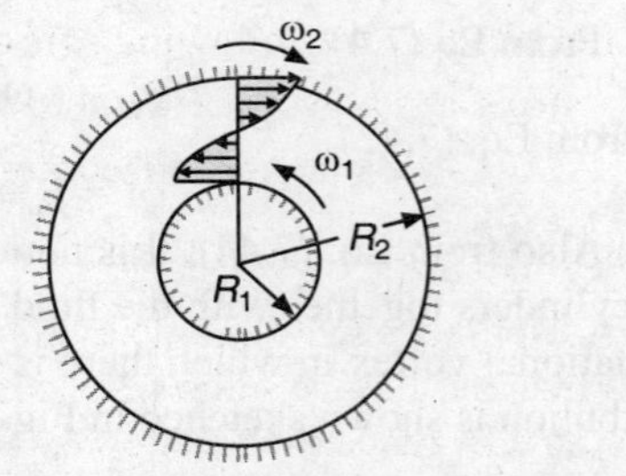

(d) CO-AXIAL ROTATING CYLINDERS

Fig. 7.7: Special Cases of Rotation (Contd.)

Case (d): When the cylinders rotate in different directions, *i.e.*, when the sense of ω_1 is opposed to the sense of ω_2, the equations still apply with appropriate signs for ω_1 and ω_2. Positive signs are used for anticlockwise rotation as viewed from the end, as shown in Fig. 7.7(d).

Case (e): If the cylinders have a narrow space between them, $(R_2 - R_1)$ is very small compared to R_1 or R_2 then

$$R_2 - R_1 = a$$

and

$$R_1 \approx R_2 \approx \frac{(R_1 + R_2)}{2} \approx R$$

Also, $R_2^2 - R_1^2 = (R_2 - R_1)(R_2 + R_1) = 2aR$

The expression for u_θ simplifies to

$$u_\theta = \frac{(\omega_2 - \omega_1)\,R}{2\,a}(r - R^2/r)$$

Shearing stresses at the inner and outer cylinders must be

$$\tau_1 = \frac{\mu R}{a}(\omega_2 - \omega_1) = \tau_2$$

If, in addition, ω, the inner cylinder is held stationary and the outer cylinder rotates at a constant rotational speed ω, the velocity is given by

$$u_\theta = \frac{\omega R}{2a}(r - R^2/r)$$

and the shearing stress is given by

$$\tau = \frac{\mu R\omega}{a}.$$

7.6 Measurement of Viscosity

It is possible to measure the viscosity of a fluid in a large number of ways., In principle,

(*i*) a viscous or laminar flow is created,

(*ii*) the equation governing the flow is identified,

(*iii*) necessary observations are taken, and

(*iv*) the viscosity of the fluid is calculated.

The apparatus used to determine the viscosity of fluid is called viscosity meter or viscometer and the science of determining the viscosity is called viscometry.

Let us study the following commonly used viscometers

(a) Capillary Tube Viscometer **(b) Coaxial Cylinder Viscometer**

(c) Falling Sphere Viscometer and **(d) Efflux Viscometers**

(a) Capillary Tube Viscometer

A capillary tube viscometer consists of a horizontal capillary tube through which the given fluid is made to flow. A large reservoir supplies the fluid at a pressure above the atmosphere at one end of the tube while the other end is open to atmosphere. The pressure in the capillary at a distance *l* upstream the open end is recorded by means of a vertical piezometer. It is necessary to leave a distance *s* between the reservoir and the piezometer because the flow at the entry to the capillary tube is of developing nature, *i.e.*, the velocity profile has yet not become the desired steady laminar velocity profile. If the head of the fluid in the piezometer is *h*, it is obvious that the pressure drop over the fully developed laminar flow length *l* of the tube must be *h*. The level of the fluid in the reservoir is maintained approximately constant so as to ensure steady flow through the tube. The rate of discharge through the tube is estimated by collecting the fluid over a specified period of time. Recalling the expression for *Q*, the Hagen-Poiseuille equation,

$$Q = -\frac{\pi D^4}{128\,\mu}\frac{dp}{dx}$$

which may be recast as

$$\mu = \frac{\pi D^4 \rho g h}{128\, Q l} \qquad ...(7.53)$$

because $$\frac{dp}{dx} = \frac{0-\rho g h}{l} = -\frac{\rho g h}{l}$$

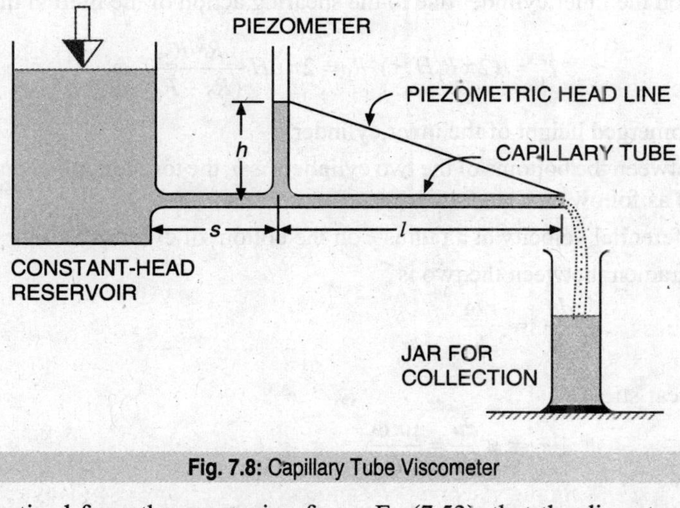

Fig. 7.8: Capillary Tube Viscometer

It may be noticed from the expression for μ, Eq.(7.53), that the diameter *D* is raised to the fourth power whereas all other variables occur as first powers. It follows that any error in *D* would substantially alter the result and hence *the diameter of the capillary tube should be measured most accurately* either by direct or by indirect methods. The viscosity of the fluid is estimated by substituting the observed set of values of ρ, *h*, *Q* and *l* in Eq. (7.53).

(b) Co-axial Cylinder Viscometer

A co-axial cylinder viscometer consists of a cylinder of radius R_1 suspended within a co-axial cylinder of radius R_2. The narrow space between the two cylinders is filled with the fluid, as shown in Fig. 7.9. The outer cylinder is arranged to rotate at a constant rotational velocity while the inner cylinder is held stationary by means of a torsional spring. The torque T acting on the inner cylinder is measured by the torsional strain caused in the spring. The measured torque on the inner cylinder must be equal and opposite to the torque applied on the outer cylinder.

The torque is applied on the outer cylinder to overcome the resistance provided by presence of the fluid in the annular space and at the bottom of the inner cylinder. Since the peripheral speed of the outer cylinder at radius R_2 is

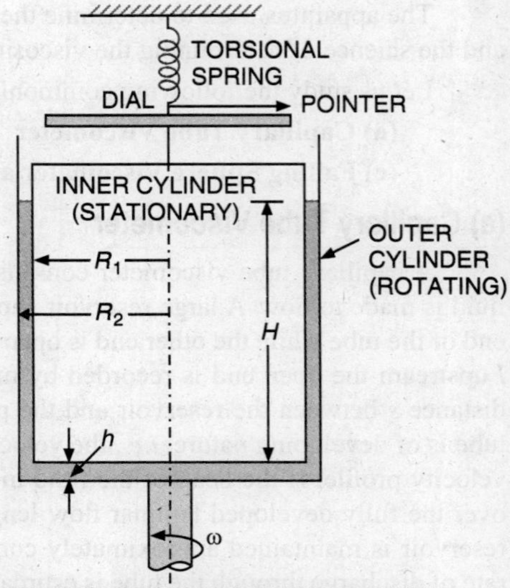

Fig. 7.9: Co-axial Cylinder Viscometer

$$u = R_2 \omega$$

the velocity gradient over the radial distance ($R_2 - R_1$) is

$$\frac{du}{dy} = \frac{R_2 \omega}{(R_2 - R_1)}$$

Invoking the Newton's law of viscosity, the shear stress may be expressed as

$$\tau = \mu \frac{du}{dy} = \frac{\mu R_2 \omega}{(R_2 - R_1)}$$

The torque on the inner cylinder due to the shearing action of the fluid in the annular space is

$$T_1 = (2\pi R_1 H . \tau). R_1 = 2\pi \mu H \frac{R_1^2 R_2}{(R_2 - R_1)} . \omega \qquad ...(7.54)$$

where H is the submerged height of the inner cylinder.

If the gap between the bottoms of the two cylinders is h, the torque applied on the inner cylinder may be calculated as follows:

The circumferential velocity at a radius r on the bottom of cylinder is $u = r\omega$ and the velocity gradient between the two is

$$\frac{du}{dy} = \frac{r\omega}{h}$$

giving rise to a shear stress

$$\tau = \mu \frac{du}{dy} = \frac{\mu r \omega}{h}$$

The elemental torque on a thin circular strip of area $2\pi r\, \delta r$ at a radius r must be

$$T = \tau\, 2\pi r\, \delta r\, r$$

$$= \frac{2\pi \mu\omega}{h} r^3 \delta r$$

The torque acting on the inner cylinder due to the shearing action of the fluid between the bottoms of the two cylinders must be

$$T_2 = \int_0^{R_1} \frac{2\pi\,\mu\,\omega}{h} r^3\, dr = \frac{\pi\,\mu\omega}{2h} R_1^4 \qquad ...(7.55)$$

The total torque acting on the inner cylinder must be

$$T = T_1 + T_2 = 2\pi\,\mu R_1^2 \left(\frac{R_2 H}{(R_2 - R_1)} + \frac{R_1^2}{4h} \right) \omega \qquad ...(7.56)$$

Finally, the dynamic viscosity may be expressed in terms of the torque T acting on the inner cylinder (measured by the strain of the torsional spring), radii R_1 and R_2 the clearance h at the bottom of the inner cylinder, the submerged height H of the inner cylinder and the rotational speed of the outer cylinder,

$$\mu = \frac{2\,(R_2 - R_1)\, hT}{\pi R_1^2 \omega\, [4HhR_2 + R_1^2 (R_2 - R_1)]} \qquad ...(7.57)$$

If the narrow radial space $(R_2 - R_1)$ is specified as a and the radii R_1 and R_2 are nearly to be equal

$$R_1 \approx R_2 \approx R$$

the expression for μ simplifies to

$$\mu = \frac{aT}{\pi R^3 \omega} \Big/ \left(2H + \frac{aR}{2h} \right) \qquad ...(7.58)$$

Further, if the clearance h at the bottom of the inner cylinder is large compared with the radial clearance a, *i.e.*,

$$\frac{a}{h} << 1$$

then, the expression for μ reduces to

$$\mu = \frac{aT}{2\pi R^3\, \omega H} \qquad ...(7.59)$$

It may also be noticed that, for the case of large clearance at the bottom of the inner cylinder, the torque contributed by the shearing action at the bottom is negligible and hence

$$T = T_1 = 2\pi\mu H \frac{R^3}{a} \omega$$

whence

$$\mu = \frac{aT}{2\pi\, R^3\, \omega H} \qquad ...(7.60)$$

which is the same expression as Eq. (7.59).

Example 7.9. *A co-axial cylinder viscometer consists of the inner stationary cylinder of diameter 10 cm and the outer drum of 10.2 cm internal diameter. When the outer drum rotates at 500 revolutions per minute, the torque registered on the torsion meter attached to the inner cylinder is 3.0 Nm. Determine the viscosity of the fluid filled in the viscometer to a depth of 20 cm.*

Solution.

In the absence of information about the clearance at the bottom of the rotating cylinder, it is assumed that the same is large and contributes negligibly to the resisting torque. The expression for the viscosity of the fluid is expressed as

$$\mu = \frac{aT}{2\pi R^3 \omega H}$$

From the data,

$$R_1 = \frac{10}{2} = 5 \text{ cm} = 0.05 \text{ m}$$

$$R_2 = \frac{10.2}{2} = 5.1 \text{ cm} = 0.051 \text{ m}$$

$$R \approx R_1 \approx R_2 = 0.05 \text{ m}$$

the radius of the cylinder on which torque is measured.

$$a = R_2 - R_1 = 0.001 \text{ m}$$

$$\omega = \frac{2\pi \times 500}{60} = 52.4 \text{ rad/s}$$

$$T = 3.0 \text{ Nm}$$

$$H = 0.2 \text{ m}$$

Substituting the same in the expression for viscosity,

$$\mu = \frac{0.001 \times 3.0}{2\pi \times (0.05)^3 \times 52.4 \times 0.2}$$

$$= \mathbf{0.365\ Ns/m^2}.$$

(c) Falling Sphere Viscometer

A falling sphere viscometer consists of a tall vertical transparent tube which is filled with the fluid as shown in Fig. 7.10. Provision is made to release and drop a small spherical ball. After a short distance of travel, the ball acquires a constant i.e., terminal velocity. The time to traverse a known vertical distance between two fixed marks on the tube is noted to calculate the terminal velocity U of the ball. The equilibrium of the ball under the steady state condition requires that the drag force and the buoyancy force must add up to balance the weight of the ball. For a ball of diameter d moving at a velocity U in a fluid of viscosity μ the drag is given by the Stokes law,

$$D = 3\pi \mu U d \quad \text{...(7.61)}$$

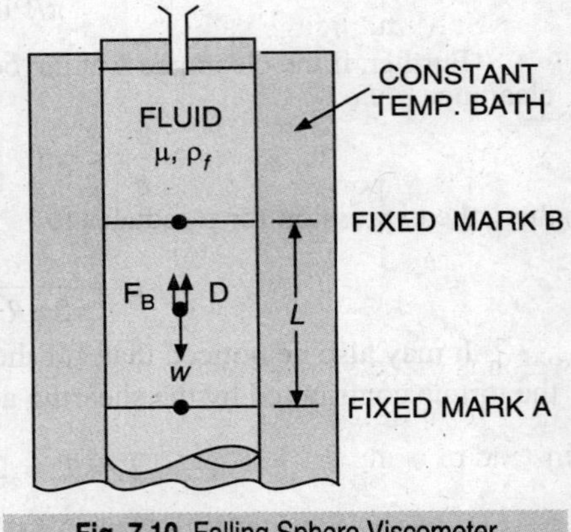

Fig. 7.10. Falling Sphere Viscometer

If the density of the solid ball is ρ_s and the density of the fluid is ρ_f the buoyancy force and the weight must be

$$F_B = \frac{\pi d^3}{6} \rho_f g$$

The weight, $$W = \frac{\pi d^3}{6} \rho_s g$$

For equilibrium,

$$3\pi \mu U d + \frac{\pi d^3}{6} \rho_f g = \frac{\pi d^2}{6} \rho_s g$$

whence

$$\mu = \frac{g d^2 (\rho_s - \rho_f)}{18 U} \quad \text{...(7.62)}$$

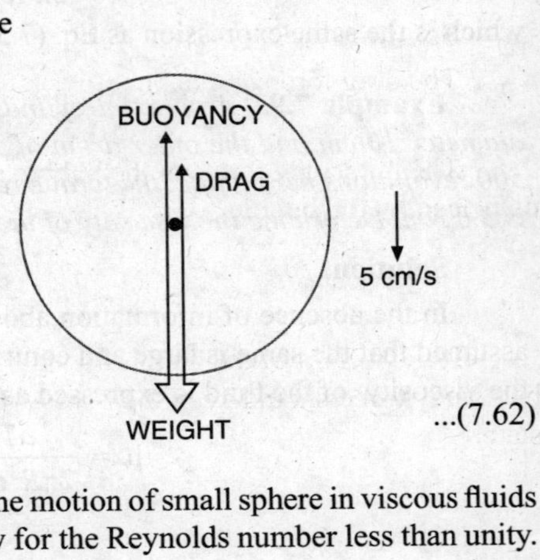

It may be noted that this relationship is true only for the motion of small sphere in viscous fluids because the expression for drag as used above is valid only for the Reynolds number less than unity.

Example 7.10. *A viscometer consisting of a long vertical tank containing a fluid at the desired ambient conditions and a small sphere is employed for determining the viscosity of an oil. If a steel sphere (specific gravity 8.0) of 1 cm diameter falls with a constant velocity of 5 cm/s through a lubricating oil (specific gravity 0.9), calculate the viscosity of the oil.*

Solution.

During the fall of the sphere with the constant, i.e., terminal velocity, it must be in equilibrium under the forces due to drag, buoyancy and its weight.

$$D = 3\pi\ \mu\ Ud,\ upward$$

$$F_B = \frac{\pi d^3}{6}\rho_f g,\ upward$$

$$\text{Weight} = \frac{\pi d^3}{6}\rho_s g,\ downward$$

where ρ_f and ρ_s are the mass densities of the fluid and sphere material respectively. For equilibrium,

$$3\pi\mu\ Ud + \frac{\pi d^3}{6}\rho_f g = \frac{\pi d^3}{6}\rho_s g$$

and

$$\mu = \frac{gd^2(\rho_s - \rho_f)}{18U}$$

Substituting the data,

$$\mu = \frac{9.81\times 0.01^2\times(8.0-0.9)\times 1000}{18\times 0.05} = \mathbf{7.74\ Ns/m^2}$$

It is important to note that the expression for drag $D = 3\pi\ \mu\ Ud$ is valid only for sub-laminar conditions, $Re < 1.0$. It is, therefore, necessary to check for this fact:

$$Re = \frac{5}{100}\times\frac{1}{100}\times\frac{0.9\times 1000}{7.74} = 0.058$$

which is indeed much less than 1.0.

If Re were greater than 1.0, we should be advised to look up the drag characteristic for a sphere instead of using the Stokes law and proceed as before.

Example 7.11. *A small air bubble of diameter 1.0 mm rises with a steady state velocity of 1.5 cm/s through an oil. Estimate the viscosity of the oil if its specific gravity is 0.9.*

Solution.

The drag force on the bubble due to the viscous resistance provided by the oil is

$$D = 3\pi\mu.\ Ud$$

The buoyancy force on the bubble equals the weight of the oil displaced by the bubble,

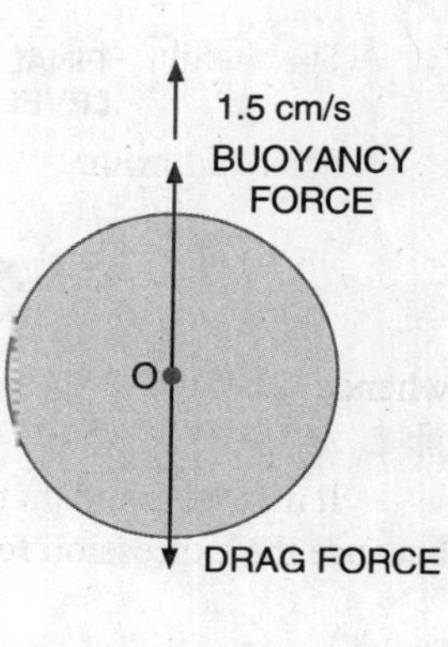

$$F = \frac{\pi d^3}{6}\rho g$$

If the weight of the bubble is neglected, the drag force and the buoyancy force should keep the bubble in equilibrium at the steady state.

$$3\pi\mu\ Ud = \frac{\pi d^3}{6}\rho g$$

whence the dynamic viscosity $\mu = \dfrac{\rho g\, d^2}{18\,U}$

For the given case,

$$d = 1.0 \text{ mm} = 0.001 \text{ m}$$
$$U = 1.5 \text{ cm/s} = 0.015 \text{ m/s}$$

Therefore, $$\mu = \frac{(1000\times 0.9)\times 9.81\times (0.001)^2}{18\times 0.015}$$

$$= \mathbf{32.4 \times 10^{-3}\ Ns/m^2}$$

The Reynolds number is given by

$$Re = \frac{Ud\rho}{\mu} = \frac{0.015\times 0.001\times 1000\times 0.9}{32.4\times 10^{-3}} = 0.41$$

which is less than unity. It is necessary to check for the Reynolds number because the expression for drag is valid only for Reynolds number less than unity.

(d) Efflux Viscometers: Redwood Viscometer

A number of efflux viscometers have been developed in different countries and at different times. An efflux viscometer consists of a container maintained at a desired temperature and has an aperture such as an orifice, a nozzle or a capillary through which a fixed volume of a liquid is drained out. The time taken to discharge the fixed volume of liquid under those standard conditions is considered as a measure of the viscosity of the liquid. Efflux viscometers can not provide the viscosity in absolute units. They may, however, be calibrated with some standard liquid and the viscosity may be expressed in terms of ratios of time taken w.r.t. the standard liquid.

In the Redwood efflux viscometer, Fig.7.11, the time in seconds for 50 ml of the liquid to flow out through an orifice outlet is a measure of the viscosity. This is expressed as `Redwood seconds'.

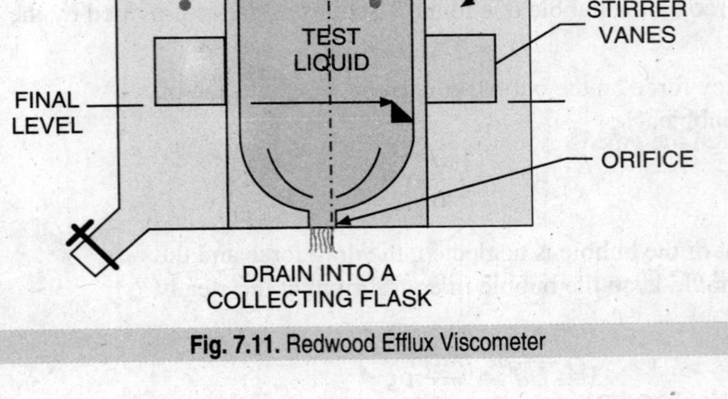

Fig. 7.11. Redwood Efflux Viscometer

SUPPLEMENTARY REVISION EXAMPLES

Example. 7.12. *A viscous liquid is flowing in the laminar regime in a 6 cm diameter circular pipe. A pitot tube at a radial distance of 2 cm from the axis indicates a velocity of 6.6 m/s. Calculate the maximum velocity, mean velocity and the discharge through the pipe.*

Solution.

In laminar flow, the velocity profile is parabolic. At a distance r from the axis, the velocity u is given by

$$u = \left[1 - \left(\frac{r}{R}\right)^2\right] K.$$

where K depends upon the viscosity and pressure gradient, which are constant across the section.

Given that

$$u = 0.6 \text{ m/s at } r = 0.02 \text{ m; } R = 0.03 \text{ m.}$$

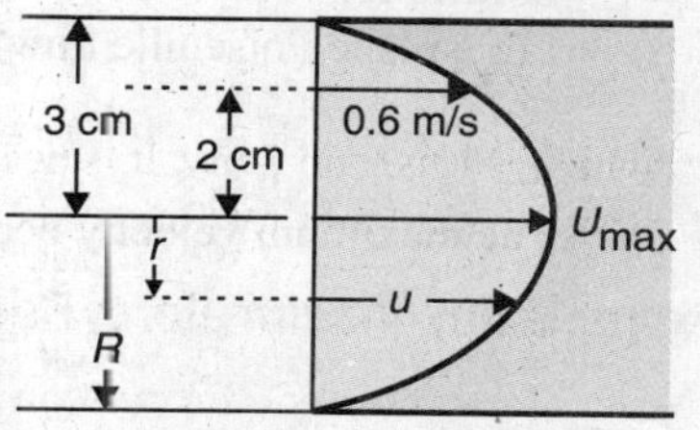

On substitution, $0.6 = \left[1 - \left(\frac{0.02}{0.03}\right)^2\right] K = 5/9\ K$

whence, $K = 1.08$ in SI units.

Velocity is maximum at the axis, $r = 0$

$$U_{max} = (1 - 0) \times 1.08 = \mathbf{1.08\ m/s}$$

Mean velocity for parabolic profile is half of maximum, *i.e.*, U_{mean} = **0.54 m/s**

Discharge = U_{mean} × area = 0.54 × π × 0.3² = **0.153 m³/s**

Example 7.13. *The maximum velocity of laminar flow in a pipe of diameter 240 mm is 2.8 m/s. Determine the mean velocity and the radius at which it occurs. Also, find the velocity at 40 mm from the wall of the pipe.*

Solution.

For fully developed laminar flow in a pipe,

$$u = u_{max}\left(1 - \left(\frac{r}{R}\right)^2\right)$$

For the data,

$$u = 2.8\left(1 - \left(\frac{r}{0.12}\right)^2\right).$$

At 40 mm for the wall, $r = 0.12 - 0.04 = 0.08$ m

$$u = 2.8\left(1 - \left(\frac{0.08}{0.12}\right)^2\right) = \mathbf{1.56\ m/s.}$$

Flow through the pipe is given by

$$Q = \int_0^R u \times 2\pi r\, dr$$

Substituting $u = u_{max}\left(1 - \left(\frac{r}{R}\right)^2\right)$ and integrating,

$$Q = \pi R^2 .\, u_{max}/2$$

Mean velocity = $Q/\pi R^2 = u_{max}/2$, *i.e.*, 50% of the maximum velocity = **1.4 m/s.**

Now, $$1.4 = 2.8\left(1-\left(\frac{r}{0.12}\right)^2\right);$$

$$1-\left(\frac{r}{0.12}\right)^2 = 0.5;$$

whence $$\frac{r}{0.12} = 0.707$$

and r = **0.085 m** from the centre of the pipe.

Example 7.14. *A liquid of relative density 0.85 flows between two parallel plates, separated by a distance of 2 mm, with a velocity of 0.75 m/s. If two pressure gauges located at a distance of 12 m apart record a difference of 50 kN/m², estimate the viscosity of the fluid.*

Solution.

For a plane Poiseuille flow between two stationary plates, h apart,

$$u = -\frac{h^2}{8\mu}\frac{dp}{dx}(1-4(y/h)^2)$$

The maximum velocity occurs at the midplane, $y = 0$

$$U_{max} = -\frac{h^2}{8\mu}\frac{dp}{dx}$$

and the average velocity

$$U_{av} = 2/3\, U_{max} = -\frac{h^2}{12\mu}\frac{dp}{dx}$$

From the data,

$$h = 0.002 \text{ m};\ U_{av} = 0.75 \text{ m/s};\ \rho = 850 \text{ kg/m}^3.$$

$$\frac{dp}{dx} = -\frac{50\times 1000}{12} = -4167 \text{ N/m}^2 \text{ per meter.}$$

By substitution,

$$\mu = -\frac{h^2}{12\times U_{av}}\cdot\frac{dp}{dx}$$

$$= \frac{0.002^2}{12\times 0.75}\times 4167 = \mathbf{1.85\times 10^{-3}\ Ns/m^2}$$

Kinematic viscosity is given by

$$\nu = \mu/\rho = 1.85\times 10^{-3}/850 = \mathbf{2.18\times 10^{-6}\ m^2/s.}$$

Example 7.15. *Hydrogen bubbles are used to visualise the streak lines in the video, 'Flow Visualisation.' A typical bubble diameter is 0.025 mm. The bubble tends to rise slowly in water and ultimately attain a terminal velocity relative to water. Assuming the local pressure p = 836 mm of mercury, temperature t = 20° C and taking the viscosity of water μ = 1 cP, estimate the terminal speed of a bubble rising in water. Assume that the Stoke's law is applicable.*

Solution.

From the data

diameter of the bubble $d = 0.025\times 10^{-3}$ m.

$$h = \frac{836}{1000} \text{ m of mercury}$$

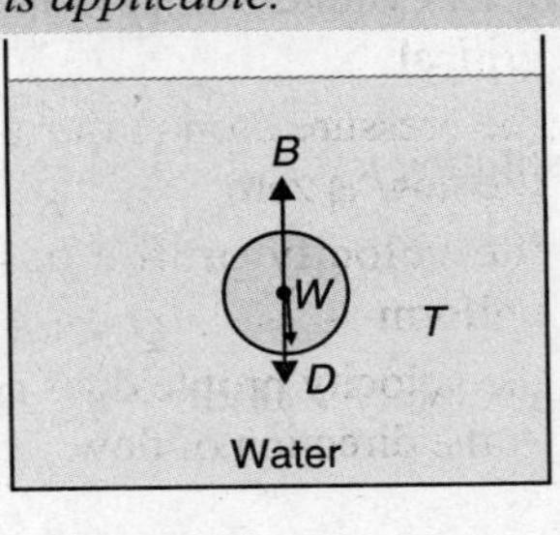

$$p = \rho g h = 13.6 \times 1000 \times 9.81 \times \frac{836}{1000}$$

$$= 111{,}536 \text{ N/m}^2.$$

$$T = 20 + 273 = 293 \text{ K}.$$

The density of hydrogen bubble must be

$$\rho = p/RT = 111536/(287 \times 293) = 1.33 \text{ kg/m}^3.$$

For the terminal (upward) speed U, the bubble must be in equilibrium;

Buoyancy B = Weight of the bubble W + Drag D.

Employing the Stokes law $D = 3\pi \mu Ud$

and substituting the viscosity $\mu = 1\ cP = 0.001$ Ns/m²

$$\frac{\pi d^3}{6} \times 1000 \times 9.81 = \frac{\pi d^3}{6} \times 1.33 \times 9.81 + 3\pi \mu Ud$$

Substituting $d = 0.025 \times 10^{-3}$ m,

$$U = \mathbf{3.4 \times 10^{-4}\ m/s = 0.34\ mm/s.}$$

ITEM BANK

Objective items of different types are given to review the concepts developed in the Chapter.

* *Select the correct response for each item.*
* *Every correct response is expected in less than a minute and carries one mark.*
* *Compare your responses with the key answers given at the end.*

Multiple-Choice Items

Attempt the following Multiple-Choice Items by ticking the correct or the most appropriate response from among the alternatives given:

1. The flow of a fluid is sure to be laminar when

(*a*) the fluid is ideal

(*b*) the fluid is viscous

(*c*) the Reynolds number is less than a critical value

(*d*) the velocity of fluid is more than a critical value

2. The term fully-developed laminar flow means that

(*a*) the Reynolds number for the flow is critical

(*b*) the pressure drop in the direction of the flow is zero

(*c*) the velocity profile has become uniform

(*d*) the velocity profile does not change in the direction of flow.

3. The flow in a capillary tube is laminar because

(*a*) the capillary tube is made of glass

(*b*) the surface tension and capillarity promote laminarisation

(*c*) only inviscid liquids can flow through a capillary tube

(*d*) the diameter of the capillary tube is very small

4. For a laminar flow through a pipe, the discharge varies

(*a*) as the square of the diameter

(*b*) as the inverse of viscosity

(*c*) inversely as the pressure gradient

(*d*) as the square of viscosity

5. The lower critical Reynolds number is

(*a*) the Reynolds number at which laminar flow changes to turbulent flow

(*b*) the Reynolds number at which turbulent flow changes to laminar flow

(c) the least Reynolds number for laminar flow ever obtained
(d) is different for different fluids.

6. The shear stress between two fixed parallel plates with a laminar flow between them
(a) is constant across the gap
(b) varies parabolically as the distance from the mid-plane
(c) varies inversely as the distance from the mid-plane
(d) varies directly as the distance from the mid-plane

7. The shear stress in a round pipe with a laminar flow in it
(a) remains constant over the cross section
(b) varies inversely as the distance from the mid-plane
(c) varies directly as the distance from the mid-plane
(d) varies parabolically across the cross-section

8. The terminal velocity of a small sphere falling under the action of gravity in a very viscous fluid varies
(a) directly as the diameter of the sphere
(b) directly as the viscosity of the fluid
(c) inversely as the diameter of the sphere
(d) inversely as the viscosity of the fluid.

9. For laminar flow through a round pipe, the shear stress
(a) remains constant over the cross section
(b) varies linearly with the radial distance
(c) must be zero at all points
(d) varies parabolically with the radial distance

10. For laminar flow in a round pipe, the energy correction factor is
(a) 0.50 (b) 1.00 (c) 1.33 (d) 2.0

11. For laminar flow in a round pipe, the momentum correction factor is
(a) 0.50 (b) 1.00 (c) 1.33 (d) 2.0

12. For the same discharge in a pipe with laminar flow, reduction of radius to half will increase the pressure gradient by a factor of
(a) 2 (b) 4 (c) 8 (d) 16.

13. The velocity distribution for laminar flow between two stationary parallel plates is
(a) linear (b) parabolic
(c) elliptic (d) logarithmic

Other Objective Items

14. Are the following statements true or false? State why?
(a) The pressure gradient for laminar flow in a pipe must be constant. T/F
(b) The average velocity for Couette flow, *i.e.*, laminar flow between two parallel plates, is half the maximum velocity. T/F
(c) The average velocity for laminar flow in a round pipe is half the maximum velocity. T/F
(d) Laminar flow cannot be irrotational. T/F

15. Identify the following statements as true or false for laminar flow:
(a) Pressure drop in a pipe is proportional to velocity. T/F
(b) Pressure drop increases with the fluid density. T/F
(c) The maximum velocity occurs at the mean radius for flow through a co-axial annulus. T/F
(d) The critical Reynolds number for pipe flow is 2000. T/F

16. Match the values in Block B with the descriptions in Block A for laminar flows:

Block A	Block B
(a) Ratio of maximum to average velocity for flow through a pipe	(k) 1.33
(b) Ratio of maximum to average velocity	(l) 1.50
(c) Kinetic energy correction factor for pipe flow	(m) 2.00
(d) Momentum correction factor for pipe flow	(n) 2.50

ANSWERS

1. (c), **2.** (d), **3.** (d), **4.** (b), **5.** (c), **6.** (d), **7.** (c), **8.** (d), **9.** (b), **10.** (e), **11.**(c), **12.**(d), **13.** (b), **14.** (T F T T), 15 (T F F T), 16 (am, bl, cm dk).

QUESTION BANK

Questions on theory and numerical problems are given for tutorial work.

* *The questions are arranged generally in order of articles in the text*
* *Every question should take 15-20 minutes to solve.*
* *Answers for the numerical problems are given alongside them.*

7.1. Calculate the power required to maintain a flow of 5.5 litres/second of an oil of viscosity 1.0 N/m² through a straight length of 1.0 km of a 75 mm diameter pipe. Assume the flow to be laminar. **[39 kW]**

7.2. Compute the head loss per kilometrs length of a pipe designed to carry oil of specific gravity 0.85 and kinematic viscosity 0.6×10^{-3} m²/s in a 15 cm diameter pipe at a rate of 0.030 m³/s **[147.67 m/km length]**

7.3. Kerosene of kinematic viscosity 0.0015 m²/s and dynamic viscosity 3.2×10^{-3} Ns/m² flows under the action of gravity under a head of 1 m through a 100 m long pipe of 6 mm diameter. Determine the discharge. **[8×10^{-7} m³/s]**

7.4. Glycerine of density 1250 kg/m² and viscosity 0.72 kg/ms flows through a pipe of 80mm diameter. If the shear stress at the wall is 300 N/m², calculate the

(*a*) pressure gradient along the flow, (*b*) average velocity in the pipe,
(*c*) rate of discharge, and (*d*) Reynolds number of the flow.

[15kN/m² per m, 4.2 m/s, 0.021 m³/s, 580]

7.5. A horizontal concentric annulus with radii 0.05 m and 0.035 m conveys a liquid of density 1100 kg/m³ and viscosity 0.1 N s/m² when maintained at pressure gradient of 4000 N/m² per metre. Calculate the volumetric flow rate. **[3×10^{-4} m³/s]**

7.6. Calculate (*a*) the discharge and (*b*) the power required to pump a liquid of specific gravity 0.8, viscosity 0.01 Ns/m² through a pipeline 2 cm diameter, 100 m long operating at Reynolds number of 500. **[0.0785 kg/s, 2.5 W]**

7.7. An oil having a viscosity of 0.048 kg/ms flows through a 50 mm diameter tube at an average velocity of 0.12 m/s. Calculate the pressure drop in 65 m of tube and the velocity at 10 mm from the tube wall. **[4.8 kN/m², 0.153 m/s]**

7.8. Calculate (*a*) the pressure gradient, (*b*) the average velocity and (*c*) the Reynolds number of operation if the maximum shear stress (at the wall) is measured as 200 N/m² for the flow of glycerine, m = 7 poise and specific gravity 1.3 in a pipe of diameter 10 cm. **[8000 N/m² per m, 3.57 m/s, 663]**

7.9. Determine (*a*) the pressure gradient along flow, (*b*) the average velocity and (*c*) the discharge for an oil of viscosity 0.02 Ns/m² flowing between two stationary parallel plates 1 m wide maintained 1 cm apart. The velocity midway between the plates is 2 m/s. **[3200 N/m² per m, 133 m/s, 0.0133 m³/s]**

7.10. Plot the

(*a*) velocity distribution and

(*b*) shear stress distribution for laminar flow between two flat plates 2 cm apart (*i*) when the upper plate is moved at 3 m/s and the lower held stationary,

(*ii*) when the upper plate is moved at 2 m/s and the lower plate at –1m/s.

The fluid between the plates is oil of specific gravity 0.9 and viscosity 0.02 Ns/m².

7.11. A thin film of oil, thickness z and viscosity μ, flows down an inclined plate. Show that the velocity profile is given by

$$u = \frac{\rho g}{2\mu}(z^2 - y^2)\sin\theta$$

where μ is the local velocity at a depth y below the free surface, θ is the plate inclination to the horizontal and ρ is the fluid density.

7.12. What must be the diameter of a mist droplet which is observed to fall with a velocity of 2 cm/s through air of viscosity 0.000 018 Ns/m² ? **[0.0257 mm]**

7.13. For laminar flow in a round tube of radius r, determine the position of the average cross-sectional velocity from the tube wall. **[0.293 × radius]**

7.14. (*a*) For laminar flow in a tube show that the mean velocity is half the centreline maximum velocity.

(*b*) The centreline velocity is measured as 3 m/s in a 0.1 m diameter tube. If the fluid flowing has a density of 1260 kg/m³ and a viscosity of 0.9 Ns/m², determine whether the flow is laminar and calculate the pressure gradient necessary.

[Re = 210; laminar; 4320 kN/m² per m]

7.15. Oil of dynamic viscosity 8.35 poise flows through two parallel plates kept 2 cm apart. If the velocity midway between the plates is 6 m/s, what is pressure gradient in the direction of flow? Determine also the average velocity. **[– 100 N/m³, 4 m/s]**

7.16. Crude oil of dynamic viscosity 1.5 poise and specific gravity 0.9 flows through a 20 mm diameter vertical pipe. Two pressure gauges have been fixed at 20 m apart. The pressure gauge fixed at higher level B reads 2 kg_f/cm^2 and the lower level A reads 6 kg/cm². Find the direction and rate of flow through the pipe. Take 1 kg = 9.81 N.

[U = 0.9 m/s from A to B; Q = 0.283 × 10⁻³ m³/s]

7.17. (*a*) What do you understand by the following terms:

(*i*) laminar flow, (*ii*) smooth turbulent flow and (*iii*) rough turbulent flow in a pipe?

(*b*) A pipe line, 30 cm in diameter and 3200 m long, is used to pump up 50 kg per second of an oil whose density is 950 kg/m³ and whose kinematic viscosity is 2.1 stokes. The centre of the pipe line at the upper end is 40 m above that at the lower end. The discharge at the upper end is atmospheric. Find the pressure at the lower end.

[Re = 1064, laminar flow, 58.15 m head of water]

7.18. A liquid flows between two parallel horizontal plates. The flow is fully developed. The x-direction is horizontal and is measured positively from left to right. Sketch the velocity profile and if it is of a simple geometric shape name it in the following three cases:

(*i*) The lower plate is at rest, the upper plate moves with a steady velocity in the positive x-direction and dp/dx is zero.

(*ii*) Both plates are at rest and dp/dx is negative.

(*iii*) The lower plate is at rest, the upper plate moves with a steady velocity in the positive x-direction and dp/dx is positive.

7.19. An oil of specific gravity 0.85 and centipoise dynamic viscosity 1.7 is flowing through a smooth pipe of 150 mm diameter at the rate of 20 litres/min. Determine whether the flow is laminar or turbulent. Draw the velocity distribution at any transverse cross-section of the pipe. What will be the pressure drop for 100 m of this pipe?

[Re = 1415; Laminar, Parabolic; U = 0.0189 m/s and 4.57 Pa]

7.20. An oil having a viscosity of 0.096 Ns/m^2 and a specific gravity of 1.59 flows through a horizontal pipe of 50 mm diameter with a pressure drop of 6 kN/sup2 per metre length of the pipe. Determine (*i*) the rate of flow in kg/s, (*ii*) the shear stress at the pipe wall, and (*iii*) the power required for 100 m length of the pipe to maintain the flow. (Consider the flow to be steady, laminar and fully developed). **[15.26 kg/s; 75 N/m^2; 5.76 kW]**

7.21. A capillary tube viscometer with a bore of 2 mm is employed to determine the viscosity of an oil of specific gravity 0.9. The head loss is 10 cm over a length of 50 cm and the time taken to collect 0.000 11m^3 of oil is 500 seconds,

(*a*) calculate the viscosity of the oil and (*b*) compute the Reynolds number.

If the same viscometer is used for an oil of twice the viscosity of this oil, what would be the corresponding observations? **[0.00315 Ns/m^2, 40]**

7.22. A lead sphere of 1.0 mm diameter acquired a terminal velocity of 7 mm/s in an oil, while falling down. Assuming that the density of lead is 11000 kg/m^3 and that of the oil is 1200 kg/m^3, compute

(*a*) the Reynolds number and (*b*) the kinematic viscosity of the oil. **[0.167, 4.19 × 10^{-5} m^2/s]**

7.23. A film of fluid, density 2400 kg/m^3 flows down a vertical plate with a free surface velocity of 0.75 m/s. If the film is 20 mm thick determine the fluid viscosity. **[1.25 Ns/m^2]**

7.24. (*a*) Explain the principle and working of a rotary viscometer with a neat sketch.

(*b*) The lower end of a vertical shaft rests in a footstep bearing. Assuming that the end of shaft and the surface of the bearing are both flat and are separated by an oil film of 0.5 mm thickness, find the torque required to rotate the shaft at 750 revolutions per minute. The diameter of the shaft is 100 mm and the dynamic viscosity of oil is 1.5 poise.

[0.232 Nm]

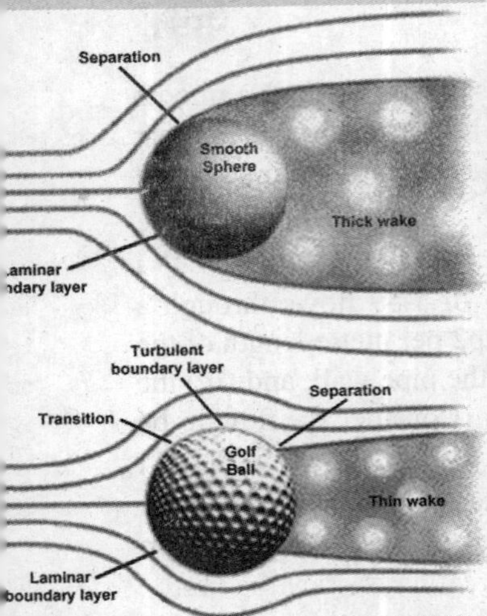

CHAPTER

8

Boundary Layer Flow

LEARNING OBJECTIVES

On successful completion of all learning activities planned and provided in the chapter, a student will be able to

- explain the development of boundary layers in external and internal flows;
- describe the rationale for the different estimates of boundary layer thickness ;
- establish the boundary layer equations for a two-dimensional incompressible flow;
- demonstrate the phenomena of turbulence, separation and recirculation of flow and suggest some means to control the separation of flow.

8.1. Introduction

When a real fluid flows past a solid surface, the velocity of the fluid at the solid surface must be the same as that of the solid surface. If the solid surface is stationary, the velocity of the fluid at the surface must be zero. As a result, there is a region close to the surface through which the velocity increases from the zero velocity at the solid surface to the velocity of the mainstream. This region in the vicinity of the solid surface is generally a narrow region where the velocity gradients are large.

The narrow region, near the solid surface, over which velocity gradients and shear stresses are large, is known as boundary layer. The subject concerned with the study of velocity gradients, shear stresses, forces and energy loss in the boundary layer is called boundary-layer theory.

The simplest case of boundary layer formation is when a free-stream approaches parallel to a sharp-edged, thin, smooth flat plate. The fluid, with a uniform free-stream velocity U, gets retarded in the vicinity of the solid surface and the boundary layer region begins at the sharp leading edge. At subsequent points downstream the leading edge, the boundary layer region increases because the retarded fluid proceeds and it is further retarded.

This is also referred to as the 'growth' of the boundary layer. Having defined the boundary layer as a region near a wall where velocity gradients and shear stresses are significant, it should be stated

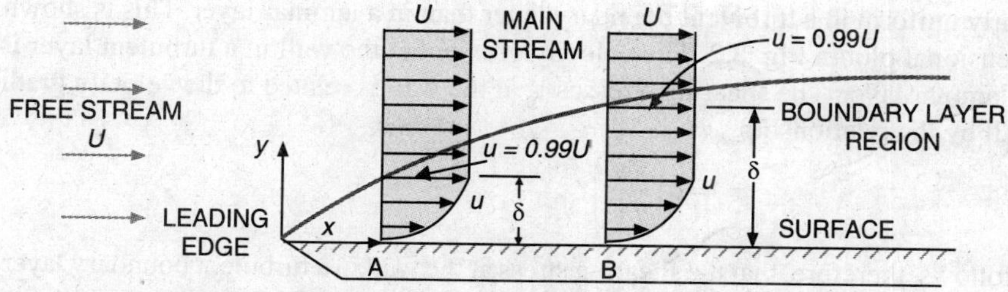

Fig. 8.1: Boundary-Layer Region over a Flat Plate (y-scale enlarged for clarity)

that the same are not so large and significant in the entire boundary layer. The velocity should level off to the free-stream uniform velocity and the shear stresses should vanish at the outer edge of the layer where the layer merges with the main stream. As a matter of fact, the velocity gradient at the outer edge of the layer has to be asymptotic and an arbitrary line of demarcation is drawn to mark the end of boundary-layer region. The limit is placed at a distance where the velocity reaches 99% of the maintstream velocity and the distance where 99% velocity is reached is called the boundary-layer thickness, usually denoted by δ. Mathematically, for the regions,

$$y < \delta \qquad 0 < u < 0.99\,U$$
$$y = \delta \qquad u = 0.99\,U$$
$$y > \delta \qquad 0.99 < u < U$$

These aspects of a boundary-layer are shown in Fig. 8.1. The boundary layer thickness at a point *A* near the leading edge is smaller than the thickness at a point *B* farther from the leading edge; the *y*-scale is enlarged for the sake of clarity.

8.2. Development of Boundary Layer

The boundary layer starts at the leading edge of a solid surface and the boundary layer thickness increases with the distance *x* along the surface. Very important and interesting changes take place in the nature of flow as the boundary-layer grows. This is referred to as the development of boundary-layer.

Near the leading edge of the solid surface, where the thickness is small, the flow in the boundary layer can be laminar and the layer is said to be laminar boundary layer. The velocity profile in the laminar layer may be approximated by a parabola. As the thickness of the layer increases in the downstream direction, the laminar layer becomes unstable and the motion within it is disturbed and irregular which leads to a transition from laminar to turbulent boundary layer. The short length over which the nature of flow changes from laminar to turbulent is called transition zone. Further downstream of the transition zone, the boundary layer is turbulent and continues to grow in thickness. The velocity profile in the turbulent boundary layer is not parabolic but follows a logarithmic law. Turbulent flow is characterised by greater interchange of mass, momentum and energy within the fluid particles than that in the case of laminar flow. The velocity profile is consequently

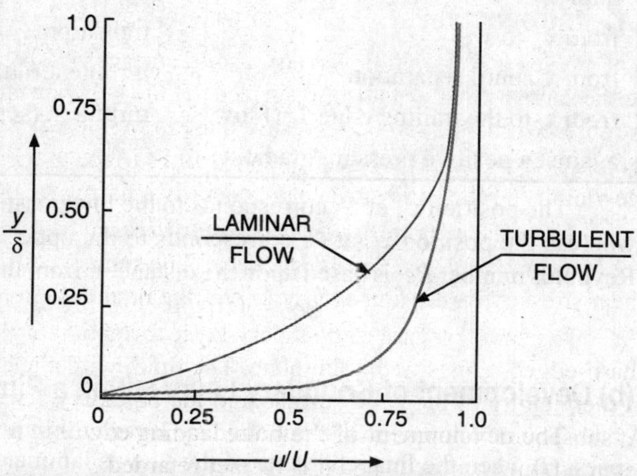

Fig. 8.2: Typical Velocity Profiles in Laminar and Turbulent Boundary Layers

more nearly uniform in a turbulent boundary layer than in a laminar layer. This is shown clearly in a non-dimensional plot in Fig. 8.2. The velocity gradient at the wall in a turbulent layer is larger than that in a laminar layer. The shearing stresses τ_0 at the wall is related to the velocity gradient $(\partial u/\partial y)$, at the wall by the relationship

$$\tau_0 = \mu\left(\frac{\partial u}{\partial y}\right). \quad ...(8.1)$$

It follows, therefore, that the shearing stress at the wall in a turbulent boundary layer is generally larger than that in a laminar layer.

(a) Development of Boundary Layer over a Flat Plate

Development of the boundary layer over the upper surface of a flat plate is shown in Fig. 8.3.(*a*). The incoming uniform flow with respect to the plate changes into non-uniform flow because the velocity of fluid at the plate must be zero. A typical velocity profile in the laminar region is parabolic as shown at section 1. As the distance *x* increases, the flow changes its character as tabulated:

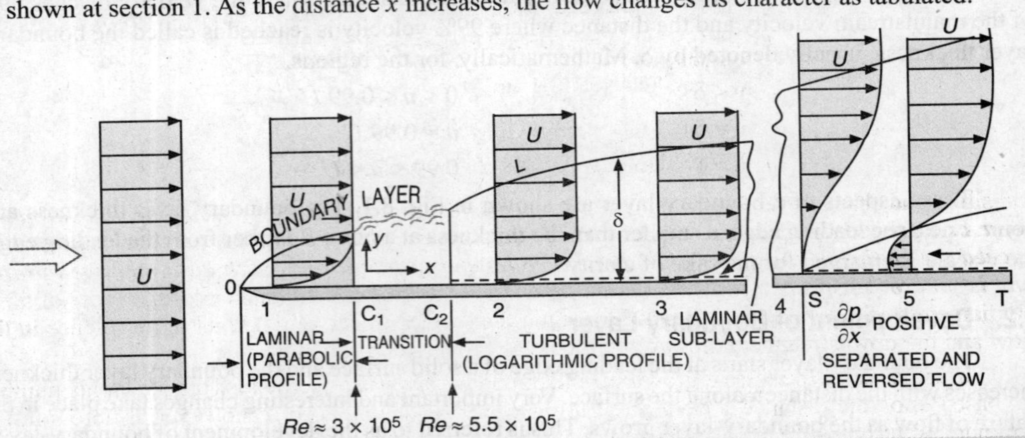

Fig. 8.3(a): Development of Flow Over a Flat Plate

Distance *x* from the leading edge O	Flow	Velocity Profile
until x_{c1}	laminar flow	parabolic as shown at 1.
from x_{c1} to x_{c2}	transition	transitory and uncertain
from x_{c2} until separation	turbulent (attached)	logarithmic as at 2 and 3.
from x_s to the trailing edge *T*: (Flow against a positive pressure gradient)	turbulent (separated)	partly reversed and forward as shown at 5.

The position x_{c1} at C_1 corresponds to the lower critical Reynolds number ≈ 3×10^5 to 3.2×10^5 whereas the position x_{c2} at C_2 corresponds to the upper critical Reynolds number ≈ 5.5×10^5. The Reynolds number Re_x is based upon the distance *x* from the leading edge as the 'characteristic length',

$$Re_x = Ux\rho / \mu$$

(b) Development of Boundary Layer within a Pipe

The development of the boundary layer within a round pipe is shown in Fig. 8.3(b) for two cases, (*i*) when the final character of the flow is laminar and (*ii*) when it is turbulent. Obviously the two cases correspond to the incoming velocity in accordance with the Reynolds number below the lower critical and above the upper critical values respectively. In case (*i*), the entire development is in the laminar region, whereas in case (*ii*), the development starts with a laminar profile, undergoes a

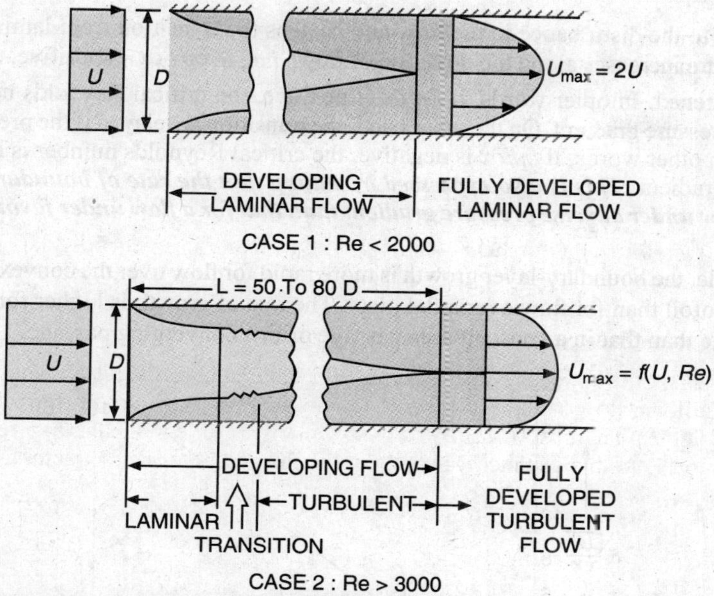

Fig. 8.3(b): Development of Flow in a Round Pipe

transition and change over to turbulent profile and stays turbulent thereafter. ***In a pipe, the boundary layer can grow only up to the centre-line of the pipe and the boundary layer thickness cannot exceed the radius of the pipe. When once it so happens, it continues with an unchanged profile and is said to constitute a fully developed flow***. The length of development is of the order of 50 to 80 diameters of the pipe and depends a lot upon the surface finish, initial level of turbulence in the flow and the downstream conditions as well as the properties of the fluid.

In either case, the rate of growth of the boundary layer is governed by the Reynolds number *Re* based upon the distance x from the leading edge. The dependence is, however, different in the laminar and turbulent regions.

For laminar layer: $$\frac{\delta}{x} = \frac{5}{\sqrt{Re}} \qquad \text{...(8.2)}$$

For turbulent layer: $$\frac{\delta}{x} = \frac{0.37}{Re^{0.2}} \qquad \text{...(8.3)}$$

It follows from the above that, for laminar flow,

$$\delta \propto x^{0.5}$$

and for the turbulent flow, $\delta \propto x^{0.8}$

The dependence of δ on *Re* is presented in Fig. 8.4, where it may be noted that δ bears a direct proportionality with $\sqrt{x}$ in the laminar range, *i.e.*, until $Re = 3.2 \times 10^5$ and the dependence on x thereafter is different.

In the light of the dependence of δ upon Re, it may be concluded that the boundary layer thickness is likely to be greater

(i) at a point downstream the flow than at a point upstream of it;

(ii) for a lower free-stream velocity U than for a higher free-stream velocity;

(iii) for a more viscous fluid than for a less viscous fluid ; and

(iv) for a lower density fluid.

In addition to the dependence upon the Reynolds number, the boundary layer growth is influenced by the roughness of the surface and the pressure gradient in the direction of flow. Roughness of a surface adds to the disturbance in the flow and hastens the transition from laminar to turbulent flow. If the pressure increases along the direction of flow, *i.e.*, if $\partial p / \partial x$ is positive, the transition to turbulence is hastened. In other words, if $\partial p/\partial x$ is positive, the critical Reynolds number is lower than that for no pressure gradient. On the other hand, the transition is delayed if the pressure decreases along the flow. In other words, if $\partial p/\partial x$ is negative, the critical Reynolds number is higher than that for no pressure gradient. ***This is also expressed by stating that the rate of boundary-layer growth is faster for a flow under adverse pressure gradient than that for a flow under favourable pressure gradient.***

For example, the boundary-layer growth is more rapid for flow over the convex shaped (upper) surface of an aerofoil than for flow over a flat plate. The rate of growth is higher for flow through a diverging passage than that in a constant area passage or in a converging passage.

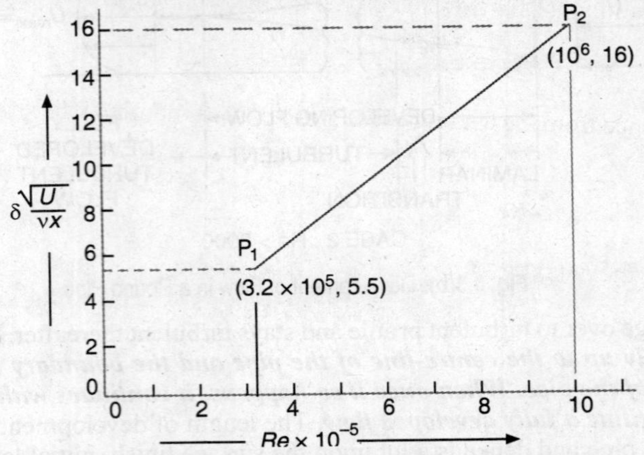

Fig. 8.4: Boundary Layer Thickness vs. Reynolds Number

Example 8.1. *Compare the rates of growth of the laminar boundary layer over a smooth flat plate in the following cases:*

(*a*) *Flat plate placed in a water stream flowing at 2 m/s.*

(*b*) *Flat plate placed in an air stream flowing at 2 m/s.*

(*c*) *Flat plate placed in an air stream flowing at 8 m/s.*

Given that the densities of water and air are 1000 and 1.2 kg/m³ respectively and the viscosities of water and air are 0.001 and 0.000 019 Ns/m² respectively.

Solution.

The rate of growth of a laminar boundary layer is given by

$$\frac{\delta}{x} = \frac{5}{\sqrt{Re}}$$

whence

$$\delta = \frac{5}{\sqrt{U}}\sqrt{\frac{\mu}{\rho}}\sqrt{x}$$

For water, $$\sqrt{\frac{\mu}{\rho}} = \sqrt{\frac{0.001}{1000}} = \mathbf{0.001\ units}$$

For air, $$\sqrt{\frac{\mu}{\rho}} = \sqrt{\frac{0.000019}{1.2}} = \mathbf{0.004\ units}$$

It follows from the above that the boundary layer thickness in air is four times as much as that in water for the same free-stream velocity.

While in the same fluid, the boundary layer is thinner at higher velocities, **it is halved when the velocity is four times.**

Of the given cases, the boundary layer grows the fastest in case (*b*), *i.e.*, in air at low velocity; half as fast in case (*c*), *i.e.*, at four times the velocity and slowest in case (*a*), *i.e.*, when the flat plate is placed in water stream at low velocity.

Example 8.2. *A smooth flat plate with a sharp leading edge is placed along a free stream of water flowing at 3 m/s. Calculate the distance from the leading edge and the boundary layer thickness where the transition from laminar to turbulent flow may commence. Assume the density of water as 1000 kg/m³ and viscosity as 1 centipoise.*

Solution.

The Reynolds number for flow over a flat plate is defined as

$$Re = \frac{Ux\rho}{\mu}$$

where x is the distance from the leading edge. The Reynolds number where transition may commence is given by

$$Re = 3.2 \times 10^5 = \frac{Ux\rho}{\mu}$$

Substituting the data, *i.e.*,

$$U = 4 \text{ m/s}, \rho = 100 \text{ kg/m}^3, \mu = 1 \text{ cP} = 0.001 \text{ Ns/m}^2$$

$$x = 3.2 \times 10^5 \times 0.001/(3 \times 1000)$$

$$= 0.11 \text{ m} = \mathbf{11 \text{ cm}}$$

From the relationship,

$$\frac{\delta}{x} = \frac{5}{\sqrt{Re}}$$

$$\delta = 5 \times 0.11/\sqrt{3.2 \times 10^{-5}} = 0.001 \text{ m} = \mathbf{0.1 \text{ cm}}$$

The growth of the laminar boundary layer would, therefore, be to the extent of 0.1 cm over a distance of 11 cm from the leading edge where transition is likely to commence.

8.3 Estimates of Boundary Layer Region

Consider a velocity profile in the vicinity of a plate as shown in Fig. 8.1. The velocity at the solid boundary is zero and it should approach the free stream uniform velocity at a large distance from the boundary,

$$u \to U \text{ or } (U - u) \to 0$$

at the outer edge, $\quad y \to \infty.$

i.e., the velocity profile is asymptotic.

For practical purposes, the boundary layer region is defined in three alternative ways:

(*a*) Boundary Layer Thickness

(*b*) Displacement Thickness

(*c*) Momentum Thickness

(*a*) Boundary Layer Thickness δ : *The distance y where u becomes 99% U, i.e., when the difference between u and U remains 1%.* This is only to approximate the region of non-uniform flow. This thickness is, however, devoid of any physical interpretation and is not commonly used in engineering. There are, therefore, other means of expressing the growth of boundary layer. These are also recognised as the boundary layer thickness parameters.

(b) Displacement thickness δ^*: *The distance y by which the external free stream is effectively displaced due to the formation of the boundary layer.*

If a free stream of velocity U is effectively displaced by δ^* the loss of flow over the displacement equals $\rho U \delta^*$ per unit time

$$L = \rho U \delta^*. \quad ...(i)$$

Loss of flow actually affected by the profile, *i.e.*, the elementary hatched area at a distance y

$$dL = \rho\,(U - u)dy$$

$$\text{Total loss of flow } L = \int_0^\infty \rho(U-u)dy \text{ per unit time} \quad ...(ii)$$

as represented by the total shaded area.

Equating (*i*) and (*ii*)

$$\rho U \delta^* = \int_0^\infty \rho(U-u)dy$$

and for incompressible flow, cancelling ρ,

$$\delta^* = \int_0^\infty \left(1 - \frac{u}{U}\right) dy \quad ...(8.4)$$

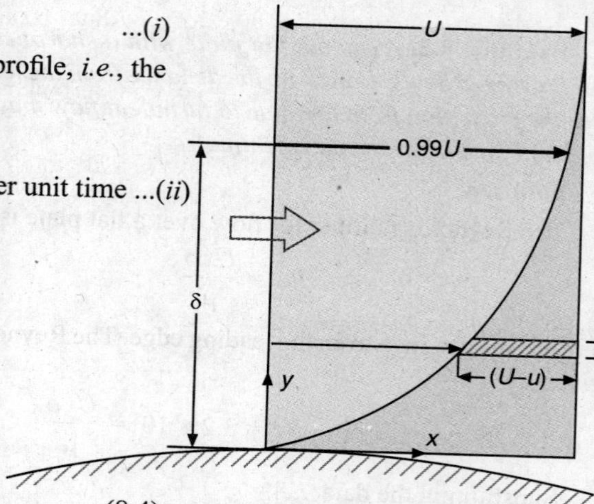

Fig. 8.5: Velocity Profile and Deficit Near a Solid Boundary.

Displacement thickness $\delta^* \approx \delta/3$ for a laminar profile

$\approx \delta/8$ for a turbulent profile

A concept closely related to the displacement thickness is the blockage correction factor for confined passages. For example, in the case of fully developed turbulent flow through pipes, the turbulent boundary layer extends from the circular wall up to the centre line. The displacement thickness concept would show that the effective area of flow corresponds to the cross-sectional area due to a radius (r - δ/8) rather than the actual radius r. Since the boundary layer thickness δ in a fully developed pipe flow must be equal to r, the effective radius is

$$r_{\text{eff}} = r - r/8 = 7/8\, r$$

The blockage correction factor is defined as the rate of effective area to the actual area, *i.e.*,

$$\frac{\pi r_{eff}^2}{\pi r^2} = (7/8)^2 = \frac{49}{64} \approx 0.8$$

The concept of blockage correction factor is used in dealing with flow through pipes, annuli, diffuser and other ducts.

(c) Momentum Thickness θ : *The distance y corresponding to the momentum defect brought about by the actual profile in relation to the uniform profile.*

For a displacement θ, the loss of momentum of the stream

$$= \rho U \theta \times U = \rho U^2 \theta \text{ per unit time} \quad ...(iii)$$

Loss of momentum associated with the loss of flow due to the elementary hatched area at a distance $y = \rho\, u(U - u)\, dy$

$$\text{Total loss of momentum} = \int_0^\infty \rho u(U-u)\, dy \quad ...(iv)$$

From equations (*iii*) and (*iv*) and for an incompressible flow, cancelling ρ

$$\theta = \int_0^\infty \frac{u}{U}\left(1-\frac{u}{U}\right)dy \qquad ...(8.5)$$

Momentum thickness $\theta \approx \frac{\delta}{8}$ for a laminar profile

$$\approx \frac{7}{72}\delta \text{ for a turbulent profile.}$$

Example 8.3. *If the velocity profile in a laminar boundary layer is approximated by a parabolic profile*

$$\frac{u}{U} = 2\left(\frac{y}{\delta}\right)-\left(\frac{y}{\delta}\right)^2$$

where u is the velocity at y and u → U as y → δ, calculate the displacement thickness and the momentum thickness.

Solution.

Displacement thickness,

$$\delta^* = \int_0^\infty\left(1-\frac{u}{U}\right)dy \approx \int_0^\delta\left(1-\frac{u}{U}\right)dy$$

$$= \int_0^\delta\left[1-2\left(\frac{y}{\delta}\right)+\left(\frac{y}{\delta}\right)^2\right]dy = \left|y-\frac{y^2}{\delta}+\frac{y^2}{3\delta^2}\right|_0^\delta = \mathbf{\frac{\delta}{3}}$$

Momentum thickness,

$$\theta = \int_0^\infty\frac{u}{U}\left(1-\frac{u}{U}\right)dy \approx \int_0^\infty\frac{u}{U}\left(1-\frac{u}{U}\right)dy$$

$$= \int_0^\delta\left[2\left(\frac{y}{\delta}\right)-\left(\frac{y}{\delta}\right)^2\right]\cdot\left[1-2\left(\frac{y}{\delta}\right)+\left(\frac{y}{\delta}\right)^2\right]dy$$

$$= \left\{\frac{\delta}{3}-\frac{\delta}{2}+\frac{3}{10}\delta\right\} = \mathbf{\frac{2}{15}}\delta$$

Example 8.4: *A uniform free stream of air at 10 m/s flows over a flat plate. Work out the growth of the boundary layer in the laminar region. Take* $\rho = 1.2$ *kg/m³ and* $\mu = 18 \times 10^{-6}$ *Ns/m².*

Solution.

From the data, $Re = \dfrac{10\times1.2x}{18\times10^{-6}} = 0.67\times10^6 x$

Assuming that the laminar boundary layer persists up to $Re = 3 \times 10^5$, the corresponding x is given by

$$x = (3 \times 10^5)/(0.67 \times 10^6) = 0.45 \text{ m}$$

Within this range, the boundary layer thickness is given by

$$\frac{\delta}{x} = \frac{5}{\sqrt{Re_x}} \text{ or } \delta = 6.1 \times 10^{-3}\sqrt{x}$$

Hence

x	0	00.1	0.2	0.4	0.45 m
δ	0	0.0013	0.002 74	0.003 87	0.0041m

Recalling that, for laminar flow,

$$\delta^* = \delta/3 \text{ and } \theta = \delta/8$$

the values of δ^* and θ are also tabulated:

δ^*	0	0.000 43	0.000 91	0.001 29	0.001 37 m
θ	0	0.000 16	0.000 34	0.000 48	0.000 51m

8.4. Boundary Layer Equations

Considering a two-dimensional steady incompressible flow over a flat plate one can simplify the Navier-Stokes equations by an 'order of magnitude' analysis as follows:

Let U be of the order of unity, written as $U = 0(1)$ and x also $0(1)$, signifying that these are not negligible quantities.

Since $\dfrac{\delta}{x} \propto \dfrac{1}{\sqrt{\text{Re}_x}}$ in general, $\delta = 0\,(\nu)$, *i.e.*, of the order of ν.

We write down the continuity equation and the Navier-Stokes equations and work out the orders of magnitude of all the terms and write underneath them:

$$\frac{\partial u}{\partial x} + \frac{\partial v}{\partial y} = 0 \qquad \text{...}(i)$$

$$1 \qquad 1$$

This is because $u = 0(1)$, $x = 0(1)$, and $\dfrac{\partial u}{\partial x} = \dfrac{0(1)}{0(1)} = 0(1)$ and the second term $\dfrac{\partial v}{\partial y}$ should be of 0(1) in order to satisfy the continuity equation. Since $y = 0(\delta)$, v must also be $0(\delta)$.

$$u\frac{\partial u}{\partial x} + v\frac{\partial u}{\partial y} = -\frac{\partial p}{\partial x} + \frac{1}{Re}\left(\frac{\partial^2 u}{\partial x^2} + \frac{\partial^2 u}{\partial y^2}\right) \qquad \text{...}(ii)$$

$$1 \times 1 \quad \delta \times \frac{1}{\delta} \qquad \delta^2 \times \left(1 \times \frac{1}{\delta^2}\right)$$

$$u\frac{\partial v}{\partial x} + v\frac{\partial v}{\partial y} = \frac{\partial p}{\partial x} + \frac{1}{Re}\left\{\frac{\partial^2 v}{\partial x^2} + \frac{\partial^2 v}{\partial y^2}\right\} \qquad \text{...}(iii)$$

$$1 \times \delta \quad \delta \times 1 \qquad \delta^2 \left\{\delta \quad \frac{1}{\delta}\right\}$$

The terms in the Navier-Stokes equations are non-dimensionalised ; p w.r.t. ρU^2; u and v w.r.t. U; x. w.r.t. a characteristic length L and y w.r.t. δ. Order of a term such as $\dfrac{\partial u}{\partial y}$ is found by writing it as $\dfrac{0(1)}{0(\delta)} = 0\left(\dfrac{1}{\delta}\right)$; of the terms $\dfrac{\partial^2 u}{\partial y^2}$ by writing it as $\dfrac{0(1)}{0(\delta)\times 0(\delta)} = 0\left(\dfrac{1}{\delta^2}\right)$; of the term $\dfrac{\partial^2 v}{\partial y^2}$ by writing it as $\dfrac{0(\delta)}{0(1)\times 0(1)} = 0(\delta)$, etc.

By observation, we conclude as follows:

(*a*) Both the terms in equation (*i*) are equally important and must be retained.

(b) The term $\frac{1}{Re}\frac{\partial^2 u}{\partial x^2}$ is $0(\delta^2)$ whereas every other term in equation (*ii*) is 0 (1): this can be dropped because of comparative smallness. Nothing being known about $\frac{\partial p}{\partial x}$ it will be retained.

(c) All the terms in equation (*iii*) are $0(\delta)$ or smaller suggesting that $\frac{\partial p}{\partial y}$ should be $0(\delta)$ or smaller. This implies that the variation of *p* in the *y*-direction in a boundary layer must be smaller. Equation (*iii*) is, therefore, dropped.

The boundary layer equations finally emerge in the well known form as:

Continuity: $$\frac{\partial u}{\partial x}+\frac{\partial v}{\partial y}=0 \qquad \text{...(8.6 a)}$$

Momentum: $$u\frac{\partial u}{\partial x}+v\frac{\partial u}{\partial y}=-\frac{1}{\rho}\frac{\partial p}{\partial x}+\frac{\mu}{\rho}\frac{\partial^2 u}{\partial y^2} \qquad \text{...(8.6 b)}$$

for a two-dimensional incompressible flow.

8.5 Drag on a Flat Plate

The drag, i.e., skin-friction on a flat plate can be evaluated from a knowledge of the velocity profile near the plate. At a distance *x* from the leading edge, the shear stress τ is given by

$$\tau_0=\mu\left(\frac{\partial u}{\partial y}\right)_{y=0}$$

and the shear force or drag over a small distance *dx* per unit width equals $\tau_0\,dx$.

This must equal the rate of change of momentum over the distance *x*.

Consider a small control volume AS_1S_2B bounded by the outer edge of the boundary layer as shown in Fig. 8.6.

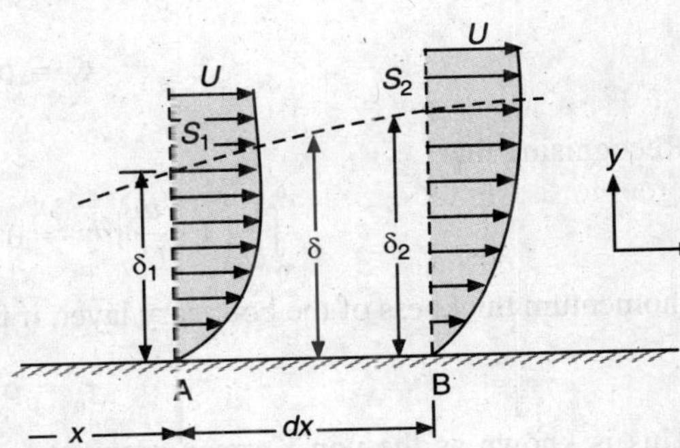

Fig. 8.6: Elementary Control Volume

Mass flow rate entering the control volume at the upstream station AS_1

$$\dot{m}_1=\int_0^\delta \rho\,u\,dy$$

per unit width of the plate.

Mass flow rate leaving the control volume at the downstream station AS_2

$$\dot{m}_2=\int_0^\delta \rho\,u\,dy+\frac{d}{dx}\left(\int_0^\delta \rho\,u\,dy\right)dx$$

Therefore, the mass flow rate entering the edge of the boundary layer S_1S_2

$$\dot{m}_2-\dot{m}_1=\frac{d}{dx}\left(\int_\delta^0 \rho\,u\,dy\right)dx$$

Momentum flux entering across $$AS_1=\int_0^\delta \rho\,u^2\,dy$$

$$\text{Momentum flux leaving across } BS_2 = \int_0^\delta \rho u^2\, dy + \frac{d}{dx}\left(\int_0^\delta \rho u^2\, dy\right)dx$$

$$\text{Momentum flux entering across } S_1S_2 = U.\frac{d}{dx}\left(\int_0^\delta \rho u\, dy\right)dx$$

The rate of change of momentum between the inflow and outflow is, therefore,

$$= \frac{d}{dx}\left(\int_0^\delta \rho u^2\, dy\right)dx - U\frac{d}{dx}\left(\int_0^\delta \rho u\, dy\right)dx$$

$$= \rho\frac{d}{dx}\left[\int_0^\delta u(u-U)\, dy\right]dx$$

This must equal the force exerted by the flat plate on the control volume. Drag force exerted by the fluid on the plate is equal and opposite to it and act along the flow:

$$D = \rho\frac{d}{dx}\left[\int_0^\delta u(u-U)\, dy\right]dx$$

$$D = \tau_0 dx$$

Equating the two expressions for D,

$$\tau_0 dx = \rho\frac{d}{dx}\left[\int_0^\delta u(u-U)\, dy\right]dx$$

or

$$\tau_0 = \rho U^2 \frac{d}{dx}\int_0^\delta \frac{u}{U}\left(1-\frac{u}{U}\right)dy$$

Recognising that

$$\int_0^\delta \frac{u}{U}\left(1-\frac{u}{U}\right)dy = \theta$$

is the momentum thickness of the boundary layer, it follows that

$$\tau_0 = \rho U^2 \frac{d\theta}{dx} \qquad \text{...(8.7 a)}$$

This is known as the von Karman momentum integral equation. For flow against pressure gradient $\frac{dp}{dx}$, the equation is

$$\tau_0 + \rho\frac{dp}{dx} = \rho U^2 \frac{d\theta}{dx} \qquad \text{...(8.7 b)}$$

Once the distribution of shear stress τ_0 is known, the drag on the length Δx of the plate per unit width is given by

$$\Delta D = \tau_0 \Delta x$$

The total drag on the plate of length l is

$$D = \int_0^l \tau_0\, dx \qquad \text{...(8.8)}$$

It may be noted that the expression for drag could have been obtained directly by applying the integral momentum equation over a control volume extending over the desired length of the plate. In that case,

$$D = \rho\left[\int_0^{\delta_2} u(U-u)\, dy - \int_0^{\delta_1} u(U-u)\, dy\right]$$

If the drag is desired on the total length of the plate from the leading edge, then the inlet velocity distribution is uniform and

$$D = \rho \int_0^\delta u(U-u)\, dy \qquad \text{...(8.9)}$$

The coefficient of drag for the plate, for laminar flow over one surface,

$$C_D = \frac{1.328}{\sqrt{Re}} \qquad \text{...(8.10)}$$

In case of turbulent boundary layers, the velocity profile is commonly assumed to follow the one-seventh power law,

$$\frac{u}{U} = \left(\frac{y}{\delta}\right)^{1/7} \qquad \text{...(8.11)}$$

With this profile and other subsidiary relations,

$$C_D = \frac{0.072}{(Re)^{0.2}} \qquad \text{...(8.12)}$$

for turbulent flow over one surface.

This relation is valid for Re between 5×10^5 and 2×10^7. For $Re > 2 \times 10^7$, the one-seventh power law is not valid and logarithmic velocity profile must be invoked. For the entire range of Reynolds number exceeding $Re = 10^5$ up to 1×10^9, however,

$$C_D = \frac{0.455}{(\log_{10} Re)^{2.58}} \qquad \text{...(8.13)}$$

Example 8.5. *A uniform free stream of air at 10 m/s flows over a flat plate. Calculate the drag coefficient and the drag for the plate, 0.5 m long and 2 m wide (Take $\rho = 1.2$ kg/m³ $\mu = 18 \times 10^{-6}$ Ns/m²)*

Solution.

From the data, $Re = \dfrac{10 \times 1.2 \times 0.5}{18 \times 10^{-6}} = 0.333 \times 10^6$

which is about the lower critical value. Assuming laminar flow over the plate,

$$C_D = \frac{1.328}{\sqrt{0.333 \times 10^3}} = 2.32 \times 10^{-3} = \mathbf{0.00232}$$

$$\text{Drag} = C_D . \frac{1}{2} \rho U^2 A$$

$$= 0.00232 \times \frac{1}{2} \times 1.2 \times 100 \times 2 \times 0.5$$

$$= \textbf{0.134 N for one surface.}$$

Example 8.6. *Calculate the drag and the power required to tow a smooth, flat plate 2 m wide 20 m long through still water (dynamic viscosity 0 001 N s/m²) at 10 m/s.*

What would be the drag and power required if the plate was half as long?

Solution.

For the flow, $Re = \dfrac{10 \times 20 \times 1000}{0.001} = 200 \times 10^6$; turbulent flow

and from tables, $\log_{10} Re = 8.3010$

$$C_D = \frac{0.455}{(8.3010)^{2.58}} = \frac{0.455}{229.1} = 0.002$$

Drag on both sides of the plate,

$$D = 2 \times 0.002 \times \frac{1}{2} \times 1000 \times 100 \times 40 = \mathbf{8000\ N}$$

Power required $\quad P = \text{Drag} \times \text{Velocity}$

$$= 8000 \times 10 = 80 \times 10^3 \text{ W} = \mathbf{80\ kW}$$

For plate half as long as this,

$$Re = 100 \times 10^6;\ \text{turbulent flow};\ \log_{10} Re = 8.0$$

$$C_D = \frac{0.455}{8.0^{2.58}} = 0.00213$$

Drag on both sides of the plate

$$D = 2 \times 0.00213 \times \frac{1}{2} \times 1000 \times 100 \times 20 = \mathbf{4250\ N}$$

This was expected to be slightly more than half of 8000 N because the drag coefficient decreases with x, the length of the plate because the flow is still turbulent in nature.

If the length of the plate is even smaller or the velocity is less, it could have been laminar flow and we could employ C_D for laminar flow.

Power required for the plate

$$P = 4250 \times 10 = 42.5 \times 10^3 \text{ W} = \mathbf{42.5\ kW}$$

Example 8.7. *Assuming that the velocity distribution in the laminar boundary layer with zero pressure gradient may be expressed as*

$$\frac{u}{U} = \sin\left(\frac{\pi}{2}.\frac{y}{\delta}\right)$$

where u is the velocity at a distance y in the boundary layer, U is the free stream velocity and δ is the boundary layer thickness. Determine the drag coefficient in terms of the Reynolds number.

Solution.

The Karman momentum integral equation for zero pressure gradient

$$\tau_0 = \rho U^2 \frac{d\theta}{dx}$$

where $\quad \theta = \text{momentum thickness}$

$$= \int_0^\delta \frac{u}{U}(1 - u/U)\,dy$$

$$= \int_0^\delta \sin\left(\frac{\pi}{2}.\frac{y}{\delta}\right)\left(1 - \sin\frac{\pi}{2}\frac{y}{\delta}\right) dy = \left(\frac{4-\pi}{2\pi}\right)\delta$$

and

$$\tau_0 = \mu \left.\frac{du}{dy}\right|_{y=0} = \frac{\mu U \pi}{2\delta}$$

Thus, we get

$$U^2 \frac{d\theta}{dx} = \frac{\mu U \pi}{2\delta\rho}$$

$$U^2\left(\frac{4-\pi}{2\pi}\right)\frac{d\delta}{dx} = \frac{\mu U\pi}{2\delta\rho}$$

Solving for δ,

$$\delta = \sqrt{\frac{2\pi^2}{(4-\pi)}}\sqrt{\frac{\mu x}{\rho U}}$$

Drag is given by $$D = \int_0^L \tau_0(x)\,dx$$

or $$D = \int_0^L \frac{\mu U\pi}{2\delta}dx$$

Simplifying for the drag-coefficient, we get the Blasius equation,

$$C_D = \frac{\text{Drag}}{\rho U^2 L} = \frac{1.310}{\sqrt{Re}} \qquad \text{...(8.14)}$$

where $$Re = \frac{\rho UL}{\mu}$$

Example 8.8. *Calculate the skin-friction drag on a plate 15 cm wide and 45 cm long placed longitudinally in a stream of oil (specific gravity 0.925 and kinematic viscosity 0.9 stokes) flowing with a free stream velocity of 6 metres per second. Also find the thickness of the boundary layer and shear stress at the trailing edge.*

Solution.

$$Re = \frac{Ux}{\nu} = \frac{6\times0.45}{0.9\times10^{-4}} = 30\,000$$

This being much less than 3.2×10^6, the flow over the plate is entirely laminar.

By the Blasius equation,

$$C_D = \frac{1.328}{\sqrt{Re}} = 0.0077$$

$$\text{Drag} = C_D\frac{1}{2}\rho U^2 A$$

$$= 0.0077 \times \frac{1}{2} \times 925 \times 6^2 \times 0.15 \times 0.45 = \mathbf{8.65\ N}$$

and $$\text{Drag} = \mathbf{17.30\ N\ for\ both\ sides.}$$

Thickness of the boundary layer in the laminar range is given by

$$\delta/x = 5/\sqrt{Re}$$

At the trailing edge,

$$\delta = 0.45 \times 5/\sqrt{30\,000}$$

$$= 0.013\ \text{m} = \mathbf{1.3\ cm}$$

Shear stress at the trailing edge,

$$\tau_0 = \frac{0.332\,\rho U^2}{\sqrt{Re}} = \frac{0.332\times925\times6^2}{\sqrt{30\,000}}$$

$$= \mathbf{63.83\ N/m^2}$$

8.6. Nature of Turbulence

Turbulence may appear in the flow around a solid object or in the flow through a passage under certain circumstances. Turbulence is a random phenomenon where the various quantities related to the flow vary randomly with time and space. The theory of turbulence is based upon statistical considerations. The instantaneous velocity u in a turbulent flow is considered to be the sum of the time-averaged-velocity $\overline{u}$ and the fluctuating component u' which may be positive or negative, as shown in Fig. 8.7.

$$u = \overline{u} + u' \qquad ...(8.15)$$

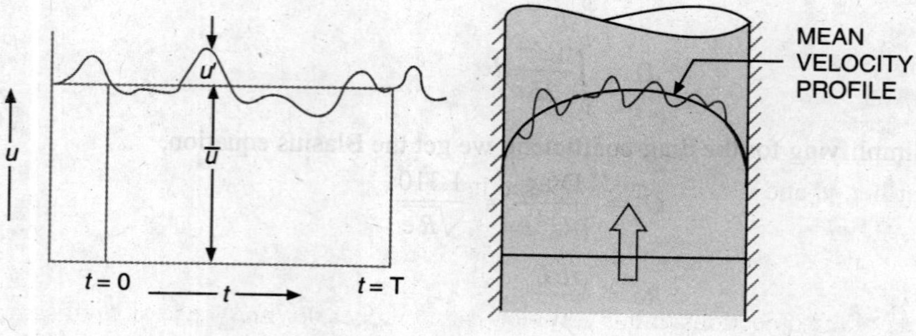

(a) TEMPORAL FLUCTUATIONS (b) INSTANTANEOUS PATTERN

Fig. 8.7: Nature of turbulent Flow

Similarly, the instantaneous v-component of velocity, pressure and density are expressed as

$$v = \overline{v} + v'$$

$$p = \overline{p} + p'$$

$$\rho = \overline{p} + p'$$

By definition, the time-averaged velocity is given by

$$\overline{u} = \frac{1}{T}\int_0^T u\,dt \qquad ...(8.16)$$

and

$$\overline{u'} = \frac{1}{T}\int_0^T u'\,dt = 0$$

since the time-averaged fluctuation must vanish.

A quantitative measure of the intensity of turbulence, I, is the *r.m.s.* value of the fluctuation. Along the direction of u, the intensity of turbulence is

$$I = \sqrt{\overline{u'^2}} \qquad ...(8.17)$$

Relative intensity of turbulence in the direction of the free stream U is $\frac{I}{u}$ and the percentage of turbulence is $\frac{I}{U}\times 100$.

The mechanism of turbulence is responsible for a large scale momentum exchange which results in apparent shear stresses,

$$\tau_t = -\rho\overline{u'v'} = \mu_t \frac{du}{dy} \qquad ...(8.18)$$

where τ_t and μ_t refer to the apparent shear stresses and turbulence-viscosity respectively.

The total viscosity in a turbulent flow may be expressed as an effective viscosity as

and $$\mu_{\text{eff}} = \mu + \mu_t \qquad ...(9.24)$$

Prandtl hypothesized that each fluctuating component is proportional to the average axial velocity gradient and the constant of proportionality is called the Prandtl mixing length l such that,

$$u' = l_1 \frac{d\bar{u}}{dy}$$

and $$v' = l_2 \frac{d\bar{u}}{dy}$$

where $$l_1 \approx l_2 \approx l$$

Further, u' and v' are always opposed in sign.

Hence $$\overline{u'\,v'} = -l_2\left(\frac{d\bar{u}}{dy}\right)^2 \qquad ...(8.19)$$

One of the objections to this equation is that it does not imply a change of sign of shear stress if $\left(\frac{d\bar{u}}{dy}\right)$ changes sign, a fact which is experimentally observed.

If $\tau_t = \rho\, l^2 \left|\frac{d\bar{u}}{dy}\right|\left(\frac{d\bar{u}}{dy}\right)$ is considered instead, it overcomes that objection. It must also be stated that this equation is also not entirely satisfactory for the shear stress distribution. Further, shear stress also equals $\mu_t \frac{d\bar{u}}{dy}$ which provides $\mu_t = \rho l^2 \left|\frac{d\bar{u}}{dy}\right|$ ...(8.20)

The fact is that most of the flows in nature and in man-made equipment are turbulent flows. Turbulence results in augmentation of mass, momentum and energy transfer between different parts of a fluid. It is desirable to promote mixing for the sake of improvement in heat transfer coefficient, better mixing and drag reduction. In nature, turbulence maintains uniformity of temperature between the earth and the environment, balances pressure differences between different locations, enables streams and rivers to transport waste from the foothills to the sea and, above all, allows human survival through quick mixing and displacement of exhaled gas by the fresh air required to be inhaled. Turbulent flows must take place through the water tubes as well as around the water tubes in a boiler plant, through a pump or a turbine, through a compressor or some other fluid machine, through pipes or open channels transporting large quantities of fluids, within combustion chambers and furnaces, around the wings of aircraft and gliders and around the other spacecraft. *In fact, there are more examples of turbulent flow than of laminar flow both in nature and in engineering practice.*

Turbulence is, however, accompanied by energy dissipation. Turbulent motion is characterised by eddies of different sizes and wave numbers. Larger eddies have smaller wave numbers. *The phenomenon of energy dissipation proceeds by way of kinetic energy being passed on from bigger eddies to smaller to still smaller eddies and eventually to the viscous dissipation action.*

8.7. Smooth and Rough Surfaces

Whether the general nature of boundary layer is laminar or turbulent, the flow in the immediate neighbourhood of a solid surface is always laminar. The thickness of the laminar sublayer is very small, approximated by

$$\delta_{sub} = 11.84 \frac{\nu}{u_*} \qquad ...(8.21)$$

A surface is said to be *hydrodynamically smooth* if the mechanical imperfections are small in comparison with δ_{sub}, the viscous sublayer thickness and it is said to be *hydrodynamically rough* if the mechanical irregularities are large and comparable with δ_{sub} as shown in Figs. 8.8 (*a*) and (*b*).

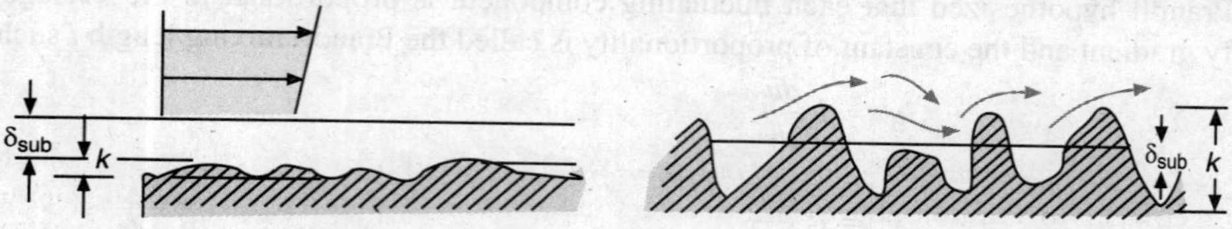

(a) SMOOTH SURFACE : $k < \delta_{sub}$ (b) ROUGH SURFACE : $k > \delta_{sub}$

Fig. 8.8. Hydrodynamic Classification of Surfaces

Quantitatively for a flat plate or a pipe of internal roughness *k*,

$$\frac{k}{\delta_{sub}} < 0.25 \text{ for smooth surface}$$

$$\frac{k}{\delta_{sub}} > 5 \text{ for rough surface} \qquad ...(8.22)$$

and $$0.25 < \frac{k}{\delta_{sub}} < 5 \text{ for transitory surface.}$$

This classification is logical on the grounds that the turbulent velocities above the viscous sublayer are not likely to penetrate a long way down into the sublayer whereas the turbulence can interact with imperfections protruding out of the mean sublayer!

8.8. Boundary Layer Separation

As the flow proceeds over a solid surface, the boundary layer thickness increases. The velocity profile changes from parabolic to logarithmic. The fluid mass in the layer close to the wall has to do work against boundary friction by consuming a fraction of its kinetic energy. This is, in turn, regained from the adjacent fluid layers through momentum exchange process. *The extent of momentum exchange between two layers in a laminar flow is considerably less than that in a turbulent flow. The boundary layer may not be able to keep sticking to the solid surface if, at a certain stage, it cannot expend energy to overcome the resistance offered by the surface.* For reasons just stated, laminar layer is in a weaker position compared to turbulent layer to withstand the expenditure of kinetic energy. *Consequently, a laminar layer tends to separate off the surface more readily than a turbulent layer.* However, measures are generally taken to change over the laminar layers into turbulent layers before separation but the stoppage of separation of the turbulent layers is an uneasy task.

Adverse pressure gradient along the flow tends to decelerate the flow and results in a rapidly thickening layer. The fluid mass in the layer close to the wall has to do both, work against friction as explained above and work against increasing or adverse pressure gradient. This causes the layers to show a tendency to separate from the surface even earlier. On the other hand, favourable pressure gradient resulting in accelerated flow provides energy to the layer-fluid enabling it to remain sticking to the solid surface. An example of this in regard to flow through diffusers and nozzles is given in Fig. 8.9.

In a diffuser, the adverse pressure gradient causes the boundary layer to separate at a section SS whereas in a nozzle flow, separation is not likely to occur.

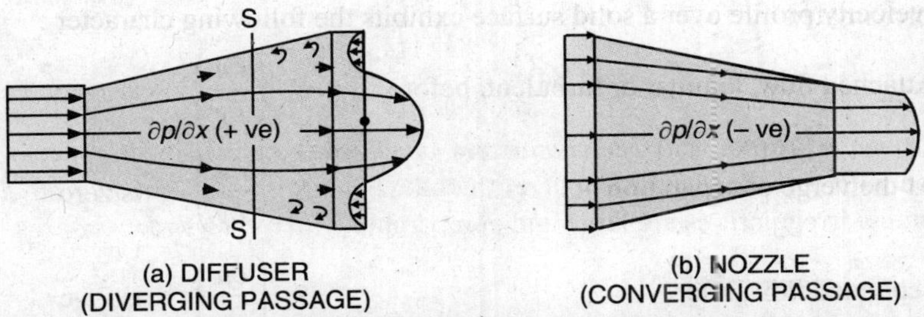

Fig. 8.9. Flow through Diffusers and Nozzles

Let us consider the flow over curved surfaces at Fig. 8.10 (*a*) and an aerofoil at (*b*). From the leading edge up to a point *P*, the flow is accelerating due to a favourable pressure gradient whereas the pressure gradient thereafter becomes adverse to cause the flow to separate at *S*, and cause eddy formation and reversal thereafter. It may also be observed that the flow on the lower surface of an aerofoil does not separate due to favourable pressure gradient.

Shown in Fig. 8.10(c) is a sudden expansion in duct.Separation of flow takes place at the corner point *S* resulting in eddying motion in the side regions as shown.

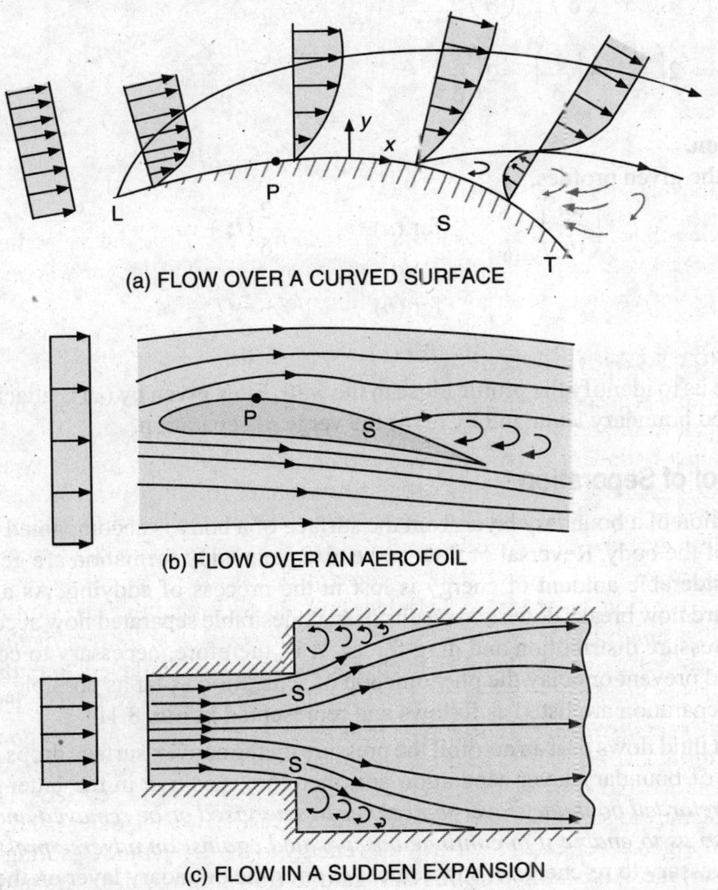

Fig. 8.10: Separation and Recirculation of Flow

The velocity profile over a solid surface exhibits the following character :

Attached flow, laminar or turbulent, before S, $\left(\frac{\partial u}{\partial y}\right)_{y=0}$ is + *ve*

At the verge of separation at S, $\left(\frac{\partial u}{\partial y}\right)_{y=0}$ is zero

Separated flow, after S, $\left(\frac{\partial u}{\partial y}\right)_{y=0}$ is – ve

It is thus possible from a given velocity profile to test whether the flow is attached, detached or at the verge of separation.

Example 8.9. *Examine the following profiles to state if the flow is attached or detached.*

(*a*) $\frac{u}{U} = 2\frac{y}{\delta} - \left(\frac{y}{\delta}\right)^2$

(*b*) $\frac{u}{U} = -2\frac{y}{\delta} + \left(\frac{y}{\delta}\right)^3 + \left(\frac{y}{\delta}\right)^4$

(*c*) $\frac{u}{U} = 2\left(\frac{y}{\delta}\right)^2 + \left(\frac{y}{\delta}\right)^3 - 2\left(\frac{y}{\delta}\right)^4$

Solution.

From the given profiles,

$\left(\frac{du}{dy}\right)_{y=0}$ for (*a*) $\frac{2}{\delta}U$; + *ve*

for (*b*) $-\frac{2}{\delta}U$; – *ve*

and for (c) 0

which enable us to identify the profile close to the wall. Flow given by (*a*) is attached, by (*b*) reversed with separated boundary layer and by (*c*) at the verge of separation.

8.9. Control of Separation

Separation of a boundary layer from the surface of a body is accompanied by reversed flow in the vicinity of the body. Reversal of flow and consequent eddy formation are generally undesirable because considerable amount of energy is lost in the process of eddying. As a matter of fact, the orderly forward flow breaks down and results in an undesirable separated flow accompanied by abrupt changes in pressure distribution and drag forces. It is, therefore, necessary to control the boundary layer flow and prevent or delay the phenomenon of separation as far as possible. Some of the means of delaying separation are listed as follows and represented in Fig. 8.11.

When a fluid flows past an aerofoil the pressure on the convex surface drops and rises thereafter. The chances of boundary layer separation are, therefore, positive in the latter part of the convex surface. *The retarded boundary layer may either be energised or be removed and replaced by high-energy fluid so as to enable it to continue unseparated against an adverse pressure gradient.* It is, therefore, necessary to re-energise, supplement or suck the boundary layer on the convex surface as is shown in Figs. 8.11 (*a*) to (*c*).

When the flow takes place round a bend, a pressure gradient is generated and there is a tendency of separation at the inner radius of the bend. Provision of guide-blades as shown at (*d*) divides the passage and brings about a well-guided flow without separation and recirculation. In order that the flow through a diffuser may continue unseparated, the adverse pressure gradient should be kept low. This is achieved by providing small divergence in the passage. The semi-angle of divergence of a diffuser is, therefore, limited to 5°. In case it is desired to provide a rapid divergence, some means such as provision of splitter blades to divide the passage into subpassages of low divergence, re-energisation by injection or removal by suction of the retarded boundary layer are necessary.

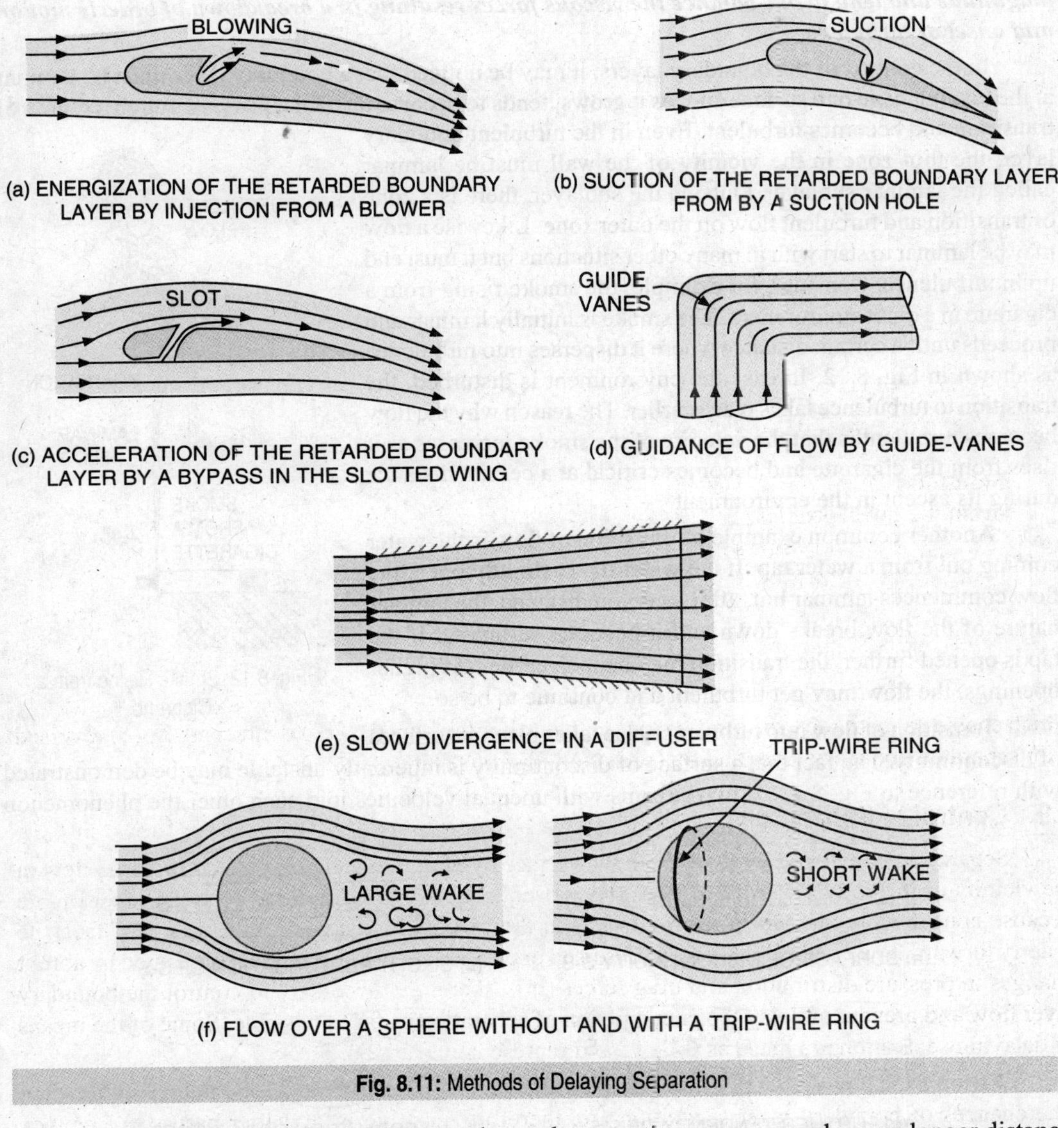

Fig. 8.11: Methods of Delaying Separation

The flow over a spherical object may be made to continue unseparated over a longer distance resulting in the reduction of wake behind it simply by installing a trip-wire ring. The boundary layer fluid is revitalised by way of increased momentum exchange and it continues to flow unseparated further as shown in Fig. 8.11 (*f*).

8.10. Mechanism of Transition

It has already been stated that a laminar flow becomes unstable at higher velocities and tends to become turbulent. In fact, the connotation of the word `turbulence' is the disorderly or irregular fluctuating motion. In a laminar flow, the shear forces due to viscosity regiment the flow and keep it orderly, *i.e.*, confined to layers of flow sliding over each other. At low velocities, inertia forces are small and the viscous forces predominate. ***At higher velocities, the inertia forces increase in magnitude and tend to out-balance the viscous forces resulting in a breakdown of orderly motion and onset of turbulence.***

In the context of the boundary layers, it may be noticed that a boundary layer must be laminar at the leading edge of a surface and, as it grows, tends to become unstable, passes through a region of transition and becomes turbulent. Even in the turbulent boundary layer, the thin zone in the vicinity of the wall must be laminar called the laminar sublayer. Outside the sublayer, there is a zone of transition and turbulent flow on the outer zone. Likewise a flow may be laminar to start with in many other situations but it must end up in turbulence. Consider, for example, the smoke rising from a cigarette in a quiet environment. The smoke is initially laminar and proceeds until a certain distance where it disperses into turbulence as shown in Fig. 8.12. In case the environment is disturbed, the transition to turbulence takes place earlier. The reason why the flow becomes turbulent is that the velocity of the smoke increases as it rises from the cigarette and becomes critical at a certain distance during its ascent in the environment.

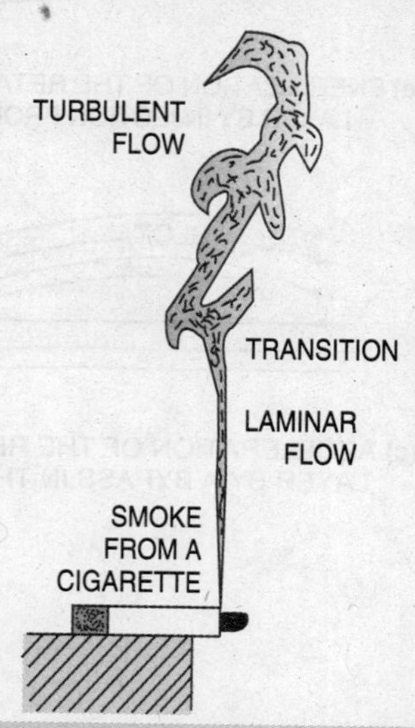

Fig. 8.12: Smoke Rising from a Cigarette

Another common example of transition of flow is the water coming out from a water tap. If the water tap is slightly open, the flow commences laminar but, after a certain distance, the laminar nature of the flow breaks down and it becomes turbulent. If the tap is opened further, the transition may happen earlier. At higher openings, the flow may get turbulent and continue to be so.

Transition of flow into turbulence also takes place at surfaces of discontinuity. The fact that a surface of discontinuity is inherently unstable may be demonstrated with reference to Fig. 8.13. If two streams with unequal velocities join each other the phenomenon

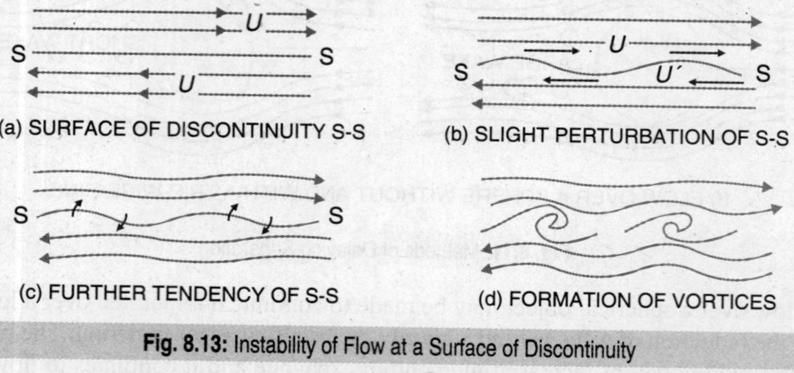

Fig. 8.13: Instability of Flow at a Surface of Discontinuity

may be represented by considering a surface of discontinuity with streams of velocity U in opposite directions on either side of it as at (a). If a slight perturbation is applied to the surface of discontinuity as shown in (b), the velocity u' at a point where the streamlines converge is greater than U, resulting in local drop of pressure. Drop of pressure above a `peak' and below a `valley' result in a moment which tends to disturb the surface further as shown at (*c*). The process of magnification of disturbance culminates in the formation of vortices as shown at (*d*).

Transition from laminar to turbulent flow in shear flows where moderate velocity gradients exist is explained by the growth of `spots' of turbulence in laminar flow. Propagation and cross-contamination of spots result in further spots which change the structure of flow from laminar to turbulent. The theory of natural transition by spot formation due to Tollmein and Schlichting assumes that there are a number of steps in the growth of spots and eventual conversion of flow from laminar to turbulent. Some controversy, however, exists as to the exact details and causes of initial spot formation and its eventual breakdown. The tendency of spot formation and the frequency increase with the Reynolds number.

SUPPLEMENTARY REVISION EXAMPLES

Example 8.10. *Calculate the velocity gradient at distance of 0, 10 and 15 cm from the boundary if the velocity profile is parabolic, given by $u = Ay^2 + By + C$ and the vertex 15 cm from the boundary where the velocity is 100 cm/s. Also calculate the shear stress at these points if the fluid has viscosity of 8.2 Poise.*

Solution.

From the data on the velocity gradient,

$$u = Ay^2 + By + C$$

$$\frac{\partial u}{\partial y} = 2Ay + B = 0 \text{ at the vertex; } y = 0.15 \text{ m}$$

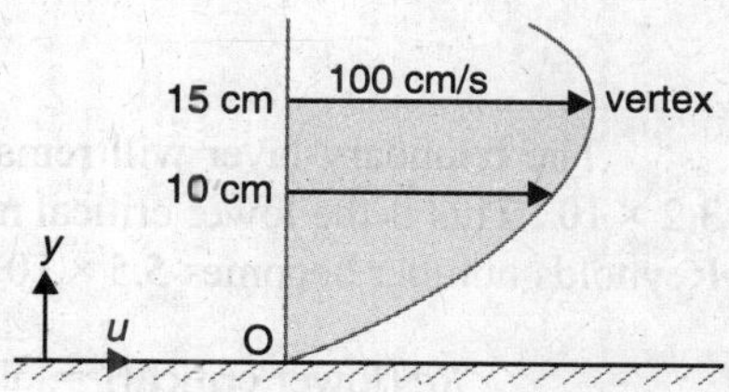

$$2A \times 0.15 + B = 0$$

$$0.3\,A + B = 0$$

Also, $u = 1$ m/s at $y = 0.15$

$$1 = A \times 0.15^2 + 0.15\,B + C$$

$$0.0225\,A + 0.15\,B + C = 1$$

We know that $u = 0$ at $y = 0$, the stationary boundary.

$$0 = 0 + 0 + C; \text{ which gives } C = 0$$

From these facts, $0.3\,A + B = 0$ ...(*i*)

$$0.0225\,A + 0.15\,B = 1 \quad ...(ii)$$

From (*i*) $B = -0.3\,A$

So, $0.0225\,A - 0.045\,A = 1;\ A = -44.4$

and $B = +13.33$

Hence, $u = -44.4y^2 + 13.33y$

Velocity gradient

$$\frac{\partial u}{\partial y} = -88.8\,y + 13.33$$

At $y = 0$, $\frac{\partial u}{\partial y}$ **= 13.33 /s**

At $y = 0.1$ m, $\frac{\partial u}{\partial y}$ **= 4.45 /s**

At $y = 0.15$ m, $\frac{\partial u}{\partial y}$ **= 0, as expected!**

Using μ = 8.2 Poise = 0.82 Ns/m².

Shear stress is given by

$$\tau = \mu\frac{\partial u}{\partial y} = 0.82(-88.8\,y + 13.33)$$

At $y = 0$ $\quad \tau =$ **10.93 N/m²**

At $y = 0.1$ $\quad \tau =$ **3.65 N/m²**

At $y = 0.15$ $\quad \tau =$ **0.**

Example 8.11. *A flat plate of 2 m width and 4 m length is kept parallel in air flowing at a velocity of 5 m/s. Determine the length of the plate over which the boundary layer is laminar and the total force on both sides of that portion of plate where the boundary layer is laminar. Take* $\rho = 1.2$ *kg/m³ and* $\nu = 1.47 \times 10^{-5}$ *m²/s.*

Solution.

From the data,

$$U = 5 \text{ m/s};\ \rho = 1.2 \text{ kg/m}^3;\ \nu = 1.47 \times 10^{-5} \text{ m}^2/\text{s}.$$

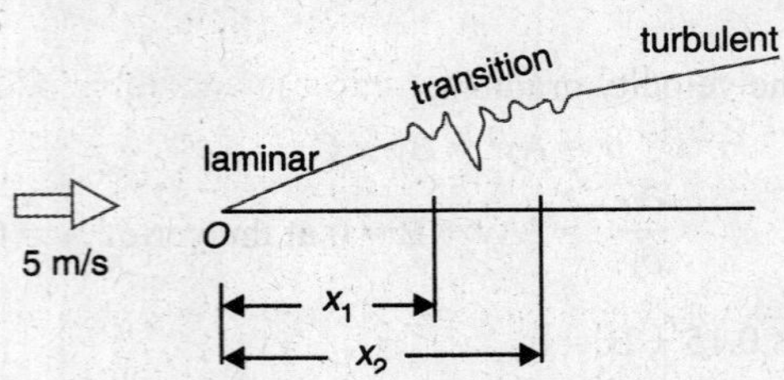

The boundary layer will remain laminar until a distance x_1, where the Reynolds number is 3.2×10^5. This is the lower critical number. It will cease to be laminar at x_2 where the upper critical Reynolds number becomes 5.5×10^5.

$$Re \text{ (lower critical)} = 3.2 \times 10^5 = \frac{U x_1 \rho}{\mu} = \frac{U x_1}{\nu}$$

$$= \frac{5\,x_1}{1.47 \times 10^{-5}}.$$

whence $\quad x_1 = 0.94$ m

Similarly $\quad x_2 = 1.47 \times 10^{-5} \times 5.5 \times 10^5/5 = 1.62$ m

The coefficient of drag for laminar flow over one surface is given by

$$C_D = \frac{1.328}{\sqrt{Re}}$$

$$= 1.328/\sqrt{3.2 \times 10^5} = 2.35 \times 10^{-3}.$$

Area of laminar part of the boundary layer on both sides of the plate is

$$A = 2 \times 0.94 \times 2 \text{ m}^2$$

$$= 3.76 \text{ m}^2$$

Drag force

$$D = \frac{1}{2} C_D . \rho U^2 A$$

$$= \frac{1}{2} \times 2.35 \times 10^{-3} \times 1.2 \times 5^2 \times 3.76$$

$$= \mathbf{0.133 \text{ N}}.$$

It is very small indeed for low velocity flow of air over a flat plate!

Shear stress τ at any location can be obtained if the velocity profile $u = f(y)$ is known close to the plate.

$$\tau = \mu\left(\frac{du}{dy}\right)_{y=0}$$

For example, if $\frac{u}{U} = 2\left(\frac{y}{\delta}\right) - f(y^2, y^3 \,..........)$

$$\frac{\partial u}{\partial y} = \frac{2}{\delta} \text{ at } y = 0$$

where $\delta = 5x/\sqrt{Re} = 5 \times 0.94/\sqrt{3.2 \times 10^5}$, etc.

Example 8.12. *Assuming the boundary layer to be laminar over the entire length of a flat plate placed at zero incidence in a stream of uniform velocity, determine the ratio of the friction drag on the leading half and on the rear half of the plate.*

Solution.

For a flat plate placed parallel to a uniform free stream of velocity U, the drag from the leading edge to any point along its length is given by

$$D = \rho\int_0^{\delta} u(U - u)dy$$

by rate of change of momentum consideration. However, it is also given by

$$D = C_D \times \frac{1}{2}\rho U^2 . ax$$

where $C_D = 1.328/\sqrt{Re}$; $Re = \frac{Ux\rho}{\mu}$

for a plate of length x.

The drag $D_x = \frac{1.328}{\sqrt{\frac{Ux\rho}{\mu}}} . \frac{1}{2}\rho U^2 a x$

For the leading half of the plate of length L, $x = L/2$

$$D_x = \frac{1.328}{\sqrt{\frac{UL\rho}{2\mu}}} . \frac{1}{2}\rho U^2 a \frac{L}{2} = \frac{1}{\sqrt{2}} . \frac{1.328}{\sqrt{\frac{UL\rho}{\mu}}} . \frac{1}{2}\rho U^2 aL$$

For the total length L of the plate, $x = L$

$$D_t = \frac{1.328}{\sqrt{\frac{UL\rho}{\mu}}} . \frac{1}{2}\rho U^2 aL;$$

It is noticed that $D_1 = \frac{1}{\sqrt{2}} . D_t = 0.707\, D_t$

Hence, for the rear half only, $D_2 = D_t - D_1 = \left(1 - \frac{1}{\sqrt{2}}\right) D_t = 0.293\, D_t$

$$D_1/D_2 = 0.707/0.293 = \mathbf{2.41.}$$

Example 8.13. *A flat plate is placed parallel to a uniform stream of air. Assuming boundary layer to be turbulent over the entire plate, work out the ratio of the skin-friction drag on the front half and the rear half of the plate.*

Solution.

In this case, when the flat plate is placed along the flow, the average drag coefficient is given by the relation.

$$C_D = 0.072/Re^{0.2} \text{ for turbulent flow}$$

and

$$\text{Drag} = C_D \cdot \frac{1}{2}\rho U^2 \text{ area of the plate}$$

for the leading half of the plate (width and length L)

$$C_D = 0.072\Big/\left(\frac{UL}{2}\frac{\rho}{\mu}\right)^{0.2}$$

and

$$D_f = 0.072\Big/\left(\frac{UL}{2}\frac{\rho}{\mu}\right)^{0.2} \cdot \frac{1}{2}\rho . U^2 . \frac{L}{2} . a$$

For the total plate

$$C_D = 0.072\Big/\left(\frac{UL\rho}{\mu}\right)^{0.2}$$

and

$$D_t = 0.072\Big/\left(\frac{UL\rho}{\mu}\right)^{0.2} \frac{1}{2}\rho U^2 La$$

$$\frac{D_t}{D_f} = \left(\frac{1}{2}\right)^{0.2} \times 2 = 1.74$$

But sum of drag on front and rear $D_t = D_f + D_r$,

$$1.74\ D_f = D_f + D_r \text{ ; } D_r = 0.74\ D_f$$

Hence

$$\frac{D_f}{D_r} = \frac{1}{0.74} = 1.35$$

Example 8.14. *A flat plate 0.3 m long and 2 m wide is placed along a 1.2 m/s flow of water at 15° C. Assume the boundary layer on each face is laminar and the velocity profile is linear for which*

$$\delta/x = 3.46 / \sqrt{Re_x}$$

Determine the drag force on the plate, taking the viscosity of water as 1.1×10^{-6} m/s and density 1000 kg/m³.

Solution.

Given that the velocity profile is linear,

Let $u/U = a + by$

At $y = 0, u = 0,$

which means $a = 0$

At $y = \delta, u = U,$

which means $b = 1/\delta$

Hence, $u/U = y/\delta.$

Given that $\delta = 3.46\ x/\sqrt{\text{Re}_x} = 3.46\ x\Big/\sqrt{\dfrac{Ux}{\nu}}$

$$= 3.46\sqrt{x}\sqrt{\frac{\nu}{U}}$$

so that $$u = \frac{U}{3.46}\sqrt{\frac{U}{\nu x}}\,y$$

whence $$\frac{du}{dy} = \frac{U}{3.46}\sqrt{\frac{U}{\nu x}}$$

and shear stress $$\tau = \mu\frac{du}{dy} = \sqrt{\nu.\rho.}\;U^{3/2}/3.46\,.\,x^{-1/2}.$$

With the given values of ρ, ν and U_1

$$\tau = 0.4\,x^{-1/2}$$

and the drag on both faces of the plate is

$$D = \left[\int_0^{0.3} \tau\,dx.2\right] \times 2 = \mathbf{0.88\ N.}$$

ITEM BANK

Objective items of different types are given to review the concepts developed in the chapter.

* *Select the correct response for each item.*

* *Every correct response is expected in less than a minute and carries one mark.*

* *Compare your responses with the key answers given at the end.*

Multiple Choice Items

Attempt the following Multiple-Choice Items by ticking the correct or the most appropriate response from among the alternatives given:

1. A boundary layer
(*a*) may exist near a solid boundary only
(*b*) may exist near any boundary or wall,
(*c*) may not exist for some fluids
(*d*) may or may not grow along the direction of flow

2. A boundary layer grows rapidly
(*a*) for laminar flow with favourable pressure gradient
(*b*) for turbulent flow with favourable pressure gradient
(*c*) for any type of flow with favourable pressure gradient
(*d*) for any flow with adverse pressure gradient

3. Separation is caused by
(*a*) reduction of pressure in the direction of flow
(*b*) reduction of the boundary layer thickness
(*c*) presence of adverse pressure gradient
(*d*) presence of favourable pressure gradient.

4. The wake is defined as a flow where
(*a*) the flow is about to separate from the boundary
(*b*) the pressure becomes very low
(*c*) the velocity gradient is very high
(*d*) the flow has separated from the boundary.

5. The law of the wall refers to
(*a*) the law for the laminar sublayer
(*b*) the law for the buffer zone
(*c*) the logarithmic velocity profile
(*d*) the entire region including the laminar sublayer and buffer zone.

6. The displacement thickness for a boundary layer
   (*a*) is greater than the boundary layer thickness
   (*b*) is less than the momentum thickness
   (*c*) represents the momentum deficit in a flow
   (*d*) represents the mass deficit in a flow

7. Nature of turbulence implies
   (*a*) unsteadiness of flow
   (*b*) non-uniformity of flow
   (*c*) unsteadiness and non-uniformity of flow
   (*d*) random component of velocity superimposed on the mean flow.

ANSWERS

1. (*b*), **2.** (*d*), **3.** (c), **4.** (*d*), **5.** (*d*), **6.** (*d*), **7.** (*d*).

QUESTION BANK

Questions on theory and numerical problems are given for tutorial work.

* *The questions are arranged generally in order of the articles in the text.*

* *Every question should take 15-20 minutes to solve.*

* *Answers for the numerical problems are given alongside them.*

8.1. The laminar boundary layer profile in a case is approximated by a cubic parabola.

$$\frac{u}{U} = \frac{3}{2}\left(\frac{y}{\delta}\right)^2 - \frac{1}{2}\left(\frac{y}{\delta}\right)^3$$

where u = velocity at a distance y from the surface and $y \to \delta$, as $u \to U$.

Calculate the displacement thickness and momentum thickness in terms of δ and work out the shear stress at the surface. **(5/8 δ, 39/280 δ; O)**

8.2. A flat plate is placed longitudinally in a turbulent stream of water flowing at 10 m/s. Locate the transition point and calculate the boundary layer thickness just before the transition. **[3.2 cm, 0.28 cm]**

8.3. Given the velocity profile

$$\frac{u}{U} = 2\frac{y}{\delta} - 2\frac{y^3}{\delta^3} + \frac{y^4}{\delta^4}$$

in a laminar boundary layer over a flat plate in a uniform stream of oil (sp. gr. 0.9 and μ = 0.08 Ns/m² at 0.2 m/s), calculate the shear stress profile and plot the velocity profile in the same sketch. **$[\tau_0 = \frac{0.032 \times 2}{\delta}, \delta = 0.065\sqrt{x}\,]$**

8.4. For a one-fifth power law profile, calculate the coefficient of drag for an oncoming free stream flow over a flat plate.

8.5. Show that the critical Reynolds number of 3×10^5 for a flat plate corresponds to the critical Reynolds number of 2200 for flow through a round pipe. Make suitable assumptions.

8.6. A uniform flow at 2 m/s enters a long horizontal pipe of diameter 0.1 m. Work out the changes in velocity profile as the flow proceeds in the following fluids:

(*a*) air	$\rho = 1.2\ kg/m^3$	$\mu = 18 \times 10^{-6}\ Ns/m^2$
(*b*) water	$\rho = 1000\ kg/m^3$	$\mu = 0.001\ Ns/m^2$
(c) oil	$\rho = 900\ kg/m^3$	$\mu = 0.1\ Ns/m^2$

At how many diameters downstream of the entrance in each case will the flow be fully developed? **[65, 65, 56]**

8.7. Determine the drag force acting on a straight segment of a pipe if at the entrance, the flow is uniform and at the exit, it is a paraboloid of revolution given by

$$u = 10(1 - r^2)$$

where u = velocity at radius r. The fluid is air, $\rho = 1.2$ kg/m^3

Calculate also the momentum and energy correction factors at the exit. **[31.5 N]**

8.8. Assuming that the velocity distribution in the laminar boundary layer is expressed as

$$\frac{u}{U} = \sin(\pi/2.\, y/\delta)$$

show that the displacement and momentum thicknesses are given by

$$\frac{\delta^*}{x} = \frac{1.741}{\sqrt{Re}} \text{ and } \frac{\theta}{x} = \frac{0.654}{\sqrt{Re}}$$

Also show also that the drag coefficient is $C_D = \dfrac{1.31}{\sqrt{Re}}$

8.9. (*a*) With neat sketches explain the development of a boundary layer along a thin flat smooth plate held parallel to a uniform flow and explain the salient features.

(*b*) What are the causes of separation of boundary layers and how can separation be controlled?

(*c*) A smooth two-dimensional flat plate is exposed to a wind velocity of 360 km/hr. If the laminar boundary layer exists up to a value of $R_x = 2 \times 10^5$, find the maximum distance from the leading edge up to which laminar boundary layer exists and the maximum thickness. The kinematic viscosity of air is 1.49×10^{-5} m^2/s.

[x = 29.8 mm, δ = 0.334 mm]

8.10. A free stream flowing at 1 m/s, density 1000 kg/m^3 and dynamic viscosity 0.01 kg/s m, flows steadily over a horizontal flat plate. Calculate the Reynolds number and the boundary layer thickness at 0.1 m from the leading edge. **[$Re = 1 \times 10^4$, δ = 0.005 m]**

8.11. (*a*) Starting with the Navier-Stokes equations of motion for two-dimensional incompressible flow, obtain the boundary layer equations.

(*b*) The local skin friction coefficient at any section x in a laminar boundary layer on a flat plate is given by

$$C_D = 0.664(Re_x)^{1/2}\}$$

Show that the average or mean skin friction coefficient based on total length L of the plate is twice the local skin friction at the end of the plate.

8.12. (*a*) Explain the significance of the terms (*i*) laminar boundary layer, (*ii*) turbulent boundary layer and (*iii*) laminar sublayer.

(*b*) A flat plate, 1.5 m × 1.5 m, moves at 50 km/hour in a stationary air of density 1.15 kg/m^3. If the coefficients of drag and lift are 0.15 and 0.75 respectively, determine (*i*) the lift force, (*ii*) the drag force, (*iii*) the resultant force and (iv) the power required to keep the plate in motion. **[(*i*) 187.4 N, (*ii*) 37.5 N, (*iii*) 191 N and (*iv*) 0.52 kW]**

8.13. (*a*) What is a boundary layer? Why does it thicken with distance from the upstream edge?

(*b*) Describe the phenomenon of boundary layer separation when the flow takes place over a curved surface.

8.14. Air at 20°C and 760 mm Hg absolute pressure flows past a smooth wind tunnel wall, with a free stream velocity of 160 km/hr. Determine the position along the wall, in the flow direction, at which the boundary layer becomes turbulent and the distance where the boundary layer thickness is 25 mm. All wall measurements may be assumed to be taken from the working section entrance and edge/corner effects may be ignored. **[162 mm]**

8.15. An open rectangular box section, sides 3 m × 30 m and 1.5 m × 20 m, is drawn submerged through still water at 20° at a velocity of 9 m/s. Determine the overall drag force, neglecting any edge effects. **[28.5 kN]**

8.16. Air at 40° C flows over a smooth flat plate of length 8 m at a velocity of 10 m/s. The plate width is 1.5 m. Assuming that air flows only on one side of the plate, calculate the force required to hold the plate stationary, considering that the flow is turbulent from the leading edge. Take the local value of turbulent drag coefficient $C_D(x) = 0.0576\ Re^{-1/5}$. At 40° C, ν (the kinematic viscosity = 17×10^{-6} m²/s and density = 1.2 kg/m³ for air) **[Re = 4.7×10^6; 2.26 N]**

CHAPTER

9

Flow Around Immersed Bodies

LEARNING OBJECTIVES

On successful completion of all learning activities planned and provided in the chapter, a student will be able to

- explain the origin of lift and drag forces for flow around an immersed body ;
- classify the types of drag ;
- describe the variation of lift and drag coefficients with variations in flow; and
- éstimate the terminal velocity for free fall of a body.

9.1. Introduction

Quite often, one is encountered with the problem of flow of a fluid over an immersed object or the motion of a body in an expanse of a fluid. In either case, the problem reduces to the study of the development of flow pattern when an object is placed in the free-stream of a real fluid. Examples of such flows are the flow of water past a bridge pier, the flow of air past a structure, motion of aeroplanes, submarines, torpedoes and other objects as well as the flow of fluids past blades in fans, blowers, compressors and turbines.

Attention is focussed on the steady flow of a real fluid past a cylinder, an aerofoil and a sphere. Development of flow and consequent fluid-dynamic forces acting on bodies are considered.

9.2. Lift and Drag

Consider a body placed in the free stream of a real fluid. The fluid exerts a force F on the body. In general, the force F may act in a direction inclined at an angle θ with the free stream as shown in Fig. 9.1.

The component of the force acting in the direction of the free stream is called Drag Force, denoted by D, such that

$$D = F \cos \theta \qquad ...(9.1)$$

The component of the force acting in a direction at right angles to the direction of the free stream is called Lift Force, denoted by L, such that

$$L = F \sin \theta \qquad ...(9.2)$$

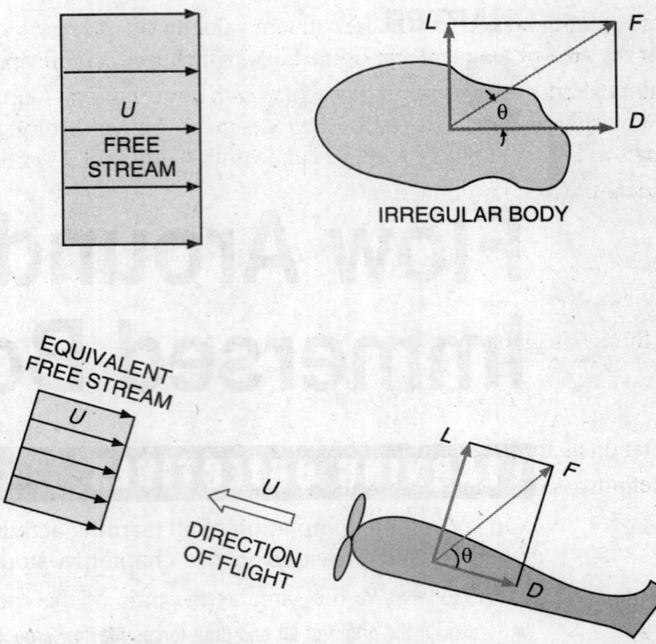

Fig. 9.1: Definitions of Lift and Drag

If the body in question is moving at a constant velocity U, the situation is equivalent to a body at rest in a free stream having a velocity equal and opposite to the velocity of the body. Consider, for example, an aeroplane **moving at a constant velocity U as also shown in Fig. 9.1.** The equivalent free stream acts in line with the direction of flight as shown. The total force acting on the aeroplane is F and the components along and normal to the direction of the free stream are the drag D and the lift L respectively.

While the force that the fluid exerts on the body along the direction of the free stream is called Drag force, the force that the body reacts on the fluid is called Resistance. The resistance is equal and opposite to drag and acts to resist the flow.

When a body has an axis of symmetry and the free stream approaches the body along the axis, the force that acts on the body is wholly along the free stream. The body is then said to be subjected to drag in the absence of lift. As a matter of fact, the production of lift requires asymmetry of flow about the direction of the free stream, whereas drag must be produced under all circumstances. *It is possible, as above, to create drag without lift but it is not possible to create lift without drag.*

The drag and lift forces are expressed in terms of coefficient of drag and lift respectively. The coefficients are defined as the ratios of corresponding forces to the dynamic forces on the projected areas or the planform area whichever is more appropriate.

The coefficient of drag, C_D is defined as

$$C_D = \frac{D}{\frac{1}{2}\rho U^2 A} \quad \text{...(9.3)}$$

such that the drag force D is given by

$$D = C_D \cdot \frac{1}{2}\rho U^2 A \quad \text{...(9.4)}$$

where ρ is the density of the fluid, U is the free stream velocity and A is the projected area normal to the free-stream or the area of the planform of the body, whichever is more appropriate.

For a circular cylinder of diameter d placed across a flow, the area A equals d times the length of the cylinder. For a sphere of diameter d, the area A is $\pi d^2/4$. For an aerofoil of chord c and span s, the area A is s times c. For a flat plate of length l and width b, the area A is l times b.

The coefficient of lift C_L is defined as

$$C_L = \frac{L}{\frac{1}{2}\rho U^2 A} \qquad ...(9.5)$$

such that the lift force L is given by

$$L = C_L \cdot \frac{1}{2}\rho U^2 A \qquad ...(9.6)$$

with the same notation as for the coefficient of drag.

From the definitions of C_D and C_L, it follows that

$$\frac{C_L}{C_D} = \frac{L}{D} \qquad ...(9.7)$$

which means that the ratio of lift to drag is the same as the ratio of the coefficient of lift to the coefficient of drag.

9.3. Classification of Drag

When a fluid flows past a body immersed in it, say a bridge pier, the following phenomena come into play in different regions as shown in Fig. 9.2.

Region (*a*) Divergence of streamlines with a consequent decrease in velocity and the occurrence of a stagnation point upstream of the body which divides the flow into two streams.

Region (*b*) Boundary layer flow; laminar boundary layer until a certain distance and turbulent boundary layer thereafter on the surfaces of the body.

Region (*c*) Separation of the boundary layers at one or more surfaces and/or rejoining of the two streams at different velocities, resulting in a surface of discontinuity which breaks up to form vortices.

Region (*d*) Shedding of vortices behind the body to constitute the wake. The size of the wake is large if the separation takes place early in the boundary layer flow over the surfaces.

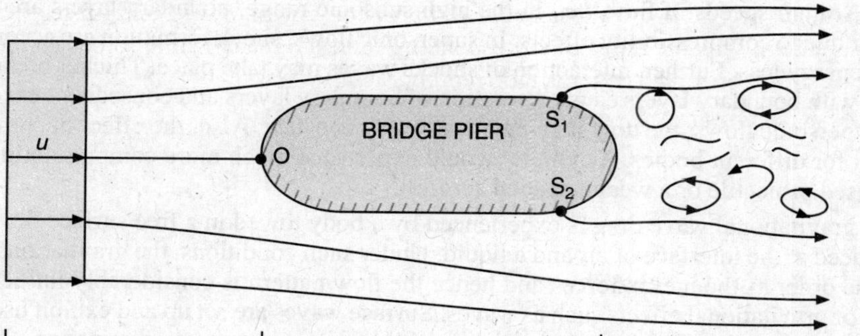

Fig. 9.2: Different Regions for Flow past a Pier

Consideration of the flow pattern around a body reveals that the fluid dynamic drag may be of different origins. Formation of the boundary layer and shear stresses at the surfaces in the region (*b*) give rise to skin-friction drag. Formation of eddies and wake in the region (*d*) result in form drag. The extent of form drag depends upon the planform of the body. The form-drag may far exceed the skin-friction drag for a badly designed or badly oriented body with respect to the flow. If a body is partly immersed in a liquid, the flow would include the formation of gravitational waves which result in wave-drag. On the other hand, if the immersed body travels at transonic and supersonic speeds, the compressibility effect would give rise to wave-drag. The skin-friction drag and the form drag are generally taken together under a single name profile drag because both of them depend upon the profile of the body in relation to the direction of flow.

The drag due to the three-dimensional nature of flow and finite size of a body is called induced drag. The induced drag is due to the induced component of velocity and vorticity distribution along the span of the body. For instance, the induced drag on an aerofoil of a finite span depends upon the lift on the aerofoil as well as on the distribution of the lift over the span of the aerofoil.

Types of drag and their dependence on different factors are shown in Table 9.1.

TABLE 9.1. Classification of Drag

TOTAL DRAG

- PROFILE DRAG
  - SKIN-FRICTION DRAG (Surface Area, shear stresses, velocity gradient, viscosity, surface roughness; streamlined bodies)
  - FORM DRAG (Volume, shape, pressure distribution, wake formation, Reynolds number; bluff bodies)
    - WAVE DRAG
      - Free-surface flow (Gravitational wave phenomena and drag; Floating bodies)
      - Compressible flow (Shock-interaction phenomena and drag; Supersonic and hypersonic flows)
- INDUCED DRAG Lift-Dependent (induced velocity, vortex formation span, volume; lifting surface)

'Compressibility drag' has its origin in the phenomena due to changes in density as the flow proceeds. At high speeds of flow, *i.e.,* in the high subsonic range, boundary layers are thicker and grow faster due to compressibility effects. In supersonic flows, shock formation are accompanied by excessive energy loss. Further, interaction of shocks waves may take place. Thicker boundary layers and shock with boundary layers cause disruption of boundary layers and consequent excessive form drag. In hypersonic flows, the drag may practically stay constant. Also, the effect of compressibility is different for different bodies. A cylinder would experience much more compressibility drag than a sharp-nosed projectile or a wedge-shaped aerofoil.

The gravitational wave drag is experienced by a body towed in a free surface flow or when a body is placed at the interface of air and a liquid. Under such conditions, the gravitational forces are of the same order as the inertia forces and hence the flow pattern is considerably influenced by the formation of gravitational effects such as waves. Surface waves are set up and exhibit oscillations of the liquid, mixing and froth formation with the air and consequent energy dissipation. Gravitational waves are formed ahead of, on the sides but more predominantly at the rear ends of boats, ships, ocean-liner and other sea-going vessels.

9.4. Streamlined and Bluff Bodies

A body is said to be streamlined if, when placed in a flow, the surfaces of the body tend to coincide with stream-surfaces. The streamlines should, therefore, conform with the boundaries of the body. The boundary layer may commence at the leading edge but the layer grows from laminar to turbulent and does not separate until the trailing edge on a streamlined body. Consequently, there is none or little eddying zone and wake formation behind a streamlined body. A body may be streamlined at low velocities but may not be so at higher velocities. A body may be streamlined when placed in a flow in a particular direction at a particular angle but it may not be so when placed in another direction. An example of streamlined body is a thin aerofoil when subjected to a flow as shown in Fig. 9.3 (*a*). A streamlined body experiences drag mainly due to the skin-friction at its surfaces. Streamlined bodies are employed to provide lift, *i.e.*, force normal to the direction of the free stream. Since the

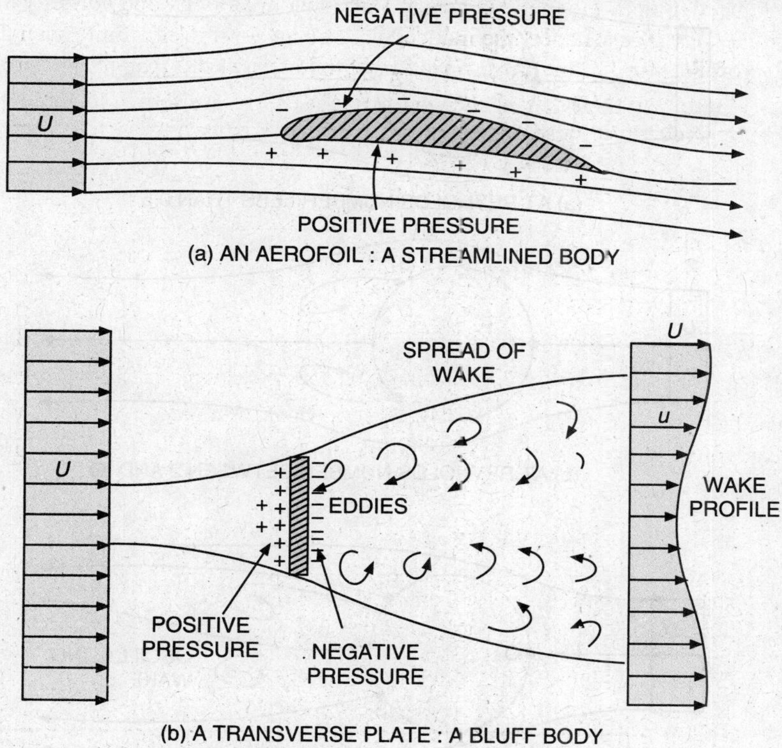

Fig. 9.3: Flow over Streamlined and Bluff Bodies

drag on a streamlined body is low, the lift/drag ratio is high. **Flight of birds is attributed to the production of high lift/drag ratio by virtue of wings attached to their bodies and their adept flying at low angles of attack.**

A body is said to be bluff if it makes the flow turn suddenly and it subdivides it by separation at or near the leading edge as shown in Fig. 9.3 (*b*). A bluff body does not allow the formation of a boundary layer. The drag on a bluff body is mainly due to eddy-formation and wake effect. Bluff bodies are used to promote turbulence and mixing of flow with other substances and to accelerate diffusion in combustion chambers.

While a streamlined body is characterised by high skin-friction drag to form-drag ratio, a bluff body is characterised by high form-drag to skin-friction drag ratio.

9.5. Flow around a Circular Cylinder

Of all the cases studied in fluid mechanics, the flow around a circular cylinder is the most representative. Detailed study of the flow of a real fluid around an infinitely long circular cylinder reveals a number of interesting phenomena which take place for flow around any immersed body. The flow pattern changes significantly with the Reynolds number.

At very low Reynolds number, *i.e., Re* < 0.5, the inertia forces are negligible in comparison with the viscous forces. The flow is, therefore, wholly laminar. The flow pattern around a circular cylinder is symmetrical about the *x* and *y* axes as shown in Fig. 9.4 (a) with stagnation points at O and S. The drag on the cylinder is wholly due to the viscous shearing at the surface of the cylinder, there being no form drag on it.

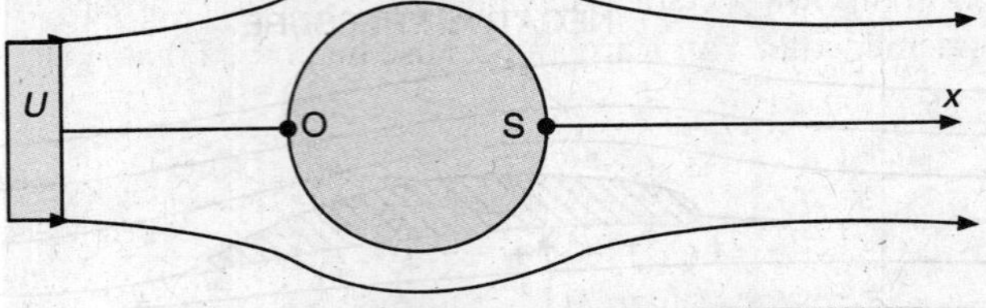

(a) AT REYNOLDS NUMBER LESS THAN 0.5

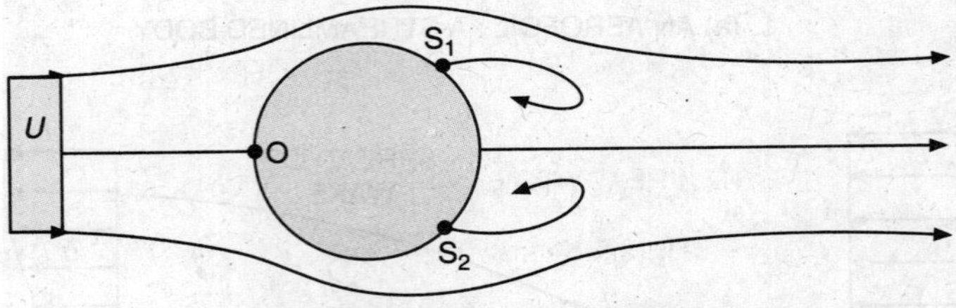

(b) AT REYNOLDS NUMBER BETWEEN 2 AND 30

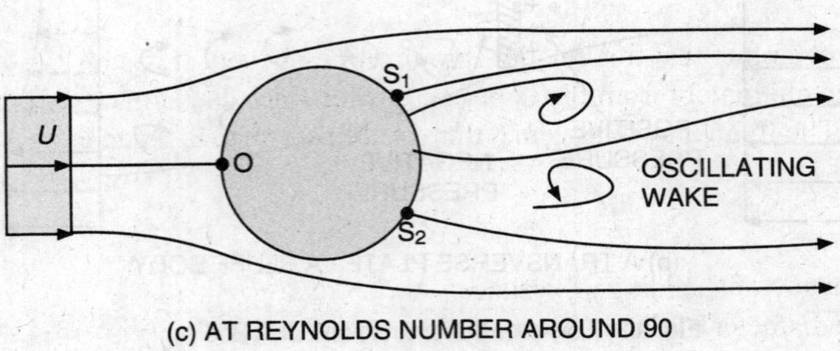

(c) AT REYNOLDS NUMBER AROUND 90

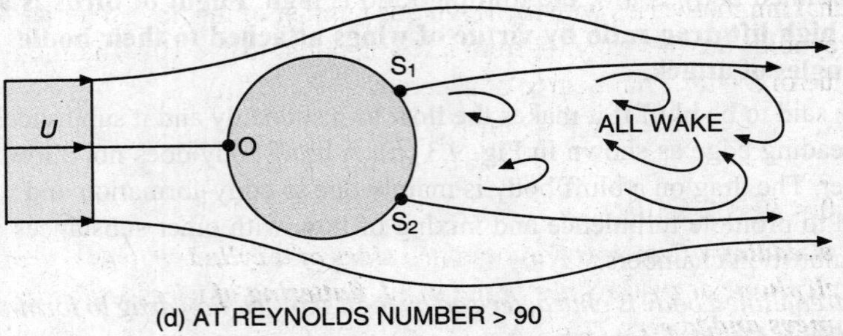

(d) AT REYNOLDS NUMBER > 90

Fig 9.4: Flow over a Circular Cylinder

At Reynolds number in the range of 2 to 30, the flow is laminar and symmetrically over the two sides until S_1 and S_2 positions where the flow separates and two eddies are formed simultaneously which rotate in opposite directions. The two eddies remain fixed in position and continue to rotate by extracting energy from the surrounding fluid. The separated boundary layers come closer to each other and join so that the length of the wake is limited to a narrow and short region as shown in Fig. 9.4 (*b*).

When the Reynolds number is more than 40, the eddying motion is no more stable as shown in Fig. 9.4 (c). The eddies elongate and eventually break off from each side of the cylinder alternately. Periodic oscillation of the wake is observed and alternate breaking and washing away of eddies is noticed at Reynolds number exceeding 90 as shown in Fig. 9.4 (d).

For flow over a range of Reynolds number in excess of 90, eddies are shed alternately from the two sides of the cylinder and, as a result, two rows of vortices are formed in the wake as shown in Fig. 9.5. The arrangement of vortices in two rows rotating in opposite directions is called a Von Karman Vortex Street. The vortex street is named after Von Karman because he proved that such an arrangement is

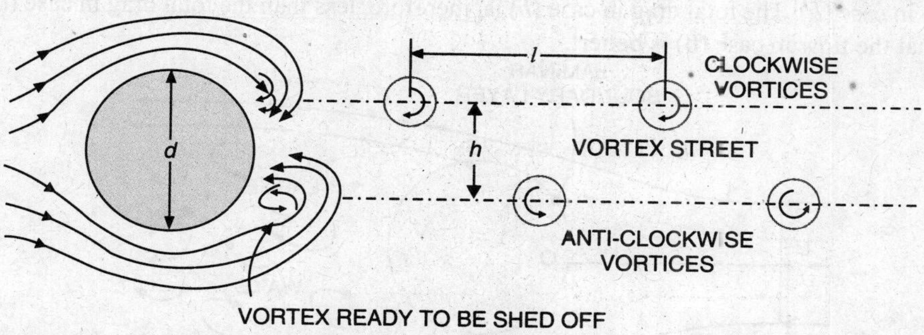

Fig. 9.5: Karman Vortex Street

unstable except when the flow pattern is such that

$$\frac{h}{l} = 0.281 \quad \text{...(9.8)}$$

It is now a proven fact that this relationship must be obeyed. It has also been confirmed that the vortices are shed alternately from the upper and lower sides causing a dynamic lateral force to act on the cylinder. The frequency with which the vortices are shed is given by the Strouhal number S defined by

$$S = \frac{nd}{U} \quad \text{...(9.9)}$$

where n is the frequency of shedding of vortices,

d is the diameter of the cylinder, and

U is the free-stream velocity.

The Strouhal number varies with the Reynolds number of the flow; S varies between 0.20 and 0.23 over the Reynolds number ranging from a few hundred to 1×10^7. The following relationship provides the value of S with a fair degree of accuracy :

$$S = \frac{nd}{U} = 0.2\left(1 + \frac{20}{Re}\right) \quad \text{...(9.10)}$$

for the range $250 < Re < 2 \times 10^7$.

Alternate shedding of vortices from the two sides of a cylinder gives rise to the phenomena of 'singing' of telephone or power lines in the wind, fluttering of wires and poles and vibration of cylindrical chimneys and towers, etc.

At Reynolds numbers higher than 1×10^7, the vortices formed downstream the cylinder are disintegrated into random turbulent motion and the Karman vortex street is not observed.The flow upstream the cylinder is gradually retarded and proceeds over the surfaces of the cylinder with the formation of the laminar boundary layer. There are two possibilities :

One, that the laminar boundary layer may separate at points S_1 and S_2 short distance downstream of the leading point O. The other, that the laminar boundary layer may change to turbulent boundary layer and by virtue of the fact that this is equivalent to energisation of the layer, the separation is delayed until far downstream for the leading point O.

The difference in the flow pattern for the two cases referred to as case (*a*) and case (*b*) is brought out in Fig. 9.6. If the laminar boundary layer separates early on the surface of the cylinder, the wake behind the cylinder is large and wide whereas if the boundary layer becomes turbulent and separates further downstream the leading point, the wake behind the body is short and narrow. The form drag in case (*a*) is much more than that in case (*b*). The skin-friction drag in case (*b*) is marginally more than that in case (*a*). The total drag in case (*b*) is, therefore, less than the total drag in case (*a*) which means that the flow in case (b) is better!.

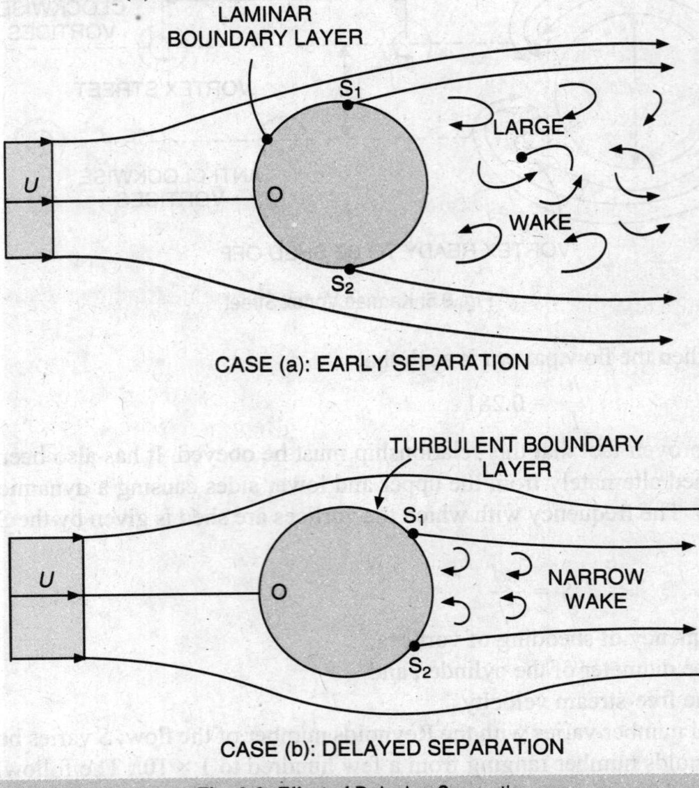

Fig. 9.6: Effect of Delaying Separation

The changeover from an early separation as in case (a) to a delayed separation as in case (b) is brought about by one or more of the following methods :

(*i*) *by increasing the Reynolds number of flow which may be achieved by increasing the velocity of the free stream.*

(*ii*) *by increasing the initial level of turbulence in the free stream.*

(*iii*) *by roughening a part of the leading face of the cylinder or*

(*iv*) *by providing a trip-wire ring in the laminar boundary layer region.*

The D'Alembert 'Paradox'

It has been shown, on the basis of the ideal-flow theory, that a cylinder placed in a free stream should experience no drag force. In fact, it can be shown that, according to ideal-flow analysis, any symmetrical body placed in a free stream along the axis to symmetry should experience no drag force.

D'Alembert discovered that if a cylinder or any other symmetrical body were placed in a real fluid, it does experience a drag force. *The paradoxial behaviour of a body is to actually experience a drag while, by ideal-flow analysis, it should not have any drag force is referred to as D'Alembert Paradox.*

The explanation for the paradox is that the actual flow past a body is not the same as the ideal-flow analysis assumes. The actual flow is influenced by the viscous effects, if the body deviates appreciably from the ideal flow pattern and the pressure distribution on the surface also differs from the distribution predicted by the ideal-flow analysis.

In reality, the flow separates from the surface of the cylinder and vortex rings are formed which are shed downstream. The reason why the flow separates from the surface of the cylinder is to do with the boundary layer formation, thickening of the layer and its inability to move against an adverse pressure gradient as discussed before.

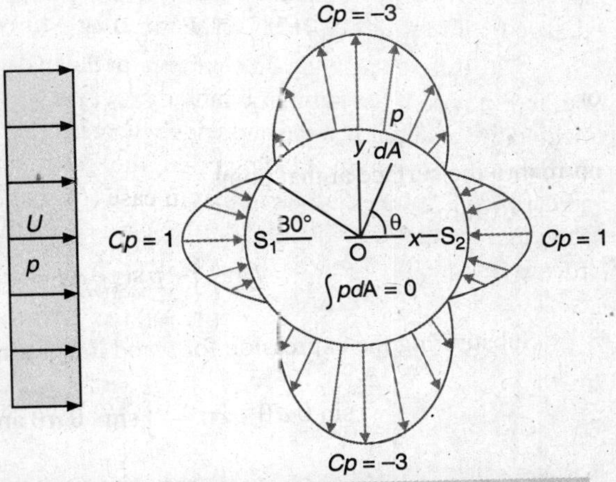

Fig 9.7: Pressure Distribution and Drag Force in an Ideal flow

The pressure distribution in a real flow is such that it deviates mainly in the downstream region.

Magnus Effect and Lift Force

If a symmetrical body is placed in a flow along the axis of symmetry, the body experiences a drag in the direction of the flow. This is obvious from the fact that the pressure distribution is symmetrical about the axis of symmetry. On the other hand, if the object is not symmetrical about the direction of flow or if a symmetrical object is rotated in a flow, the pressure distribution results in a force, a component of which is perpendicular to the direction of flow, *i.e.*, a lift force as shown in Fig. 9.8. *The phenomenon of lift production by a rotating object placed in a free stream is known as the Magnus effect.*

It may be noted that the drag on a cylinder without rotation may be high due to separation and wake effect but it is considerably reduced when the cylinder is rotated. In other words, real-flow effects are less important when the flow past a rotating cylinder is considered. We shall, therefore, confine ourselves to the ideal flow analysis for the computation of the lift force. The velocity at any point on the surface of a cylinder placed in a free stream with circulation Γ associated with its rotation is given by

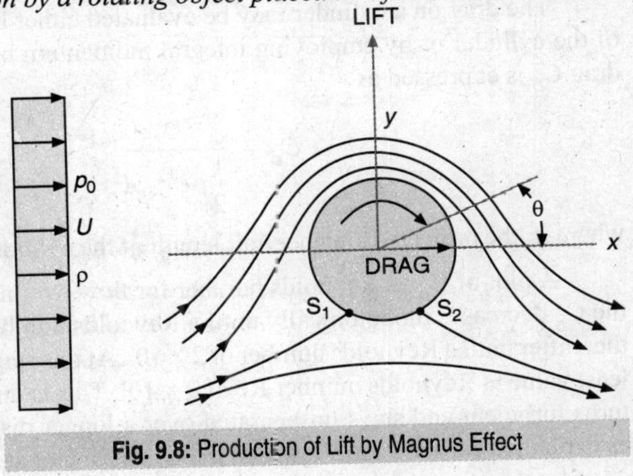

Fig. 9.8: Production of Lift by Magnus Effect

$$u_\theta = -\left(2U\sin\theta + \frac{\Gamma}{2\pi R}\right)$$

$$u_r = 0$$

representing a purely circulatory motion.

Assuming no energy loss between the free stream and a point of interest; applying the Bernoulli equation,

$$p_0 + \frac{1}{2}\rho U^2 = p + \frac{1}{2}\rho u_\theta^2$$

or
$$p = p_0 + \frac{1}{2}\rho\left[U^2 - \left(2U\sin\theta + \frac{\Gamma}{2\pi R}\right)^2\right]$$

normal to the surface at that point.

Lift force

$$L = \int_s -p\sin\theta\, dA = -\int_0^{2\pi} p\, R\sin\theta\, d\theta$$

Substituting the expression for p and integrating, bearing in mind that

$$\int_0^{2\pi} \sin\theta\, d\theta = 0 = \int_0^{2\pi} \sin^3\theta\, d\theta \text{ and } \int_0^{2\pi} \sin^2\theta\, d\theta = \pi$$

$$L = \rho U \Gamma \text{ per unit span}$$

It follows from this expression that

(*a*) Lift force upwards as shown by L for clockwise rotation of the cylinder placed in free stream flowing from left to right.

(*b*) Lift force is directly proportional to the fluid density p, free stream velocity U and the circulation Γ around the cylinder.

The ultimate significance of the expression $L = pU\Gamma$ is due to the fact that this is valid for flow over an object of any cross-section as long as Γ is produced around it *whether by rotating it or otherwise*. In general, Γ is not zero for most of the unsymmetrical bodies such as aerofoil at a small angle of attack. The lift force is given by the same expression irrespective of the origin of circulation and the shape of the body.

The drag on a cylinder may be evaluated either by integrating the pressure over the total area of the cylinder or by employing integral momentum balance or by other means. The coefficient of drag C_D is expressed as

$$C_D = \frac{D}{\frac{1}{2}\rho U^2 . A} \qquad ...(9.11)$$

where A, the projected area per unit length of the cylinder, which equals the diameter of the cylinder.

A plot of C_D vs Reynolds number for flow over a cylinder is shown in Fig. 9.9. It is seen that the C_D decreases monotonically until a Reynolds number of 2000, and remains practically constant thereafter until a Reynolds number of 2×10^5. At this stage, the C_D drops steeply by a factor of 5 to the least value at Reynolds number Re of 3×10^5. This is due to the fact that the laminar boundary layer turns turbulent and stays unseparated over a longer distance, thus reducing the wake considerably as explained earlier. Also shown on the C_D *vs Re* plot are the variations of the C_D for a flat plate held across the flow as a bluff body and for an aerofoil. The drag for the flat plate stays constant irrespective of the Reynolds number because it offers no chance for the formation and development of boundary layers. The drag for an aerofoil is considerably less than that for other bodies.

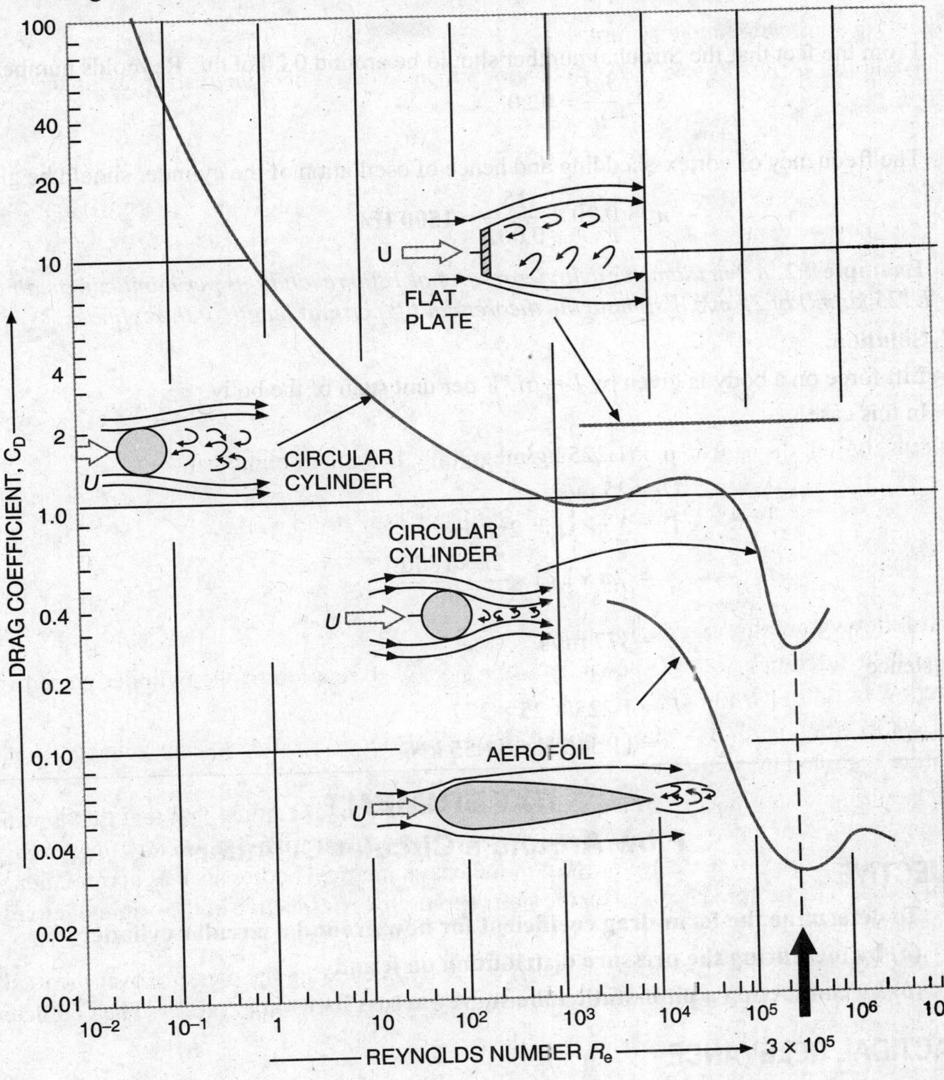

Fig. 9.9: Coefficient vs Reynolds Number for Two-Dimensional Bodies

Example 9.1. *Determine the frequency of oscillation produced by a 15 m/s wind at standard sea-level conditions blowing across a 2 mm diameter taut wire across it.*

Solution.

Given that

$$U = 15 \text{ m/s and } d = 0.002 \text{ m}$$

The density and viscosity of air at the standard sea-level conditions are

$$\rho = 1.225 \text{ kg/m}^3 \text{ and } \mu = 1.8 \times 10^{-5} \text{ Ns/m}^3$$

From the data,

$$Re = \frac{Ud\rho}{\mu} = \frac{15 \times 0.002 \times 1.225}{1.8 \times 10^{-5}} = 2041.7$$

From the fact that the Strouhal number should be around 0.20 at this Reynolds number,

$$S = \frac{nd}{u} = 0.20$$

The frequency of vortex-shedding and hence of oscillation of the cylinder should be given by

$$n = 0.20 \times \frac{15}{0.002} = \mathbf{1500\ Hz}$$

Example 9.2. *A 2 m diameter cylinder rotates at 1800 revolutions per minute in an air stream* (ρ = *1.225 kg/m³*) *of 25 m/s. Estimate the theoretical lift per unit length of the cylinder.*

Solution.

Lift force on a body is given by $L = \rho U \Gamma$ per unit span of the body.

In this case

$$\rho = 1.225 \text{ kg/m}^3$$

$$U = 25 \text{ m/s}$$

$$\Gamma = 2\,\pi r.\, u_\theta = 2\pi r.\, r\omega$$

$$= 2\pi \times 1 \times 1 \times \frac{2\pi \times 1800}{60}$$

$$= 377 \text{ m}^2/\text{s}$$

Hence

$$L = 1.225 \times 25 \times 377$$

$$= 11\,546 \text{ N} = \mathbf{11.55\ kN}$$

EXPERIMENT

Flow Around a Circular Cylinder

OBJECTIVE

To determine the form-drag coefficient for flow around a circular cylinder

(*a*) by measuring the pressure distribution on it and

(*b*) by conducting a pitot-static tube survey across its wake.

PRACTICAL RELEVANCE

Flow around cylindrical objects with circular cross-section is often encountered in engineering applications. Flow of gases over heat exchanger tubes, flow of air past transmission wires, chimneys, and other structures, flow of steam over boiler tubes and condenser tubes are a few examples where such flows are encountered. Designers of such systems require an understanding of the flow around a circular cylinder with particular reference to the variation of drag with the nature of flow, the Reynolds number and other parameters. Studies in terms of non-dimensional parameters are helpful in predicting the drag for similar situations.

The study of flow over a circular cylinder has an academic significance too. The names of D'Alembert with reference to his paradox, Stokes for creeping flows, Von Karman for the shedding of vortices and many other researchers are associated with the flow over a circular cylinder. The academic importance and the relative simplicity of the flow pattern justify detailed studies of flow over circular cylinders.

BACKGROUND INFORMATION

When a circular cylinder is placed in a uniform flow of a real fluid, it is found that the nature of the flow in the immediate vicinity of the surface and behind the cylinder in the wake region

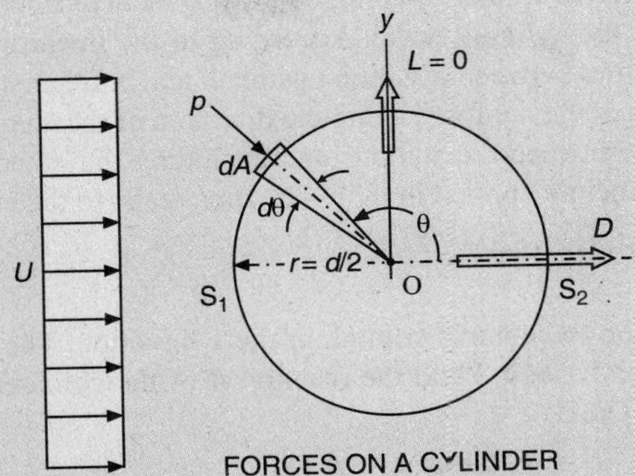

FORCES ON A CYLINDER

depends upon a number of factors such as the viscosity and density of the fluid, the velocity and the laminar or turbulent nature of the flow. At very low speeds, the flow may be laminar without a trace of separation from the surface. At higher speeds, the flow may continue to be laminar over a substantial part of the surface but would separate near the 90° point and eddies would be formed in the wake region. At still higher speeds, the flow in the boundary layer may become turbulent and due to enhanced momentum exchange, the eddy-formation may be delayed and suppresed. Experimental evidence indicates that a non-dimensional parameter called the Reynolds number, $Re = Ud\rho/\mu$, governs the transition from laminar to turbulent flow over the surface of the cylinder. Transition is found to occur at a value of $Re > 2 \times 10^5$ for a cylinder and at a slightly higher value for a sphere.

Considering the pressure p over an elementary area dA on the surface of the cylinder, as shown the drag force on it must be $pdA \cos\theta$. For a cylinder of length l, the total drag force is then given by

$$D = \int_A p \cos\theta \, dA = \int_0^{2\pi} p \cos\theta \, l \, r \, d\theta$$

$$= r\, l \int_0^{2\pi} p \cos\theta \, d\theta$$

Alternately, on defining a drag coefficient

$$C_D = \frac{D}{\frac{1}{2}\rho U^2 l \, d}$$

one gets

$$C_D = \int_0^{2\pi} \frac{p}{(\rho U^2/2)} \cos\theta \, d\theta = \int_0^{2\pi} \frac{p}{(\rho U^2/2)} \, d(\sin\theta).$$

The above expression for C_D may be interpreted as the area under the $p/(\rho u^2/2)$ *vs* $\sin\theta$ plot between the limits of $\theta = 0$ and $\theta = 2\pi$. This is obtained from the measured distribution of the pressure p over the circumference of the cylinder from the leading edge or the front stagnation point S_1 to the trailing edge or the hind stagnation point S_2. It may be noted that the pressure p is measured relative to the static pressure in the wind tunnel, *i.e.*,

$$p = p_{\text{cylinder}} - p_{\text{tunnel}}$$

The term $p/(\rho u^2/2)$ is often called the pressure coefficient C_p. The problem of determining the drag coefficient is thus reduced to an evaluation of the pressure coefficient at different angular position θ on the cylindrical surface.

The total drag and the coefficient of drag could also be determined by conducting a pitot-static tube survey across the cylinder wake. According to the linear momentum principle, the rate of change of momentum between a station upstream and another station downstream of the immersed object must equal the net force along the direction of the change of momentum on the control volume enclosed between the stations and excluding the object. With reference to the figure, the momentum principle applied in the x-direction leads to the expression

$$\int_{A_2} u_2 . \rho u_2 \, dA_2 - \int_{A_1} u_1 . \rho u_1 \, dA = F$$

where F is the net force on the control volume in the x-direction. The pressure forces exerted by the isolated fluid beyond 1 and 2 and the reaction R of the cylinder on the control volume contribute to make up F. Thus

$$F = p_1 A_1 - p_2 A_2 + R$$

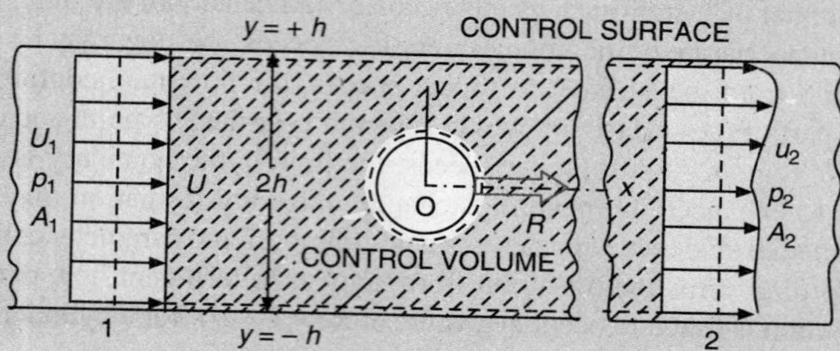

CONTROL VOLUME INCLUDING THE CYLINDER

if it is assumed that the static pressures p_1 and p_2 are constant across the stations 1 and 2 respectively. Further, assuming that the upstream velocity is uniform,

$$R = \rho \int_{A_2} u_2^2 \, dA_2 - \rho U^2 A_1 + p_2 A_2 - p_1 A_1$$

The drag force D exerted by the fluid on the cylinder must be equal and opposite to R.

Hence $$D = \rho U^2 A - \rho \int_A u_2^2 \, dA + (p_1 - p_2) A$$

if the areas of cross-section at the upstream and downstream stations are equal, *i.e.*,

$$A_1 = A_2 = A$$

PROCEDURE

1. Record the atmospheric pressure and the air temperature in the wind tunnel and look up the corresponding values of the density and viscosity of air from standard tables.
2. From the knowledge of the value of the transition Reynolds number, compute the two values of the free-stream velocities at which the tests would be conducted. Note that the tests should be conducted such that at one free-stream velocity, the boundary layer flow around the cylinder is in the laminar flow regime and that at the other value of free-stream velocity, it is in the turbulent flow regime.

3. Start the fan of the wind tunnel and adjust the damper to create a steady flow at the higher of the two free-stream velocities.
4. Connect the static pressure tap on the surface of the cylinder to the u-tube manometer and turn the cylinder manually until the static pressure tap is at the forward stagnation point S_1; this occurs when the pressure reading is at its maximum. Note the reading on the protractor.
5. Turn the cylinder manually clockwise until the pressure on the cylinder is zero and note the reading on the protractor. Now, turn the cylinder manually anticlockwise until the pressure is again zero. Again note the reading on the protractor. The mid-position between these two settings should coincide with the positions at which the maximum static pressure occurs, *i.e.*, the stagnation point S_1. In case the two positions do not coincide, the average position between the two may be used as the reference position.
6. Turn the cylinder in steps of 15° and read the static pressure on the cylinder referred to the static pressure in the wind tunnel. Complete the table of observations for θ between 0° and 360°.
7. Record the static pressure and the total pressure, sufficiently ahead of the cylinder, on the centreline of the tunnel. As a check, note that the total pressure as measured with the pitot tube should be the same as the reading of the cylinder pressure tap when it is positioned at the forward stagnation point S_1, *i.e.*, at the reference position located in step 5.
8. Traverse the pitot static tube in a vertical plane cross the wake of the cylinder, at a section fairly far downstream of the cylinder. Record the static pressure and the total pressure at this plane at intervals of approximately 0.5 cm.
9. Stop the fan of the wind tunnel and return to the initial conditions.

RESULTS

1. Plot C_p *vs* sin θ for θ varying from 0° to 360° for each condition of flow. Observe that the area under the plot equals the integral

$$\int_0^{2\pi} \frac{p}{(\rho U^2/2)} d(\sin\theta)$$

and gives the value of C_D. The magnitude of C_D could also have been obtained by plotting $C_p \cos\theta$ *vs* θ over the same range of θ and then by computing the area under the curve.

2. For ideal fluid flow around a circular cylinder, the theoretical pressure distribution is represented by the expression

$$C_p = 1 - 4\sin^2\theta:$$

Plot C_p *vs* sin θ for the ideal-fluid flow case on the experimental real-flow plot. Observe that the area under the ideal-fluid plot must be zero.

3. Calculate the values of C_D and D from the pressure coefficient plots and compare with the standard values.
4. Calculate the drag on the cylinder by applying the linear momentum principle. This involves the evaluation of the term

$$\rho \int_A u_2^2\, dA = \rho l \int_{-h}^{+h} u_2^2\, dh$$

by graphical integration of the plot of u^2 *vs* *h*.

5. Make a polar plot of the real and ideal flow pressure distribution around the cylinder. In such a plot the convention employed states *that a radial line directed on the surface*

represents positive pressure at that point and a radial line directed away from the surface represents negative pressure at that point.

POINTS FOR DISCUSSION

1. Explain the difference between an ideal-fluid flow and a real-fluid around an object with particular reference to a circular cylinder.
2. At what angular locations is a zero pressure observed on the cylinder surface and why so?
3. Locate the points of maximum and minimum velocities and pressures. Indicate the regions of favourable and adverse pressure gradients.
4. Locate the points of separation of flow from the cylinder surface and state what happens to the flow between the separation point and the downstream station selected for the pressure and velocity traverse.
5. Where is the static pressure on the cylinder equal to the total pressure in the free-stream and why ?
6. Indicate whether it is better to measure the static pressure distribution on the cylinder surface by rotating one pressure tap over 360° or by means of a large number of static taps at different angular locations.
7. Why is it necessary to locate the downstream station for the velocity and pressure traverse far downstream of the cylinder ? Can the upstream station be as close to the cylinder as practicable ? Discuss these points with reference to the streamline pattern for the flow.
8. Discuss the nature of the curve C_D *vs* Re for flow around a cylinder commenting specifically on the drop in C_D at Re of the order of 2×10^5. Would the C_D *vs* Re curve for a spherical object be expected to coincide, be below or above the corresponding curve for a circular cylinder ?

Any other observation or comment?

Example 9.3. *A television transmitter antenna consists of a vertical pipe 20 cm diameter and 30 m high on top of a tall structure. Determine the total drug on the antenna and the bending moment about the base in a 30 m/s wind at NTP.*

(Given that the density of air is 1.22 kg/m³ and viscosity 1.79 × 10⁻⁵ Ns/m² at NTP)

Solution.

The Reynolds number is given by

$$Re = \frac{Ud\rho}{\mu} = \frac{30 \times 0.2 \times 1.22}{1.79 \times 10^{-5}} = 4.1 \times 10^5$$

The value of C_D may be taken as 0.2 at this Reynolds number of flow. The drag is given by

$$D = C_D . \frac{1}{2} \rho U^2 . A$$

$$= 0.2 \times \left(\frac{1}{2} \times 1.22 \times 30^2\right) \times (0.2 \times 30)$$

$$= 659 \text{ N}$$

The moment about the base is given by

$$M = 659 \times \frac{30}{2} = 9885 \text{ Nm} = \mathbf{9.9 \text{ kNm}}$$

9.6. Flow Around an Aerofoil

An aerofoil is a streamlined body designed to produce a streamlined flow pattern when placed in a free stream. The terminology for an aerofoil is as follows :

Aerofoils may be symmetrical or unsymmetrical about the chord line 1-2 drawn from the leading edge to the trailing edge as shown in Fig. 9.10. The camber line is straight and coincides with the chord line for a symmetrical aerofoil whereas it is curved for an unsymmetrical aerofoil. The width of the aerofoil normal to the plane of the cross-sectional view is called the span, S.

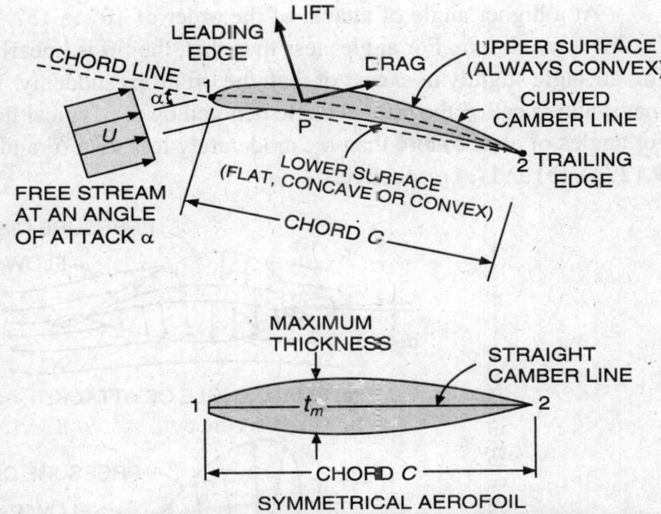

Fig. 9.10: Nomenclature for an Aerofoil

The ratio of the span S to the chord C of the aerofoil is called its aspect ratio,

$$AR = \frac{S}{C} \qquad ...(9.12)$$

If the span is large compared to the chord, the aspect ratio is large. The flow over an aerofoil is two dimensional if its aspect ratio is infinite. If the aspect ratio is large, the end flow effects may then be neglected.

The flow around an aerofoil placed in the free stream of a real fluid varies with the angle of attack, *i.e.*, the angle between the free stream and the chord. The pressure distribution accordingly changes. At a certain angle of attack, *i.e.*, zero for a symmetrical aerofoil and negative for an asymmetrical aerofoil as shown in Fig. 9.11(*a*) and (*b*), the pressure distribution is such that the total lift is zero.

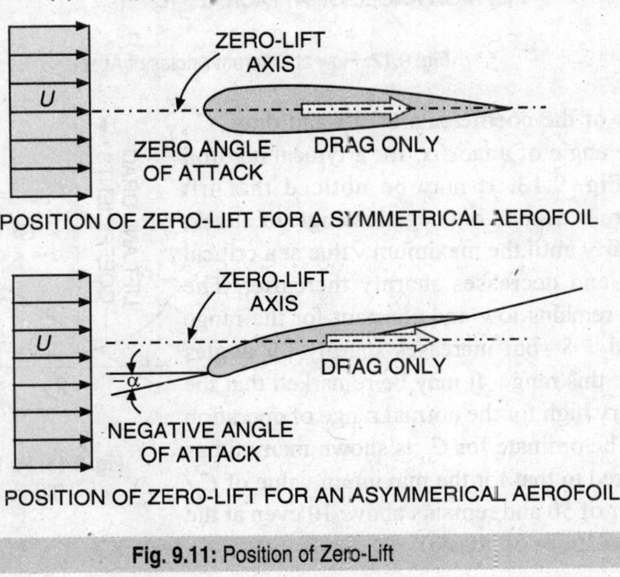

Fig. 9.11: Position of Zero-Lift

At a higher angle of attack, of the order of 10° to 15°, the pressure distribution is such that the lift is maximum. For angles less than that, the lift is linearly related with the angle of attack and for an angle slightly in excess of that, the lift drops suddenly. The angle of attack at which the lift is maximum is called the critical angle denoted by α_c. Typical flow patterns and pressure distribution for angles of attack more than α_c, moderately less than α_c and greater than α_c, are sketched in Fig. 9.12 (*a*), (*b*) and (*c*) respectively.

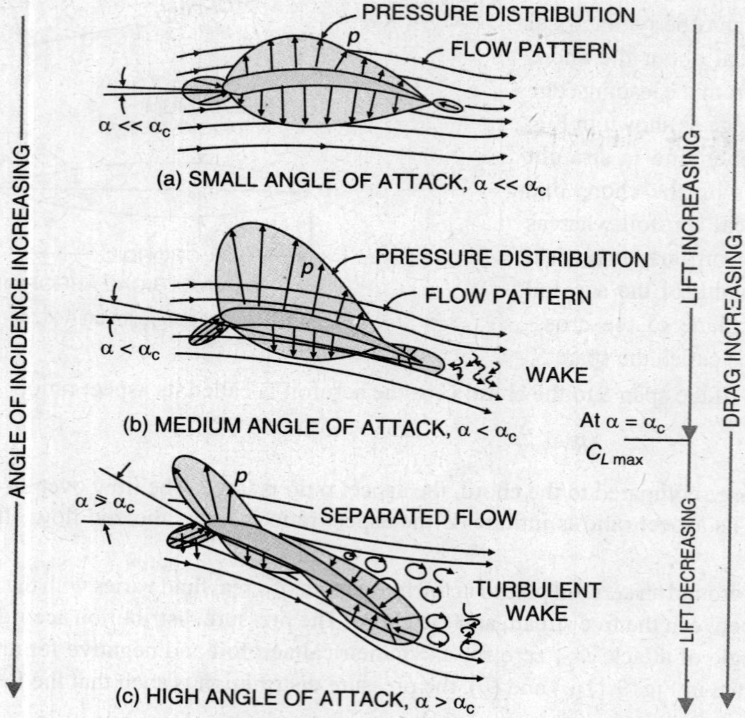

Fig. 9.12: Flow at Different Angles of Attack

Variations of the coefficients of lift and drag, C_L, and C_D with the angle of attack α, for a typical aerofoil are shown in Fig. 9.13. It may be noticed that lift coefficient is zero at a small but negative angle of attack, it increases linearly until the maximum value at a critical angle of attack and decreases sharply thereafter. The drag coefficient remains low and constant for the range between -5° and + 8° but increases sharply for angles of attack outside this range. It may be remarked that the ratio C_L/C_D is very high for the normal range of operation of an aerofoil. The ordinate for C_D is shown many times enlarged compared to that for the maximum value of C_L/C_D is of the order of 50 and remains above 10 even at the position of critical angle of attack !

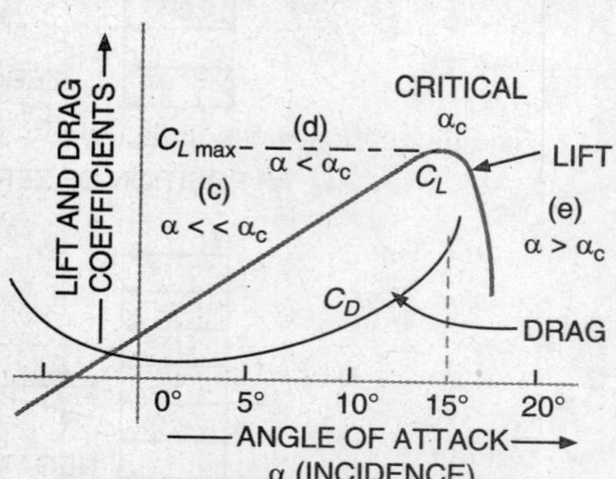

Fig. 9.13: Performance of an Aerofoil at Different Angles of Attack

The critical angle of attack is also known as the stalling angle or stalling incidence because the aerofoil is said to be stalled at that orientation. Inclination with the free stream slightly in excess of the stalling angle renders the aerofoil to be a bluff body and hence inoperative because the flow separates from the convex surface and a large turbulent wake is formed in the separated region.

Example 9.4. *Experiments were conducted in a wind tunnel with a wind speed of 50 kilometre per hour on a flat plate 1 metre long and 1 metre wide. The specific weight of air is 1.15 kilogram per cubic metre. The plate is kept at such an angle that the coefficients of lift and drag are 0.75 and 0.15 respectively. Determine (i) the lift force, (ii) the drag force, (iii) the resultant force, and (iv) power exerted by the air stream on the plate.*

Solution.

From the data,

$$U = 50 \text{ km/hour} = \frac{50 \times 1000}{60 \times 60} = 13.9 \text{ m/s}$$

$$l = 1 \text{ m and } b = 1 \text{ m}$$

$$\rho = 1.5 \text{ kg/m}^3;$$

$$C_L = 0.75;\ C_D = 0.15$$

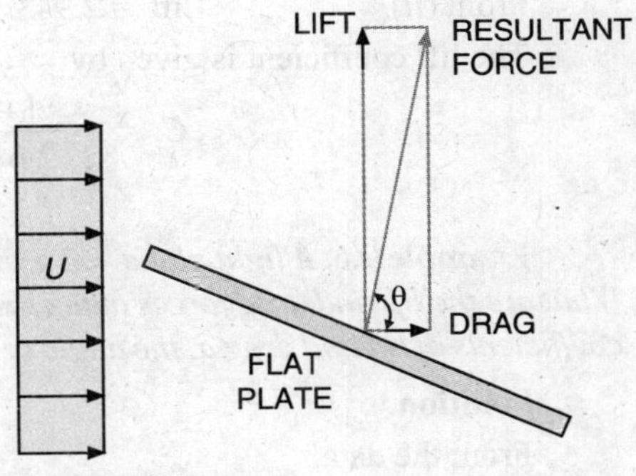

(*i*) Lift force $L = C_L \times \frac{1}{2}\rho U^2 A$

$$= 0.75 \times \frac{1}{2} \times 1.15 \times 13.9^2 \times 1 \times 1$$

$$= \mathbf{83\ N}$$

(*ii*) Drag force $D = C_D \times \frac{1}{2}\rho U^2 A$

$$= 0.15 \times \frac{1}{2} \times 1.5 \times 13.9^2 \times 1 \times 1$$

$$= \mathbf{16.7\ N}$$

(*iii*) Resultant force

$$\mathbf{F} = \sqrt{83^2 + 16.7^2}$$

$$= \mathbf{84.7\ N}$$

The inclination θ which the direction of the free stream is given by

$$\tan\theta = \frac{83}{16.7} = 5$$

and $\theta = \mathbf{78.7°}$

(*iv*) Power exerted by the air stream on the plate

= free stream velocity × resultant force along the free stream

$= 13.9 \times 84.7 \times \cos 78.7° = 13.9 \times 16.7$ W

$= 230.7$ W $=$ **0.231 kW**

Example 9.5. *A kite has an effective area of 0.5 m² and mass 0.3 kg. It experiences a drag of 12 N in a wind speed of 40 km/hour, (a) Determine the tension in the cord if it makes an angle of 45° with the horizontal, (b) Estimate the lift coefficient for the kite.*

Solution.

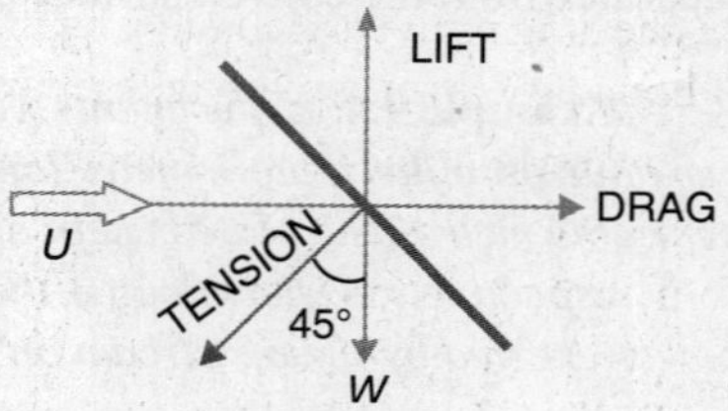

From the data, the wind speed is

$$U = 40 \times 1000/3600 = 11.11 \text{ m/s}$$

and the weight of the kite is

$$W = 0.3 \times 9.81 = 2.943 \text{ N}$$

$$\text{Drag} = 12 \text{ N}$$

The kite is in equilibrium under the forces shown in the sketch. For equilibrium,

$$\text{Drag} - T\cos 45° = 0 \quad -(i)$$

$$\text{Lift} - W - T\sin 45° = 0 \quad -(ii)$$

From (*i*), $T = 12/\cos 45° = 12/0.707 = \mathbf{16.97\ N}$

From (*ii*), $\text{Lift} = 2.943 + 16.97 \sin 45 = \mathbf{14.94\ N}$

The lift coefficient is given by

$$C_L = \frac{\text{Lift}}{1/2\,\rho U^2.A} = \frac{14.94}{1/2 \times 1.2 \times (11.11)^2 \times 0.5} = 0.404$$

Example 9.6. *A light plane has a wing span of 10 m and a chord of 2 m for the aerofoil section. What are the lift and drag forces at the instant of take-off at a speed of 120 miles/hour. Given that the coefficients of lift and drag at the angle of attack of 10° during take-off are 0.95 and 0.08 respectively.*

Solution.

From the data,

$$U = 120 \text{ miles/hr} = \frac{120 \times 1.609 \times 1000}{3600} = 53.6 \text{ m/s}$$

A, the projected area of the wing = 10 × 2 = 20 m²

ρ, the density of air = 1.2 kg/m³

The lift is given by

$$L = C_L.\frac{1}{2}\rho U^2 A$$

$$= 0.95 \times \frac{1}{2} \times 1.2 \times 53.6^2 \times 20$$

$$= \mathbf{32\,752\ N \approx 32.8\ kN}$$

The drag is given by

$$D = C_D.\frac{1}{2}\rho U^2 A$$

$$= 0.08 \times \frac{1}{2} \times 1.2 \times 53.6^2 \times 20$$

$$= \mathbf{2758\ N \approx 2.76\ kN}$$

The drag force could also be determined from the lift force by the relationship

$$D = \frac{C_D}{C_L} L$$

such that

$$D = \frac{0.08}{0.95} \times 32.8 = \mathbf{2.76\ kN}$$

9.7 Flow around Axisymmetric and Three-Dimensional Bodies

The same principles and concepts which apply to the flow around two-dimensional bodies also apply to the flow around axisymmetric and three-dimensional bodies. At very low Reynolds number *Re,* the drag is given accurately by the viscous flow analysis. At high Reynolds number the phenomena of boundary layer growth and separation and consequent vortex shedding etc. come into play.

Consider the flow around a sphere. For an ideal fluid flow, the pressure coefficient is given by

$$C_p = \frac{p - p_0}{\frac{1}{2}\rho U^2} = 1 - \frac{9}{4}\sin^2\theta$$

where θ is the angle measured *w.r.t.* the lagging stagnation point.

For laminar flow over the sphere, the pressure distribution over the surface of the sphere is given by

$$p - p_0 = -\frac{3}{2}\mu U \frac{x}{R^2} \qquad \text{...(9.13)}$$

where x is the distance of the point along the x-axis and R is the radius of the sphere. The pressure at the leading stagnation point is given by substituting $x = -R$,

$$p - p_0 = \frac{3}{2}\mu U / R \qquad \text{...(9.13 a)}$$

The pressure at the trailing stagnation point is given by substituting $x = R$,

$$p - p_0 = -\frac{3}{2}\mu U / R \qquad \text{...(9.13 b)}$$

The total force exerted by the viscous fluid on the sphere is obtained by integrating the component of pressure along the direction of the free stream over the surface area. Stokes proved that

$$D = 3\pi\,\mu\,Ud \qquad \text{....(9.14)}$$

for very low Reynolds number, Re < 0.2.

This relationship is called the Stokes law for drag. The coefficient of drag C_D may be obtained by employing the definition

$$C_D = \frac{D}{\frac{1}{2}\rho U^2 A}$$

Substituting the expression for D from Eq. 9.14, and realising that the projected area $A = \pi d^2/4$ and the Reynolds number $Re = Ud\rho/\mu$,

$$C_D = \frac{3\pi\mu U d}{\frac{1}{2}\rho U^2 \pi d^2 / 4}$$

or

$$C_D = \frac{24}{Re} \text{ for } Re < 0.5 \qquad \text{...(9.15)}$$

which is also known as the Stokes law. The law is valid for very small Reynolds number, *i.e., Re* less than 0.2 where the flow is completely laminar and viscous forces predominate.

Oseen improved upon the Stokes' theoretical expression in the light of the experimental data over the range of Reynolds number between 0 and 2 and suggested the following expression for C_D:

$$C_D = \frac{24}{Re}\left(1+\frac{3}{16}Re\right) \text{ for } Re < 2.0 \qquad ...(9.16)$$

At higher Reynolds number, the flow is such that the boundary layer grows and separates sooner or later depending upon whether it is laminar or turbulent. ***Later the separation, smaller the size of the wake and hence less is the total drag on the sphere.*** As a matter of fact, increase of Reynolds number triggers the laminar boundary layer into turbulent boundary layer before separation and the point of separation is delayed. A sudden reduction in drag is, therefore, observed when the Reynolds number exceed a critical value of 3×10^5. The variation of C_D *vs Re* is shown in Fig. 9.14 for flow

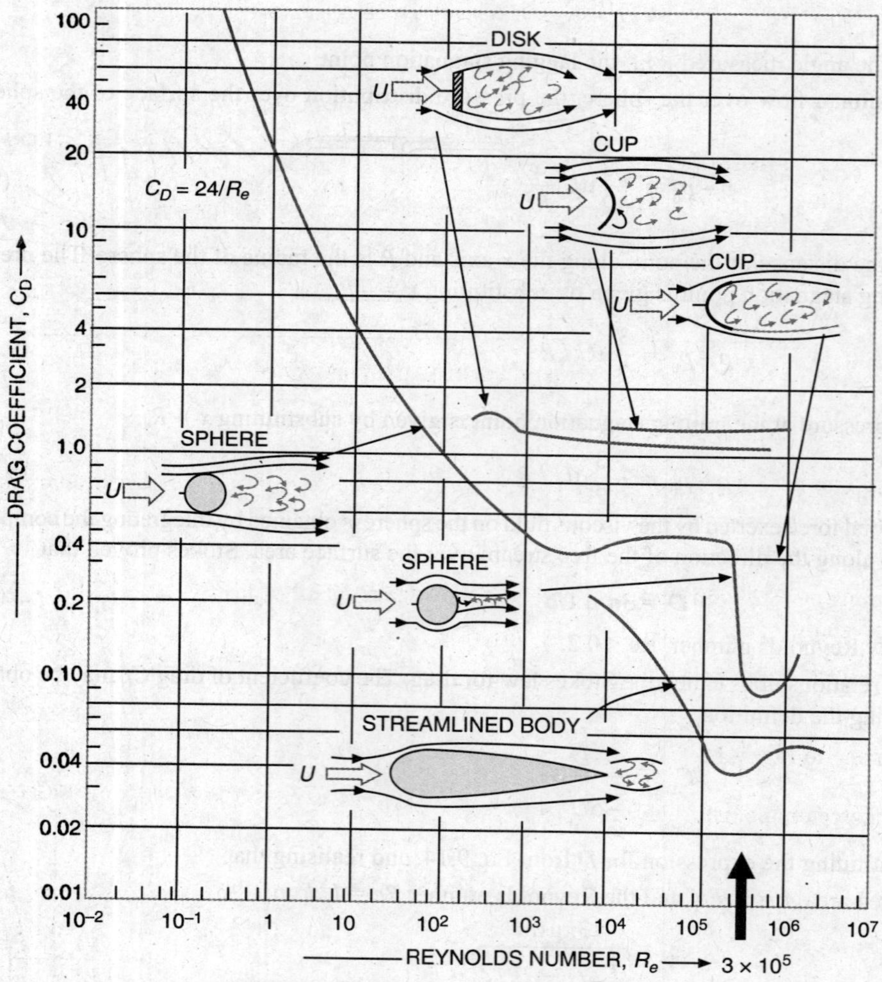

Fig. 9.14: Drag Coefficient vs Reynolds Number for Axisymmetric Bodies

over a sphere as well as for a disk and cups as bluff bodies and an aerofoil as a streamlined body. While the drag for bluff bodies remains practically the same for different Reynolds number, the drag on the aerofoil is considerably less and it is the least at $Re = 3 \times 10^5$. Pressure distributions on typical three-dimensional bodies are shown in Fig. 9.15.

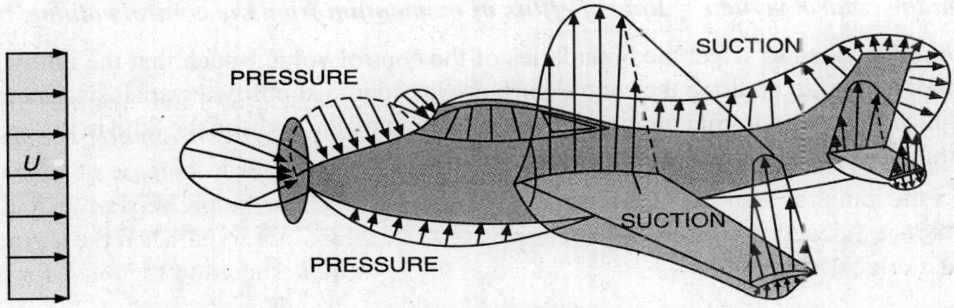

(a) AN AEROPLANE IN A LEVEL FLIGHT

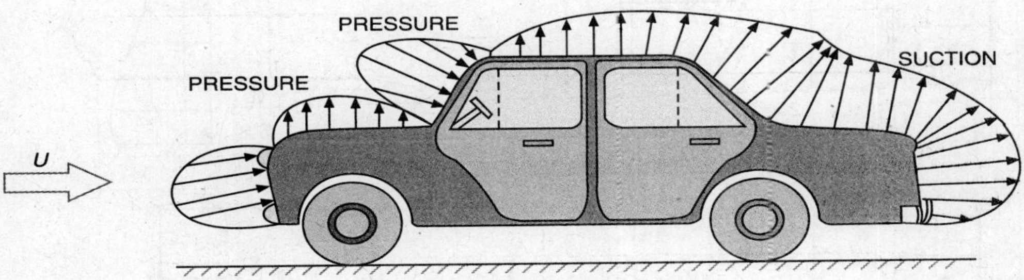

(b) A VEHICLE ON THE ROAD

Fig. 9.15 : Pressure Distributions on 3-d Bodies

Example 9.7. *A 7.5 cm diameter ball is just supported at rest in a vertical stream of air at 10 m/s. Calculate the mass of the ball.*

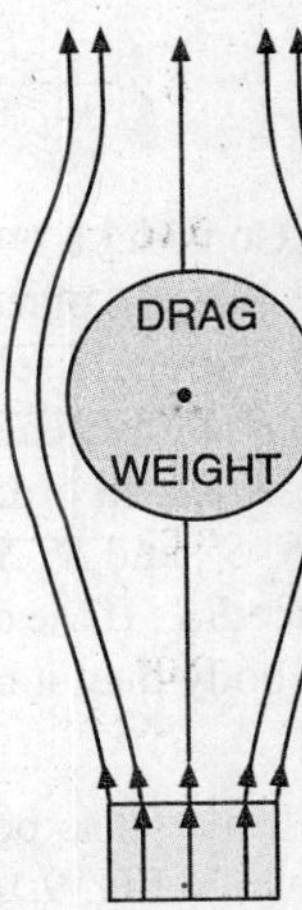

Solution.

Assuming $\rho = 1.2$ kg/m³ and $\mu = 18 \times 10^{-5}$ Ns/m² for air, the Reynolds number is given by

$$Re = 10 \times \frac{7.5}{100} \times \frac{1.2}{1.8 \times 10^{-5}} = 5 \times 10^4$$

Referring to Fig. 9.14, $C_D = 0.4$

Drag force on the ball $D = C_D . \frac{1}{2} \rho U^2 . A$

$$= 0.4 \times \frac{1}{2} \times 1.2 \times 10^2 \times \frac{\pi (0.075)^2}{4}$$

$$= 0.106 \text{ N}$$

Since the ball is in equilibrium, the weight of the ball should be equal to the drag force,

Mass of the ball $m = \frac{0.106}{9.81} = \mathbf{0.0108\ kg}$

9.8 Forces on Immersed Bodies : Integral Momentum Equation

The force exerted by a fluid flowing over a body may be determined by considering a suitable control volume and by applying the integral momentum equation to it. By the integral momentum equation,

Force on the control volume = Rate of efflux of momentum from the control volume

Care should be taken to select the boundaries of the control volume such that the inflow and outflow are either normal to or along the control surfaces in order to simplify the analysis. The body should be excluded from the control volume so that the reaction by the body on the fluid in the control volume constitutes the desired force on the control volume.

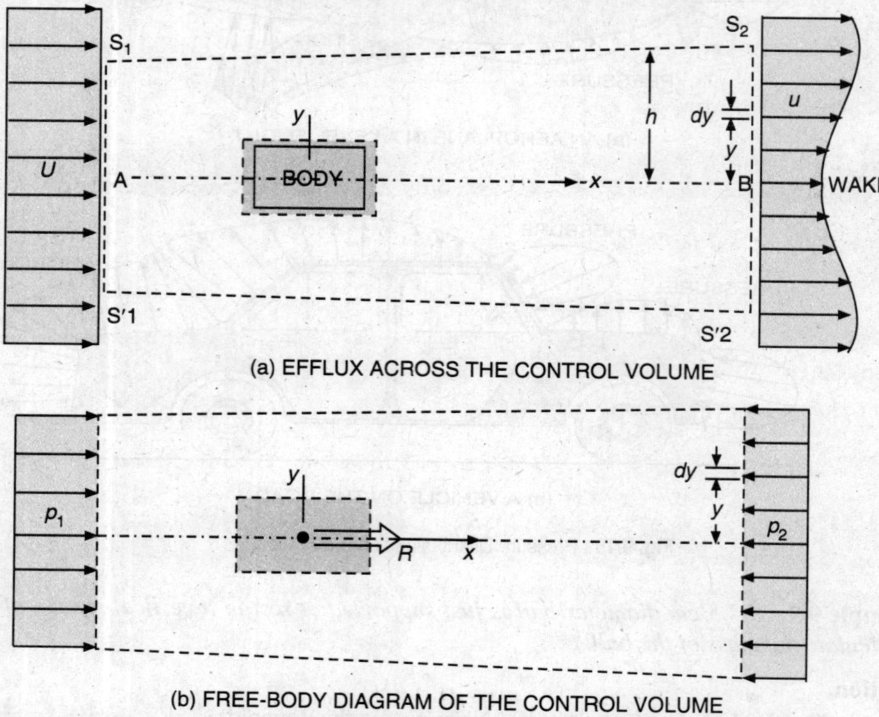

Fig. 9.16: Control Volume to Determine the Drag

Consider a two-dimensional body such as a cylinder, an aerofoil or some other body (*i.e.* with infinite length) placed in a uniform stream of velocity U as shown in Fig. 9.16 (*a*). A control volume, shown dotted, is selected such that the velocity across $S_1\, S'_1$ is the uniform velocity U, the velocity across $S_2\, S'_2$ varies in the y direction in the wake of the body. The distance h is chosen until the end of the wake. In order to ensure that no flow leaves across the control surfaces, $S_1 S_2$ and $S_1{}'\, S'_2$, the lines $S_1\, S'_2$ and $S'_1\, S'_2$ should be streamlines. This fact requires that the distance AS_1 is less than the distance BS_2. If the distance h is chosen to be sufficiently large compared to the transverse dimension of the body then it can be assumed that

$$AS_1 \approx BS_2 = h \qquad ...(9.17)$$

It may also be noted that the control volume just excludes the body. The free-body diagram, as drawn in Fig. 9.16 (*b*), depicts that the forces acting on the control volume are due to pressure p_1 on the upstream control surface $S_1\, S'_1$, pressure p_2 on the down stream control surface $S_2\, S'_2$ and the reaction R by the body on the control volume. The net force is, therefore, given by

$$p_1 .\, 2h - p_2 .\, 2h + R \qquad ...(i)$$

per unit width of the body.

The rate of mass outflow through a strip of height dy at a distance y from the axis is $\rho\, u\, dy$ and the rate of momentum outflow through the same strip is $\rho\, u^2\, dy$. The momentum inflow at the upstream associated with the same mass flow rate must be $\rho\, u\, dy.\ U$ and the net rate of efflux of momentum is given by

$$\rho\, u^2 dy - \rho\, u\, dy\,.U = \rho\, u(u - U)dy \qquad ...(ii)$$

The net rate of efflux of momentum over the entire control surface $S_2\, S'_2$ is

$$\int_{-h}^{+h} \rho\, u(u - U)\, dy$$

Equating the expression (*i*) and (*ii*), according to the integral momentum equation,

$$(p_1 - p_2)2h + R = \int_{-h}^{+h} \rho\, u(u - U)dy$$

whence R, the reaction by the body on the control volume, is given by

$$R = \int_{-h}^{+h} \rho\, u(u - U)dy - (p_1 - p_2)\,.\,2h \qquad ....(9.18)$$

This is also the resistance force offered by the body to the flow of the fluid. The drag D exerted by the fluid on the body must be equal and opposite to R.

$$D = -R$$

Hence,
$$D = \int_{-h}^{+h} \rho\, u(U - u)\, dy + (p_1 - p_2).\, 2h \qquad ...(9.19)$$

The expression for the drag *D* shows that it consists of two components,

(*a*) *due to the velocity defect, (U – u) between the free stream and the wake, and*

(*b*) *due to the pressure drop from the upstream to the downstream.*

If the body is placed in the atmosphere or in a mass of fluid subjected to atmospheric pressure or if the upstream and downstream are identical, then

$$p_1 = p_2$$

and the pressure drag, *i.e.*, the component of drag due to the pressure drop vanishes.

Thus,
$$D = \int_{-h}^{+h} \rho\, u(U - u)dy \qquad ...(9.20)$$

If the body is not symmetrically placed about the axis of the free-stream, there must be a change of momentum in the y-direction resulting in a force across the flow. In that case, the wake will have two components of velocity u and v, along and normal to the free stream respectively. The drag D is determined by applying the integral momentum equation along the x-axis as before and the lift L may be determined by applying the integral momentum equation in the y-direction. The net efflux of momentum in the y-direction is given by

$$\int_{-h}^{+h} \rho\, u\, dy.\, v \qquad ...(9.21)$$

because the inflow has no momentum in the y-direction and the outflow $\rho u\, dy$ through an elementary height of dy leaves with the component of velocity v. The lift force L on the body is consequently given by

$$L = \int_{-h}^{+h} \rho\, u\, dy.\, v \qquad ....(9.22)$$

since the contribution by the pressure drop is negligible if the upstream and downstream stations are chosen far from the body where the pressures tend to be equal.

Example 9.8. *The wake of the cylinder of diameter 10 cm placed in a wind tunnel of free stream 20 m/s is such that the velocity at the axis is 10 m/s and increases to 20 m/s linearly over two diameters on either side of the axis. If the difference in pressure between an upstream station and the wake is recorded as 100 N/m², calculate the drag on the cylinder per unit length.*

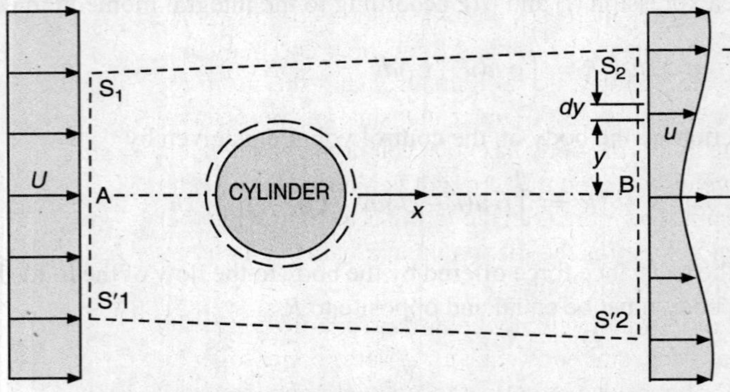

SELECTED CONTROL VOLUME FOR THE ANALYSIS

Solution.

Consider the control volume $S_1S_2S'_1S'_2$ such that $S'_2\ S'_1$ extends from $y = -h$ to $y = +h$ over the entire wake and $S_1\ S_2$ as well as $S'_1S'_2$ are streamlines such that the'inflow through $S_1S'_1$ equals the outflow through $S_2\ S'_2$.

The width of the wake is 40 cm compared to the diameter of the cylinder 10 cm.

Let us assume that $AS_1 \approx BS_2$

The drag on the cylinder is given by

$$D = \int_{-h}^{+h} \rho\, u\,(U-u)dy + (p_1 - p_2).\,2h$$

From the data,

$$h = 2 \times 0.1 = 0.2 \text{ m}$$

$$U = 20 \text{ m/s}$$

$$u = 10 + \frac{20-10}{0.2}y = 10 + 50y$$

above the x-axis and a symmetrical distribution below it.

$$p_1 - p_2 = 100 \text{ N/m}^2$$

Substituting the data in the expression for drag,

$$D = 2\int_0^h 1.2 \times (10+50y)(20-10-50y)\,dy + 100 \times 0.4$$

$$= 2.4 \times \left|\left(100y - 2500\frac{y^3}{3}\right)\right|_0^{0.2} + 40$$

$$= 32 + 40 = \mathbf{72\ N}$$

9.9 Terminal Velocity of a Body

When a body is allowed to drop freely in a fluid, ultimately it acquires a velocity U such that it remains constant thereafter. For example, if a body is dropped, starting from rest, in the atmosphere it will accelerate under the action of its own weight, *i.e.,* with an acceleration due to gravity. As the speed of the body increases, the body will experience a drag force which tends to oppose the motion of the body. The drag force increases with the velocity of fall. At the terminal or equilibrium state, the upward drag force and buoyancy would just balance the downward weight force and the net force acting on the body would be zero. Under this state of 'no external force' acting on the body, the body has no acceleration and hence it attains a constant velocity called the terminal velocity. The velocity of a falling body cannot be more than the terminal velocity. In other words, the terminal velocity is the maximum attainable velocity by a freely falling body in an environment.

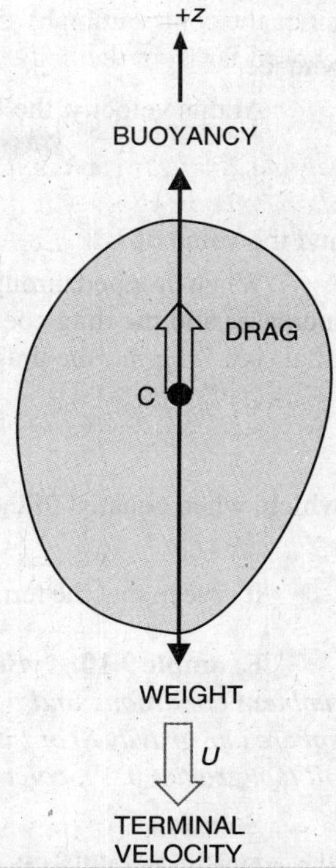

Fig. 9.17: Concept of Terminal Velocity

The concept of terminal velocity holds for a simple body such as a sphere or for a composite body such as 'a paratrooper together with the parachute'. It may also be realised that the terminal velocity is low for high-drag bodies and high for low-drag bodies. ***If a coin and a feather were dropped from a great height, (such as the Leaning Tower of Pisa or the Qutab Minar) the terminal velocity of the coin would be much more than that of the feather because the drag on the coin per unit weight is much less than the drag on the feather per unit weight.***

At the instant the body has acquired the terminal velocity, the body must be in equilibrium, *i.e.,* the net force on the body must be zero. *The weight of the body, the drag force and the buoyancy force must add to zero.* This is shown in Fig. 9.17

Example 9.9. *Determine the terminal velocity of a spherical steel ball of diameter 50 mm when dropped in a large mass of water. Assume that the specific gravity of steel is 8 and that, the C_D is 0.4.*

Solution.

The forces on the steel ball are as follows :

$$\text{Weight} = \frac{4}{3}\pi\left(\frac{0.05}{2}\right)^3 \times 8000 \times 9.81 = 5.14 \text{ N}$$

$$\text{Buoyancy} = \frac{4}{3}\pi\left(\frac{0.05}{2}\right)^3 \times 1000 \times 9.81 = 0.64 \text{ N}$$

$$\text{Drag} = C_D.\frac{1}{2}\rho U^2.A$$

$$= 0.4 \times \frac{1}{2} \times 1000 \times U^2 \times \frac{\pi(0.05)^2}{4}$$

$$= 0.39\, U^2$$

Note that the projected area A for a sphere is $\pi d^2/4$.

For equilibrium,

$$\text{Weight} = \text{Buoyancy} + \text{Drag}$$

$$5.14 = 0.64 + 0.39\ U^2$$

whence $\quad U = \mathbf{3.4\ m/s}$

At this velocity, the Reynolds number is given by

$$Re = \frac{Ud\rho}{\mu} = \frac{3.4 \times 0.05 \times 1000}{0.001} = 170000 = 1.7 \times 10^5$$

and the value of $\quad C_D = 0.4$ is appropriate.

When dropped through the atmosphere, the buoyancy force is negligible. The Reynolds number increases and the drag coefficient decreases to a value of the order of 0.1 to 0.15. Since the density of air is 1.2 kg/m³, the drag is given by

$$\text{Drag} = 0.1 \times \frac{1}{2} \times 1.2 \times U^2 \times \frac{\pi(0.05)^2}{4}$$

which, when equated to the weight, *i.e.*, 5.14 N, provides the terminal velocity

$$U = \mathbf{207\ m/s}$$

It is seen that the terminal velocity of a body is much higher in air as compared to that in water.

Example 9.10. *A viscometer consisting of a long vertical tank containing the fluid at the desired ambient conditions and a small sphere is employed to determine the viscosity of an oil. If a steel sphere (sp. gravity 8) of 1 cm diameter falls with a constant velocity of 0.05 m/s through a lubricating oil (sp. gravity 0. 9), calculate the viscosity of the oil.*

Solution.

During the fall of the sphere, it must be in equilibrium under the forces due to drag, buoyancy and its weight.

For very low velocity in a highly viscous fluid,

$$D = 3\ \pi\ \mu\ Ud$$

$$F_B = \frac{\pi d^3}{6} \rho_f . g$$

$$\text{Weight} = \frac{\pi d^3}{6} \rho_s . g$$

where p_f and p_s the mass densities of the fluid and material of the sphere respectively. For equilibrium,

$$3\pi\mu\ Ud + \frac{\pi d^3}{6} \rho_f . g - \frac{\pi d^3}{6} \rho_s . g = 0$$

and

$$\mu = \frac{gd^2(\rho_s - \rho_f)}{18\,U}$$

Substituting the numerical figures,

$$\mu = \frac{9.81 \times 0.01^2 \times (8.0 - 0.9) \times 1000}{18 \times 0.05} = \mathbf{7.74\ Ns/m^2}$$

It is important to note that the expression for drag $D = 3\pi\ \mu\ Ud$ is valid only for sub-laminar conditions, $Re < 0.1$. It is, therefore, necessary to check for this fact :

$$Re = \frac{5}{100} \times \frac{1}{100} \times \frac{0.09 \times 100}{7.74} = 0.058$$

which is less than 0.1.

If *Re* were greater than 0.1, we would be advised to look up the drag characteristic for a sphere instead of using the Stokes law and proceed as above.

Example 9.11. *A man of mass 60 kg descends to the ground from an aeroplane with the help of a parachute of mass 20 kg against the resistance of air. The shape of the parachute with which it comes down is hemispherical with 2m in diameter. Find his velocity.*

Assume that $C_D = 1.5$ *and* $\rho_{air} = 1.25$ kg/m³

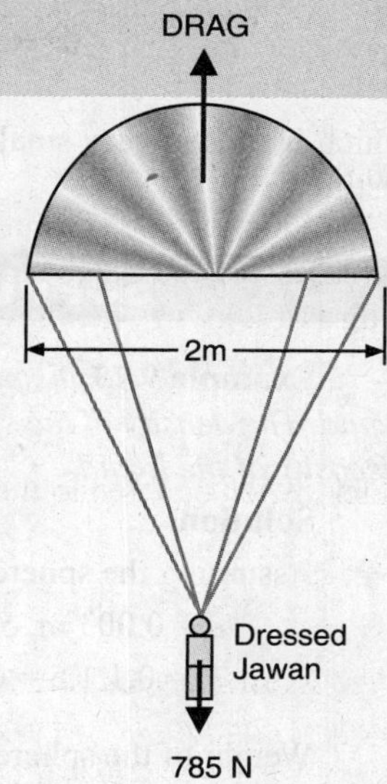

Solution.

The drag on the mass and the parachute is given by

$$D = C_D.\frac{1}{2}\rho U^2.A$$

$$= 1.5\times\frac{1}{2}\times 1.25U^2\pi.\frac{2^2}{4}$$

$$= 2.95\ U^2$$

At the steady state condition, the drag should equal the weight,

$$2.95\ U^2 = (60 + 20) \times 9.81 = 785\text{ N}$$

whence $U = \mathbf{16\ m/s}$

One may like to check the Reynolds number

$$Re = \frac{Ud\rho}{\mu} = \frac{16\times 2}{1.15/10^4} = 2.1\times 10^6$$

For this very high Reynolds number the given value of C_D is appropriate.

Example 9.12. *Determine the final steady rate at which an air bubble, half a milimetre diameter, rises in a jar of honey with specific gravity 1.3 and dynamic viscosity 0.50 Ns/m².*

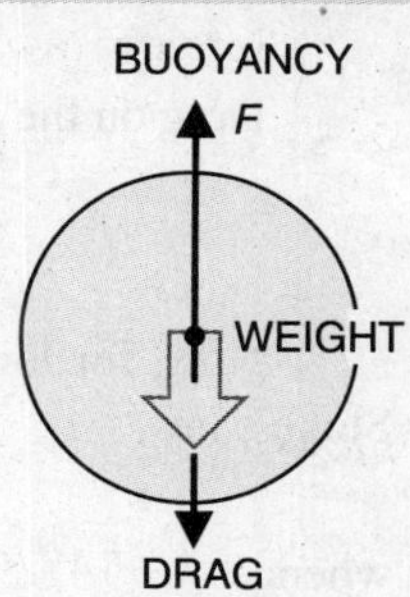

Solution.

The steady-state rise of the bubble takes place when it is in equilibrium under its weight, buoyancy force and drag.

This buoyant force on it equals the weight of the honey displaced.

$$F = \frac{\pi d^3}{6}\rho hg$$

The weight of the air bubble is

$$Wt = \frac{\pi d^3}{6}\rho\, a \times g$$

The drag force on the bubble for very slow movement, *i.e.,* very small Reynolds number is

$$D = 3\pi\,\mu Ud$$

Under equilibrium,

$$\pi d^3/6.\ \rho hg = \pi d^3/6.\ \rho\, a\, g + 3\,\pi\,\mu\, U\, d$$

$$\pi d^3/6.\ (\rho h - \rho a)\, g = 3\,\pi\,\mu\, U\, d$$

From the data, $\rho_h = 1.3 \times 1000 = 1300$ kg/m³, $\rho_a = 1.2$ kg/m³

$\mu = 0.5$ Ns/m² and $d = 0.5/1000 = 0.0005$ m

Substituting these values in the above equation,

$$U = 354 \times 10^{-6}\text{ m/s} = \mathbf{0.354\ mm/s}$$

Checking for the Reynolds number,

$$Re = \frac{Ud\rho}{\mu} = 354 \times 10^{-6} \times 0.0005 \times 1300 / 0.50$$

$$= 460 \times 10^{-6}$$

which is indeed very small ; much smaller than the value of 0.1 for which the Stokes law is well justified.

SUPPLEMENTARY REVISION EXAMPLES

Example 9.13. *A sphere of diameter 3 mm falls steadily 100 mm in 1.5 seconds in a viscous liquid. The density of the sphere is 7000 kg/m³ and of the liquid is 800 kg/m³. Find the coefficient of viscosity of the liquid.*

Solution.

Assuming the sphere to be falling with a constant (terminal) velocity U,

$d = 0.003$ m, $S = 0.1$ m, $t = 1.5$ seconds.

$U = 0.1/1.5 = 0.0667$ m/s

Weight of the sphere $$W = \frac{4}{3}\pi\left(\frac{0.003}{2}\right)^3 \times 7000 \times 9.81$$

$$= 7.77 \times 10^{-3} \text{ N.}$$

Buoyancy by the liquid $$B = \frac{4}{3}\pi\left(\frac{0.003}{2}\right)^3 \times 800 \times 9.81$$

$$= 8.876 \times 10^{-4} \text{ N}$$

Drag on the sphere $D =$ Weight – Buoyancy

$$= 7.77 \times 10^{-3} - 8.876 \times 10^{-4}$$

$$= 6.88 \times 10^{-3} \text{ N.}$$

Now, for low Re, which it should be for such a small diameter sphere and low velocity, by Stoke's law,

$$D = 3\pi\mu Ud.$$

whence $$\mu = 6.88 \times 10^{-3}/(3 \times \pi \times 0.0667 \times 0.003)$$

$$= \mathbf{3.65 \text{ Ns/m}^2}$$

Example 9.14. *A metallic ball 3 mm in diameter having a sp. gravity of 12 falls in a liquid of sp. gravity 0.84 and viscosity 1.5 Ns/m². Determine*

(a) *the drag on the ball*

(b) *the pressure drag and skin drag and*

(c) *terminal velocity of the ball in the fluid.*

Solution.

From the data,

Diameter $d = 0.003$ m, Volume $V = \frac{4}{3}\pi\left(\frac{0.003}{2}\right)^3 = 1.41 \times 10^{-8}$ m³.

Weight of the ball $W = mg = 12 \times 1000 \times 1.41 \times 10^{-8} \times 9.81 = 1.66 \times 10^{-3}$ N.

Buoyancy on the ball $B = 1.41 \times 10^{-8} \times 0.84 \times 1000 \times 9.81 = 1.16 \times 10^{-4}$ N.

Assuming the ball to be falling with a constant velocity U, it is in equilibrium:

Drag + Buoyancy = Weight

Hence Drag on the ball $D = 1.66 \times 10^{-3} - 1.16 \times 10^{-4} = \mathbf{1.544 \times 10^{-3}\ N}$.

$$Re = \frac{Ud\rho}{\mu} = \frac{0.003 \times 840}{1.5} U = 1.68\, U.$$

It should be very low Re number of flow, where Stokes law applies,

$$D = 3\pi\, \mu\, U\, d$$

$$1.544 \times 10^{-3} = 3\pi \times 1.5 \times U \times 0.003$$

whence $U = \mathbf{0.036\ m/s}$, the terminal velocity

showing that Re is 0.06, very low indeed!

whence $C_D = 24/Re = 24/0.06 = 400$

$$\text{Skin friction Drag} = C_D \times \frac{1}{2}\rho U^2 A = 400 \times \frac{1}{2} \times 840 \times 0.036^2 . \frac{\pi(0.003)^2}{4}$$

$$= \mathbf{1.54 \times 10^{-3}\ N}$$

which shows that the total drag is almost wholly skin friction drag.

Example 9.15. *A sphere has a projected area of 1m². Compare the drag force in water and in air when travelling at a speed of 30 km per hour.*

Solution.

Projected area $A = 1\text{m}^2 = \dfrac{\pi d^2}{4}$

whence $d = 1.13$ m

Given that $v = 30$ km/hour $= 30 \times 1000/3600$ m/s $= 8.33$ m/s

For water	For Air
$\rho = 1000$ kg/m³	$\rho = 1.2$ kg/m³.
$\mu = 0.001$ Ns/m²	$\mu = 0.000018$ Ns/m²
$Re = \dfrac{v\, d\rho}{\mu}$	$Re = \dfrac{v\, d\rho}{\mu}$
$= \dfrac{8.33 \times 1.13 \times 1000}{0.001}$	$= \dfrac{8.33 \times 1.13 \times 1.2}{0.000018}$
$= 9.4 \times 10^6$	$= 0.63 \times 10^6$

We, therefore, notice that the Reynolds number is beyond the critical value of 3×10^5 in water as well as in air. From the Re-Drag diagram,

C_D for travelling in water or in air.

$$C_D \approx 0.1$$

$$\text{Drag} = C_D . \frac{1}{2}\rho U^2 A$$

$$\text{Drag in water} = 0.1 \times \frac{1}{2} \times 1000 \times 8.33^2 . 1 = \mathbf{3470\ N}$$

$$\text{Drag in air} = 0.1 \times \frac{1}{2} \times 1.2 \times 8.33^2 . 1 = \mathbf{4.2\ N}$$

Drag in water is more than 800 times the drag in air !

Example 9.16. *A spherical balloon containing helium is tied to the ground by a cable. What would be the inclination of the cable when a wind of 10 km/hr blows horizontally past the balloon. Given that*

Diameter of balloon = 3 m

Weight of balloon = 135 N

Density of air = 1.2 kg/m³

Density of helium = 0.22 kg/m³

Drag coefficient = 0.2.

Solution.

The balloon must be in equilibrium under the action of four concurrent forces (Weight, Buoyancy, Drag and Tension in the cable) acting on it.

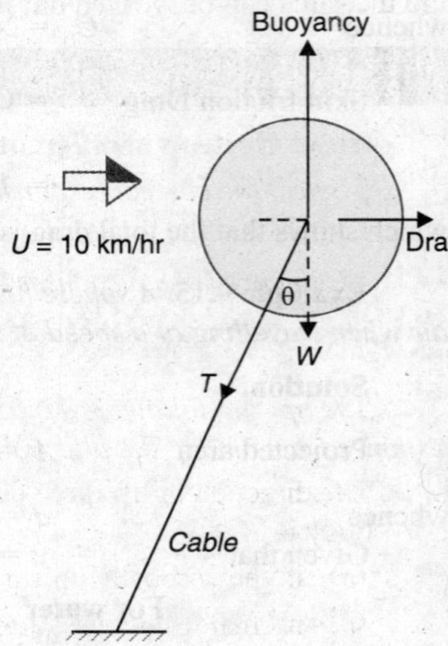

Drag $D = C_D \cdot \frac{1}{2} \rho U^2 \cdot A$

$$= 0.2 \times \frac{1}{2} \times 1.2 \times \left(\frac{10,000}{3600}\right)^2 \times \frac{\pi \times 3^2}{4}$$

$$= 6.5 \text{ N}$$

Buoyancy by the air

$$B = \frac{4}{3} \pi (3/2)^3 \times 1.2 \times 9.81 = 166.4 \text{ N}.$$

Weight of the balloon, $W = 135$ N.

Tension T in the cable is inclined at an angle θ.

Hence, $T \cos \theta = 166.4 - 135 = 31.4$ N

$T \sin \theta = 6.5$ N

whence, $\tan \theta = 0.21$; $\theta =$ **11.7°.**

Example 9.17. *A high tension cable 50 mm in diameter is strung out between two towers. Calculate the drag per unit length of the cable when the wind velocity is 80 km/hr across it. Given that $\rho = 1.2$ kg/m³, $\mu = 0.000018$ Pas for air.*

Solution.

Given that

$$d = 50/1000 = 0.05 \text{ m}$$

$$U = 80 \text{ km/hr} = \frac{80 \times 1000}{3600} = 22.2 \text{ m/s}$$

The Reynolds number of flow is

$$Re = \frac{Ud\rho}{\mu} \approx \frac{22.2 \times 0.05 \times 1.2}{0.000\,018}$$

$$\approx 74074 \approx 7.4 \times 10^4.$$

which shows that the flow is turbulent and the coefficient of drag

$C_D \approx 0.5$, as looked up from the $Re - C_D$ diagram.

The drag per meter length is given by

$$D = C_D . \frac{1}{2} \rho U^2 \times A$$

$$= 0.5 \times \frac{1}{2} \times 1.2 \times 22.2^2 \times 0.05 \times 1$$

$$= 7.4 \text{ N/m}$$

(The drag per metre is indeed small. But such cables tend to vibrate due to alternate vortices shed by them. The frequency of shadding of vortices n is governed by the Strouhal number

$$S = \frac{nd}{U} \approx 0.20$$

and the same can be worked out for a given set of conditions)

ITEM BANK

Objective items of different types are given to review the concepts developed in the Chapter.

* *Select the correct response for each item.*

* *Every correct response is expected in less than a minute and carries one mark.*

* *Compare your responses with the key answers given at the end.*

Multiple-Choice Items

Attempt the following Multiple-Choice Items by ticking the correct or the most appropriate response from among the alternatives given :

1. The direction of lift force on an immersed body is
(*a*) in the vertically upward direction.
(*b*) normal to the direction of motion of the body
(*c*) normal to the zero-lift axis
(*d*) normal to the longitudinal axis of the body

2. A streamlined body is such that
(*a*) it produces no drag for flow around it
(*b*) the surface of the body coincides with the stream surface
(*c*) the flow is laminar around it
(*d*) it is symmetrical about the axis along the free stream.

3. The critical Reynolds number for flow over a circular cylinder implies that, beyond that number,
(*a*) the drag increases at a fast rate
(*b*) the coefficient of drag increases
(*c*) the coefficient of drag decreases
(*d*) the flow changes from laminar to turbulent over the entire cylinder

4. The critical angle of attack of an aerofoil is that where
(*a*) the lift becomes zero
(*b*) the drag becomes zero
(*c*) the drag begins to rise
(*d*) the lift begins to drop

5. The terminal velocity of a body in a stationary fluid is such that
(*a*) the body acquires a constant acceleration
(*b*) the weight of the body equals the buoyancy force acting on it
(*c*) the net force acting on the body acts in the vertical direction
(*d*) the net vertical force acting on the body equals zero

6. A trip-wire is mounted near the leading edge of a body
(*a*) to increase the lift
(*b*) to decrease the lift
(*c*) to advance the point of separation
(*d*) to delay the point of separation

Other Objective Items

7. Identify the following statement as true or false:
   (*a*) Lift coefficient cannot exceed unity. (T/F)
   (*b*) Drag coefficient cannot exceed unity. (T/F)
   (*c*) For flow over a cylinder, delayed separation results in less drag than that for early separation. (T/F)
   (*d*) For flow over a sphere, delayed separation result in less drag than that for early separation (T/F)
8. State the following statements as true or false:
   (*a*) Karman vortex trail is formed for flow over a sphere. (T/F)
   (*b*) Karman vortex trail produces induced drag. (T/F)
   (*c*) Form drag is always more than profile drag. (T/F)
   (*d*) An aerofoil is stalled at low angle of attack. (T/F)
9. Are the following statements true or false ?
   (*a*) Wave drag is associated with wave phenomena. (T/F)
   (*b*) Profile drag is the same as skin-friction drag. (T/F)
   (*c*) Induced drag depends upon the lift. (T/F)
   (*d*) Larger the wake, greater the drag. (T/F)
10. Match the predominant type of drag from Block B with the description of- flow in Block A.

Block A	Block B
(*a*) Descent of a parachute	(*k*) profile drag
(*b*) Flight of an aeroplane	(*l*) form drag
(*c*) Vortex formation	(*m*) wave drag
(*d*) Supersonic jet	(*n*) induced drag
(*e*) Rise of gas bubbles in oil	
(*f*) Fall of mist droplets in air	

11. Arrange the following bodies in order of increasing value of the drag coefficient at high Reynolds numbers :
    (*a*) a cylinder (*b*) a sphere (*c*) an aerofoil (*d*) a flat plate normal to flow
    (*e*) a racing car
12. Match the drag coefficient for high Reynolds numbers in Block B with the bodies in Block A.

Block A	Block B
(*a*) a cylinder	(*k*) 0.20
(*b*) a sphere	(*l*) 0.30
(*c*) an aerofoil	(*m*) 0.80
(*d*) a bluff body	(*n*) 1.60
(*e*) a racing car	(*o*) 0.10

13. Match the three consequences in Block B with the three orientations of a flat plate placed in the uniform stream of a fluid in Block A.

Block A	Block B
(*a*) plate normal to the flow	(*k*) low form drag
(*b*) plate along the flow	(*l*) creation of lift
(*c*) plate inclined to the flow	(*m*) high profile drag

14. Match the parameters in Block B with the consequences in Block A

Block A	Block B
(*a*) increase in lift coefficient	(*k*) Karman vortices
(*b*) decrease in drag coefficient	(*l*) higher angle of attack
(*c*) drop in form drag	(*m*) delay of separation
(*d*) vibration of taut wires in flow	(*n*) higher Reynolds number

ANSWERS

1. (*b*), **2.** (*b*), **3.** (*c*), **4.** (*d*), **5.** (*d*), **6.** (*d*), **7.** (*F, F, T, T*), **8.** (*T, T, F, F*), **9.** (*T, F, T, T*), **10.** (*a l, b k, c n, d m, e k, f k*), **11.** (*c, b, e, a, d*), **12.** (*a m, b k, c o, d n, e l*), **13.** (*a m, b k, c l*), **14.** (*a l, b n, c m, d k*)

QUESTION BANK

Questions on theory and numerical problems are given for tutorial work.

* *The questions are arranged generally in order of the articles in the text.*
* *Every question should take 15–20 minutes to solve.*
* *Answers for the numerical problems are given alongside them.*

9.1 The chimney of a boiler house is 50 m tall and has an outside diameter of 3 m. Compute the overturning moment about the base if a 30 m/s wind blows past it at the standard atmospheric conditions. **[411.75 kN m]**

9.2 A truck having a projected area of 6.5 square metres travelling at 70 km/hr has a total resistance of 2 kN. Of this, 20 per cent is due to rolling friction and 10 per cent due to surface friction. The rest is due to form drag. Calculate the coefficient of form drag. **[C_D = 0.95]**

9.3 A kite of dimensions 0,6 m × 0.6 m, weighing 5 N flies in equilibrium when it makes an angle of 10° with the horizontal in a wind of speed 4 km/hr and density 1.2 kg/m³. If the string attached to it makes an angle of 45° with the horizontal and experiences a tension of 20 N, calculate the life and drag coefficients. **[C_L = 0.72, C_D = 0.53]**

9.4 An aeroplane of mass 500 kg and wing area of 10 m² is flying in a straight and level flight at a speed of 45 m/s near the sea-level. Estimate the lift coefficient of the aerofoil wing. **[0.40]**

9.5 A wing of a small aeroplane is rectangular in plan having a span of 10 m and a chord of 1.2 m. In straight and level flight at 240 km/hour the total aerodynamic force acting on the wing is 20 kN. If the lift/drag ratio is 10, calculate the coefficient of life and the total weight the aeroplane can carry. Assume air density to be 1.2 kg/m³. **[0.622]**

9.6 An aircraft weighing 1000 kN when empty has a wing area of 226 m². It is to take off at a velocity of 300 km per hour and a 20° angle of attack. Determine the allowable weight of cargo and power required for the engine. Assume density of air as 1.2 kg/m³. Assume coefficient of lift for the wing at 20° angle of attack as 1.42 and coefficient of drag as 0.17. **[337 kN and 13.33 MW]**

9.7 What is the terminal velocity of a 10 mm diameter hailstone in air at the standard atmospheric conditions ? Assume that the hailstone has a specific gravity of 0.9. **[25.45 m/s]**

9.8 A 1.5 m diameter spherical balloon filled with helium of density one-fifth that of air is let go from a meteorological observatory. If the weight of the empty balloon is 50 N, determine its steady velocity of ascent under normal atmospheric conditions. Take $C_D = 0.8$. **[6.27 m/s]**

9.9 Determine the maximum load that a 6 m diameter parachute may carry if the maximum permissible velocity is 10 m/s. The coefficient of drag may be assumed as 1.2 **[2 kN]**

9.10 A paratrooper and parachute together weigh 1 kN. Estimate its rate of descent if the parachute diameter is 7 m and the drag coefficient is assumed to be 1.2. Given that the density of air is 1.22 kg/m^3. **[6 m/s]**

9.11 Determine the frequency of vortex shedding from a 5 mm telephone wire in a wind speed of 30 m/s at standard conditions of density 1.2 kg/m^3 and viscosity 1.8×10^{-5} N s/m^2. **[1.2 kHz]**

9.12 A submarine periscope is 0.15 m in diameter and is travelling at 15 km/hour. What is the frequency of the alternating vortex shedding and the force per unit length of the periscope. Take density of water as 1030 kg/m^3 and kinematic viscosity as 1.25 mm^2/s.

[5.5 Hz, 805 N/m]

CHAPTER

10 Flow Through Pipes

LEARNING OBJECTIVES

On successful completion of all learning activities planned and provided in the chapter, a student will be able to

- derive an expression for the frictional head loss in a pipe flow;
- estimate minor losses in pipe transitions ;
- draw the energy gradient and hydraulic lines for flow through a pipe
- calculate the flow characteristics through pipes connected in series and in parallels and
- work out power transmission in a pipe.

10.1 Introduction

A fluid is conveyed through a confined passage when it is necessary to maintain its pressure different from the atmospheric pressure. Round pipes are most frequently employed to carry fluids over small or large distances. In view of the fact that pipe-flow constitutes the largest class of flows of engineering importance, some basic aspects of pipe-flow are studied in this chapter.

Pipes are used to convey water for water distribution networks, oil for oil transportation, air for pneumatic systems and gases or steam for commercial and domestic requirements. Pipes are also used for power transmission when high-pressure fluids are transmitted from one point to another. Large pipes called penstocks used to lead high pressure water to hydraulic turbines, interconnecting pipes in oil servo-mechanisms and pneumatic control systems are some such examples.

Pressure in a pipe may be above or below the atmospheric pressure as desired. While there is no upper limit on the positive pressure or gauge pressure, there is a definite restriction to the negative pressure or vacuum being dropped below a certain level. If the pressure drops below the saturation vapour pressure at that temperature, liquids tend to evaporate and constitute liquid-vapour mixtures which may result in vapour-lock and obstruction to flow. Besides, there are other phenomena such as cavitation which may be detrimental to the system. It is, therefore, necessary to ensure that pipes convey fluids at pressures in excess of the vapour pressure. It may also be noted that when a pipe discharges a fluid to the atmosphere the pressure at the exit must be atmospheric. Similarly, when a pipe draws in a fluid from the environment, the pressure at the entry must be the ambient pressure.

Flow through a confined passage is said to be fully developed if the velocity profile does not alter along the flow. Fully developed flows can occur only in constant area symmetrical passages such as pipes, annuli and between parallel plates. It may also be understood that a fully developed flow can be laminar or turbulent depending upon the the Reynolds number of operation.

In a fully developed pipe-flow, the pressure drops linearly along the length of the pipe-line. In other words, the pressure, gradient along the flow remains constant. It is customary to refer to the average velocity of flow in a pipe as 'the velocity' and base all further analysis on this assumption of uniform velocity at each cross section.

The velocity profile for fully developed turbulent flow through a long, straight and round pipe of radius R is adequately described by the one-seventh power law,

$$u = \left(1 - \frac{r}{R}\right)^{1/7} U_{max} \quad \text{...(10.1 } a)$$

where U_{max} is the velocity at the centreline.

This relationship is applicable for high Reynolds numbers and hydraulically smooth surfaces. In that case, the average velocity in the pipe is given by

$$U_{av} = 49/60\ U_{max} \quad \text{...(10.1 } b)$$

which shows that the average velocity is slightly in excess of 80% of the maximum velocity. The momentum-correction factor is 1.06 and the energy-correction is 1.02. It follows, therefore, that the percentage error in estimating the kinetic energy by assuming the flow through a pipe as one dimensional *i.e.*, with uniform velocity over any cross-section is of the order of 2 per cent.

10.2. Friction Loss in Pipe Flow : Darcy-Weisbach Formula

When a fluid flows steadily through a pipe of constant diameter, the average velocity at each cross section remains the same. This is necessary from the condition of continuity since the velocity U is given by

$$U = \frac{Q}{A}$$

The static pressure p drops along the direction of flow because the stagnation pressure drops due to loss of energy in overcoming friction as the flow occurs.

Darcy equation relates the head loss due to frictional or turbulent flow through a pipe to the velocity of the fluid U, friction factor f and the diameter d of the pipe. This is derived by considering fluid-control-volume in a section of the pipe between stations 1 and 2 as shown in Fig. 10.1.

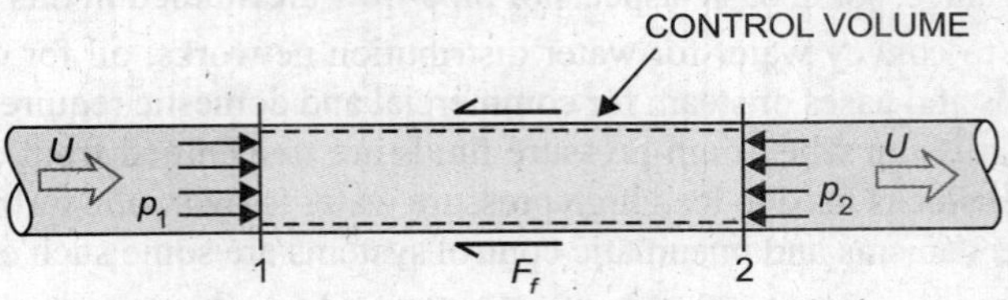

Fig 10.1: Flow through a Pipe

Pressure force along the flow, $F_p = (p_1 - p_2).\ \pi \frac{d^2}{4}$ ...(*i*)

Frictional force due to shearing at the pipe wall resisting the flow

$$F_f = C_D.\frac{1}{2}\rho U^2.\pi\ dl \quad \text{...(}ii\text{)}$$

Equating (*i*) and (*ii*) for equilibrium,

$$(p_1 - p_2).\pi\frac{d^2}{4} = C_D.\frac{1}{2}\rho U^2.\pi\, dl$$

Noting that $(p_1 - p_2)/\rho g$ is the head loss due to friction and h_f or '4 C_D' equals the friction factor for the pipe roughness, it becomes the Darcy-Weisbach formula,

$$h_f = \frac{f\,l}{d}.\frac{U^2}{2g} \quad ...(10.2)$$

In terms of the discharge Q.

$$h_f = \frac{f\,l}{d}.\frac{Q^2}{2g}.\frac{16}{\pi^2 d^2} = \frac{f\,lQ^2}{12\,d^5} \quad ...(10.3)$$

It has been shown that for laminar flow,

$$f = \frac{64}{Re}$$

For turbulent flow, relations such as

$$f = \frac{0.316}{Re^{1/4}} \text{ for } 4 \times 10^3 < Re < 1 \times 10^5 \quad ...(10.4)$$

are available but a reference to f v/s Re plot as in Fig. 10.2 is very convenient. The carpet plot takes

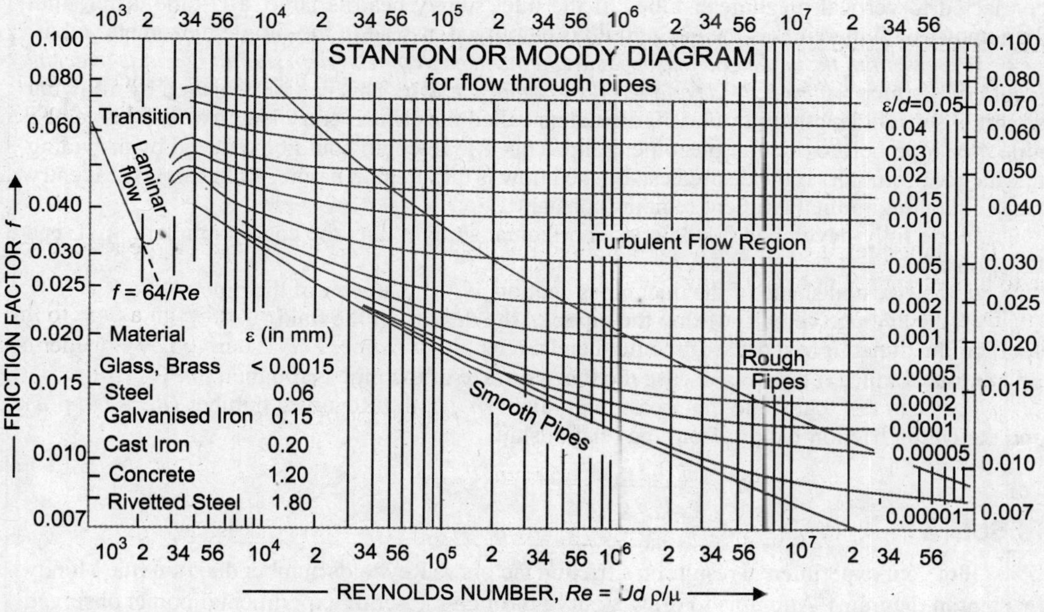

Fig. 10.2: Friction Factor for Pipe Flow

into account the variation of f with respect to Re as well as the relative roughness parameter of the pipe. In fact, these plots, basically due to Nikuradse, have been complied after extensive experimentation. These plots are frequently referred to as the Stanton or Moody diagram.

EXPERIMENT
Fully Developed Flow Through Pipes

OBJECTIVE

To study the frictional characteristics of fully-developed laminar and turbulent flow through pipes.

PRACTICAL RELEVANCE

Fluids are variously transported through pipes. Distribution of water and gas to domestic consumers through conduits, supply of steam through pipes in thermal power plants and of gases in process plants, off-shore pumping of oil, etc. are some of the examples of transportation of fluids through pipes. In order that the designer of such systems may provide for adequate pumping requirements, it is necessary to study the friction-flow characteristics for fully developed flow through pipes. The prediction of friction loss for flow through a pipeline enables the designer to estimate the power consumption and hence the type and size of pumps required for a given application. Although the flow is generally turbulent in nature, the characteristics of flow over the entire domain of flow should be studied with greater emphasis on turbulent flow.

PROCEDURE

The experiment is performed by using a number of long horizontal pipes of different diameters, connected to a common supply of water at a constant head and discharging into a common sump. Pressure tappings are provided on each pipe at suitable distances apart and connected to vertical piezometer tubes. If the inlet supply head is large, a U-tube differential manometer with mercury as the gauge fluid is employed instead of the simple piezometer tubes.

It is essential that the head loss measurements in each pipe system should be made over the length of the pipe where fully-developed flow actually exists. This can be achieved by allowing an initial straight length of around 50 diameters called the critical entry length after a valve or a pipe transition before the first piezometer tap. This initial length could be reduced by providing turbulence promotors if the desired established flow is turbulent in nature. After this critical entry length, the longitudinal gradient remains constant.

In the fully-developed flow along a horizontal straight duct the energy gradient, h_f /L is a function of :

a, the size and shape of the duct cross-section, *e.g.*, diameter of the pipe ;
U, the average velocity within the pipe ; ρ, the density of the fluid ;
μ, the dynamic viscosity of the fluid and
ε, the surface roughness in case the inner surface of the pipe is rough.

For each test, calculate the pressure gradient (h_f/L), the Reynolds number ($Re = Ud\,\rho/\mu$) and the Darcy friction factor / from the relationship,

$$h_f = \frac{f\,L}{d}\frac{U^2}{2g}$$

RESULTS

Plot your experimental results on a friction factors vs Reynolds number diagram (the Moody or Stanton diagram). Attempts to draw a curve through the set of experimental points observed by employing more than one pipe may not provide a smooth curve because the friction factors for different pipes would not be the same.

POINTS FOR DISCUSSION

For a laminar flow, $f = 64/Re$. How do your results for laminar flow compare with this equation?

1. What is the significance of the upper and lower critical Reynolds numbers and wha are their values ?

2. What is the effect of ε/d on pipe friction ?
3. Can you use the Moody or Stanton diagram for non-circular ducts and for compressible flow ? Comment.
4. Comment on the importance of the Reynolds number as a single dynamical parameter which characterises the incompressible flow of any fluid in a pipe.
5. How would you calculate the power required to pump a given fluid through a long pipeline with a known diameter made of a specified material ?
6. How do the wall shear stress, the longitudinal pressure gradient and the shape of the velocity profile vary in the entrance region of a pipe and in the fully-developed region for turbulent flow ? What is the effect of an upstream disturbance and roughness on the 'critical entry distance' for a pipe flow.

ANY OTHER OBSERVATION OR COMMENT?

10.3. Minor Losses in Pipe Flow

Apart from the head loss in a pipe flow due to friction, there are other sources of loss of head due to enlargements, contractions, entrance and exit as well as bends. These are considered here.

In a Sudden Enlargement such as that in Fig. 10.3(*a*) the head loss occurs due to the separation of flow at the periphery of the smaller pipe and consequent eddying motion in the corner region. The pressure drop or head loss from 1 to 2 can be evaluated by taking a fluid-control-volume $S_1S_1S_2S_2$, and applying the continuity and integral momentum equations.

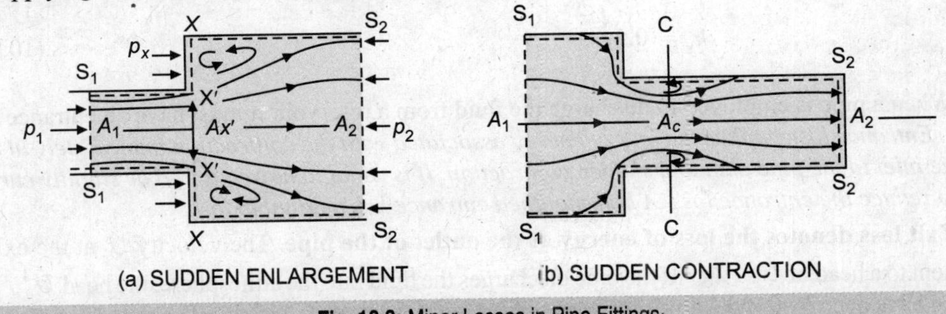

Fig. 10.3: Minor Losses in Pipe Fittings

By continuity,

$$A_1U_1 = A_2U_2 = Q \quad \text{...}(i)$$

for incompressible flow.

By momentum conservation

$$p_1A_1 + p_x(A_2 - A_1) - p_2A_2 = \rho Q(U_2 - U_1)$$

Assuming that p_x the pressure in the stagnant zone just after expansion equals p_1, because there cannot be an abrupt change of pressure across the same section *xx*.

$$(p_1 - p_2)A_2 = \rho Q(U_2 - U_1)$$

or

$$(p_1 - p_2)/\rho = U_2(U_2 - U_1) \quad \text{...}(ii)$$

By the Bernoulli equation applied between 1 and 2, accounting for the head drop due to enlargement,

$$p_1/\rho g + U_1^2/2g = p_2/\rho g + U_2^2/2g + h_1$$

whence $$h_l = (p_1 - p_2)/\rho g + (U_1^2 - U_2^2)/2g \qquad ...(iii)$$

Substituting (*ii*) in (*iii*), the head loss is

$$h_l = [U_2(U_2 - U_1) + (U_1^2 - U_2^2]/2g = (U_1 - U_2)^2/2g \qquad ...(10.5)$$

or $$h_l = \left(1 - \frac{A_1}{A_2}\right)^2 \frac{U_1^2}{2g} \qquad ...(10.6)$$

For a tapered enlargement, the head loss is considerably less

$$h_l = k\left(1 - \frac{A_1}{A_2}\right)^2 \frac{U_1^2}{2g} \quad \text{where } k < 1.$$

For a sudden contraction as in Fig. 10.3 (*b*), the head loss occurs only after the vena contracta at CC. This is because the flow upto this section is accelerating and the boundary layer separation does not occur. Let the area at cc be A_c, then h_l the head loss in contraction equals

$$h_l = k\left(1 - \frac{A_c}{A_2}\right)^2 \frac{U_c^2}{2g} = (U_c - U_2)^2/2g \qquad ...(10.7)$$

The ratio of areas A_c/A_2 is called the coefficient of contraction C_c. Although the value of C_c depends upon the ratio of the areas A_2/A_1, it may be taken as 0.62 on an average. In that case,

$$U_c = U_2/0.62 \qquad ...(10.8)$$

With these substitutions, the expression for the head loss in a sudden contraction reduces to

$$h_l = (1 - 0.62)^2 \times \frac{U_2^2}{(0.62)^2 \times 2g} = 0.38\,\frac{U_2^2}{2g} \qquad ...(10.9\ a)$$

The expression for head loss over a sudden contraction is frequently assumed to

$$h_l = 0.5\,\frac{U_2^2}{2g} \qquad ...(10.9\ b)$$

When a pipe is employed to discharge the fluid from a reservoir, it must have an entrance and an exit. *Entrance loss implies the loss of energy associated with the contraction immediately at and after the inlet to the pipe akin to a sudden contraction. It is usual to have a faired or slightly curved entry to reduce the entrance loss. A bell-mouthed entrance has negligible loss.*

Exit loss denotes the loss of energy at the outlet of the pipe. The velocity U_e at the exit is equivalent to a head of $U_e^2/2g$. If the pipe discharges the fluid into the atmosphere, the head $U_e^2/2g$ is said to have been lost. If the pipe discharges into a tank, the head $U_e^2/2g$ is dissipated in diffusion, sound and heat. The kinetic head must be taken into account in all cases except those where the pipe is very long and the friction loss is far in excess of the exit loss.

When a fluid flows round a bend, the streamlines are curved and a pressure gradient is created transversely such that the pressure at the outer radius is more than the pressure at the inner radius of the bend. The pressure at the inner radius decreases further with the contraction of streamlines to constitute a vena contracta and separation of the flow at the inner wall. The expansion following the contraction is accompanied by the formation of eddies and consequent head loss of the same order as that at the entrance to a pipe. The head loss for flow through a bend may be reduced by making a bend of larger radius which would eliminate the phenomena of contraction and expansion.

The pressure and velocity variations over the radial line AB are shown plotted in Fig. 10.4 alongside the nature of flow through a bend.

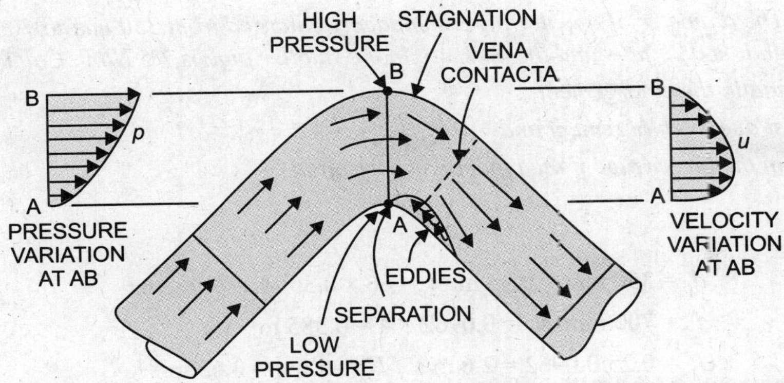

Fig. 10.4: Flow Through a Pipe Bend

It is customary to refer the losses of all the fittings and transitions in a pipe to **inlet** velocity head by the specifying constants *Ks* for these fittings etc. such that

$$\text{head loss} = K\frac{U^2}{2g}$$

It is equally customary to refer the head loss due to fittings etc. in terms of equivalent length *l* of the pipeline so that :

$$\text{head loss} = \frac{K\,U^2}{2g} = \frac{f\,l'}{d}\frac{U_2^2}{2g} \text{ or } K = f\left(\frac{l'}{d}\right) \quad \text{...(10.10)}$$

The advantage of doing so is that the equivalent lengths of the fittings and transitions can be added to the physical length *l* of the pipe to get the gross length

$$\text{Gross length} = \text{Physical length } l + l'_1 + l'_2 + l'_3 \ldots\ldots$$

and then apply the Darcy's equation with the gross length instead of the physical length.

Values of the constant *K* for different transitions are listed in Table 7.1.

Table 10.1

Transition/Fitting	K	Ref. Velocity
Sudden enlargement	$\left(1-\frac{A_1}{A_2}\right)^2$	U_1
Sudden contraction	0.5	U_2
Entrance (sharp edged)	0.5	U
Entrance (bell mouthed)	0 – 0.05	U
Elbow	1	U
Tee	2	U
Gate Valve (Fully open)	0.2	U
Globe Valve (Fully open)	1	U
Exit	1	U

Example 10.1. *The diameter of a water pipe is suddenly enlarged from 350 mm to 700 mm. The rate of flow through it is 0.25 m³/s and the pressure in the smaller pipe is 7.5 N/m². Calculate*

(*a*) *the loss of head in the enlargement*

(*b*) *the power lost due to enlargement and*

(*c*) *the pressure in the larger pipe if the pipeline is horizontal*

Solution.

From the data,

$$d_1 = 350 \text{ mm} = 0.35 \text{ m}; A_1 = 66 \times 0.35^2/4 = 0.962 \text{ m}^2$$

$$d_2 = 700 \text{ mm}; A_2 = 0.0962 \times 4 = 0.385 \text{ m}^2$$

$$U_1 = 0.25/0.0962 = 2.6 \text{ m/s} \quad U_2 = 2.6/4 = \mathbf{0.65 \text{ m/s}}$$

(*a*) The head loss is given by

$$h_1 = (1 - A_1/A_2)^2 U_1^2/2g$$

$$= (1 - 0.0962/0.385)^2 . U_1^2/(2 \times 9.81)$$

$$= 0.563 \times 0.345 = \mathbf{0.194 \text{ m head of water}}$$

(*b*) The power lost due to enlargement is

$$P = \rho g Q h_1$$

$$= 1000 \times 9.81 \times 0.25 \times 0.194 = \mathbf{476 \text{ W}}$$

(*c*) Applying the Bernoulli equation between the points across the enlargement,

$$p_1/\rho g + U_1^2/2g = p_2/\rho g + U_2^2/2g + h_1$$

whence, $$\frac{7.5}{1000 \times 9.81} + \frac{2.6^2}{2 \times 9.81} = \frac{p_2}{1000 \times 9.81} + \frac{(0.65)^2}{2 \times 9.81} + 0.194$$

or $$0.000\,764 + 0.3445 = \frac{p_2}{\rho g} + 0.021\,53 + 0.194$$

and $$p_2 = \mathbf{1272.7 \text{ N/m}^2}$$

This could be obtained alternatively by applying the momentum equation :

$$(p_1 - p_2)A_2 = \rho A_2 U_2 (U_2 - U_1)$$

$$(p_1 - p_2) = \rho U_2 (U_2 - U_1)$$

or $$7.5 - p_2 = 1000 \times 0.65 \times (0.65 - 2.6).$$

whence $$p_2 = \mathbf{1275 \text{ N/m}^2}$$

Example 10.2. *A pipe of diameter D_1 suddenly enlarged to a diameter D_2. Find the ratio between D_1 and D_2 in order that the rise of pressure shall be a maximum for a given discharge through the pipe.*

Solution.

For a sudden enlargement, the loss of head is given by

$$h_1 = \left(1 - \frac{A_1}{A_2}\right)^2 \frac{U_1^2}{2g}$$

Applying the energy equation between a point just before enlargement and a point just after enlargement,

$$\frac{U_1^2}{2g}+\frac{p_1}{\rho g} = \frac{U_2^2}{2g}+\frac{p_2}{\rho g}+\left(1-\frac{A_1}{A_2}\right)^2\frac{U_1^2}{2g}$$

By continuity, $U_1A_1 = U_2A_2 = Q,$

$$U_2 = U_1\frac{A_1}{A_2}$$

Then, $$\frac{p_2-p_1}{\rho g} = \frac{U_1^2}{2g}-\frac{U_1^2}{2g}\left(\frac{A_1}{A_2}\right)^2-\left(1-\frac{A_1}{A_2}\right)^2\frac{U_1^2}{2g}$$

or $$\frac{\Delta p}{\rho} = \left[1-\left(\frac{A_1}{A_2}\right)^2-\left(1-\frac{A_1}{A_2}\right)^2\right]\frac{U_1^2}{2}$$

$$\frac{\Delta p}{\rho} = \left[-2\left(\frac{A_1}{A_2}\right)^2+2\left(\frac{A_1}{A_2}\right)\right]\frac{U_1^2}{2}$$

Let $$\frac{A_1}{A_2} = r$$

Then $$\frac{\Delta p}{\rho} = [-r^2+r]\,U_1^2$$

For a given pipe and a given discharge, U_1 is constant. The pressure rise would be maximum when

$$\partial\left(\frac{\Delta p}{\rho}\right)/\partial r = 0$$

$$-2r+1 = 0$$

or $$r = \frac{1}{2}$$

Since $$\frac{A_1}{A_2} = \left(\frac{D_1}{D_2}\right)^2 = \frac{1}{2}$$

we get, $$\frac{D_2}{D_1} = \sqrt{2} = \mathbf{1.414}$$

The maximum pressure rise is given by

$$\frac{\Delta p}{\rho} = \left(-\frac{1}{4}+\frac{1}{2}\right)U_1^2 = \frac{1}{4}U_1^2$$

The maximum pressure head may be expressed as

$$h = \frac{\Delta p}{\rho g} = \frac{1}{2}.\frac{U_1^2}{2g}$$

which is half of the available velocity head in the pipe.

EXPERIMENT

Head Loss in a Pipe Transition

OBJECTIVE

To determine the head loss in a pipe transition in relation to the velocity head of the fluid.

PRACTICAL RELEVANCE

It is always necessary to instal bends, elbows, tees, reducers and valves, etc. in pipelines in order to convey the fluid from one point to another. Flow through such ' fittings, called transitions, is invariably accompanied by loss in head. Knowledge of head loss through transitions enables a designer to select the types of fittings and optimise the number as also to work out the head loss across all of them in advance. Pumping power and discharge through pipe systems is, thus, worked out realistically.

PROCEDURE

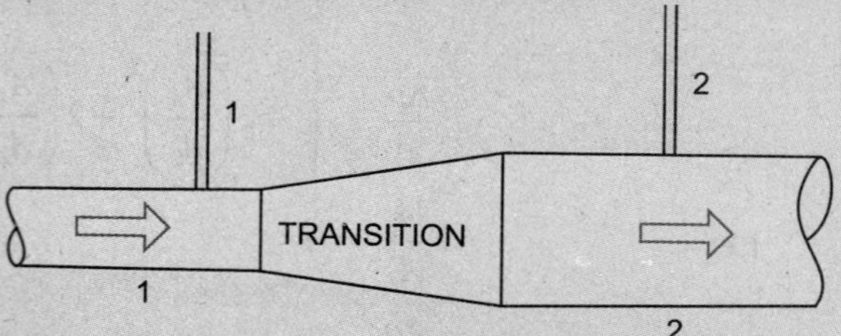

Install piezometers or an inverted U-tube manometer across the transition selected for study.

Adjust the flow to be steady at a valve setting. Measure the discharge Q gravimetrically or volumetrically and measure the difference of static pressure, $p_1 - p_2$, across the transition. From the knowledge of the cross section areas A_1 and A_2 at the two locations, work out the corresponding velocities U_1 and U_2.

Total head at 1 $\quad h_1 = p_1 + U_1^2/2g$

and total head at 2 $\quad h_2 = p_2 + U_2^2/2g$

Head loss $\quad h_1 = h_1 - h_2$

From the knowledge that the head loss is a function of the velocity head, $U_1^2/2g$,

$$h_1 = k.\ U_1^2/2g$$

the value of k may be determined.

The experiment is repeated at different rates of flow and a mean value of k is obtained

RESULTS

The value of k for the given transition may be compared with the expected value for a standard transition. Comment on the difference between the two.

QUESTIONS FOR DISCUSSION

* Why is the head loss in a transition proportional is the velocity head ?
* If the head loss is proportional to the square of velocity of the fluid, why don't we run the pipe at lower velocities only ?
* Compare the head loss across the given transition with the head loss across other alternative transitions, e.g., sudden expansion vs inverted reducer, elbow vs bend and state which of the either fittings you would prefer.
* Does the static pressure always drop across a transition ?
* Draw the hydraulic gradient line and the total energy line across the given transition.

ANY OTHER OBSERVATION OR COMMENT?

Example 10.3. *Water is discharged from a tank maintained at a constant head of 5 m above the exit of a straight pipe 100 m long, 15 cm diameter. Estimate the rate of flow if the friction coefficient for the pipe is given as 0.01.*

Solution.

Since the free surface of water in the tank is 5 m above the exit of the pipe discharging into the atmosphere, the entire available head of 5 m is lost in overcoming the friction for flow through the pipe and as exit loss.

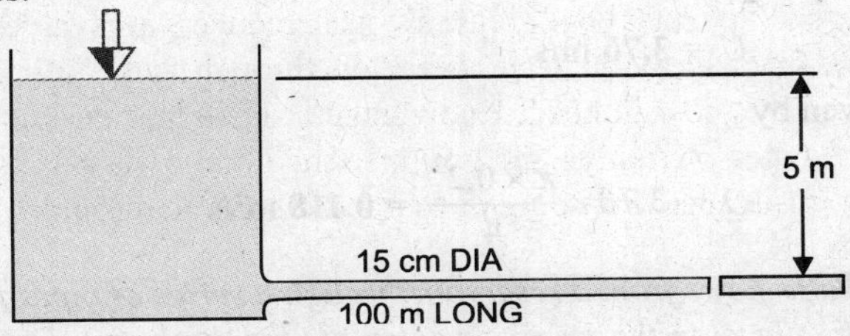

To check if minor losses are negligible

$$\text{Entrance loss} = 0.5 \times \frac{U^2}{2g}$$

This is not negligible compared to the exit loss $U^2/2g$ and thus cannot be neglected. Employing the relationship,

$$h_f = \frac{f\,l\,U^2}{d\;2g}$$

$$5 = \frac{0.01 \times 100}{0.15} \frac{U^2}{2 \times 9.81} + \frac{U^2}{2 \times 9.81} + \frac{0.5U^2}{2 \times 9.81}$$

whence $U = 3.466$ m/s

and discharge $Q = 3.466 \times \frac{\pi}{4}(0.15)^2$

$= \mathbf{0.061\ m^3/s}$

Example 10.4. *The difference in water surface elevations in two reservoirs A and B is 10 m and the gauge pressure of the air space in A is 50 kN/m². They are connected by a single pipe 200 m long and 20 cm in diameter. If the friction factor for the pipe is 0.02, Calculate the discharge.*

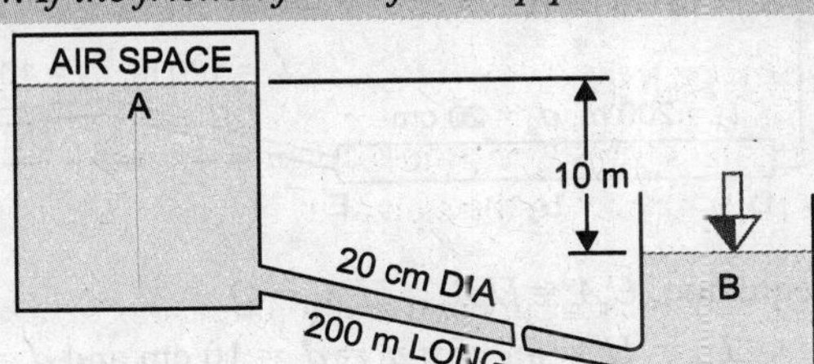

Solution.

The piezometric head difference between the two tanks is

$$H = 10 + \frac{50 \times 1000}{1000 \times 9.81} = 15.1 \text{ m}$$

Since the entire available head must be lost in overcoming the pipe friction and in creating the exit velocity,

$$15.1 = \frac{f\,l}{d}\frac{U^2}{2g} + \frac{U^2}{2g} = \left(\frac{f\,l}{d} + 1\right)\frac{U^2}{2g}$$

$$= \left(\frac{0.02 \times 200}{0.2} + 1\right) \times \frac{U^2}{2 \times 9.81}$$

whence $U = \mathbf{3.76\ m/s}$

and the discharge is given by

$$Q = 3.76 \times \frac{\pi \times 0.2^2}{4} = \mathbf{0.118\ m^3/s}$$

Example 10.5. *Water flows from a reservoir through a series of pipes joined as shown. If the exit is 20 m below the free level in the reservoir calculate the discharge. Find the percentage error in discharge if the minor losses are neglected.*

Solution.

In addition to the major loss due to friction, there are minor losses at A, B, C, D, E and F due to obvious reasons. Working in terms of K factors of the fittings and transitions, the available head of 20 m is used up in providing the exit discharge and overcoming the head losses all through the flow.

Total available head = Sum of losses at A, B, pipe 1, C, D, pipe 2, E, pipe 3 and F.

$$20 = \left[0.5\,U_1^2 + 1U_1^2 + \frac{f_1\,l_1}{d_1}U_1^2 + 1U_1^2 + \left(1 - \frac{A_1}{A_2}\right)^2 U_1^2 + \frac{f_2\,l_2}{d_2}U_2^2 + 0.5\,U_2^2 + \frac{f_3\,l_3}{d_3}U_3^2 + U_3^2\right] / 2g \quad \text{...(i)}$$

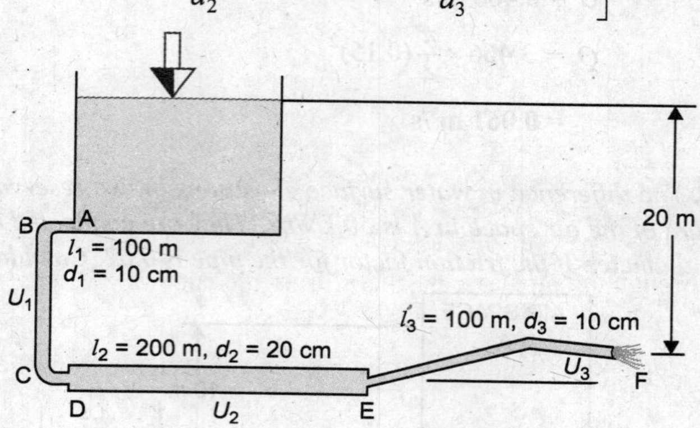

By the continuity equation, $U_1A_1 = U_2A_2 = U_3A_3 = Q$

Hence, $U_1 = U_3 = 4U_2$ for $d_1 = d_3 = 10$ cm and $d_2 = 20$ cm

The friction factor depends upon the Reynolds number but the velocity is not yet known. Avoiding an iterative procedure, engineers prefer to assume a reasonable value of f for the pipe material and roughness. For G.I. pipes, let $f = 0.02$,

Substitution in (*i*) provides

$$20 = \frac{U_3^2}{2g}(0.5 + 1.0 + 20.0 + 1.0 + 0.56 + 1.25 + 0.03 + 20.0 + 1.0)$$

$$= 45.34 \frac{U_3^2}{2g}$$

Hence $\quad U_3 = 2.94$ m/s

and the discharge $\quad Q = U_3 A_3 = 2.94 \times \frac{\pi \times 0.1^2}{4} = 0.0231$ m³/s

Neglecting the minor losses

$$20 = \frac{f_1 l_1}{d_1} U_1^2 + \frac{f_2 l_2}{d_2} U_2^2 + \frac{f_3 l_3}{d_3} U_3^2$$

$$= (20.0 + 1.25 + 20.0)\frac{U_3^2}{2g} = 41.25 \frac{U_3^2}{2g}$$

$$U_3 = 3.08 \text{ m/s}$$

$$Q = 3.08 \times \pi \times 0.1^2/4 = 0.0242 \text{ m}^3/\text{s}$$

Percentage error

$$e = \frac{0.0242 - 0.0231}{0.0231} = \mathbf{4.8\%}$$

Since the largest percentage of head loss is due to the pipe-friction effects, particularly in long pipe lines, engineers tend to ignore the minor losses such as in the entrance, elbows, exit, etc.

Example 10.6. *A special pipe of diameter D carrying a flow rate Q at its upstream is arranged to discharge q per unit length uniformly along its length. Develop an expression for the head loss over a length L and hence determine the head loss for the entire length of the pipe.*

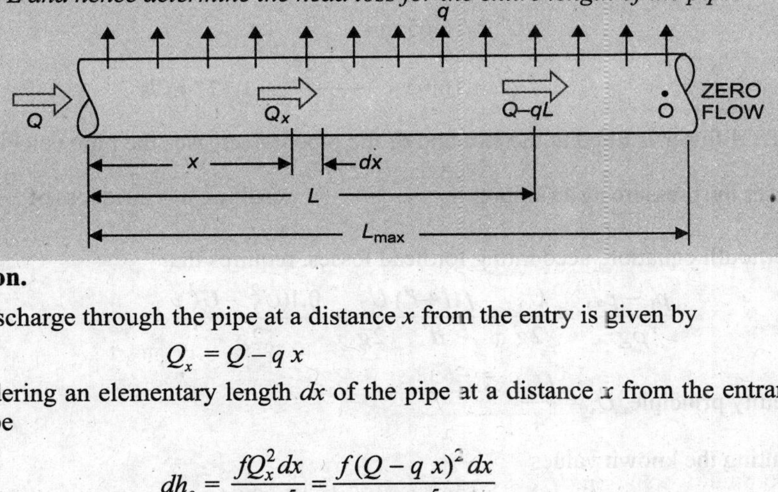

Solution.

The discharge through the pipe at a distance x from the entry is given by

$$Q_x = Q - q\,x$$

Considering an elementary length dx of the pipe at a distance x from the entrance, the head loss should be

$$dh_f = \frac{f Q_x^2 dx}{12\,D^5} = \frac{f(Q - q\,x)^2\,dx}{12 D^5}$$

The head loss over a length L from the entrance must be

$$h_f = \int_0^L dh_f = \frac{f}{12D^5}(Q^2 x + q^2 x^3/3 - Qqx^2)\Big|_0^L$$

$$= \frac{f\,LQ^2}{12D^5}(1 + q^2L^2/3Q^2 - qL/Q)$$

The length of the pipe until the flow through it becomes zero is

$$L_{max} = Q/q.$$

Substituting it for L,

$$h_f = \frac{f\,L_{max}Q^2}{12D^5}(1 + 1/3 - 1) = 1/3.\frac{f\,L_{max}Q^2}{12D^5}$$

or

$$h_f = \frac{f\,Q^3}{36\,q\,D^5}$$

Example 10.7. *Calculate the exit velocity and discharge of air flowing from a tank under 3 cm WG to the atmosphere through a 30 cm diameter 10 m long pipe.*

If a short diffuser of area ratio 4 is fitted at the exit end and it works at 90% efficiency, will the discharge through the system be effected ?

Assume well rounded entry with equivalent length of five diameters and $f = 0.02$.

Solution.

Applying the Bernoulli equation at the entrance and exit of the pipe, and accounting for the losses

$$\frac{p_1}{\rho g} + \frac{U_1^2}{2g} + z_1 = \frac{p_2}{\rho g} + \frac{U_2^2}{2g} + z_2 + \text{head losses}$$

For a horizontal pipe,

$$\frac{p_1 - p_2}{\rho g} = \frac{U_2^2}{2g} + \frac{f(l + l')}{d}\frac{U_2^2}{2g}$$

$$0.03 \times \frac{1000}{1.2} = \left(1 + \frac{0.02(10 + 5 \times 0.3)}{0.3}\right)\frac{U_2^2}{2 \times 9.81}$$

whence

$$U_2 = 16.65 \text{ m/s}$$

and

$$Q = 16.65 \times \frac{\pi \times 0.3^2}{4} = \mathbf{1.177\ m^3/s}$$

Now, if a diffuser is fitted at the exit end of the pipe it decreases the pipe velocity from U_2 to U_3 and increases the pressure by an amount $\frac{0.9\,(U_2^2 - U_3^2)}{2g}$ resulting in a head loss of $\frac{0.1\,(U_2^2 - U_3^2)}{2g}$

The Bernoulli equation, accounting for head losses, requires that

$$\frac{p_1 - p_3}{\rho g} = \frac{U_3^2}{2g} + \frac{f(l + l')}{d}\frac{U_2^2}{2g} + \frac{0.1(U_2^2 - U_3^2)}{2g}$$

By the continuity principle, $U_3 = \frac{U_2}{4}$

Substituting the known values

$$U_2^2 = 0.03 \times \frac{100}{1.2} \times \frac{2 \times 9.81}{\left(0.062 + \frac{11.5 \times 0.02}{0.3} + 0.094\right)}$$

$$U_2 = 23.06 \text{ m/s}$$

Hence,

$$Q = 23.06 \times \frac{\pi}{4}(0.3)^2 = \mathbf{1.63\ m^3/s}$$

This is 38.5% more than that without a diffuser at the exit.

The mode of action of the diffuser is to decrease the exit velocity head loss in comparison with the friction loss through the pipe. Fitting of a diffuser at the end of a *long* water line would not augment the discharge because the exit head loss in a long water pipe is already negligible compared to the friction loss in the pipe.

Example 10.8. *A reservoir 10 m × 5 m cross section contains water upto a depth of 3 m. The water level is 25 m above the ground. It is desired to empty the tank through a 15 cm diameter opening near the bottom of the reservoir. Is it faster to empty it through the opening as an orifice or through a straight vertical pipe reaching the ground. Take Cd per the orifice as 0.6, and f for the pipe as 0.02.*

Solution.

For discharge through an orifice, whose diameter is small enough compared to the head

$$t = \int dt = -\int_H^0 \frac{Adh}{C_d a\sqrt{2gh}} = \int_0^H \frac{Adh}{C_d a\sqrt{2gh}}$$

$$= \frac{2\times10\times5\times\sqrt{3}}{0.6\times\pi\dfrac{(0.15)^2}{4}\sqrt{2g}}$$

$$\mathbf{= 3687.4\ s = 61.46\ minutes}$$

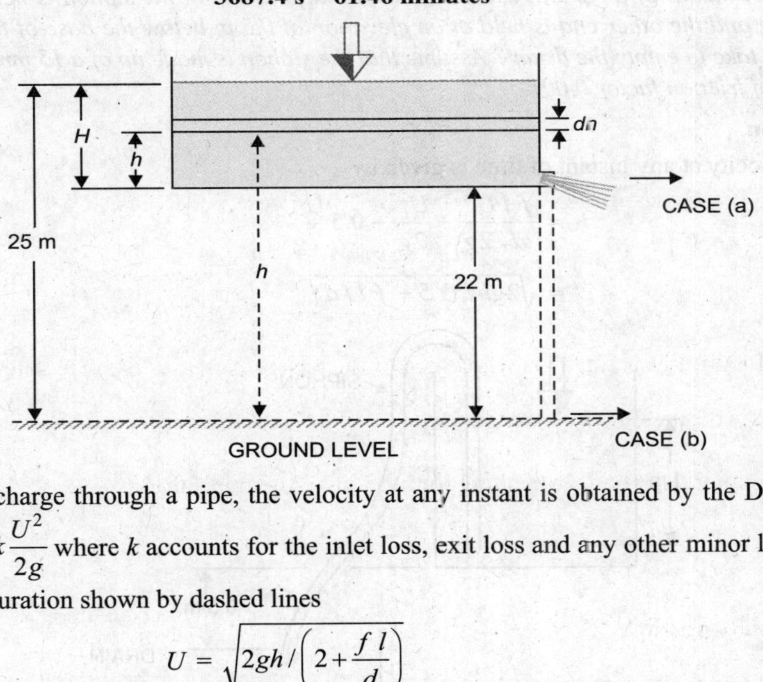

For discharge through a pipe, the velocity at any instant is obtained by the Darcy equation $h = \frac{fl}{d}\frac{U^2}{2g} + k\frac{U^2}{2g}$ where k accounts for the inlet loss, exit loss and any other minor loss. Let $k = 2$ for the configuration shown by dashed lines

$$U = \sqrt{2gh\Big/\left(2+\frac{fl}{d}\right)}$$

Considering the flow in a short time dt when the level falls by dh and integrating,

$$t = \int dt = \left|\frac{2A\sqrt{2+fl/d}}{a\sqrt{2g}}h^{1/2}\right|_{22}^{25}$$

$$t = \frac{2\times10\times5\sqrt{2+\dfrac{0.02\times22}{0.15}}}{\pi\times\dfrac{(0.15)^2}{4}\times(4.43)}\left(\sqrt{25}-\sqrt{22}\right)$$

$$\mathbf{t = 879.5\ s = 14.66\ minutes}$$

It is much faster to drain with a long vertical pipe than with an orifice.

The time of emptying the tank by a vertical pipe is considerably less than that by a simple orifice for the following reasons :

(*a*) The head causing the flow for an orifice varies between 3 and 0 meters whereas that for a vertical pipe varies between 25 and 22 meters. The values of the $\sqrt{h}$ under these limits bear a factor of

$$\frac{\sqrt{3}-0}{\sqrt{25}-\sqrt{22}} = \frac{1.732}{5-4.69} = 5.6$$

which is the major reason of time reduction.

(*b*) Pipes run full whereas the orifices are accompanied by a reduction of the area of flow resulting in $C_d < 1$. However, pipes have frictional loss !

Example 10.9. *A drum of 0.6 m diameter which contains water to a depth of 1 m is emptied into a drain by means of a siphon, as shown, such that one end of the siphon is held in the drum near the base and the other end is held at an elevation of 0.5 m below the base of the drum. How long would it take to empty the drum ? Assume that the siphon is made up of a 15 mm diameter, 3.5 m long tube of friction factor 0.001.*

Solution.

The velocity at any instant of time is given by

$$h = \frac{f\,l}{d}\frac{U^2}{2g} + \frac{U^2}{2g} + 0.5\frac{U^2}{2g}$$

$$U = \sqrt{2gh/(1.5 + f\,l/d)} \qquad \text{...(i)}$$

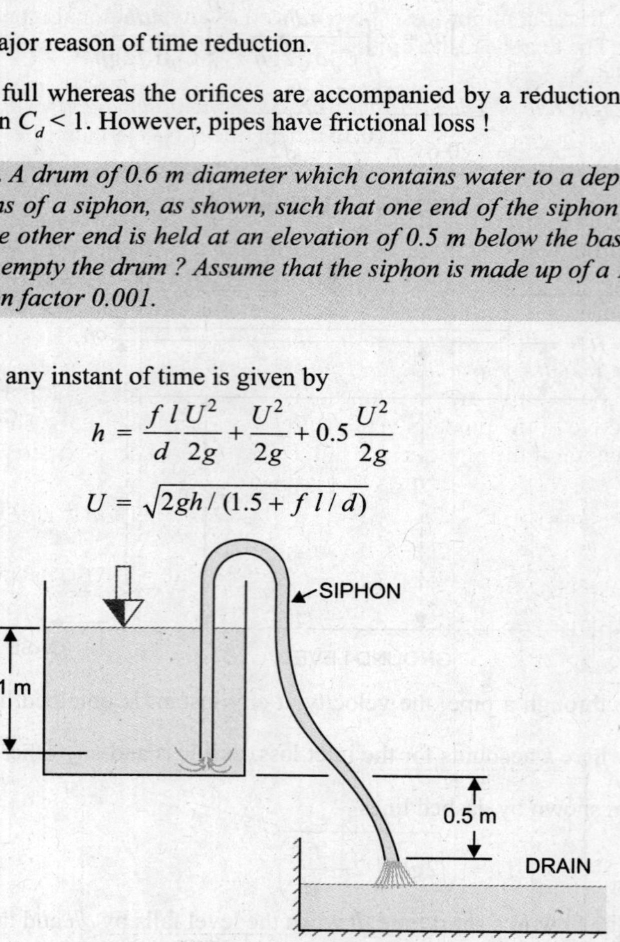

Consider the flow over a short time dt while the level in the tank drops by a height dh. The volume of water discharged must be

$$dV = U\frac{\pi \times (15/1000)^2}{4}dt = -\frac{\pi \times 0.6^2}{4}dh$$

whence
$$dt = -\frac{1600}{U}dh$$

Substituting for U from (*i*) in (*ii*) and integrating between the limits of 1.5 and 0.5 m,

$$t = \int dt = \left| \frac{2\times1600}{\sqrt{(2\times9.81)/\left(1.5+\frac{0.001\times3.5}{0.15}\right)}} h^{1/2} \right|_{0.5}^{1.5}$$

$$= 487.78 \text{ s} = \textbf{8.13 minutes}$$

10.4. Energy Line and Hydraulic Gradient Line

The energy-line diagram for a pipe system is a longitudinal display of the total head at all salient sections of the pipe. The energy line E E -----, therefore, represents the variation of energy along the flow due to friction, minor losses, etc. as well as any additional input or output by means of pumps or turbines. The total head at a cross section consists of the pressure head $p/\rho g$, the velocity head $U^2/2g$ and the datum head z.

Hydraulic gradient line is a display of the pressure and datum heads, i.e., piezometric head at all salient sections of a pipe line. The hydraulic gradient line H H-----, therefore, lies below the energy line by a magnitude equal to the local velocity head $U^2/2g$. Obviously, if a segment of the pipe lies above the hydraulic gradient line, there must be negative pressure or partial vacuum at all sections in that segment. Such a segment is said to constitute a siphon. The absolute pressure at any point in a siphon should not drop below a critical value of 3 m of water below which separation may occur.

The energy line and the hydraulic gradient line show an abrupt and coincident rise or drop at the location of a pump, fan, turbine or some other rotodynamic machine. In the absence of these machines, *the energy line will continue to drop along the flow whereas the hydraulic gradient line, though always below the energy gradient line, may show an upward or downward trend.* These lines have been shown for some simple pipe systems in Fig. 10.5. It may also be noted that these are straight sloping lines irrespective of the pipeline being straight or curved since the slopes of these lines are referred to per unit length of the pipe and not unit length in any specified direction.

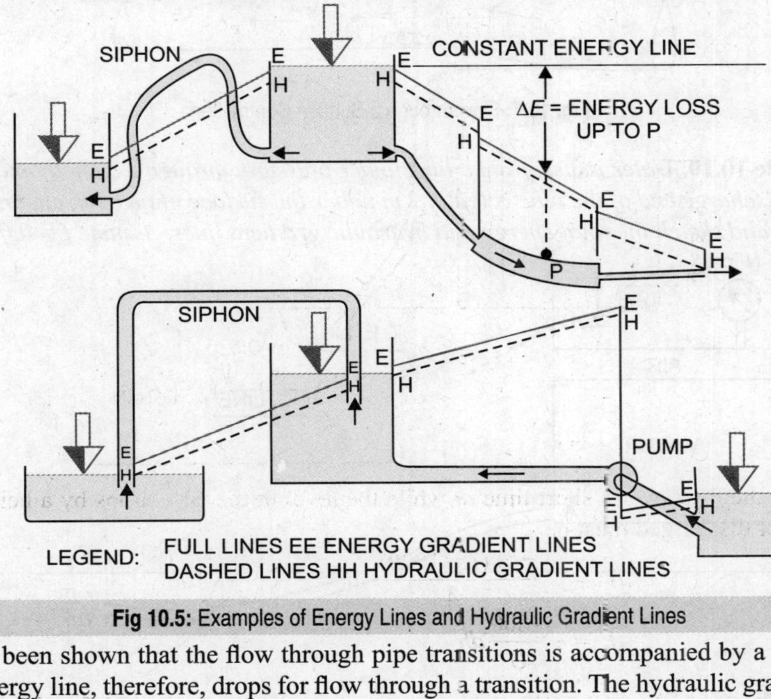

Fig 10.5: Examples of Energy Lines and Hydraulic Gradient Lines

It has been shown that the flow through pipe transitions is accompanied by a loss in energy. The total energy line, therefore, drops for flow through a transition. The hydraulic gradient line may drop or rise depending upon the change in velocity. Typical plots for flow through a sudden expansion and a sudden contraction are shown in Figs. 10.6 and 10.7 respectively.

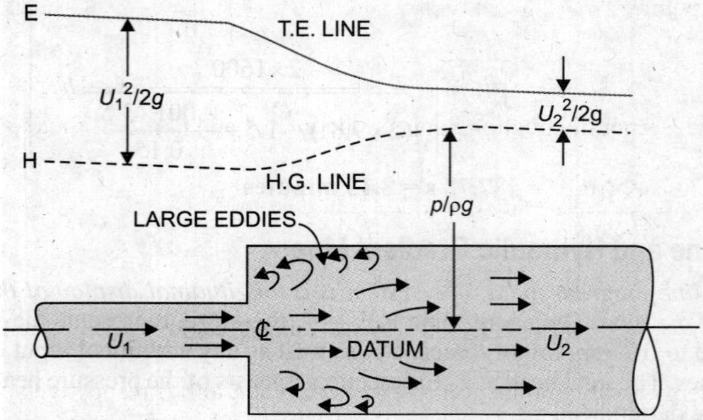

Fig. 10.6: Flow through a Sudden Expansion

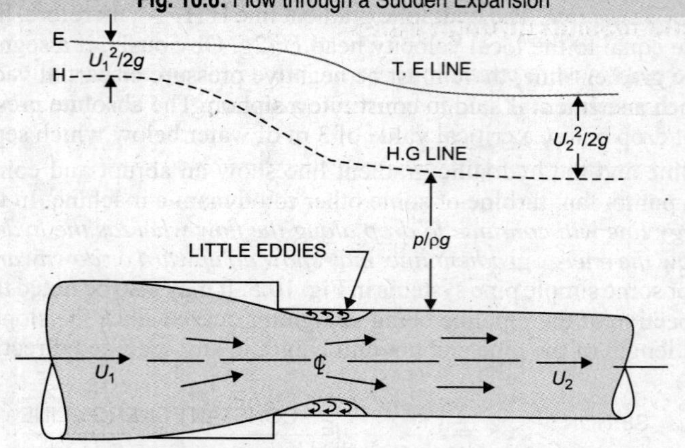

Fig. 10.7: Flow through a Sudden Contraction

Example 10.10. *Water issues, from a tank under pressure, through a 5 cm diameter pipe, 500 m long. The discharge end of the pipe is left at 2 m under the surface of an open channel. Calculate the discharge and sketch the total energy and hydraulic gradient lines. Assume $f = 0.02$.*

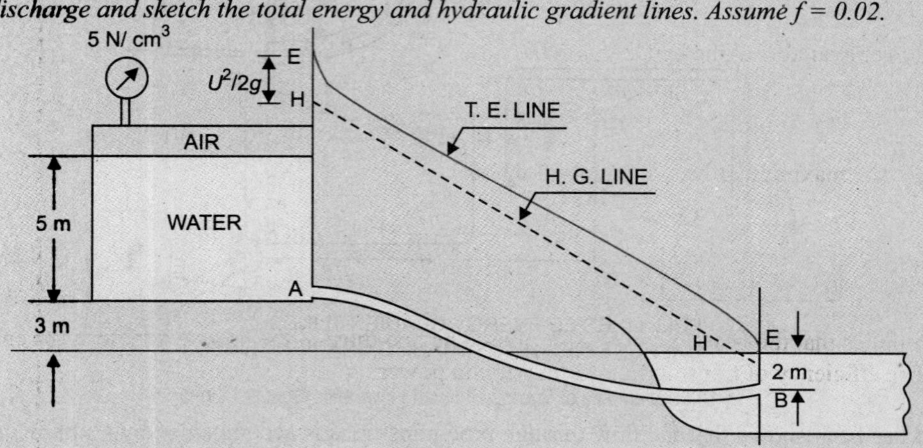

Solution.

The minor losses like entrance losses, exit losses, etc. are negligible compared to the friction losses in a long pipeline.

$$h_f = \frac{fl}{d}\frac{U^2}{2g}$$

Applying the Bernoulli equation between the inlet A and exit B of the pipe and considering A as the datum.

$$5 + \frac{5 \times 10^4}{10^3 \times 9.81} = -3 + \frac{U^2}{2g} + \frac{fl}{d}\frac{U^2}{2g}$$

the velocity $U = \mathbf{1.3\ m/s}$

and the discharge $Q = 1.13 \times \frac{\pi \times 0.05^2}{4} = \mathbf{0.0022\ m^3/s}$

The total energy and hydraulic gradient lines start at E and H respectively which are above the level of water in the tank by an equivalent column of water due to 5 N/cm² pressure.

10.5. Power Transmission through Pipes

Pipes may be employed to transmit hydraulic power in conjunction with hydraulic machines. Consider a pipe of diameter d and length l conveying discharge Q from a head H_1 to a head H_2.

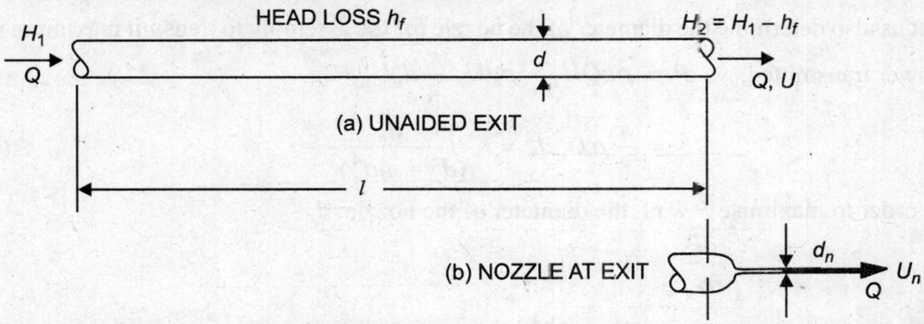

Fog. 10.8: Transmission of Fluid Power

If h_f is the head loss in the transmission,

$$H_2 = H_1 - h_f = H_1 - \frac{f\,l}{d}\frac{U^2}{2g} = H_1 - \frac{f\,l\,Q^2}{12\,d^5} \quad ...(10.11)$$

Power available at the exit $P = \rho g Q H_2$

$$= \rho g\left(H_1 Q - \frac{f\,lQ^2}{12\,d^5}\right) \quad ...(10.12)$$

For the maximum power available at the exit,

$$\frac{\partial P}{\partial Q} = 0 = \rho g\left(H_1 - 3\frac{f\,l}{12 d^5}Q^2\right) = \rho g(H_1 - 3h_f)$$

and

$$h_f = \frac{H_1}{3} \quad ...(10.13)$$

which implies that the head lost in friction should be one-third of the head available at the entry.

The efficiency of transmission for maximum power

$$\eta = \frac{H_2}{H_1}$$

$$= \frac{H_1 - h_f}{H_1} = \frac{H_1 - H_1/3}{H_1} = \mathbf{66.67\%} \quad ...(10.14)$$

If, at the exit, a nozzle discharging into the atmosphere is provided, the exit head H_2 will equal the exit velocity head $\frac{U_n^2}{2g}$ only

$$H_2 = \frac{U_n^2}{2g} \quad ...(10.15)$$

For maximum power transmission,

$$h_f = \frac{H_1}{3} \text{ and } \eta = 66.67\%$$

The exit velocity U_n is evaluated from the energy conservation

$$H_1 = h_f + \frac{U_n^2}{2g} = \frac{f\,l}{d}\frac{U^2}{2g} + \frac{U_n^2}{2g} \quad ...(10.16)$$

and by mass conservation $U d^2 = U_n d_n^2$

Combining these two relations,

$$U_n^2 = 2gH_1 / \left(1 + \frac{f\,l}{d}.\frac{d_n^4}{d^4}\right) \quad ...(10.17)$$

Let us also determine the diameter of the nozzle for the assembly to transmit maximum power.

Power transmitted, $P = \rho g Q H_2 = \rho g U_n . \frac{\pi}{4} d_n^2 U_n^2 / 2g$

$$= \frac{\pi}{8} \rho U_n^2 d_n^2 = K \frac{d_n^2}{(d^5 + fld_n^4)^{3/2}} \quad ...(10.18)$$

In order to maximise P w.r.t. the diameter of the nozzle, d_n

$$\frac{\partial P}{\partial d_n} = 0$$

whence

$$d_n = \left(\frac{d^5}{2f\,l}\right)^{1/4} \quad ...(10.19)$$

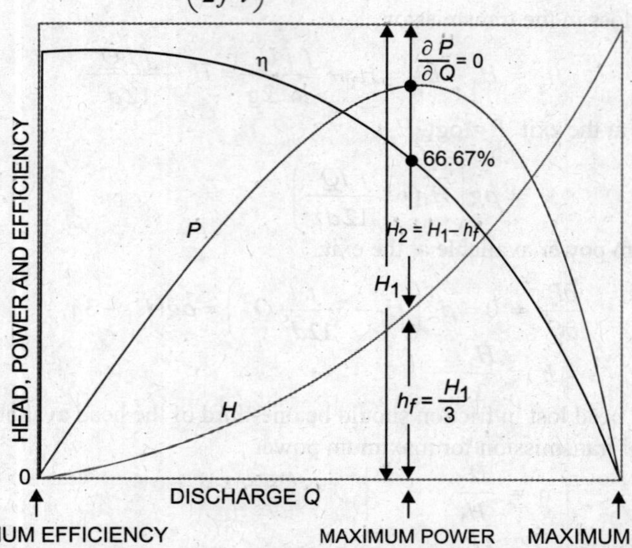

Fig. 10.9: Criteria of Hydraulic Transmission

Example 10.11. *A power transmission pipe 10 cm diameter and 500 m long is fitted with a nozzle at the exit. The inlet is from a lake with water level 60 m above the discharge nozzle.*

Taking $f = 0.03$, calculate the maximum power which can be transmitted, and the diameter of the nozzle required.

Solution.

By the conditions of maximum power transmitted,

$$h_f = \frac{H_1}{3} = 20 \text{ m}$$

and $$H_2 = 60 - 20 = 40 \text{ m} = U_n^2/2g$$

or $$U_n = 28 \text{ m/s}$$

From $$h_f = 20 = \frac{f\,l\,U^2}{d\;2g}$$

$$U = 1.62 \text{ m/s}$$

and $$Q = 1.62 \times \frac{\pi}{4} \times 0.1^2 = 0.0127 \text{ m}^3\text{/s}$$

Maximum power transmitted = $\rho g Q H_2$

$= 1000 \times 9.81 \times 0.0127 \times 40 = 4983.5$ W = **4.98 kW**

Diameter of the nozzle d_n may be determined by the continuity equation

$$1.62 \times 0.1^2 = 28 \times d_n^2$$

whence $$d_n = 0.024 \text{ m} = \textbf{2.4 cm}$$

Alternatively, using Eq. 10.19,

$$d_n = \left(\frac{(0.1)^5}{2 \times 0.03 \times 500}\right)^{1/4} = 0.024 \text{ m} = \textbf{2.4 cm}$$

Example 10.12. *Compare the cost of pumping the same fluid at the same flow rate through a 150 mm pipe and through a 200 mm pipe both having the same roughness factor $f = 0.03$. Assume negligible minor losses.*

Solution.

Let the flow rate be Q. The frictional head loss is given by

$$h = f\,l\,Q^2/12d^5$$

From the data,

$$d_1 = 150 \text{ mm} = 0.15 \text{ m}; \; d_2 = 0.20 \text{ m}$$

The ratio of head loss in the two pipes

$$\frac{h_1}{h_2} = \frac{f\,l\,Q^2}{12d_1^5}\frac{d_2^5}{f\,lQ^2} = (d_2/d_1)^5$$

$$= (0.20/0.15)^5 = \textbf{4.2}$$

The cost of pumping varies as the power required for pumping which, in turn, depends directly upon the frictional head loss. The cost of pumping the same fluid at the same discharge through a 150 mm pipe is, therefore, **4.2 times** the cost of pumping it through a 200 mm pipe.

10.6. Pipes in Series and in Parallels

If a pipeline is joined to one or more pipelines in continuation directly or indirectly, these are said to constitute pipes *in series*, as in Fig. (10.10 a). In this case,

$$Q = Q_1 = Q_2 = Q_3 \quad \text{...(10.20 a)}$$

by the continuity principle and the total head loss is given by

$$h_f = h_{f1} + h_{f2} + h_{f3} = h_{fAB} \quad \text{...(10.20 b)}$$

neglecting minor transition losses.

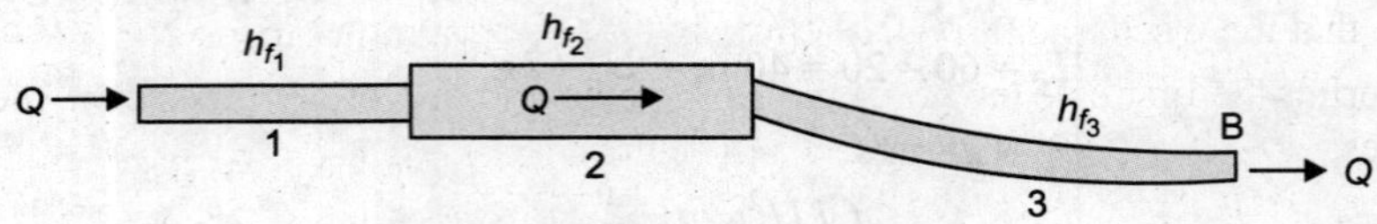

(a) PIPES IN SERIES

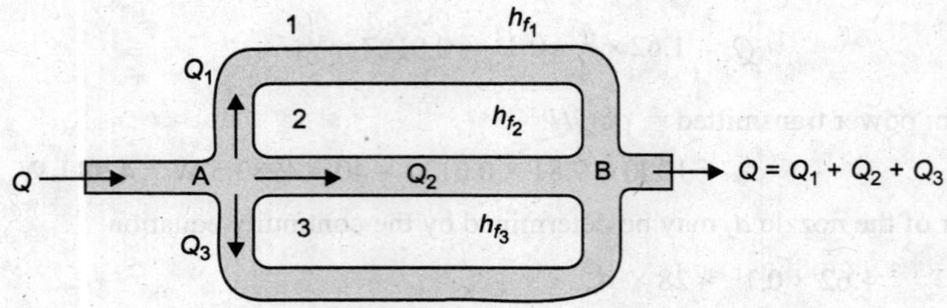

(b) PIPES IN PARALLELS

Fig. 10.10: Flow through Connected Pipes

If two or more pipes emanate from, and join to form, a single pipe, these are said to be in *parallels*. In Fig. (10.10 *b*), pipes 1, 2 and 3 are in parallels between *A* and *B*. For this case, by continuity principle,

$$Q = Q_1 + Q_2 + Q_3 \quad \text{...(10.21 a)}$$

and h_f across *A* and *B* is

$$h_f = h_{f1} = h_{f2} = h_{f3} = h_{fAB} \quad \text{...(10.21 b)}$$

i.e., the discharge is shared by the pipes and the head loss through each of them between the two common points is equal.

Example 10.13. *A pipe of 0.6 m diameter is 1.5 km long. In order to augment the discharge, another line of the same diameter is introduced parallel to the first in the second half of the length. Neglecting minor losses, find the increase in discharge if f = 0.04. The head at inlet is 30 m.*

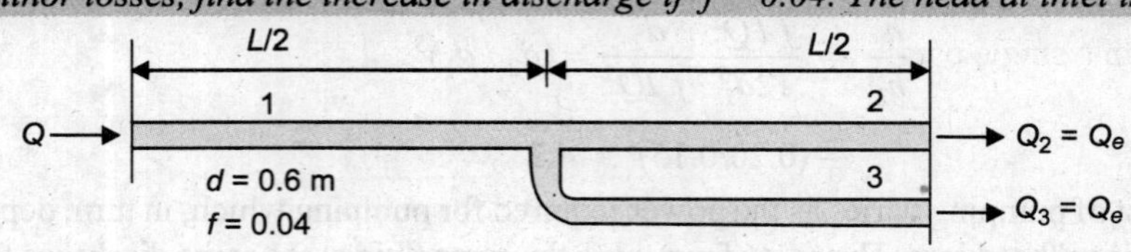

Solution.

Initially, the discharge is calculated from the relationship,

$$h_f = \frac{f\,l\,U^2}{d\;2g} = \frac{f\,l\,Q^2}{12 d^5}$$

whence $$Q^2 = \frac{12 \times 0.6^5 \times 30}{0.04 \times (1.5 \times 1000)} = 0.466$$

and $$Q = 0.683 \text{ m}^3/\text{s}$$

When another pipe line is added over the second half of the length, let the discharge through each be Q_e so that the discharge through the first half of the common pipe is $2Q_e$.

Considering the first half length of the single pipe and one of the pipes over the second half length in series,

$$30 = \frac{f.l/2}{d}\frac{U_1^2}{2g} + \frac{f.l/2}{d}\frac{U_2^2}{2g}$$

$$= \frac{f.l/2.Q_1^2}{12d^5} + \frac{f.l/2.Q_2^2}{12\,d^5} = \frac{f.l/2}{12d^5}(4Q_e^2 + Q_e^2)$$

$$= \frac{0.04 \times (1.5 \times 1000/2)}{12 \times 0.6^5} \times 5\,Q_e^2$$

$$= 160.75\,Q_e^2$$

whence $$Q_e = 0.432 \text{ m}^3/\text{s}$$

and the total discharge through this system is

$$2\,Q_e = 0.864 \text{ m}^3/\text{s}$$

Increase in discharge by this arrangement

$$2Q_e - Q = 0.864 - 0.683 = 0.181 \text{ m}^3/\text{s}$$

which is $\frac{0.181}{0.683} \times 100$ **or 26.5% increase.**

Example 10.14. *A pipeline conveys 0.00833 m³/s water from an overhead tank to a building. The pipe is 2 km long and 0.15 m diameter; f = 0.03. It is desired to increase the discharge by 30% by installing another pipeline in parallel with this over half the length. Suggest a suitable diameter of the pipe to be installed. Is there an upper limit on discharge augmentation by this arrangement?*

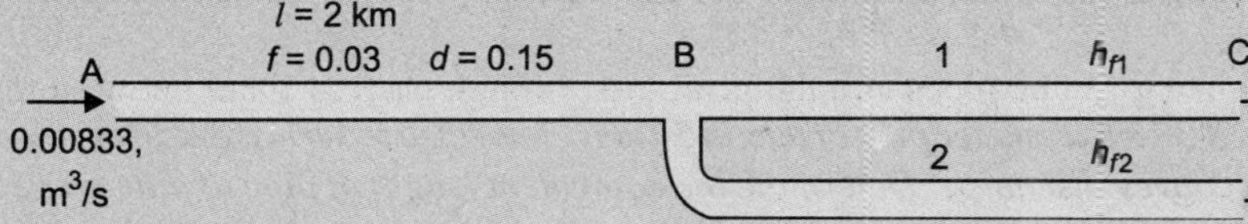

Solution.

The height h of the overhead tank above the building can be determined from the initial conditions with a single pipe.

$$h = h_f = \frac{0.03 \times 2000 \times (0.008\,33)^2}{12 \times (0.15)^5} = 4\,57 \text{ m}$$

assuming negligible losses in the fittings, as the pipe is very long.

In the new plan as shown in the figure,

$$h_f = 4.57 = h_{fAB} + h_{fBC} \quad \text{... (i)}$$

and $$h_{fBC} = h_{f1} = h_{f2} = \frac{f_1 l_1 Q_1^2}{12 d_1^5} = \frac{f_2 l_2 Q_2^2}{12 d_2^5}$$

Assuming $f_1 = f_2 = f$

taking $l_1 = l_2 = 1000$ m

$$\left(\frac{Q_1}{Q_2}\right)^2 = \left(\frac{d_1}{d_2}\right)^5 \qquad \text{... (ii)}$$

By continuity

$$Q_1 + Q_2 = Q$$

and using (*i*) $$\frac{1000 \times f(Q_1+Q_2)^2}{12 d_1^5} + \frac{1000 \times f Q_1^2}{12 d_1^5} = 4.57$$

$$\frac{(Q_1+Q_2)^2}{d_1^5} + \frac{Q_1^2}{d_1^5} = 1.828 \qquad \text{...(iii)}$$

Taking $Q = 1.3 \times 0.008\ 33 = 0.0108\ m^3/s$

From (*iii*) $Q_1^2 = 1.828 \times 0.15^5 - 0.0108^2$

and $Q_1 = 0.004\ 71\ m^3/s$

From (*ii*) $Q_2 = 0.0108 - 0.004\ 71 = 0.006\ 09\ m^3/s$

From (*i*), $$d_2 = \left(\frac{0.006\ 09}{0.00471}\right)^{2/5} \times 0.15$$

$$= 0.166 \text{ m}$$

It can be observed from (*iii*) that

$$Q^2 = 1.828 \times 0.15^5 - Q_1^2$$

$$= 0.000\ 139 - Q_1^2$$

The total discharge Q can only be a maximum of

$$\sqrt{0.000\ 139} = \mathbf{0.0118\ m^3/s}$$

which is **41.66% more** than the initial condition. For this case $Q_1 = 0$; $Q_2 = 0.0118\ m^3/s$, $d_2 \to \infty$; an infinitely large pipe.

Example 10.15. *A 30 cm diameter pipe of length 30 m is connected in series to a 20 cm diameter pipe of length 20 m to convey discharge. Determine the equivalent length of pipe of diameter 25 cm, assuming that the friction factor remains the same and the minor losses are negligible.*

Solution.

From the data,

$$d_1 = 0.3 \text{ m}, l_1 = 30 \text{ m } d_2 = 0.2 \text{ m}, l_2 = 20 \text{ m}$$

If U_1 is the velocity in the first pipe, $U_2 = (0.3/0.2)^2 \times U_1 = 2.25\ U_1$

The head loss in the given pipes in series is

$$h = \frac{f \times 30}{0.3}\frac{U_1^2}{2g} + \frac{f \times 20}{0.2}\frac{U_2^2}{2g}$$

$$= (100 + 506)\ f\ U_1^2 / 2g = 606\ f\ U_1^2 / 2g$$

For equivalence, the discharge and the frictional head loss in the new pipe should be the same as in the original system.

$$Q = \pi/4\,(0.25)^2 \times U = \pi/4\,(0.3)^2 \times U_1$$

whence, $U = (0.3/0.25)^2 . U_1 = 1.44\, U_1$

$$h = \frac{f \times l}{0.25}\frac{(1.44\, U_1)^2}{2g} = 606\, fU_1^2/2g$$

whence, **length l = 73 m**

Example 10.16. *Two pipes of the same material and of equal lengths are available for connection to an overhead tank which can supply 0.085 m/s of 'water. The diameters of the pipes are 30 cm and 15 cm respectively. Determine the ratio of the head loss when the pipes are connected in series to the head loss when they are connected in parallels. Neglect minor losses.*

Solution.

Let the length of each pipe be l. When connected in series, the discharge through each pipe is 0.085 m³/s and the total head loss

$$h_f = \frac{f\,l}{12}.\frac{(0.085)^2}{(0.3)^5} + \frac{f\,l}{12}.\frac{0.085^2}{0.15^2}$$

$$= \frac{f\,l}{12} \times (2.973 + 95.14) = 98.11 \times \frac{f\,l}{12}$$

When connected in parallels.

$$Q = 0.085 = Q_1 + Q_2$$

and head loss in each pipe is the same

$$h_{f1} = \frac{f\,l}{12}\frac{Q_1^2}{(0.3)^5} = h_{f2} = \frac{f\,l}{12}\frac{Q_2^2}{(0.15)^3}$$

whence $Q_1 = 5.66\, Q_2$

From these equations,

$$Q_1 = 0.0722 \text{ m}^3/\text{s} \,;\, Q = 0.0128 \text{ m}^3/\text{s}$$

and $$h_{f1} = \frac{f\,l}{12} \times 0.0722^2/0.3^5 = 2.15 \times \frac{f\,l}{12} = h_{f2}$$

Therefore, the ratio of heads lost is 98.11/2.15, *i.e.*, **45.6.**

Example 10.17. *Determine the length of an equivalent pipe of diameter 20 cm and friction factor 0.02 for a given pipe system discharging 0.1 m³/s consisting of*

(a) a 10 m line of 20 cm diameter with f = 0.03 with

(b) three 90° bends, K = 0.5 for each,

(c) two sudden expansions ; 20-30 cm diameter,

(d) a 15 m line of 30 cm diameter with f = 0.025 and

(e) a globe value, fully open, K = 10.

Solution.

The equivalence of two flow systems requires that the frictional head loss across them and the discharge through them should be the same. For the required discharge, 0.1 m³/s, the velocities in the 20 cm pipe and 30 cm pipe are :

$$U_1 = 0.1/(\pi \times 0.2^2/4) = 3.18 \text{ m/s}$$

$$U_2 = 3.18 \times (0.2/0.3)^2 = 1.41 \text{ m/s}$$

For (*a*), equating the frictional head loss $\dfrac{f\,l_e}{d}\dfrac{U^2}{2g}$ in the two pipes,

$$l_e = \frac{0.03 \times 10}{0.02} = \mathbf{15\ m}$$

For (*b*), equating the head loss across the bends to the frictional head loss in the required pipe,

$$\frac{0.02.\,l_e}{0.2}\ \frac{3.18^2}{2g} = 0.5 \times \frac{3.18^2}{2g} \times 3$$

whence, $l_e = \mathbf{15\ m}$

For (*c*), the sudden expansions

$$\frac{0.02\,l_e}{0.2} \times \frac{3.18^2}{2g} = \left[\left(1 - \left(\frac{0.2}{0.3}\right)^2\right)^2 \times \frac{3.18^2}{2g}\right] \times 2$$

$$l_e = \mathbf{6.17\ m.}$$

For (*d*), the 30 cm dia pipeline, $f = 0.025$

$$\frac{f_1 l_e}{d_1} \cdot \frac{u_1^2}{2g} = \frac{f_2 l_2}{d_2} \cdot \frac{u_2^2}{2g}$$

$$\frac{0.02 l_e}{0.2} \times \frac{3.18^2}{2g} = \frac{0.025 \times 15}{0.3} \times \frac{1.41^2}{2g}$$

whence $l_e = \mathbf{2.5\ m.}$

For (*e*), the globe valve,

$$\frac{0.02 l_e}{0.2} \times \frac{3.18^2}{2g} = 10 \times \frac{1.41^2}{2g}$$

$$l_e = \mathbf{19.66\ m}$$

Total equivalent length of the 20 cm dia pipe for the given system is

$$l_e = 15 + 15 + 6.17 + 2.5 + 19.66 \text{ m} = \mathbf{58.33\ m.}$$

Example 10.18. *Two similar pipes of the same diameter of lengths l_1 and l_2 are placed in parallel. Calculate the equivalent length of a single pipe of the same diameter. What would be the equivalent length if the two given pipes were equal in length ?*

Solution.

For the pipes connected in parallel as shown,

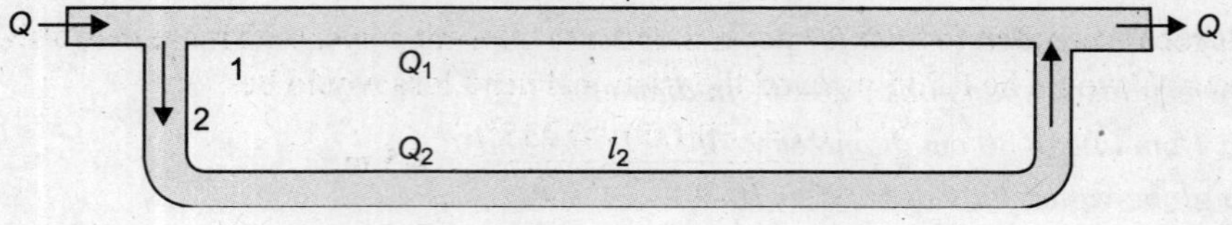

By continuity $Q = Q_1 + Q_2$

and for equal head loss across them,

$$h_f = h_{f1} = h_{f2}$$

or
$$h_f = \frac{f\,l_1Q_1^2}{12d_1^5} = \frac{f\,l_2Q_2^2}{12d_2^5};\ \frac{Q_2}{Q_1} = \sqrt{l_1/l_2}$$

If the length of the equivalent pipe is l_e, then
$$h_f = \frac{f\,l_eQ^2}{12d^5}$$

whence,
$$l_eQ^2 = l_1\,Q_1^2 = l_2\,Q_2^2$$

But
$$Q_1 = \frac{Q}{(1+\sqrt{l_1/l_2})}$$

Therefore,
$$l_e = \frac{l_1}{(1+\sqrt{l_1/l_2})^2}$$

For the special case when $l_1 = l_2 = l$
$$l_e = \frac{l}{(1+1)^2} = l/4$$

Example 10.19. *It is required to deliver 20 kW hydraulic power by a 25 cm pipe over a distance of 10 km with a minimum efficiency of 90%. The pressure at the delivery end is maintained at 80 m of water. Calculate the number of parallel pipes required if the friction factor f is assumed as 0.04.*

Solution.

The power available at the delivery end is 90% of 20 kW, i.e., 18 kW. Since the delivery head is 80 m

$$\rho g Q \times 80 = 18000$$

Taking $\rho = 1000$ kg/m³,

We get $Q = 0.023$.

If there are *n* parallel pipes, the velocity, in each is
$$U = \frac{1}{n} \times 0.023/(\pi \times 0.25^2/4) = 0.47/n \text{ m/s.}$$

The frictional head loss in each pipe must be 10% of the head supplied, *i.e.*
$$h_f = \frac{1}{10}(h_f + 30)\,;\ h_f = 8.9 \text{ m}$$

also,
$$h_f = \frac{0.04 \times 10{,}000}{0.25} \times \frac{0.47^2}{n^2 \times 2 \times 9.81}$$

whence $n = 1.42$. (more than 1!)

It is recommended to install 2 pipes in order to transmit power with more than 90% efficiency. The velocity *U* would be 0.235 m/s and the frictional head loss would be
$$h_f = \frac{0.04 \times 10{,}000}{0.25} \times \frac{0.235^2}{2g} = 4.5 \text{ m}$$

This corresponds to an 84.5 m head at the entrance and hence an efficiency of 80/84.5, *i.e.*, **95%.**
Installation of just one pipeline would have meant a head loss of
$$h_f = \frac{0.04 \times 10{,}000}{0.25} \times \frac{(0.47)^2}{2g} = 18\text{m}$$

which would have corresponded to an efficiency of 80/98, *i.e.*, 82%.

Example 10.20. *Three pipes are connected in series with the following details.*

	Diameter	*Length*	*Friction coefficient*
Pipe 1	*10 cm*	*500 m*	*0.018*
Pipe 2	*15 cm*	*750 m*	*0.018*
Pipe 3	*20 cm*	*1000 m*	*0.018*

Determine the diameter of the single equivalent pipe with the same friction coefficient.

Solution.

Since the three pipe are connected in series, the same discharge Q is passing through each of them; and Q should also pass through the single equivalent pipe of diameter d and length equal to the sum of the three pipes, *i.e.*, 2250 m.

Using the Darcy-Weisbach equation,

$$h_f = \frac{f\,l\,Q^2}{12\,d^5}$$

For the three pipes connected in series,

$$h_f = \frac{fQ^2}{12}\left[\frac{500}{(0.1)^5} + \frac{750}{(0.15)^5} + \frac{1000}{(0.2)^5}\right]$$

$$= \frac{fQ^2}{12} \times 63 \times 10^6.$$

For the equivalent pipe,

$$h_f = \frac{fQ^2}{12}\left(\frac{2250}{d^5}\right)$$

Equating the two expressions for the head loss, for equivalence

$$\frac{2250}{d^5} = 63 \times 10^6$$

whence, $d = 0.13 \text{ m} = \mathbf{13\ cm.}$

Example 10.21. *An overhead tank of 5 m × 10 m cross section in a village is used to supply water to another village across a hill. For the configuration shown, calculate the discharge conveyed. Is there an upper limit on the discharge conveyed by the pipe ? Given, length AT = 2 km and TB = 1 km.*

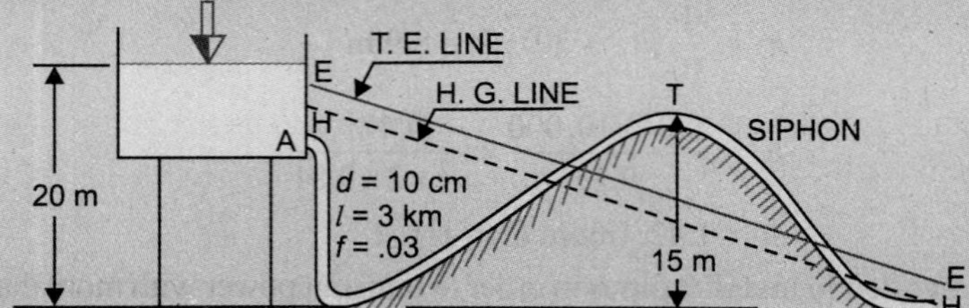

Solution.

$$h_f = \frac{f\,l}{d}\frac{U^2}{2g}$$

or $$20 = \frac{0.03 \times 3000}{0.1} \times \frac{U^2}{2 \times 9.81}$$

whence $U = 0.66$ m/s

and $$\frac{U^2}{2g} = 0.022 \text{ m}$$

In order to check on this velocity in the light of the maximum feasible velocity, apply the Bernoulli equation, accounting for the head losses between A and T, considering the ground as datum,

$$\left(\frac{p_A}{\rho g} + Z_A\right) + \frac{U_A^2}{2g} = \frac{p_T}{\rho g} + \frac{U_T^2}{\rho g} + Z_T + h_{f_1}$$

$$10.33 + 20 + 0 = h_T + 0.0027 + 15 + (2/3 \times 20)$$

$$h_T \approx 2 \text{ m of water}$$

This is less than the separation limit (of 3 m head of water) at T, The pipe line appears to have been designed without regard to siphon considerations. The separation at T could have been avoided by choosing a bigger diameter of the pipe or a pipe material of lower friction coefficient. What can be done at this stage is to limit the discharge so as not to let h_T drop below 3 m of water. By the Bernoulli equation,

$$10.33 + 0 + 20 = 3 + \frac{U_T^2}{2g} + 15 + h'_f$$

whence $$h'_f + \frac{U_T^2}{2g} = 12.33 \text{ m}$$

or $$\frac{U_T^2}{2g}\left(\frac{0.03 \times 2000}{0.1} + 1\right) = 12.33$$

and $$U_T = 0.634 \text{ m/s}$$

This corresponds to a discharge of

$$Q = 0.634 \times \frac{\pi \times 0.1^2}{4} = \mathbf{0.005\ m^3/s}$$

Example 10.22. *A siphon of diameter 20 cm connects two reservoirs having a difference in elevation of 20 m. The total length of the siphon is 800 m and the summit is 5 m above the water level in the upper reservoir. If the separation takes place at 2.8 m of water absolute, find the maximum length of the siphon from the water level in the upper reservoir to the summit. Take f = 0.004 for the siphon and the atmospheric pressure as 10.3 m of water. Neglect minor losses.*

Solution.

Flow through the siphon is due to the 20 m difference in elevation ; it must be equal to the head loss accompanying the flow.

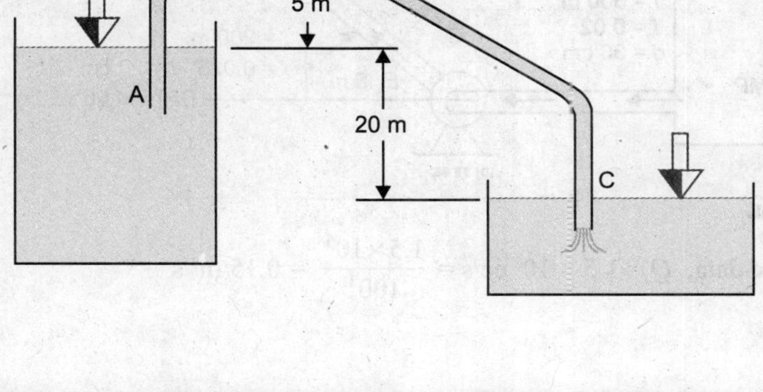

$$\frac{f\,l}{d}\times\frac{U^2}{2g} = 20$$

or $$\frac{0.004\times 800\times U^2}{0.2\times 2\times 0.81} = 20;\ U = 4.952 \text{ m/s}$$

Now, consider the head loss in the section of siphon *AB*,

$$h_{\text{fAB}} = \frac{f\,l}{d}\frac{U^2}{2g}$$

By energy balance

$$10.3 = 5 + 2.8 + h_{\text{fAB}} + \frac{U^2}{2g}$$

$$l_{\text{AB}} = \frac{2.5\times 0.2\times 2\times 9.81}{0.004\times(4.952)^2} = \mathbf{50\ m}$$

Let us check this result. The head loss in the remaining 700 m length of the siphon should be such that

$$5 + 20 + 2.8 + 12.5 - 10.3 \equiv \frac{0.004\times 750}{0.2}\times\frac{4.952^2}{2\times 9.81}$$

The expression on each side reduces to 18.75 m which shows that our solution is correct.

Example 10.23. *A piping system is shown. Considering all losses, determine the pressure head in N/m² on the suction and discharge side of the pump and the power of motor needed to run the pump if the pump and motor efficiencies are each 70%. Note that the head loss for the nozzle is*

$$h_L = \frac{(I - C_v^2)}{C_v^2}\frac{V_j^2}{2g}$$

C_v = *coefficient of velocity and*

V_j = *velocity through the nozzle.*

Draw the energy and hydraulic grade lines approximately to scale and table the segments.

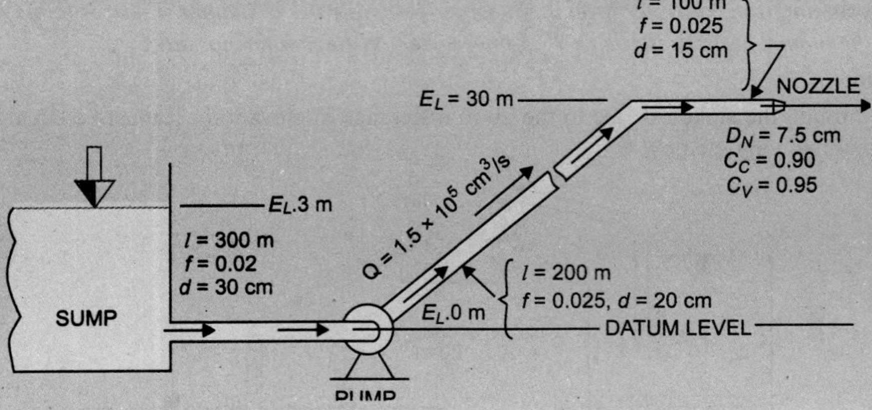

Solution.

From the data, $Q = 1.5\times 10^5 \text{ cc/s} = \dfrac{1.5\times 10^5}{100^3} = 0.15 \text{ m}^3/\text{s}$

$$U_1 = \frac{0.15 \times 4}{\pi \times 0.3^2} = 2.12 \text{ m/s}$$

$$U_2 = 2.12 \times \left(\frac{0.3}{0.2}\right)^3 = 4.77 \text{ m/s}$$

$$U_3 = 2.12 \times \left(\frac{0.3}{0.15}\right)^2 = 8.48 \text{ m/s}$$

Head loss at entrance to the suction pipe is

$$0.5U_1^2/2g = 0.5 \times 2.12^2/(2 \times 9.81) = 0.115 \text{ m}$$

Head loss due to friction in the suction length of 300 m is

$$h_{f_1} = \frac{f_1 l_1}{d_1}\frac{U_1^2}{2g} = \frac{0.02 \times 300}{0.3} \times \frac{2.12^2}{2 \times 9.81} = 45.8 \text{ m}$$

The head of water at the suction side of the pump is, therefore,

$$3 - 0.115 - 4.58 = -1.695 \text{ m w.r.t. atmosphere}$$

$$= -1.695 \times 1000 \times 9.81 = -16630 \text{ N/m}^2$$

$$= \mathbf{16.63 \text{ kN/m}^2 \text{ vacuum}}$$

Head loss due to friction in the discharge line from the pump upto the bend is

$$h_{f_2} = \frac{0.025 \times 200}{0.2} \times \frac{4.77^2}{2 \times 9.81} = \mathbf{29 \text{ m}}$$

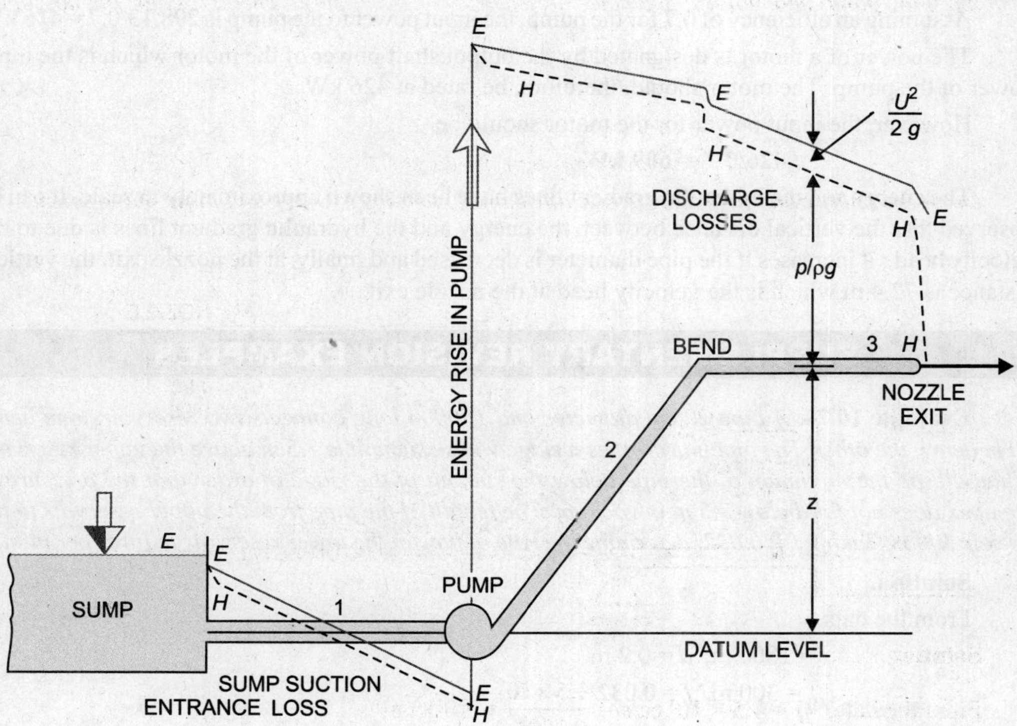

and the head loss due to friction in the horizontal section of the discharge line is

$$h_{f_3} = \frac{0.025\times 100}{0.15}\times\frac{8.48^2}{2\times 9.81} = \mathbf{61.1\ m}$$

Head loss in the bend may be taken as small as the head loss in a contraction, *i.e.*,

$$0.5U_2^2/2g = \frac{0.5\times 4.77^2}{2\times 9.81} = 0.58\ \text{m}$$

Head loss in the nozzle flow is

$$\frac{1-C_v^2}{C_v^2}\frac{V_j^2}{2g} = \frac{1-0.95^2}{0.95^2}\times\frac{37.7^2}{2\times 9.81} = \mathbf{7.83\ m}$$

Since $C_v = 0.95$

and $$V_j = \frac{0.15\times 4}{\pi\times 0.075^2\times 0.90} = 37.7\ \text{m/s}$$

The velocity head available at the exit is

$$\frac{V_j^2}{2g} = \frac{37.7^2}{2\times 9.81} = \mathbf{72.4\ m}$$

The pressure head at the exit of the nozzle being atmospheric, the pressure at the discharge end of the pump for the head

$$72.4 + 7.83 + 61.1 + 0.58 + 29 + 30 = 200.9\ \text{m}$$

must be $p = \rho gh = 209.0\times 1000\times 9.81 = 1\,970\,930\ \text{N/m}^2 = \mathbf{1.97\ MN/m^2}$

The water-power delivered is

$$0.15\times[1\,970930-(-16\,630)] = 298\,130\ \text{W} = 298.13\ \text{kW}$$

Assuming an efficiency of 0.7 for the pump, the input power to the pump is 298.13/0.7 = **426 kW**

The power of a motor is designated by the output shaft power of the motor which is the input power of the pump. The motor should, therefore, be rated at 426 kW.

However, the input power for the motor should be

$$426/0.7 = \mathbf{609\ kW}$$

The energy and the hydraulic gradient lines have been shown approximately to scale. It can be observed that the vertical distance between the energy and the hydraulic gradient lines is due to the velocity head ; it increases if the pipe diameter is decreased and finally, at the nozzle exit, the vertical distance is 72.4 m which is the velocity head at the nozzle exit.

SUPPLEMENTARY REVISION EXAMPLES

Example 10.24. *A pipe 20 cm diameter and 1800 m long connects two reservoirs one being 30 m below the other. The pipeline crosses a ridge whose summit is 7.5 m above the upper reservoir. What will be the minimum of the pipe below the summit of the ridge in order that the pressure at the apex does not fall below 7.5 m in vacuum. The length of the pipe from the upper reservoir to the apex is 300 m. Taking f = 0.032, determine the rate of flow in the lower reservoir in litres per minute*

Solution.

From the data,

$$l = 1800\ \text{m},\quad d = 0.2\ \text{m}$$

$$l_{AB} = 300\ \text{m},\quad f = 0.032$$

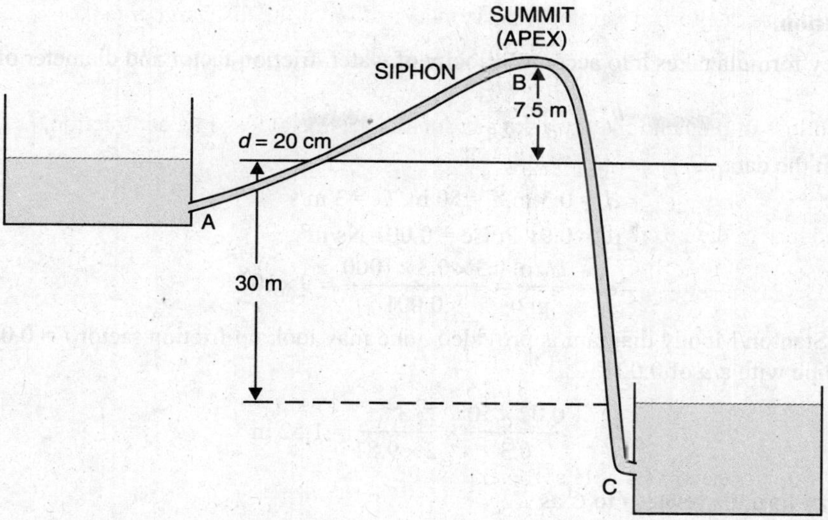

Flow through the pipe ABC is due to the 30 m head difference between A and C. Using the equation,

$$h_f = 30 = \frac{f\,l\,U^2}{d\;2g} = \frac{0.032 \times 1800}{0.02}\cdot\frac{U^2}{2g}$$

$$U = 0.45 \text{ m/s}$$

$$Q = UA = 0.45 \times \pi\frac{0.2^2}{4} = 0.0142 \text{ m}^3\text{/s} = 0.0142 \times 1000 \text{ litres/s}$$

$$= 14.2 \text{ litres per second.}$$

Let us check the pressure at the apex, the summit of the ridge:

$$h_f\text{AB} = \frac{f.l_{AB}}{d}.\frac{U^2}{2g} = \frac{0.032 \times 300}{0.02}.\frac{0.45^2}{2 \times 9.81}$$

$$= 4.98 \text{ m}$$

It could have been obtained as 300/1800 times 30 m, *i.e.,* 5 m head loss from A to B.

It means that p_B is 5 cm vacuum.

The question requires us to find out the pipe lengths if the pressure at the apex must not fall below 7.5 m vacuum. For that condition,

$$\frac{f.l'\,U^2}{d\;\;2g} = 7.5 = \frac{0.032 \times l'}{0.02}.\frac{0.45^2}{2 \times 9.81}$$

whence l' = **450 m.**

The minimum length of the pipe below the summit on the other side, *i.e.,* towards the lower reservoir must be (1800 – 450), *i.e.,* **1350 m.**

Example 10.25. *Find the head lost due to friction in a pipe of diameter 300 mm and length 50 m through which water is flowing at a velocity of 3 m/s by using (i) Darcy formula (ii) Chezy formula, taking C = 60.*

Take μ for water 0.01 Poise.

Solution.

Darcy formula takes into account velocity of water, friction factor and diameter of pipe,

$$h_f = \frac{f\,l}{d}.\frac{U^2}{2g}.$$

From the data,

$$d = 0.3 \text{ m}, l = 50 \text{ m}, U = 3 \text{ m/s}$$

$$\mu = 0.01 \text{ Poise} = 0.001 \text{ Ns/m}^2.$$

$$Re = \frac{Ud\rho}{\mu} = \frac{3\times 0.3\times 1000}{0.001} = 9\times 10^5.$$

If a Stanton/Moody diagram is provided , one may look up friction factor $f = 0.02$ for rough cast iron pipe with ε/d of 0.001.

$$h_f = \frac{0.02\times 50}{0.3}\times\frac{3^2}{2\times 9.81} = 1.52 \text{ m}$$

Chezy formula relates f to C as

$$C = \sqrt{\frac{8g}{f}};\ f = \frac{8g}{C^2} = \frac{8\times 9.81}{60^2} = 0.022$$

Then

$$h_f = \frac{0.022\times 50}{0.3}\times\frac{3^2}{2\times 9.81} = 1.67 \text{ m}$$

Example10.26. *At a sudden enlargement of a water main from 240 mm to 480 mm diameter, the hydraulic gradient rises by 10 mm. Estimate the rate of flow.*

Solution.

Total energy $E = z + \frac{U^2}{2g} + \frac{p}{\rho g}$ in terms of 'head'.

The hydraulic gradient refers to $H = z + \frac{p}{\rho g}$.

In a sudden enlargement, change in hydraulic gradient from 1 to 2 is

$$\left(z+\frac{p_2}{\rho g}\right)-\left(z+\frac{p_1}{\rho g}\right) = \frac{p_2-p_1}{\rho g} = 0.1 \text{ m, rise.}$$

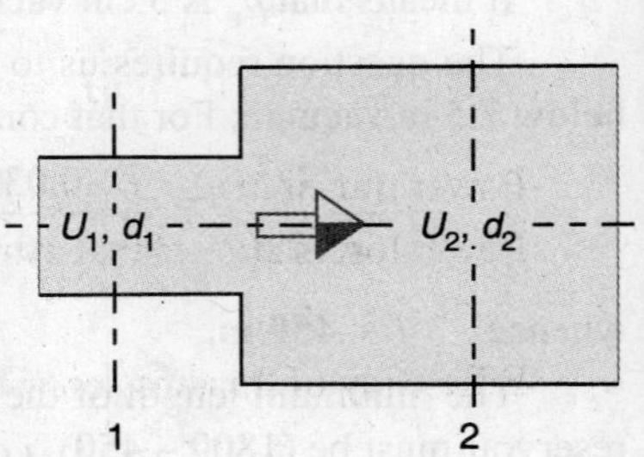

Total energy must at 1 and 2 is such that

$$E_1 = E_2 + h_l$$

$$= E_2 + \left(1-\frac{A_1}{A_2}\right)^2\frac{U_1^2}{2g}$$

$$z+\frac{U_1^2}{2g}+\frac{p_1}{\rho g} = z+\frac{U_2^2}{2g}+\frac{p_2}{\rho g}+\left(1-\frac{A_1}{A_2}\right)^2\frac{U_1^2}{2g}$$

Since $d_2 = 2d_1$, $A_2 = 4\,A_1$ and $U_1 = 4\,U_2$.

$$\frac{U_1^2}{2g}-\frac{1}{16}\frac{U_1^2}{2g}-\left(1-\frac{1}{4}\right)^2\frac{U_1^2}{2g} = 0.1.$$

whence, $U_1 = \mathbf{2.29\ m/s}$

Discharge $Q = U_1 A_1 = 2.29 \times \dfrac{\pi\, 0.24^2}{4} = \mathbf{0.104\ m^3/s}$

Example 10.27. *Two pipes of lengths 2500 m each and diameters 80 cm and 60 cm respectively, are connected in parallels. The coefficient of friction for each pipe is 0.006 and the total flow is 250 litres per second. Find the rate of flow in each pipe.*

Solution.

Head loss through each pipe, placed in parallel, must be the same

$$h_f = \frac{f\,l\,Q_1^2}{12\ d_1^5} = \frac{f\,l\,Q_2^2}{12\ d_2^5}$$

whence
$$\left(\frac{Q_1}{Q_2}\right)^2 = \left(\frac{d_1}{d_2}\right)^5 = \left(\frac{0.8}{0.6}\right)^5 = 4.21$$

$$\frac{Q_1}{Q_2} = 2.05.$$

Given that, $Q_1 + Q_2 = 250\ \text{litres/s} = 0.25\ \text{m}^3\text{/s}$

$$2.05\,Q_2 + Q_2 = 0.25,$$

$$Q_2 = 0.25/3.05 = \mathbf{0.082\ m^3/s}$$

$$Q_1 = 0.25 - 0.082 = \mathbf{0.168\ m^3/s.}$$

Example 10.28. *A pipeline 2100 m long is used to transmit 103 kW power. The pressure at the inlet is 392.4 N/cm². If the efficiency of transmission is 80% and f = 0.005, find the diameter of the pipe.*

Solution.

From the data, $L = 2100$ m, Power transmitted = 103 kW.

$$p_1 = 392.4 \times 100 \times 100\ \text{N/m}^2$$

$$\frac{p_1}{\rho g} = \frac{392.4 \times 100 \times 100}{1000 \times 9.81} = 400\ \text{m} = H_1$$

$$H_2 = H_1 - \frac{f\,l\,Q^2}{12\ d^5}$$

Power transmitted = $\rho\, g\, Q.\, H_2$ = 103 × 1000 = 103,000 W.

Power lost is 20% of that available at the entry; *i.e.* = 103,000 × 100/80 – 103,000 = 25,750 W.

= **25.75 kW**

Efficiency of transmssion = $\eta = H_2/H_1 = 0.80$.

$$\left(H_1 - \frac{f\,l\,Q^2}{12\ d^5}\right) / H_1 = 0.8 \qquad ...(i)$$

$$\rho g\, Q\left(H_1 - \frac{f\,l\,Q^2}{12\ d^5}\right) = 103{,}000 \qquad ...(ii)$$

Substituting $H_1 = 400$ m, $f = 0.005$, $L = 2100$ m,

Dividing (*ii*) by (*i*)

$$\rho g Q H_1 = \frac{103{,}000}{0.8} = 128{,}750$$

$$1000 \times 9.81 \times Q \times 400 = 128{,}750$$

$$Q = 0.033 \text{ m}^3/\text{s}$$

From (*i*), solve for d ; d = **10.3 cm.**

Example 10.29. *A fluid of mass density 1790 kg/m³ and viscosity 2.1 Pas flows at a velocity of 3 m/s in a 6 cm diameter pipe. Estimate the head loss over a length of 12 m pipe.*

Solution.

From the data $\rho = 1790 \text{ kg/m}^3$, $\mu = 2.1$ *Pas.*

$u = 3$ m/s, $d = 0.06$ m, $l = 12$ m

The Reynolds number of the flow is

$$Re = \frac{ud\rho}{\mu} = \frac{3 \times 0.06 \times 1790}{2.1} = 153.$$

Re is much less than 2000, which shows it is a laminar flow of a viscous fluid.

For laminar flow,

$$f = \frac{64}{Re} = \frac{64}{153} = 0.42$$

Head loss

$$h_f = \frac{f\,l}{d}\frac{U^2}{2g}$$

$$= (0.42 \times 12 \times 3^2)/(0.06 \times 2 \times 9.81)$$

= **38.4 m head** of the fluid.

Example 10.30. *A round pipe 10 cm in diameter has 2 m length which is porous. In this porous section the velocity at the exit is constant. If the velocities at the entry and exit are 2.0 m/s and 1.2 m/s respectively, estimate (a) the discharge emitted through the porous wall of the pipe and (b) the average velocity of the emitted discharge.*

Solution.

Cross section area of the pipe $= \pi \times \frac{0.1^2}{4} = 0.00785 \text{ m}^2$

Q_1 at entry $= 2 \times 0.00785 = 0.0157 \text{ m}^3/\text{s}$

Q_2 at exit $= 1.2 \times 0.00785 = 0.0094 \text{ m}^3/\text{s}$

Q, the discharge emitted from the porous pipe is

$$Q = Q_1 - Q_2$$

$$= 0.0157 - 0.0094 = 0.0063 \text{ m}^3/\text{s}$$

Peripheral area of the porous pipe through which the discharge has been emitted is

$$\pi \times 0.1 \times 10 = 3.142 \text{ m}^2$$

Average velocity of the discharge through the porous pipe is

$$v = \frac{0.0063}{3.142} = \mathbf{0.002 \text{ m/s}}.$$

Example 10.31. *If two pipes of diameters D and d and equal length L are arranged in parallel. The loss of head for a flow of Q is h. If the same pipes are arranged in series, the loss of head for the same flow is H. If d = 0.5 D, find the percentage of total flow through each pipe when placed in parallel. Also, find the ratio H/h. Neglect minor losses and assume f to be constant.*

Solution.

When the two pipes are placed in parallel over an equal length L, the loss of head in each pipe is h and the flow is shared as Q_1 and Q_2, totalling to Q. Hence,

$$h = \frac{f\,L\,Q_1^2}{12D^5} = \frac{f\,L\,Q_2^2}{12d^5}.$$

which requires

$$\frac{Q_1}{Q_2} = \left(\frac{D}{d}\right)^{5/2} = \left(\frac{D}{0.5D}\right)^{5/2}$$

$$= 2^{5/2} = \mathbf{5.66}$$

Percentages of flow in the pipes are given by 5.66 : 1, *i.e.*, 5.66/6.66 × 100 and 1/6.66 × 100 *i.e.*, 85% and 15%.

When pipes are placed in series, loss of head

$$H = \frac{f\,L\,Q^2}{12D^5} + \frac{f\,L\,Q^2}{12\,d^5}$$

$$H = \frac{f\,L\,Q^2}{12}\left(\frac{1}{D^5} + \frac{1}{d^5}\right)$$

If $$d = 0.5\,D,$$

$$H = \frac{f\,L\,Q^2}{12\,D^5}\left(1 + \frac{1}{(0.5)^5}\right) = 33\frac{f\,L\,Q^2}{12D^5}$$

Also, $$h = \frac{f\,L\,Q_1^2}{12\,D^5} = \frac{f\,L(0.85\,Q)^2}{12D^5} = 0.722\frac{f\,L\,Q^2}{12D^5}$$

Then, $$\frac{h}{H} = \frac{0.722}{33} = \mathbf{0.02};$$ ***i.e.*, one fiftieth!**

Example 10.32. *A pumping plant forces water through a 600 mm diameter main, with 27 m head loss. If another pipe, similar the existing one in every respect is laid parallel in the existing pipe to reduce the friction loss to 9.6 m only, what should be the diameter of the new pipe?*

Solution.

We know that, $$h_f = \frac{f\,l\,Q^2}{12d^5}.$$

For the given pipe, $$27 = \frac{f\,l}{12}\frac{Q^2}{(0.6)^5};\quad \frac{f\,l\,Q^2}{12} = 2.1$$

When the new pipe of diameter D is placed in parallel, head loss in each pipe will become 9.6 m. So, in the existing pipe, the discharge will become q, such that

$$9.6 = \frac{f\,l}{12}\frac{q^2}{(0.6)^5}$$

which means, $q^2/Q^2 = 9.6/27 = 0.3555$

and $q/Q = 0.6.$

Implying that the new pipe should carry only 40% of the water.

So, $$9.6 = \frac{f\,l}{12}\frac{(0.4\,Q)^2}{D^5}$$

$$D^5 = 2.1 \times \frac{0.16}{9.6} = 0.035$$

$$D = \mathbf{0.51\ m\ diameter.}$$

Example 10.33. *An oil pipeline 60 cm in diameter and roughness height 0.00005 m carries a flow of 0.55 cumecs across a country. The pumping stations are located at every 80 km. If the pump efficiency is 85%, determine the power input required at each station when the kinematic viscosity and specific gravity of the oil are 2 × 10⁻⁶ m²/s and 0.88 respectively.*

Solution.

From the data,

$$l = 80\text{ km} = 80{,}000\text{ m};\ d = 0.6\text{ m}$$

$$Q = 0.55\text{ cumecs} = 0.55\text{ m}^3/\text{s}$$

$$\nu = 2 \times 10^{-6}\text{ m}^2/\text{s};\ \rho = 0.88 \times 1000 = 880\text{ kg/m}^3.$$

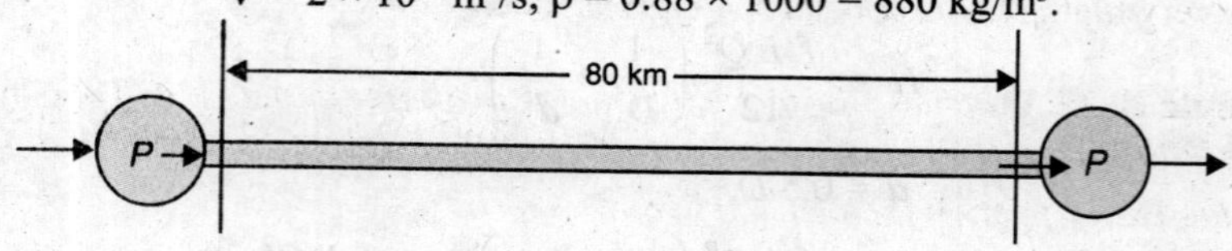

Velocity in pipe $= Q/A = \dfrac{0.55 \times 4}{\pi \times (0.6)^2} = 1.95$ m/s

Reynold's number $= Re = \dfrac{u d \rho}{\mu} = 1.95 \times 0.6/(2 \times 10^{-6})$

$= 0.585 \times 10^6$

Looking at the Stanton (or Moody) diagram for smooth pipes (because given roughness is very small), $f = 0.013$.

Head loss for energy 80 km is

$$h_f = \frac{0.013 \times 80{,}000}{0.6} \times \frac{1.95^2}{2 \times 9.81} = 336\text{ m}$$

Power input every 80 km is

$$P = \rho g Q h_f = 880 \times 9.81 \times 0.55 \times 336 = 1.6 \times 10^6\text{ W}$$

$$= \mathbf{1600\ kW.}$$

Example 10.34. *A fluid of viscosity 0.833 Pas and sp. gravity 1.26 is pumped along a horizontal pipe of 65 m length and 10 cm diameter at a flow rate of 0.18 m³/s. Determine the Reynolds number and calculate the head loss in the pipe if the flow is assumed laminar.*

Solution.

From the data,

$$Q = 0.18\text{ m}^3/\text{s},\ \mu = 0.883\text{ Pas}$$

$$A = \frac{\pi \times 0.1^2}{4} = 0.00785 \text{ m}^2$$

$$U = \frac{Q}{A} = \frac{0.18}{0.00785} = 22.9 \text{ m/s}$$

$$Re = \frac{U\, d\rho}{\mu} = \frac{22.9 \times 0.1 \times 1.26 \times 1000}{0.883}$$

$$= 3270$$

Assuming the flow to be laminar,

$$f = \frac{64}{Re} = \frac{64}{3270} \approx 0.02$$

The head loss is given by

$$h_f = \frac{f\, l}{d} \cdot \frac{U^2}{2g} = \frac{0.02 \times 65}{0.1} \times \frac{22.9^2}{2 \times 9.81}$$

$$= \mathbf{340 \text{ m}}$$

This is a very significant loss! From the value of viscosity, it appears to be a highly viscous oil and the diameter of the pipe is too small for pumping through a long distance of 65 m. The required discharge is very high, resulting in such a high head loss!

Example 10.35. *Calculate the maximum power available at the far end of a hydraulic pipeline 4.8 km long, 200 mm in diameter when water at 6.9 MPa pressure is fed into it at the near end. Take the Darcy friction factor f = 0.028.*

Solution.

For the maximum power available at the exit, the friction loss must be one third of the head available at the inlet,

$$h_f = H_1/3$$

From the data $\quad p_1 = 6.9 \times 10^6 \text{ Pa} = \rho g H_1$

Hence $\quad H_1 = 6.9 \times 10^6/(1000 \times 9.81) = 703.4$ m of water.

$$h_f = 703.4/3 = 234.47 \text{ m}$$

$$= \frac{f.l.Q^2}{12\, d^5}$$

whence $\quad Q^2 = 6.7 \times 10^{-3};$

$$Q = 0.082 \text{ m}^3\text{/s.}$$

Available head at the far end

$$H_2 = 2/3\, H_1 = 469 \text{ m of water}$$

Available power at the far end

$$p = \rho.\, g.Q.\, H_2$$

$$= 1000 \times 9.81 \times 0.082 \times 469$$

$$= 377272 \text{ W} = \mathbf{377.272 \text{ kW.}}$$

Example 10.36. *A liquid of sp. gravity 1.5 is discharged from a tank through a siphon whose summit point is 1.2 m above the liquid level in the tank. The siphon pipe has a diameter of 10 cm and it discharges the liquid into atmosphere at 100 kPa. If the vapour pressure of the liquid is 28 kPa, how far below the liquid level in the tank can the outlet be safely located? Estimate the discharge, neglecting loss of head.*

Solution.

Let us take the datum level as the level in the tank; at point 1 in the pipe $z = 0$.

At point 2, $z = 1.2$ m;

and at point 3, $z = -h$.

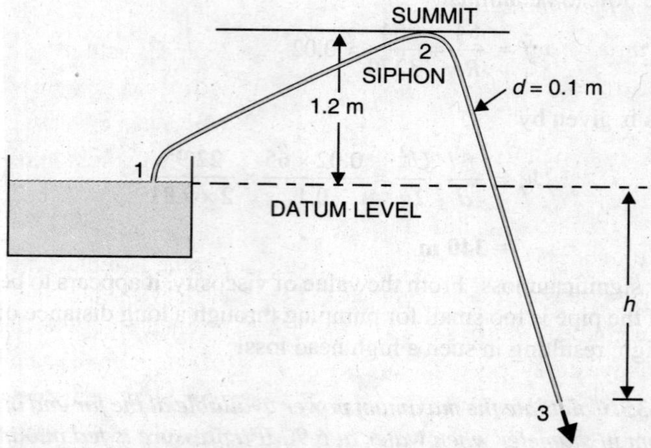

The pressure at point 2 cannot be below 28 kPa;

$$p_2 = 28 \text{ kPa}.$$

In the absence of all head losses, we apply the Bernoulli's equation between points 1, 2 and 3.

$$\frac{p_1}{5\rho g} + \frac{U_1^2}{2g} = \frac{p_2}{\rho g} + \frac{U_2^2}{2g} + 1.2 = \frac{p_3}{\rho g} + \frac{U_3^2}{2g} - h$$

For the pipe 2-3

$$U_2 = U_3 = U;$$

$$p_1 = p_3 = 100 \text{ kPa};$$

and $\rho = 1500$ kg/m³.

$$\frac{100 \times 1000}{1500 \times 9.81} + 0 = \frac{28 \times 1000}{1500 \times 9.81} + \frac{U^2}{2 \times 9.81} + 1.2$$

$$= \frac{100 \times 1000}{1500 \times 9.81} + \frac{U^2}{2 \times 9.81} - h$$

whence $h = 3.7$ m (from the equation between 2 and 3)

and $U = 8.52$ m/s. (from the equation between 1 and 3)

Discharge $Q = 8.52 \times 0.1^2 \times \frac{\pi}{4} = \mathbf{0.067\ m^3/s}$

$= 0.067 \times 1.5 \times 1000 = \mathbf{100\ kg/s.}$

ITEM BANK

Objective items of different types are given to review the concepts developed in the Chapter.

* Select the correct response for each item.
* Every correct response is expected in less than a minute and carries one mark.
* Compare your responses with the key answers given at the end.

Multiple-Choice Items

Attempt the following Multiple-Choice Items by ticking the correct or the most appropriate response from among the alternatives given :

1. Friction loss through a pipe now implies
(*a*) loss of energy due to static coefficient of friction
(*b*) loss of energy due to dynamic coefficient of friction
(*c*) loss of flow rate in a pipe due to surface roughness
(*d*) loss of energy due to surface roughness

2. The hydraulic gradient line is
(*a*) always above the energy gradient line
(*b*) always sloping downward in the direction of flow
(*c*) velocity head below the energy gradient line
(*d*) velocity head above the energy gradient line.

3. For two pipes arranged in series,
(*a*) the flow may be different in the two pipes
(*b*) the head loss per unit length must be more in the smaller pipe
(*c*) the velocity must be the same in both pipes
(*d*) the head loss must be the same in both pipes.

4. In parallel-pipe systems,
(*a*) the pipes be placed geometrically parallel to each other
(*b*) the flow must be equal in all pipes
(*c*) the head loss per unit length must be the same in all pipes
(*d*) the head loss across each of the parallel pipes must be the same.

5. A pipe is said to be a siphon if
(*a*) it can operate without external power
(*b*) it can discharge a liquid at a fast rate
(*c*) it has subatmospheric pressure in it
(*d*) the exit is at a lower level than the inlet.

6. A pipe is replaced by two parallel pipes, each with half the cross section of the original pipe. The discharge will
(*a*) remain the same
(*b*) increase by more than 10%
(*c*) decrease by more than 10%
(*d*) change by less than 5%.

7. If it is required to convey the same discharge by replacing a pipe of diameter D by two equal parallel pipes of diameter d then d/D should be
(*a*) 0.37 (*b*) 0.42
(*c*) 0.50 (*d*) 0.76

8. A pipeline is said to be equivalent to another if, in both,
(*a*) length and discharge are the same
(*b*) velocity and diameter are the same
(*c*) discharge and frictional head loss are the same
(*d*) length and diameter are the same.

9. The head loss in a sudden pipe expansion from area A_1 to area A_2 and from velocity U_1 to velocity U_2 is given by
(*a*) $(1 - A/A_2)^2 . U_1^2/2g$
(*b*) $(1 - A_1/A_2)^2 . U_2^2/2g$
(*c*) $(1 - A_2/A_1)^2 . U_2^2/2g$
(*d*) $(1 - A_2/A_1) . U_1^2/2g$

10. For pipes (1, 2,...) connected in series,

(*a*) $U = U_1 + U_2 + ...$

(*b*) $f = f_1 + f_2 + ...$

(*c*) $h_f = h_{f_1} + h_{f_2} + ...$

(*d*) $Q = Q_1 + Q_2 + ...$

11. For pipes (1, 2,...) connected in parallels,

(*a*) $U = U_1 + U_2 + ...$

(*b*) $f = f_1 + f_2 + ...$

(*c*) $h_f = h_{f_1} + h_{f_2} + ...$

(*d*) $Q = Q_1 + Q_2 + ...$

12. The condition for maximum power transmission through a pipe is that the frictional head loss should be the following fraction of the available head :

(*a*) Zero (*b*) 0.333

(*c*) 0.50 (*d*) 0.667

Other Objective Items

13. Are the following statements true or false?

(*a*) The pressure gradient in a fully developed pipe flow is constant (T/F)

(*b*) Friction factor for turbulent flow is more than that in laminar flow (T/F)

(*c*) Minor losses in pipe fittings can be expressed as equivalent length of pipe (T/F)

(*d*) Energy line and hydraulic gradient line cannot intersect each other (T/F)

14. Assertion *A* : Total energy line and the hydraulic gradient line for a pipe flow cannot intersect each other.

Reason *R* : The vertical difference between these two lines must remain equal to velocity head. Which one of the following is right ?

(*a*) *A* is true ; *R* is true and *R* explains *A*.

(*b*) *A* is true ; *R* is true but *R* does not explain *A*.

(*c*) *A* is true ; *R* is false

(*d*) *A* is false ; *R* is true.

15. Assertion *A* : Head loss for a sudden expansion is more than the head loss for a sudden contraction for the same diameter ratio.

Reason *R* : Head loss varies as the square of the velocity in a pipe. Which one of the following is right ?

(*a*) *A* is true ; *R* is true and *R* explains *A*.

(*b*) *A* is true ; *R* is true but *R* does not explain *A*.

(*c*) *A* is true ; *R* is false.

(*d*) *A* is false ; *R* is true.

16. Match the *K* factors in block *B* with the type of transitional fitting in block *A*.

Block A	Block B
(*a*) sudden contraction	(*k*) 0.05
(*b*) Bellmonth entrance	(*l*) 0.20
(*c*) Elbow	(*m*) 0.50
(*d*) Open gate valve	(*n*) 1.00
	(*o*) 2.00

ANSWERS

1. (d), **2.** (c), **3.** (b), **4.** (d), **5.** (c), **6.** (c), **7.** (d), **8.** (c), **9.** (a), **10.** (c), **11.** (d), **12.** (b), **13.** (T, F, T, T), **14.** (a), **15.** (b), **16.** (a m, b k, c n, d l)

QUESTION BANK

Questions on theory and numerical problems are given for tutorial work.

* *The questions are arranged generally in order of the articles in the text.*

* *Every question should take 15-20 minutes to solve.*

* *Answers for the numerical problems are given along side them.*

10.1 Calculate the drop in pressure accompanying the flow of 10 m^3/s of air at standard conditions through a duct 0.5 m diameter, 100 m long. Take $f = 0.02$. **[6225 N/m^2]**

10.2 Two reservoirs are connected by three pipes in series

	l	d
first pipe	300 m	30 cm
second pipe	250 m	25 cm
third pipe	200 m	20 cm

Assuming $f = 0.02$, 0.025 and 0.03 respectively, calculate the discharge through them if the elevation difference of the levels is in the reservoirs is 20 m. Neglect minor losses.

[0.0936 m³/s]

10.3 Calculate (*a*) the discharge and (*b*) the power necessary to run the pump for a 5 cm diameter pipe line supplying water from a river to a factory reservoir level which is 20 m above the river free surface. The pipe line ($f = 0.02$ m) is 200 m long and incorporates 5 bends ($K = 0.3$ each), one strainer ($K = 3$), one non-return valve ($K = 2$) and one gate valve ($K = 0.2$). The efficiency of the pump is around 40% at this load. **[0.004 m³/s, 2.0 kW]**

10.4 Two reservoirs are connected by three pipes in series constituting a siphon operating at a head differential of 10 m.

	l	d
first pipe	500 m	20 cm
second pipe	200 m	10 cm
third pipe	100 m	10 cm

Calculate the discharge, neglecting minor losses. Draw also the total energy and hydraulic gradient lines for the system. Take $f = 0.02$. **[0.014 m³/s]**

10.5 Two reservoirs are connected by three pipes in parallels.

	l	d
first pipe	100 m	20 cm
second pipe	200 m	20 cm
third pipe	100 m	10 cm

The total discharge is 0.2 m/s³. Calculate (*a*) the difference in elevations of the free surfaces in the reservoirs if $f = 0.02$ for each pipe, (*b*) The slope of the hydraulic gradient line between the reservoirs placed 100 m apart. **[5.82 m, 0.0582]**

10.6 A single pipe 30 cm diameter, 300 m long conveys a discharge of 0.1 m³/s. What length of another 40 cm diameter pipe is required to be placed in parallel with the existing pipe in order to augment the discharge by 30% ? Take f for each pipe = 0.015 and neglect the minor losses.

[137.24 m]

10.7 A power-transmission pipe line, 30 cm diameter, 500 m long operates between a reservoir maintained at 60 m head and a Pelton-turbine nozzle. Assuming $f = 0.03$, calculate

(*a*) the maximum power which can be transmitted and

(*b*) the nozzle diameter to do so. **[77.3 kW, 9.5 cm]**

10.8 A pipe, 30 cm diameter 3000 m long conveys a maximum of 0.20 m³/s from one tank to another with a surface elevation difference of 30 m. Calculate the increase in maximum flow rate if another similar pipe of the same size is looped parallel to it and connected to it over

(*a*) one third the length,

(*b*) two third the length and

(*c*) the entire length. **[15.5%, 41.4%, 100%]**

10.9 A 20 cm diameter pipe 200 m long discharges oil from a tank into atmosphere. At the midpoint of the pipe length, the gauge pressure is one and a half times the atmospheric pressure. The specific gravity of the oil is 0.9 and viscosity 0.02 Ns/m². Calculate

(*a*) the discharge of oil and

(*b*) the gauge pressure in the tank at the inlet of the pipe. **[0.18 m³/s, 303 kN/m²]**

10.10 Calculate the time for a liquid to fall in a tank from height of 9 m to a height of 4 m above the level of the exit end of an outlet pipe with the following data :
Tank diameter 1 m, pipe diameter 0.02 m, f = 0.032, pipe length = 25 m. **[2.04 hours]**

10.11 (*a*) Derive an expression for the loss due to sudden enlargement in pipe flow and therefrom deduce the loss due to sudden contraction.

(*b*) At a sudden enlargement of a water line from 240 mm to 480 mm diameter pipe, the hydraulic gradient rises by 10 mm. Estimate the rate of flow.

(*c*) Show how a pipe bend can be used as a flow measuring device. **[0.025 m³/s]**

10.12 A pipe 7.5 cm in diameter and 250 m long has a nozzle fitted at the discharge end. The total at the inlet to the pipe is 40 m. Taking f = 0.01 for the pipe determine the maximum power transmitted. **[732.74 kW]**

10.13 Three pipes of 400 mm, 200 mm and 300 mm diameters and having lengths of 400 m, 200 m and 300 m, respectively are connected in series to make a compound pipe. The ends of this compound pipe are connected with two tanks whose difference of water levels is 16 m. If the friction factor f for these pipers is same and equals to 0.02, determine the discharge through the compound pipe considering both the major and minor losses (take coefficient of contraction C_c = 0.6). **[0.108 m³/s]**

10.14 A pipe system consists of three pipes connected in series (*i*) 300 m long, 15 cm in dia (*ii*) 150 cm long, 10 cm in dia (*iii*) 250 m long 20 cm in dia. Determine the equivalent length of 12.5 cm diatmeter pipe. Take f = 0.01 for each pipe. **[602 m]**

10.15 Determine the difference in water levels of two reservoirs both connected by a pipe system consisting of three pipes laid in series, each 700 metre long, but of diameters 45, 30 and 60 cm respectively-starting from the end of the pipe system connected to the higher reservoir. The discharge through the pipe is 150 litres/s. The minor losses may be neglected. Assume f = 0.01. **[6.224 m]**

10.16 Two reservoirs are connected by a long pipe 30 cm diameter carrying 0.14 m³/s of water. If another parallel pipe of the same material is laid beside this one to carry 0.28 m³/s of water, find its diameter. Neglect minor losses. **[0.387 m]**

10.17 Two reservoirs having water level difference of 20 m are connected by a pipe 300 mm in diameter and 4000 m long. At a distance 1500 m from the supply reservoir, a leakage of 10 litres/s is noticed. Calculate the discharge to the lower reservoir. Assume f = 0.02.

CHAPTER

11

Flow Through Open Channels

LEARNING OBJECTIVES

On successful completion of all learning activities planned and provided in the chapter, a student will be able to

- develop an expression for the discharge for uniform flow in an open channel;
- obtain and employ the condition for maximum discharge;
- draw and interpret the specif c energy diagram for a free-surface flow; and
- describe the formation of a hydraulic jump and estimate the variation of Froude number and other parameters across a jump.

11.1. Uniform Flow: Chezy Equation

A liquid flowing with a free surface exposed to a constant pressure throughout constitutes an important class of flows with numerous applications in engineering. This is known as open-channel flow irrespective of the fact whether the channel is open or closed as long as its surface is exposed to constant pressure. Examples of free-surface flows are natural water courses, canals, water falls, irrigation drains and sewer lines.

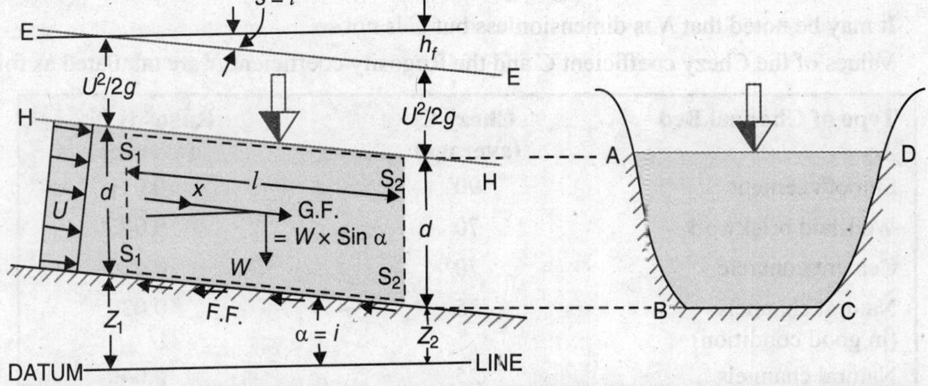

Fig. 11.1: Uniform Flow through an Open Channel

The motive force causing a free surface flow is the gravity force due to a drop in the total head, often by a slope of the bed of the channel. Although the actual flow is three-dimensional, non-uniform and unsteady in character, a fair approximation is possible if the flow is considered to be one-dimensional and uniform over each cross section and steady with respect to time. Under these assumptions, consider the flow represented in Fig. 11.1. The flow is considered fully developed, i.e., uniform over each cross-section as well as along the direction of flow. *The total energy line EE is parallel to the free surface at a distance of $U^2/2g$ above it* and the slope *s of* the total energy line is equal to the bed slope *i* for uniform flow. *The hydraulic gradient line HH is the free surface itself, whatever be the nature of flow.*

The flow can be in equilibrium if the gravity force along the flow in a control volume $S_1S_1S_2S_2$ balances the frictional forces due to shearing at the bed and the walls. The wetted perimeter *P* over which shearing occurs is *ABCD.* The hydrostatic forces on the control surface S_1S_1 and S_2S_2 cancel off mutually.

Weight of the fluid in the control volume.

$$W = \rho g.Al$$

Gravity Force, *G.F.,* parallel to the channel bed.

$$G.F. = W \sin \alpha = \rho g.Al \sin \alpha \quad \text{... (11.1)}$$

Frictional force

$$F.F. = \frac{f}{4}.Pl. \frac{\rho U^2}{2} \quad \text{... (11.2)}$$

Equating the two

$$U = C\sqrt{mi} \quad \text{... (11.3)}$$

where

$$C = \sqrt{\frac{8g}{f}} = \text{Chezy coefficient}$$

$$m = \frac{A}{P} = \text{hydraulic mean depth}$$

$i = \tan \alpha \approx \sin \alpha \approx \alpha$, the slope of the bed

f = friction coefficient

P = wetted perimeter of the cross section

l = length of the section of the channel

and A = Area of cross section of flow.

The Chezy coefficient *C* can be evaluated with reference to an expression due to Manning.

which states that

$$C = \frac{1}{N} m^{\frac{1}{6}} \quad \text{...(11.4)}$$

where N = Rugosity coefficient

It may be noted that *N* is dimensionless but *C* is not so.

Values of the Chezy coefficient *C* and the Rugosity coefficient *N* are tabulated as follows :

Type of Channel Bed	Chezy C (average)	Rugosity N (average)
Smooth cement	90	0.011
Well-laid brickwork	70	0.015
Cement concrete	70	0.015
Natural channels (in good condition)	35	0.030
Natural channels (in bad condition)	25	0.040

Likewise, the Manning's formula for velocity in an open channel is expressed as

$$U = \frac{1}{N} m^{2/3} i^{1/2}$$

The discharge through it is given by

$$Q = \frac{A}{N} m^{2/3} i^{1/2}$$

which must be the same as $Q = AC\sqrt{mi}$

Since the value of Chezy's constant is more difficult to determine than the rugosity coefficient N, the Manning's formula is more commonly employed. The value of N for a channel is selected through inspection and experience by taking into account surface roughness, vegetation, irregularities silting and scouring, etc.

Now, for a constant area of flow, A, and for a given slope i of the channel bed, the discharge Q is a monotonic function of the hydraulic mean depth m. Maximum Q corresponds to maximum m.

Since $$m = \frac{A}{P} \propto \frac{1}{P}$$

it shows that the maximisation of discharge requires the minimisation of the wetted perimeter of the channel for a given area of flow.

It is interesting to note that the perimeter P is a function of the geometrical variable of the channel cross-section. Minimisation of P implies that

$$\frac{\partial P}{\partial g_1} = 0 = \frac{\partial P}{\partial g_2} = \frac{\partial P}{\partial g_3} \ .....$$

where g_1, g_2 .... are the geometric variable parameters defining the cross-section. The cross-section corresponding to the minimum perimeter for a given flow area A is called the most hydraulically 'economical' the 'most efficient", or the 'best' section.

In the foregoing analysis it was assumed that the flow is uniform at each cross-section. In practice, it may not be so. The velocity profile in a fully-developed turbulent flow may nearly be uniform in the core region but the velocity at the base and at the walls must be zero. The velocity at the surface is influenced by the air resistance and may be governed by the wind velocities. A typical actual velocity profile in a turbulent flow is shown in Fig. 11.2. It may also be noted that the roughness of the surface influences the velocity profile. In general, the velocity profile is more nearly uniform for a rough surface than for a smooth surface. The assumption of uniform velocity is, therefore, not far from reality for most purposes.

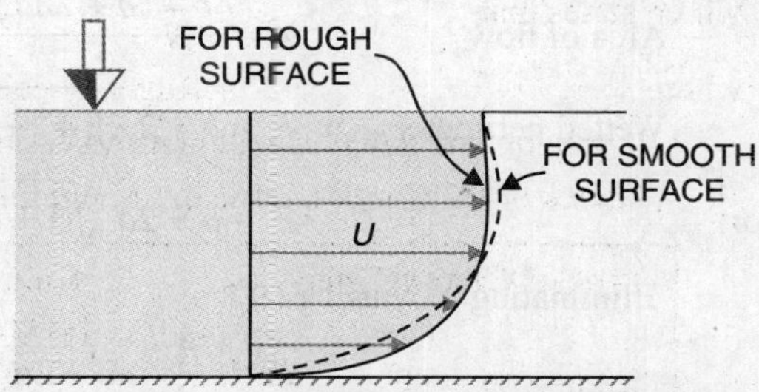

Fig. 11.2: Velocity Profiles for Rough and Smooth Surfaces

The fact that the flow is turbulent and not laminar may be established by verifying that the Reynolds number exceeds a critical value. Recalling that the Reynolds number is based upon the diameter D of the pipe in a pipe flow. For a pipe running full, the area of flow is

$$A = \frac{\pi D^2}{4}$$

and the wetted perimeter is

$$P = \pi D$$

The hydraulic mean depth m for a pipe running full is, therefore,

$$m = \frac{A}{P} = \frac{\pi D^2/4}{\pi D} = \frac{D}{4}$$

or $$D \equiv 4\,m$$

The diameter may, therefore, be replaced by four times the hydraulic mean depth of a channel;

$$Re = \frac{UD\rho}{\mu} = \frac{U(4m)\rho}{\mu}$$

Equating it to 2000, it is observed that the value of the critical Reynolds number on the basis of the hydraulic mean depth is 500;

$$Re = \frac{Um\rho}{\mu} = 500$$

for critical flow through open channels.

Consider, for example, a small channel 1 m wide, 0.5 m deep with water flowing at 2 m/s. The hydraulic mean depth is given by

$$m = \frac{1 \times 0.5}{(1 + 2 \times 0.5)} = 0.25 \text{ metre}$$

the Reynold number is given by

$$Re = \frac{2 \times 0.25 \times 1000}{0.001} = 500 \times 10^3$$

which is far more than the critical value of 500.

Flow through open channels is turbulent except for very small shallow channels and models for laboratory tests. Artificial roughness is often provided in models to create turbulence for the sake of similarity.

Example 11.1. *Develop the following set of conditions for the maximum discharge through a channel of trapezoidal section for a given cross-sectional area and a fixed bed slope.*

$$d\sqrt{(1+n^2)} = \frac{b + 2nd}{2}$$

and $$n = \frac{1}{\sqrt{3}} \text{ or } m = \frac{d}{2}$$

Interpretand Simplify these conditions for a rectangular channel and for a triangular channel.

Solution.

For a trapezoidal section as shown,

Area of flow $$A = \frac{b + (b + 2nd)}{2} \cdot d = (b + nd)d \quad \text{...(i)}$$

Wetted perimeter $$P = b + 2\sqrt{(d^2 + n^2 d^2)}$$

or $$P = b + 2d\sqrt{(1+n^2)} \quad \text{...(ii)}$$

Eliminating b by using (*i*)

$$P = \frac{A}{d} - nd + 2d\sqrt{(1+n^2)}$$

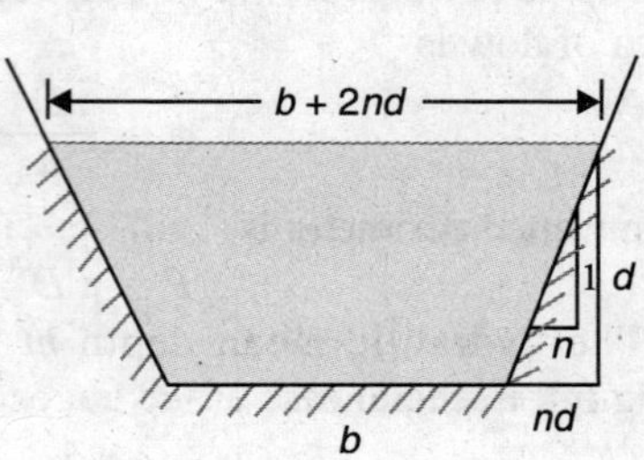

Discharge $$Q = AU = AC\sqrt{mi}$$

$$= AC\sqrt{\frac{A}{P}i} \propto \sqrt{\frac{1}{P}}$$

if A, C and i are otherwise fixed, as is implied in the problem.

Maximum Q corresponds to minimum P.

This condition requires that

$$\frac{\partial P}{\partial d} = 0 \text{ and } \frac{\partial P}{\partial n} = 0$$

(*a*) Firstly,

$$\frac{\partial P}{\partial d} = 0 = -\frac{A}{d^2} - n + 2\sqrt{(1+n^2)}$$

Substituting $A = (b + nd)\,d$ and simplifying,

$$d\sqrt{(1+n^2)} = \frac{b+2nd}{2} \qquad \text{... (iii)}$$

which is interpreted by saying that 'the sloping side equals half the top width'.

Further, for this section,

$$m = \frac{A}{P} = \frac{(b+nd)d}{b+2d\sqrt{(1+n^2)}}$$

or

$$m = \frac{(b+nd)d}{b+b+2nd} = \frac{d}{2} \qquad \text{...(iv)}$$

implying that the hydraulic mean depth equals half the actual depth.

(*b*) Now,

$$\frac{\partial P}{\partial n} = 0$$

for constant or optimum depth, requires that

$$d + 2d\,.\,\frac{1}{2}\left(1+n^2\right)^{1/2} \times 2n = 0$$

or

$$2n = -\sqrt{1+n^2}$$

or

$$4n^2 = 1 + n^2;\ n = \frac{1}{\sqrt{3}} \qquad \text{...(v)}$$

which interprets to a 'side slope of 60° with the horizontal'.

An important comment here is that the ultimate maximum discharge through a trapezoidal channel demands the fulfilment of both the above conditions but in practice, only one of the two conditions may be invoked, the other being assigned by the design and feasibility considerations.

For a Rectangular Channel

It is easy to visualise a rectangular channel as a special case of the trapezoidal channel when $n = 0$

From conditions (*iii*) and (*iv*) for a rectangular channel,

$$d = \frac{b}{2}$$

and

$$m = \frac{d}{2}$$

which means that the depth should be 50% of the width and the hydraulic mean depth is half of the depth.

For a Triangular Channel

Another special case of the trapezoidal channel is a triangular channel when

$$b = 0$$

Then

$$A = nd^2$$

and $$P = 2d\sqrt{1+n^2} = 2\sqrt{A}\sqrt{\frac{1+n^2}{2}}$$

For minimum P $$\frac{\partial P}{\partial \theta} = 0$$

or $$\frac{\sec\theta\tan\theta}{\sqrt{\tan\theta}} - \frac{\sec^3\theta}{2(\tan\theta)^{1.5}} = 0$$

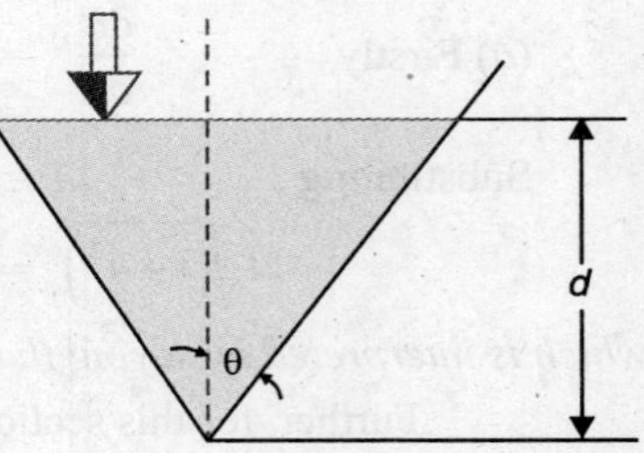

Rearranging and simplifying the terms,

$$2\tan^2\theta = \sec^2\theta$$

or $$\sin\theta = \frac{1}{\sqrt{2}}$$

and $$\theta = 45°$$

for the maximum discharge through a triangular channel.

It means that the maximum discharge occurs in a right-angled triangular channel.

Example 11.2. *Compute the dimensions of the 'most economical' trapezoidal section with 1:1 side slope to carry 10 m³/s of water on a cement concrete bed slope of 1 in 1600:*

Solution.

From the data, $n = 1;\ i = \frac{1}{1600};\ Q = 10\ \text{m}^3/\text{s}$

By condition (*iii*), above example,

$$\sqrt{2}d = \frac{b+2d}{2}$$

or $$b = 0.83\,d \text{ and } A = 1.83\,d^2$$

By condition (*iv*),

$$m = \frac{A}{P} = \frac{d}{2}$$

By the Chezy equation,

$$Q = C\sqrt{mi} \times A$$

where $$C = \frac{1}{N}m^{1/6}$$

Taking $N = 0.015$ for a cement concrete bed,

$$Q = \frac{1}{0.015} \times \left(\frac{d}{2}\right)^{2/3} \times \left(\frac{1}{1600}\right)^{1/2} \times 1.83d^2$$

$$10 = 1.92\,d^{8/3} \text{ or } d = \mathbf{1.86\ m}$$

and $$\mathbf{b = 0.83\,d = 1.54\ m}$$

These dimensions are in accordance with the maximisation of the discharge by the minimisation of the wall shear forces or the wetted perimeter for a required area of flow and the given slope of the bed material.

Example 11.3. *Determine the expressions for the most economical depths of flow of water in terms of the diameter of a channel of circular cross-section (a) for maximum velocity and (b) for maximum discharge. Comment on their relative importance.*

Solution.

Consider an open channel of diameter D and radius R with a depth Z of water in it.

The area of flow is given by

$$A = \text{area of sector } AOBPA + \text{area of triangle } AOB$$
$$= \frac{1}{2}R^2 . 2\theta + \frac{1}{2}.2R\sin(180° - \theta) . R\cos(180° - \theta)$$
$$= R^2\theta + R^2 \sin\theta \cos\theta$$
$$= R^2 (\theta + \frac{1}{2}\sin 2\theta)$$

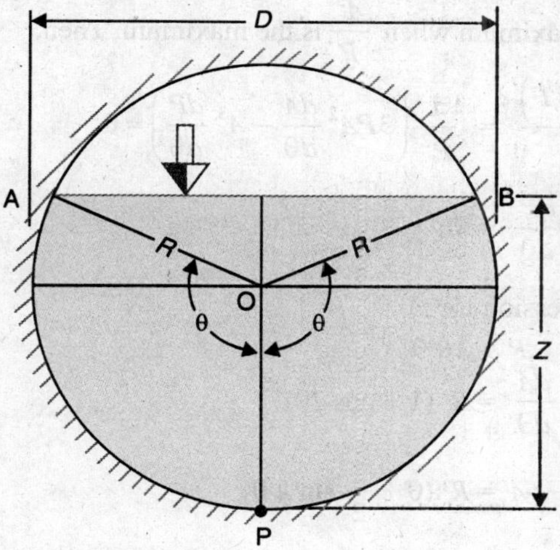

The wetted perimeter is given by

$$P = 2R.\theta$$

and the hydraulic mean depth is

$$m = \frac{A}{P} = \frac{R\left(\theta + \frac{1}{2}\sin 2\theta\right)}{2\theta}$$

(*a*) The condition for maximum velocity is that the hydraulic mean depth *m* should be the maximum;

which means that $\frac{\partial m}{\partial \theta} = 0$

$$= \frac{1}{4\theta^2}\left[2\theta.(R + \frac{R}{2}.2.1\cos 2\theta) - R\,\theta + \frac{1}{2}\sin 2\theta).2\right] = 0$$

whence $\theta \cos 2\theta - \frac{1}{2}\sin 2\theta = 0$

or $\tan 2\theta = 2\theta$

which provides the answer.

$$\theta = \mathbf{128.75°}$$

The corresponding depth of flow is given by

$$Z = R + R\cos(180 - \theta)$$
$$= R + R\cos(180 - 128.75)°$$
$$= R(1 + \cos 51.25°) = 1.63\,R$$
$$= \mathbf{0.81\,D}$$

The optimum depth of flow for a circular open channel is, therefore, 81% of its diameter as far as the maximisation of velocity is concerned.

(*b*) The condition for maximum discharge is that the expression for discharge, *i.e.*,

$$Q = AC\sqrt{mi}$$

should be maximised

Since $$m = \frac{A}{P}$$

the discharge will be the maximum when $\frac{A^3}{P}$ is the maximum. Then,

$$\frac{d\left(A^3/P\right)}{d\theta} = \frac{1}{P^2}\left(3PA^2\frac{dA}{d\theta} - A^3\frac{dP}{d\theta}\right) = 0$$

or $$3P\frac{dA}{d\theta} - A\frac{dP}{d\theta} = 0$$

Substituting the expression for *P*,

$$P = 2R\,\theta$$

$$\frac{dA}{d\theta} = R^2\,(1 + \cos 2\theta)$$

$$A = R^2\,(\theta + \frac{1}{2}\sin 2\,\theta)$$

and $$\frac{dP}{d\theta} = 2R$$

the condition becomes,

$$6\,R^3\theta\,(1 - \cos 2\theta) - 2\,R^3\,(\theta - \frac{1}{2}\sin 2\theta) = 0$$

or $$4\,\theta - 6\theta\cos 2\theta + \sin 2\theta = 0$$

which provides the answer

$$\boldsymbol{\theta = 154°}$$

The corresponding depth of flow is given by

$$Z = R[1 + \cos(180 - 154)°]$$

$$= R\,(1 + 0.90) = 1.90\,R$$

$$= \mathbf{0.95\,D}$$

The optimum depth of flow for a circular open channel is, therefore, 95% of its diameter as far as the maximisation of discharge is concerned.

Comments. In a channel of circular cross section, the area of flow and wetted perimeter increase with the depth of flow in a manner such that

(*i*) as the level of free surface rises, a point is reached beyond which the wetted perimeter increases very rapidly in comparison with the area of flow. The velocity and the hydraulic mean depth are the maximum at that state and decrease thereafter.

(*ii*) as the level rises further, another point is reached beyond which the increase in area is more than offset by the reduction in the mean velocity. The discharge is the maximum at that state and decreases thereafter.

The hydraulic mean depth for the maximum discharge is given by substituting

$$\boldsymbol{\theta = 154° = 2.69\ \text{rad}}$$

in the expression for m ;

$$m = \frac{R^2 \,(2.69 + \theta \sin 2\theta)}{2R \times 2.69} = 0.573R$$

The hydraulic mean depth for the channel flowing full would be

$$m = \frac{\pi R^2}{2\pi R} = 0.5\, R$$

The area of flow at the state of maximum discharge is given by

$$A = R^2 \left(2.69 + \frac{1}{2}\sin 2\theta\right) = 3.084R^2$$

and the area of the section for the channel flowing full is given by

$$A = \pi\, R^2 = 3.1416\ R^2$$

$$\frac{\text{Discharge when running full}}{\text{Maximum discharge}} = \frac{3.1416\sqrt{0.5}}{3.084\sqrt{0.573}} = 0.95$$

The interesting fact brought out as above may be stated as follows:

A circular channel, flowing full carries only 95% of the discharge that it would carry when running up to a depth of 95% of diameter of the channel!

11.2. Specific Energy and Momentum Relations

The energy per unit width at a section of a free surface flow with respect to the bed of the channel is called the specific energy. If the depth of the liquid in a wide rectangular channel is d and the velocity of flow is U, then the specific energy is given by

$$E_s = d + \frac{U^2}{2g} = E_p + E_k \qquad ...(11.5)$$

where E_p = potential energy, due to the depth of the water referred to the channel bed d and the kinetic energy

$$E_k = \frac{U^2}{2g} = \frac{q^2}{2g\,d^2}$$

where q = discharge per unit width of the reectangular channel.

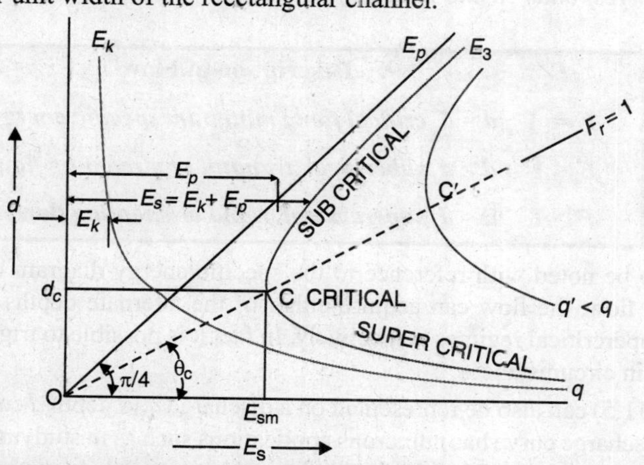

Fig. 11.3: Specific Energy Diagram

The expression for E_S becomes

$$E_S = d + \frac{U^2}{2g} = d + \frac{q^2}{2g\,d^2} \qquad \text{... (11.5)}$$

A plot of the specific energy E_s *vs* depth d is shown in Fig. 11.3. For minimum E_s,

$$\frac{\partial E_S}{\partial d} = 0 = 1 - \frac{q^2}{gd^2} = 1 - \frac{U^2}{gd} \qquad \text{... (11.6)}$$

Critical depth for the minimum E_s from Eq. (11.6)

$$d_c = \sqrt{\frac{q^2}{g}} \qquad \text{... (11.7a)}$$

and the corresponding critical velocity, also from Eq (11.6),

$$U_c = \sqrt{gd_c} \qquad \text{...(11.7b)}$$

The minimum specific energy E_{sm} is given by

$$E_{sm} = d_c + \frac{U_c^2}{2g} = d_c + \frac{gd_c}{2g} = \frac{3}{2}d_c \qquad \text{...(11.8)}$$

It also follows that, at the critical conditions,

$$E_p = \text{potential energy} = \frac{2}{3}E_{sm}$$

$$E_k = \text{Kinetic energy} = \frac{1}{3}E_{sm}$$

$$d_c = \frac{U_c^2}{g} = 2 \times \frac{U_c^2}{2g} = \text{twice the critical kinetic energy.}$$

A reference to the dimensional analysis would show that the Froude number F_r is defined as

$$F_r = \frac{U}{\sqrt{gd}}$$

It may be noted that the critical depth d_c corresponds to $F_r = 1$. It is also clear that this condition corresponds to the speed of propagation of a small disturbance in channel of depth d.

F_r	Description of Flow
$F_r = 1$	$d = d_c$ *critical flow; minimum specific energy*
$F_r < 1$	$d > d_c$ *subcritical, tranquil or streaming flow*
$F_r > 1$	$d < d_c$ *supercritical, rapid or shooting flow*

It may also be noted with reference to the specific energy diagram that for a given specific energy in a fluid flow, the flow can acquire either of the alternate depths d_1 and d_2 and be in the subcritical and supercritical regimes respectively. In fact it is possible to trigger from one type to the other under certain circumstances.

Equation (11.5) can also be represented on a discharge q *vs* depth d curve for constant specific energy E_s. The discharge curve has numerous applications such as in studying the flow under a sluice gate at different openings and in studying the flow over a mound. Assuming no loss of energy, the specific energy is maintained constant at different openings of the sluice gate. Expressing Eq. (11.5) as

$$q^2 = 2g(E_s d^2 - d^3) \qquad \text{...(11.9)}$$

a plot of q *vs* d for constant E_s appears as shown in Fig. 11.4.

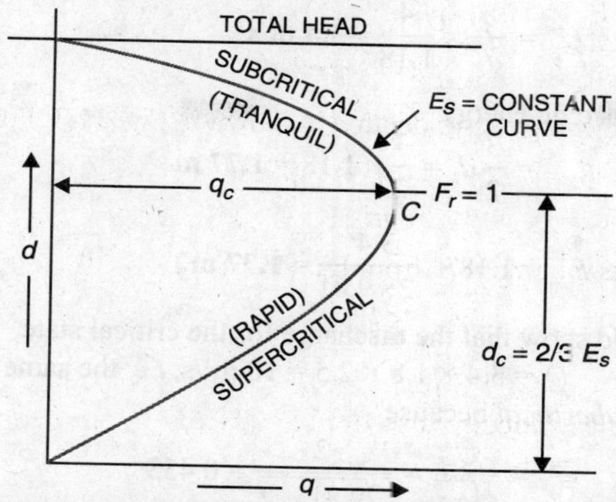

Fig. 11.4: Discharge Curve

For maximum discharge through a sluice gate, $\dfrac{\partial q}{\partial d} = 0$

or $$2q\,\frac{\partial q}{\partial d} = 2g\left|(2E_s d - 3d^2)\right|\; d = d_c$$

where U_c is the critical velocity.

Example 11.4. *Calculate the specific energy, critical depth and velocity for the flow of 10 m³/s in a cement lined rectangular channel 2.5 m wide with 2 m depth of water. Is the given flow subscritical or supercritical?*

Solution.

From the data, the discharge per unit width

$$q = \frac{10}{2.5} = 4 \text{ m}^3\text{/s per metre}$$

$$U = \frac{10}{2.5\times 2} = 2 \text{ m/s}$$

$$\frac{U^2}{2g} = \frac{2^2}{2\times 9.81} = 0.204 \text{ m}$$

Specific energy $$E_s = d + \frac{U^2}{2g} = 2 + 0.204 = \mathbf{2.204\ m}$$

Critical depth $$d_c = \left(\frac{q^2}{g}\right)^{1/3}$$

$$= \left(\frac{4^2}{9.81}\right)^{1/3} = \mathbf{1.18\ m}$$

and the critical velocity $$U_c = \sqrt{gd_c}$$

$$= \sqrt{(9.81\times 1.18)} = \mathbf{3.4\ m/s}$$

alternatively determined as

$$U_c = \frac{q}{d_c} = \frac{4}{1.18} = \mathbf{3.4\ m/s}$$

E_{sm}, the minimum specific energy

$$= \frac{3}{2}d_c = \frac{3}{2} \times 1.18 = \mathbf{1.77\ m}$$

alternatively determined as $E_{sm} = 1.18 + \dfrac{3.4^2}{2 \times 9.81} = \mathbf{1.77\ m}$

A quick check would show that the discharge for the critical state,

$Q = 3.4 \times 1.8 \times 2.5 = 10\ m^3/s$, *i.e.* the same as for the given state.

The given state is subcritical because

$$F_r = \frac{U}{\sqrt{gd}} = \frac{2}{\sqrt{9.81 \times 2}} = 0.453$$

which is less than 1

or $d = 2$ m; greater than 1.18 m, the critical depth.

11.3. Hydraulic Jump

If an incoming flow is supercritical as emerging from a sluice gate in Fig. 11.5 (*a*) *or* the slope of the bed provided is inadequate for a stable supercritical flow to proceed as in Fig. 11.5 (*b*), the flow may transform abruptly from supercritical to subcritical. This is in accordance with a triple arrowed line d_1d_2 instead of a vertical line d_1d_2' as shown in Fig. 11.6. This is to allow an accompanying loss of energy ΔE from d'_2 due to real-flow effects, *e.g.*, eddying and turbulence.

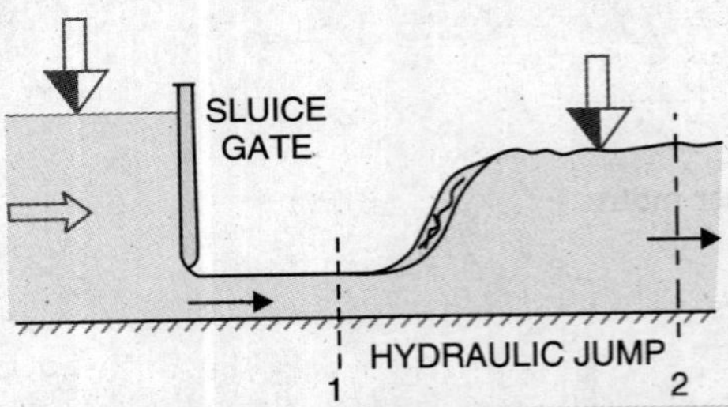

Fig. 11.5(a): Flow under a Sluice Gate

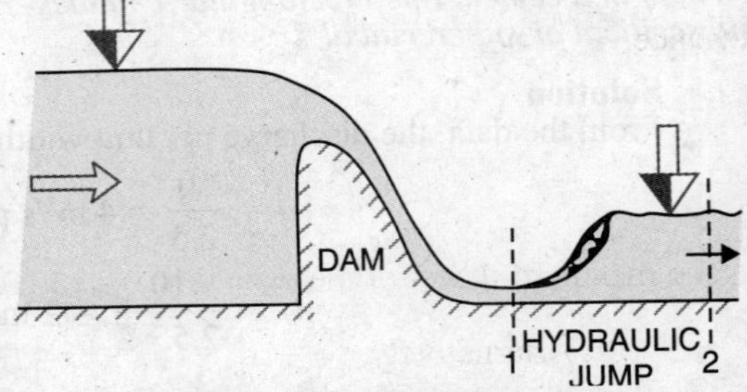

Fig. 11.5(b): Flow over a Dam

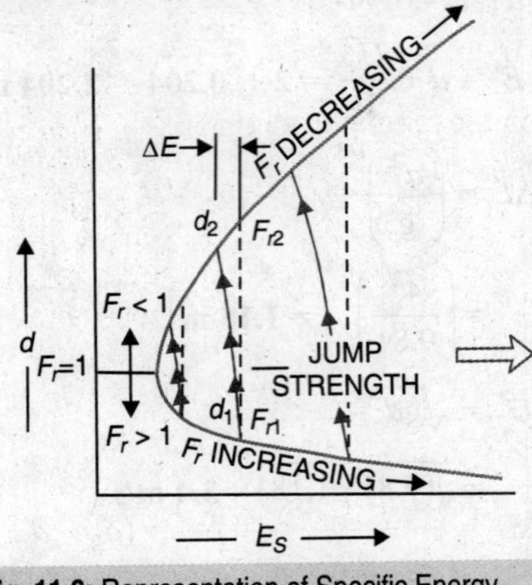

Fig. 11.6: Representation of Specific Energy

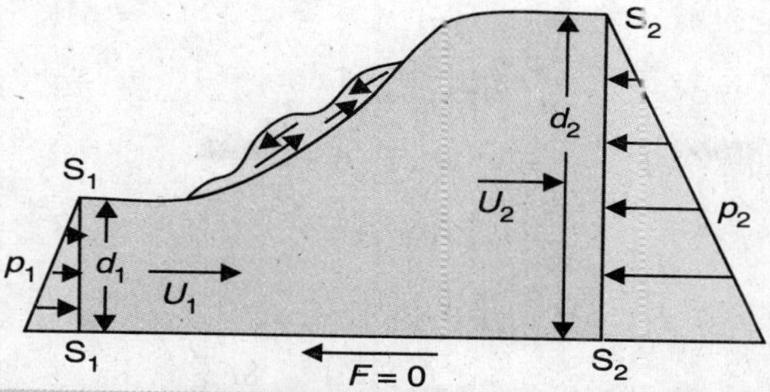

Fig. 11.7: Control Volume for Analysis

In either situation, consider a control volume $S_1S_2S_2S_1$ as shown in Fig. 11.7. for the analysis. By the continuity equation,

$$U_1d_1 = U_2d_2 = q \quad ...(i)$$

the flow per unit width.

By the momentum equation,

$$\rho g d_1^2/2 - \rho g d_2^2/2 = (U_2 - U_1)\,\rho q \quad ...(ii)$$

Solving equations (*i*) and (*ii*) for d_2 as follows,

$$d_1^2 - d_2^2 = \frac{2q^2}{g}\left(\frac{1}{d_2} - \frac{1}{d_1}\right)$$

or $$d_2^2\, d_1 + d_2\, d_1^2 - \frac{2q^2}{g} = 0$$

whence $$d_2 = -\frac{d_1}{2} + \sqrt{\left(\frac{d_1^2}{4} + \frac{2q^2}{gd_1}\right)} \quad ...(11.10\ a)$$

or $$d_2 = -\frac{d_1}{2} + \sqrt{\left(\frac{d_1^2}{4} + \frac{2d_1U_1^2}{g}\right)} \quad ...(11.10\ b)$$

as the meaningful root of the equation.

Alternatively, $$\frac{d_2}{d_1} = \frac{1}{2}(\sqrt{1 + 8F_{r1}^2} - 1) \quad ...(11.10\ c)$$

By the energy equation,

$$E_1 = d_1 + \frac{U_1^2}{2g} = d_2 + \frac{U_2^2}{2g} + \Delta E$$

where ΔE is the energy loss in the jump phenomenon;

$$\Delta E = \frac{(U_1^2 - U_2^2)}{2g} - (d_2 - d_1)$$

or $$\Delta E = \frac{q^2}{2g}\left(\frac{1}{d_1^2} - \frac{1}{d_2^2}\right) - (d_2 - d_1)$$

Invoking Eq (11.10 *a*), $$\frac{q^2}{2g} = \frac{(d_1 + d_2)d_1d_2}{4}$$

$$\Delta E = \frac{(d_1 + d_2)(d_2^2 - d_1^2)}{(4d_1d_2)} - (d_2 - d_1)$$

or $$\Delta E = \frac{(d_2 - d_1)^3}{4d_1 d_2} \quad ...(11.11)$$

Since $$F_{r1} = \frac{U_1}{\sqrt{gd_1}} \text{ and } F_{r2} = \frac{U_2}{\sqrt{gd_2}}$$

$$\frac{F_{r1}^2}{F_{r2}^2} = \left(\frac{d_2}{d_1}\right)^3$$

which, by using Eq. (11.10 *c*), reduces to $F_{r2}^2 = \dfrac{8F_{r1}^2}{(\sqrt{2 + 8F_{r1}^2} - 1)^3}$

It can be observed that :

(*i*) If $F_{r1} < 1$, $F_{r2} > 1$, but this requires an increase of available energy showing a decrease in entropy of a system which is in violation of the second law of thermodynamics. Therefore, no jump can be formed for a flow where $F_{r1} < 1$.

(*ii*) If $F_{r1} = 1$, $F_{r2} = 1$. This implies the continuation of a critical flow ; no jump !

(*iii*) If $F_{r1} > 1$, $F_{r2} < 1$. This corresponds to a real jump in conformity with the laws of Physics. It may also be noted that higher the value of F_{r1}, lower is the resulting F_{r2} after the jump and 'stronger' is the jump!

The hydraulic jumps are classified as shown in Fig. 11.8 and tabulated alongside.

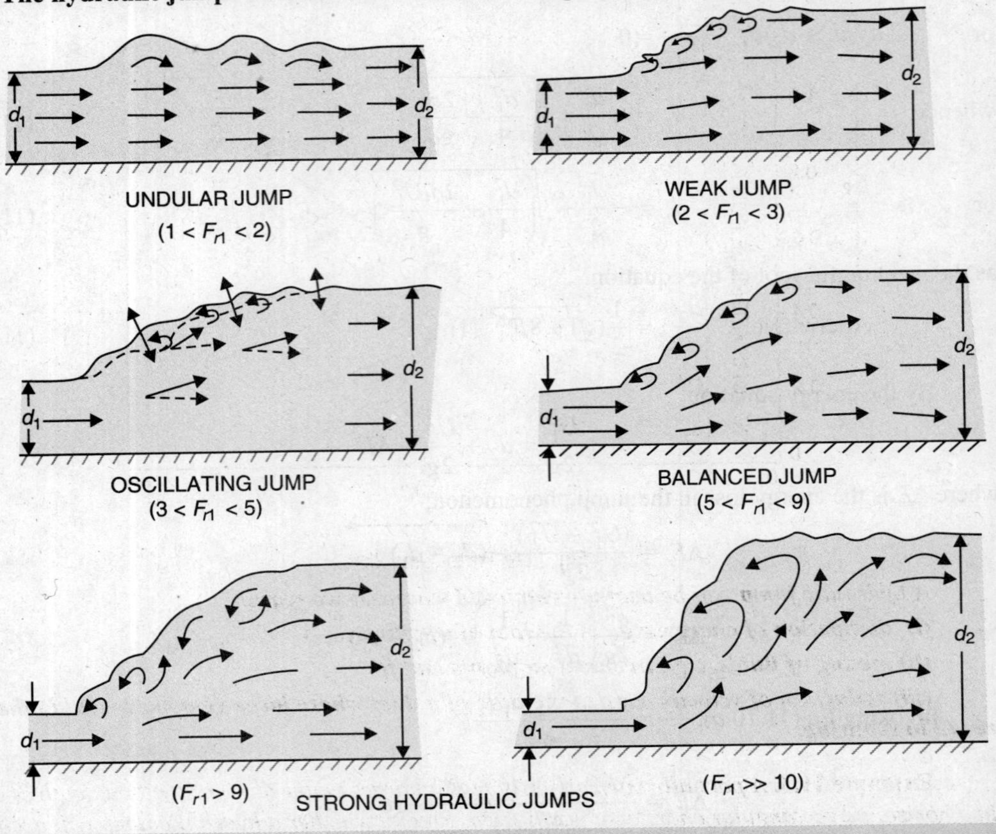

Fig. 11.8: Classification of Hydraulic Jumps

CLASSIFICATION OF HYDRAULIC JUMPS

Range of F_{r1},	Type of Jump	Features of the Jump
$1 < F_{r1} < 2$	Undular jump	Standing wave of a small height
$2 < F_{r1} < 3$	Weak jump	Smooth and quiet
$3 < F_{r1} < 5$	Oscillating jump	Transition range unpredictable and excessively large bursts of jumps
$5 < F_{r1} < 9$	Balanced jump	Good jump with well-defined features accompanied by over 50% energy loss
$F_{r1} > 9; >> 10$	Strong jump	Rough nature ; losing up to 90% of energy

Let us summarise the conclusions from the one-dimensional analysis of a hydraulic jump, as shown in Fig. 11.9,

(*a*) A jump can occur only if $F_{r1} > 1$; greater the F_{r1}, stronger the jump.

(*b*) After the jump, F_{r2} must be less than 1; greater the F_{r1}, less is F_{r2}

(*c*) the depth of water across the jump must increase and d_2/d_1 increase with F_{r1}

(*d*) the jump strength, defined by $\frac{d_2}{d_1}$ or $\frac{d_2 - d_1}{d_1}$ or $\frac{F_{r1}}{F_{r2}}$ increases with F_{r1}

(*e*) stronger the jump, greater the energy loss ΔE resulting in dissipation.

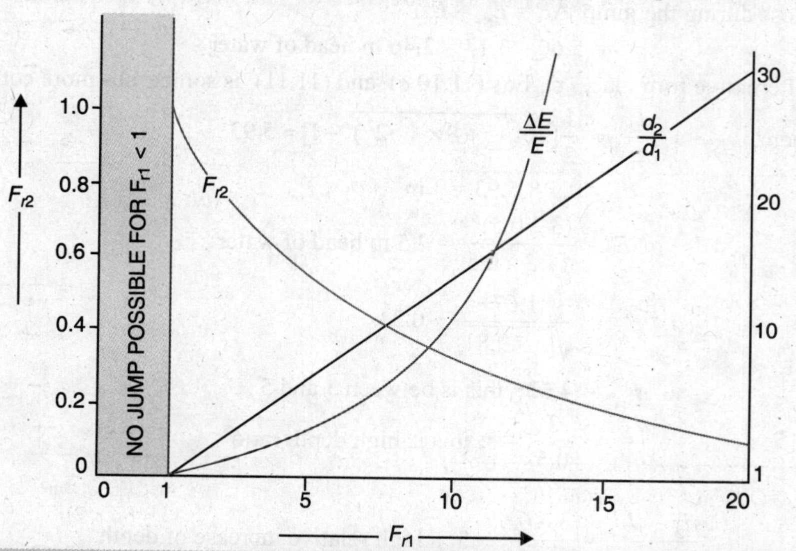

Fig. 11.9: Trans-Jump Plots

A hydraulic jump can be usefully employed wherever we require

(i) dissipation of energy, e.g., at the foot of a spillway,

(ii) mixing of fluids, e.g., in chemical plants and for

(iii) reduction of velocity, e.g., at the base of a dam where large velocities would otherwise result in scouring.

Example 11.5. *A partially open sluice gate discharges water at 10 m/s with a depth of 50 cm in a horizontal rectangular channel of width 10 m. Predict whether a hydraulic jump will occur and if so, calculate the salient features of the jump.*

Solution.

The discharge per unit width,

$$q = 10 \times 0.5 = 5 \text{ m}^2\text{/s per metre}$$

$$F_{r1} = \frac{10}{\sqrt{(0.5 \times 9.81)}} = 4.52$$

which is greater than unity.

It shows that a hydraulic jump is likely to occur on a horizontal bed.

Employing Eq. (8.10 *b*).

$$d_2 = -\frac{0.5}{2} + \left(\frac{0.5^2}{4} + \frac{2 \times 0.5 \times 10^2}{9.81}\right)^{1/2} = 3 \text{ m}$$

$$U_2 = \frac{q}{d^2} = \frac{5}{3} = 1.67 \text{ m/s}$$

Height of the jump $H = d_2 - d_1 = 3 - 0.5 = 2.5$ m

Now $$E_{S1} = 0.5 + \frac{10^2}{2 \times 9.81} = 5.60 \text{ m head water}$$

and $$E_{S2} = 3 + \frac{1.67^2}{2 \times 9.81} = 3.14 \text{ m head of water}$$

Energy loss during the jump $\Delta E = E_{S1} - E_{S2}$

$$= 5.60 - 3.14 = 2.46 \text{ m head of water}$$

Use of alternative formulae, *i.e.*, Eqs (11.10 *c*) and (11.11) is sometimes more convenient :

Using them, $$\frac{d_2}{d_1} = \frac{1}{2}[20\sqrt{(1 + 8 \times 4.52^2)} - 1] = 5.93$$

$$d_2 = 0.5 \times 5.93 = 3 \text{ m}$$

$$\Delta E = \frac{(3 - 0.5)^3}{4 \times 3 \times 0.5} = 2.5 \text{ m head of water}$$

$$F_{r2} = \frac{1.67}{\sqrt{(3 \times 9.81)}} = 0.31$$

F_{r1} = **4.52;** this is between 3 and 5

$\frac{d_2}{d_1} = \frac{3}{0.5}$ = **6;** this is high depth ratio

$\frac{d_2 - d_1}{d_1} = \frac{3 - 0.5}{0.5}$ = **5;** a high relative increase of depth

$\frac{F_{r1}}{F_{r2}} = \frac{4.52}{0.31}$ = **14.6** ; a large ratio of Froude number

$\frac{\Delta E}{E_1} = \frac{2.46}{5.6}$ = **44%;** a substantial loss of energy

It is apparent from either of the above related results that the data refers to fairly strong oscillating jump :

Example 11.6. *During an experiment on a hydraulic jump, in rectangular open channel 0.5 m wide, the depth of water changes from 0.2 m to 0.5 m Calculate (a) discharge in the channel, (b) the initial and final Froude number and (c) the loss of head.*

Solution.

(*a*) For a hydraulic jump,

$$d_2/d_1 = 1/2(\sqrt{(1+8Fr_1^2)}-1)$$

Using the data,

$$0.5/0.2 = 1/2(\sqrt{(1+8Fr_1^2)}-1)$$

whence $Fr_1 = \mathbf{2.09} = U_1/\sqrt{gd_1}$

and $U_1 = 2.09 \times \sqrt{9.81 \times 0.2} = 2.93$ m/s

The discharge is given by

$$Q = U_1 \times A_1 = 2.93 \times 0.5 \times 0.2 = \mathbf{0.293\ m^3/s}$$

(*b*) The initial Froude number Fr_1 is 2.09; the final Froude number is given by

$$(Fr_2/Fr_1)^2 = (d_1/d_2)^3$$

and $Fr_2^2 = (0.2/0.5)^3 \times (2.09)^2 = 0.28$

whence $Fr_2 = \mathbf{0.529}$ and $U_2 = 0.529 \times \sqrt{9.81 \times 0.5} = 1.17$ m/s

(*c*)The loss of head is given by

$$\Delta E = (d_1 + U_1^2/2g) - (d_2 - U_2^2/2g)$$

$$= [0.2 + 2.93^2/(2 \times 9.81)] - [0.5 + 1.17^2/(2 \times 9.81)]$$

$$= 0.637 - 0.57 = \mathbf{0.067\ m\ head\ of\ water}$$

Alternatively, the head loss is expressed as

$$\Delta E = (d_2 - d_1)^3/4d_1d_2$$

$$= (0.5 - 0.2)^3/(4 \times 0.2 \times 0.5) = \mathbf{0.067\ m\ head\ of\ water}$$

EXPERIMENT

Study of a Hydraulic Jump

OBJECTIVE

To study the salient features of a hydraulic jump in a horizontal rectangular open channel and to compare the experimental results with the prediction based on a one-dimensional flow analysis.

PRACTICAL RELEVANCE

A hydraulic jump is formed if a supercritical free-surface flow is caused to change abruptly into a subcritical flow as a result of an obstruction placed in the passage or by a change in the slope of the channel bed. The phenomenon of a hydraulic jump is accompanied by a sudden rise in the depth of the liquid surface, increase in the level of turbulence, pronounced mixing and dissipation of energy. Hydraulic jumps are, therefore, employed to achieve

(*i*) dissipation of energy at the foot of a spillway and in many other hydraulic structures wherein large velocities would result in scouring of the river bed.

(*ii*) mixing of fluids in an industry where a jump is formed in the primary liquid and a secondary liquid or a granulated solid is dropped from above.

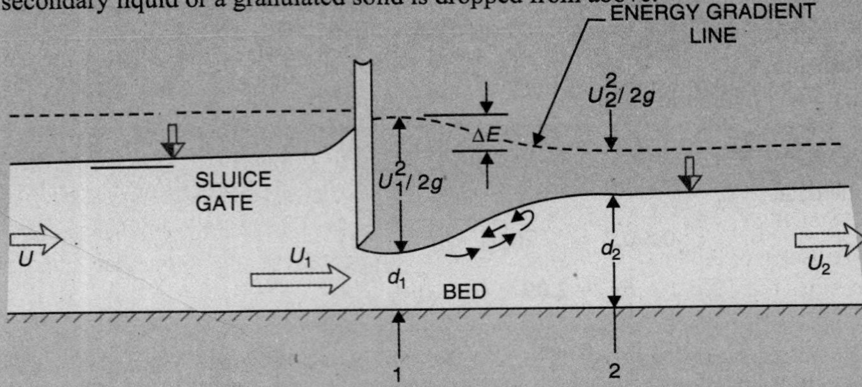

HYDRAULIC JUMP FORMATION

Study of the jump phenomenon in a laboratory is necessary to discover the interesting features in a systematic way.

PROCEDURE

1. Check that the floor of the channel or flume is horizontal. Adjust the 'zero' of the depth gauge so that it corresponds to the level of the bed of the channel or flume.
2. Start the flow and regulate it to obtain the maximum possible flow. Be careful to avoid spillage of water over the side-walls of the channel or flume.
3. Adjust the height of the upstream sluice gate to obtain a depth d_1 just downstream of the gate such that $F_{r1} > 1$.
4. Adjust the height of the end obstruction such that a hydraulic jump occurs in the channel.
5. When conditions have become relatively steady, measure :
   (*a*) the depth, d_1 just in front of the hydraulic jump;
   (*b*) the depth, d_2 behind the hydraulic jump at a section where the flow is approximately uniform;
   (*c*) the rate of flow, Q
6. Keeping the flow rate constant as adjusted in step 2, repeat steps 3 to 5 for six different values of F_{r1}. Use as wide a range of values as possible, always keeping $F_{rl} > 1$. F_{rl} should be adjusted by varying d_1 by changing the opening of the sluice gate.
7. Verify that no jump can occur for d_1 slightly larger than or equal to d_c.
8. Observe the jump by suitably injecting fine potassium permanganate crystals.

RESULTS

1. For each run, calculate the initial and final velocities U_1 and U_2 in the channel, the upstream and downstream Froude numbers F_{rl} and F_{r2}, the upstream and down stream specific energies E_{sl} and E_{s2}, the experimental and theoretical energy loss ΔE, and d_2/d_1.
2. Plot the theoretical curves of d_2/d_1 *vs* F_{rl}. $\Delta E/E_{sl}$ *vs* F_{rl}. Superimpose the experimental points on these curves.
3. Plot F_{r2} against F_{rl} and compare the experimental result with the theoretical result.
4. Plot the specific energy $E_s = d + U^2/2g$ as the abscissa against d as the ordinate for the particular value of q used in the experiment. Calculate the energy loss for each case from this graph.

POINTS FOR DISCUSSION

1. Why is the initial Froude number F_{r1} considered to be significant ?
2. Why cannot a jump occur if the initial Froude number is equal to or less than unity ?
3. Is it essential that a stronger jump must be accompanied by greater energy loss ?
4. Discuss the variation of the depth ratio with the Froude number. At what value of F_{r1} do the experiments indicate a value of unity for d_2/d_1 ? When should it be unity theoretically ? Comment on the difference.
5. Comment on the discrepancy between the experimental and theoretical curves.
6. Why is the energy loss taken as the difference of specific energies rather than of the mechanical energies as is usually the case ?
7. Is it possible to reduce the velocity of a supercritical flow with less energy loss than that which is encountered in a jump ? If so, why is a hydraulic jump invariably used in a stilling basin ?
8. Is it possible to bring about mixing of two fluids with less energy loss than that which is needed for a jump ? When would the jump be preferred to other methods ?

ANY OTHER OBSERVATION OR COMMENT?

ITEM BANK

Objective items of different types are given to review the concepts developed in the Chapter.

* *Select the correct response for each item.*

* *Every correct response is expected in less than a minute and carries one mark.*

* *Compare your responses with the key answers given at the end.*

Multiple-Choice Questions

Attempt the following Multiple-Choice Items by ticking the correct or the most appropriate response from among the alternatives given :

1. In an open-channel flow, the free surface, the hydraulic gradient and energy lines are such that

(*a*) the three of them coincide

(*b*) the first two coincide

(*c*) the last two must remain parallel

(*d*) the first and the last must remain parallel

2. The corss-section of an open channel is said to be '*the best*' if

(*a*) the roughness coefficient is the least

(*b*) the section has a maximum area for the flow

(*c*) the section has the least perimeter for a given area of flow

(*d*) the hydraulic mean depth is the maximum

3. A hydraulic jump must occur when

(*a*) the flow is supercritical

(*b*) the depth is less than the critical depth

(*c*) the slope is mild or level

(*d*) the bed slope changes from steep to mild.

4. The strength of a jump is indicated by

(*a*) the upstream velocity of the liquid

(*b*) the downstream velocity of the liquid

(*c*) the upstream Froude number

(*d*) the upstream depth of flow

5. The depth of flow in an open channel where the specific energy is minimum is called
(*a*) critical depth
(*b*) conjugate depth
(*c*) normal depth
(*d*) minimum depth.

6. Uniform flow in a channel is characterised by
(*a*) uniform flow across the section.
(*b*) total energy constant along the channel.
(*c*) specific energy constant along the channel.
(*d*) total energy rising along the channel

7. A 3 metre wide rectangular channel flowing at its normal depth of 0.8 m carries a discharge of 5.5 m^3/s. The bed slope of the channel is
(*a*) Steep (*b*) Critical
(*c*) Mild (*d*) Adverse.

8. If the Froude Number characterizing flow in an open channel is less than unity, an increase in channel width causes the water surface elevation to
(*a*) form ripples (*b*) remain the same
(*c*) decrease (*d*) increase

9. A rectangular channel of 4 m width conveys water at 8 m^3/sec under critical condition. Specific energy for this flow is :
(*a*) 1.1123 m (*b*) 1.4830 m
(*c*) 0.3703 m (*d*) 0.7416 m

Other Objective Items

10. Are the following statements true or false for flow through an open channel ?
(*a*) The most economical cross section is that for which the area to perimeter ratio is the maximum. T/F
(*b*) A hydraulic jump cannot be formed if the upstream Froude number is less than unity. T/F
(*c*) The energy loss in a jump is proportional to the upstream Froude number. T/F
(*d*) Greater the upstream Froude number, stronger the hydraulic jump. T/F

11. Match the type of hydraulic jump in block A with the initial Froude number in block B.

Block A	Block B
(a) Oscillating jump	(k) less than 2
(b) Strong jump	(l) between 2 and 3
(c) Weak jump	(m) around 4
(d) undulac jump	(n) around 10

ANSWERS

1. (b), **2.** (c), **3.** (d), **4.** (c), **5.** (a), **6.** (c), **7.** (a), **8.** (c), **9.** (a), **10.** (aF, bT, cF, dT)
11. (*a m, bn, cl, dk*)

QUESTION BANK

Questions on theory and numerical problems are given for tutorial work.

* The questions are arranged generally in order of the articles in the text.
* Every question should take 15-20 minutes to solve.
* Answers for the numerical problems are given alongside them.

11.1 Estimate the rate of flow through a rectangular channel 2 m wide laid at a slope of 5 in 10 000 for 1 m depth of water when the bed is (*a*) natural ($N = 0.035$) and (*b*) cement-concrete ($N = 0.015$) **[0.805, 1.88 m^3/s]**

11.2 The breadth of a rectangular channel is twice its depth. Assuming the Chezy's C to be 55, find the cross-sectional dimensions of the channel and the slope to satisfy the following two conditions:
(*a*) the discharge when flowing full should be 0.8 m^3/s
(*b*) the velocity when flowing half full should be 0.6 m/s **[$d = 0.74$ m, $b = 1.48$ m, $i = 1/20731$]**

11.3 A V-shaped open channel of included angle 90° conveys a discharge of 0.05 m^3/s when the depth of water at the centre is 22.5 cm. Assuming that $C = 50\ m^{1/2}/s$ in the Chezy equation, calculate the slope of the channel. **[1 in 204]**

11.4 Calculate the dimensions of the rectangular cross section of an open channel which requires minimum area to convey 10 m^3/s, the slope being 1 in 1500. Take the Kutter constant N as 1/60. **[d = 1.85 m, b = 3.7 m]**

11.5 A semicircular channel is required to convey 1.5 m^3/s when flowing full, the gradient being 1 in 1000. Taking the Chezy C as 60, calculate the diameter of the channel. **[1.74 m]**

11.6 Compare the depths of excavation for a rectangular channel required to convey the same discharge laid at the same slope for two different bed materials with the Manning's constants in the ratio of 1 to 2.

11.7 Determine the average shear stress over the periphery of a rectangular channel 10 m wide laid at a slope of 1 in 1000, conveying a discharge of 10 m^3/s with a depth of water 1 m everywhere. **[8.175 N/m²]**

11.8 A rectangular channel 3 m wide takes a discharge of 5 m^3/s at a water depth of 1 m. If it is converted to a trapezoidal channel of side slopes 1 : 2, would there be an appreciable saving in the power lost ? Take C, the Chezy's constant = 50 in SI units for each case. **[No]**

11.9 Calculate the minimum slope of an open channel required to convey a discharge of 10 m^3/s with an average velocity of 1 m/s if a trapezoidal section of sides with slopes of 1 to 1 is preferred. Take C = 50 in *SI* units. **[1 in 2922]**

11.10 Calculate the size of a sewage pipeline required to carry 2 m^3/s discharge when running 90% full. The pipe is to be laid to a slope of 1 in 1000. Take C = 50 in *SI* units.

[d = 1.68 m, {$\theta = 1.9\ \pi$ radians }]

11.11 A trapezoidal channel has a bottom width of 6.1 m and side slopes of 2 horizontal to 1 vertical. When the depth of water is 1.07 m, the flow is 10.47 m^3/s. (*i*) What is the specific energy of flow? (*ii*) Is the flow tranquil or rapid? **[1.14 *m* ; Fr < 1, tranquil]**

11.12 In a hydraulics laboratory, a flow of 0.412 m^3/s was measured from a rectangular channel flowing 1.22 m wide and 0.64 m deep. If the slope of the channel was 0.0004, find its roughness factor using Manning's formula. **[m = 0.312 ; N = 0.0174]**

11.13 During an experiment on a hydraulic jump in a rectangular open channel 0.5 m wide, the depth changes from 0.2 m to 0.5 m. Calculate (*a*) the discharge, (*b*) the initial and final Froude numbers and (*c*) the loss of head. **[0.293 m³/s; 2.1, 0.53; 0.0675 m]**

11.14 Show that an excavated channel of a desired geometry flowing full has the best cross section when the hydraulic mean depth equals half the height of the channel.

11.15 An open channel has a U-section, semicircular at the bottom with vertical sides. It is 2 m wide and is laid at 1 in 5000 slope. Calculate the depth of flow if it should carry 1 m^2/s discharge. Take C = 60 in SI units. **[1.02 m]**

11.16 Deduce and specify the geometrical conditions for the form of the best trapezoidal channel flowing full,

(*a*) if the side slopes are fixed and (*b*) if the side slopes can also be optimised.

11.17 A horizontal rectangular channel of constant width has a sluice gate installed in it. At a position of 1 m opening, the velocity of water is 10 m/s. Determine whether a jump can occur, and if so,

(*a*) the height downstream, (*b*) the loss of energy per unit discharge, and
(*c*) the ratio of Froude number across it.

[Fr_1 = 3.2, a jump can occur, 4.043 m, 1.742 m, 0.123]

11.18 Prove that the velocity of flow in a circular culvert of diameter d is the maximum when the depth h of flow at the centre is 0.81 times the diameter.

$$\left[\frac{dm}{d\theta} = 0, \theta = 128.7°, h = 0.81\, d\right]$$

11.19 A sewer pipe of diameter 0.6 m has been laid at a bed slope of 1 in 300. Determine the maximum velocity of flow and the corresponding discharge through it. Take $C = 50$ in SI units. **[0.989 m/s, 0.156 m³/s]**

11.20 In a rectangular channel 0.6 m wide, a hydraulic jump occurs where the Froude number is 3. The depth after the jump is 0.6 m. Estimate the total loss of head and the power dissipated due to the jump. **[0.743 m, 0.912 kN]**

11.21 A sluice gate across a channel 6 m wide discharges a stream 1.2 m deep. What will be the flow if the upstream depth is 6 m ? **[78 m³/s]**

11.22 The stream issuing from beneath a vertical sluice gate is 0.3 m deep at the vena contracta. Its mean velocity is 6 m/s. A standing wave is created on the level bed downstream the sluice gate. Find the height of the jump, the loss of head and the power dissipated per unit width of the sluice gate. **[1.05 m, 0.7 m, 12.4 kW]**

11.23 A trapezoidal channel carries a discharge of 2.5 m³/s. If the bed slope is 1 in 2000 and sides of channel slope at 1 horizontal to 0.5 vertical, design the most economical section for this channel. Assume $C = 55$ in Chezy's formula. **[$d = 1.0617$ m, $b = 0.5$ m]**

11.24 Determine the depth flow after the jump in a horizontal rectangular channel and consequent loss in total head when the sluice gate discharges water into the channel with a velocity of 20 m/s and depth of 2 m. **[11.81 ; 10 m]**

CHAPTER

12

Compressible Flow

LEARNING OBJECTIVES

On successful completion of all learning activities planned and provided in the chapter, a student will be able to

- differentiate between various processes of expansion and compression ;
- determine the sonic velocity in a fluid ;
- describe the effect of Mach number in flow ;
- analyse a normal shock ;
- describe the flow through a converging-diverging nozzle; and
- estimate a supersonic velocity by employing a pitot-static probe.

12.1 Introduction

Flow of fluids where density variations become too significant to be ignored is termed as compressible flow. Compressibility effects become increasingly important as the fluid velocity approaches the sonic velocity and thereafter. The compressible flow studied here is restricted to one-dimensional variations, *i.e.,* variations along the flow, Fig. 12.1. It is considered uniform over any cross-section. The fluid is assumed to behave as a perfect gas. However severe these assumptions may appear to be, this analysis has numerous practical applications in aerodynamics, pneumatics and fluid machines.

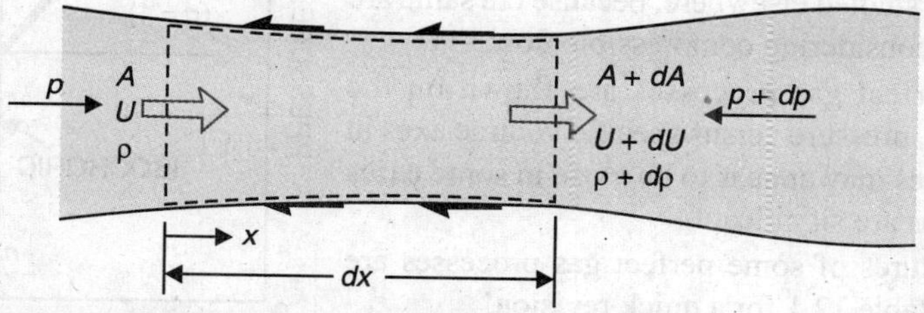

Fig. 12.1: One-dimensional Compressible Flow

Recalling the behaviour of a perfect gas and the processes,

The equation of state is $p = \rho RT,$ ... (12.1)

Change in internal energy $\delta i = C_v \delta T,$ ... (12.2)

and change in enthalpy $\delta h = C_p \delta T,$ ... (12.3)

where T is the absolute temperature in K, degrees kelvin

such that $T = t$ (in degree Celsius) + 273.15 K

and R is the characteristic gas constant, 287 J/kg K

Standard Atmospheric Temperature, Pressure and Density referred to as STP at the sea level are:

Temperature	**15° C or 288.15 K**
Pressure	**101.325 kN/m^2 or 10.33 m water gauge**
Density	**1.225 kg/m^3**
Sonic velocity	**340.29 m/s in air**

The universal gas constant is 8314 J/mole K. The characteristic gas constant R for a gas may be obtained by dividing the universal gas constant with molecular weight of the gas. For air, the mean molecular weight is 28.966. Therefore, for air,

$$R = 8314/28.966 = 287 \text{ J/kg K}$$

and C_p and C_v are the specific heats at constant pressure and constant volume respectively.

$$C_p / C_v = \gamma :$$

$$C_p = \gamma / \gamma - 1) R$$

and $$C_v = (1/ \gamma - 1) R$$

In a one-dimensional compressible flow, continuity of flow requires that

$$\rho AU = \text{Constant}$$

where ρ is the density and U is the velocity at a cross section where the area is A. On logarithmic differentiation, it becomes

$$d\rho / \rho + dA / A + dU/ U = 0 \quad ...(12.4)$$

The equation of motion for the one-dimensional steady flow reduces to

$$UdU + d\rho/\rho = 0 \quad ...(12.5)$$

The Bernoulli equation for a compressible flow undergoing adiabatic process takes the form,

$$\frac{\gamma}{\gamma - 1} \cdot \frac{p}{\rho} + \frac{U^2}{2} + gz = \text{constant} \quad ...(12.6)$$

Perfect Gas Processes

Let us revisit the perfect gas processes of expansion and compression, studied elsewhere, because the same are often relevant in considering compressible flow.

Several perfect gas processes are shown on the p - v diagram, i.e., pressure versus specific volume axes in Fig. 12.2. The lines may appear to be close in some cases but the differences are significant.

Salient features of some perfect gas processes are also tabulated in table 12.1 for a quick revision!

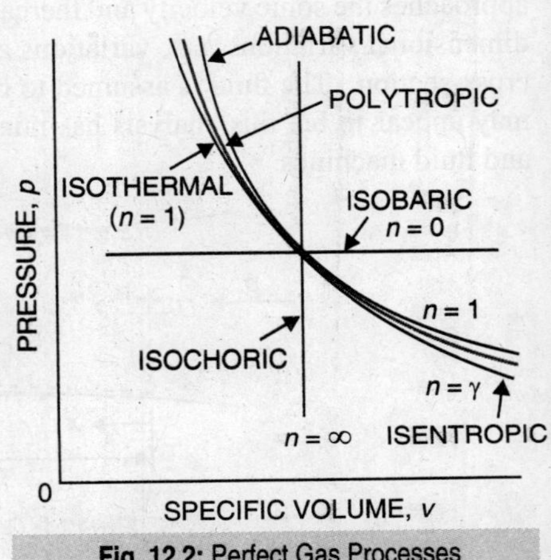

Fig. 12.2: Perfect Gas Processes

Table 12.1. **Salient Features of Perfect Gas Processes**

Process	**Isobaric** $(p_1 = p_2 = p)$	**Isometric** $(\rho_1 = \rho_2 = \rho)$	**Isothermal** $(T_1 = T_2 = T)$	**Isentropic** $(s_1 = s_2 = s)$ *Adiabatic	**Polytropic** n = constant
Index n in p/ρ^n = const.	$n = 0$	$n = \infty$	$n = 1$	$n = \gamma = \frac{C_P}{C_v}$	n
p, ρ and T relationship	$\rho_2 T_2 = \rho_1 T_1 = \frac{p}{R}$ (Charles law)	$\frac{p_2}{T_2} = \frac{p_1}{T_1} = \rho R$	$\frac{p_2}{\rho_2} = \frac{p_1}{\rho_1} = RT$ (Boyles law)	$\frac{p_2}{p_1} = \left(\frac{\rho 2}{\rho 1}\right)^{\gamma}$ $\frac{T_2}{T_1} = \left(\frac{\rho_2}{\rho_1}\right)^{\gamma-1} = \left(\frac{p_2}{p_1}\right)^{\frac{\gamma-1}{\gamma}}$	$\frac{p_2}{p_1} = \left(\frac{\rho_2}{\rho_1}\right)^{n}$ $\frac{T_2}{T_1} = \left(\frac{\rho_2}{\rho_1}\right)^{n-1} = \left(\frac{p_2}{p_1}\right)^{\frac{n-1}{n}}$
Heat added	$C_p (T_2 - T_1)$	$C_v (T_2 - T_1)$	$\frac{p_1}{\rho_1} \log_e \frac{\rho_1}{\rho_2}$	0	$C_n (T_2 - T_1)$ where $C_n = C_v \left(\frac{\gamma - n}{1 - n}\right)$
Process work done $\int_1^2 p dv$; $v = \frac{1}{\rho}$	$p_1(v_2 - v_1)$	0	$\frac{p_1}{\rho_1} \log_e \frac{\rho_1}{\rho_2}$	$\frac{p_1 v_1 - p_2 v_2}{\gamma - 1}$	$\frac{p_1 v_1 - p_2 v_2}{n - 1}$

* An adiabatic process is that where the index n equals γ but the entropy is not constant. This is because frictional effects are there although heat exchange with the surroundings is zero.

12.2 Sonic Velocity

The velocity of sound is the propagation of 'a small pressure disturbance'. It may be determined by the application of the continuity and momentum principles to a fluid in a long tube as shown in Fig. 12.3. An infinitesimal disturbance wave front moves at speed c and we locate our coordinates on it moving with it so that the wave front appears stationary and the undisturbed fluid appears to be moving at speed c while that on the other side at $c + \delta U$. By continuity,

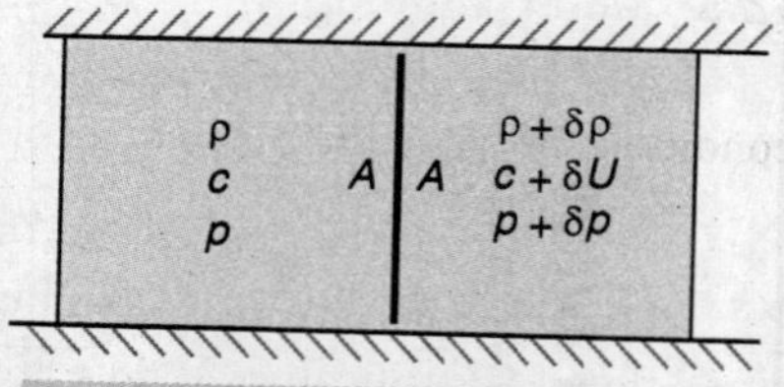

Fig. 12.3: Sonic Velocity in a Fluid

$$\rho\, c A = (\rho + \delta\rho) \,.\, (c + \delta U) A = \rho\, c A + \rho\, \delta U A + \delta\rho\, c A + \delta\rho\, \delta U A$$

Dividing each term by $\rho\, c\, A$ and neglecting the higher order last term,

$$d\rho \,/\, \rho = - dU/c$$

By the momentum principle,

$$(p - (p + \delta p))\, A = \rho\, c A\, (c + \delta\, U - c)$$

or

$$dp = - \rho\, c\, dU$$

Combining the continuity and momentum relations,

$$c^2 = dp/d\rho \;\; ; \qquad c = \sqrt{dp \,/\, d\rho} \qquad \text{...(12.7)}$$

In terms of the bulk modulus of elasticity K for a liquid,

$$c = \sqrt{K \,/\, \rho} \qquad \text{...(12.8)}$$

where

$$K = dp \,/\, (d\rho \,/\, \rho)$$

Eq. (12.7) and Eq. (12.8) reveal that an infinitesimal disturbance propagates at a velocity equal to the velocity of sound in a medium.

For an insentropic process in a gas,

$$dp \,/\, d\rho = \gamma\, p/\rho = \gamma\, RT$$

where R is the gas constant, *e.g.*, 287 *J/kg K* for air,

Hence,

$$c = \sqrt{\gamma R T}$$

It may be observed from this equation that, for a given gas, the sonic velocity in a fluid increases with a rise in its temperature.

Example 12.1. *Find the sonic velocity for the following fluids :*
(i) Crude oil. sp. gr. 0.8, bulk modulus 1.5 × 10⁶ kN/m² (ii) Water having a bulk-modulus of 2.07 × 10⁶ kN / m² and (iii) Air at a temperature of 280 K

Solution.

The sonic velocity c in a liquid is given by $c = \sqrt{K \,/\, \rho}$

Employing SI units for case (*i*) $\quad c = \sqrt{1.5 \times 10^9 \,/\, 1000} =$ **1225 m/s**

and for case (*ii*) $\quad c = \sqrt{2.07 \times 10^9 \,/\, 1000} =$ **1439 m/s**

The sonic velocity in a gas is given by $\quad c = \sqrt{\gamma R T}$

Threfore, for case (*iii*) $\quad c = \sqrt{1.4 \times 287 \times 280} =$ **335 m/s**

It may be noted from the example that the sonic velocity in a liquid is over four times the sonic velocity in a gas. Higher the bulk modulus of compressibility K of a fluid, more is the velocity of sound in it. In an ideal fluid, i.e., incompressible (and inviscid), the value of K and hence the value of the sonic velocity must be infinite.

12.3 Mach Number

The ratio of the fluid velocity U to the velocity of sound c, in that fluid under the existing conditions is called the Mach number M,

$$M = U/c \qquad ...(12.10)$$

The Mach number may vary from a very low value at low value to more than 7 for very high speed flows. Classification of flow with Mach number is shown in Table 12.2.

Table 12.2

Mach Number	Classification of Flow
$M < 0.3$	Incompressible Flow
$0.3 < M < 1$	Subsonic Flow
$0.9 < M < 1.1$	Transonic Flow
$M = 1$	Sonic Flow
$1 < M < 7$	Supersonic Flow
$M > 7$	Hypersonic Flow

The Mach number also stands for the ratio of the inertia force to the compressibility force, as shown under the chapter on Dimensional Analysis. Mechanics of a compressible flow has, therefore, an important bearing on its Mach number.

It may also be shown that the condition for 'incompressible flow' corresponds to very low Mach number. Let us see how. The continuity equation for steady flow along the streamwise coordinates is given by

$$U\frac{\partial \rho}{\partial s} + \rho\frac{\partial U}{\partial s} = 0$$

and the Euler's equation for steady flow is $\rho U\dfrac{\partial U}{\partial s} = -\dfrac{\partial p}{\partial s}$

The condition for incompressible flow is stated as

$$U\frac{\partial \rho}{\partial s} << \rho\frac{\partial U}{\partial s} << \left|\frac{1}{U}\frac{\partial p}{\partial s}\right|$$

whence $$\frac{U^2}{\partial p/\partial \rho} << 1$$

Introducing the bulk modulus $$K = \rho\frac{\partial p}{\partial \rho}$$

the condition becomes $$\frac{U^2}{K/\rho} << 1$$

or $$U^2/c^2 < 1;$$
$$U/c << 1$$

which means $$M << 1.$$

i.e., the flow must be at very low Mach number!

The speed of sound c in air at STP is about 340 m/s. Air velocity of 100 m/s corresponds to a Mach number 0.294 and $M^2 = 0.087$ which is much less than 1. Therefore, flow of air up to say 100 m/s may be considered as incompressible flow. Similarly, taking the speed of sound in water as 1400 m/s, the flow of water may be considered incompressible up to a speed of 400 m/s; beyond this speed, the flow of water must be considered as compressible flow!.

The continuity equation for compressible flow is given by

$$\nabla.\rho U + \frac{\partial \rho}{\partial t} = 0$$

For steady flow, it reduces to $\nabla.\rho U = 0$

or $$U.\nabla\rho + \rho\nabla.U = 0$$

It may be noted that the sonic wave front in a stationary fluid must be spherical, emanating from the source of sound. If the source of sound is moving or if the fluid is set in motion at a velocity *U*, the wave front would no longer appear to be spherical to a stationary observer. The nature of wave front would depend upon the ratio of *U* to *c*. Change of the shape of wavefront in subsonic and sonic flows is shown in Fig. 12.4 where the circles labelled 1, 2 and 3 represent the wave fronts with the passage of time.

For supersonic flow, the wave front would be conical with the semi-cone angle, also called Mach angle, given by

$$\sin\alpha = \frac{ct}{Ut} = c/U = 1/M \qquad \text{...(12.11)}$$

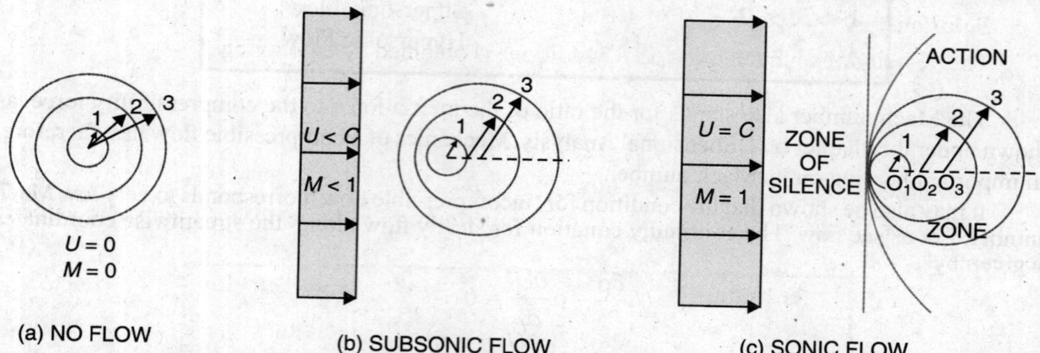

Fig. 12.4: Wave fronts until Sonic Flow

If a series of disturbances were created at the same location in a fluid flow, an observer outside the Mach cone would never hear the disturbance, hence the region outside Mach cone is called the silence zone, in contrast to the action zone within the cone as shown in Fig. 12.5. An example of this is in 'not being able to hear' a supersonic jet plane by a ground observer until it has passed over him/her. Higher the supersonic velocity, more the time lag.

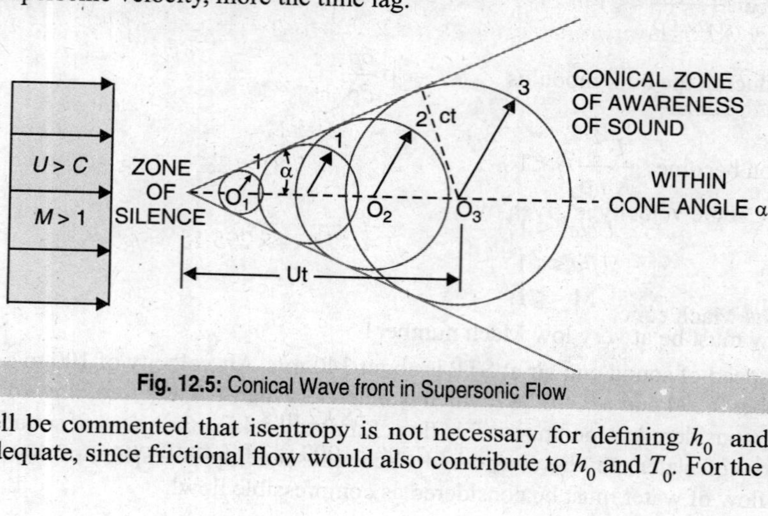

Fig. 12.5: Conical Wave front in Supersonic Flow

It may well be commented that isentropy is not necessary for defining h_0 and T_0 : only adiabaticity is adequate, since frictional flow would also contribute to h_0 and T_0. For the stagnation pressure p_0 and density ρ_0 concepts, isentropic retardation to rest is necessary. *For the same reason, constancy of the stagnation enthalpy or the stagnation temperature implies either isentropic or adiabatic flow, whereas constancy of the stagnation pressure or the stagnation density refers only to isentropic flow.*

Stagnation properties are also known as 'reservoir' or 'total' conditions in contrast to the static conditions in a flow; the differences between the stagnation and static values are termed as 'dynamic' values.

It may also be observed that the Bernoulli equation

$$\int dp/\rho + U^2/2 = \text{constant}$$

for a compressible flow undergoing isentropic changes

$$p = k\rho^\gamma$$

reduces to $\gamma/(\gamma - 1)\, p/\rho + U^2/2 = \text{constant} = \gamma/(\gamma - 1)\, p_0/\rho_0$ ...(12.12)

which is also consistent with the definition of the stagnation properties.

Example 12.2. *Determine the sonic velocity in air at STP conditions. How does it change with the altitude in the standard atmosphere ?*

Solution.

The velocity of sound in air at STP conditions is obtained by employing

$$c = \sqrt{\gamma RT}$$

where $\gamma = 1.4$, $R = 287$ J/kg K and $T = 288.15$ K

Therefore $c =$ **340.3 m/s**

At higher altitudes, until tropopause, the temperature T decreases. Hence the sonic velocity decreases

Altitude	T	Sonic Velocity
1 km	281.7 K	**336.4 m/s**
5 km	255.7 K	**320.5 m/s**
10 km	223.3 K	**299.5 m/s**
11 km	216.7 K	**295.1 m/s**
20 km	216.7 K	**295.1 m/s**

Example 12.3. *Determine the velocity of a bullet fired in the air if the Mach angle is observed to be 30°. Given that the temperature of the air is 22°C, density 1.2 kg/m³ Take $\gamma = 1.4$ and R = 29.3 m/K.*

Solution.

From the data,

$$T = 273.15 + 22 = 295.15 \text{ K}$$

$$R = 29.3 \text{ m}/K = 29.3 \times 9.81 = 287.4 \text{ J/kg K}$$

The sonic velocity is given by

$$c = \sqrt{\gamma RT} = \sqrt{1.4 \times 287.4 \times 295.15} = 344.6 \text{ m/s}$$

For the Mach cone,

$$\sin 30° = 1/M = 0.5$$

whence $M = 2.0 = U/c$

and $U = 2.0 \times 344.6 = 689.2$ m/s

$= 689.2 \times 3600 / 1000 =$ **2481 km/hour.**

Example 12.4. *An observer on the ground hears the sonic boom of a plane 15 km above when the plane has gone 20 km ahead of him. Estimate the speed of flight of the plane.*

Solution.

The observer hears the sound after the plane has passed over his head because he must be in the zone of silence before he hears it.

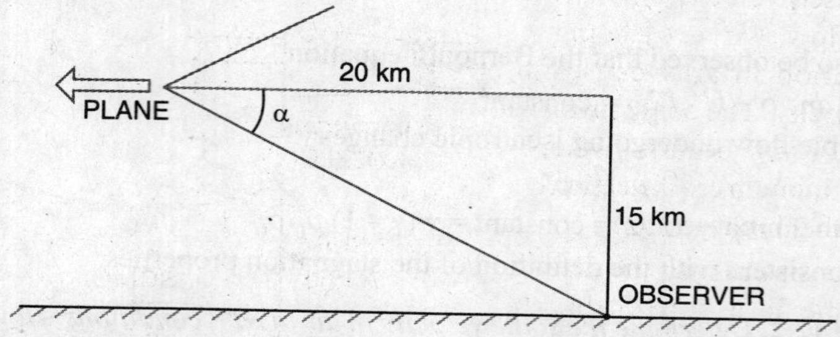

As soon as he becomes within the conical zone of awareness, he hears it. In that position,

$$\sin \alpha = c/U = 1/M$$

From the data, $\tan \alpha = 15/20 = 0.75 : \alpha = 36.87°$

Therefore. $\sin 36.87° = 0.6 = 1/M$ whence $M = \mathbf{1.67}$

The plane must be flying at a supersonic speed corresponding to a local Mach number of 1.67.

Example 12.5. A *Mach cone of Mach angle π/6 radian is observed for a fighter aircraft at an altitude where the temperature is 280 K. If the Mach angle is estimated within ± 5%, what is the range of velocity of the fighter ?*

Solution.

Velocity of sound $c = \sqrt{\gamma RT} = \sqrt{1.4 \times 287.4 \times 295.15} = 336$ m/s

$$1/M = c/U = \sin \alpha = \sin (\pi/6) = 0.5$$

whence, $M = 2$ and $U = 2 \times 336 = \mathbf{772\ m/s}$

Now, $U \sin \alpha = c = \text{constant}$

differentiating w.r.t. α, $U \cos \alpha + dU/d\alpha \sin \alpha = 0$

At $\alpha = \pi/6$ rad and $U = 772$ m/s.

$$dU/d\alpha = -772\ (0.866/0.30) = -2240 \text{ m/s per rad}$$

Error in α is $\pm\ \pi/6 \times 0.05 = \pm\ 0.0261$ rad

Therefore, error in U is $\pm 2240 \times 0.0261 = \mp$ **58 m/s**

12.4 Isentropic Flow

Isentropic flow implies that there is no heat transfer and no dissipation and hence '*adiabatic and frictionless*' flow.

By the continuity equation $d\rho/\rho + dU/U + dA/A = 0$

By the momentum equation $U\,dU + dp/\rho = 0$

or $U\,dU + dp/d\rho \,.\, d\rho/\rho = 0$

or $U\,dU + c^2 d\rho/\rho = 0$

Substituting for $d\rho/\rho$ from Eq. (12.11) and simplifying,

$$dA/A = dU/U\ (M^2 - 1) \quad ...(12.13)$$

This result brings out at a glance a very important difference between the subsonic and supersonic flow behaviour. In subsonic flow, $(M^2 - 1)$ being negative, reduction in area results in increased velocity whereas in supersonic flow, $(M^2 - 1)$ being positive, reduction in area causes reduction in velocity. The sonic flow, $M = 1$, with $dA/A = 0$ occurs at the location of the minimum cross-section which is called the throat section.

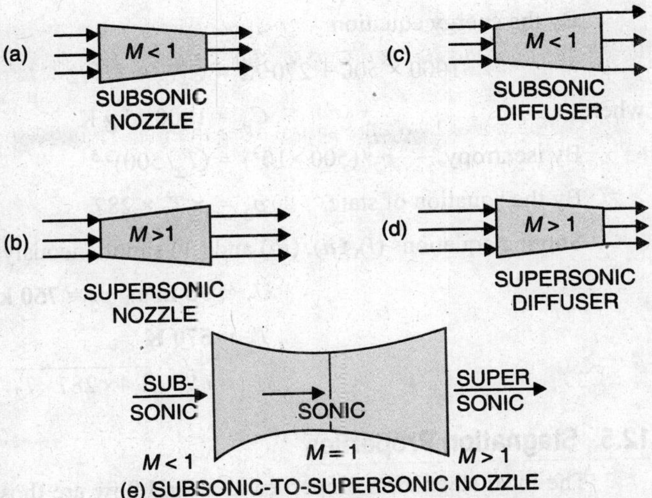

Fig. 12.6: Effect of Cross-sectional Area in Flow

A nozzle operating in the subsonic regime is a converging passage as in Fig. 12.6 (*a*) whereas it is a diverging passage in supersonic flow as shown in Fig. 12.6(*b*). *It follows that a subsonic-to-supersonic nozzle should consist of a converging-diverging passage shown in Fig. 12.6 (e).* Likewise, a diffuser is diverging passage in subsonic flow and a converging passage in supersonic regime, as shown in Fig. 12.6 (*c*) and (*d*).

Variation of different parameters with the geometry of the passage is shown in Table 12.3.

Table 12.3. Flow through Converging and Diverging Passages

	Converging Passage		*Diverging Passage*	
	Subsonic $M < 1$	*Supersonic* $M > 1$	*Subsonic* $M < 1$	*Supersonic* $M > 1$
Velocity, U	increases	decreases	decreases	increases
Pressure, p	decreases	increases	increases	decreases
Density, ρ	decreases	increases	increases	decreases

Example 12.6. *A diffuser of area ratio 2 : 1 operates at the inlet conditions,*

$$p_1 = 500 \text{ kN/m}^2,\ T_1 = 500 \text{ K},\ \ M_1 = 0.6 \text{ and } \gamma = 1.4.$$

Estimate the following at the exit :

(a) velocity, (b) pressure, (c) temperature and (d) Mach number.

Solution.

From the data,

$$c_1 = \sqrt{\gamma R T_1} = \sqrt{1.4 \times 287 \times 500} = 450 \text{ m/s}$$

$$U_1 = 0.6 \times 450 = 270 \text{ m/s}$$

$$\rho_1 = p_1/RT_1 = 500 \times 10/287 \times 500 = 3.5 \text{ kg/m}^3$$

By continuity,

$$\rho_1 U_1 A_1 = \rho_2 U_2 A_2$$

or

$$\rho_2 U_2 = 3.5 \times 270^2/2 = 470 \text{ kg/m}^2\text{s} \qquad ...(i)$$

By the energy equation,

$$1000 \times 500 + 270^2/2 = C_p T_2 + U_2^2/2$$

where $C_p = 1000$ J/kg K ...(ii)

By isentropy, $p_2/(500 \times 10^3) = (T_2/500)^{3.5}$ ...(iii)

By the equation of state, $p_2 = \rho_2 T_2 \times 287$ ...(iv)

Solving equations (i), (ii), (iii) and (iv) simultaneously; by hit and trial,

$$U_2 = \mathbf{95\ m/s} \quad p_2 = \mathbf{750\ kN/m^2}$$

$$T_2 = \mathbf{570\ K}$$

$$M_2 = U_2/\sqrt{1.4 \times 287 \times T_2} = \mathbf{0.2}$$

12.5 Stagnation Properties

The local stagnation properties in a fluid flow are those which would be reached if the fluid were brought to rest isentropically either actually or hypothetically.

Stagnation enthalpy, $h_0 = h + U^2/2$

Stagnation temperature $T_0 = 1 + U^2/2\,C_p$ ...(12.14)

or $T_0/T = 1 + U^2/2\,C_p T = 1 + (\gamma - 1)/2)\,M^2$ ...(12.15)

using $C_v = (\gamma/\gamma - 1)\,R,\ c^2 = \gamma RT$

and $M = U/c,$

the stagnation pressure, by a hypothetical isentropic flow process,

$$p_o/p = [1 + (r - 1)/2\,M^2]^{\gamma/\gamma - 1} \quad ...(12.16)$$

Likewise, the stagnation density ρ_0 is given by

$$\frac{\rho_0}{\rho} = \left(1 + \frac{r-1}{2} M^2\right)^{\frac{1}{r-1}} \quad ...(12.17)$$

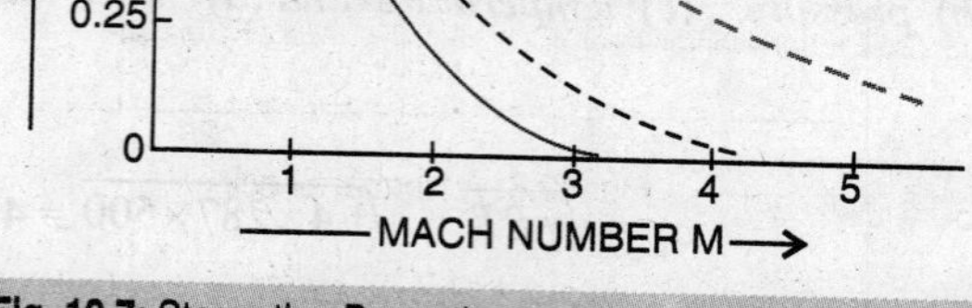

Fig. 12.7: Stagnation Properties versus Mach Number

In terms of stagnation values, the local values of temperature, pressure and density are related as follows :

$$\frac{T}{T_0} = \left(1 + \frac{r-1}{2} M^2\right)^{-1};$$

$$\frac{p}{p_0} = \left(1 + \frac{r-1}{2} M^2\right)^{\frac{-r}{r-1}} ;$$

$$\frac{\rho}{\rho_0} = \left(1 + \frac{r-1}{2} M^2\right)^{\frac{-r}{r-1}}$$

All these relations are shown plotted in the figure 12.7.

Example 12.7. *An air stream with a velocity of 300 m/s at 10 kN/m² vacuum and temperature of 320 K passes through a section where its gauge pressure is 20 kN/m². Compute its stagnation properties. The atmospheric pressure is 100 kN/m² and R = 287 J/kg K.*

Solution.

From the data, $p_1 = 100 - 10 = 90 \text{ kN/m}^2, \gamma = 1.4$

$U_1 = 300$ m/s,

$R = 287$ J/kg K

$T_1 = 320$ K,

$C_p = \gamma / (\gamma - 1) R = 1.4 \times 287/0.4 = 1004.5$

Now, $c_1 = \sqrt{\gamma R T_1} = \sqrt{1.4 \times 287 \times 320} = 358$ m/s

Therefore, $M_1 = U_1 / c_1 = 300/358 = 0.84$

and $T_0 = T_1 (1 + (\gamma - 1)/2) M_1^2) = 320(1 + 0.2 \times 0.84^2) = 320 \times 1.14 = \mathbf{365\ K}$

$p_0 = 90 \times 1.14^{3.5} = \mathbf{142\ kN/m^2}$

12.6. Normal Shock

For supersonic flow in a passage or around a body, the downstream pressure and geometrical conditions may require an abrupt reduction of velocity and consequent changes of the flow properties. Such abrupt changes take place through a 'shock'. Normal shocks are substantially perpendicular to the flow and oblique shocks inclined at other angles. Shock formation is possible for confined flows as well as for external flows. A shock may be attached or detached with respect to the body over which it may occur. Normal and oblique shocks may mutually interact to make a shock pattern.

Considering a normal shock, as visualised in Fig. 12.8 for perfect gas undergoing one-dimensional steady adiabatic flow in a constant area duct, $A_1 = A_2 = A$.

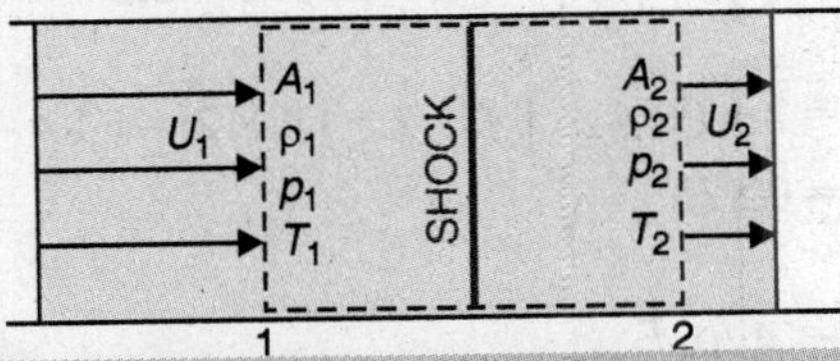

Fig. 12.8: A Normal Shock

By continuity, $\rho_1 U_1 = \rho_2 U_2$ ...(i)

By the momentum principle, $p_1 - p_2 = \rho_1 U_1 (U_2 - U_1)$ ...(ii)

and by energy conservation, $C_p T_1 + U_1^2/2 = C_p T_2 + U_2^2/2$ ...(iii)

Substituting (i) in (ii) $p_1 - p_2 = \rho_2 U_2^2 - \rho_1 U_1^2$

Introducing $U = c M = \sqrt{\gamma . p / \rho} . M$

$$p_1 - p_2 = \gamma(\rho_2 M_2^2 - \rho_1 M_1^2)$$

whence, $$p_2/p_1 = (1+\gamma M_1^2)/(1+\gamma M_2^2) \qquad ...(12.18)$$

Finally, we can state that in general

$$p_2/p_1 = f(M_1, M_2, \gamma) \qquad ...(12.19)$$

Shock waves are accompanied by dissipation of energy and can only be considered adiabatic, i.e., not isentropic. Equating expressions for p_2/p_1 from Eqs. (12.18) and (12.19), and neglecting the trivial solution $M_1 = M_2$ for no shock at all,

$$M_2^2 = \frac{2+(\gamma-1)M_1^2}{2\gamma M_1^2 - (\gamma-1)} \qquad ...(12.20)$$

With this substitution,

$$p_2/p_1 = 1 + 2\gamma/(\gamma+1).(M_1^2 - 1) \qquad ...(12.21)$$

The following conclusions are drawn from the one-dimensional analysis of a normal shock:

(a) A normal shock can occur only if M_1 is greater than 1

(b) M_2 must be less than 1 for a normal shock

(c) Compression must occur across it; p_2/p_1 rises sharply with M_1

(d) T_2/T_1 increases with M_1

(e) Shock strength defined by $p_2 - p_1/p_1$ or M_1/M_2 increases with an increase in M_1.

(f) Entropy rise across a shock increases abruptly with M_1 showing that a stronger shock is associated with greater loss of energy. In other words, the efficiency of compression across a shock decreases with an increase in the inlet Mach number.

Example 12.8. *Calculate the downstream pressure and temperature, shock strength and the energy efficiency of a normal shock wave observed to occur in an air nozzle at*

$$M_1 = 2, p_1 = 20\ kN/m^2, T_1 = 300\ K$$

Solution.

By Eq. (12.21),

$$p_2 = 20 \times 10^3 [1 + (2 \times 1.4/(1+1.4)(2^2 - 1)]$$
$$= \mathbf{90 \times 10^3\ N/m^2 = 90\ kN/m^2}$$

The shock strength $S = (p_2 - p_1)/p_1 = (90 - 20)/20 = \mathbf{3.5}$

By Eq. (12.20),

$$M_2^2 = 2 + (1.4 - 1)2^2/2 \times 1.4 \times 2^2 - (1.4 - 1) = 0.33$$

and $$M_2 = 0.575$$

By the equation of state,

$$p_1/\rho_1 T_1 = p_2/\rho_2 T_2$$

or $$T_2/T_1 = p_2/p_1 \times \rho_1/\rho_2 = 4.5/2.67 = 1.7$$

whence $$T_2 = 1.7 \times 300 = \mathbf{510\ K}$$

If the process took place isentropically, from pressure p_1, to pressure p_2 the final temperature T_2 would have been given by

$$T_2/T_1 = (p_2/p_1)^{\gamma - 1/\gamma} = (4.5)^{1.4 - 1/1.4} = 1.53$$
$$T_2 = 300 \times 1.53 = \mathbf{459\ K}$$

The energy efficiency of the shock wave is thus

$$\eta = (459 - 300)/(510 - 300) = 0.756 \text{ or } \mathbf{75.6\%}$$

12.7 Flow through a Converging-Diverging Nozzle

A converging-diverging nozzle is usually employed to expand a flow from a given reservoir conditions to a controlled back-pressure condition. Let the reservoir conditions be specified by T_0, p_0 and ρ_0 with an inlet velocity U. The back-pressure conditions are, in general T, p and ρ. It is interesting to observe the flow, *i.e.*, variation in pressures and velocity through the nozzle as the back-pressure p is reduced from the initial value of p_a to a low value of p_f gradually as shown in Fig. 12.9

1. *For p/p_0 close to unity, the entire nozzle operates subsonic.*
2. *Further lowering of p/p_0 to p_c/p_0 will cause the sonic velocity at the throat which drops off to subsonic flow at the exit and choking just occurs.*
3. *Very low back-pressure p_f causes the flow to be supersonic after the throat up to the exit isentropically.*
4. *Low pressure such as p_d and p_e causes the flow to continue supersonic up to a certain distance from the throat and then a shock wave occurring within the diverging portion renders the exit condition subsonic.*
5. *If the back-pressure is lowered below p_f, a series of oblique expansion waves are observed to occur outside the nozzle.*

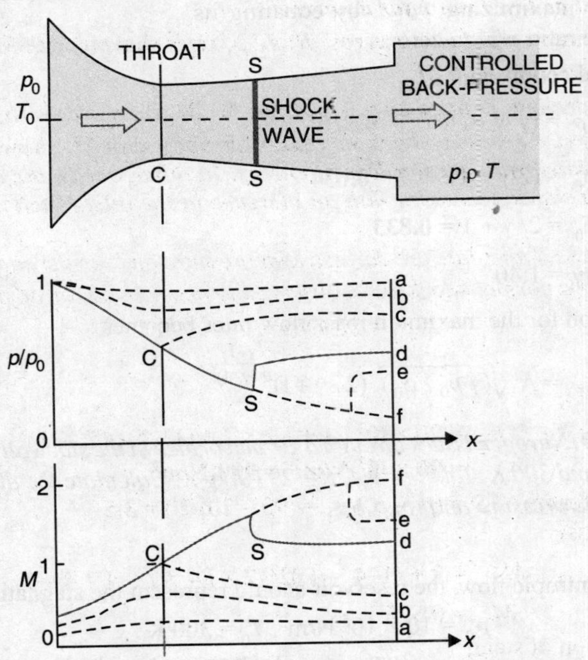

Fig.12.9: Flow through a Nozzle

By the energy conservation principle,

$$C_pT_0 = C_pT + U^2/2$$

Invoking $$C_p = \gamma/(\gamma-1)R$$

and $$p/\rho = RT$$

$$\gamma/(\gamma-1)\, p_0/\rho_0 = \gamma/(\gamma-1)\, p/\rho + U^2/2$$

$$U^2 = 2\gamma/(\gamma-1)\,(p_0/\rho_0 - p/\rho)$$

$$= 2\gamma/(\gamma-1)\, p_0/\rho_0 \cdot (1 - p/p_0 \cdot \rho_0/\rho)$$

Employing $\quad p_0/\rho_0^{\gamma} = p/\rho^{\gamma}$ for isentropic flow

$$U = [2\gamma/(\gamma-1)\, p_0/\rho_0 \cdot (1-(p/p_0)^{\gamma-1/\gamma}]^{1/2} \qquad ...(12.22)$$

The mass flow rate per unit area

$$\dot{m}/A = \rho U = (p/p_0)^{1/\gamma}\, \rho U$$

or

$$\dot{m}/A = [2\gamma/(\gamma-1)\, p_0/\rho_0\, (p/p_0)^{2/\gamma}\, (p/p_0)^{\gamma+1/\gamma}]^{1/2} \qquad ...(12.23)$$

With the reduction of p/p_0 the mass flow rate $\dot{m}/A$ increases until a maximum value is reached, then drops off to zero at $p/p_0 = 0$ as shown in Fig. 12.10. In actual practice the flow represented by the first part of the curve up to C is observed and after the maximum m/A, the flow rate remains constant even in the face of reduction of p/p_0. This is explained by the fact that the critical condition *e.g.*, sonic velocity is reached at C and any further lowering of the exit pressure is not felt upstream, it being in the zone of 'silence'. Evaluation of p_c/p_0 is made by maximizing $\dot{m}/A$ by equating its derivative w.r.t. the ratio p/p_0 to zero.

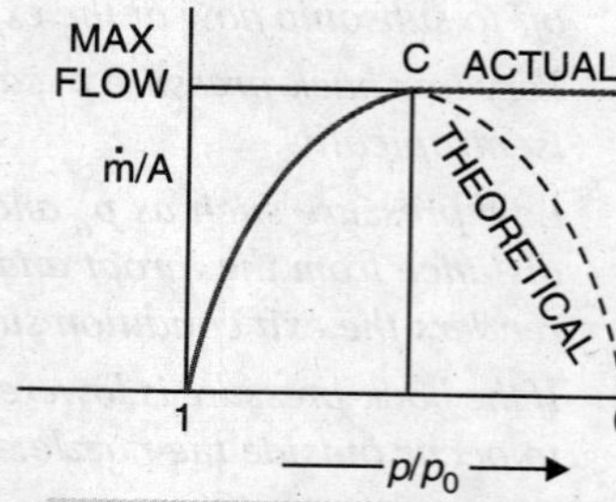

Fig. 12.10

At the critical condition,

$$p_c/p = (\gamma/\gamma+1)^{\gamma/\gamma-1} = \mathbf{0.528}$$

$$\rho_c/\rho_0 = (2/\gamma+1)^{1/\gamma-1} = \mathbf{0.634}$$

$$T_c/T_0 = 2/\gamma+1 = \mathbf{0.833} \qquad ...(12.24)$$

For air, $\quad \gamma = 1.40$

The expression for the maximum mass flow rates becomes

$$\dot{m}_{max} = A\sqrt{(\gamma\, p_0/\rho_0)}\ (2/\gamma+1)^{\frac{\gamma+1}{2(\gamma-1)}} \qquad ...(12.25)$$

Example 12.9. *Nitrogen issues through a 10 mm orifice in the sidewall of a reservoir maintained at 10 bar pressure and 300 K. Take γ = 1.4, R = 297 J/kg K. Calculate the discharge in kg/s if the exit is at (a) atmospheric pressure and (b) 6 bar.*

Solution.

Assuming isentropic flow, the reservoir should represent the stagnation conditions.

$$p_0 = 10 \times 10^5\ \text{N/m}^2,\ T_0 = 300\ \text{K}$$

$$\rho_0 = p_0/RT_0 = 10 \times 10^5/297 \times 300 = 11.4\ \text{kg/m}^2$$

In such problems, it should be ascertained whether the critical conditions govern the flow or the actual conditions do so. In order to do so, let us work out.

$$p_c/p_0 = (2/(\gamma+1))^{\gamma/\gamma-1} = \mathbf{0.528}$$

Case (a): $\quad p_2/p_0 = 1/10 = 0.1$; less than the critical ratio.

The orifice exit, in effect, becomes the critical section.

$$T_c/T_0 = 2/(\gamma+1) = \mathbf{0.833},$$

whence $\quad T_c = 300 \times 0.833 = 250\ \text{K}$

Velocity at the orifice exit $c = \sqrt{\gamma RT_c} = \sqrt{(1.4 \times 297 \times 250)} = 322$ m/s

$$\rho_c = p_c / RT_c = \frac{0.528 \times 10 \times 10^5}{297 \times 250} = 7.2 \text{ kg/m}^3$$

At the critical conditions $\dot{m} = \rho U A$

$= 7.2 \times 322 \times \pi \times 0.01^2/4 = \mathbf{0.182 \text{ kg/s}}$

Case (b): $p/p_0 = 6/10 = 0.6$; more than the critical ratio.

Here, the orifice velocity is going to be less than the sonic velocity.

$$T/T_0 = (0.6)^{(\gamma - 1)/\gamma} = (0.6)^{(1.4 - 1)/1.4} = 0.864$$

$$T = 300 \times 0.864 = 259 \text{ K}$$

$$\rho = p/RT = 0.6 \times 10 \times 10^5/297 \times 259 = 7.9 \text{ kg/m}^3$$

Mass flow rate is given by

$$\dot{m}/A = \left[\frac{2 \times 1.4}{1.4 - 1} \times 10 \times 10^5 \times 11.4\left\{(0.6)^{2/1.4} - (0.6)^{(1.4+1)/1.4}\right\}\right]^{1/2}$$

whence $\dot{m} = \mathbf{0.179 \text{ kg/s}}$ for $A = \dfrac{\pi \times 0.01^2}{4}$ m²

If the exit pressure is raised further the discharge would decrease until zero value at the exit pressure of 10 bar, i.e., the same as at the entry.

Example 12.10. *A venturimeter with inlet area 100 cm² and throat 25 cm² is employed to measure the discharge of carbon dioxide supplied at two bar and 290 K. The U-tube upright differential manometer connected between the inlet and throat shows 20 cm deflection of the water column. Estimate the mass flow rate. Given that*

$$\gamma = 1.3, R = 188 \text{ J/kg K}, p_{atm} = 200 \text{ kN/m}^2.$$

Solution.

$$p_1 = 200 \text{ kN/m}^2;$$

$$T_1 = 290 \text{ K}$$

$$\rho_1 = 200 \times 1000/290 \times 188 = 3.67 \text{ kg/m}^3$$

By Bernoulli equation,

$$(p_1 - p_2)/\rho g + (z_1 - z_2) = h_m\left(\frac{\rho_m}{\rho} - 1\right)$$

$$= 0.2\,(100/3.67 - 1)$$

$$= 52.2 \text{ m head of gas.}$$

Neglecting the datum difference $(z_1 - z_2)$

$$p_2 = 200 - 52.2 \times 3.67 \times 9.81/1000 = 198 \text{ kN/m}^2.$$

By isentropy, $p_1/p_2 = (\rho_1/\rho_2)^\gamma$

$$\frac{200}{198} = \left(\frac{3.67}{\rho_2}\right)^{1.3}$$

whence, $\rho_2 = 3.65 \text{ kg/m}^3$

By the Bernoulli equation, for compressible flow,

$$\gamma/(\gamma - 1) \cdot p_1/\rho_1 + U_1^2/2 = \gamma/(\gamma - 1) \cdot p_2/\rho_2 + U_2^2 \quad \text{....(i)}$$

By the continuity equation,

$$\rho_1 A_1 U_1 = \rho_2 A_2 U_2 = \dot{m} \quad \text{...(ii)}$$

From (*ii*),

$$3.67 \times 100 \times U_1 = 3.65 \times 25 \times U_2$$

$$U_1 = 0.249\, U_2$$

From (*i*), by substitution,

$$1.3/(1.3 - 1) \times 200 \times 1000/3.67 + 0.249^2 \times U_2^2/2$$

$$= 1.\, 3/(1.\, 3 - 1) \times 198 \times 1000/3.65 + U_2^2/2$$

whence $U_2 = 186.49$ m/s

and $U_1 = 46.44$ m/s

Mass flow rate $\dot{m} = 3.65 \times 25 \times 186.49/10000 =$ **1.70 kg/s**

12.8 Pitot-Static Probe in a Compressible Flow

The theory and use of a pitot-static problem in an incompressible flow has already been discussed under the applications of the dynamical equations governing the flow of fluids. The pitot-static problem may also be used to estimate the velocity of flow of a compressible fluid whether it is flowing at subsonic or at supersonic conditions.

Let us first consider the flow at subsonic compressible conditions and arrive at a correction factor which must be applied to estimate the velocity.

By the energy equation,

$$T_0/T = (1 + (\gamma - 1)/2)M^2$$

and for an isentropic process,

$$p_0/p_1 = (1 + (\gamma - 1)/2\, M_1^2)^{\gamma/\gamma-1}$$

This equation is not straightforward for use. Instead, at least for subsonic flow up to Mach number of 0.4, an alternalive technique is employed. This is by the introduction of a compressibility correction faclor (*CCF*) as follows : Expanding the r.h.s. terms for $M_l < 1$,

$$p_0/p_1 = 1 + (\gamma/2)\, M_1^{\,2} + (\gamma/8)\, M_1^{\,4} + [\gamma\,(2 - \gamma)/48]\, M_1^{\,6} + ...$$

$$p_0 - p_1 = \gamma\, p_1 / 2\, M_1^2\, (1 + \; M_1^2\, /4 + (2 - \gamma)/24\; M_1^4\; + ...)$$

This equation has the same form as for incompressible flow. The bracketed term constitutes the Compressibility Correction Factor,

$$CCF = (1 + \; M_1^2\, /4 + (2 - \gamma)\, /24\; M_1^4\; +....) \quad ...(12.26)$$

which reduces to unity for low subsonic flows.

The compressibility correction factor is shown plotted as a function of M in Fig. 12.11.

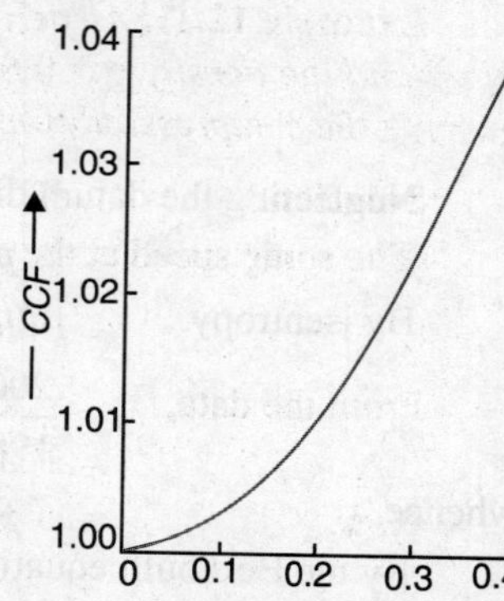

Fig. 12.11: Compressibility Correction Factor

If a pitot-static problem is placed in a supersonic flow, then a detached shock, approximately a normal stock, appears in front of the leading tip. The process of pressure rise is two-fold, *i.e.*, (*i*) through the normal shock and (*ii*) by virtue of the stagnation point of the leading tip. The pitot-static tube remains wholly in the subsonic region. This is shown in Fig. 12.12.

If the velocity indicator on the pitot-static probe is calibrated according to the incompressible flow,

$$p_0 - p_1 = \frac{1}{2}\rho U_i^2$$

then $$\left(\frac{U_i}{U}\right)^2 = 1 + M_1^2/4 + \ldots\ldots$$

whence $$\frac{U_i}{U} = 1 + M_1^2/8 + \ldots\ldots$$

It shows that the indicated speed *Ui* is higher than the true speed by a factor of $(1 + M_1^2/8)$.

At low Mach numbers, the error is indeed negligible. At $M_1 = 0.1$, *i.e.* at 10% the speed of sound, the factor is 1.001. At $M_1 = 0.2$, the factor is 1.005. The error becomes 10% at a Mach number exceeding 0.28.

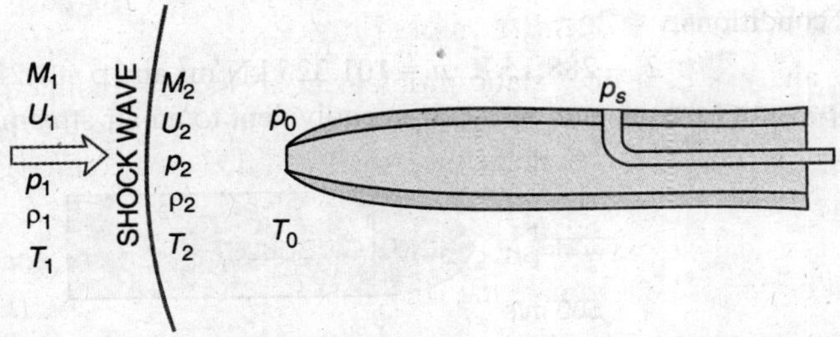

Fig. 12.12: Pitot - Static Probe in Supersonic Flow

Pressure rise through the normal shock is governed by

$$p_2/p_1 = 1 + 2\gamma/(\gamma+1)\,.\,(M_1^2 - 1)$$

Further pressure rise from p_2 to p_0 in the subsonic flow must be given by

$$p_0/p_2 = (1 + (\gamma - 1)/2\, M_2^2)^{\gamma/\gamma - 1}$$

From these two equations,

$$p_0/p_1 = p_0/p_2 \times p_2/p_1 \quad \ldots.(12.27)$$

Example 12.11. *An aeroplane flies at 1000 km/hour at an altitude where the pressure p is 50 kN/m² and the density ρ = 0.6 kg/m². Calculate the pitot speed indicctor reading if it is calibrated ignoring the compressibility effects.*

Solution.

The sonic speed at the ambient conditions is given by

$$c = \sqrt{\gamma p/\rho} = \sqrt{1.4 \times 50 \times 10^3/0.6} = 342 \text{ m/s}$$

From the data, $$U = 1000 \times 1000/60 \times 60 = 278 \text{ m/s}$$

$$M = U/c = 278/342 = 0.81$$

Now, $$p_0 = 50 \times (1 + (1.4 - 1)/2 \times 0.81^2)^{1.4/1.4 - 1}$$

$$= 77 \text{ kN/m}^2$$

Neglecting compressibility,

$$p_0 - p_1 = \rho U^2/2$$

or $$U = \sqrt{2(77 - 50) \times 10^3/0.6} = 300 \text{ m/s}$$

The pitot probe, uncorrected for compressibility, is likely to indicate 300 m/s which is in excess of the true value of 278 m/s by

$$(300 - 278)/278 \times 100 = \mathbf{7.9\%}$$

Alternatively, the correction factor,

$$CCF = (1 + M^2/4 + ...)$$
$$\approx 1 + 0.81^2 \approx 1.164$$

This is a correction on the square of velocity. Correction to the velocity is $\sqrt{1.164} = 1.079$, *i.e.*, **7.9%.**

The indicated speed of 300 m/s is, therefore, corrected to 300/1.079, i.e., 278 m/s true speed!

Example 12.12. *Calculate the air pressure at the nose of a bullet passing through air at 200 m/s velocity. Assume STP conditions. What happens if the velocity of the bullet is doubled ?*

Solution.

At the STP conditions,

$$T = 288.15 \text{ K}, p = 101.325 \text{ kN/m}^2 \text{ and } \rho = 1.225 \text{ kg/m}^3$$

The bullet passing through stationary air is equivalent to an air stream with 500 m/s flowing past the bullet.

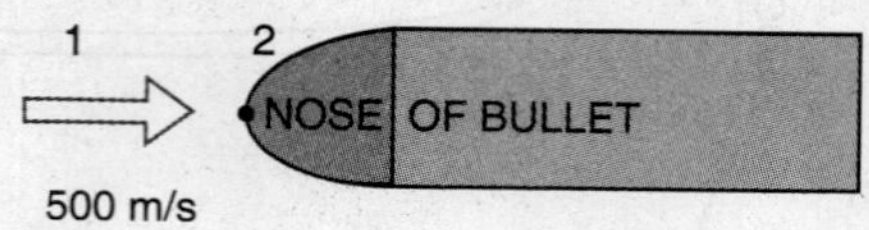

The sonic velocity in air is

$$c = \sqrt{\gamma RT} = \sqrt{1.4 \times 287 \times 288.15} = 340.3 \text{ m/s}$$

and the Mach number of flow is

$$M = U/c = 200/340.3 = 0.59.$$

For isentropic compression from 1 to the nose of the bullet 2, a point of stagnation,

$$\frac{p_0}{p_1} = (1 + \frac{\gamma - 1}{2} M_1^2)^{\frac{\gamma}{\gamma - 1}}$$

$$= (1 + \frac{1.4 - 1}{2} \times 0.59^2)^{3.5} = 1.266$$

Hence $\quad p_0 = 1.266 \times 101.325 = \mathbf{128.24 \text{ kN/m}^2}$

Alternatively, one could determine the pressure at the nose by employing the relation

$$p_0 - p_1 = \frac{\rho_1 U_1^2}{2}.(CCF)$$

where *CCF* is the compressibility correction factor,

$$CCF = (1 + \frac{M_1^2}{4} + ...) = 1 + \frac{0.59^2}{4} = 1.087$$

Therefore,

$$p_0 - p_1 = \frac{1.225 \times 200^2}{2} \times 1.087$$

and $\quad p_0 = 101325 + 26\,632 = 127\,597 \text{ N/m}^2$

$$= \mathbf{127.96 \text{ kN/m}^2}$$

This is indeed an approximation to the value of 128.4 kN/m² determined earlier.

If the velocity of the bullet is 400 m/s

$$M = 400/340.2 = 1.176$$

When supersonic flow approaches the bullet, a shock wave is formed ahead of it. The pressure across the normal shock is given by

$$p_2 = \left(\frac{2\gamma}{\gamma+1}M_1^2 - \frac{\gamma-1}{\gamma+1}\right)p_1$$

$$= \left(\frac{2\times1.4}{1.4+1}\times1.176^2 - \frac{1.4-1}{1.4+1}\right)\times101.325$$

$$= 146.6 \text{ kN/m}^2$$

The Mach number across the normal shock is given by

$$M_2^2 = \frac{2+(\gamma-1)M_1^2}{2\gamma M_1^2 - (\gamma-1)}$$

$$= \frac{2+0.4\times1.176^2}{2\times1.4\times1.176^2-0.4} = 0.735$$

and $M_2 = 0.86$

The pressure rise from this point until the tip of the nose is given by

$$\frac{p_0}{p_2} = (1+\frac{\gamma-1}{2}M_2^2)\frac{\gamma}{\gamma-1}$$

$$= \left(1+\frac{0.4}{2}\times0.86^2\right)^{3.5} = 1.62$$

whence $p_0 = 146.5\times1.62 = \mathbf{237.4\ \ kN/m^2}$

12.9 Transonic and Supersonic Flows

When an aerofoil shaped body is subjected to compressible flow, different types of flow patterns develop in the vicinity of the body.

As shown in Fig. 12.13, for low subsonic free stream M_1, the flow may remain subsonic

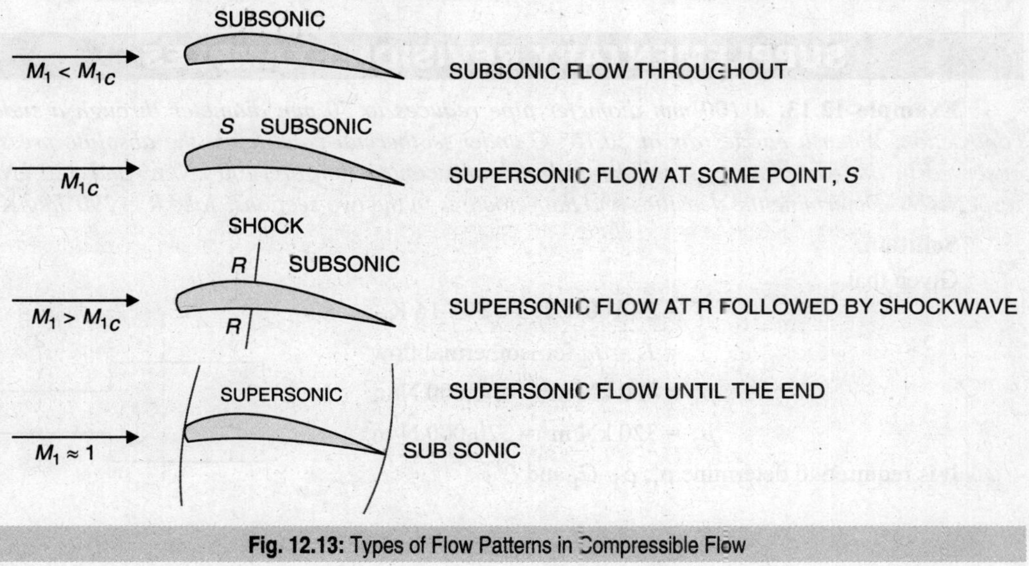

Fig. 12.13: Types of Flow Patterns in Compressible Flow

throughout. At higher subsonic free stream values, the flow may appear to be similar in nature. At a critical free stream velocity, the flow at some point on the body may become sonic; the free stream Mach number M_{1C} is said to have become critical. The value of M_{1C} is always less than unity. For a thick aerofoil, M_{1C} may be 0.8 and for a thin aerofoil, it may be 0.9. For free stream Mach numbers beyond M_{1C}, the flow must become supersonic in some region or the other. It may change to subsonic in the downstream region. It is sure to become subsonic in the tail region, because the obstruction vanishes downstream of the body. For free stream Mach number becoming unity, the flow is supersonic throughout the region close to the body; it becomes subsonic downstream of it.

If the free stream flow is supersonic, $M_1 > 1$, a shock wave, called bow shock, is formed ahead of the body as shown in Fig. 12.14. The Mach number drops across the shock, but it may remain partially supersonic and partially subsonic over the body. There is bound to be a tail shock wave as well.

Presence of 'standing' bow shock ahead of the body results in a great deal of drag force as also loss of energy. A normal aerofoil designed for subsonic flow is, therefore, not suitable in supersonic flow. It is for this reason that supersonic aerofoils are sharp-edged or diamond-edge thin aerofoils. Supersonic flow over such sharp-edged aerofoils is accompanied by oblique shocks, which result in significantly less loss of energy than normal shocks do.

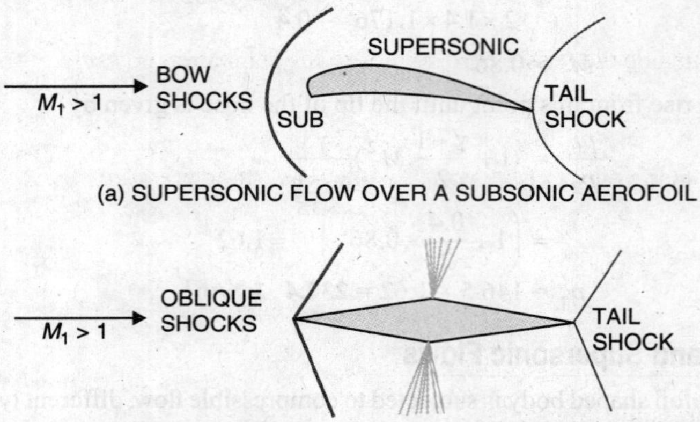

Fig. 12.14: Supersonic Flow past Aerofoils

SUPPLEMENTARY REVISION EXAMPLES

Example 12.13. *A 100 mm diameter pipe reduces to 50 mm diameter through a sudden contraction. When it carries air at 20.16° C under isothermal conditions, the absolute pressure observed in the two pipes, just before and after the contraction are 400 kN/m² and 320 kN/m² respectively. Determine the densities and the velocities in the two sections. Take R = 290 J/kg K.*

Solution.

Given that

$$T = 20.160 + 273 = 293.16 \text{ K}$$

$$= T_1 = T_2 \text{ for isothermal flow}$$

$$p_1 = 400 \text{ kN/m}^2 = 400{,}000 \text{ N/m}^2$$

$$p_2 = 320 \text{ kN/m}^2 = 320{,}000 \text{ N/m}^2.$$

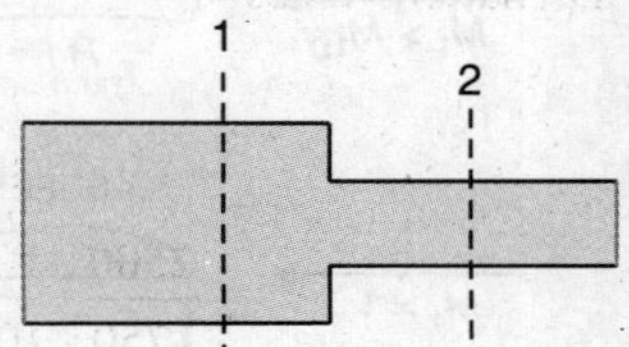

It is required to determine ρ_1, ρ_2, U_1 and U_2.

From the equation of state,

$$p = \rho RT$$

$$\rho_1 = \frac{400{,}000}{290 \times 293.16} = \mathbf{4.70\ kg/m^3}$$

$$\rho_2 = \frac{320{,}000}{290 \times 293.16} = \mathbf{3.76\ kg/m^3}$$

By continuity, $\rho_1 A_1 U_1 = \rho_2 A_2 U_2$

$$\frac{U_2}{U_1} = \frac{4.7 \times \pi\, 0.1^2/4}{3.76 \times \pi\, 0.05^2/4} = \frac{4.7 \times 0.1^2}{3.76 \times 0.05^2} = \mathbf{5.0}$$

The velocity at section 2 is five times the velocity at section 1.

Example 12.14. *A large vessel contains air at a pressure of 2500 kN/m² and temperature 20° C. It is fitted with nozzle. If the pressure at the outlet of the nozzle is 1750 kN/m², find the velocity of the air flowing at the outlet of the nozzle.*

Solution.

The pressure inside the vessel corresponds to the stagnation pressure

$$p_0 = 2500 \times 10^3\ \text{N/m}^2.$$

Likewise $T_o = 20 + 273 = 293$ K.

The pressure at the outlet of the nozzle is the local static pressure

$$p = 1750 \times 10^3\ \text{N/m}^2$$

The equation of state is

$$p = \rho RT$$

whence $$\rho = \frac{2500 \times 1000}{287 \times 293} = 30\ \text{kg/m}^3.$$

In an incompressible flow,

$$p_0 = p + \frac{\rho u^2}{2}.$$

and u would be given by

$$u = \sqrt{\frac{2(p_0 - p)}{\rho}} = \sqrt{\frac{2(2500 - 1750) \times 10^3}{30}}$$

$$= 223.6\ \text{m/s}.$$

Approximate sonic velocity $= \sqrt{\gamma RT} = \sqrt{1.4 \times 287 \times 293} = 343$ m/s

Compared to which u of 223.6 m/s is not so small; the flow is not subsonic and incompressible!.

Now, that we know that the flow is compressible, let us assume it adiabatic and use the stagnation values

$$\frac{p_0}{p} = \left[1 + \frac{\gamma - 1}{2} M^2\right]^{\frac{\gamma}{\gamma - 1}}$$

$$\frac{2500 \times 10^3}{1750 \times 10^3} = \left[1 + \frac{1.4 - 1}{2} M^2\right]^{\frac{1.4}{0.4}} = \left[1 + 0.2\, M^2\right]^{3.5}$$

whence $1 + 0.2\,M^2 = (1.429)^{1/3.5} = 1.107,\ M^2 = 0.534$

and $M = 0.73.$

We can now find the temperature of the air leaving the nozzle,

$$\frac{T_0}{T} = 1 + \frac{\gamma - 1}{2} M^2 = 1 + 0.2\,M^2$$

$$T = \frac{293}{1 + 0.2 \times 0.534} = 265\ K.$$

Sonic velocity $c = \sqrt{\gamma R T} = \sqrt{1.4 \times 287 \times 265}$

$= 326$ m/s.

Velocity of the air at the exit is given by

$$u = c\,M = 326 \times 0.73 = \mathbf{238\ m/s.}$$

This is a much better estimate than the initial 223.6 m/s by assuming incompressible flow.

Example 12.15. *Oxygen flows in a conduit at an absolute pressure of 170 kN/m². If the absolute pressure and temperature at the nose of a small object in the stream are 200 kN/m² and 70.16° C respectively, determine the velocity of oxygen in the conduit. Assume γ = 1.4 and R = 281.43 J/kg K.*

Solution.

When a small object is placed in a stream of a gas, the pressure and temperature at the nose are the stagnation values.

$$p_0 = p\left[1 + \frac{\gamma - 1}{2} M^2\right]^{\frac{\gamma}{\gamma - 1}}$$

$$T_0 = T\left[1 + \frac{\gamma - 1}{2} M^2\right].$$

Given that $p = 170 \times 10^3$ N/m²; $\gamma = 1.4$

and $p_0 = 200 \times 10^3$ N/m²

$$200 \times 10^3 = 170 \times 10^3\left[1 + 0.2\,M^2\right]^{3.5}$$

whence $1 + 0.2\,M^2 = 1.176^{0.286} = 1.047$

$0.2\,M^2 = 0.047$

$M = 0.485$

We determine the temperature T from

$$(70.16 + 273) = T\,[1 + 0.2\,M^2]$$

$$T = \frac{343.16}{1 + 0.047} = \mathbf{327.8\ K}$$

The sonic velocity is given by

$$c = \sqrt{\gamma R T}$$

$$= \sqrt{1.4 \times 281.43 \times 327.8}$$

$$= 359.3 \text{ m/s}$$

Velocity of the oxygen in the conduit is estimated to be

$$U = 0.485 \times 359.3$$

$$= \mathbf{174.3\ m/s}$$

Example 12.16. *An aeroplane moving in an atmosphere with pressure 0.45 bar (abs) and density 0.65 kg/m³ records a stagnation pressure of 0.75 bar (abs). Estimate the speed of the plane and the stagnation temperature, with suitable assumptions.*

Solution.

The relation between stagnation pressure and static pressure is

$$\frac{p_0}{p} = \left[1+\frac{\gamma-1}{2}M^2\right]^{\frac{\gamma}{\gamma-1}}$$

Taking $\gamma = 1.4$,

$$0.75/0.45 = 1.667 = (1 + 0.2\,M^2)^{3.5}$$

whence $M = 0.887$

By the equation of state,

$$p = \rho R T$$

whence $T = \dfrac{0.45 \times 1.013 \times 10^5}{0.65 \times 287} = 244\text{ K.}$

The stagnation temperature T_0 is given by

$$\frac{T_0}{T} = 1+\frac{\gamma-1}{2}M^2$$

$$T_0 = 244(1 + 0.2 \times 0.887^2) = \mathbf{282.4\ K.}$$

Sonic velocity $c = \sqrt{\gamma RT} = \sqrt{1.4 \times 287 \times 244}$

$$= 313\text{ m/s}$$

The speed of the aeroplane must be

$$U = 0.887 \times 313 = \mathbf{278\ m/s.}$$

Example 12.17. *A convergent, divergent nozzle, with supersonic flow at exit, has a throat area of 500 mm² and exit area of twice as much. Air enters the nozzle with a stagnation pressure of 1 MPa and stagnation temperature of 360 K. Determine the maximum flow rate that the nozzle can pass.*

Solution.

Entry conditions are

$$p_0 = 1\text{ MPa} = 1000\text{ kPa};\ T_0 = 360\text{ K}$$

Let us take $\gamma = 1.4$ and $R = 287$ J/kg K for air.

The maximum flow rate occurs when the critical conditions are reached at the throat and the velocity becomes sonic.

$$\frac{p^*}{p_0} = \left(\frac{2}{\gamma+1}\right)^{\frac{\gamma}{\gamma-1}} = \mathbf{0.528}$$

and

$$\frac{T^*}{T_0} = \frac{2}{\gamma+1} = \mathbf{0.833}$$

Therefore, $p^* = 0.528 \times 1000 = 528$ kPa

$T^* = 0.833 \times 360 = 300$ K.

and $\rho^* = p^*/(RT^*) = \dfrac{528 \times 1000}{287 \times 300} = 6.13$ kg/m³.

Sonic velocity (at the throat) $c = \sqrt{\gamma RT^*} = \sqrt{1.4 \times 287 \times 300}$

$= 347$ m/s $= U^*$

Mass flow rate for the critical conditions is the maximum possible flow rate:

$$U_{max} = \rho^* A^* U^*$$

$$= 6.13 \times 500 \times 10^{-6} \times 347$$

$$= \mathbf{1.064\ kg/s}$$

The exit area plays no role in the above calculations but one can notice that increase in area in supersonic flow results in increase in velocity.

Example. 12.18. *A vessel contains air at a pressure of 135 kPa and temperature 27° C. The local barometer pressure is 100 kPa. Air discharges out of the vessel into atmosphere through a converging nozzle. Determine the exit velocity and the discharge. The cross-sectional area of the nozzle is 500 mm².*

Solution.

From the data,

$$p_0 = 135 \text{ kPa}, T_0 = 27 + 273 = 300 \text{ K}.$$

$$\rho_0 = p_0/RT_0 = \frac{135 \times 10^3}{287 \times 300} = 1.568 \text{ kg/m}^3.$$

It discharge into atmosphere at 100 kPa;

$$\frac{p}{p_0} = \frac{100}{135} = 0.74$$

This value is higher than the critical pressure ratio of 0.528; the velocity is going is be less than the sonic velocity. For isentropic flow,

$$U^2 = \frac{2\gamma}{\gamma - 1}\frac{p_0}{\rho_0}.\left(1 - \frac{p}{p_0}\right)^{\frac{\gamma-1}{\gamma}}$$

and $$U^2 = \frac{2 \times 1.4}{0.4}\frac{135 \times 1000}{1.568}(1 - 0.74^{0.286}) = 49730 \text{ (m/s)}^2$$

whence $U = \mathbf{223\ m/s}$

Discharge $Q = \rho UA$

$$= \left(\frac{p}{p_0}\right)^{\frac{1}{\gamma}} . \rho_0 \; 223 \times 500 \times 10^{-6}$$

$$= (0.74)^{0.714} \times 1.568 \times 223 \times 500 \times 10^{-6}$$

$$= \mathbf{0.141\ kg/s.}$$

ITEM BANK

Objective items of different types are given to review the concepts developed in the Chapter.

* *Select the correct response for each item.*
* *Every correct response is expected in less than a minute and carries one mark.*
* *Compare your responses with the key answers given at the end.*

Multiple-Choice Items

Attempt the following Multiple-Choice Items by ticking the correct or the most appropriate response from among the alternatives given :

1. An isentropic process is

(*a*) reversible isothermal (*b*) reversible adiabatic

(*c*) irreversible isothermal (*d*) irreversible adiabatic.

2. The sonic velocity in a medium is given by

(*a*) $\sqrt{p/\rho}$ (*b*) $\sqrt{\rho/p}$ (*c*) $\sqrt{\gamma\rho/p}$ (*d*) $\sqrt{\gamma p/\rho}$

3. In supersonic flow, a diverging passage results in

(*a*) increase in velocity and pressure (*b*) decrease in pressure and density

(*c*) increase in velocity and density (*d*) decrease in velocity and pressure.

Other Objective Items

4. State whether the following statements are true or false :

(*a*) A frictionless adiabatic flow is called an isentropic flow. (T/F)

(*b*) Across a normal shock, stagnation pressure must remain constant. (T/F)

(*c*) A nozzle for $M_1 > 1$ consists of converging and diverging (T/F)

(*d*) A normal shock occurs ahead of a pilot probe placed in a compressible flow. (T/F)

5. Identify the following statements as true or false :

(*a*) Entropy cannot increase in adiabatic flow. (T/F)

(*b*) Higher the velocity, smaller the angle of the Mach cone. (T/F)

(*c*) A shock wave must occur at the throat of a converging-diverging passage. (T/F)

(*d*) The maximum flow through a converging-diverging passage corresponds to flow with sonic velocity at the throat. (T/F)

6. Match the value of the '*index of compression*' in Block B with the process of a perfect gas in Block A.

Block A	**Block B**
(*a*) Isothermal process	(*k*) 0
(*b*) Isentropic process	(*l*) 0.5
(*c*) Isochoric process	(*m*) 1.0
(*d*) Isobaric process	(*n*) γ
(*e*) Polytropic process	(*o*) n
(*f*) Adiabatic process	(*p*) ∞

7. Match the value of Mach number from Block B with the type of flow in Block A.

Block A	Block B
(*a*) Ideal fluid flow	(*k*) $M = 0.0$
(*b*) Incompressible flow	(*l*) $M < 0.2$
(*c*) Subsonic flow	(*m*) $0.9 < M < 1.1$
(*d*) Supersonic flow	(*n*) $M < 1.0$
(*e*) Hypersonic flow	(*o*) $M = 1.0$
(*f*) Transonic flow	(*p*) $M > 1.0$
	(*q*) $M > 7.0$

8. **Assertion A :** In a supersonic flow, the velocity increases as the area increases.

Reason R : Conservation of mass, momentum and energy require that dA/A should increase with dM/M for M greater than unity.

Which of the following is correct ?

(*a*) A is true ; R is true and it explains A ;
(*b*) A is true ; R is true but it does not explain A
(*c*) A is true ; R is false
(*d*) A is false ; R is true.

ANSWERS

1. (*b*), **2.** (*d*), **3.** (*b*), **4.** (T, F, T, F), **5.** (F, T, F, T), **6.** (*am,bn,cp,dk,eo,fn*), **7.** (*al, bl, cnq, dp, eqp, fm*), **8.** (*a*)

QUESTION BANK

Questions on theory and numerical problems are given for tutorial work.

* *The questions are arranged generally in order of the articles in the text.*
* *Every question should take 15-20 minutes to solve.*
* *Answers for the numerical problems are given alongside them.*

12.1. From the following data at the sea level, calculate the pressure and density of air at a height of 7.6 km. Assume that the temperature diminishes with the height at a rate of 6.55° K/km and $R = 287$ J/kg K.

Sea-level data : Pressure 101.3 kN/m^2, temperature 15° C and density 1.23 kg/m^3.

[38 kN/m^2, 0.55 kg/m^3]

12.2. Calculate the Mach number of an object travelling at 300 m/s in an environment of air at 200 K. Estimate also the corresponding stagnation temperature. **[1.06, 244.8 K]**

12.3. The speed of the first Indian Satellite Aryabhatta at an altitude of 8000 m was 1000 m/s. (Take $R = 287$ J/kg K, $\gamma = 1.4$). Calculate

(*a*) the Mach number for local conditions at 240 K, (*b*) the angle of the Mach cone and (*c*) the stagnation temperature. **[3.22, 18.1°, 738 K]**

12.4. A jet fighter travels at 500 m/s in air at 280 K. Calculate

(*a*) the Mach cone angle and (*b*) the stagnation properties.

[42.13°, $T_0 = 404.5$ *K*, $p_0 = 344$ kN/m^2, $\rho_o = 2.96$ kg/m^3)]

12.5. At what air speed should a 1 : 100 model of an aircraft be tested if the laboratory temperature is 20°C and the aircraft is expected to fly at 500 m/s in an environment at – 20° C.

[538 m/s]

12.6. An orifice, 10 cm diameter, discharges air from a reservoir maintained at a pressure of 10 atmospheres and a temperature of 300 K. Calculate

(*a*) the exit velocity, (*b*) the mass flow rate and (*c*) the temperature of the ar stream.

[540 m/s, 14.6 kg/s, 155.4 K]

12.7. A horizontal venturimeter, inlet 5 cm and throat 2 cm, is used to meter air under adiabatic conditions. The absolute pressure at the inlet and the throat are 15 N/cm² and 12 N/cm² respectively. Assuming $C_d = 0.95$ and the mass density of air at entry is 1.3 kg/m³. Calculate the mass flow rate. **[0.069 kg/s]**

12.8. Compute the discharge of air through a 10 cm diameter pipe if the differential water manometer connected across a 5 cm diameter orifice meter installed in the pipe shows 10 cm head drop. The density of air initially may be taken as 1.2 kg/m³ and that of water 1000 kg/m³.

[0.011 kg/s]

12.9. During a normal shock in a constant area duct containing air, the initial conditions are

$p_1 = 10$ N/cm² absolute

$T_1 = 0°$ C

$U_1 = 1000$ m/s

Calculate the corresponding conditions and the percentage change in density across the shock if $R = 287$ J/kg K.

[Ans. $p_2 = 1.06 \times 10^6$ N/m², $T_2 = 741$ K, $U_2 = 254$ m/s, 289 %]

12.10. Calculate the correction factor for a pitot-static tube used to measure the velocity of air-stream at a Mach number of 0.9. **[1.2]**

CHAPTER

13

Dimensional Analysis and Similitude

LEARNING OBJECTIVES

On successful completion of all learning activities planned and provided in the chapter, a student will be able to

- state and apply the Buckingham Pi theorem to arrange a set of given variables into dimensionless groups ;
- employ the Rayleigh's indicial method to determine the functional form of a phenomenon in terms of dimensionless groups ;
- state the advantages of the dimensional method, dimensionless groups and dimensional homogeneity in engineering; and
- describe the principle of similitude and apply it for model testing.

13.1. Introduction

The technique of analysing fluid flow problems based upon similitude study is a practical approach to the solution of more complex problems in engineering Need for a practical approach arises from the mathematical complexity of writing the basic fluid flow equations and solving them for the basic variables for a real-flow situation. This complexity is symbolically shown in Fig. 13.1.

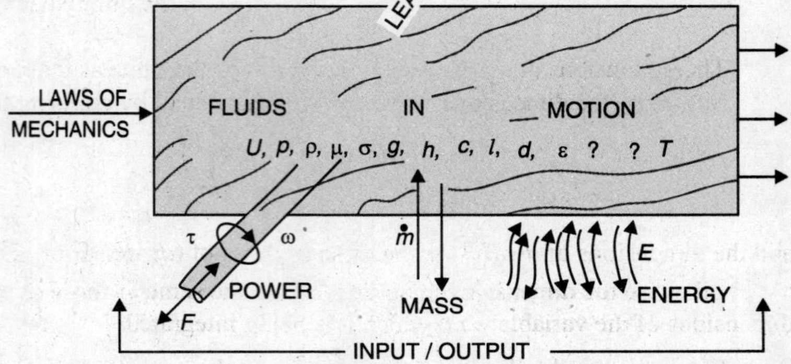

Fig. 13.1: Complexity of Solutions: Need for Dimensional Analysis

The similitude technique consists of a study in *three stages:*

The first stage is concerned with the recognition of the more predominant variables for the problem and the application of the dimensional analysis methods for grouping the variables to form meaningful dimensionless groups.

In the second stage, the criteria for similitude is decided upon in view of the relative importance of the dimensionless groups of variables for the problem. The importance of the geometric, kinematic and dynamic similarity should be realised and steps be taken to achieve the complete similarity as closely as possible.

The third stage consists of the actual execution of the similarity criteria in making a geometric model, deciding the kinematic conditions, carrying out the testing process and predicting the behaviour of the prototype in view of the dynamic parameters involved. The chapter on Dimensional Analysis and Similitude is likewise divided into parts dealing with each stage and a number of engineering examples have been cited.

13.2. Dimensions of Physical Quantities

All physical quantities are expressed by magnitudes and units. For example, the acceleration of a body may be expressed as 10 m/s² and the force on a body may be denoted as 20 N. *In the SI system, the primary physical quantities which are assigned base units are the mass, length, time and temperature besides current, luminous intensity and amount of substance.* Of these, the first four, used in Fluid Mechanics, are assigned the dimensions of *M, L, T* and *t* respectively. Since all other physical quantities may be expressed in terms of these primary physical quantities, their dimensions are likewise expressible in terms of *M, L, T* and *t.*

For example, the dimensions of velocity are determined by expressing it as

$$\text{Velocity} = \text{Distance/Time}$$

$$\text{Dimensions of velocity} = \text{Dimensions of distance/Dimensions of time}$$

$$= L/T \text{ or } LT^{-1}$$

Similarly, the dimensions of acceleration are determined by expressing the acceleration in the form

$$\text{Acceleration} = \text{velocity/time}$$

$$\text{Dimensions of acceleration} = \text{Dimensions of velocity/Dimensions of time}$$

$$= LT^{-1}/T = LT^{-2}$$

Likewise, the dimensions of derivatives, partial or total, are determined by realising the physical meaning of the terms. For example, the dimensions of dy/dx are determined by realising that

$$\frac{dy}{dx} \text{ implies } \frac{\delta y}{\delta x} \text{ in the limit when } \delta x \text{ tends to zero.}$$

The dimensions of dy/dx should, therefore, be the same as those of $\delta y/\delta x$ or of y/x. Similarly, the dimensions of d^2y/dx^2 are determined by realising that

$$\frac{d^2 y}{dx^2} = \frac{d}{dx}\left(\frac{dy}{dx}\right) = \frac{\delta\left(\frac{dy}{dx}\right)}{\delta x}$$

and the dimensions of d^2y/dx^2 are the same as those of $(dy/dx)/\delta x$ or of $(y/x.\ 1/x)$ or of y/x^2.

Likewise the dimensions of an integral are the same as those of the integral multiplied by the dimensions of the variable w.r.t. which it is being integrated.

For example, the dimensions of $\int y\, dx$ are the same as those of $y\ x$ and the dimensions of $\iint p\, dx\, dy$ are the same as those of $p\ x\ y$.

Dimensions of various physical quantities useful in fluid mechanics are listed in Table 13.1

Table 13.1 : Dimensions of Physical Quantities

Physical Quantity	Dimensions	Physical Quantity	Dimensions
Mass	M	Weight	MLT^{-2}
Length	L	Pressure ; Stress	$ML^{-1}T^{-2}$
Time	T	Angular momentum	ML^2T^{-1}
Temperature	t	Moment ; Torque	ML^2T^{-2}
Diameter ; Radius	L	Stream function	L^2T^{-1}
Area	L^2	Velocity potential	L^2T^{-1}
Volume	L^3	Circulation	L^2T^{-1}
Speed ; Velocity	LT^{-1}	Work ; Energy	ML^2T^{-2}
Angular speed	T^{-1}	Power	ML^2T^{-3}
Acceleration	LT^{-2}	Frequency	T^{-1}
Angular acceleration	T^{-2}	Surface tension	MT^{-2}
Density	ML^{-3}	Dynamic viscosity	$ML^{-1}T^{-1}$
Momentum	MLT^{-1}	Kinematic viscosity	L^2T^{-1}
Force ; Thrust	MLT^{-2}		

It may be noted that the dimensions of physical quantities are independent of the system of units employed. All dimensions are expressed in terms of M, L, T and *t*. For example, irrespective of the fact whether the unit of mass is kg, gram, pound or some other, its dimension is M. It is for this reason that the dimensional representation of physical quantities is preferred to the specification of units and the technique of dimensional analysis is employed in engineering problems.

Strain, efficiency and the coefficients of velocity, contraction and discharge as well as correction factors are non-dimensional. Dimensions of pressure drop Δp are the same as those of the pressure. Dimensions of the moduli of elasticity of materials and bulk modulus of compression of fluids are the same as the dimensions of stress. Piezometric head, the head lost in a flow and the head differential between two points have the dimensions of length.

13.3. Dimensional Homogeneity

Two quantities may be added or subtracted if and only if they have the same dimensions.

For example, it is correct to add or subtract 10 newton and 6 newton forces, each with dimension MLT^{-2}, whereas it is incorrect to add or subtract 10 newton force and 6 watt power.

The requirement that the dimensions of each term in an equation should be the same is known as the principle of dimensional homogeneity. The principle holds for equations of all types, *i.e.*, algebraic scalar, or vector, linear or non-linear, differential or integral equations. For example,

$$v = u + at$$

is dimensionally homogeneous or correct equation because the dimensions of *v*, *u* and *at* are the same.

The equation

$$\text{‘ } s = ut + \frac{1}{2}\, at \text{ ’}$$

is dimensionally incorrect equation because the dimensions of *at* are not the same as those of *s* and *ut*. It needs to be corrected to

$$s = ut + \frac{1}{2}\, at^2.$$

The equation

$$u\frac{\partial u}{\partial x} + v\frac{\partial u}{\partial y} = -\frac{1}{\rho}\frac{\partial p}{\partial x}$$

is dimensionally correct equation because the dimensions of each term are the same.

13.4. Dimensionless Groups ; Buckingham π Theorem

The study of dimensional analysis starts with a recognition of the important variables in a given fluid-flow problem and by shortlisting them. General guidelines for this step are :

(*a*) these variables should be the most important physical quantities in the problem,

(*b*) all the dimensions should be included in the variables selected,

(*c*) non-repeating independent variables must be listed,

(*d*) there should be geometric, kinematic and dynamic variables in the list, and

(*e*) the relevant fluid properties must be included.

The study proceeds with a knowledge of the dimensions of the different variables. Reference to the dimensions of the physical quantities presented in Table 13.1 may, therefore, be made and off-hand thoroughness with the dimensions should be acquired.

A set of physical quantities may always be arranged to form an infinite number of dimensionless groups of which only a few can be independent. *Buckingham π theorem states n physical quantities with r base dimensions can generally be arranged to provide only (n – r) independent dimensionless parameters also referred to as π s.*

For example, if U, d, ρ, μ and g were to be placed in non-dimensional groups, the answers would be $\frac{Ud\rho}{\mu}, \frac{U}{\sqrt{gd}}, \frac{U^3 d\rho}{\mu g d}, \log\left(\frac{Ud\rho}{\mu}\right), \frac{U^3 d\rho}{\mu g^2 d^2}$, etc. which make an infinite number. However, there can be only (5 – 3), i.e., 2 mutually independent non-dimensional such as $\left(\frac{Ud\rho}{\mu} \text{ and } \frac{U^2}{gd}\right)$ or $\left(\frac{U}{\sqrt{gd}} \text{ and } \log\left(\frac{Ud\rho}{\mu}\right)\right)$ or $\left(\frac{Ud\rho}{\mu} \text{ and } \frac{U}{\sqrt{gd}}\right)$.

There are two different methods of dimensional analysis either of which may be followed to group the recognised set of variables into a functional form of the non-dimensional parameters.

13.5. The Group Method

In the method of grouping the variables, it is necessary to recognise the common or repeating variables, **three in number,** which are common to all the groups desired to be formed. The number of groups is, of course, governed by the Buckingham π Theorem. The non-repeating variables are taken one by one together with the three repeating variables and the method is applied to determine the dimensionless parameters for each of the sets of four variables. The three repeating variables are generally, the velocity U, the density ρ and the characteristic length l. The non-repeating variables are those among viscosity μ, gravitational acceleration g, surface tension σ, etc. The reason for this is that U, ρ and l are representative of inertial behaviour of the fluid whereas the selection of μ pertains to the presence of viscous forces, selection of g for the gravitational forces and σ for surface tension forces, etc. which need to be taken one at a time to generate the corresponding dimensionless numbers.

Consider the force F on the propeller of an aircraft. It is known to depend upon the forward speed of the aircraft U, the density of air ρ, the viscosity of air μ, diameter of the propeller D and the speed of rotation of the propeller N. The general form of the relationship is, therefore,

$$F = f(U, \rho, \mu, D, N) \qquad \ldots (i)$$

In order to apply the method of grouping, the three repeating variables are chosen as U, D and ρ and the non-repeating variables as F, μ and N. The groups of variables are thus,

(1) U, D, ρ and F ;

(2) U, D, ρ and μ and

(3) U, D, ρ and N

The total number of variables is 6. According to the Buckingham π theorem, only (6 – 3), *i.e.*, 3 non-dimensional parameters can be formed. The three desired π parameters (π_1, π_2 and π_3) are constituted by the groups of variables at (1), (2) and (3).

For π_1, the dimensions of the variables in the group 1 are written as a dimensional matrix

	U	D	ρ	F
M	0	0	1	1
L	1	1	–3	1
T	–1	0	0	–2

and let
$$\pi_1 = U^{x1} D^{y1} \rho^{z1} F$$

In terms of their dimensions,
$$(M^\circ L^\circ T^\circ) = (LT^{-1})^{x1} . (L)^{y1} . (ML^{-3})^{z1} . (MLT^{-2})$$

or
$$(M^\circ L^\circ T^\circ) = M^{z_1+1} . L^{x_1+y_1-3z_1+1} . T^{-x_1-2}$$

For dimensional homogeneity, the dimensions of M, L and T on both sides should be equated:
$$z_1 + 1 = 0$$
$$x_1 + y_1 - 3z_1 + 1 = 0$$
$$-x_1 - 2 = 0$$

These equations provide
$$x_1 = -2 \quad y_1 = -2 \quad z_1 = -1$$

Substituting these values in the expression for π_1,
$$\pi_1 = U^{-2} D^{-2} \rho^{-1} F$$
$$\pi_1 = \frac{F}{\rho U^2 D^2}$$

which is the first dimensionless parameter.

Similarly for π_2, the dimensions of the variables in the group 2 are written :

	U	D	ρ	μ
M	0	0	1	1
L	1	1	–3	–1
T	–1	0	0	–1

Now let
$$\pi_2 = U^{x_2} D^{y_2} \rho^{z_2} \mu$$

In terms of their dimensions,
$$(M^\circ L^\circ T^\circ) = (LT^{-1})^{x_2} . (L)^{y_2} . (ML^{-3})^{z_2} . (ML^{-1} T^{-1})$$

or
$$M^\circ L^\circ T^\circ = M^{z_2+1} L^{x_2+y_2-3z_2-1} T^{-x_2-1}$$

By equating the indices of M, L and T respectively on either side of the equation
$$z_2 + 1 = 0$$
$$x_2 + y_2 - 3z_2 - 1 = 0$$
$$-x_2 - 1 = 0$$

whence
$$x_2 = -1 \quad y_2 = -1 \quad z_2 = -1$$

and by substituting the same in the expression for π_2,
$$\pi_2 = U^{-1} D^{-1} \rho^{-1} \mu = \frac{\mu}{UD\rho}$$

In practice, the inverse of this parameter, well known as the Reynolds number

$$\pi_2 = Re = \frac{UD\rho}{\mu}$$

is taken as the dimensionless parameter.

Similarly, the expression for π_3 may be obtained as

$$\pi_3 = \frac{ND}{U}$$

Finally, the functional form (*i*) may be expressed as

$$\frac{F}{\rho U^2 D^2} = f\left(\frac{UD\rho}{\mu}, \frac{ND}{U}\right) \quad \text{... (ii)}$$

It is interesting to see what would have happened if the initial grouping of variables was not done carefully. For example, if no control was exercised on the choice of repeating and non-repeating variables, one could have expressed

$$\pi_1' = U, \rho, \mu, N$$

$$\pi_2' = F, \rho, \mu, D$$

and

$$\pi_3' = F, U, N, \rho$$

In that case, the solution would have provided the three parameters as follows:

$$\pi_1' = \frac{\rho U^2}{\mu N}$$

$$\pi_2' = \frac{F\rho}{\mu^2}$$

$$\pi_3' = \frac{FN^2}{\rho U^4}$$

The parameters are a correct set of non-dimensional numbers but unfortunately, they lack physical interpretation. However, all sets of three independent dimensional numbers must be mutually equivalent. It can, therefore, be shown that these three parameters can be reduced to the three obtained earlier or that these parameters are made up by simple operations from the three parameters π_1, π_2, and π_3 obtained earlier. It is seen

$$\pi_1' = \frac{\rho U^2}{\mu N} = \frac{UD\rho}{\mu} \cdot \frac{U}{ND} = \frac{\pi_2}{\pi_3}$$

$$\pi_2' = \frac{F\rho}{\mu} = \frac{F}{\rho U^2 D^2} \cdot \frac{U^2 D^2 \rho^2}{\mu^2} = \pi_1 \pi_2^2$$

$$\pi_2' = \frac{FN^2}{\rho U^4} = \frac{F}{\rho U^2 D^2} \cdot \frac{N^2 D^2}{U^2} = \pi_1 \pi_3^2$$

and hence the new set may be replaced by the earlier set of parameters.

13.6. The Rayleigh's Method of Indices

The alternative method, called Indicial method, consists of dealing with all the variables together, setting up a set of 3 indicial equations and solving them for the purpose of obtaining the dimensionless groups in the desired functional form.

Consider *again* the force F acting on the propeller of an aircraft which depends upon the variables U, ρ, μ, D and N, as before

$$F = f(U, \rho, \mu, D, N)$$

The general relationship may be expanded as the sum of an infinite series of terms giving

$$F = A.(U^{a_1}\ \rho^{b_1}\ \mu^{c_1}\ D^{d_1}\ N^{e_1}) + B.(U^{a_2}\ \rho^{b_2}\ \mu^{c_2}\ D^{d_2}\ N^{e_2}) + \ldots \qquad \ldots (13.1)$$

where *A, B,*..... are numerical constants and *a, b, c, d* and *e* are unknown indices.

The principle of dimensional homogeneity requires that the dimensions of each term on the right hand side of the equation must be equal to the dimensions of the term *F* on the left hand side. Recalling the dimensions of the variables and placing them in the form of a dimensional matrix,

	F	U	ρ	μ	D	N
M	1	0	1	1	0	0
L	1	1	–3	–1	1	0
T	– 2	– 1	0	– 1	0	–1

and substituting the dimensions of the variables in the equation

$$F = K.(U^a\ \rho^b\ \mu^c\ D^d\ N^e)$$

$$(\text{MLT}^{-2}) = (\text{LT}^{-1})^a.\ (\text{ML}^{-3})^b.(\text{ML}^{-1}\ \text{T}^{-1})^c.(\text{L})^d.(\text{T}^{-1})^e \qquad \ldots (iii)$$

or

$$\text{MLT}^{-2} = \text{M}^{b+c}\ .\ \text{L}^{a-3b-c+d}\ .\ \text{T}^{-a-c-e}$$

Equating the indices of M, L and T, the following indicial equations are obtained

$$b + c = 1$$

$$a - 3b - c + d = 1$$

$$-a - c - e = -2$$

These are three equations for five unknowns. Therefore, one may at best express three of them in terms of the other two unknowns. For example,

$$b = 1 - c$$

$$a = 2 - c - e$$

$$d = 2 - c + e$$

Substituting these values in (*iii*),

$$F = K.U^{2-c-e}\ .\ \rho^{1-c}\ .\ \mu^e\ .\ d^{2-c+e}\ .\ N^e$$

Regrouping the variables,

$$F = K.\rho U^2 D^2.\left(\frac{\mu}{UD\rho}\right)^c\left(\frac{ND}{U}\right)^e$$

or

$$\frac{F}{\rho U^2 D^2} = K.\left(\frac{\mu}{UD\rho}\right)^c\left(\frac{ND}{U}\right)^e$$

Instead of this form where *K, c,* and *e* are unknown quantities, the above equation is written in the functional form

$$\frac{F}{\rho U^2 D^2} = f\left(\frac{UD\rho}{\mu}, \frac{ND}{U}\right) \qquad \ldots (13.2)$$

where the right hand side represents a function of the Reynolds number $UD\rho/\mu$ and the speed parameter ND/U, as before!

13.7. Important Force-Ratio Dimensionless Numbers

The following force-ratio numbers are frequently involved in fluid mechanics:

Reynolds Number

Froude Number

Euler Number

Mach Number

Weber Number and

Lift Coefficients

Let us look at them in some detail:

(i) Reynolds Number Re **= Inertia force/viscous force**

Inertia force = mass × acceleration = $\rho l^3 \dfrac{U}{t} = \rho l^2 \, Ul/t = \rho l^2 U^2$

and viscous force = area × shear stress = $l^2 \, \mu \dfrac{U}{l} = \mu l U$

Hence, $$Re = \frac{\rho l^2 U^2}{\mu l U} = Ul\rho/\mu \quad \text{... (13.3)}$$

where l refers to a characteristic length, say diameter of a pipe, or chord of an aerofoil. Reynolds number is considered the only number for dynamic similarity if the viscous forces predominate over others such as gravity and elastic ; for example in the following situations :

(a) incompressive flow through pipes and venturimeters,

(b) flow around aeroplanes and submarines,

(c) flow around structures and bodies and

(d) flow through low speed rotodynamic machines.

(ii) Froude Number Fr **= (Inertia force/gravity force)½**

Gravity force = mass × acceleration due to gravity = $\rho l^3 g$

and inertia force = $\rho l^2 U^2$

Hence, Froude number $$Fr = \sqrt{(\rho l^2 U^2 / \rho l^3 g)} = \frac{U}{\sqrt{lg}} \quad \text{... (13.4)}$$

Froude number alone dictates the dynamic similarity of flow in case the forces other than gravitational force are negligible such as in

(a) flow in open channels,

(b) flow of liquid jets from orifices,

(c) surface wave motion created by boats, and

(d) flow over notches and weirs.

(iii) Euler Number E **= Pressure force/inertia force**

Pressure force = pressure × area = pl^2

and inertia force = $\rho l^2 U^2$

Hence, Euler number $$E = \frac{pl^2}{\rho l^2 U^2} = \frac{p}{\rho U^2} \quad \text{... (13.5)}$$

This is also expressed alternatively as $E = \dfrac{\Delta p}{\rho U^2}$

where Δp = pressure variation in the flow.

It is also known as the **Pressure Coefficient** in the study of confined flows.

The Euler number E or the pressure coefficient assumes importance when the pressure changes in a flow are predominant. Some examples are :

(a) flow through long pipes,

(b) flow through penstocks leading water to hydraulic turbines and

(c) flow over submerged bodies.

(iv) Mach Number M **= (Inertia force/elastic force)**$^{1/2}$

Elastic force $= El^2 = \rho c^2 l^2$

where c = speed of sound in the fluid

Hence, Mach number $M = (\rho l^2 U^2 / \rho c^2 l^2)^{1/2} = \frac{U}{c} = \frac{\text{speed of fluid}}{\text{speed of sound in that fluid}}$ ... (13.6)

Mach number effects are considered only when M exceeds 0.4 and when it is the only factor baffecting the motion in

(*a*) transonic flow when M ranges between 0.9 and 1.1,

(*b*) supersonic flow when $1 < M < 7$ and

(*c*) hypersonic flow when $M > 7$

(v) Weber Number W **= Inertia force/surface tension force**

The surface tension force is expressible as $\sigma\, l$,

Hence, Weber number $W = \rho l^2 U^2 / \sigma l = \rho l U^2 / \sigma$ ... (13.7)

Some cases where the Weber number assumes importance are :

(*a*) capillary tube flows,

(*b*) human blood flow studies, and

(*c*) thin sheet flows.

(vi) **Lift Coefficient** $L/\rho U^2 l^2$ ... (13.8)

Drag Coefficient $D/\rho U^2 l^2$ ... (13.9)

Torque Coefficient $T/\rho\omega^2 d^5$ ... (13.10)

and many other such coefficients are automatically dimensionless parameters and acquire a high status whenever they occur. A summary of such numbers together with their significance is presented in a tabulated form in Table 13.2.

Table 13.2. Summary of the Dimensionless Groups

Name	Symbol	Definition	Significance	Example of Use
Reynolds number	Re	$\frac{Ud\rho}{\mu}$	$\frac{\text{Inertia force}}{\text{Viscous forces}}$	Confined viscous flow
Froude number	Fr	$\frac{U}{\sqrt{gd}}$	$\frac{\text{Inertia force}}{\text{Gravitational force}}$	Free surface flow
Mach number	M	$\frac{U}{c}$	$\frac{\text{Inertia force}}{\text{Elastic force}}$	Compressible flow
Drag or lift Coefficient	C_D	$\frac{F}{\rho l^2 U^2}$	$\frac{\text{Drag or lift force}}{\text{Inertia force}}$	Aerodynamics
Grashof number	F_r	$\frac{l^3 \rho g \beta \Delta T}{\mu^2}$	$\frac{\text{Buoyancy force}}{\text{Viscous force}}$	Free convection
Prandtl number	P_r	$\frac{c_p \mu}{k}$	$\frac{\text{Momentum diffusivity}}{\text{Thermal diffusivity}}$	Convection
Nusselt number	Nu	$\frac{hd}{k}$	$\frac{\text{Total heat transfer}}{\text{Conductivity heat transfer}}$	Heat transfer

Name	Symbol	Definition	Significance	Example of Use
Weber number	W	$\frac{\rho l U^2}{\sigma}$	$\frac{\text{Inertia force}}{\text{Surface tension force}}$	Capillary and sheet flow
Euler number	E	$\frac{p}{\rho U^2}$	$\frac{\text{Pressure force}}{\text{Inertia force}}$	Pressure variations in flow
Strouhal number	S	$\frac{nd}{U}$	Frequency of vortices	Shedding of vortices
Efficiency	η	$\frac{P_o}{P_i}$ or $\frac{FU}{TW}$	$\frac{\text{Useful output}}{\text{Total input}}$	System analysis
Gamma	γ	c_p/c_v	$\frac{\text{Sp. heat at const. pressure}}{\text{Sp. heat at const. volume}}$	Compressible flow
Knudsen number	K_n	$\frac{\lambda}{l}$	$\frac{\text{Mean free path}}{\text{Characteristic dimension}}$	Rarefied gas flow
Speed parameter	S	$\frac{ND}{U}$	$\frac{\text{Peripheral speed}}{\text{Free stream speed}}$	
Flow coefficient	Q_c	$\frac{Q}{Nd^3}$	$\frac{\text{Discharge}}{\text{Characteristic discharge}}$	Incompressible flow through fluid machines such as cenrifugal pumps, mixed flow pumps, wind-mills, axial flow pumps and other rotodynamic equipment such as fluid couplings, fluid torque converters Pumps and turbines and Fluid dynamometers
Head coefficient	H_c	$\frac{gH}{N^2 d^2}$	$\frac{\text{Head}}{\text{Characteristic head}}$	
Power coefficient	P_c	$\frac{P}{\rho N^3 d^5}$	$\frac{\text{Power}}{\text{Characteristic power}}$	
Specific speed	N_s	$\frac{N\sqrt{Q}}{(gH)^{3/4}}$ or $\frac{N\sqrt{P/\rho}}{(gH)^{5/4}}$	$\frac{\text{Operating speed}}{\text{Characteristic speed}}$	
Torque coefficient	T_c	$\frac{T}{\rho D^5 N^2}$	$\frac{\text{Torque}}{\text{Characteristic torque}}$	
Pressure coefficient	P_c	$\frac{\Delta p}{\rho U^2/2}$	$\frac{\text{Pressure drop}}{\text{Dynamic pressure}}$	Confined flows
Roughness coefficient	ε'	$\frac{\varepsilon}{d}$	$\frac{\text{Roughness}}{\text{Characteristic dimension}}$	Surface roughness simulation.

13.8. Applications of the Dimensional Method

Let us outline the three major applications of the dimensional analysis and similitude:

(A) If a number of variables such as U, ρ, d, μ are involved in the study of an effect, say drag D, on a body, then one way of studying the effect of each variable is to vary only one variable at a time keeping all others constant. This would result in a large number of curves requiring a number of fluids and sizes of articles for the laborious sets of experimentation. Instead, if we group the variables into dimensionless groups like $Ud\rho/\mu$, and $D/\rho U^2d^2$, then the experimenter needs to plot a single curve requiring very little pains of varying only one variable, say U only. This is demonstrated in Fig. 13.2. *In other words, the effect of variation of d, p and u on the drag can be predicted without ever studying the variation experimentally with respect to these variables. This is indeed the greatest advantage of the dimensional method.*

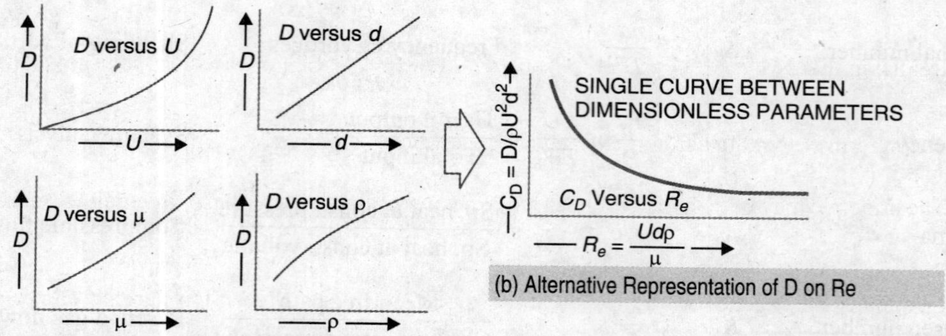

(a) Dependence of D on Individual Variables

(b) Alternative Representation of D on Re

Fig. 13.2: Advantage of the Dimensionless Parameters

(B) A course in fluid mechanics reveals that only the relatively simple flows can be studied theoretically leaving large grounds for experimental studies. Lift and drag on a complete aeroplane, performance of a rotodynamic machine and many other practical problems would have remained unsolved but for the dimensional techniques. *Applying the criteria of similitude, smaller models are constructed and tested in the laboratory and fair predictions for the behaviour of prototypes are made possible.*

(C) Dimensional homogeneity principle aids in *checking the terms of an aquation and dimensional analysis sets the relative magnitude of each term before the complicated equations are solved.*

For example, if we were to check the dimensional homogeneity of the following equation,

$$\frac{p}{\rho} + \frac{U^2}{2g} + z = \text{constant}$$

we write for the dimensions of each term,

$$p/\rho \qquad (ML^{-1}T^{-2})/(ML^{-3}) = L^2T^{-2}$$

$$\frac{U^2}{2g} \qquad (LT^{-1})^{-2}/(LT^{-2}) = L$$

$$z \qquad = L$$

which shows that the dimensions of the second and third terms are length whereas the first term has L^2T^{-2}. *A* little reflection will show that it requires a division by LT^{-2}, an acceleration term, namely g. The dimensionally correct equation is then discovered to be

$$\frac{p}{\rho g} + \frac{U^2}{2g} + z = \text{constant}$$

Since the SI units are coherent, we could have as well written the SI units of each term and compared as :

$$\frac{p}{\rho} \quad \frac{N}{m^2} \cdot \frac{m^3}{kg} = \frac{Nm}{kg}$$

$$\frac{U^2}{2g} \quad \frac{m^2}{s^2} \cdot \frac{s^2}{m} = m$$

$$z \qquad m$$

Realising that $N \equiv \frac{kg\ m}{s^2}$, the SI units of p/ρ are $\frac{kg\ m}{s^2 m^2} \cdot \frac{m^3}{kg} = \frac{m^2}{s^2}$ which also shows that an acceleration term $\frac{m}{s^2}$ in the denominator has been left out.

The principle may be extended to check any type of equation, differential or integral; for example,

$$\frac{\partial u}{\partial x} + \frac{\partial v}{\partial y} = 0$$

is dimensionally homogeneous because the dimensions or SI units of each term are the same.

Term	Dimensions	SI Units
$\frac{\partial u}{\partial x}$	$\frac{LT^{-1}}{L} = T^{-1}$	$\frac{m}{s}\frac{1}{m} = \frac{1}{s}$
$\frac{\partial v}{\partial y}$	$\frac{LT^{-1}}{L} = T^{-1}$	$\frac{m}{s}\frac{1}{m} = \frac{1}{s}$

It may be noted that dimensionally homogeneous equation may not necessarily be the correct equation. For example, a term may be missing or a sign may be wrong. On the other hand, a dimensionally incorrect equation must be wrong. *The dimensional check is therefore a negative check and not a positive check on the correctness of an equation.*

An equation may, in the first instance, appear to be dimensionally incorrect but it may actually be correct because the empirical constants may have dimensions. For example, the drag D experienced by a body varies with the velocity U and it is expressed as

for laminar flow and as $\quad D_1 = k_1 U$

and for turbulent flow. $\quad D_2 = k_2 U^2$

These equations are dimensionally correct because the constants k_1 and k_2 have dimensions such that

D_1 has the same dimensions as $k_1 U$

and $\quad D_2$ has the same dimensions as $k_2 U^2$

In such equations, care must be taken to substitute D and U in the same units in which the value of k is prescribed.

13.9. Similitude

Two flows are said to be mechanically similar if the flows are similar (*a*) geometrically, (*b*) kinematically and (*c*) dynamically as shown in Fig 13.3.

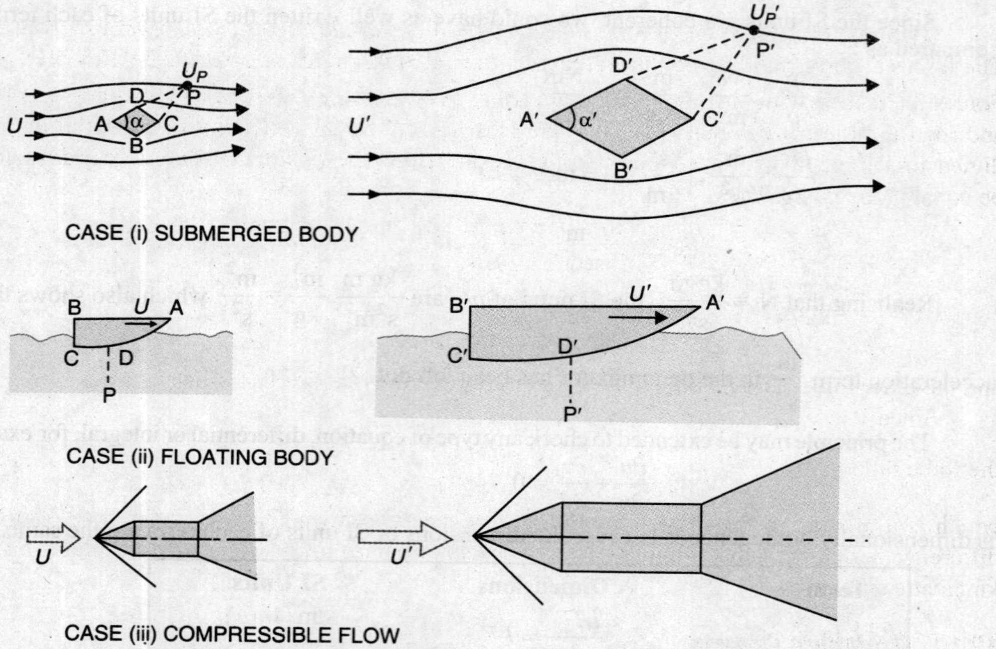

Fig. 13.3: Similitude between Model and Prototype

Geometric Similarity implies that *the corresponding geometric dimensions in the flows are related through the same geometrical ratio.* It refers to the similarity of the form. The geometric similarity also demands that the corresponding areas are related by the square of the scale factor and the corresponding volumes by the cube of the scale factor, etc.

Thus, for geometric similarity,

$$\frac{AB}{A'B'} = \frac{BC}{B'C} = \frac{AC}{A'C'} = \frac{l}{l'} = \text{scale ratio} \qquad \text{... (13.11)}$$

$$\frac{\text{Area } ABCD}{\text{Area } A'B'C'D} = \left(\frac{l}{l'}\right)^2 = (\text{scale ratio})^2 \qquad \text{... (13.12)}$$

$$\frac{\text{Volume } V}{\text{Volume } V'} = \left(\frac{l}{l'}\right)^3 = (\text{scale ratio})^3 \qquad \text{... (13.13)}$$

Kinematic Similarity requires that *the direction of velocity and acceleration at corresponding points in the two flows should be the same.* The velocities at the corresponding points would bear a constant ratio called the velocity ratio. The streamline patterns will, therefore, appear similarly in the two flows. Similarly the accelerations at the corresponding points are related through a constant acceleration ratio. Thus, for kinematic similarity.

$$\frac{U}{U'} = \frac{U_p}{U_p'} = \ldots = \text{velocity ratio} \qquad \text{... (13.14)}$$

$$\frac{a}{a'} = \frac{a_p}{a_p'} = \ldots = \text{acceleration ratio} \qquad \text{... (13.15)}$$

As a corrollary of the geometric and kinematic similarities the discharge ratio,

$$\frac{Q}{Q''} = \text{velocity ratio x (scale ratio)}^2.$$

Dynamic Similarity *means the similarity of forces at the corresponding points in the flows.* The forces at corresponding points are, therefore, related through a constant called the force ratio. Forces interacting with the motion are of a variety of origins, *e.g.*, viscous, gravity, pressure, elastic and inertia forces. It has been seen that the ratios of the forces constitute some of the important dimensionless numbers. For dynamic similarity, therefore, these dimensionless force-ratios should be equal in the flows. Thus,

$$Re = \frac{\text{Inertia force}}{\text{Viscous force}}$$

$$F_r = \frac{\text{Inertia force}}{\text{Gravity force}}$$

and the other force-ratios should have the same value for dynamically similar flows.

Since $F' = m'\,a'$ and $F = m\,a,$

the force ratio $\dfrac{F}{F'} = \dfrac{m}{m'} \cdot \dfrac{a}{a'}$

which is a product of the mass ratio and the acceleration ratio. The mass ratio arises out of the difference in densities and volumes of the two fluids whereas the acceleration ratio stems from the kinematic similarity.

13.10. Modelling Criteria

Geometric, kinematic and dynamic similarities are mutually independent.

Existence of one does not imply the existence of another similarity. We should appreciate that the geometric similarity is complete when the surface roughness profiles are also in the scale ratio. Since it is not possible to prepare a 1/20 scale model to 20 times better surface finish, it is unlikely that complete geometric similarity will be achieved. The kinematic similarity is even more difficult because the flow patterns around small objects tend to be qualitatively different from those around large objects viz. laminar flow at low speeds and turbulent flow at high speeds.

Complete dynamic similarity is practically impossible. Let us demonstrate this for a case where Reynolds number and Froude number are equated for the model and the prototype.

Equality of Re means

$$\frac{Ud\rho}{\mu} = \frac{U_m d_m \rho_m}{\mu_m}$$

or

$$\frac{U_m}{U} = \left(\frac{\rho}{\mu} \cdot \frac{\mu_m}{\rho_m}\right)\frac{d}{d_m} = C_1 \frac{d}{d_m} \qquad \text{...}(i)$$

The bracketed term is unity for the same fluid used in the two flows, but it can be another constant in accordance with the properties of the two fluids ; let it be C_1.

Equality of F_2 means

$$\frac{U}{\sqrt{gd}} = \frac{U_m}{\sqrt{gd_m}} \quad \text{or} \quad \frac{U_m}{U} = \sqrt{\frac{d_m}{d}} \qquad \text{...}(ii)$$

Conditions (*i*) and (*ii*) are entirely different showing that the *Reynolds number and the Froude number cannot be equated simultaneously*. Similarly it can be shown that the other force-ratios require different kinematic conditions which are mutually inconsistent.

For reasons described above, complete similitude is ruled out. However, engineers make the following concessions and employ the technique gainfully :

(*a*) Assume the surface roughness profile similarity.

(*b*) Assume the real flow effects to be negligible.

(*c*) Equate only the most predominant dimensionless force ratio number for obtaining the kinematic condition for flow over the model. Consider, for example, the flow over an immersed body. At low Reynolds numbers, viscous forces predominate but at large Reynolds numbers, the viscous forces are negligible. Care should, therefore, be taken to recognise the important force-parameters for a given situation.

Table 13.4

Low Reynolds Number	High Reynolds Number
($Re < 1000$)	($Re > 2000$)
Viscous Flow	Turbulent Flow
Inertia forces are negligible	Viscous forces are negligible

(*d*) Neglect the effect of the other parameters, but if desired, the same may be considered in addition. *Dependence on the next predominant dimensionless number is called the scale effect.*

Sometimes, engineers resort to distorted scale models, where the geometrical scale ratio is different in different directions. For example, the true geometric model of a river may require a very thin layer of fluid prone to surface tension and excessive shear effects. If the vertical scale-ratio is made several times the horizontal scale ratio, the model study becomes more meaningful.

A model may not always be smaller than the prototype. For example, the model of the first stage steam turbined blade row is invariably bigger than the actual blade row in order to observe and measure the flow parameters more effectively.

Let us examine the modelling criteria and the prediction procedure of the performance of a pump or a fan for incompressible flow.

The measure-of-performance variables are power input P, specific energy gH and efficiency η. For a given geometric design, the independent variables influencing the performance are ρ, d, μ, N and Q. The problem reduces to determining the functional forms

$$P = f_1 (\rho, d, \mu, N, Q)$$

$$gH = f_2 (\rho, d, \mu, N, Q) \text{ and}$$

$$\eta = f_3 (\rho, d, \mu, N, Q)$$

Starting with the first ; the six variables should be expressible in (6 – 3), *i.e.*, 3 dimensionless groups. Let these be

(*a*) $\pi_1 = \rho^{x_1} N^{y_1} d^{z_1} P$

(*b*) $\pi_2 = \rho^{x_2} N^{y_2} d^{z_2} \mu$

and (*c*) $\pi_3 = \rho^{x_3} N^{y_3} d^{z_3} Q$

For group (*a*) to be dimensionless,

$$(ML^{-3})^{x_1} (T^{-1})^{y_1} L^{z_1} ML^2T^{-3} = 0$$

whence $x_1 + 1 = 0, -3x_1 + z_2 + 2 = 0$ and $-y_1 - 3 = 0$

giving $x_1 = -1, y_1 = -3$ and $z_1 = -5$

meaning that π_1 should be $P/\rho N^3 d^5$

Similarly working out π_2 and π_3 we obtain,

$$P/(\rho N^3 d^5) = f_1(\rho N d^2/\mu, Q/Nd^3) \quad \text{... (13.16)}$$

Proceeding in a similar fashion for the head and efficiency,

$$gH/(N^2d^2) = f_2(\rho Nd^2/\mu, Q/Nd^3) \quad \text{... (13.17)}$$

and

$$\eta = f_3(\rho Nd^2/\mu, Q/Nd^3) \quad \text{... (13.18)}$$

For pumps and fans running at high speeds the flow is turbulent and independent of viscosity effects, so the Reynolds number $\rho Nd^2/\mu$, $= \rho d(Nd)/\mu$ is unimportant. Discarding it, the criterion of dynamic similarity should be the flow parameter, Q/Nd^3.

Geometric model of a pump or a fan should, therefore, operate at a speed and discharge to keep Q/Nd^3 constant ; the power parameter, head parameter and η also being the same respectively for the model and the prototype.

It is customary to define and use the term 'specific speed' in relation to rotodynamic machines. The specific speed is defined as the speed of a geometrically similar machine operating under a unit head and unit discharge. This is obtained by dividing the square of $\left(\frac{Q}{Nd^3}\right)$ by the cube of $\left(\frac{gH}{N^2d^2}\right)$ in order to eliminate d,

$$\frac{Q^2}{N^2d^6} \div \frac{(gH)^3}{N^6d^6} = \frac{N^4Q^2}{(gH)^3}$$

The fourth root of this is $\frac{N\sqrt{Q}}{(gH)^{3/4}}$; the well known *specific speed* parameter. One could have arrived at this parameter by grouping the variables, N, Q, g and H directly and by proceeding to get a dimensionless parameter.

The following functional forms of the dimensionless numbers with usual notation are employed in different situations :

(*i*) Thrust of a propeller

$$F/(\rho U^2d^2) = f\left(\frac{Nd}{U}, \frac{Ud\rho}{\mu}\right) \quad \text{... (13.19)}$$

(*ii*) Friction loss in a pipe flow and ducts

$$\frac{\Delta h}{l} = f\left(\frac{Ud\rho}{\mu}\right) \quad \text{... (13.20)}$$

(*iii*) Resistance experienced by a ship

$$\frac{R}{\rho U^2 l^2} = f\left(\frac{U^2}{lg}, \frac{Ul\rho}{\mu}\right) \quad \text{... (13.21)}$$

(*iv*) Drag on a submerged body such as an aeroplane or a submarine or the resistance to confined flows such as flow through a pipe.

$$\frac{D}{\rho U^2 l^2} = f\left(\frac{Ul\rho}{\mu}, \frac{\varepsilon}{l}\right) \quad \text{... (13.22)}$$

(*v*) Lift on an aerofoil or some other submerged body

$$\frac{L}{\rho U^2 A} = f\left(\frac{Ul\rho}{\mu}, \frac{\varepsilon}{l}\right) \quad \text{... (13.23)}$$

(*vi*) Torque of a propeller or a rotodynamic machine

$$\frac{T}{\rho N^2 d^5} = f\left(\frac{Nd}{U}, \frac{\rho Nd^2}{\mu}\right) \quad \text{... (13.24)}$$

(*vii*) Discharge over a spillway

$$\frac{Q}{bg^{1/2}H^{3/2}} = f\left(\frac{b}{H}, \frac{U^2}{gH}\right) \quad \text{... (13.25)}$$

The terms in the parentheses have been arranged in order of importance. Thus, for the resistance experienced by a ship, the kinematic condition is decided by equating the Froude number ; and then the Reynolds number effects are either ignored or taken care of additionally by way of determining the scale effects with reference to pre-drawn curves, formulae or charts.

Example 13.1. *A model of an air duct operating with water produces a pressure drop of 10 kN/m² over 10 m length. If the scale ratio is 1/50, ρ_w = 1000 kg/m³, ρ_a = 1.2 kg/m² and μ_w = 0.001 Ns/m²* and μ_a = *0.000 02 Ns/m², estimate the corresponding drop in a 20 m long air duct.*

Solution.

For a confined flow such as that in a duct under pressure, the Reynolds number is the most predominant parameter. Hence

$$\left(\frac{Ud\rho}{\mu}\right)_{water} = \left(\frac{Ud\rho}{\mu}\right)_{air}$$

and

$$\frac{U_a}{U_w} = \frac{d_w}{d_a} \cdot \frac{\rho_w}{\rho_a} \cdot \frac{\mu_a}{\mu_w} = \frac{1}{50} \times \frac{100.0}{1.2} \times \frac{0.00002}{0.001} = 0.33$$

which shows that the water model must have a velocity three times that in the air duct.

Equating the Euler number or the pressure drop coefficient,

$$\left(\frac{\Delta p}{\rho U^2}\right)_{water} = \left(\frac{\Delta p}{\rho U^2}\right)_{air}$$

Corresponding to a 10 kN/m² pressure drop in the water flow model 10 m long,

$$\Delta p \text{ in air} = 10 \times \frac{1.2}{1000} \times 0.33^2 = 0.001\ 31 \text{ kN/m}^2$$

$$= \mathbf{1.31\ N/m^2}$$

The Δp in air duct over 20 m length is therefore **2.62 N/m²**. It may be appreciated, by observing the magnitudes of pressure drops in air and water ducts, why water duct models are used for pressure drop studies.

Example 13.2. *It is desired to estimate the drag force on a 10 m long wing with a frontal projected area of 20 cm × 10 m subjected to 20 m/s wind by undertaking a model study in the laboratory. What should be the sizes of the aerofoils tested with the following methods :*

(a) in an air stream at 40 m/s in a wind tunnel ?

(b) in a water stream at 10 m/s ?

If the drag forces measured in (a) and (b) are 420 N and 1200 N respectively, calculate (c) the drag force on the wing. Which of the two methods is preferred ?

(Given that μ_w = 0.001 Ns/m² μ_a = 0.00002 Ns/m², ρ_w = 1000 kg/m³ and ρ_a = 1.2 kg/m³)

Solution.

For a fully submerged object such as an aerofoil, the Reynolds number should be equal for the model and the prototype,

$$\left(\frac{Ud\rho}{\mu}\right)_p = \left(\frac{Ud\rho}{\mu}\right)_m$$

For case (*a*), Scale ratio $= \dfrac{d_m}{d_p} = \dfrac{20}{40} = \dfrac{1}{2}$

which suggests that the frontal projected area of the model aerofoil

$$a = \left(20 \times \frac{1}{2}\right) \text{cm} \times \left(10 \times \frac{1}{2}\right) \text{m}$$

or $= 10 \text{ cm} \times 5\text{m}$

and the length of the model is $10/2 = \mathbf{5\ m}$

For case (*b*), Scale ratio $= \dfrac{d_m}{d_p} = \dfrac{20}{10} \times \dfrac{0.001}{0.00002} \times \dfrac{1.2}{1000} = 0.12$

The frontal projected area of the model is, therefore,

$$a = (20 \times 0.12) \text{ cm} \times (10 \times 0.12) \text{ m}$$

or $= 2.4 \text{ cm} \times 1.2 \text{ m}$

and the length of the model $= 10 \times 0.12 = \mathbf{1.2\ m}$

(*c*) Equating the drag coefficient,

$$\left(\frac{F}{\rho U^2 A}\right)_m = \left(\frac{F}{\rho U^2 A}\right)_p$$

For the model (*a*), $F_m = 420$ N

Hence, $F_p = 420 \times 1 \times \left(\dfrac{20}{40}\right)^2 \times 2^2$

$= 420$ N

For the model (*b*), $F_m = 1200$ N

and hence, $F_p = 1200 \times \left(\dfrac{20}{10}\right)^2 \times \left(\dfrac{1}{0.12}\right)^2 \times \dfrac{1.2}{1000}$

$= \mathbf{400\ N}$

The preference is in favour of (b) i,e. testing under water because the model is smaller and the forces involved are more, allowing greater measuring accuracy. Availability of a wind tunnel or a water tunnel is the next consideration for the preference,

Example 13.3. *A 10 cm diameter plate is to be rotated in water and a 25 cm diameter plate in air under dynamically similar conditions. Find the ratio of the torques required. Given that,*

	Water	*Air*
mass density ρ (*kg/m*³)	*1000*	*1.2*
dynamic viscosity μ (*Ns/m*²)	*0.001*	1.8×10^{-6}

Solution.

For dynamic similarity of fully immersed bodies, the Reynolds number should be equated, *i.e.*,

$$Re = Ud\rho/\mu = \text{constant} \quad ... (i)$$

However, for kinematic similarity,

$$\text{Speed parameter} = Nd/u = \text{constant} \quad ... (ii)$$

From (*i*) and (*ii*), by multiplication,

$$Nd^2\rho/\mu = \text{Constant}$$

Substituting the data for water and air,

$$N_1 \times 0.1^2 \times 1000/0.001 = N_2 \times (0.25)^2 \times 1.2/1.8 \times 10^{-6}$$

whence $N_1/N_2 = \mathbf{4.17}$

Now, by equating the torque coefficient for the two,

$$\frac{T_1}{\rho_1 N_1^2 d_1^5} = \frac{T_2}{\rho_2 N_2^2 d_2^5}$$

$$\frac{T_1}{T_2} = \left(\frac{\rho_1}{\rho_2}\right) \times \left(\frac{N_1}{N_2}\right)^2 \times \left(\frac{d_1}{d_2}\right)^5$$

$$= \left(\frac{1000}{1.2}\right) \times (4.17)^2 \times \left(\frac{0.1}{0.25}\right)^5$$

$$= (833.33) \times (17.39) \times 0.010\ 24$$

$$= \mathbf{148.38}$$

Example 13.4. *Estimate (a) the speed of rotation, (b) the thrust produced, (c) the torque developed and (d) the efficiency of propulsion by a 3 m diameter propeller to cruise at 10 m/s if a 1/16 scale model provided the following results :*

U = 5 m/s, N = 750 revolutions per minute

F = 50 N, T = 10 Nm

Solution.

(*a*) It is seen that the dynamic similitude requires Nd/U to be equated for the model and the prototype,

$$\left(\frac{Nd}{U}\right)_m = \left(\frac{Nd}{U}\right)_p$$

The speed of rotation, $N_p = 750 \times \frac{10}{5} \times \frac{1}{10}$ = **150 revolutions per minute**

(*b*) Equating the thrust parameter $F/\rho U^2 d^2$, taking ρ the same.

The thrust produced, $F_p = 50 \times \left(\frac{10}{5}\right)^2 \times \frac{100}{1000}$ = **20 kN**

(*c*) Equating the torque parameter $T/\rho N^2 d^5$, taking ρ the same,

The torque developed $T = 10 \times \left(\frac{150}{750}\right)^2 \times \frac{10^5}{1000}$ = **40 kNm**

(*d*) Propulsive efficiency

$$n_p = \left(\frac{FU}{T\omega}\right) = \frac{20 \times 10 \times 60}{40 \times 2\pi \times 150}$$

$$= 0.318 = \mathbf{31.8\%}$$

$$\eta_m = \left(\frac{FU}{T\omega}\right)_m = \frac{50 \times 5 \times 60}{10 \times 2\pi \times 750}$$

$$= 0.318 = \mathbf{31.8\%}$$

which is equal for the model and the prototype thereby confirming our calculations of the thrust and torque parameters to be correct!

Example 13.5. *A 1/20 scale model of a spillway studied in the laboratory requires 5 m³/s discharge and a hydraulic jump formed therein dissipates 500 W. Calculate*

(*a*) *the discharge in the spillway, neglecting viscous and surface tension effects,*

(*b*) *the power lost in the spillway jump, and*

(*c*) *the velocity ratio between the two flows.*

Solution.

It this case, b/H should be equated for geometric and kinematic similarity,

$$\left(\frac{b}{H}\right)_m = \left(\frac{b}{H}\right)_p$$

$$\frac{H_p}{H_m} = \frac{b_p}{b_m} = 20$$

(*a*) Equating the discharge parameter,

$$\left(\frac{Q}{bg^{1/2} \cdot H^{3/2}}\right)_p = \left(\frac{Q}{bg^{1/2} \cdot H^{3/2}}\right)_m$$

$$Q_p = 5 \times 20 \times 20^{3/2} = \mathbf{8944.27\ m^3/s}$$

(*b*) From the fact that P_p = Power lost in the prototype jump = $8950 \times H_p \rho g$ and P_m = Power lost in the model jump = $5 \times H_\omega \rho g = 500$ W

the power lost is $P_p = \frac{8950}{5} \times 20 \times 500 = 17.89 \times 10^5$ W = **17.89 MW**

(*c*) Equating the Froude number $Fr = \sqrt{U/gH}$ for the model and prototype,

the velocity ratio $\frac{U_m}{U_p} = \frac{\sqrt{gH_m}}{\sqrt{gH_p}} = \sqrt{1/20} = \mathbf{0.224}$

Example 13.6. *A hydraulic prime mover, when tested in a laboratory at 200 revolutions per minute under 10 m head, develops 50 kW power. Estimate the power potential, size ratio and the specific speed of a similar machine operating under a head of 30 m and running at the same speed.*

Solution.

The Reynolds number effects are unimportant in a well designed hydraulic prime mover because, above a certain Reynolds number ($> 10^5$), dependence on viscous effects is negligible. The other parameters are, therefore, equated.

$$\left(\frac{gH}{N^2 d^2}\right)_p = \left(\frac{gH}{N^2 d^2}\right)_m$$

$$\left(\frac{d_m}{d_p}\right)^2 = \frac{10}{30} \times 1 = \frac{1}{3}, \frac{d_p}{d_m} = 1.73$$

Equating the power parameter, $P/(\rho N^3 d^5)$, taking ρ the same,

$$P_P = 50 \times 1 \times 1.73^5 = \mathbf{779.42\ kW}$$

When an idealised model turbine operating under a unit head develops unit power, the speed at which it does so is called the specific speed N_s a non-dimensional parameter,

$$N_s = \frac{N\sqrt{P/\rho}}{(gH)^{5/4}}$$

The specific speed should also be equal for the model and the prototype. This fact could have been used to obtain the power straightaway. However, let us calculate the specific speed as a check.

For the prototype $N_p = \frac{2\pi}{60} \times 200 \times \sqrt{\frac{780 \times 100}{1000}} \Big/ (9.81 \times 30)^{5/4}$

$$= \mathbf{0.48}$$

For the model $N_m = \frac{2\pi}{60} \times 200 \times \sqrt{\frac{50 \times 100}{1000}} \Big/ (9.81 \times 10)^{5/4}$

$$= \mathbf{0.48}$$

Equal specific speeds for the two confirm the accuracy of our calculations for the size and the power required!

Example 13.7. *A 1/30 model of a ship with 900 m² wetted area towed in water at 2 m/s experiences a resistance of 20 N. Calculate*

(a) the corresponding speed of the ship, (b) the wave making drag on the ship, (c) the skin-friction drag if the skin-drag coefficient for model is 0.004 and for prototype 0.015 (d) the total drag on the ship and (e) the power to propel the ship.

Solution.

In view of the high wave-making drag in comparison with the skin-friction drag for a partially submerged body, the Froude number should be equated,

$$\left(\frac{U}{\sqrt{lg}}\right)_m = \left(\frac{U}{\sqrt{lg}}\right)_p$$

(*a*) The corresponding speed $U_p = \sqrt{30}\, U_m$

$$= \sqrt{30} \times 2 = \mathbf{10.95\ m/s}$$

Area ratio $a_r = \dfrac{a_m}{a_p} = \left(\dfrac{1}{30}\right)^2 = \dfrac{1}{900};\ a_m = \dfrac{900}{900} = 1\ \text{m}^2$

Resistance $R = 20 = \dfrac{1}{2} \times 1000 \times 2^2 \times 0.004 + R_{wm} = 8 + R_{wm}$

Hence, R_{wm} = wave making resistance of the model = 12 N

Equating the resistance coefficient $\left(\dfrac{R_w}{\rho U^2 l^2}\right)$ due to wave making for the model and the prototype,

For (*b*), $R_{wp} = 12 \times 1 \times 30 \times 900 = 324 \times 10^3\ \text{N} = \mathbf{324\ kN}$

For (*c*), R_{sp} = skin-friction resistance of the prototype

$$= \frac{1}{2} \times 1000 \times 10.95^2 \times 0.015 \times 900$$

$$= 810 \times 10^3\ \text{N} = \mathbf{809\ kN}$$

Total drag $R = 324 + 809 = \mathbf{1133\ kN}$

Propulsive power required $= 1133 \times 10.95 = \mathbf{12410\ kW = 12.41\ MW}$

13.11 Distorted Models

It is often necessary to make *models which may not be geometrically similar to the prototype.* River and harbour models, for example, cannot be made so large as to have sufficient depth of water in them. From the cost point of view, it is desirable to make smaller models but the depth of water may reduce to a few millimetres on the basis of geometric similarity. Also, it may not be possible to reduce the surface roughness with the criterion of geometric similarity on the one hand and to be able to create a turbulent flow on the other. It is, therefore, necessary that distorted models, *i.e.,* with different scales-factors for horizontal and vertical directions be made and tested. It is typical to adopt a vertical scale of 1/100 and a horizontal scale of 1/200 to 1/500.

The dynamic criterion for model testing is the Froude number which should be equated for the model and the prototype,

$$Fr_m = Fr_p$$

or $$\frac{U_m}{\sqrt{gd_m}} = \frac{U_p}{\sqrt{gd_p}}$$

whence $$\frac{U_m}{U_p} = \sqrt{\frac{d_m}{d_p}}$$

If the breadths of the model and prototype are b_m and b_p, the areas of flow are

$$A_m = b_m d_m \text{ and } A_p = b_p d_p$$

whence $$\frac{A_m}{A_p} = \frac{b_m d_m}{b_p d_p}$$

The discharge through the model and the prototype are related by

$$\frac{Q_m}{Q_p} = \frac{U_m A_m}{U_p A_p} = \sqrt{d_m / d_p} \cdot \frac{b_m d_m}{b_p d_p}$$

or $$\frac{Q_m}{Q_p} = \frac{b_m}{b_p} \cdot \left(\frac{d_m}{d_p}\right)^{3/2}$$

If the horizontal scale is 1 : x and the vertical scale is 1 : y

$$\frac{b_m}{b_p} = \frac{1}{x} \text{ and } \frac{d_m}{d_p} = \frac{1}{y}$$

$$\frac{Q_m}{Q_p} = \frac{1}{x}\left(\frac{1}{y}\right)^{3/2}$$

Example 13.8. *A 1 : 50 scale model of a hydroelectric power station tailrace is proposed to be built. If the design bypass flow rate is 600 m³/s and the allowable discharge is 800 m³/s, calculate the corresponding flow rates for model testing.*

Solution.

A free-surface model of a tailrace should be tested under dynamically similar conditions by equating the Froude numbers.

$$Fr = \frac{U}{\sqrt{lg}}$$

Equating the Froude number for the model and the prototype,

$$\frac{U_m}{\sqrt{l_m g}} = \frac{U_p}{\sqrt{l_p g}}$$

or $$\frac{U_m}{U_p} = \sqrt{l_m / l_p}$$

This gives the ratio of the velocities. The ratio of the discharges may be obtained by considering that

$$\frac{Q_m}{Q_p} = \frac{U_m A_m}{U_p A_p}$$

Since $$\frac{A_m}{A_p} = \left(\frac{l_m}{l_p}\right)^2$$

it follows that

$$\frac{Q_m}{Q_p} = \left(\frac{l_m}{l_p}\right)^{2.5}$$

Substituting the data,

$$Q_m = 600 \times \left(\frac{1}{50}\right)^{2.5} = 0.034 \text{ m}^3/\text{s}$$

at the design point.

The maximum discharge required for the model tested is given by

$$0.034 \times \frac{800}{600} = \mathbf{0.0453\ m^3/s}$$

Example 13.9. *The drag on a small submarine hull is desired when it is moving far below the surface of water. A 1/10 scale model is to be tested. What dimensionless group should be duplicated between model and prototype and why ? If the drag of the prototype at 1 knot is desired, at what speed should the model be moved to give the drag to be expected by the prototype ? Would this result still be true if the prototype were to be moved close to the surface ? Explain.*

Solution.

A submarine is a wholly submerged body in a large mass of fluid at rest. The Reynolds number must be duplicated for the model and the prototype. If the prototype were to move close to the surface then the gravity-aided wave making action and the consequent drag would also contribute to the resistance. Corrections due to 'scale effect' would then be necessary.

Equating the Reynolds number for dynamic similarity

$$\frac{U_1 l_1 \rho_1}{\mu_1} = \frac{U_2 l_2 \rho_2}{\mu_2}$$

If $\qquad l_1 = 1/10\, l_2,\ \rho_1 = \rho_2$ and $\mu_1 = \mu_2$

for the same fluid, *i.e.*, water for the model and the prototype,

$$U_1 = U_2 l_2 / l_1$$

$$= 1 \times 10 = \mathbf{10\ knots}$$

The drag coefficient $D/\rho U^2 l^2$ should be equated to evaluate the drag experienced by the prototype.

$$D_2 = D_1 \frac{\rho_2 U_2^2 l_2^2}{\rho_1 U_1^2 l_1^2}$$

$$= \frac{1}{10^2} \times 10^2 \times D_1 = D_1$$

It shows that if the 1/10 model is moved at 10 knots the drag experienced by it would be the same as that experienced by the prototype at 1 knot speed.

Example 13.10. *Calculate the speed of rotation, the torque produced and the power of a windmill of 5 m diameter in a wind speed of 30 km/hr from the performance of a 1/10 scale geometrical model in a wind tunnel with 10 m/s free stream. Tlie model rotated at 1200 revolutions per minute produced a torque of 2.5 Nm.*

Solution.

Equating the speed parameter for the model and the prototype,

$$\frac{N_1 D_1}{U_1} = \frac{N_2 D_2}{U_2}$$

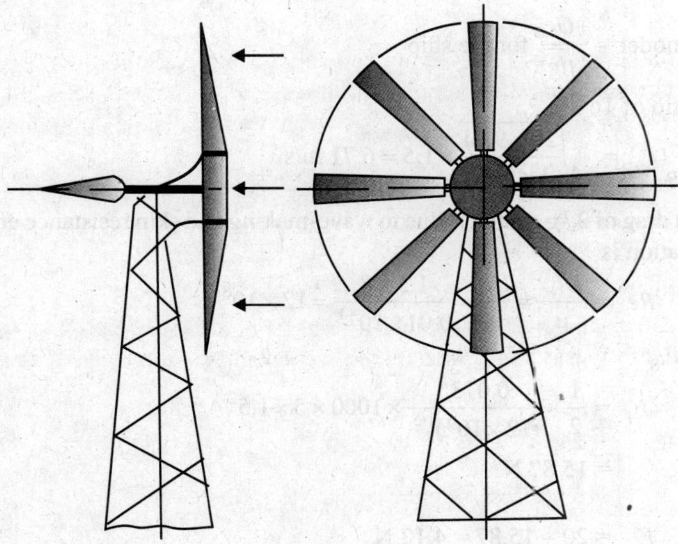

Since $D_1/D_2 = 0.1$ and $U_1 = 10$ m/s

and $U_2 = 30$ km/hr $= \dfrac{30 \times 1000}{3600} = 8.33$ m/s

$N_1 = 1200$ revolutions per minute

Therefore, $N_2 = 1200 \times 0.1 \times 8.33/10$

$= \mathbf{100}$ **revolutions per minute**

which corresponds to $\omega_2 = 100 \times 2\pi/60 =$ **10.5 rad/s**

Equating the torque coefficient,

$$\frac{T_1}{\rho N_1^2 D_1^5} = \frac{T_2^2}{\rho N_2^2 D_2^5}$$

The torque expected of the windmill

$$T_2 = 2.5 \times \left(\frac{100}{1200}\right)^2 \times 10^5 = \mathbf{1736.1\ Nm}$$

Power of the windmill

$$P = T_2\omega_2 = 1736 \times 10.5$$

$$= 18\,230 \text{ W} = \mathbf{18.23\ kW}$$

Example 13.11. *A 1 : 20 model of a naval 'ship having a submerged surface area of 5 m² and length 8 m has a total drag of 20 N when towed through water at a velocity of 15 m/s. Calculate the total drag on the prototype when moving at the corresponding speed. Use relation* $F_1 = \frac{1}{2} C_f \rho A u^2$ *for calculating the skin friction resistance. The value of* C_f *is given by*

$$C_f = 0.0735/Re^{1/5}$$

Take the kinematic viscosity of water as 0.01 stokes.

Solution.

Model of a ship is tested under dynamically similar conditions based upon the Froude number, whence, $\dfrac{U}{\sqrt{lg}}$ for the model = $\dfrac{U}{\sqrt{lg}}$ for the ship

Substituting the scale ratio of 1 : 20,

the corresponding velocity, $U_{ship} = \sqrt{\dfrac{20 \times 9.81}{1 \times 9.81}} \times 1.5 = 6.71$ m/s

The model experiences a drag of 2 kg_t which is due to wave-making and skin resistance effects. The Reynolds number of operation is

$$Re = \frac{Ul\rho}{\mu} = \frac{Ul}{\nu} = \frac{1.5 \times 8}{0.01 \times 10^{-4}} = 12 \times 10^6$$

and the skin friction resistance is

$$F_t = \frac{1}{2} \times \frac{0.0735}{(12 \times 10^6)^{1/5}} \times 1000 \times 5 \times 1.5^2$$

$$= 15.87 \text{ N}$$

leaving the wave making drag

$$F_w = 20 - 15.87 = 4.13 \text{ N}$$

The ship experiences a total drag made up of two similar components. The skin friction resistance can be calculated by the same relation as given. The Reynolds number is

$$Re = \frac{Ul}{\nu} = \frac{6.71 \times (8 \times 20)}{0.01 \times 10^{-4}} = 1074 \times 10^6$$

and the skin friction resistance, $F_t = \dfrac{1}{2} \times \dfrac{0.0735}{(1074 \times 10^6)^{1/5}} \times 1000 \times (5 \times 20^2) \times 6.71^2$

$$= 51\,710 \text{ N} = \mathbf{51.71 \text{ kN}}$$

The wave-making drag is calculated by similarity with the model,

$$\frac{F_p}{\rho A U^2} \text{ for the ship} = \frac{F_w}{\rho A U^2} \text{ for the model}$$

$$F_w \text{ for the ship} = \left(\frac{20}{1}\right)^2 \times \left(\frac{6.71}{1.50}\right)^2 \times 4.13$$

$$= 33100 \text{ N} = \mathbf{33.1 \text{ kN}}$$

The total drag on the prototype ship is thus 51.71 + 33.1 = **84.8 kN**

ITEM BANK

Objective items of different types are given to review the concepts developed in the Chapter.

* *Select the correct response for each item.*
* *Every correct response is expected in less than a minute and carries one mark.*
* *Compare your responses with the key answers given at the end.*

Multiple-Choice Items

Attempt the following Multiple-Choice Items by ticking the correct or the most appropriate response from among the alternatives given :

1. The following physical variables are dimensionless :

(*a*) Stream function and acceleration

(*b*) Angular displacement and velocity

(*c*) Reynolds number and angular momentum

(*d*) Velocity ratio and Reynolds number

2. Dimensional homogeneity principle implies that
   (*a*) Only the variables with the same dimensions may be multiplied
   (*b*) Only the variables with the same dimensions may be added or subtracted
   (*c*) Dimensions of the two sides of an equation may not be the same
   (*d*) Only the variables with the same dimensions may be divided.
3. The repeating variables in a dimensional analysis should be such as to
   (*a*) exclude the dependent variables
   (*b*) include the unimportant set of factors
   (*c*) include the important set of factors
   (*d*) have two variables with same dimensions
4. The Reynolds number is defined as the ratio of
   (*a*) gravity force to viscous force
   (*b*) viscous force to inertia force
   (*c*) inertia force to viscous force
   (*d*) gravity force to inertia force
5. In which one of the following situations are the inertia forces *unimportant* ?
   (*a*) flow through a rectangular notch
   (*b*) flow through a sluice gate
   (*c*) flow through a sudden contraction in a pipe
   (*d*) flow through a capillary tube.
6. A dimensionless group formed with the variables ρ, *w*, μ, and *D* is
   (*a*) $\rho\omega\mu/D^2$ (*b*) $\rho\omega D^2/\mu$
   (*c*) $\mu D^2\rho\omega$ (*d*) $\rho\omega\mu D$
7. Laminar flow between closely spaced parallel plates is governed by the consideration of the following pair of most important forces:
   (*a*) pressure and inertial forces
   (*b*) gravity and inertial forces
   (*c*) viscous and inertial forces
   (*d*) pressure and viscous forces
8. The time scale ratio for a model based on Froude Law criterion in terms of Length Scale ratio *Lr* is :
   (*a*) L_r (*b*) $\sqrt{L_r}$
   (*c*) $1/\sqrt{L_r}$ (*d*) $L_r^{1.5}$
9. In Hagen-Poiseuille flow of viscous liquid, one of the following pairs of forces is equated:
   (*a*) Inertial and Viscous forces
   (*b*) Pressure and Viscous forces
   (*c*) Gravity and Viscous forces
   (*d*) Inertial and Gravity forces.
10. Water turbines may be classified in the decreasing order of specific speeds as :
   (*a*) Propeller turbine, Reaction turbine, Impulse turbine.
   (*b*) Pelton wheel, Francis turbine, Kaplan turbine.
   (*c*) Reaction turbine, Impulse turbine, Propeller turbine.
   (*d*) None of the above.
11. Which one of the following is dimensionless ?
   (*a*) $\frac{\partial p}{\partial x}\cdot\frac{D^4}{\mu Q^2}$ (*b*) $\frac{\partial p}{\partial x}\cdot\frac{D^3}{\mu Q}$
   (*c*) $\frac{\partial p}{\partial x}\cdot\frac{D^4}{\mu Q}$ (*d*) $\frac{\partial p}{\partial x}\cdot\frac{\mu Q}{D^4}$
12. If there are *n* physical quantities then the number of independent dimensionless parameters with *r* base dimensions are
   (*a*) $n+r$ (*b*) $n-r$
   (*c*) $n-(n-r)$ (*d*) $n+(n-r)$
13. Which one of the following sets of variables should be selected as the repeating variables in a fluid flow problem?
   (*a*) U, l, ρ (*b*) U, g, ρ
   (*c*) U, ρ, μ (*d*) ρ, μ, g

Other Objective Items

14. Are the following statements true or false ?

(*a*) Reynolds number is the ratio of inertia to viscous forces in a flow. T/F

(*b*) Froude number is the ratio of inertia to gravity forces. T/F

(*c*) Multiplication of two dimensionless numbers must be dimensionless. T/F

(*d*) There can be more than one dimensionless numbers for the same set of four physical variables. T/F

15. Match the dimensionless parameters from Block B with the phenomena in Block A.

Block A	Block B
(*a*) Flow over a notch	(*k*) Reynolds number
(*b*) Hydraulic jump	(*l*) Froude number
(*c*) Shock waves	(*m*) Mach number
(*d*) Flow through a pipe	
(*e*) Flow past a cylinder	
(*f*) Vortex in a basin.	

16. Match the dimensionless parameters from Block B with the phenomena in Block A.

Block A	Block B
(*a*) Confined flows	(*k*) Knudsen number
(*b*) Rarified gas flow	(*l*) Prandtl number
(*c*) Convection currents	(*m*) Euler number
(*d*) Vortex shedding	(*n*) Strouhal number
	(*o*) Froude number
	(*p*) Weber number

17. **Assertion A :** Reynolds number must be the same for the model and the prototype immersed in a subsonic flow.

Reason R : A model should be geometrically similar to the prototype.

Which one of the following is correct ?

(*a*) A is true; R is true and explains the assertion

(*b*) *A* is true, R is true but does not explain the assertion

(*c*) A is true; R is false

(*d*) A is false ; R is true

ANSWERS

1. (*d*), **2.** (*b*), **3.** (*c*), **4.** (*c*), **5.** (*d*), **6.** (*b*), **7.** (*d*), **8.** (*b*), **9.** (*b*), **10.** (*b*), **11.** (*c*), **12.** (*b*), **13.** (*a*), **14.** (T, F, T, T), **15.** (al, bl, cm, dk, ek, fl), **16.** (am, bk, cl, dn), **17.** (*b*)

QUESTION BANK

Questions on theory and numerical problems are given for tutorial work.

* *The questions are arranged generally in order of the articles in the text.*

* *Every question should take 15-20 minutes to solve.*

* *Answers for the numerical problems are given alongside them.*

13.1. Check that the specific speed of a pump or a turbine defined as

$$\frac{N\sqrt{Q}}{(gH)^{3/4}} \text{ or } \frac{N\sqrt{P/\rho}}{(gH)^{3/4}}$$

is dimensionless and has no units for the data in SI units, *i.e.*, when N is in rad/s, g in m/s², Q in m³/s, ρ in kg/m³, P in W and H in metres

13.2. Check if the following equations are dimensionally homogeneous. If not, what conclusion would you draw ?

(*a*) $C_D = \frac{1.328}{\sqrt{Re}}$ (*b*) $\text{Drag} = C_D \cdot \frac{1}{2}\rho U^2 . A$

(c) $F = \dot{m}\,(U_2 - U_1) - (p_{g_1} A_1 - p_{g_2} A_2)$ (*d*) $Q = \frac{8}{15}\tan\frac{\theta}{2}\sqrt{2g}H^{5/2}$

(*e*) $T = (r_1 u_{\theta_1} - r_1 u_{\theta_1})^m$ (*f*) $E = \frac{U^2}{2} + gz + \frac{p}{\rho} = \text{constant}$

13.3. In a static fluid, the pressure p at a point is known to depend upon the specific weight ρ*g* and its depth z below the free surface. By dimensional method, determine the form of the hydrostatic law. **[$p = \rho g z$]**

13.4. If the shear stress τ varies with the strain rate *du/dy,* determine the form of the Newton's law of viscosity by dimensional analysis. **[$\tau = \mu\, du/dy$]**

13.5. Determine the form of the Mach number, M, if it is known to depend upon the pressure p, density ρ, specific heat ratio γ and velocity U of a fluid in a situation. $\left(M = U\Big/\sqrt{\frac{\gamma p}{\rho}}\right)$

13.6. Predict the torque and the rotational speed of a propeller to operate in water at 5 m/s if a 1 : 20 model rotating at 100 r.p.m. tested in a wind tunnel at a speed of 20 m/s requires a torque of 10 Nm. **[4.166 MNm, 0.131 rad/s]**

13.7. The overall head losses in a water piping system are planned to be determined on 1 : 5 model with air flowing through it. Calculate the air velocity and mass rate required at the standard conditions to simulate the flow of water at 0.5 m/s in a 6 m diameter pipe. Take ρ for water = 1000 kg/m³; μ for air = 0.000 018 Ns/m² and μ for water = 0.001 Ns/m².

[37.5 m/s, 50.89 kg/s]

13.8. By dimensional consideration, obtain an equation for the discharge through an orifice if the variables on which it depends are the fluid properties μ, ρ and p as well as d and g.

Determine the pressure drop across an orifice installed in a 30 cm diameter pipe for dynamic similarity to exist between this and another orifice in 10 cm diameter pipe where a pressure drop of 10 cm of mercury is registered across a 4 cm diameter orifice. The fluid in each case is water. **[30 cm of mercury]**

13.9. A water turbine running at 200 r.p.m. under a head of 15 m produces an output of 100 kW. Determine the scale ratio and the speed of a similar turbine to generate 500 kW running under a head of 30 m under similar circumstances. **[Ans. 1.33, 212.7 r.p.m.]**

13.10. A propeller model 0.2 m diameter advancing at 1 m/s and rotating at 200 r.p.m. experiences a thrust of 50 N and torque of 5 Nm. Determine the speed of rotation, thrust and torque of another similar propeller 2 m diameter advancing at 3 m/s. **[60 r.p.m., 45 kN, 45 kN m]**

13.11. Calculate the thrust required to run a motor-boat 5 m long at 100 m/s in a lake if the force required to tow a 1 : 30 model is 0.5 N. Neglect the viscous resistance due to water and air in comparison with the wave-making resistance. **[13.5 kN]**

13.12. A torpedo-shaped object 900 mm in diameter is to move in air at 60 m/s and its drag has to be estimated from tests in water on a half scale model. Determine the necessary speed of the model and the drag of the full scale object if that of the model is 1140 N.

(viscosity of air : 1.86×10^{-5} Pas ; viscosity of water : 1.01×10^{-3} Pas. Density of air : 1.20 kg/m^3; Density of water : 1000 kg/m^3 values are quoted at the standard temperature at which both the model and prototype work.) **[7.82 m/s ; 322 N]**

13.13. In a 1 in 20 model of a stilling basin, the height of the hydraulic jump in the model is observed to be 0.2 m. What is the height of hydraulic jump in the prototype ? If the energy dissipated in the model is 0.1 kW, what is the corresponding value in the prototype ?

[4 metre and 3577.71 kW]

13.14. Define the term dynamical similarity as used in model studies of hydraulic structures.

A spillway model is to be built to a geometrically similar scale of 1/50 across a flume of 60 cm width. The prototype is 15 m high and the maximum head on it is expected to be 1.5 m. (*i*) What height of model and what head on the model should be used ? (*ii*) If the flow over the model at a particular head is 12 litres/second, what flow per metre width of the prototype is expected ? (*iii*) If the negative pressure in the model is 20 cm, what is the negative pressure in the prototype ? Is it practicable ?

[(*i*) 0.3 and 0.03 m (*ii*) 7.07 m^3/s per metre width; (*iii*) 10 m negative head of water which is impracticable]

13.15. At the inlet to a pump, a 3.2 m diameter pipe has to be reduced to a diameter of 0.3 m and the most convenient way of doing this is to have a sudden reduction in area as shown in the figure. Sketch the streamline pattern when both pipes are flowing full of water.

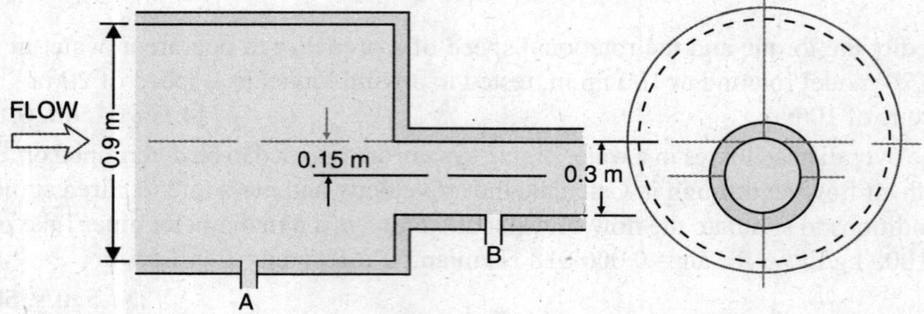

When a 1/10 scale model of the contraction is tested in air with a velocity of 5 m/s in the larger pipe, a water U-tube manometer connected between the tapping points A and B has a difference in levels of 0.22 m. Assuming that the air density in the pipes remains constant at 1.23 kg/m^3, estimate the rate at which the fluid energy is being dissipated in the contraction.

If the flow conditions in the model and prototype are similar, calculate the pressure difference between the points corresponding to A and B on the prototype when the water flow rate is 0.5 m^3/s.

13.16. (*a*) Explain the principle of dimensional homogeneity and its use.

(*b*) The thrust F of a supersonic plane during flight can be considered as dependent upon the length of aircraft l, velocity V, air viscosity μ, air density ρ, and bulk modulus of air K. Express the functional relationship between the variables and the thrust.

13.17. (*a*) State the Buckingham π–theorem.

(*b*) The efficiency η of a fan depends on density ρ, dynamic viscosity μ of the fluid, angular velocity ω, diameter D, and discharge Q. Express η in terms of dimensionless parameters.

(*c*) The resistance in fresh water of 1 in 8 model ship when towed at 2.5 m/sec is 20 N. Calculate the corresponding velocity and resistance in the prototype in sea water.

[7.07 m/s and 1024 kN]

13.18. An overflow structure 30 m long discharges water at the rate of 750 m^3/s under a head of 8 m. A model test of this structure is to be carried out in a laboratory with a maximum supply available at 50 litres per second. Recommend a suitable scale ratio for the model. **[0.02]**

13.19. A ship is to be built to have a wetted hull area of 2500 m^2 and for a cruising speed of 12 m/s. A 1/40 scale model is tested at the corresponding speed and the resistance is 32 N. From separate tests, the skin function was found to be related to velocity as 3.7 $U^{1.95}$ N per unit wetted area for the model and it is estimated to be related to velocity as 2.9 $U^{1.8}$ N per unit area for the ship where U is the velocity in m/s. The model was tested in fresh water whereas the ship would operate in sea water of density 1020 kg/m^3. Determine

(*a*) the speed of towing the model,

(*b*) the skin-friction and wave-making drag on the model, and

(*c*) the total on the ship. **[1.9 m/s ; 20.2 and 11.8 N ; 1406.4 kN]**

CHAPTER

14

Fluid Machines

LEARNING OBJECTIVES

On successful completion of all learning activities planned and provided in the chapter, a student will be able to

- classify various types of fluid machines ;
- calculate the efficiency of centrifugal pumps and hydraulic turbines;
- describe the performance characteristics of centrifugal pumps and of hydraulic turbines; and
- calculate the head developed by centrifugal pumps operating in series and in parallels.

14.1 Classification on the Basis of General Features

Fluid Machines deal with conversion of power from fluid power to shaft power or from shaft power to fluid power as shown in Fig. 14.1. A fluid machine which operates to convert fluid

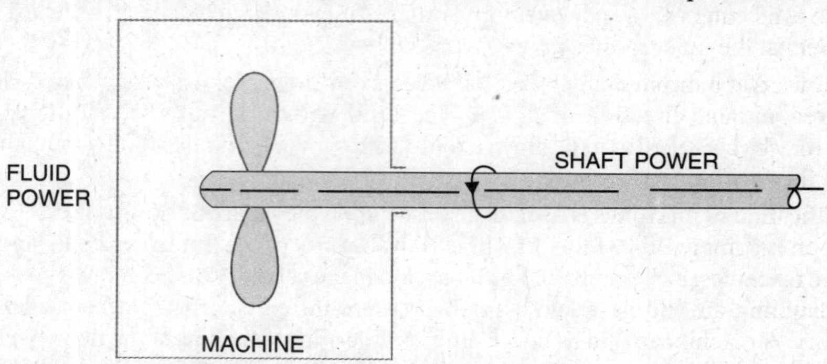

Fig. 14.1: Concept of a Fluid Machine

power into shaft power is called a turbine or an expander ; a fluid machine which operates to convert shaft power into fluid power is called a pump, a fan, a blower or a compressor. The exact nomenclature of a fluid machine depends upon its purpose, the fluid-dynamic action as well as the fluid handled.

Turbines are classified as follows :

Fluid	Type of Turbine
Water	Hydraulic Turbine
Steam	Steam Turbines
Gas or air	Gas Turbines
Wind	Windmills

Similarly, fluid machines which convert shaft power to fluid power by raising the energy content per unit mass of the fluid are classified as follows :

Fluid	Type of Machine
Water and Other Liquids	Pumps
Air and Other Gases (with slight pressure rise)	Fans, Blowers and Propellers
Air and Other Gases (with higher pressure rise)	Compressors

Fluid machines are recognised by their geometrical features as follows:

Axial Flow Machines if the fluid flows through the machine axially with respect to it.

Radial-Flow Machines/ Centrifugal Machines if the fluid flows through the machine radially with respect to it.

Mixed-Flow Machines if the fluid flow is arranged to conform to a relative path which is neither axial nor radial. A mixed flow machine is a cross between axial radial machines.

The names 'Inward Flow Machine' and 'Outward Flow Machine' are used to imply whether the flow takes place with decreasing or increasing racial distance respectively. In an inward flow machine, the fluid enters at the periphery and proceeds inward whereas in an outward flow machine, the fluid enters at the inner radius and proceeds outward.

It follows that a turbine may be classified as axial-flow, radial-flow or mixed-flow depending upon the predominant direction of flow of the fluid within the turbine. Similarly, a pump or a compressor may be classified as axial-flow, radial-flow or mixed-flow depending upon the predominant direction of flow within the machine.

Classification of machines is also made based upon the nature of flow within it. A fluid machine is said to be an Incompressible-Flow Machine if the density of the fluid does not change appreciably. All hydraulic machines, *e.g.*, pumps and turbines, fall in the category of incompressible-flow machines. Machines handling air and gases with small pressure differences across them also belong to the same category. A machine is said to be a Compressible-Flow Machine if the density change for flow through it is appreciable. Compressible-flow machines may be classified as Subsonic-Flow Machines, Transonic Flow-Machines and Supersonic-Flow Machines depending upon the range of Mach number.

Mach Number	Classification
$M > 1$	Supersonic-Flow Machines
$1 < M < 1$	Transonic-Flow Machines
$M < 1$	Subsonic-Flow Machines
$M \approx 0; < 0.2$	Incompressible-Flow Machines

14.2 Specific Speed : Type Number in SI Units

The Specific speed or Type number of a fluid machine is defined by a non-dimensional combination of its speed of rotation, head across the machine and the discharge or power for the machine under the optimum operating conditions :

$$N_s = \frac{N\sqrt{Q}}{(gH)^{3/4}} \text{ (based on discharge } Q) \qquad ...(14.1\ a)$$

and

$$N_s = \frac{N\sqrt{P/\rho}}{(gH)^{5/4}} \text{ (based on power } P) \qquad ...(14.1\ b)$$

The specific speed N_s is a non-dimensional number. It may be verified with reference to the dimensions of the variables.

From Eq. 14.1 (*a*),

$$\text{dimensions of } N_s = \frac{T^{-1}.(L^3T^{-1})^{1/2}}{(LT^{-2}.L)^{3/4}} = \frac{L^{3/2}.T^{-3/2}}{L^{3/2}T^{-3/2}} = 1$$

From Eq. 14.1 (*b*),

$$\text{dimensions of } N_s = \frac{T^{-1}(ML^2T^{-3}M^{-1}L^3)^{1/2}}{(LT^{-2}L)^{5/4}}$$

$$= \frac{L^{5/2}T^{-5/2}}{L^{5/2}T^{-5/2}} = 1$$

The SI units of the quantities used should be as follows:

N, the speed of rotation in rad/s;

Q, the discharge through the machine in m³/s;

P, the power for the machine in W;

H, the head across the machine in m of the fluid and

ρ, the density of the fluid is in kg/m³.

It may also be noticed that the discharge-based and power-based specific speeds are nearly equal ;

$$\frac{N\sqrt{P/\rho}}{(gH)^{5/4}} \approx \frac{N\sqrt{Q}}{(gH)^{3/4}} \qquad ...(14.2)$$

This can be shown by expressing the power *P* in terms of discharge *Q* and head *H* across the machine :

$$P = \rho g Q H$$

Then,

$$N\frac{\sqrt{P/\rho}}{(gH)^{5/4}} = \frac{N\sqrt{g\,Q\,H}}{(g\,H)^{5/4}} = \frac{N\sqrt{Q}}{(g\,H)^{3/4}}$$

In case the power referred to is the output power of a power generating machine operating with an efficiency η, then the power is written as

$$P = \rho g Q H.\eta$$

and

$$\frac{N\sqrt{P/\rho}}{(g\,H)^{5/4}} = \frac{N\sqrt{Q}}{(g\,H)^{3/4}} \times \sqrt{\eta}$$

which is seen to be $\sqrt{\eta}$ times the discharge-based specific speed. Since the efficiency is of the order of 90%. $\sqrt{\eta}$ is of the order of $\sqrt{0.09}$ or 0.95.

Engineers find it convenient to employ the discharge-based expression for specific speed for pumps, compressors, fans and blowers where discharge rating is specified and to employ the power-based expression for specific speed for all types of turbines where the power output is the main criterion. *Theoretically, either expression is adequate to work out the specific speed of a machine.*

The concept of specific speed or type number is useful in referring to a family of similar machines. The specific speed may alternatively be defined as a non-dimensional number, proportional to the speed at which a geometrically similar fluid machine would rotate when subjected to a unit head across it and when a unit discharge or a unit power is being dealt with at the optimum operating conditions. Let us obtain an expression for the specific speed of a fluid machine on the basis of this criterion. The discharge Q through a machine may be expressed as

$$Q \propto D\,bU \quad \text{...(14.3)}$$

where D = diameter of the rotating element, runner or impeller.

b = width of the rotating element, and

U = velocity of flow across the area of flow

Since $U \propto \sqrt{2\,gH}$ ...(*i*)

$\propto$ circumferential velocity of the rotating element

$\propto ND$

it follows that $D \propto \dfrac{\sqrt{gH}}{N}$ ...(*ii*)

where H = head across the machine

and N = speed of rotation of the machine

From the fact, $b \propto D$

it follows that $b \propto \dfrac{\sqrt{gH}}{N}$ ...(*iii*)

Substituting (*i*), (*ii*) and (*iii*) in Eq. (14.3)

$$Q \propto \frac{\sqrt{gH}}{N}\cdot\frac{\sqrt{gH}}{N}\sqrt{gH} \text{ or } \propto \frac{(gH)^{3/2}}{N^2}$$

By definition,

$N = kN_s$

if $H = 1$ m and $Q = 1\ \text{m}^3/\text{s}$

Hence the specific speed.

$$N_s = \frac{N\sqrt{Q}}{(gH)^{3/4}}$$

The expression for the power-based specific speed may be obtained by considering that the power developed should be given by

$$P = \rho\, g\, Q\, H$$

$$= \rho\, g.\frac{(gH)^{3/2}}{N^2}.\,H = \frac{\rho(gH)^{5/2}}{N^2}$$

By definition, $N = kN_s$

if $H = 1$ m and $P = 1$ W

Hence the specific speed. $N_s = \dfrac{N\sqrt{P/\rho}}{gH^{5/4}}$,

The author proposes the use of non-dimensional specific speed because

(i) N_s is dimensionless regardless of the units of measurement;

(ii) it applies to all fluid machines, e.g., pumps and turbines, irrespective of the fluid used and

(iii) it takes care of the variations in g which may not appear to be important in the first instance.

The trend in the past has been to preserve only the form of these equations and write specific speed as

$$n_s = \frac{N\sqrt{Q}}{H^{3/4}} \text{ pumps, fans, blowers and compressors} \quad \text{...(14.4 } a)$$

$$n_s = \frac{N\sqrt{P}}{H^{5/4}} \text{ for turbines} \quad \text{...(14.4 } b)$$

The essential merits of the authentic definition of N_s are lost but engineers appear to have been content with this practice.

If the units of the variables are:

N	revolutions per minute	Q	m³/s or litres/second
H	m	P	kW metric horse power

as has been the practice in the past, the specific speed n_s appears in a clumsy unit unworthy of mention. Such a discrepancy in the era of SI units is unfortunate. *We should, therefore, accept the definition of N_s on the non-dimensional basis.*

14.3 Classification on the Basis of Specific Speed

The specific speed of a machine is indicative of the type of the machine. *This is why it is also called the type number*. Machines are, therefore, classified as low, medium and high specific-speed machines.

It is customary to refer to the rotating element in a fluid machine on the basis of non-dimensional specific speed as follows :

Fluid Machine	Specific Speed
Low specific speed	less than 0.3
Medium specific speed	0.3 to 3.0
High specific speed	3.0 to 30

It is noticed from the definition of specific speed that it is inversely, though non-linearly, proportional to the head across the machine. Low specific speed, therefore, corresponds to high head across it. For hydraulic turbines, the following classification is frequently employed :

Hydraulic Turbine	Head
High head turbines	more than 250 m
Medium head turbines	60 m to 250 m
Low head turbines	less than 60 m

It may also be noticed that the specific speed is directly proportional to the square root of discharge through the machine. Low specific speed, therefore, refers to low discharge through the machine. The passage provided for the fluid to flow through a machine is in accordance with the extent of discharge desired to pass through it. In general, radial-flow machines allow low discharge and axial-flow machines allow high discharge to pass through them. Mixed-flow machines allow medium discharge in the intermediate range.

A complete range of fluid machines classified on the basis of specific speed and head is shown in Fig. 14.2.

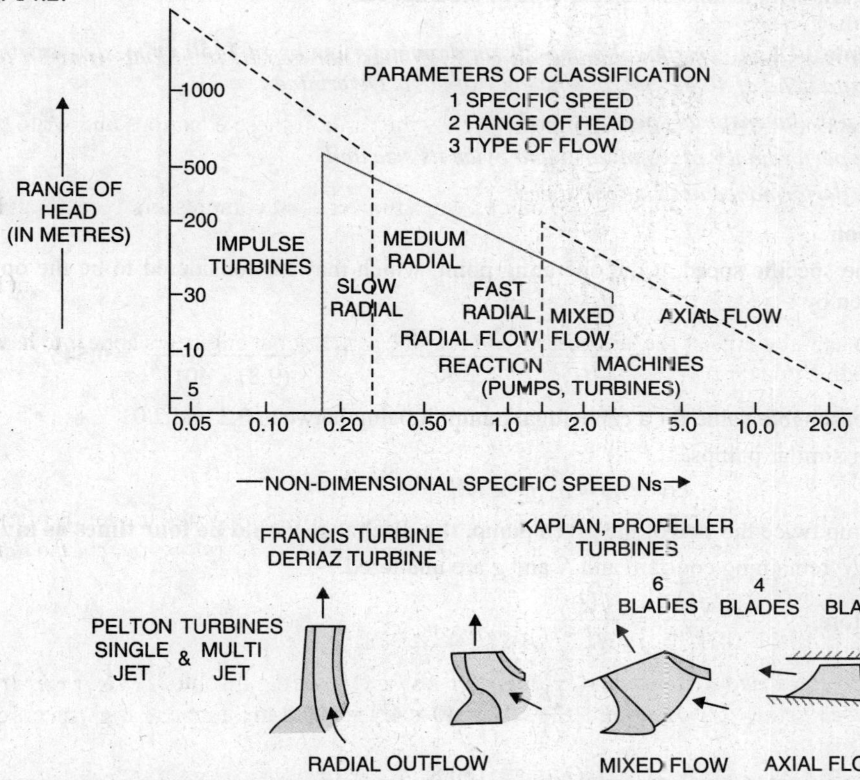

Fig. 14.2: Classification of Fluid Machines

14.4 Impulse and Reaction Principles

The action of a fluid in relation to the machine may be one of impulse or of reaction.

In an impulse machine, a fluid jet of high velocity impinges on the blades of the rotor and exerts force on it by impulsive action. There is no pressure change across the rotor blades ; the entire fluid machine operating on impulse principle may remain at the atmospheric pressure.

Typical examples of impulse machines are the Pelton water turbine and the conventional wind-mill because there is no pressure difference across them ; they operate under atmospheric pressure conditions.

In a reaction machine, a part of the force exerted by the fluid in contact with blades of the machine is due to the pressure change across the rotor blades. In a reaction machine, there may be a contribution by the impulse action whereas an impulse machine operates purely on the impulse action. In a reaction machine, a part of the pressure change occurs in the rotor and remaining in the stator. Degree of reaction, R of a reaction machine is defined as the pressure change that occurs in the rotor as a fraction of the pressure change across the rotor and stator stage.

$$R = \frac{(\Delta p)_{\text{rotor}}}{(\Delta p)_{\text{rotor and stator stage}}} \qquad \text{...(14.5)}$$

Clearly, the degree of reaction may be anything between 0 and 1. A machine with zero degree of reaction is synonymous with an impulse machine. Degree of reaction of unity is not possible in practice. Reaction machines with a degree of reaction of 0.4 to 0.6 are most frequently employed.

Typical examples of reaction machines are the Francis and Kaplan water turbines and the centrifugal, axial-flow and mixed-flow pumps and compressors. The degree of reaction for such machines differs with details of design, type of blading, etc.

Example 14.1. *A centrifugal pump 20 cm diameter running at 1450 r.p.m. delivers 0.1 m³/s water against a head of 40 m with an efficiency of 90%. Determine :*

(a) its non-dimensional specific speed,

(b) the performance of a similar pump twice its size and

(c) the performance at twice the speed.

Solution.

(*a*) The specific speed at the operating point, which may be considered to be the optimum point, is given by :

$$N_s = \frac{N\sqrt{Q}}{(gH)^{3/4}} = \frac{2\pi \times 1450}{60} \times \sqrt{0.1} \times \frac{1}{(9.81 \times 40)^{3/4}} = \mathbf{0.545}$$

which is a reasonable value for a centrifugal pump; it being between 0.3 and 2.0

(*b*) For similar pumps,

$$Q \propto D \times b; \; Q \propto D^2,$$

and for a pump twice the size of the given pump, **the discharge would be four times as much.**

Now, N_s remaining constant and N and g are unaltered.

Therefore, $$\frac{\sqrt{Q_1}}{H_1^{3/4}} = \frac{\sqrt{Q_2}}{H_2^{3/4}}$$

whence $$H_2 = H_1 \cdot \left(\frac{Q_2}{Q_1}\right)^{2/3} = 40 \times 4^{2/3} = \mathbf{100.8\ m}$$

and $$P_2 = \frac{H_2 Q_2}{H_1 Q_1} \cdot P_1 = \frac{100.8}{40} \times 4 \times P_1 = \mathbf{10.1\ P_1}$$

showing that the **power required to run it is ten times** that for the smaller pump at the same efficiency.

(*c*) From the fact that

$$\frac{Q}{ND^2} = \text{constant}$$

$$Q \propto N$$

the discharge would be doubled at twice the speed.

Similarly, judging from

$$\frac{H}{N^2 D^2} = \text{constant},$$

$$H \propto N^2$$

the head would become four times at twice the speed.

The power input will, therefore, be **8 times**

Hence $$Q = 2 \times 0.1 = \mathbf{0.2\ m^3/s}$$

and $H = 4 \times 40 = \mathbf{160\ m}$

Let us check it also from the constancy of specific speed

$$N_s = \frac{2\pi \times 2900}{60} \times \frac{\sqrt{0.2}}{(9.81 \times 160)^{3/4}} = \mathbf{0.545}$$

which is the same as before.

Power input $P_2 = \dfrac{160 \times 0.2}{40 \times 0.1} \times P_1 = \mathbf{8\ P_1}$

which is 8 times that at the initial speed as already shown!

14.5 Energy Transfer : Radial Flow Machines

Consider the flow of a fluid in a typical fluid machine. The fluid enters the machine at a radius r_1 and leaves it at a radius r_2. The flow may be outward or inward depending upon the type of the machine. In practice, the fluid is guided by a number of blades and the transfer of energy takes place by the turning effect of the blades. All bladed channels are identical and provide the same guiding effect. For the purpose of simple analysis, it is assumed that there are an infinite number of blades

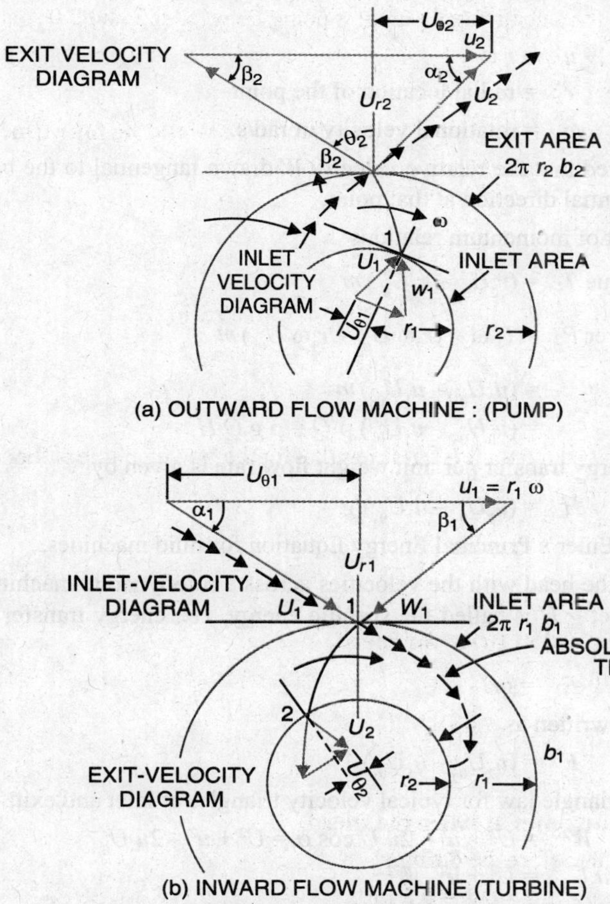

Fig. 14.3: Velocity Diagrams for Fluid Machines

of zero-thickness. This assumption is equivalent to stipulating axisymmetry of flow. In other words, the fluid variables do not change in the circumferential direction ;

$$\frac{\partial}{\partial \theta}\,(\text{variables}) \equiv 0$$

It is also assumed that the flow is one-dimensional and that the fluid variables change only in the direction of flow.

Typical velocity diagrams at inlet radius r_1 and exit radius r_2 for an outward-flow machine and an inward-flow machine are shown in Fig. 14.3. Variations of velocity occur only along the radial direction. It may be observed that the relative velocities are tangential to the blades at entry and at exit. In fact the shape of the blade conforms to the relative flow through the machine. The absolute velocities are obtained by adding vectorially the local peripheral velocities to the relative velocities. Thus, the absolute velocity U equals the relative velocity W plus the peripheral velocity u.

The absolute velocity U at a typical location may be resolved into two components:

(*a*) the radial component $U_r = U \sin \alpha$, and

(*b*) the circumferential component

$$U_\theta = U \cos \alpha$$

where α is the angle between the absolute velocity and the peripheral velocity,. The peripheral velocity u is such that

$$u = r\,\omega$$

where r = radial location of the point

ω = rotational velocity in rad/s.

It may also be observed that the relative velocity W drawn tangential to the blade makes an angle β with the circumferential direction at that point.

Recalling the moment of momentum relations

$$\text{Torque } T = (r_2 U_{\theta_2} - r_1 U_{\theta_1})\,\dot{m}$$

$$\text{Power } P = T.\,\omega = (r_2 \omega\, U_{\theta_2} - r_1 \omega\, U_{\theta_1})\,\dot{m}$$

$$= (u_2 U_{\theta_2} - u_1 U_{\theta_2})\,\dot{m}$$

$$= (u_2 U_{\theta_2} - u_1 U_{\theta_1})\,\rho\, Q = \rho\, g\, Q\, H \quad \text{....(14.6)}$$

Hence, the rate of energy transfer per unit weight flow rate is given by

$$H = (u_2 U_{\theta_2} - u_1 U_{\theta_1})/g \quad \text{....(14.7)}$$

This is known as the Euler's Principal Energy Equation for fluid machines.

The equation relates the head with the velocities across a rotodynamic machine of any type. Recognising that the product $g\,H$ is called the specific energy, *i.e.*, energy transfer per unit mass flow rate,

$$E = gH$$

the Euler's equation is also written as

$$E = (u_2 U_{\theta_2} - u_1 U_{\theta_1}) \quad \text{....(14.8)}$$

Invoking the cosine triangle law for typical velocity triangles at inlet and exit,

$$W^2 = U^2 + u^2 - 2u\, U \cos \alpha = U^2 + u^2 - 2u\, U_\theta$$

or

$$2\, u\, U_\theta = U^2 + u^2 - W^2$$

At the inlet,

$$u_1 U_{\theta_1} = \frac{U_1^2 + u_1^2 - W_1^2}{2} = \left(\frac{U^2 + u^2 - W^2}{2}\right)_1$$

and at the exit,

$$u_2 U_{\theta_2} = \frac{U_2^2 + u_2^2 - W_2^2}{2} = \left(\frac{U^2 + u^2 - W^2}{2}\right)_2$$

The Euler's Principal Equation may, therefore, be rewritten as

$$H = \left(\frac{U^2 + u^2 - W^2}{2g}\right)_2 - \left(\frac{U^2 + u^2 - W^2}{2g}\right)_1$$

$$= \text{Rate of energy transfer per unit weight of flow}$$

Hence, $$H = \frac{U_2^2 - U_1^2}{2g} + \frac{u_2^2 - u_1^2}{2g} + \frac{W_1^2 - W_2^2}{2g} \quad ...(14.9)$$

Let us interpret the significance of the three terms on the right hand side:

Term 1 : due to change of absolute kinetic energy in the impeller ; impulse effect.

Term 2 : due to rotation of the fluid mass ; centrifugal effect.

Term 3 : due to acceleration or retardation of fluid ; diffusion or reaction effect.

14.6 Centrifugal Machines

Let us now consider the application of the Eulcr's Principal Equation to centrifugal machines. For a centrifugal machine, it is a common practice to ignore the circumferential velocity at the smaller radius ; the angle *a* at that radius is assumed to be 90°. For an outward-flow centrifugal machine such as a pump or a compressor,

$$\text{Torque } T = r_2 U_{\theta_2} . \dot{m}$$

$$\text{Power } P = T.\omega = \rho\, g\, Q\, H$$

$$\text{Head } H = \frac{u_2 . U_{\theta_2}}{g} = \frac{u_2(u_2 - U_{r2} \cot\beta_2)}{g} \quad ...(14.10)$$

$$\text{Specific energy } E = u_2 U_{\theta_2} = u_2(u_2 - U_{r_2} \cot \beta_2) \quad ...(14.11)$$

where the mass flow rate $\dot{m} = 2\pi\, r_1 . b_1 . U_{r_1} . \rho_1 = 2\pi\, r_2 . b_2 . U_{r_2} . \rho_2$ ...(14.12)

since the radial velocity U_r is identical with the velocity of flow U_f

Assuming incompressible flow

$$\rho_1 = \rho_2 = \rho$$

and $$U_1 = U_{r_1}$$

the mass flow rate is given by

$$\dot{m} = \rho\, Q = 2\pi\, r_1 . b_1 . U_r . \rho = 2\pi\, r_2 . b_2 . U_{r_2} . \rho$$

Another point which deserves attention is that the outlet blade angle has a marked influence on the performance of a centrifugal pump.

$$H = \frac{u_2 U_{\theta_2}}{g} = \frac{u_2}{g}(u_2 - U_{r_2} \cot \beta_2)$$

$$= \frac{u_2}{g}\left(u_2 - \frac{Q}{2\pi r_2 b_2}\cot\beta_2\right)$$

or $$H = A + BQ \qquad \text{...(14.13)}$$

where $A = u_2^2 / g$,

a positive constant for a particular speed

and $B = -u_2 \cot\beta_2 / 2\pi r_2 b_2 g$,

also a constant for a given speed and fixed outlet angle β_2.

Shape of Blades	β_2	Slope B	H vs Q curve	Efficiency
Blades curved backward w.r.t. rotation	< 90°	– ve	falling	high
Radial blades	= 90°	zero	level	moderate
Blades curved forward w.r.t. rotation	> 90°	+ ve	rising	low

It is also seen that the shut-off head (for $Q = 0$) should be equal to u_2^2 / g, the centrifugal head. which shows that H *vs* Q curve must be a straight line and that the slope of the curve is as under:

This is represented in Figs. 14.4 (*a*) and (*b*). These curves are theoretical curves based upon the Euler's Principal Equation.

Let us also examine the change in the velocity triangle, for a typical centrifugal pump with backward curved vanes, when the discharge Q is changed. For a partial closure of the delivery valve, the discharge is reduced to Q'. If the pump continues to run at the same speed, the peripheral velocity u_2 remains the same. Since the direction of the relative velocity cannot change for the same vanes, the angle β_2 would remain the same. The radial velocity component U'_{r_2} is directly proportional to the discharge

$$\frac{U'_{r_2}}{U_{r_2}} = \frac{Q'}{Q} = q \qquad \text{...(14.14)}$$

where q is the discharge ratio.

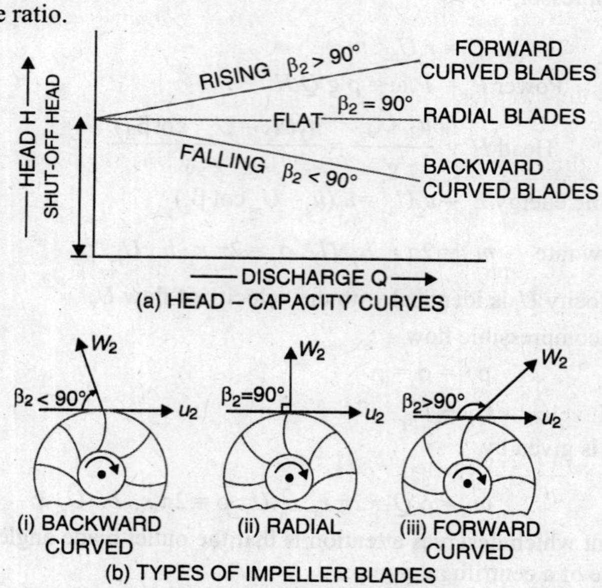

Fig. 14.4: Effect of the Shape of the Blades

From the similarity of the right-angled triangles OAB and OA′ B′ with the common vertex O,

$$\frac{U'_{r_2}}{U_{r_2}} = \frac{u_2 - U'_{\theta_2}}{u_2 - U_{\theta_2}} = q$$

and

$$\frac{W'_2}{W_2} = \frac{U'_{r_2}}{U_{r_2}} = q \qquad \text{...(14.15)}$$

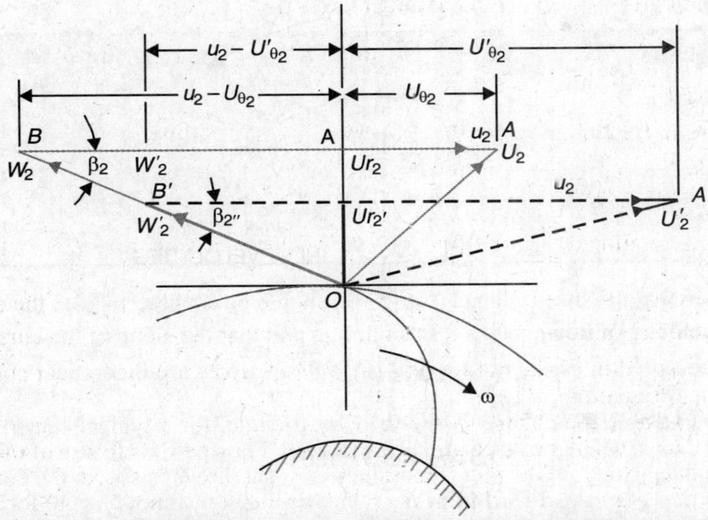

Fig. 14.5: Change in the Velocity Triangle for Reduced Discharge

Example 14.2. *A centrifugal pump delivers salt water against a net head of 15 m at a speed of 1000 r.p.m. The vanes are curved backward at 30° with the periphery. Obtain the discharge for an impeller diameter of 30 cm and outlet width of 5 cm at a manometric efficiency of 90%.*

Solution.

At the exit of the impeller,

$$u_2 = \pi \times 0.3 \times 1000/60 = 15.7 \text{ m/s}$$

Manometric efficiency implies the ratio of the net head or manometric head to the ideal head $\frac{U_{\theta_2} u_2}{g}$ so that,

$$0.90 = \frac{15 \times 9.81}{U_{\theta_2} \times 15.7};$$

whence $U_{\theta_2} = 10.4$ m/s

From the exit velocity triangle,

$$\frac{U_{r_2}}{u_2 - U_{\theta_2}} = \tan 30°$$

$$U_{r_2} = 0.577 \times (15.7 - 10.4)$$
$$= 3.05 \text{ m/s}$$

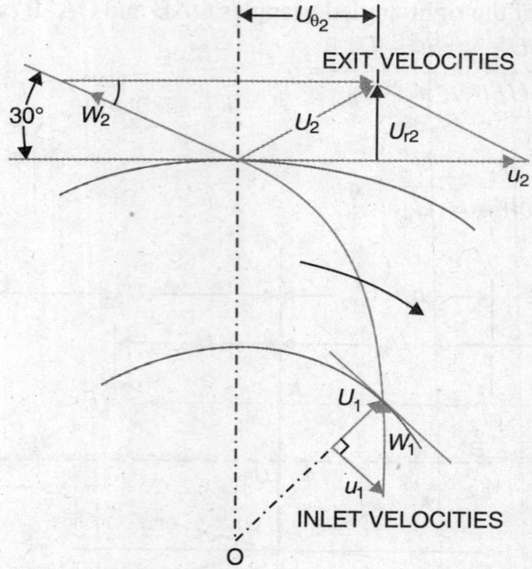

Since the velocity of flow

$$U_{f_2} = U_{r_2} = 3.05 \text{ m/s}$$

$$\text{Discharge } Q = U_{f_2} \cdot \pi . D_2 . b_2$$

$$= 3.05 \times \pi \times 0.3 \times 0.05$$

$$\mathbf{= 0.144 \ m^3/s}$$

14.7. Departure from the Ideal Flow

The actual flow in a fluid machine differs from the ideal fluid flow model assumed in the derivation of the Euler's Principal Equation. The following factors account for the departure of the actual flow from the ideal flow :

1. ***Existence of viscosity ; viscous resistance, boundary layer formations, separation of flow, etc.***
2. ***Finite number of blades ; flow through bladed channels, secondary flows in the bladed passages and leakage effects.***
3. ***Existence of compressibility ; density variations, temperature rise and shock phenomena.***
4. ***Off-design operating conditions ; velocities and angles different from the design values.***

Existence of viscosity and surface roughness results in the formation of a boundary layer and separation which are accompanied by skin-friction drag and form-drag on the blades. Head loss due to frictional effect are expressible as

$$h_f = kQ^2 \qquad ...(14.16)$$

where h_f = head lost due to friction and shearing,

Q = volumetric flow rate and

k = a constant.

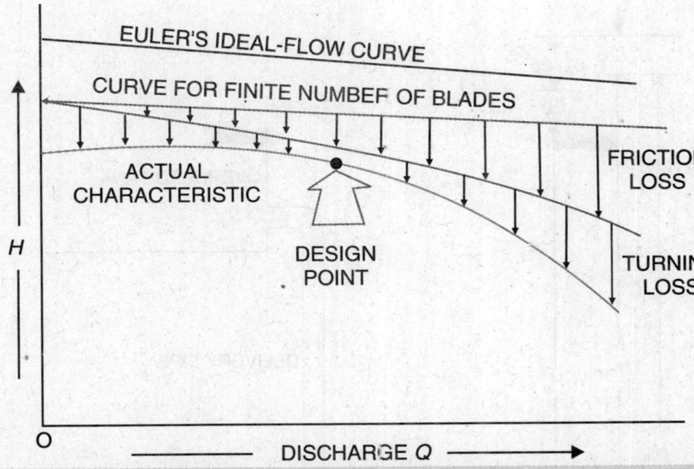

Fig. 14.6: Actual and Euler's Characteristics

Losses due to improper fluid incidences on the blades at the inlet and lack of complete guidance by the blades are called turning losses. Turning losses are expressed as

$$h_t = C(Q - Q_n^2) \qquad \text{...(14.17)}$$

where h_t = head lost in turning due to improper incidence

Q = volumetric flow rate

Q_n = designed flow rate

and C = a constant

The effect of all losses on the head-discharge characteristic of a typical impeller pump is shown in Fig. 14.6. The actual characteristic is lower than the Euler's characteristic at all values of the discharge. The difference between them is the largest at very high and very low discharge.

14.8. Efficiency of Centrifugal Pumps

Consider a typical installation of a centrifugal pump as shown in Fig. 14.7. The liquid enters at the bottom of the suction pipe through a strainer and a foot valve and the pump operates so as to discharge the liquid to an overhead tank by means of a delivery pipe. The symbols used have the following meaning :

H_s = Suction lift .
= vertical height from the level in the sump to the centre-line of the pump.

H_d = Delivery lift
= vertical height from the centre-line of the pump to the level in the overhead tank.

H = Total static head = $H_s + H_d$

h_{fs} = Head loss in friction in the suction pipe

h_{fd} = Head loss in friction in the delivery pipe

U_s = Velocity in the suction pipe

U_d = Velocity in the delivery pipe

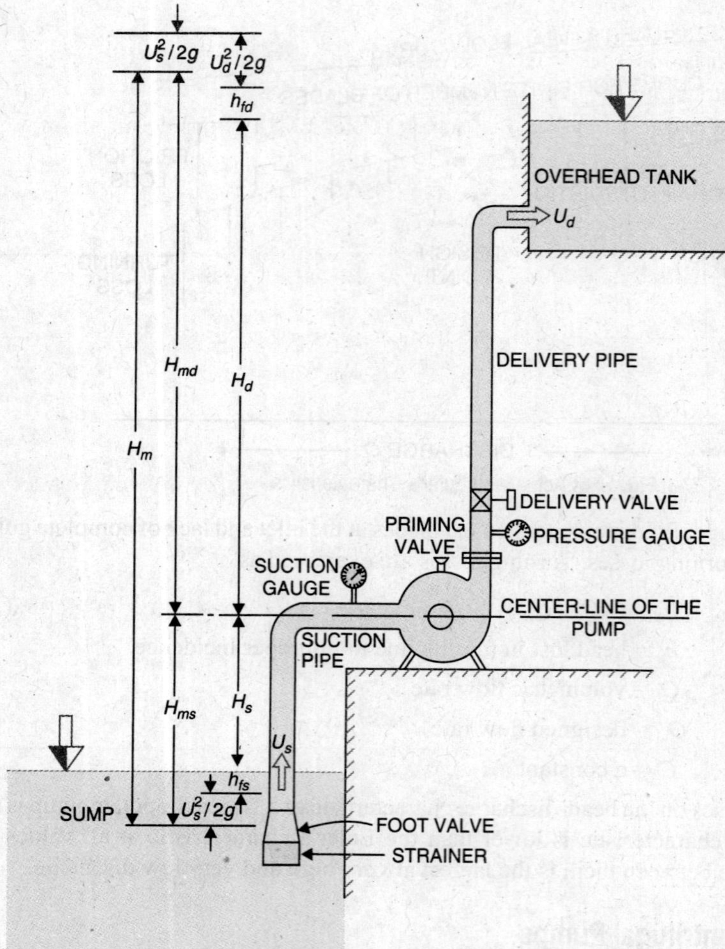

Fig. 14.7: A Typical Installation of a Centrifugal Pump

$$H_{ms} = \text{Manometric suction head}$$

$$= H_s + h_{fs} + \frac{U_s^2}{2g}$$

$$H_{md} = \text{Manometric delivery head}$$

$$= H_d + h_{fd} + \left(\frac{U_d^2}{2g} - \frac{U_s^2}{2g}\right) \quad ...(14.18)$$

$$H_m = \text{Manometric head across the pump}$$

$$= H_{ms} + H_{md}$$

$$= H_s + H_d + h_{fs} + h_{fd} + \frac{U_d^2}{2g}$$

$$= H + (h_{fs} + h_{fd}) + \frac{U_d^2}{2g} \quad ...(14.19)$$

If the discharge through the pump is Q and the density of the liquid is ρ, the manometric power is defined as the liquid power associated with the manometric head ;

Manometric power $= \rho g Q H_m$ ...(14.20)

The net output power is only that which is associated with the static head *H*, whence the net system Output power $= \rho g Q H$ ...(14.21)

The power available at the impeller is that which is obtained by considering the flow to be ideal.

Available impeller power $= \rho Q u_2 U_{\theta 2}$ ...(14.22)

The various efficiencies are defined in Table 14.1.

Table 14.1. Efficiencies and Sources of Inefficiency in a Pump

	← Available Impeller Power, $\rho Q u_2 U_{\theta_2}$ →		
	← Manometric Power $= \rho g Q H_m$ →		
← Mechanical Losses →	← Hydraulic Losses →	← Pipe Losses →	← System Output Waterpower →
Disc Friction, Bearing Glands and Windage	Fluid Friction, Turbulence, Shocks and Leakage	Pipe Friction and Transitions	NET PUMP POWER $\rho g Q H$
← INPUT SHAFT POWER →			

$$\eta_{mech} = \frac{\text{Available Impeller Power}}{\text{Input Shaft Power}} \qquad \eta_{vol} = \frac{\text{Delivery Discharge}}{\text{Suction Discharge}} = \frac{Q}{Q + \delta Q}$$

$$\eta_{mano} = \frac{\text{Manometric Power}}{\text{Available Impeller Power}} \qquad \eta_{overall} = \frac{\text{Manometric Power}}{\text{Input Shaft Power}}$$

$$\eta_{system} = \frac{\text{System Output Waterpower}}{\text{Input Shaft Power}} \qquad = \eta_{mech} \times \eta_{mano} \times \eta_{vol}$$

Example 14.3. *The impeller of a centrifugal pump has a diameter of 0.3 m and runs at 1440 revolutions per minute. The pressure gauges on suction and delivery sides show a difference of 25 m head of water. If the blades are curved backward at an angle of 30° to tangent at the outlet tips of the blades and the velocity of flow through the impeller is constant at 2.5 m/s, find the manometric efficiency of the pump. If frictional losses in the impeller amount to 2 m, find the fraction of total energy which is converted into pressure energy by the impeller. Also, determine the pressure rise in the casing of the pump.*

Solution.

At the outlet tip of a blade, the peripheral velocity is given by

$$u_2 = \frac{\pi DN}{60} = \frac{\pi \times 0.3 \times 1440}{60} = 22.6 \text{ m/s}$$

From the outlet velocity triangle drawn approximately,

$$U_{\theta_2} = u_2 - U_{r2} \cot 30°$$

$$= 22.6 - 2.5 \times 1.732 = 18.27 \text{ m/s}$$

Rate of work done per kg per second is given by

$$u_2 \frac{U_{\theta_2}}{g} = \frac{22.6 \times 18.27}{9.81} = 42.1 \text{ m}$$

$$\text{Manometric head} = \frac{\text{Manometric head}}{(u_2 U_{\theta_2} / g)}$$

$$= \frac{25}{42.1} = 0.594 = \mathbf{59.4\ \%}$$

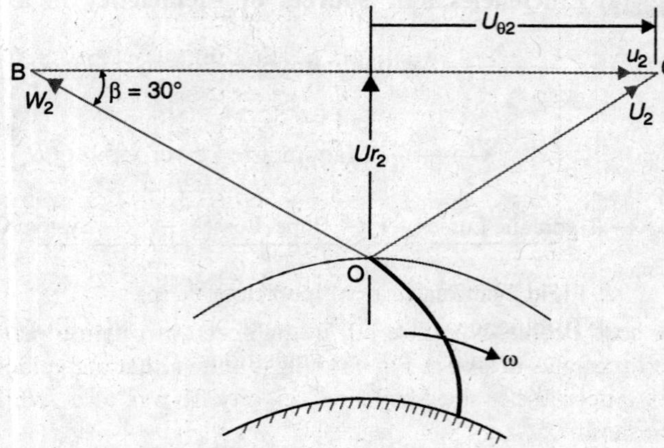

The rate of work done per kg per second in accordance with the velocity triangle should provide for the change in kinetic energy, rise of pressure across the pump and the losses in the impeller. Therefore,

$$42.1 = \frac{U_2^2 - U_1^2}{2g} + \text{pressure rise} + 2.0$$

Assuming that U_1^2 is small compared with U_2^2,

$$\text{pressure rise} = 40.1 - \frac{U_2^2}{2g}$$

$$= 40.1 - \frac{U_{r2}^2 + U_{\theta_2}^2}{2g}$$

$$= 40.1 - \frac{2.5^2 + 18.27^2}{2 \times 9.81}$$

$$= \mathbf{22.8\ m}$$

The fraction of total energy which is converted into pressure energy by the impeller equals

$$\frac{22.8}{42.1} = 0.54 = \mathbf{54\%}$$

Pressure rise in the casing must be the difference between the head rise in the pump and that in the impeller only, *i.e.,*

$$25 - 22.8 = \mathbf{2.2\ m}$$

14. 9. Performance of Centrifugal Pumps

The characteristics of a centrifugal pump are plotted as a set of constant speed lines for

(a) Head of the fluid vs discharge

(b) Power input vs discharge

(c) Overall efficiency vs discharge.

A typical set of characteristics is shown in Fig. 14.8 where the arrows indicate the trend for the speed increasing above the design speed. The trend is reversed for speeds decreasing below the design speed.

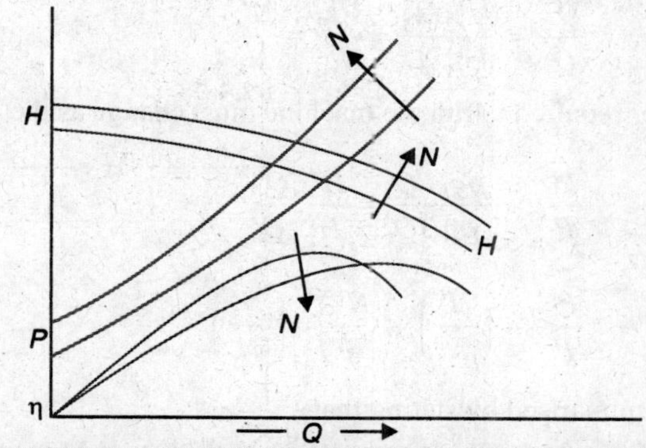

Fig. 14.8: Performance of Centrifugal Pumps

The fact that the head decreases with a fall in the speed of rotation may be explained with reference to the velocity triangles drawn in Fig. 14.9. It is shown that the velocity triangle changes from OBC to OB′C′ for a decrease in the peripheral velocity from u_2 to u'_2 while the blade and the blade angles remain the same.

From the definitions amd velocity triangles,

$$u_2 = \frac{\pi D_2 N}{60}$$

$$u'_2 = \frac{\pi D_2 N'}{60}$$

whence

$$\frac{u_2'}{u_2} = \frac{N'}{N}$$

and

$$\frac{U_{\theta_2'}}{U_{\theta_2}} = \frac{OC'}{OC} = \frac{B'C'}{BC} = \frac{u_2'}{u_2} = \frac{N'}{N} \qquad \text{...(14.23)}$$

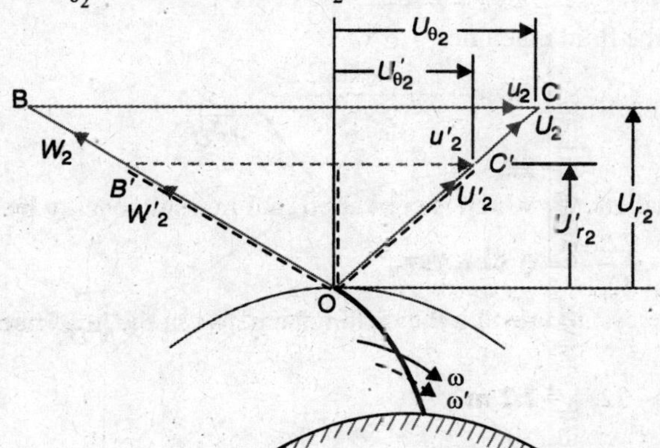

Fig. 14.9: Effect of a decrease in the rotational speed

It follows that

$$\frac{H'}{H} = \frac{u_2'}{u_2}\frac{U'_{\theta_2}}{U_{\theta_2}} = \left(\frac{N'}{N}\right)^2 \quad ...(14.24)$$

The discharge through the machine changes in direct proportion to the speed of rotation ;

$$\frac{Q'}{Q} = \frac{U'_{r2}A_2}{U_{r2}A_2} = \frac{U'_{r_2}}{U_{r_2}} = \frac{N'}{N} \quad ...(14.25)$$

Finally, the power required to run the machine must change as the product of the head and the discharge

$$\frac{P'}{P} = \frac{\rho g H'Q'}{\rho g HQ} = \frac{H'}{H}.\frac{Q'}{Q}$$

or

$$\frac{P'}{P} = \left(\frac{N'}{N}\right)^2 . \left(\frac{N'}{N}\right) = \left(\frac{N'}{N}\right)^3 \quad ...(14.26)$$

The results are summarised by stating that

* ***the discharge Q varies in direct proportion to the speed N,***
* ***the head H varies as the square of N, and***
* ***the power P varies as the cube of N.***

It may also be shown, by way of drawing velocity triangles, that the performance of a centrifugal pump changes as the diameter of the impeller is changed. The results are summarised by stating that

* ***the discharge Q varies as the cube of the diameter D,***
* ***the head H varies as the square of D, and***
* ***the power P varies as the fifth power of D.***

In fact, one. should consider the non-dimensional parameters ;

Flow coefficient $\frac{Q}{ND^3}$

Head coefficient $\frac{gH}{N^2D^2}$

Power coefficient $\frac{P}{\rho N^3 D^5}$

and the efficiency η of the fluid machine.

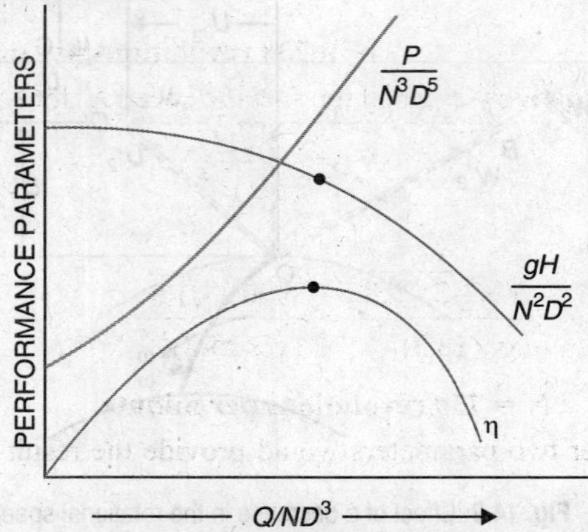

Fig. 14.10: Non-dimensional Performance Curves

Performance curves can be plotted for a family of geometrically similar pumps operating under dynamically similar conditions. Such curves, based upon non-dimensional parameters $\frac{Q}{ND^3}, \frac{gH}{N^2D^2}, \frac{P}{\rho N^3 D^5}$ and η are shown in Fig. 14.10.

Such performance curves are referred to as 'unit' curves. The unit quantities refer to unit size and unit speed. Thus, a knowledge of the 'unit' curve of head *vs.* discharge enables a designer to figure out the head and discharge for any size of a family of pumps operating at any speed. Unit head multiplied by N^2D^2 yields the actual head and unit discharge multiplied by ND^3 provides the actual discharge at a speed N for an impeller of diameter D in a family of pumps.

Example 14.4. *A centrifugal pump having an impeller of 30 cm diameter was tested in a laboratory and gave the best efficiency when running at 900 revolutions per minute and delivering 1.8m³/minute against a head of 15.2 m. If an exactly similar design of pump is required to deliver twice the discharge against a head of 18.3 m, determine the diameter of the impeller and the speed at which it should be run.*

Solution.

For similar pumps, the non-dimensional coefficients, such as flow coefficient may be equated.

Equating $\frac{Q}{ND^3}$ for the pumps,

$$\frac{2 \times 1.8}{ND^3} = \frac{1.8}{900 \times 0.3^3}$$

whence

$$ND^3 = 48.6 \qquad ...(i)$$

Equating $\frac{gH}{N^2D^2}$ for the pumps,

$$\frac{g \times 18.3}{N^2D^2} = \frac{g \times 15.2}{900^2 \times 0.3^2}$$

whence

$$N^2D^2 = 87,768$$

and

$$ND = 296.2 \qquad ...(ii)$$

Dividing (*i*) by (*ii*) and solving for D,

$$D = 0.405 \text{ m} = \mathbf{40.5\ cm}$$

and from (*i*) or (*ii*),

$$\mathbf{N = 730\ revolutions\ per\ minute.}$$

One could have alternatively equated the specific speed N_s for the two pumps,

$$N_s = \frac{N\sqrt{Q}}{(gH)^{3/4}}$$

Substituting the data,

$$\frac{N \times \sqrt{2 \times 1.8}}{(g \times 18.3)^{3/4}} = \frac{900 \times \sqrt{1.8}}{(g \times 15.2)^{3/4}}$$

which provides **N = 730 revolutions per minute.**

Equating of the other two parameters would provide the result for diameter of the impeller required.

EXPERIMENT
Performance Characteristics of a Centrifugal Pump

OBJECTIVE

To study the performance characteristics of a single-stage centrifugal pump.

PRACTICAL RELEVANCE

Performance curves show the behaviour of a centrifugal pump under different operating conditions. For large centrifugal pumping units, the performance characteristics of a dynamically similar model are determined and then the theory of similitude is used to predict the performance characteristics of the prototype.

A centrifugal pump lifts the liquid to a higher level as a result of a modification of the hydraulic gradient caused by the centrifugal action and change in angular momentum. This is in contrast to a positive displacement pump wherein lifting action is due to pushing in a confined space.

The main advantage of a centrifugal pump vis-a-vis positive displacement pump is that its discharge capacity is much greater and it can be used to pump highly viscous liquids. Also, it can be operated at high speeds with less danger of separation and cavitation and its maintenance requirements are low. However, it cannot build up pressures as high as those that can be built up by reciprocating pumps.

PROCEDURE

A centrifugal pump is initially primed, *i.e.,* the suction pipe, casing of the pump and the portion of the delivery pipe up to the delivery valve are completely filled with the liquid to be pumped. With the delivery valve closed, the impeller is made to rotate. As a result a forced vortex is developed which imparts a centrifugal head to the liquid. Simultaneously, the angular momentum is changed resulting in an increase of the liquid pressure. When the delivery valve is opened, the liquid is forced to flow in an outward radial direction thereby leaving the vanes of the impeller at

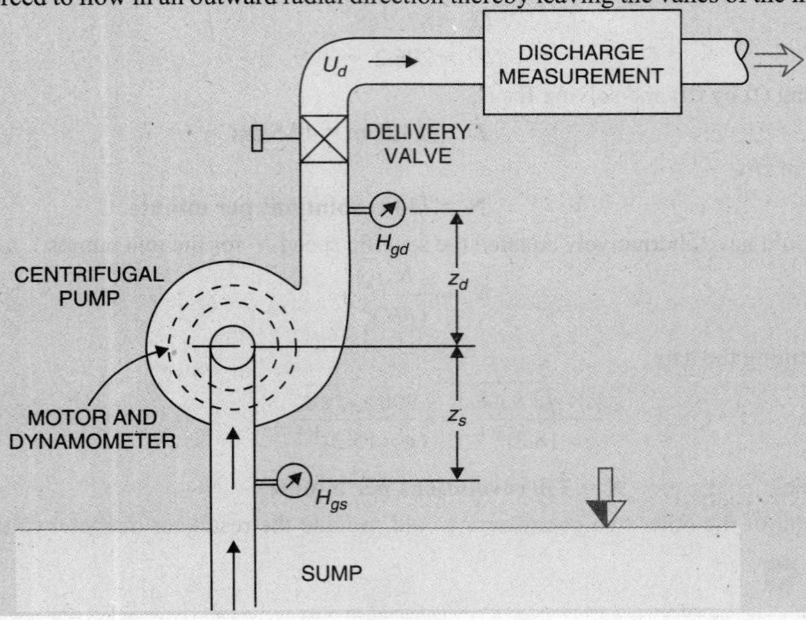

the outer circumference with high velocity and pressure. The high pressure of the liquid leaving the impeller enables the liquid to rise to a high level. This action is a continuous process because the eye of the impeller is continuously supplied with replacement liquid from the sump as a result of the pressure gradient in the suction pipe (a partial vacuum exists at the eye of the impeller and the liquid in the sump is at the atmospheric pressure). The high absolute velocity at the outlet of the vanes is converted to useful pressure energy by shaping the casing such that the liquid flows through expanding passage.

While studying a performance characteristic of a centrifugal pump, its speed is kept constant throughout. The head *H*, the input power P_i and the output power P_o are then determined as functions of the volume rate of flow *Q*.

The head *H* is defined as the total effective head generated by the pump and it is given by the expression :

$$H = \left(H_{gd} + z_d + \frac{U_d^2}{2g}\right) - \left(H_{gs} + z_s + \frac{U_s^2}{2g}\right)$$

where H_{gd} and H_{gs} are the delivery and suction gauge readings in meter of water, respectively, z_d and z_s are the vertical distances of the delivery and suction gauges from the pump axis respectively and $U_d^2/2g$ and $U_s^2/2g$ are the average kinetic energy per unit weight of the pumped liquid in the delivery and suction pipes respectively, at the sections where the gauges are fitted.

The input power P_i is the motor power determined from the dynamometer or otherwise.

The output power P_o is the water power determined from the expression

$$P_o = \rho g Q H$$

where ρ is the mass density of the liquid.

The overall efficiency η of the pump is then given by

$$\eta = \frac{P_o}{P_i}$$

RESULTS

The following graphs may be drawn for operation at a constant speed :

1. Total head generated *H vs* discharge *Q*.
2. Power input P_i *vs* discharge *Q*.
3. Overall efficiency η *vs* discharge *Q*.

Students may prefer to draw the curves on non-dimensional basis and interpret their curves.

POINTS FOR DISCUSSION

1. Discuss the nature of the curves plotted and comment on the expected shapes of the curves if the experiment is performed at different speeds.
2. Can cavitation take place in a pump ? If so, where and why ?
3. Calculate the specific speed of the pump for any one observation and comment on its value in relation to the design point specific speed of the pump.

ANY OTHER OBSERVATION OR COMMENT?

14.10. Diffusers for Centrifugal Pumps

A partial increase in head occurs in the diffuser surrounding the impeller. A diffuser operates on the principle of increasing the pressure energy by decreasing the kinetic energy in a free vortex or spiral flow. The energy conversion can be achieved by one or more of the methods tabulated and shown in Fig. 14.11.

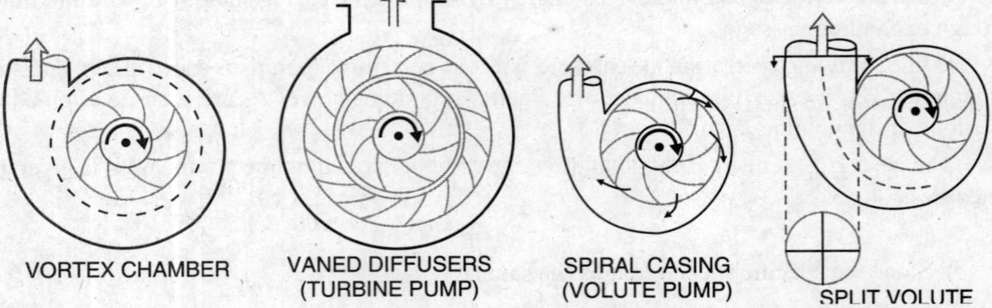

Fig. 14.11: Types of Diffusers for Centrifugal Pumps

Vortex Chamber	Vaned Diffuser	Volute Casing	Split Volute
Parallel sided passage based on the free-vortex principle. The velocity drops inversely with the radius. Used before the volute casing or by itself.	A series of vanes around the impeller for aerodynamic diffusion. Used on centrifugal compressors, super chargers and on turbine-type centrifugal pumps.	A spiral passage uniformly increasing in area enclosing the impeller periphery. Used for medium head pumps.	A spiral passage partitioned so as to provide a number of longitudinal paths, decreasing the effective angle and increasing the diffusion effect. Used for large pumps.

A comparison of different types of pumps, shown in Fig. 14.12, reveals that the volute and the split-volute casing centrifugal pumps have the highest efficiency.

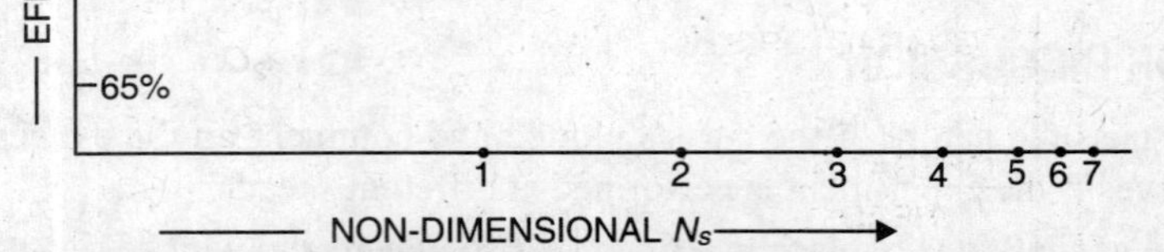

Fig. 14.12: Comparison of Different Pumps

Example 14.5. *An exhaust fan delivers 1 m³/s against a static pressure of 10 cm WG at a speed of 1450 revolutions per minute. Calculate*

(a) the speed of rotation in r.p.m.

(b) the specific speed of the machine, and

(c) the power of the motor required for delivering 50% more discharge.

Given that the impeller diameter is 0.5 m and the overall efficiency is 33%.

Solution.

For (*a*) Equating the discharge parameter for the given and required conditions,

$$\frac{Q_1}{N_1 D_1^3} = \frac{Q_2}{N_2 D_2^3}$$

$$N_2 = \frac{1.5\, Q_1}{Q_1} \times 1450 \text{ for } D_1 = D_2$$

$$= \mathbf{2715 \text{ r.p.m.}}$$

(*b*) The specific speed of either machine is given by

$$N_s = \frac{N\sqrt{Q}}{(gH)^{3/4}} = \frac{2\pi \times 1450}{60} \times \frac{\sqrt{1}}{\left(9.81 \times \frac{10}{100} \times \frac{1000}{1.2}\right)^{3/4}} \approx 1.0$$

(*c*) Since the specific speed remains the same,

$$H_2 = H_1 \left(\frac{Q_2}{Q_1}\right)^{2/3}$$

$$H_2 = 10 \times 1.5^{2/3} \text{ cm WG} = 13.1 \text{ cm WG.}$$

$$\text{Power} = \rho\, g\, Q\, H = 1.2 \times 9.81 \times 1.5 \times \left(\frac{13.1}{100} \times \frac{1000}{1.2}\right)$$

$$= \mathbf{1930 \text{ W} = 1.93 \text{ kW}}$$

This is the power input to the fluid. The power of the motor required to run the fan with an overall efficiency of 33% is over 3 times this power. Power of the motor is, therefore, 6 kW.

14.11. Operating Point of a Pump

The operating point of a centrifugal pump in a given installation is determined by considering the equilibrium between the generation of head by the pump and loss of head in system. The point of intersection of the pump and the system characteristics, therefore, locates the actual operating point of the pump when installed in a given system.

A system characteristic between the head required, H and the discharge to be maintained, Q is generally expressed as a parabolic equation,

$$H = k_1 + k_2\, Q^2 \qquad \text{... (14.27)}$$

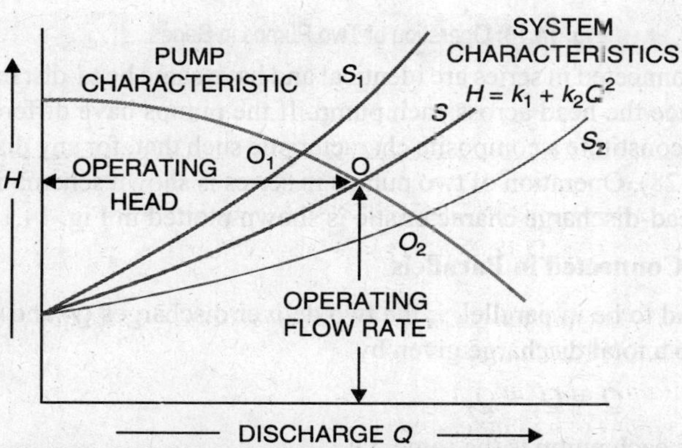

Fig. 14.13: Operation of a Pump in a System

$$H = k_1 + k_2 Q^2 \quad ...(14.27)$$

where k_1 and k_2 are constants for the system. The system characteristic rises with the discharge; it must intersect the head discharge characteristic of a pump at a point, say O, as shown in Fig. 14.13. The same pump may operate at different points such as O_1 and O_2 when it is installed in different systems.

14.12. Centrifugal Pumps in Parallels and in Series

Two pumps may be connected in series or in parallels if the pumps are not capable ol providing the required duty singly. Let us consider both cases:

(Case a): Pumps Connected in Series

Two pumps are said to be placed in series if the discharge of one is passed through the other such that the head raised by the former is augmented by the second and the total head is given by

$$H = H_1 + H_2 \quad ...(14.28a)$$

and $$Q = Q_1 = Q_2 \quad ...(14.28b)$$

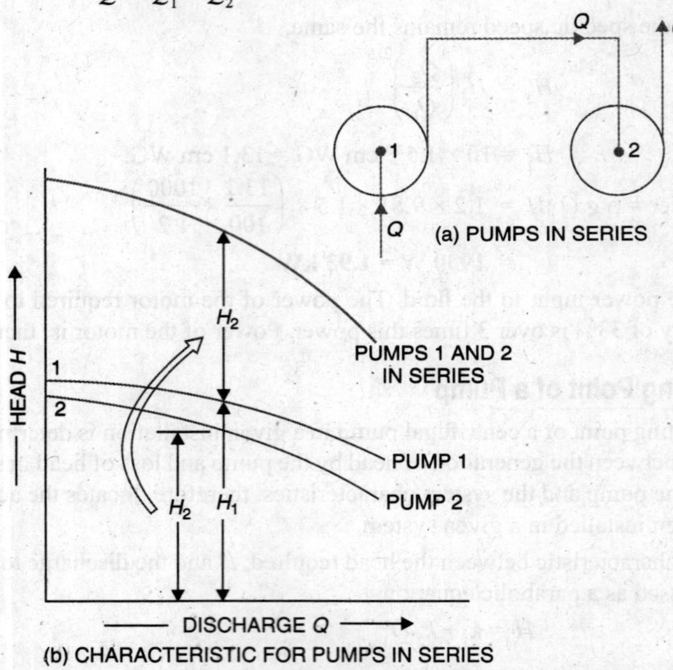

Fig. 14.14: Operation of Two Pumps in Series

If the two pumps connected in series are identical and have same head-discharge characteristic, then the total head is twice the head across each pump. If the pumps have different characteristics, the two may be added to constitute a composite characteristic such that, for any discharge Q, the total head is given by Eq. (14.28). Operation of two pumps in series is shown schematically in Fig. 14.14 (*a*) and the composite head-discharge characteristic is shown plotted in Fig. 14.14 (*b*).

(Case b): Pumps Connected in Parallels

Two pumps are said to be in parallel, if the two deliver discharges Q_1 and Q_2 and supply into a common main to pump a total discharge given by

$$Q = Q_1 + Q_2 \quad ...(14.29a)$$

such that the head across each pump is the same, i.e.

$$H = H_1 = H_2 \qquad \text{... (14.29b)}$$

If the two pumps connected in parallel are identical, then the total discharge delivered is $2Q$ and the head is the same. If the pumps have different characteristics, the two may be added to constitute a composite characteristic such that for any head H, the total discharge is given by Eq. (14.29). Operation of two pumps in parallel is shown schematically in Fig. 14.15 (a) and the composite head-discharge characteristic is shown plotted in Fig 14.15 (b).

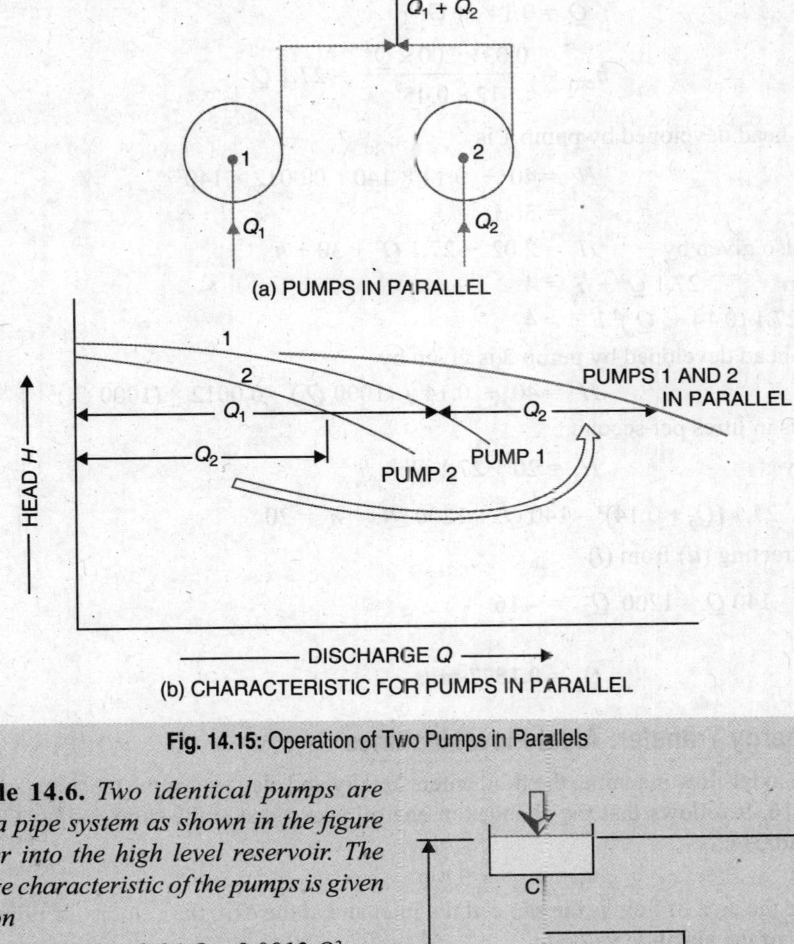

Fig. 14.15: Operation of Two Pumps in Parallels

Example 14.6. *Two identical pumps are connected to a pipe system as shown in the figure to pump water into the high level reservoir. The head-discharge characteristic of the pumps is given by the equation*

$$H = 40 + 0.14\,Q - 0.0012\,Q^2$$

where Q is the discharge in litres/second and H is the head developed by the pump in m.

The pipe AB is 45 cm in diameter and 100 m long. The pipe BC is 45 cm in diameter and 200 m long. For both the pipes the friction factor can be taken as 0.03. V is a regulating valve on the pipeline BC. The static lifts against which the pumps are working are 20 m and 30 m. For a given position of the valve V, the pump operating against a static lift of 30 m develops a discharge of 140 litre/second. Determine the discharge of the second pump and the loss coefficient of the valve.

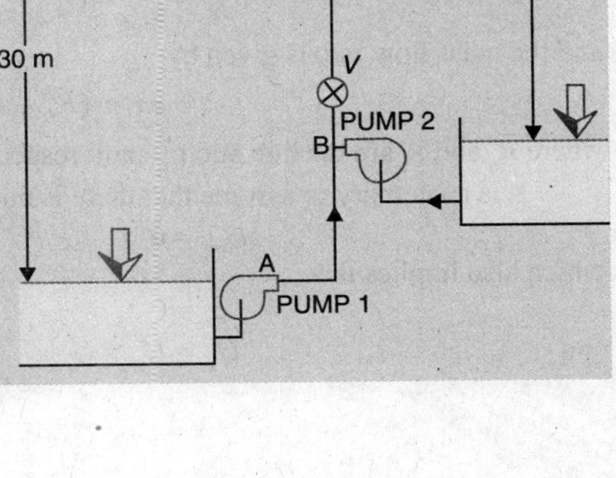

Solution.

For the given arrangement of pumps,

$$h_{fAB} = \frac{0.03 \times 100 \times 0.14^2}{12 \times 0.3^6} = 2.02 \text{ m}$$

If the total discharge in pipe BC is Q

where $$Q = 0.14 + Q_2$$

$$h_{fBC} = \frac{0.03 \times 200 \times Q^2}{12 \times 0.45^5} = 27.1\, Q^2$$

The head developed by pump 1 is

$$H_1 = 40 + 0.14 \times 140 - 0.0012 \times 140^2$$
$$= 36.1$$

which is also given by $$H_1 = 2.02 + 27.1\, Q^2 + 30 + h_v$$

Then $$27.1\, Q^2 + h_v = 4$$

or $$27.1\,(0.14 + Q_2)^2\, h_v = 4 \qquad \text{... } (i)$$

The head developed by pump 2 is given by

$$H_2 = 40 + 0.14 \times (1000\, Q_2) - 0.0012 \times (1000\, Q_2)^2$$

where Q_2 is in litres per second.

It gives $$H_2 = 20 + 27.1\, Q^2 + h_v$$

or $$27.1\,(Q_2 + 0.14)^2 - 140\, Q_2 + 1200\, Q_2^2 + h_v = 20 \qquad \text{... } (ii)$$

Subtracting (*ii*) from (*i*)

$$140\, Q_2 - 1200\, Q_2^2 = -16$$

we get $$\mathbf{Q_2 = 0.1877\ m^3/s}$$

14.13. Energy Transfer: Axial-Flow Machines

In an axial-flow machine, the fluid enters axially and also leaves the machine axially as shown in Fig. 14.16. It follows that the changes in energy take place at the same radius. Considering the flow at a radius r,

$$u_1 = u_2 = u = r\,\omega \qquad \text{...(14.30)}$$

Since the area of flow is the same at the inlet and at the exit, the velocity of flow, *i.e.*, the axial component of the absolute velocity,

$$U_{f1} = U_{f_2} = U_f. \qquad \text{...(14.31)}$$

and the mass flow rate is given by

$$= \dot{m}\ \pi\,(R_t^2 - R_h^2)\, U_f \rho \qquad \text{...(14.32)}$$

where R_h and R_t are the hub and tip radii respectively.

It is customary to assume that there is no prewhirl at entry

$$U_{\theta 1} = 0 \qquad \text{...(14.33 } a)$$

which also implies that

$$\alpha = 0$$

and $$U_1 = U_{f1} = U_f \qquad \text{...(14.33 } b)$$

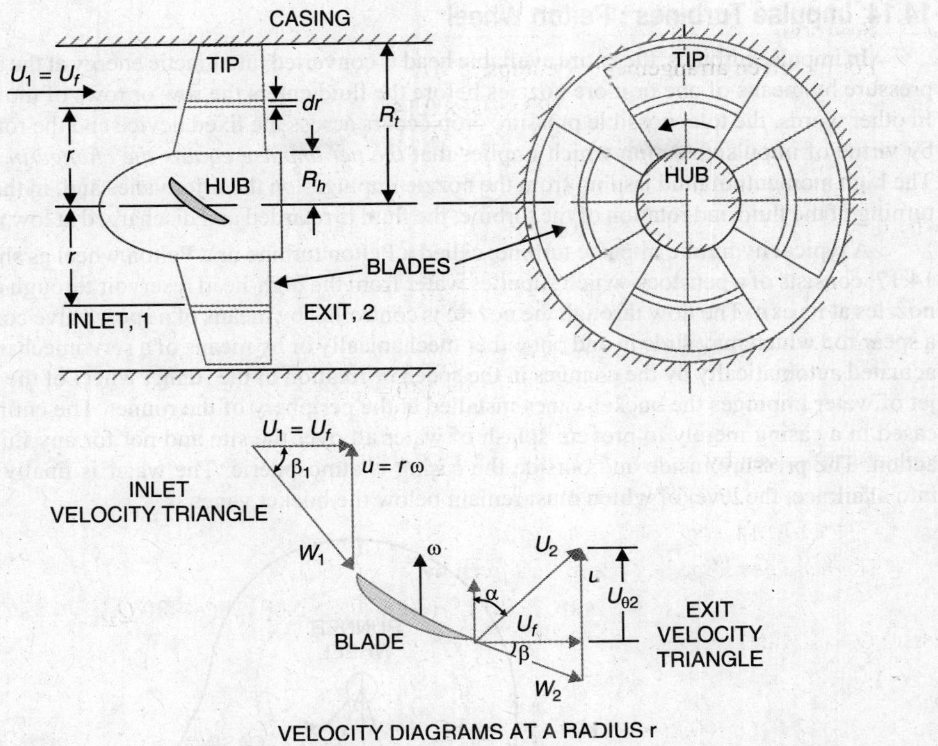

Fig 14.16: Flow Through an Axial-Flow Machine

It follows that, at a radial location r,

Head $$H = \frac{u\,U_{\theta 2}}{g} \quad \text{...(14.34)}$$

and specific energy $$E = u\,U_{\theta 2} = r\,\omega\,U_{\theta 2} \quad \text{...(14.35)}$$

One may assume that the velocities at a mean radius R given by

$$R = \frac{R_h + R_t}{2} \quad \text{...(14.36)}$$

are representative for the fluid machine and work out the specific energy and power on that basis. It is more accurate to apply the equations for any radius r and to integrate over the radius, from R_h to R_t, to obtain the gross behaviour of the machine. The mass flow rate through a strip of radial distance dr at a radius r is given by

$$d\dot{m} = 2\pi r.\,dr\,U_f\rho$$

The total rate of energy transfer would, therefore, be

$$\text{Power } P = \int E\,dm$$

whence $$P = \int r\omega\,U_{\theta 2}.\,2\pi r\,U_f\rho\,dr$$

and, therefore, $$P = 2\pi\rho\omega\,U_f \int_{R_t}^{R_h} r^2\,U_{\theta 2}\,dr \quad \text{...(14.37)}$$

14.14 Impulse Turbines : Pelton Wheel

In impulse turbines, the entire available head is converted into kinetic energy at the atmospheric pressure by means of one or more nozzles before the fluid enters the row or rows of moving blades. In other words, the total possible pressure drop occurs across the fixed device and the rotor is driven by virtue of impulsive action which implies that *the net impulse equals the change in momentum.* The high momentum fluid issuing from the nozzle impinges on the rotor vanes and, in the process of turning of the fluid and rotation of the turbine, the fluid is retarded and discharged at low momentum.

A typical hydraulic impulse turbine, called a Pelton turbine or a Pelton wheel as shown in Fig. 14.17, consists of a penstock which supplies water from the high-head reservoir through one or more nozzles at its exit. The flow through the nozzle is controlled by means of a spear valve constituted by a spear rod which may slide in and out either mechanically or by means of a servomechanism device actuated automatically by the changes in the speed of rotation of the runner wheel of the turbine, the jet of water impinges the bucket-vanes installed at the periphery of the runner. The entire turbine is cased in a casing merely to prevent splash of water all over the site and not for any fluid-dynamic action. The pressure inside and outside the casing is atmospheric. The water is finally discharged into a tailrace, the level of which must remain below the bucket vanes.

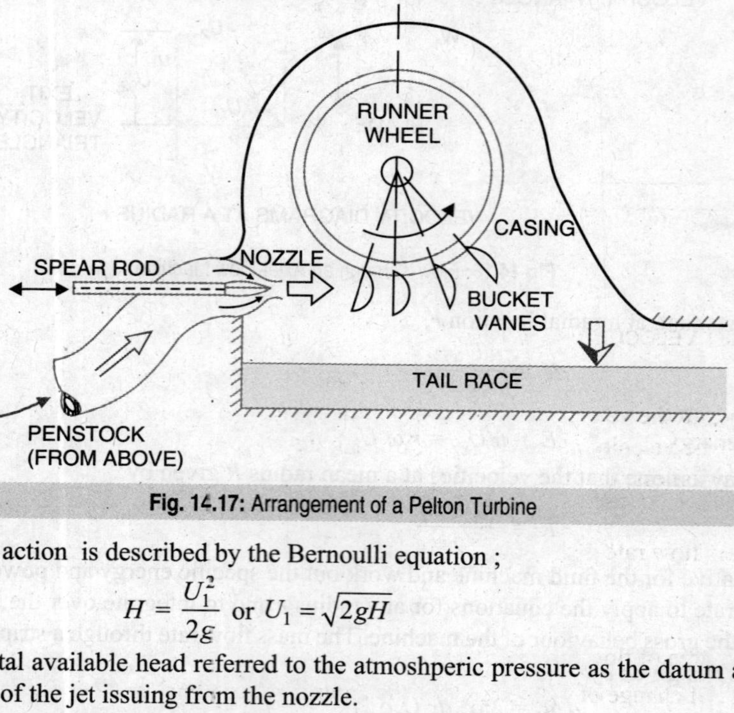

Fig. 14.17: Arrangement of a Pelton Turbine

The nozzle action is described by the Bernoulli equation ;

$$H = \frac{U_1^2}{2g} \text{ or } U_1 = \sqrt{2gH} \qquad ...(14.38)$$

where H is the total available head referred to the atmoshperic pressure as the datum and U_1 is the absolute velocity of the jet issuing from the nozzle.

If the coefficient of velocity for the nozzle is C_v, which is of the order of 0.98, then the expression for the actual jet velocity is

$$U_1 = C_v . \sqrt{2gH} \qquad ...(14.39)$$

It follows that the head lost in the nozzle is

$$h_n = H(1 - C_v^2)$$

and the efficiency of the nozzle is

$$\eta_n = \frac{U_1^2 / 2g}{H} = \frac{C_v^2 . H}{H} = C_v^2 \qquad ...(14.40)$$

From the inlet velocity triangle as shown in Fig. 14.18 (*a*),

$$U_{\theta 1} = U_1$$

and $$W_1 = U_1 - u \quad ...(14.41)$$

where W_1 is the relative velocity of the jet with respect to the vane moving with a peripheral velocity u given by

$$u = \frac{\pi DN}{60} \quad ...(14.42)$$

for the wheel of diameter D rotating at N revolutions per minute.

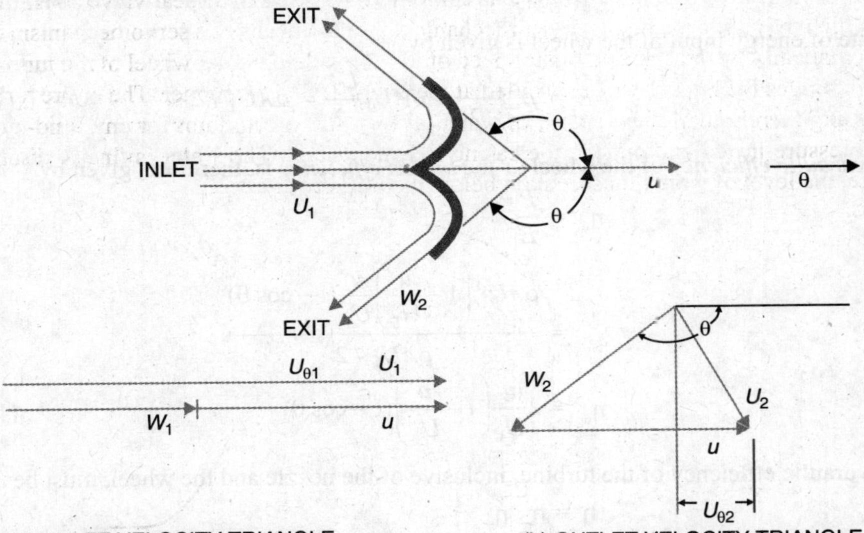

(a) INLET VELOCITY TRIANGLE (b) OUTLET VELOCITY TRIANGLE

Fig. 14.18: Velocity Triangles for the Bucket-Vanes

Referring to the velocity triangle at the outlet as shown in Fig. 14.18 (*b*), the circumferential component of the velocity relative to the vane is reduced to $W_2 \cos \theta$ assuming that the surface of the vane is frictionless,

$$|W_2| = |W_1| \quad ...(14.43)$$

The mass flow rate is given by

$$\dot{m} = \rho Q = \rho (U_1 A) \quad ...(14.44)$$

where A is the area of flow through the nozzle.

The rate of change of momentum for the control volume enclosed between the nozzle and the vane must equal the reaction by the vane on the control volume ;

$$R = \rho Q(U_{\theta 2} - U_{\theta 1}) = \rho Q(u - W_2 \cos\theta - U_1)$$
$$= \rho Q(u - W_1 \cos \theta - U_1)$$

The force acting on the bucket-vane must be equal and opposite to the reaction by the vane on the control volume

$$F = -R = \rho Q[U_1 - (u - W_1 \cos \theta)] \quad ...(14.45)$$

Substituting $$W_1 = U_1 - u$$

$$F = -\rho Q(U_1 - u)(1 - \cos \theta) \quad ...(14.46)$$

It has been assumed in the foregoing analysis that the entire water issuing from the nozzle is utilised in the turbine. This is indeed the case because the vanes follow in quick succession to receive the water. The vanes are *undercut* so as to allow the next vane to receive the water as soon as the vane ahead of it has done its job.

The rate of work done by the water on the wheel is the hydraulic power of the wheel ;

$$P = F.u$$

$$= \rho\, Q(U_1 - u)(1 - \cos\theta).\, u$$

or $$P = \rho A U_1^2 \left(1 - \frac{u}{U_1}\right)\frac{u}{U_1}(1 - \cos\theta) \quad ...(14.47)$$

The rate of energy input of the wheel is given by

$$E = \rho Q \frac{U_1^2}{2} = \rho A U_1 \frac{U_1^2}{2} = \rho A U_1^3 / 2 \quad ...(14.48)$$

The *hydraulic efficiency* of the wheel or the *bucket efficiency* is, therefore, given by

$$\eta_w = \frac{P}{E}$$

$$= \frac{\rho A U_1^3 \left(1 - \frac{u}{U_1}\right)\frac{u}{U_1}(1 - \cos\theta)}{\rho A U_1^3 / 2}$$

or $$\eta_w = 2\frac{u}{U_1}\left(1 - \frac{u}{U_1}\right)(1 - \cos\theta) \quad ...(14.49)$$

The hydraulic efficiency of the turbine, inclusive of the nozzle and the wheel, must be

$$\eta = \eta_n . \eta_w$$

or $$\eta = 2C_v^2 . \frac{u}{U_1}\left(1 - \frac{u}{U_1}\right)(1 - \cos\theta) \quad ...(14.50)$$

The condition for the maximum hydraulic efficiency for a given wheel may be arrived at by employing the fact that

$$\frac{\partial \eta}{\partial u} = 0$$

which provides that $$u = \frac{U_1}{2} \quad ...(14.51)$$

i.e., when the peripheral velocity is 50% of the jet velocity issuing from the nozzle. The expression for the maximum efficiency becomes

$$\eta = \frac{(1 - \cos\theta)}{2} . C_v^2 \quad ...(14.52)$$

The maximum possible efficiency of a Pelton turbine is achieved when the turning is maximum;

$$\theta = 180° \quad ...(14.53\ a)$$

and $$\cos\theta = -1$$

which happens for hemispherical bucket-vanes :

$$\eta = C_v^2 \quad ...(14.53\ b)$$

If $$C_v = 1$$

$$\eta = 1, \textit{i.e.}, 100\ \%$$

For flat-plate vanes,

$$\theta = 90° \quad \text{...(14.54 a)}$$

and $\cos\theta = 0$

so that

$$\eta = 0.5\ C_v^2 \quad \text{...(14.54 b)}$$

which may at best be 0.5, *i.e.*, 50% for $C_v = 1$.

The maximum efficiency for hemispherical bucket-vanes is, therefore, twice as much as for flat-plate vanes.

The expressions for the power and efficiency of an impulse turbine may be written as follows :

$$P = \rho A U_1^3 \phi(1-\phi)(1-\cos\theta) \quad \text{...(14.55)}$$

and

$$\eta = 2C_v^2 . \phi\,(1-\phi)(1-\cos\theta) \quad \text{...(14.56)}$$

where ϕ, defined as the speed factor, is the ratio of the peripheral speed to the jet speed

$$\phi = \frac{u}{U_1} \quad \text{...(14.57)}$$

For most efficient turbine operation, the value of ϕ has been found to depend upon the specific speed of the turbine, *i.e.*,

$$\phi = \phi(N_s) \quad \text{...(14.58)}$$

However, in theory, for ideal flow,

$$\phi = 0.5$$

as proved earlier, but in practice ϕ varies between 0.42 and 0.47 depending upon the specific speed of the turbine.

It may be mentioned that there are other variations of the impulse turbine. Some of them are Turgo impluse wheel which is an axial-flow machine, Banki impulse turbine which is a radial-flow machine and the Wind turbines which go by different names.

A general comment on the impulse machines is that they are low-speed machines. In view of the low specific speeds, they are suitable for high head and low discharge as already discussed under the classification of fluid machines.

Example 14.7. *A Pelton turbine is required to develop 9000 kW when working under a head of 300 metre. The impeller may rotate at 500 revolutions per minute. Assuming the jet ratio as 10 and overall efficiency of 85%, calculate the following :*

(a) quantity of water required,

(b) diameter of the wheel,

(c) number of jets and

(d) number and size of bucket-vanes on the runner.

Solution.

(*a*) Since the output power is 9000 kW and the overall efficiency is 85%, the power supplied by the water must be given by

$$P = \rho g Q H = \frac{9000 \times 10^3}{0.85} = 10\,590 \times 10^3 \text{ W}$$

and

$$Q = \frac{10\,590 \times 10^3}{1000 \times 9.81 \times 300} = \mathbf{3.6\ m^3/s}$$

(*b*) The velocity of the jet is given by

$$U_1 = \sqrt{2gH}$$

$$= \sqrt{2 \times 9.80 \times 300} = 76.7 \text{ m/s}$$

Assuming the speed parameter as 0.48

$$u = 0.48 \times 76.7 = 36.82$$

From the relationship

$$u = \frac{\pi DN}{60}$$

the diameter of the impeller is obtained as

$$D = \frac{60 \times 36.82}{\pi \times 500} = \mathbf{1.41\ m}$$

(*c*) Diameter of the jet $d = \dfrac{D}{10} = \dfrac{14.1}{10} = 0.14$ m

Discharge through one jet is given by

$$q = \frac{\pi \times (0.14)^2}{4} \times 76.7 = 1.18 \text{ m}^3/\text{s}$$

Since the total discharge required is

$$Q = 3.6 \text{ m}^3/\text{s}$$

the number of jets is

$$n = \frac{3.6}{1.18} = 3.04, \text{ say } \mathbf{3\ jets}$$

It is necessary to recalculate the diameter of the nozzle in view of the rounding off to 3 jets,

Discharge per nozzle $q = \dfrac{3.6}{3} = 1.2 \text{ m}^3/\text{s}$

$$\frac{\pi d^2}{4} \times 76.7 = 1.2$$

whence $d = 0.141 \text{ m} = \mathbf{14.1\ cm}$

(*d*) The number of bucket-vanes is determined from an empirical relationship.

$$N = \frac{1}{2}\left(\frac{D}{d}\right) + 15$$

noted from a handbook where D/d is the ratio of the wheel and jet diameters. For the problem in hand,

$$\frac{D}{d} = 10$$

Therefore, $N = \dfrac{1}{2} \times 10 + 15 = 20$

It is recommended to instal **20 bucket vanes.** The size of buckets is also decided empirically.

Width of bucket $\approx 5\,d = 5 \times 14.1 \approx 70$ cm

and depth of bucket $\approx 1.2\,d = 1.2 \times 14.1 \approx 17$ cm

Example 14.8.

(a) Plot a curve showing the variation of efficiency of a Pelton wheel with the speed parameter and hence locate the point of maximum efficiency.

(b) Determine the output power from a Pelton wheel when the flow is reduced by 20% by means of a throttle valve before the nozzle if the wheel developed 400 kW while operating at the maximum efficiency, the speed of the wheel and the nozzle opening remain the same before and after the reduction of flow.

Solution.

(*a*) Referring to expression for the efficiency of a Pelton turbine,

$$\eta = 2\phi(1-\phi)(1-\cos\theta)$$

For a given turbine, θ is constant

whence, $$(1-\cos\theta) = k$$

Hence $$\eta = 2k(\phi - o^2)$$

which implies a parabolic dependence of η on ϕ. The result is shown plotted. It is observed that the efficiency is the maximum at

$$\phi = 0.5$$

Alternatively, at the point of maxima,

$$\frac{\partial\eta}{\partial\phi} = 0$$

or $$2k(1-2\phi) = 0$$

whence $$\phi = 0.5$$

(*b*) The hydraulic power of the impeller is given by

$$P = \rho A U_1^2 o(1-\phi)(1-\cos\theta)$$

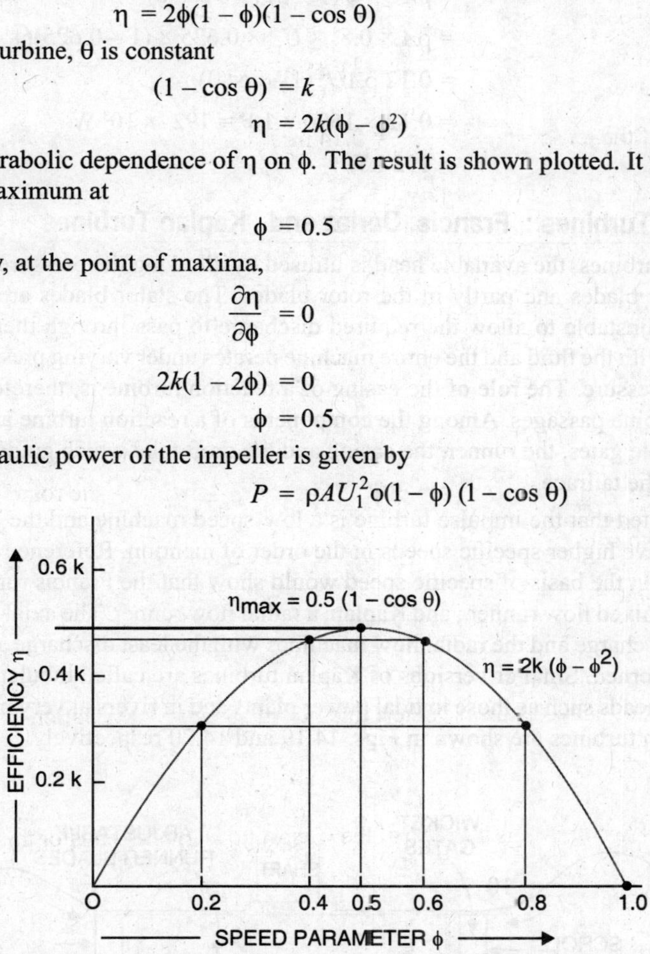

Fig. Ex. 14.8: Variation of the Efficiency of a Pelton Wheel with Speed Parameter

For operation at the maximum efficiency,

$$\phi = 0.5 = \frac{u}{U_1}$$

whence $$u = 0.5\,U_1$$

and the power $$P = 0.25\,\rho A U_1^3(1-\cos\theta) = 1600 \times 10^3 \text{ W} = \textbf{1600 kW}$$

After the reduction of flow

$$Q_1' = 0.8\, Q_1$$

$$U_1' = 0.8\, U_1$$

$$u' = u$$

$$\phi' = \frac{u'}{U_1'} = \frac{u}{0.8U_1} = \frac{0.5}{0.8} = 0.625$$

The power output is given by

$$P' = \rho A U_1'^3 \phi'(1 - \phi)\,(1 - \cos\theta)$$

$$= \rho A \times 0.8^3 \times U_1^3 \times 0.625 \times (1 - 0.625)(1 - \cos\theta)$$

$$= 0.12\, \rho A U_1^3 (1 - \cos\theta)$$

$$= 0.12 \times 1600 \times 10^3 = 192 \times 10^3 \text{ W}$$

$$= \mathbf{192\ kW}$$

14.15 Reaction Turbines : Francis, Deriaz and Kaplan Turbines

In reaction turbines, the available head is utilised by allowing the pressure of the fluid to drop partly in the stator blades and partly in the rotor blades. The stator blades are also called wicket gates which are adjustable to allow the required discharge to pass through them. All passages are completely filled with the fluid and the entire machine perates under varying pressures different from the atmospheric pressure. The role of the casing of a reaction turbine is, therefore, to maintain the pressure in the turbine passages. Among the components of a reaction turbine are the penstock, the scroll, the adjustable gates, the runner, the casing and the draft tube which guides the fluid from the exit the runner to the tailrace.

It may be noted that the impulse turbine is a low-speed machine and the Francis, Deriaz and Kaplan turbines have higher specific speeds in the order of mention. Reference to the classification of fluid machines on the basis of specific speed would show that the Francis runner is a radial flow runner ; Deriaz, a mixed flow runner ; and Kaplan, a radial flow runner. The axial-flow machines deal with the highest discharge and the radial flow machines with the least discharge as far as the reaction machines are concerned. Smaller versions of Kaplan turbines are called tubular turbines which are used for very low heads such as those in tidal power plants and in rivers at very modest falls. Typical Francis and Kaplan turbines are shown in Figs. 14.19 and 14.20 respectively.

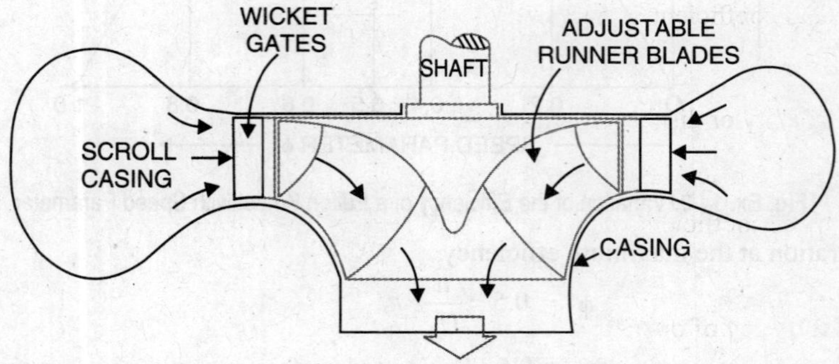

Fig. 14.19: A Typical Francis Turbine

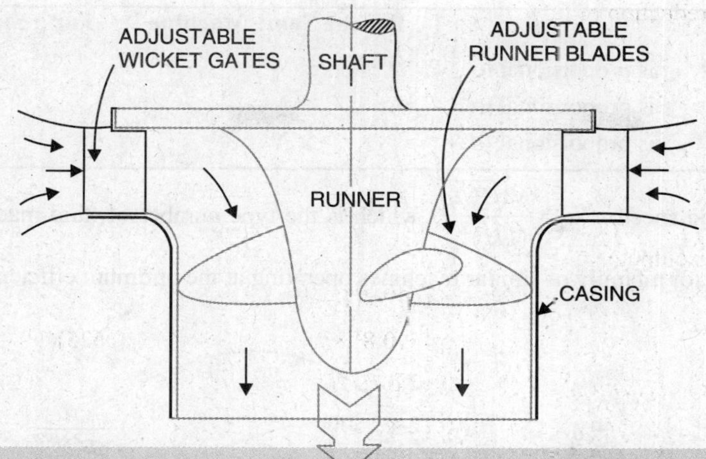

Fig. 14.20: A Typical Kaplan Turbine

The velocity triangles at inlet and outlet of the blades may be constructed in accordance with the principles and practice already outlined. A typical Francis turbine is a radial entry and axial-outflow machine whereas a typical turbine is an axial-inflow and axial-outflow machine. In a mixed flow machine, such as a Deriaz turbine, the flow may enter with axial and radial components of velocity and may leave axially.

Francis turbines are further classified as low, medium and high specific speed-Francis turbines alternatively called as slow, normal and fast runner turbines respectively. As the specific speed increases, the discharge becomes more axial at the tip of the runner, the discharge, in all cases, is axial through the exit to the tailrace.

Another common feature of most reaction machines is that the runner blades are adjustable during operation. This is to allow the runner blades to operate at optimum efficiency. The Francis runner has fixed blades on it.

Reaction turbines are installed appreciably above the tailrace level without sacrificing the power and efficiency. This is achieved by means of installing draft tubes as discussed later.

14.16 Performance of Hydraulic Turbines

Recapitulating the relations developed earlier by the dimensional analysis,

Head coefficient or parameter, $\dfrac{gH}{N^2D^2}$

Capacity or discharge parameter, $\dfrac{Q}{ND^3}$ or $\dfrac{Q}{D^2\sqrt{gH}}$

Power coefficient, $\dfrac{P}{\rho N^3 D^5}$ or $\dfrac{P}{\rho D^2 H^{3/2}}$

Efficiency of operation, η

The laws of performance of turbines are stated as follows from the fact that each of these parameters should remain constant for the same machine or for a family of similar machines :

Variables		For the Same Machine	For a Similar Machine
Discharge Q	is proportional to	N	D^3
Head H	is proportional to	N^2	D^2
Power P	is proportional to	N^3	D^5

The specific speed $N_s = \dfrac{N\sqrt{P/\rho}}{(gH)^{5/4}}$ which is the type number of fluid machines must also remain the same for a family of similar machines operating at the optimum efficiency.

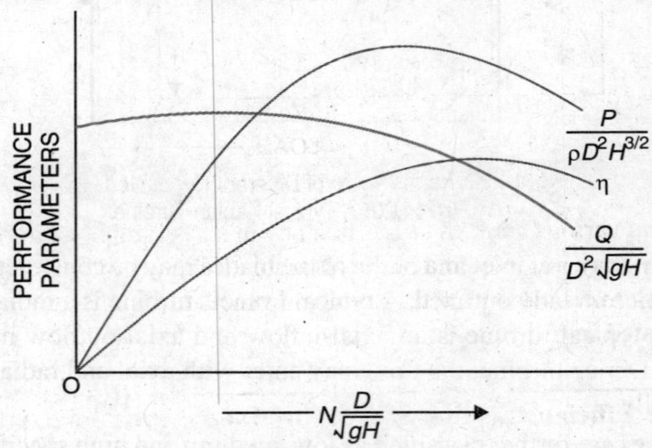

Fig. 14.21: Dimensionless Turbine Curves

The terms 'unit speed', 'unit power' and 'unit discharge' are frequently used to express the operational features of hydraulic turbines. The unit quantities are theoretical features for a head of 1 metre for the same or similar turbine. Sometimes, the unit quantities are referred to a head of 1 metre for a turbine of 1 metre diameter.

Unit Values		For Unit Head	For Unit Head and Unit Diameter
Unit power	P_u	$\dfrac{P}{H^{3/2}}$	$\dfrac{P}{D^2 H^{3/2}}$
Unit discharge	Q_u	$\dfrac{Q}{\sqrt{H}}$	$\dfrac{Q}{D^2\sqrt{H}}$

These definitions follow from the definitions of the head, discharge and power coefficients respectively.

The characteristic curves for hydraulic reaction turbines are usually plotted in terms of unit quantities *vs* 'unit' speed. A set of performance curves are shown plotted in Fig. 14.21. A comparative set of curves for different types of turbines is shown in Fig. 14.22. It can be seen that the Deriaz and Kaplan turbines have the highest efficiency in the entire load range. This is due to the fact that the runner blades of these two types of turbines are adjustable during operation. Consequently the flow is efficient and well-guided by the runner blades at all flow conditions unlike the other turbines where the rotor-vane adjustability is not provided. The guide vanes of each of reaction turbines are adjusted for varying the discharge. It may also be mentioned that the vane-angle alterations are automatically effected by a governor which controls the flow through the turbine with the variation of load, attempting to keep the speed constant which is required by the A.C. generator.

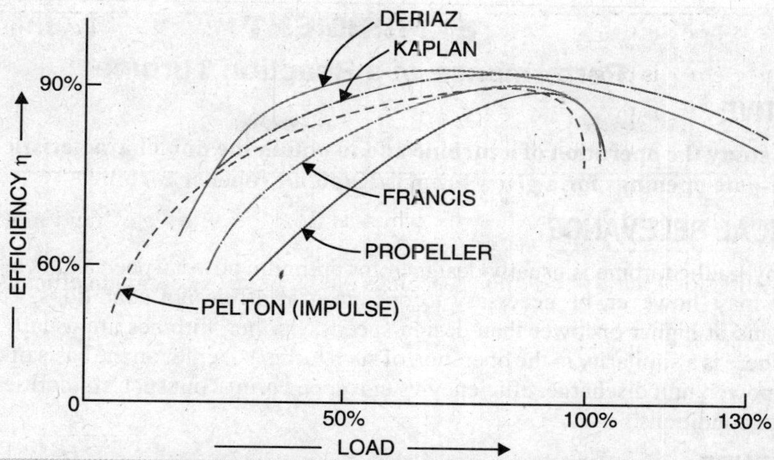

Fig. 14.22: Performance of Different Hydraulic Turbines

Efficiencies and various sources of inefficiency in a hydraulic turbine are presented in a tabular form in Table 14.2. It may be noted that the hydraulic efficiency of a hydraulic turbine may be high but the overall or system efficiency for the hydro-power plant installation is very low due to the large number of factors listed in the table.

Table 14.2. Efficiencies and Sources of Inefficiency in a Hydraulic Turbine

	← Net Available Power for Turbine $\rho Q g H$ →		
		← Water Power of the Runner = $\rho Q\, U_{\theta 1}\, u_1$ →	
← Penstock Losses →	← Hydraulic Losses →	← Mechanical losses →	← Net Output Brake power →
Penstock, Inlet loss Friction loss through the penstock	Fluid Friction, Turbulence Shock-in-Vanes, Draft Tube and Leakage	Disk Friction, Bearing, Glands, Windage	← NET SHAFT POWER →
← GROSS AVAILABLE POWER AT RESERVOIR →			

$$\eta_{vol} = \frac{\text{Useful Runner Discharge}}{\text{Supply Discharge}} = \frac{Q - \delta Q}{Q} \qquad \eta_{mech} = \frac{\text{Net Shaft Power Output}}{\text{Water Power Developed by Runner}}$$

$$\eta_{hyd} = \frac{\text{Water Power Developed by Runner}}{\text{Net Available Power for Turbine}} \qquad \eta_{overall} = \frac{\text{Net Shaft Power Output}}{\text{Net Available Power}}$$

$$\eta_{system} = \frac{\text{Net Output Shaft Brake Power}}{\text{Gross Available Power at Reservoir}} = \eta_{mech} \times \eta_{hyd} \times \eta_{vol}$$

EXPERIMENT

Performance of a Reaction Turbine

OBJECTIVE

To study the operation of a turbine and to obtain the unit characteristic curves at full and part-gate openings for a given Francis/Kaplan/Propeller turbine.

PRACTICAL RELEVANCE

A hydraulic turbine is usually designed for optimum power, speed and head at the 'design point'. It may, however, be necessary to operate it at off-design conditions, *i.e.,* at part-gate opening and at higher or lower than design speeds. Further, turbines are usually designed as a family. There is a similarity in the operation of such turbines. A plot on the basis of unit quantities, *i.e.,* unit power, unit discharge, efficiency vs unit speed brings out such similarities at design and off-design conditions.

PROCEDURE

Ensure that the arrangements to measure the discharge, speed and power output are adequate. The power output is measured by a prony brake or a hydraulic dynamometer. Observe the zero errors and least counts as also the sources of error.

At full gate opening, perform the test by varying the parameters for a selected value of the head H.

CALCULATIONS AND CURVES

The unit quantities, *i.e.,* speed, discharge and power in terms of unit head are given by

$$N_u = N/\sqrt{H}, Q_u = Q\sqrt{H}, P_u = P/H^{3/2}$$

Input water power is estimated as

$$P_{input} = \rho g Q H$$

and the output power on a prony brake of radius R,

$$P_{output} = T.\,\omega = (W_1 - W_2)\,R.\,2\pi N/60$$

where W_1 and W_2 are the spring balance readings.

The efficiency of the turbine is given by

$$\eta = P_{output}/P_{input}$$

The discharge Q is usually measured by creating a flow over a V-notch.

$$Q = 8/15\sqrt{2g}.\,C_d.h^{5/2}$$

POINTS OF DISCUSSION

1. Why do we prefer unit characteristics to be plotted for a turbine ?
2. Determine the specific speed as a dimensionless parameter,

$$N_s = \frac{N\sqrt{Q}}{(gH)^{3/4}} = \frac{N\sqrt{P/\rho}}{(gH)^{5/4}}$$

What does this number imply and why ?

3. How would the curves look if plotted in terms of discharge parameter, power coefficient and specific speed ?

ANY OTHER OBSERVATION OR COMMENT?

Example 14.9. *A turbine is to operate under a head of 25 m at 200 r.p.m. The available discharge is 9 m³/s. Assuming an efficiency of 90%, determine*

(a) its specific speed,

(b) power generated,

(c) performance under a head of 20 m and

(d) the type of turbine.

Solution.

(*a*) The specific speed of a fluid machine is expressed as

$$N_s = \frac{N\sqrt{Q}}{(gH)^{3/4}}$$

$$N_s = \frac{2\pi \times 200}{60} \times \frac{\sqrt{9}}{(9.81 \times 2.5)^{3/4}} = 1.0, \text{ dimensionless.}$$

It is conventional, for turbines to use the dimensionless expression

$$\frac{N\sqrt{P/\rho}}{(gH)^{5/4}} = \frac{N\sqrt{10\, gQH \times \eta}}{(gH)^{5/4}} = \frac{N\sqrt{Q}}{(gH)^{3/4}} \times \sqrt{\eta}$$

which is $\sqrt{\eta}$ times the former ; substantially the same.

For this case it would be $1 \times \sqrt{0.9} = \mathbf{0.95}$, dimensionless.

(*b*) Power generated $P = \rho\, g\, Q\, H\,.\, \eta$

$$= 1000 \times 9.81 \times 9 \times 25 \times 0.9$$

$$= 2 \times 10^6 \text{ W} = \mathbf{2\ MW}$$

(*c*) Under 20 m head, the speed is determined by the head parameter ;

$$\frac{gH}{N^2D^2} = \text{constant or } N \propto \sqrt{H}$$

Hence

$$N = \sqrt{\frac{20}{25}} \times 200 = \mathbf{179\ r.p.m.}$$

The power parameter $\dfrac{P}{\rho D^2 H^{3/2}}$ remains constant.

Hence $P \propto H^{3/2}$

and the power generated $P = 2 \times 10^2 \times \left(\dfrac{20}{25}\right)^{3/2} = 1.43 \times 10^3$ W

$$= \mathbf{1.43\ KW}$$

(*d*) Judging from the non-dimensional specific speed of 1 or 0.95, the turbine should be **Francis or Deriaz type.** In a Francis turbine only the guide vanes are adjustable whereas in a Deriaz turbine the runner vanes are also adjustable during operation. Consequently, the Deriaz turbine should result in better off-design performance.

Example 14.10. *An inward-flow reaction turbine operates under a head of 25 m running at 200 r.p.m. The peripheral velocity of the runner is 20 m/s and the radial velocity at the runner exit is 5 m/s. If the hydraulic losses are 20% of the available head, calculate*

(a) the guide-vane exit angle,

(b) the runner-vane angles,

(c) the runner diameter,

(d) the specific speed, if the width of the runner at the periphery is 30 cm and

(e) the power produced by the turbine.

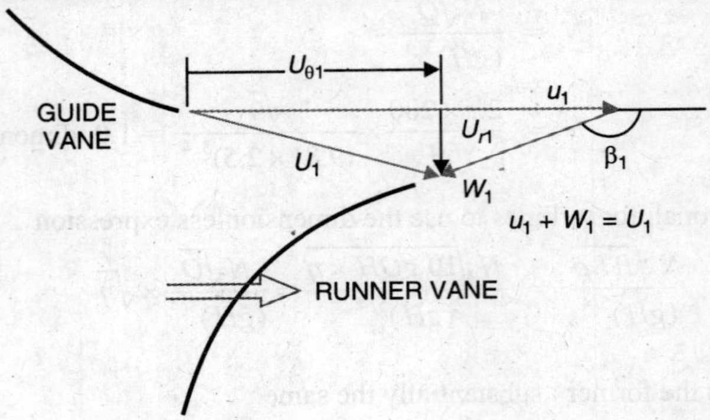

Solution.

From the data, $$\frac{U_1^2}{2g} = 25 \times \frac{100 - 20}{100} = 20 \text{ m},$$

$$U_1 = \sqrt{2 \times 9.81 \times 20} = 19.8 \text{ m/s}$$

$$\frac{U_{\theta 1} u_1}{g} = 20 \text{ m}$$

whence $$U_{\theta 1} = \frac{20 \times 9.81}{20} = 9.81 \text{ m/s}$$

$$U_{r1} = 5 \text{ m/s}$$

$$\tan \alpha = \frac{5}{9.81} = 0.51 ; \alpha = \mathbf{27°}$$

$$\tan(180° - \beta_1) = \frac{5}{20 - 9.81} = \frac{5}{10.19} = 0.49$$

$$180° - \beta_1 = 26°; \beta_1 = \mathbf{154°}$$

(*a*) the guide-vane exit angle is 27°

(*b*) the runner-vane inlet angle is 154°. The runner-vane exit angle is 90°, assuming the discharge to be radial in the absence of any other information

For (*c*) $$u_1 = \frac{\pi D \times 200}{60}$$

The runner diameter, $D = \mathbf{1.91\ m}$

For (*d*) $$N_s = \sqrt{\eta}\, \frac{N\sqrt{Q}}{(gH)^{3/4}}$$

Taking the velocity of flow $U_{f1} = U_{r1} = 5$ m/s

$$Q = \pi \times 1.91 \times 0.3 \times 5 = 9 \text{ m}^3/\text{s}$$

$$N_s = \sqrt{\eta}\,\frac{2\pi \times 200}{60} \times \frac{\sqrt{9}}{(9.81 \times 25)^{3/4}} = \mathbf{0.9} \text{ dimensionless}$$

For (*e*) $\quad P = \rho g Q H.\,\eta = 1000 \times 9.81 \times 9 \times 25 \times 0.8 = 1.76 \times 10^6 \text{ W} = \mathbf{1.76\ MW}$

14.17 Draft Tubes for Reaction Turbines

A draft tube is a confined passage provided for the flow leaving the turbine runner until it reaches the tailrace. Three different types of draft tubes are shown in Fig. 14.33.

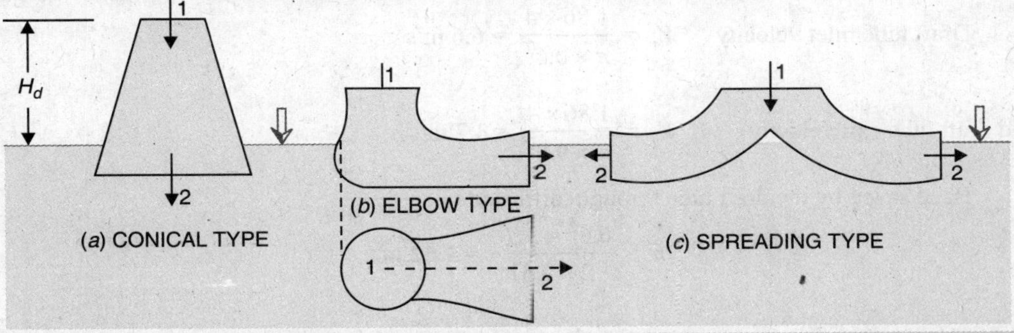

Fig. 14.23: Types of Draft Tubes for Reaction Turbines

The functions of a draft tube are as follows:

(a) To recover part of the kinetic energy otherwise going waste at the exit.

(b) To allow the recovery of head between the level of the turbine runner and the tailrace level

(c) To facilitate the installation of a turbine above the tailrace level without loss of head.

In order to maintain the continuity of flow without vaporisation, the pressure at any place should not be allowed to drop below the vapour pressure of the liquid.

Example 14.11. *A vertical reaction turbine operates under 60 m head at 400 r.p.m. The area and diameter of the runner at inlet are 0.7 m² and 1 m respectively. The absolute and relative velocities of the fluid entering the runner are at 15° and 60° to the tangential direction. The swirl-free entry to the draft tube is at 2 m above the tailrace. If inlet and exit diameters are 0.6 m and 0.8 m respectively, calculate*

(a) the hydraulic efficiency

(b) the head saved by the draft tube through diffusion.

Solution.

From the data $\quad u_1 = \dfrac{\pi \times 1 \times 400}{60} = 21$ m/s

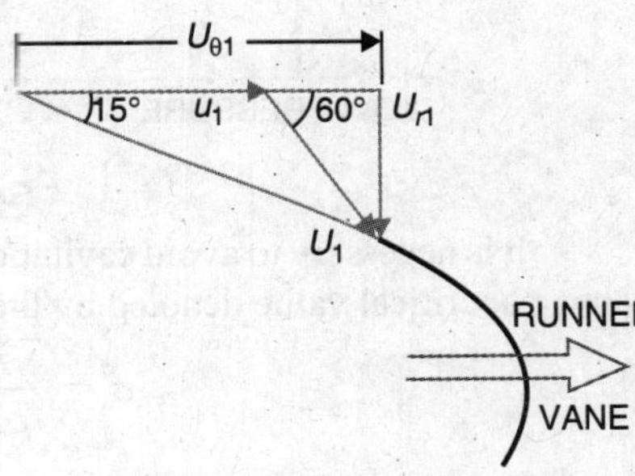

(*a*) From the velocity triangle at runner inlet,

$$\frac{U_1}{\sin 120°} = \frac{u_1}{\sin 45°}$$

$$U_1 = \frac{21 \times 0.866}{0.707} = 25.8 \text{ m/s}$$

$$U_{\theta 1} = 25.8 \cos 15° = 24.9 \text{ m/s}$$

Centrifugal head $= U_{\theta 1} \times \dfrac{u_1}{g} = \dfrac{24.9 \times 21}{9.81} = 53.3 \text{ m}$

Hence, hydraulic efficiency $\eta = \dfrac{53.3}{60} = \mathbf{89\%}$

(*b*) We have $U_{r1} = 25.8 \sin 15° = 6.2$ m/s

and the discharge $Q = 0.3 \times 6.2 = 1.86$ m³/s

Velocity of flow $U_{f1} = U_{r1} = 6.2$ m/s

Draft tube inlet velocity $V_1 = \dfrac{1.86 \times 4}{\pi \times 0.6^2} = 6.6$ m/s

and draft tube exit velocity $V_2 = \dfrac{1.86 \times 4}{\pi \times 0.8^2} = 3.7$ m/s

Head saved by the draft tube through diffusion

$$h = \frac{6.6^2 - 3.7^2}{2 \times 9.81} = \mathbf{1.52 \text{ m}}$$

14.18 Cavitation Phenomenon

A problem commonly encountered in the operation of machines dealing with liquids is cavitation which manifests itself in erosion, noise, vibration and los of energy. Cavitation is caused as follows, as demonstrated in Fig. 14.24.

(*a*) vaporisation of the liquid and/or release of dissolved air at low pressure,

(*b*) movement of the vapour/gas into a high pressure region,

(*c*) collapse of the vapour/gas cavities when subjected to high pressure, and

(*d*) release of energy and pressure wave of high intensity resulting in

(i) pitting and erosion of metal surfaces

(ii) noise and vibration of the machine

(iii) energy loss and drop in efficiency.

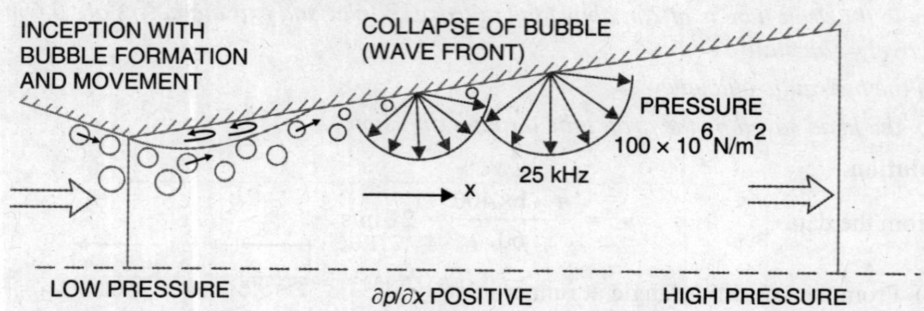

Fig. 14.24: Mechanism of Cavitation

It is necessary to avoid cavitation which can be done by limiting the minimum suction pressure above a critical value denoted by the Thoma Cavitation Parameter,

$$\sigma_c = \frac{p}{\rho H} \qquad \text{...(14.59)}$$

where p = local pressure

ρ = fluid density

and H = operating head

Alternatively, $$\sigma_c = \frac{p - p_v}{\frac{1}{2}\rho U^2} \quad \text{... (14.60)}$$

where U = reference velocity, *i e.,* through the runner impeller

and p_v = vapour pressure of the water at that temperature

Vapour pressure and density of water at different temperatures is as follows:

Temperature (Celcius)	Vapour pressure of water kN/m²	Density kg/m³
15	1.71	999
20	2.33	998
25	3.16	997
30	4.21	996
35	5.61	994
40	7.36	992
45	9.55	990
50	12.31	988
100	101.30	958

It is observed that the efficiency of a machine deteriorates rapidly if the cavitation parameter σ is allowed to drop below the critical value of σ_r. This is shown in Fig. 14.25 where the efficiency of a pump, varied by changing the inlet pressure while keeping head and discharge constant, is plotted against the cavitalion parameter.

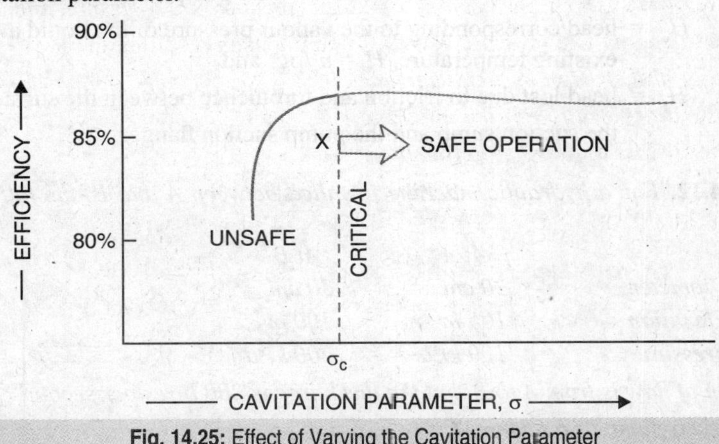

Fig. 14.25: Effect of Varying the Cavitation Parameter

The parameter σ_c is related to the specific speed of the machine. For example,

At N_s	σ_c for Safe Operation
0.5	0.15
1.0	0.30
2.0	0.60

Cavitation is more likely to take place in the following places :

(a) Suction side of the pumps,

(b) Draft tube of the hydraulic turbines,

(c) Venturi or the minimum area section of passages,

(d) Boundaries in the vicinity of high velocity flow and

(e) Siphon pipes with low pressure.

It is sometimes preferred to express the cavitation parameter as the 'suction specific speed'

$$S = \frac{N\sqrt{Q}}{(gH)^{3/4}} \quad \text{...(14.61)}$$

instead of σ_c.

Minimum Safe *S*	For
3.0	Pumps
4.0	Turbines

The current design practice is to allow the formation of bubbles but to stop the collapse of bubbles until downstream of the runner and impeller passage. These are referred to as the supercavitating conditions in flow. Axial flow pumps are generally designed with reference to the supercavitating principle.

A term commonly used to express the margin of cavitation in the centrifugal pumps is the 'Net Positive Suction Head' written as *NPSH*. This is the total suction head at the pump centreline corrected for the local vapour pressure.

$$NPSH = H_p - H_z - H_v - H_f \quad \text{...(14.62)}$$

where H_p = absolute pressure on the surface in the suction sump ;
For standard atmospheric conditions take H_p = 10.32 m.

H_z = depth of the fluid surface below the pump centreline;

H_v = head corresponding to the vapour pressure of the liquid at the existing temperature, $H_v = p_v/\rho g$; and

H_f = head lost due to friction and turbulence between the surface in the suction sump and the pump suction flange.

Example 14.12. *For a hydraulic machine installed between A and B, the following data is available:*

	At A	*At B*
Diameter	*20 cm*	*30 cm*
Elevation	*105 m*	*100 m*
Pressure	*100 kPa*	*200 kPa*

The direction of flow is from A to B and the discharge is 200 litres per second. Is the machine a pump or a turbine ?

Solution. Discharge $Q = \frac{200}{1000}$ m^3/s = 0.2 m^3/s

Let us proceed in the tabular form, as given, and find out the total head at A and at B.

	At A	**At B**
X-section area	$\pi \times 0.2^2/4$ = 0.0314 m^2	$\pi \times 0.3^2/4$ = 0.0707 m^2

Velocity U	0.2/0.0314 = 6.37 m/s	0.2/0.0707 = 2.83 m/s
Velocity head $U^2/2g$	2.07 m	0.408 m
Elevation z	105 m	100 m
Pressure head $p/\rho g$:	10.19 m	20.39
Total head $U^2/2g + z + p/\rho g$	**117.26 m**	**120.8 m**

Since the total head at B is greater than the total head at A, it must be pump which is adding energy to the fluid!

ITEM BANK

Objective items of different types are given to review the concepts developed in the chapter.

* *Select the correct response for each item.*

* *Every correct response is expected in less than a minute and carries one mark.*

* *Compare your responses with the key answers given at the end.*

Multiple-Choice Items

Attempt the following Multiple-Choice Items by ticking the correct or the most appropriate response from among the alternatives given:

1. Specific speed of a fluid machine
(*a*) is specific to the particular machine
(*b*) refers to the speed of a machine of unit dimensions
(*c*) depends only upon the head under which the machine operates
(*d*) is a type-number representative of its performance.

2. The degree of reaction of a machine is defined as the ratio of the
(*a*) static pressure changes in the rotor to that in the stator
(*b*) static pressure change in the rotor to that in the stage
(*c*) static pressure change in the stator to that in the rotor
(*d*) total pressure changes in the rotor to that in the stage.

3. An impeller with backward curved blades
(*a*) exhibits a rising head-discharge characteristic
(*b*) exhibits a drooping head-discharge characteristic
(*c*) is an unbalanced rotor with flat characteristic
(*d*) has a flat characteristic for reverse rotation.

4. For a given centrifugal pump
(*a*) the discharge varies directly as the speed
(*b*) the head varies inversely as the speed
(*c*) the power varies as the square of the speed
(*d*) the discharge varies as the square of the speed.

5. The operating point of a pump installed in a pipeline is decided by
(*a*) the speed of the pump
(*b*) the opening of the delivery valve
(*c*) the length of the pipeline
(*d*) the system characteristic.

6. The efficiency of an impulse turbine
(*a*) may never be beyond 50% even theoretically
(*b*) may approach 100% for frictionless vanes
(*c*) may exceed 50% with inclined flat-plate vanes
(*d*) may approach 100% for hemispherical bucket vanes.

7. The runner vanes of reaction turbines are made adjustable
   (*a*) to permit the turbines to operate at optimum efficiency at part-loads
   (*b*) to allow the operation under all temperature and pressure conditions
   (*c*) to allow running at different speeds of rotation
   (*d*) to reduce the wear and tear of the runner.
8. Chances of occurrence of cavitation are high if
   (*a*) the pressure becomes very high
   (*b*) the temperature becomes low
   (*c*) the thoma cavitation parameter becomes high
   (*d*) the pressure falls below the vapour pressure

Other Objective Items

9. Match the items in the following blocks :

Block A	Block B
(*a*) Kaplan turbine	(*l*) works at atmospheric pressure
(*b*) Pelton Wheel	(*m*) has high part-load efficiency
(*c*) Axial flow pumps	(*n*) gradually converts the velocity head to pressure
(*d*) Draft Tube	(*o*) creates pressure head recovery
(*e*) Volute Casing	(*p*) has high values of N_s

10. State 'TRUE' or 'FALSE' for the following :
   (*a*) Priming is necessary in a centrifugal pump but not in a reciprocating pump.
   (*b*) All pumps or turbines of the same shape have the same specific speed regardless of size.
   (*c*) Unit quantities are helpful in predicting the performance of geometrically similar turbines.
   (*d*) A high head turbine has high specific speed and speed of rotation.
   (*e*) For a centrifugal pump, the discharge varies inversely as the speed.
   (*f*) The suction lift of a centrifugal pump can be up to 10 m head of water

ANSWERS

1. (d), **2.** (b), **3.** (b), **4.** (a), **5.** (d), **6.** (d), **7.** (a), **8.** (d), **9.** (am, bl, cp, do, en)
10. (aT, bT, cT, dF, eF, fF)

QUESTION BANK

Questions on theory and numerical problems are given for tutorial work.

* *The questions are arranged generally in order of articles in the text*

* *Every question should take 15-20 minutes to solve.*

* *Answers for the numerical problems are given alongside them*

14.1. A centrifugal pump, 30 cm impeller diameter running at 1450 r.p.m. delivers 0.2 m^3/s discharge of water against a head of 50 m with an efficiency of 90%. Estimate
   (*a*) its non-dimensional specific speed,
   (*b*) the performance of a pump half its size, *i.e.,* with 15 cm impeller diameter and
   (*c*) the performance if it were coupled to a motor running at 2900 r.p.m.

14.2. Calculate the speed and power input for an exhaust fan to deliver twice the discharge if during a test, it yields

$$Q = 1 \text{ m}^3/\text{s}, \quad R = 15 \text{ cm WG at}$$
$$N = 1450 \text{ revolutions per minute}$$

Assume that the fan operates at 40% efficiency in each case.

14.3. Calculate the discharge of a centrifugal pump with an impeller 40 cm diameter and outlet width 5 cm if the blades are

(*a*) curved backward at 30° (*b*) curved forward at 30° (*c*) radial type.

The pump runs at 1450 r.p.m. and delivers water against a net head of 20 m.

14.4. Recommend the type of turbine to operate under a head of 30 m intended to run at 200 r.p.m. if the available discharge is 10 m³/s. Assume the efficiency to be 90%.

Calculate its power output and sketch the salient features.

14.5. An inward flow reaction turbine operates under a head of 30 m running at 200 r.p.m. The runner diameter is 1 m and the radial velocity at runner exit is 6 m/s. Assuming the hydraulic losses as 25% of 'the available head, determine the

(*a*) guide vane exit angle, (*b*) runner vane angles and (*c*) power produced by the turbine.

14.6. A hydroelectric plant is planned for a discharge of 300 m³/s and available head of 5 m. Assuming a speed of 60 revolutions per minute determine the type and the number of hydraulic turbines required.

14.7. Recommend the power requirement and speed of a motor intended to double the discharge through a ventilation fan currently operating at 1450 r.p.m., delivering 5 m³/s air with a consumption of 2 kW.

14.8. Select a suitable turbine and determine its location above the tailrace level to develop 5 MW power operating at 100 r.p.m. under a head of 20 m. Assume $\sigma_c = 0.15$ for safe operation.

14.9. An inward flow reaction turbine is supplied with water at the rate of 600 litres per second with a velocity of 2 m/s. The velocity of periphery and velocity of whirl at inlet is 24 m/s and 18 m/s. Assuming the discharge to be radial at outlet and the velocity of flow to be constant, find (*i*) vane angle at inlet and (*ii*) head of water on the wheel. **(Ans. 18.5, 44.2 m)**

14.10. An outward flow reaction turbine has internal and external diameters of 2.4 metres and 3.0 metres respectively. The turbine has a radial discharge of 6 m³/s, and is running at 200 r.p.m. The total head on the turbine is 40 metres and width of the wheel at inlet and outlet is 30 cm. Neglecting the thickness of the vanes, find the velocities of

(*i*) flow at inlet, (*ii*) flow at outlet and (*iii*) whirl at inlet.

(Ans. 2.65, 2.12, 15.5 m/s)

14.11. A Kaplan turbine, operating under a net head of 20 metres, develops 36 775 kW power with an overall efficiency of 86 per cent. The speed ratio is 2.0 and flow ratio is 0.60. The hub diameter of the wheel is 0.35 times the outside of the wheel. Find the diameter and speed of the turbine. **(Ans. 5.47 m, 14.52 rad/s)**

14.12. A centrifugal pump having 30 cm diameter impeller is discharging water at the rate of 200 1/s. while running at 1200 r.p.m. The discharge area at outlet is 0.1 m² and vanes are bent backward at 35° to the tangent at outlet. The diameters of the suction and delivery pipes are 30 cm and 22.5 cm respectively.

The gauges fixed on the suction and delivery pipes, close to the pump, show heads of 3.5 metres below and 18.5 metres above atmosphere. If the power required for the pump is 71.6 kW, find the (*a*) overall efficiency and (*b*) manometric efficiency. **(Ans. 63.6%, 74.5%)**

14.13. A quarter-scale turbine model is tested under a head of 12.0 metres. The full scale turbine is required to work under a head of 30 metres and to run at 428 r.p.m. At what speed must the model be run ?

If the model develops 100 kW and uses 1100 litres per second water at this speed, what power will be obtained from the full scale turbine, assuming that the efficiency is 3% better than the model ? **(Ans. 113.7 rad/s, 4150 kW)**

14.14. A hydel power plant is to be installed in a natural stream where a stable discharge of 23 m^3/s and a head of 114 m are available. If the overall efficiency of the water turbine to be used is 87%, determine its type. Assume the speed of the runner as 300 r.p.m.

[0.73 Non-dimensional; Francis]

14.15. The impeller of a centrifugal pump is 0.3 m in diameter and runs at 1450 rpm. The pressure gauges on suction and delivery side show the difference of 25 m. If the blades are curved back to an angle of 30° to tangent at outlet tips of blades and velocity of flow through the impeller, being constant, equal 2.5 m/s, find the manometric efficiency of the pump. If the frictional losses in the impeller amount to 2 m, find the fraction of total energy which is converted into pressure energy by the impeller. **[58.4%; 54.8%]**

Useful Appendices

APPENDIX 1

REVIEW OF SI UNITS

SI BASE UNITS AND SUPPLEMENTARY UNITS

Base Unit	Unit Symbol	Dimensions	Physical Quantity
kilogram	kg	M	mass
metre	m	L	length
second	s	T	time
degree kelvin	K	t	temperature
ampere	A	I	current
candela	cd		luminous intensity
mole	mol		amount of substance
Supplementary Unit			
radian	rad	—	plane angle
steradian	sr	—	solid angle

SI DERIVED UNITS WITH NEW NAMES

Derived Unit	Unit Symbol and Meaning	Physical Quantity
newton	N $= \text{kg m/s}^2$	force, weight
joule	J $= \text{N m} = \text{kg m}^2/\text{s}^2$	energy, work, heat
watt	W $= \text{J/s} = \text{N m/s} = \text{kg m}^2/\text{s}^3$	power, rate of work
pascal	Pa $= \text{N/m}^2 = \text{kg/m/s}^2$	pressure, stress
hertz	Hz $= \text{s}^{-1}$	frequency
ampere	A	current
volt	V $= \text{W/A}$	potential
coulomb	C $= \text{A s}$	charge
ohm	$\Omega = \text{V/A} = \text{W/A}^2$	resistance
farad	F $= \text{A s /V} = \text{A}^2\text{ s/W}$	capacitance
weber	Wb $= \text{V s} = \text{W s/A}$	magnetic flux
henry	H $= \text{V s/A} = \text{W s/A}^2$	inductance
siemen	S $= \text{V/A} = \text{W/A}^2$	conductance
tesla	T $= \text{Wb/m}^2 = \text{V s/m}^2 = \text{W s/A/m}^2$	magnetic induction
lumen	lm $= \text{cd sr}$	luminous flux
lux	$\text{lm/m}^2 = \text{cd sr/m}^2$	illuminance

RULES TO WRITE SI UNITS

1. **Leave at least one blank space between the number and the units, *e.g.,* 32 kg, 205 N/m^2.**
2. **Do not use full stops, dots, dashes or plurals for units. Write Nm, not N.m, N-m etc., e.g., write 125 Nm, not 125 Nms.**
3. **For numbers less than unity, write 0 (zero) on the left of the decimal, *e.g.,* 0.37 W, 0.003 m**
4. **For numbers exceeding five digits, leave one blank space or put a comma every three digits to show thousands or thousandth and millions or millionth as the case may be, e.g. write as 713 290 or 713,290 and 0.625 10, etc.**
5. **For numbers with exponentials, use 10^3, 10^6, 10^9, etc. and 10^{-3}, 10^{-6}, 10^{-9}, etc. followed by units; e.g., 7.02×10^6 N and 0.02×10^{-3} m or 20×10^{-6} m.**

SOME BASIC CONVERSION FACTORS

To convert the following	*Into SI units*	*Multiply by*
inches	cm	2.5400
feet	m	0.3048
miles	km	1.6093
gallons	m^3	4.546×10^{-3}
pints	m^3	0.5683×10^{-3}
litres	m^3	1.0×10^{-3}
pounds	kg	0.4536
pounds/sq. inch	kPa	6.895
gallons/minute	m^3/s	75.8×10^{-6}
cusecs	m^3/s	0.0283
poise	Ns/m^2	0.1
stoke	m^2/s	0.1×10^{-3}

APPENDIX 2

REVIEW OF VECTOR OPERATIONS

1. Vector entities possess magnitude, direction and sense and obey the commutative law of vector addition. Thus, for vectors A and B

$$\boldsymbol{A} = A_x\,\boldsymbol{i} + A_y\,\boldsymbol{j} + A_z\,\boldsymbol{k}$$
$$\boldsymbol{B} = B_x\,\boldsymbol{i} + B_y\,\boldsymbol{j} + B_z\,\boldsymbol{k}$$
$$\boldsymbol{A} + \boldsymbol{B} = \boldsymbol{B} + \boldsymbol{A} = (A_x + B_x)\,\boldsymbol{i} + (A_y + B_y)\,\boldsymbol{j} + (A_z + B_z)\,\boldsymbol{k}$$

where i, j and k are the unit vectors along the three Cartesian coordinate axes.

2. $\boldsymbol{A}\,.\,\boldsymbol{B} = A\,B\cos\alpha = A_xB_x + A_yB_y + A_zB_z$

3. $\boldsymbol{A}\times\boldsymbol{B} = A\,B\sin\theta\,\boldsymbol{n} = \begin{vmatrix} \boldsymbol{i} & \boldsymbol{j} & \boldsymbol{k} \\ A_x & A_y & A_z \\ B_x & B_y & B_z \end{vmatrix} = -\,\boldsymbol{B}\times\boldsymbol{A}$

where $\boldsymbol{n}$ is the unit vector normal to the plane containing $\boldsymbol{A}$ and $\boldsymbol{B}$ such that $\boldsymbol{A}$, $\boldsymbol{B}$ and $\boldsymbol{n}$ form a right-handed triad.

4. 'Del' operator : ∇ is defined as follows in the chosen coordinate system :

Cartesian : $\nabla \equiv \boldsymbol{i}\dfrac{\partial}{\partial x} + \boldsymbol{j}\dfrac{\partial}{\partial y} + \boldsymbol{k}\dfrac{\partial}{\partial z}$

Cylindrical : $\nabla \equiv \boldsymbol{e}_r\dfrac{\partial}{\partial r} + \boldsymbol{e}_\theta\dfrac{1}{r}\dfrac{\partial}{\partial \theta} + \boldsymbol{e}_z\dfrac{\partial}{\partial z}$

where e_r, e_θ, e_z are the unit vectors along the cylindrical coordinate axes.

5. Gradient of a scalar, say of pressure p, ∇p is a vector extending in a direction in which the space derivative of p is a maximum and ∇p is that maximum value

$$\nabla p = \boldsymbol{i}\frac{\partial p}{\partial x} + \boldsymbol{j}\frac{\partial p}{\partial y} + \boldsymbol{k}\frac{\partial p}{\partial z}$$

6. Divergence of a vector A, *i.e.*. $\nabla.\boldsymbol{A}$ is a scalar quantity depicting the net rate of flux across a control surface.

Cartesian : $\nabla.\boldsymbol{A} = \dfrac{\partial A_x}{\partial x} + \dfrac{\partial A_y}{\partial y} + \dfrac{\partial A_z}{\partial z}$

Cylindrical : $\nabla.\boldsymbol{A} = \dfrac{\partial A_r}{\partial r} + \dfrac{1}{r}\dfrac{\partial A_\theta}{\partial \theta} + \dfrac{\partial A_z}{\partial z} + \dfrac{A_r}{r}$

7. Curl of vector ***A*** *i.e.,* $\nabla \times \boldsymbol{A}$ or 'curl $\boldsymbol{A}$' is vector,

Cartesian : $$\nabla \times \boldsymbol{A} = \begin{vmatrix} \boldsymbol{i} & \boldsymbol{j} & \boldsymbol{k} \\ \frac{\partial}{\partial x} & \frac{\partial}{\partial y} & \frac{\partial}{\partial z} \\ A_x & A_y & A_z \end{vmatrix}$$

Cylindrical : $$\nabla \times A = \frac{1}{r} \begin{vmatrix} e_r & re_\theta & e_z \\ \frac{\partial}{\partial r} & \frac{\partial}{\partial \theta} & \frac{\partial}{\partial z} \\ A_r & rA_\theta & A_z \end{vmatrix}$$

8. The Laplacian operator, ***i.e.,*** ∇^2 **or** $(\nabla . \nabla)$**, is defined as**

Cartesian : $$\nabla^2 \equiv \frac{\partial^2}{\partial x^2} + \frac{\partial^2}{\partial y^2} + \frac{\partial^2}{\partial z^2}$$

Cylindrical : $$\nabla^2 \equiv \frac{\partial^2}{\partial r^2} + \frac{1}{r}\frac{\partial}{\partial r} + \frac{1}{r^2}\frac{\partial^2}{\partial \theta^2} + \frac{\partial^2}{\partial z^2}$$

9. Operator 'following the motion' or the 'total' or 'substantial derivative' of a vector,

$$\frac{D}{Dt} \equiv \frac{\partial}{\partial t} + (U.\nabla)$$

Cartesian : $$\frac{D}{Dt} \equiv \frac{\partial}{\partial t} + u\frac{\partial}{\partial x} + v\frac{\partial}{\partial y} + w\frac{\partial}{\partial z}$$

Cylindrical : $$\frac{D}{Dt} \equiv \frac{\partial}{\partial t} + u_r\frac{\partial}{\partial r} + \frac{u_\theta}{r}\frac{\partial}{\partial \theta} + u_z\frac{\partial}{\partial z}$$

where the term $\frac{\partial}{\partial t}$ refers to the temporal or local rate of change and the other three terms refer to the convective rate of change.

10. Some Vector Identities

$$\nabla(\phi\psi) = \phi\nabla\psi + \psi\nabla\phi$$

$$\nabla.(\phi A) = \phi\nabla \times A + \nabla\phi \times A$$

$$\nabla \times (\phi A) = \phi\boldsymbol{\nabla} \times \boldsymbol{A} + \nabla\phi \times \boldsymbol{A}$$

$$\boldsymbol{\nabla} \times (\boldsymbol{\nabla} \times A) = \nabla(\nabla.A) - \nabla^2.A$$

APPENDIX 3

PLANE GEOMETRICAL FIGURES

z_c	*Area*	I_{cc}	Figure
$\frac{d}{2}$	bd	$\frac{bd^3}{12}$	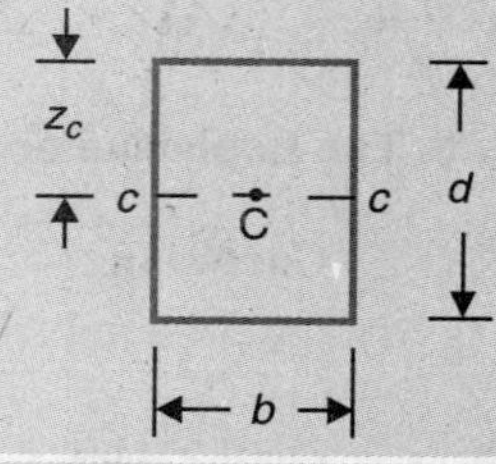
$\frac{2d}{3}$	$\frac{bd}{2}$	$\frac{bd^3}{36}$	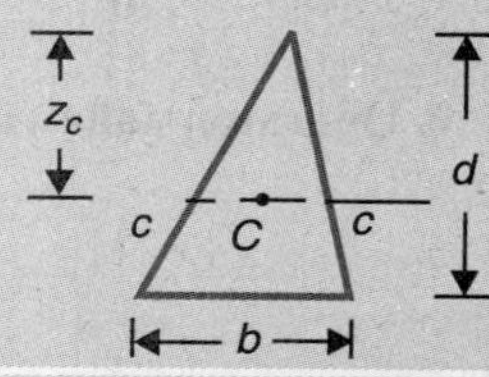
$\frac{d}{2}$	$\frac{\pi d^2}{4}$	$\frac{\pi d^4}{64}$	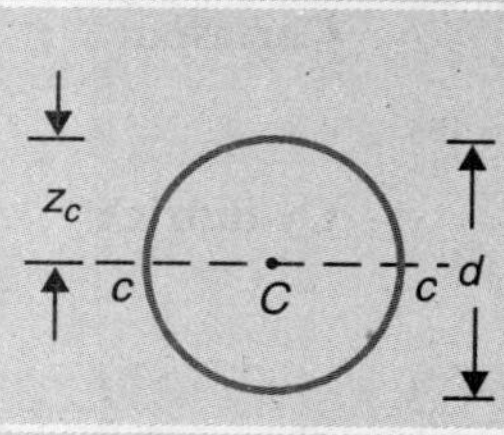
0.212 d	$\frac{\pi d^2}{8}$	0.0069 d^4	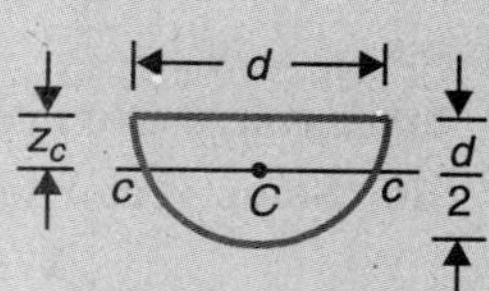
0.212 d	$\frac{\pi d^2}{16}$	0.0044 d^4	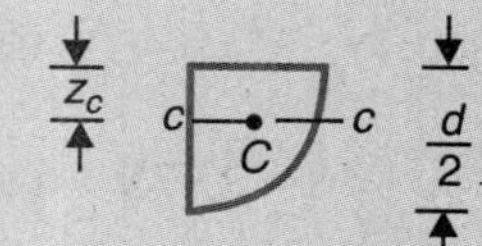
$\frac{d}{2}$	$\frac{\pi bd}{4}$	$\frac{\pi bd^3}{64}$	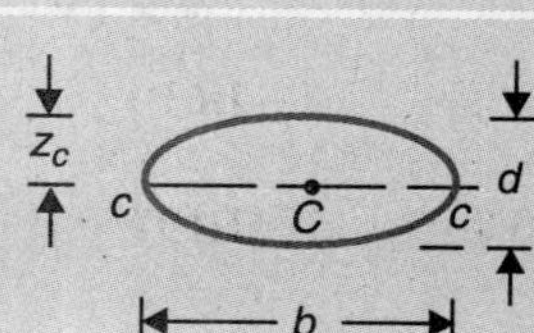

APPENDIX 4

INTERNATIONAL STANDARD ATMOSPHERE

STP CONDITIONS: **Altitude 0** **Pressure 101.325 kN/m^2**
Temperature 288.15 K **Density 1.225 kg/m^3**

Altitude above sea level (m)	Absolute pressure (kN/m^2)	Absolute temperature (K)	Velocity of sound (m/s)	Mass density, ρ (kg/m^3)	Kinematic viscosity, ν (m^2/s)	Dynamic viscosity, μ (Ns/m^2)
0	101.325	288.15	340.3	1.2250	1.461×10^{-5}	17.90×10^{-6}
1000	89.88	281.7	336.4	1.1117	1.581×10^{-5}	17.58×10^{-6}
2000	79.50	275.2	332.5	1.0066	1.715×10^{-5}	17.26×10^{-6}
4000	61.66	262.2	324.6	0.8194	2.028×10^{-5}	16.62×10^{-6}
6000	47.22	249.2	316.5	0.6602	2.416×10^{-5}	15.95×10^{-6}
8000	35.65	236.2	308.1	0.5258	2.904×10^{-5}	15.27×10^{-6}
10 000	26.50	223.3	299.5	0.4134	3.525×10^{-5}	14.57×10^{-6}
11 500	20.98	216.7	295.1	0.3375	4.213×10^{-5}	14.22×10^{-6}
14 000	14.17	216.7	295.1	0.2279	6.239×10^{-5}	14.22×10^{-6}
16 000	10.35	216.7	295.1	0.1665	8.540×10^{-5}	14.22×10^{-6}
18 000	7.565	216.7	295.1	0.1216	11.686×10^{-5}	14.22×10^{-6}
20 000	5.529	216.7	295.1	0.088 92	15.989×10^{-5}	14.22×10^{-6}
22 000	4.047	218.6	296.4	0.064 51	22.201×10^{-5}	14.32×10^{-6}
24 000	2.972	220.6	297.7	0.046 94	30.743×10^{-5}	14.43×10^{-6}
26 000	2.188	222.5	299.1	0.034 26	42.439×10^{-5}	14.54×10^{-6}
28 000	1.616	224.5	300.4	0.025 08	58.405×10^{-5}	14.65×10^{-6}
30 000	1.197	226.5	301.7	0.018 48	80.134×10^{-5}	14.75×10^{-6}
32 000	0.889	228.5	303.0	0.013 56	109.620×10^{-5}	14.86×10^{-6}

APPENDIX 5

PROPERTIES OF LIQUIDS

(At STP unless otherwise stated)

Liquid	Density ρ (kg/m³)	Viscosity μ (Ns/m²)	Surface Tension σ (N/m)	Melting Point (K)	Boiling Point (K)	Thermal Conductivity (W/m K)	Specific Heat (J/kg K)
Acetic Acid	1049	1.219×10^{-3}	27.6×10^{-3}	290	391	0.180	1960
Acetone	780	0.329×10^{-3}	23.7×10^{-3}	178	330	0.161	2210
Benzene	879	0.647×10^{-3}	28.9×10^{-3}	279	353	0.140	1700
Chloroform	1489	0.569×10^{-3}	27.1×10^{-3}	210	334	0.121	960
Crude Oil	860	1.379×10^{-3}	...	...	...	0.155	...
Ethyl Alcohol	789	1.197×10^{-3}	22.3×10^{-3}	156	353	0.177	2500
Glycerine	1260	1.495×10^{-3}	63×10^{-3}	293	562	0.270	2400
Mercury	13,456	1.552×10^{-3}	4.72×10^{-3}	234	630	7.96	140
Olive Oil	920	85×10^{-3}	32×10^{-3}	267	570	0.17	1970
Paraffin Oil	800	1000×10^{-3}	26×10^{-3}	...	...	0.15	2130
Sea-water	1020	1.02×10^{-3}	70×10^{-3}	264	380	...	3900
Turpentine	870	1.49×10^{-3}	27×10^{-3}	263	429	0.136	1760
Water	998 (Assumed as 1000)	1.00×10^{-3}	72.7×10^{-3}	273	373	0.591	4190

APPENDIX 5 (Contd.)

PROPERTIES OF LIQUIDS

(At STP unless otherwise stated)

Latent Heat (J/kg)	Heat of Vaporization (J/kg)	Coefficient of Cubical Expansion	Refractive Index	Bulk Modulus (GPa)	Sonic Velocity (m/s)	Liquid
181×10^3	390×10^3	1070×10^{-6}	1.3718	2.49	1510	**Acetic Acid**
82×10^3	520×10^3	1430×10^{-6}	1.3620	0.8	1000	**Acetone**
127×10^3	400×10^3	1220×10^{-6}	1.5011	1.10	1090	**Benzene**
79×10^{-3}	250×10^{-3}	1270×10^{-6}	1.4467	1.10	830	**Chloroform**
...	...	900×10^{-6}	...	...	...	**Crude Oil**
104×10^3	850×10^3	1080×10^{-6}	1.3610	1.32	1260	**Ethyl Aclochol**
199×10^3	830×10^3	470×10^{-6}	1.4730	4.03	1760	**Glycerine**
117×10^3	290×10^3	182×10^{-6}	1.73	26.2	1370	**Mercury**
...	...	700×10^{-6}	1.47	1.60	1300	**Olive Oil**
...	...	900×10^{-6}	...	...	...	**Paraffin Oil**
330×10^3	...	...	1.343	2.275	1494	**Sea-Water**
...	290×10^3	970×10^{-6}	1.47	1.28	1180	**Turpentine**
334×10^3	2260×10^3	210×10^{-6}	1.333	2.05	1410	**Water**

APPENDIX 6

PROPERTIES OF VAPOURS AND GASES

(At STP unless otherwise stated)

Name	Formula	Density ρ (kg/m³)	Viscosity μ (Ns/m²)	Heat of Vaporization (J/kg)	Melting Temperature (K)	Boiling Temperature (K)	Thermal Conductivity (W/mk)
Acetylene	C_2H_2	1.175	9.35×10^{-6}	...	192.4	189	18.9×10^{-3}
Ammonia	NH_3	0.971	9.18×10^{-6}	1371×10^{3}	195.4	240	21.8×10^{-3}
Carbon Dioxide	CO_2	1.977	14.0×10^{-6}	364×10^{3}	216	195	14.5×10^{-3}
Carbon Monoxide	CO	1.250	16.6×10^{-6}	211×10^{3}	66	81	23.2×10^{-3}
Chlorine	Cl_2	3.214	12.9×10^{-6}	281×10^{3}	171.7	238	7.2×10^{-3}
Hydrogen	H_2	0.090	8.35×10^{-6}	453×10^{3}	13.7	20.35	168.4×10^{-3}
Methane	CH_4	0.717	10.4×10^{-6}	511×10^{3}	90.1	109	30.2×10^{3}
Nitrogen	N_2	1.250	16.7×10^{-6}	209×10^{3}	62.7	77	24.3×10^{-3}
Oxygen	O_2	1.429	19.2×10^{-6}	243×10^{-3}	54	90	24.4×10^{-3}
Water Vapour (at 273 K)	H_2O	0.800	8.7×10^{-6}	2261×10^{3}	272	373	15.8×10^{-3}
Freon-12	CCl_2F_2	...	26.0×10^{-6}	166×10^{3}	118.2	242.6	41.0×10^{-3}
Freon–22	$CHClF_2$	...	23.0×10^{-6}	234×10^{3}	113.2	229.5	52.0×10^{-3}
Air	78.09 N_2 20.95 O_2; 0.96 Misc.	1.225	18.0×10^{-6}	214×10^{3}	...	83	25.3×10^{-3}

APPENDIX 6 (Contd.)

PROPERTIES OF VAPOURS AND GASES

(At STP unless otherwise stated)

Gas Constant (J/kg K)	Specific Heat (J/kg K) at 300 K	Ratio of Specific Heats	Refractivity	Sonic Velocity at S.T.P. (m/s)	Critical Pressure (MPa)	Critical Temperature (K)	Critical Volume (m^3/mol)	Name
320	1590	1.26	606×10^{-6}	341	6.14	309	113×10^{-6}	**Acetylene**
489	2190	1.310	376×10^{-6}	430	11.3	405	72.5×10^{-6}	**Ammonia**
189	845	1.304	451×10^{-6}	266	7.38	304	94.0×10^{-6}	**Carbon Dioxide**
297	1040	1.404	338×10^{-6}	346	3.50	134	93.1×10^{-6}	**Carbon Monoxide**
115	478	1.355	773×10^{-6}	215	7.71	417	124×10^{-6}	**Chlorine**
4127	14300	1.410	132×10^{-6}	1262	12.93	3.3	65.5×10^{-6}	**Hydrogen**
518	2200	1.313	444×10^{-6}	443	4.62	191	98.7×10^{-6}	**Methane**
297	1040	1.404	297×10^{-6}	347	3.39	126	90.1×10^{-6}	**Nitrogen**
260	918	1.401	272×10^{-6}	324	5.08	154	78.0×10^{-6}	**Oxygen**
462	1890 (373 K)	1.3	254×10^{-6}	416	22.12	647	56.8×10^{-6}	**Water Vapour (at 273 K)**
68.64	...	...	1.13	175	4.1	388.5	217×10^{-6}	**Fereon-12**
96.1	...	...	1.18	190	4.9	396.2	165×10^{-6}	**Fereon-22**
287	993	1.402	292×10^{-6}	340	3.77	132	...	**Air**

APPENDIX 7

BASIC PROPERTIES OF AIR AND WATER

Properties of Air at Standard Atmospheric Pressure, 101.325 kN/m²

Temperature, T (°C)	Density, ρ (kg/m³)	Dynamic Viscosity, μ (Ns/m²)	Kinematic Viscosity, ν (m²/s)
– 20	1.39	1.56×10^{-5}	1.13×10^{-5}
– 10	1.34	1.62×10^{-5}	1.21×10^{-5}
0	1.29	1.68×10^{-5}	1.30×10^{-5}
20	1.20	1.80×10^{-5}	1.49×10^{-5}
40	1.12	1.91×10^{-5}	1.70×10^{-5}
60	1.06	2.03×10^{-5}	1.92×10^{-5}
80	0.99	2.15×10^{-5}	2.17×10^{-5}
100	0.94	2.28×10^{-5}	2.45×10^{-5}

Properties of Water at Standard Atmospheric Pressure, 101.325 kN/m²

Temperature (°C)	Density, ρ (kg/m³)	Viscocity, μ (Ns/m²)	Kinematic Viscosity, ν (m²/s)	Surface Tension, σ (N/m)	Vapour-Pressure Head (m)	Bulk Modulus of Elasticity (MN/m²)
0	999.9	1.792×10^{-3}	1.792×10^{-6}	7.62×10^{-2}	0.06	2040
4	1000.0	1.519×10^{-3}	1.519×10^{-6}	7.54×10^{-2}	0.09	2060
10	999.7	1.308×10^{-3}	1.308×10^{-6}	7.48×10^{-2}	0.12	2110
15	999.1	1.140×10^{-3}	1.141×10^{-6}	7.41×10^{-2}	0.17	2140
20	998.2	1.005×10^{-3}	1.007×10^{-6}	7.36×10^{-2}	0.25	2200
25	997.1	0.894×10^{-3}	0.897×10^{-6}	7.26×10^{-2}	0.33	2220
30	995.1	0.801×10^{-3}	0.804×10^{-6}	7.18×10^{-2}	0.44	2230
35	994.1	0.723×10^{-3}	0.727×10^{-6}	7.10×10^{-2}	0.58	2240
40	992.2	0.656×10^{-3}	0.661×10^{-6}	7.01×10^{-2}	0.76	2270
45	990.2	0.599×10^{-3}	0.605×10^{-6}	6.92×10^{-2}	0.98	2290
50	988.1	0.549×10^{-3}	0.556×10^{-6}	6.82×10^{-2}	1.26	2300
55	985.7	0.506×10^{-3}	0.513×10^{-6}	6.74×10^{-2}	1.61	2310
60	983.2	0.469×10^{-3}	0.477×10^{-6}	6.68×10^{-2}	2.03	2280
65	980.6	0.436×10^{-3}	0.444×10^{-6}	6.58×10^{-2}	2.56	2260
70	977.8	0.406×10^{-3}	0.415×10^{-6}	6.50×10^{-2}	3.20	2250
75	974.9	0.380×10^{-3}	0.390×10^{-6}	6.40×10^{-2}	3.96	2230
80	971.8	0.357×10^{-3}	0.367×10^{-6}	6.30×10^{-2}	4.86	2210
85	968.5	0.336×10^{-3}	0.347×10^{-6}	6.20×10^{-2}	5.93	2170
90	965.3	0.317×10^{-3}	0.328×10^{-6}	6.12×10^{-2}	7.18	2160
95	961.9	0.299×10^{-3}	0.311×10^{-6}	6.02×10^{-2}	8.62	2110
100	958.4	0.284×10^{-3}	0.296×10^{-6}	5.94×10^{-2}	10.33	2070

NOMENCLATURE

Symbol	Meaning in Relation to the Text, as Appropriate
A	area ; a constant
a	acceleration in fluid flow ; area
B	breadth ; centre of buoyancy ; a constant
b	breadth or width
C	centroid ; a constant
C	degrees Celsius
C_c	coefficient of contraction
C_D	coefficient of drag
C_d	coefficient of discharge
C_f	skin friction coefficient
C_L	coefficient of lift
C_p	specific heat at constant pressure
C_v	specific heat at constant volume ; coefficient of velocity
CM	metacentric height
c	chord length ; celerity or speed of sound
D	depth ; diameter ; drag force
d	depth of liquid; diameter
d_n	diameter of nozzle
E	energy ; Euler number ; modulus of elasticity ; specific energy
e	energy per unit mass ; error
F	force ; thrust
F_r	Froude number
F_B	body force
F_S	surface force
f	friction factor; skin friction
g	gravitational acceleration ; ≈ 9.81 m/s^2
H	head of fluid ; rate of energy transfer
H_s	suction head
H_m	manometric head
H_d	delivery head
h	head of fluid ; specific enthalpy, distance
h_f	head loss due to friction
h_t	head loss due to transition
I	intensity of turbulence ; moment of inertia
i	specific internal energy ; slope of the channel bed
K	bulk modulus of elasticity ; a constant ; loss coefficient
L	lapse rate ; length ; lift force
l	length ; mixing length
l'	equivalent length
M	Mach number ; mass ; moment
M_B	moment due to body force

Symbol	Meaning in Relation to the Text, as Appropriate
M_S	moment due to surface force
m	mass ; hydraulic mean depth
m	mass flow rate
N	Kutter constant ; Newton number ; rotational speed
N_s	specific speed of fluid machines
n	index ; Manning's constant ; a number
P	centre of pressure ; power ; wetted perimeter
p	pressure
Q	heat transfer to a system ; volumetric discharge
q	discharge per unit width ; heat transfer per unit mass
R	characteristic gas constant ; radial distance ; radius ; reaction force
Re	Reynolds number
r	radial distance ; radius
r_{max}	radius for maximum velocity
r.p.m.	revolution per minute
S	span; specific gravity
s	distance ; slope ; specific entropy ; streamwise direction
T	temperature ; thrust ; time ; torque
t	thickness ; time
U	velocity along x-axis ; free stream velocity ; absolute velocity
U_{av}	average velocity or speed of flow
U_{max}	maximum speed of flow
U_{res}	resultant speed of flow
u, v, w	rectangular components of velocity
u_r, u_θ, u_z	cylindrical components of velocity
u	local velocity along x-axis ; rotational speed
u', v', w'	velocity fluctuations in turbulent flow
$\overline{u'v'}$	Reynolds stresses in turbulent flow
V	volume ; velocity along y axis
v	specific volume
W	weight ; Weber number ; work done ; relative velocity
W_f	rate of flow work done by the system
W_s	rate of shaft work done by the system
WG	water guage head of the fluid
x, y, z	rectangular coordinates
x	distance along x-axis; horizontal
y	distance along y-axis ; along incline; vertical
z, Z	vertical distance ; depth ; datum head
$\overline{z}$	vertical depth of centroid, C
z_p	vertical distance of the centre of pressure P

GREEK SYMBOL

α (alpha)	angle of attack ; semi Mach-cone angle ; kinetic energy correction factor
β (beta)	momentum correction factor; angle
Γ (Gamma)	circulation ; line integral of velocity
γ (gamma)	adiabatic index ; ratio of C_p and C_v
Δ, δ(delta)	increment or change; boundary layer thickness
∇ (del)	vector operator
δ^* (delta star)	displacement tickness of boundary layer
ε (epsilon)	absolute roughness ; eccentricity ; shear strain rate
η (eta)	efficiency ; y/δ in boundary layer flow
θ (theta)	angle ; momentum thickness ; angle of contact
μ (mue)	absolute viscosity
ν (nue)	kinematic viscosity
ξ (xi)	dimensionless parameter in boundary layer flow
π (pi)	constant ; 3.141592654 or 22/7 ; dimcnsionless parameter
ρ (rho)	mass density
σ (sigma)	surface tension ; cavitation parameter ; normal stress
τ (tau)	shear stress
τ_ω, τ_0	shear stress at wall or at a boundary
ϕ (phi)	velocity potential function ; angle
ψ (psi)	stream function
Ω (Omega)	vorticity, angular velocity
ω (omega)	rotation, angular velocity

SOME ADDITIONAL SUFFIXES

0	For free stream ; stagnation value
1	At inlet ; input or at first station
2	At exit ; output or at second station
abs	absolute value
at	atmospheric value
eff	effective value
c	critical value; at sonic state
d	delivery
g	guage value
l	lower
m	mean, maximum or minimum
n	normal
ref	reference value
r	radial component; radial
θ	tangential component; circumferential
s	static value; specific; suction
t	total or stagnation value
t	throat value
u	upper ; unit value
x	component along *x*-axis
y	component along *y*-axis
z	component along *z*-axis

SUBJECT INDEX

A

B

C

D

E

F

G

H

N

O

P